Lasers

GRADUATE TEXTS IN PHYSICS

Graduate Texts in Physics publishes core learning/teaching material for graduate- and advanced-level undergraduate courses on topics of current and emerging fields within physics, both pure and applied. These textbooks serve students at the MS- or PhD-level and their instructors as comprehensive sources of principles, definitions, derivations, experiments and applications (as relevant) for their mastery and teaching, respectively. International in scope and relevance, the textbooks correspond to course syllabi sufficiently to serve as required reading. Their didactic style, comprehensiveness and coverage of fundamental material also make them suitable as introductions or references for scientists entering, or requiring timely knowledge of, a research field.

Series Editors

Professor Richard Needs
Cavendish Laboratory
JJ Thomson Avenue
Cambridge CB3 0HE, UK
E-mail: rn11@cam.ac.uk

Professor William T. Rhodes
Florida Atlantic University
Imaging Technology Center
Department of Electrical Engineering
777 Glades Road SE, Room 456
Boca Raton, FL 33431, USA
E-mail: wrhodes@fau.edu

Professor H. Eugene Stanley
Boston University
Center for Polymer Studies
Department of Physics
590 Commonwealth Avenue, Room 204B
Boston, MA 02215, USA
E-mail: hes@bu.edu

K. Thyagarajan · Ajoy Ghatak

Lasers

Fundamentals and Applications

Second Edition

Springer

Prof. Dr. K. Thyagarajan
Department of Physics
Indian Institute of Technology
Hauz Khas
New Delhi 110 016, India
ktrajan@physics.iitd.ac.in

Prof. Dr. Ajoy Ghatak
Department of Physics
Indian Institute of Technology
Hauz Khas
New Delhi 110 016, India
ajoykghatak@yahoo.com

ISSN 1868-4513   e-ISSN 1868-4521
ISBN 978-1-4614-2694-3   ISBN 978-1-4419-6442-7 (eBook)
DOI 10.1007/978-1-4419-6442-7
Springer New York Dordrecht Heidelberg London

Printed on acid-free paper

Springer is part of Springer Science+Business Media (www.springer.com)

To: Abhinandan, Amitabh, Arjun, Divya, Kalyani, Kamayani and Krishnan

Preface

It is exactly 50 years since the first laser was realized. Lasers emit coherent electro-magnetic radiation, and ever since their invention, they have assumed tremendous importance in the fields of science, engineering, and technology because of their impact in both basic research as well as in various technological applications. Lasers are ubiquitous and can be found in consumer goods such as music players, laser printers, scanners for product identification, in industries like metal cutting, welding, hole drilling, marking, in medical applications in surgery, and in scientific applications like in spectroscopy, interferometry, and testing of foundations of quantum mechanics. The scientific and technological advances have enabled lasers spanning time scales from continuous operation up to as short as a hundred attoseconds, wavelengths spanning almost the entire electromagnetic spectrum up to the X-ray region, power levels into the terawatt region, and sizes ranging from tiny few tens of nanometers to lasers having a length of 270 km. The range of available power, pulse widths, and wavelengths is extremely wide and one can almost always find a laser that can fit into a desired application be it material processing, medical application, or in scientific or engineering discipline. Laser being the fundamental source with such a range of properties and such wide applications, a course on the fundamentals and applications of lasers to both scientists and engineers has become imperative.

The present book attempts to provide a coherent presentation of the basic physics behind the working of the laser along with some of their most important applications and has grown out of the lectures given by the authors to senior undergraduate and graduate students at the Indian Institute of Technology Delhi.

In the first part of the book, after covering basic optics and basic quantum mechanics, the book goes on to discuss the basic physics behind laser operation, some important laser types, and the special properties of laser beams. Fiber lasers and semiconductor lasers which are two of the most important laser types today are discussed in greater detail and so is the parametric oscillator which uses optical non-linearity for optical amplification and oscillation and is one of the most important tunable lasers. The coverage is from first principles so that the book can also be used for self study. The tutorial coverage of fiber lasers given in the book is unique and should serve as a very good introduction to the subject of fiber amplifiers and lasers.

Toward the end of the first part of the book we discuss quantization of electromagnetic field and develop the concept of photons, which forms the basic foundation of the field of quantum optics.

The second part of the book discusses some of the most important applications of lasers in spatial frequency filtering, holography, laser-induced fusion, light wave communications, and in science and industry. Although there are many more applications that are not included in the book, we feel that we have covered some of the most important applications.

We believe that the reader should have some sense of perspective of the history of the development of the laser. One obvious way to go about would be to introduce the reader to some of the original papers; unfortunately these papers are usually not easy to read and involve considerable mathematical complexity. We felt that the Nobel lectures of Charles H Townes, Nicolai G Basov, and A M Prokhorov would convey the development of the subject in a manner that could not possibly be matched and therefore in the third part of the book we reproduce these Nobel Lectures. We have also reproduced the Nobel lecture of Theodor W Hansch who in 2005 was jointly awarded the Nobel Prize for developing an optical *"frequency comb synthesizer,"* which makes it possible, for the first time, to measure with extreme precision the number of light oscillations per second. The frequency comb techniques described in the lecture are also offering powerful new tools for ultrafast physics.

Numerical examples are scattered throughout the book for helping the student to have a better appreciation of the concepts and the problems at the end of each chapter should provide the student with gaining a better understanding of the basics and help in applying the concepts to practical situations. Some of the problems are expected to help the reader to get a feel for numbers, some of them will use the basic concepts developed in the chapter to enhance the understanding and a few of the problems should be challenging to the student to bring out new features or applications leading perhaps to further reading in case the reader is interested. This book could serve as a text in a course at a senior undergraduate or a first-year graduate course on lasers and their applications for students majoring in various disciplines such as Physics, Chemistry, and Electrical Engineering.

The first edition of this book (entitled LASERS: Theory & Applications) appeared in 1981. The basic structure of the present book remains the same except that we have added many more topics like Erbium Doped Fiber Lasers and Amplifier, Optical Parametric Oscillators, etc. In addition we now have a new chapter on Semiconductor Lasers. A number of problems have now been included in the book which should be very useful in further understanding the concepts of lasers. We have also added the Nobel Lecture of Theodor Hansch. Nevertheless, the reader may find some of the references dated because they have been taken from the first edition.

We hope that the book will be of use to scientists and engineers who plan to study or teach the basic physics behind the operation of lasers along with their important applications.

New Delhi, India   K. Thyagarajan
                                                                                Ajoy Ghatak

Acknowledgments

At IIT Delhi we have quite a few courses related to Photonics and this book has evolved from the lectures delivered in various courses ranging from *Basics of Lasers* to *Quantum Electronics*, and our interaction with students and faculty have contributed a great deal in putting the book in this form. Our special thanks to Professor M R Shenoy (at IIT Delhi) for going through very carefully the chapter on Semiconductor Lasers and making valuable suggestions and to Mr. Brahmanand Upadhyaya (at RRCAT, Indore) for going through the chapter on Fiber Lasers and for his valuable suggestions. We are grateful to our colleagues Professor B D Gupta, Professor Ajit Kumar, Professor Arun Kumar, Professor Bishnu Pal, Professor Anurag Sharma, Professor Enakshi Sharma, and Dr. Ravi Varshney for continuous collaboration and discussions. Our thanks to Dr. S. V. Lawande (of Bhabha Atomic Research Center in Mumbai) for writing the section on laser isotope separation.

We are indebted to various publishers and authors for their permission to use various figures appearing in the book; in particular, we are grateful to American Institute of Physics, American Association of Physics Teachers, Institute of Physics, UK, Optical Society of America, SPIE, Oxford University Press, IEEE, Laser Focus World and Eblana Photonics for their permissions. Our sincere thanks to Elsevier Publishing Company for permitting us to reproduce the Nobel lectures. We are grateful to Dr. A.G. Chynoweth, Professor Claire Max, Professor Gurbax Singh, Dr. H Kogelnik, Dr. T.A. Leonard, Dr. D. F. Nelson, Dr. R.A. Phillips, Dr. R.W. Terhune, Dr. L.A. Weaver, Ferranti Ltd., and the United States Information service in New Delhi for providing some of the photographs appearing in the book.

One of the authors (AG) is grateful to Department of Science and Technology, Government of India, for providing financial support.

Finally, we owe a lot to our families – particularly to Raji and Gopa – for allowing us to spend long hours in preparing this difficult manuscript and for their support all along.

K. Thyagarajan
Ajoy Ghatak

Contents

Milestones in the Development of Lasers and Their Applications

1917: A Einstein postulated stimulated emission and laid the foundation for the invention of the laser by re-deriving Planck's law

1924: R Tolman observed that "molecules in the upper quantum state may return to the lower quantum state in such a way to reinforce the primary beam by *"negative absorption"*

1928: R W Landenberg confirmed the existence of stimulated emission and negative absorption through experiments conducted on gases.

1940: V A Fabrikant suggests method for producing population inversion in his PhD thesis and observed that *"if the number of molecules in the excited state could be made larger than that of molecules in the fundamental state, radiation amplification could occur"*.

1947: W E Lamb and R C Retherford found apparent stimulated emission in hydrogen spectra.

1950: Alfred Kastler suggests a method of "optical pumping" for orientation of paramagnetic atoms or nuclei in the ground state. This was an important step on the way to the development of lasers for which Kastler received the 1966 Nobel Prize in Physics.

1951: E M Purcell and R V Pound: In an experiment using nuclear magnetic resonance, Purcell and Pound introduce the concept of negative temperature, to describe the inverted populations of states usually necessary for maser and laser action.

1954: J P Gordon, H J Zeiger and C H Townes and demonstrate first MASER operating as a very high resolution microwave spectrometer, a microwave amplifier or a very stable oscillator.

1956: N Bloembergen first proposed a three level solid state MASER

1958: A Schawlow and C H Townes, extend the concept of MASER to the infrared and optical region introducing the concept of the laser.

1959: Gordon Gould introduces the term LASER

1960: T H Maiman realizes the first working laser: Ruby laser

1960: P P Sorokin and M J Stevenson Four level solid state laser (uranium doped calcium fluoride)

1960: A Javan W Bennet and D Herriott invent the He-Ne laser

1961: E Snitzer: First fiber laser.

1961: P Franken; observes optical second harmonic generation

1962: E Snitzer: First Nd:Glass laser

1962: R. Hall creates the first GaAs semiconductor laser

1962: R W Hellwarth invents Q-switching

1963: Mode locking achieved

1963: Z Alferov and H Kromer: Proposal of heterostructure diode lasers

1964: C K N Patel invents the CO_2 laser

1964: W Bridges: Realizes the first Argon ion laser

1964: Nobel Prize to C H Townes, N G Basov and A M Prochorov *for fundamental work in the field of quantum electronics, which has led to the construction of oscillators and amplifiers based on the maser-laser principle*

1964: J E Geusic, H M Marcos, L G Van Uiteit, B Thomas and L Johnson: First working Nd:YAG laser

1965: CD player

1966: C K Kao and G Hockam proposed using optical fibers for communication. Kao was awarded the Nobel Prize in 2009 for this work.

1966: P Sorokin and J Lankard: First organic dye laser

1966: Nobel Prize to A Kastler *for the discovery and development of optical methods for studying Hertzian resonances in atoms*

1970: Z Alferov and I Hayashi and M Panish: CW room temperature semiconductor laser

1970: Corning Glass Work scientists prepare the first batch of optical fiber, hundreds of yards long and are able to communicate over it with crystal clear clarity

1971: Nobel Prize: D Gabor *for his invention and development of the holographic method*

1975: Barcode scanner

1975: Commercial CW semiconductor lasers

1976: Free electron laser

1977: Live fiber optic telephone traffic: General Telephone & Electronics send first live telephone traffic through fiber optics, 6 Mbit/s in Long Beach CA.

1979: Vertical cavity surface emitting laser VCSEL

1981: Nobel Prize to N Bloembergen and A L Schawlow *"for their contribution to the development of laser spectroscopy"*

1982: Ti:Sapphire laser

1983: Redefinition of the meter based on the speed of light

1985: Steven Chu, Claude Cohen-Tannoudji, and William D. Phillips develop methods to cool and trap atoms with laser light. Their research is helps to study fundamental phenomena and measure important physical quantities with unprecedented precision. They are awarded the Nobel Prize in Physics in 1997.

1987: Laser eye surgery

1987: R.J. Mears, L. Reekie, I.M. Jauncey, and D.N. Payne: Demonstration of Erbium doped fiber amplifiers

1988: Transatlantic fiber cable

1988: Double clad fiber laser

1994: J Faist, F Capasso, D L. Sivco, C Sirtori, A L. Hutchinson, and A Y. Cho: Invention of quantum cascade lasers

1996: S Nakamura: First GaN laser

1997: Nobel Prize to S Chu, C Cohen Tannoudji and W D Philips *"for development of methods to cool and trap atoms with laser light"*

1997: W Ketterle: First demonstration of atom laser

1997: T Hansch proposes an octave-spanning self-referenced universal optical frequency comb synthesizer

1999: J Ranka, R Windeler and A Stentz demonstrate use of internally structured fiber for supercontinuum generation

2000: J Hall, S Cundiff J Ye and T Hansch: Demonstrate optical frequency comb and report first absolute optical frequency measurement

2000: Nobel Prize to Z I Alferov and H Kroemer *"for developing semiconductor heterostructures used in high-speed- and opto-electronics"*

2001: Nobel Prize to E Cornell, W Ketterle and C E Wieman *"for the achievement of Bose-Einstein condensation in dilute gases of alkali atoms, and for early fundamental studies of the properties of the condensates"*

2005: H Rong, R Jones, A Liu, O Cohen, D Hak, A Fang and M Paniccia: First continuous wave Raman silicon laser

2005: Nobel Prize to R J Glauber *"for his contribution to the quantum theory of optical coherence"* and to J L Hall and T H Hansch *"for their contributions to the development of laser-based precision spectroscopy, including the optical frequency comb technique"*

2009: Nobel Prize to C K Kao *"for groundbreaking achievements concerning the transmission of light in fibers for optical communication"*

Ref: Many of the data given here has been taken from the URL for Laserfest: http://www.laserfest.org/lasers/history/timeline.cfm

Part I
Fundamentals of Lasers

Chapter 1
Introduction

An atomic system is characterized by discrete energy states, and usually the atoms exist in the lowest energy state, which is normally referred to as the ground state. An atom in a lower energy state may be excited to a higher energy state through a variety of processes. One of the important processes of excitation is through collisions with other particles. The excitation can also occur through the absorption of electromagnetic radiation of proper frequencies; such a process is known as stimulated absorption or simply as absorption. On the other hand, when the atom is in the excited state, it can make a transition to a lower energy state through the emission of electromagnetic radiation; however, in contrast to the absorption process, the emission process can occur in two different ways.

(i) The first is referred to as spontaneous emission in which an atom in the excited state emits radiation even in the absence of any incident radiation. It is thus not stimulated by any incident signal but occurs spontaneously. Further, the rate of spontaneous emissions is proportional to the number of atoms in the excited state.

(ii) The second is referred to as stimulated emission, in which an incident signal of appropriate frequency triggers an atom in an excited state to emit radiation. The rate of stimulated emission (or absorption) depends both on the intensity of the external field and also on the number of atoms in the upper state. The net stimulated transition rate (stimulated absorption and stimulated emission) depends on the difference in the number of atoms in the excited and the lower states, unlike the case of spontaneous emission, which depends only on the population of the excited state.

The fact that there should be two kinds of emissions – namely spontaneous and stimulated – was first predicted by Einstein (1917). The consideration which led to this prediction was the description of thermodynamic equilibrium between atoms and the radiation field. Einstein (1917) showed that both spontaneous and stimulated emissions are necessary to obtain Planck's radiation law; this is discussed in Section 4.2. The quantum mechanical theory of spontaneous and stimulated emission is discussed in Section 9.6.

K. Thyagarajan, A. Ghatak, *Lasers*, Graduate Texts in Physics,
DOI 10.1007/978-1-4419-6442-7_1, © Springer Science+Business Media, LLC 2010

The phenomenon of stimulated emission was first used by Townes in 1954 in the construction of a microwave amplifier device called the maser,[1] which is an acronym for *m*icrowave *a*mplification by *s*timulated *e*mission of *r*adiation (Gordon et al. 1955). At about the same time a similar device was also proposed by Prochorov and Basov. The maser principle was later extended to the optical frequencies by Schawlow and Townes (1958), which led to the device now known as the laser. In fact "laser" is an acronym for *l*ight *a*mplification by *s*timulated *e*mission of *r*adiation. The first successful operation of a laser device was demonstrated by Maiman in 1960 using ruby crystal (see Section 11.2). Within a few months of operation of the device, Javan and his associates constructed the first gas laser, namely, the He–Ne laser (see Section 11.5). Since then, laser action has been obtained in a large variety of materials including liquids, ionized gases, dyes, semiconductors. (see Chapters 11–13).

The three main components of any laser device are the active medium, the pumping source, and the optical resonator. The active medium consists of a collection of atoms, molecules, or ions (in solid, liquid, or gaseous form), which acts as an amplifier for light waves. For amplification, the medium has to be kept in a state of population inversion, i.e., in a state in which the number of atoms in the upper energy level is greater than the number of atoms in the lower energy level. The pumping mechanism provides for obtaining such a state of population inversion between a pair of energy levels of the atomic system. When the active medium is placed inside an optical resonator, the system acts as an oscillator.

After developing the necessary basic principles in optics in Chapter 2 and basic quantum mechanics in Chapter 3, in Chapter 4 we give the original argument of Einstein regarding the presence of both spontaneous and stimulated emissions and obtain expressions for the rate of absorption and emission using a semiclassical theory. We also consider the interaction of an atom with electromagnetic radiation over a band of frequencies and obtain the gain (or loss) coefficient as the beam propagates through the active medium.

Under normal circumstances, there is always a larger number of atoms in the lower energy state as compared to the excited energy state, and an electromagnetic wave passing through such a collection of atoms would get attenuated rather than amplified. Thus, in order to have amplification, one must have population inversion. In Chapter 5, we discuss the two-level, three-level, and four-level systems and obtain conditions to achieve population inversion between two states of the system. It is shown that it is not possible to achieve steady-state population inversion in a two-level system. Also in order to obtain a population inversion, the transition rates of the various levels in three-level or four-level systems must satisfy certain conditions. We also obtain the pumping powers required for obtaining population inversion in three- and four-level systems and show that it is in general much easier to obtain inversion in a four-level system as compared to a three-level system. In Chapter 6

---

[1] A nice account of the maser device is given in the Nobel lecture of Townes, which is reproduced in Part III of this book.

Fig. 1.1 A plane parallel resonator consisting of a pair of plane mirrors facing each other. The active medium is placed inside the cavity. One of the mirrors is made partially transmitting to couple out the laser beam

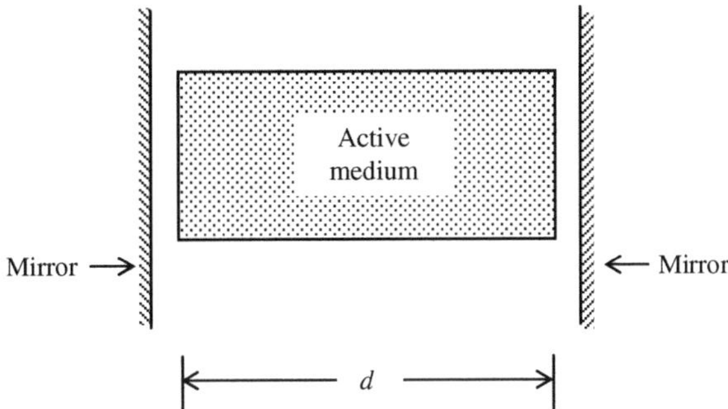

we give the semiclassical theory of laser operation and show that the amplification process due to stimulated transitions is phase coherent – i.e., an electromagnetic wave passing through an inverted medium gets amplified and the phase of the wave is changed by a constant amount; the gain depends on the amount of inversion.

A medium with population inversion is capable of amplification, but if the medium is to act as an oscillator, a part of the output energy must be fed back into the system.[2] Such a feedback is brought about by placing the active medium between a pair of mirrors facing each other (see Fig. 1.1); the pair of mirrors forms what is referred to as an optical resonator. The sides of the cavity are, in general, open and hence such resonators are also referred to as open resonators. In Chapter 7 we give a detailed account of optical resonators and obtain the oscillation frequencies of the modes of the resonator. The different field patterns of the various modes are also obtained. We also discuss techniques to achieve single transverse mode and single longitudinal mode oscillation of the laser. In many applications one requires pulsed operation of the laser. There are primarily two main techniques used for operating a laser in a pulsed fashion; these are Q-switching and mode locking. Chapter 7 discusses these two techniques and it is shown that using mode locking techniques it is indeed possible to achieve ultrashort pulses in the sub picosecond time scale.

Because of the open nature of the resonators, all modes of the resonator are lossy due to the diffraction spillover of energy from the mirrors. In addition to this basic loss, the scattering in the laser medium, the absorption at the mirrors, and the loss due to output coupling of the mirrors also lead to losses. In an actual laser, the modes that keep oscillating are those for which the gain provided by the laser medium compensates for the losses. When the laser is oscillating in steady state, the losses are exactly compensated by the gain. Since the gain provided by the medium depends on the amount of population inversion, there is a critical value of population inversion beyond which the particular mode would oscillate in the laser. If the population

---

[2] Since some of the energy is coupled back to the system, it is said to act as an oscillator. Indeed, in the early stages of the development of the laser, there was a move to change its name to loser, which is an acronym for light oscillation by stimulated emission of radiation. Since it would have been difficult to obtain research grants on losers, it was decided to retain the name laser.

inversion is less than this value, the mode cannot oscillate. The critical value of population inversion is also called the threshold population inversion. In Chapters 4 and 6 we obtain explicit expressions for the threshold population inversion in terms of the parameters of the laser medium and the resonator.

The quantum mechanical theory of spontaneous and stimulated emission is discussed in Chapter 9; the necessary quantum mechanics is given in Chapter 8. Chapter 9 also discusses the important states of light, namely coherent states and squeezed states. The emission from a laser is in the form of a coherent state while squeezed states are non-classical states of light and find wide applications. We also discuss the properties of a beam splitter from a quantum mechanical perspective and show some interesting features of the quantum aspects of light.

The onset of oscillations in a laser cavity can be understood as follows. Through some pumping mechanism one creates a state of population inversion in the laser medium placed inside the resonator system. Thus the medium is prepared to be in a state in which it is capable of coherent amplification over a specified band of frequencies. The spontaneous emission occurring inside the resonator cavity excites the various modes of the cavity. For a given population inversion, each mode is characterized by a certain amplification coefficient due to the gain and a certain attenuation coefficient due to the losses in the cavity. The modes for which the losses in the cavity exceed the gain will die out. On the other hand, the modes whose gain is higher than the losses get amplified by drawing energy from the laser medium. The amplitude of the mode keeps on increasing till non-linear saturation depletes the upper level population to a value when the gain equals the losses and the mode oscillates in steady state. In Chapter 5 we study the change in the energy in a mode as a function of the rate of pumping and show that as the pumping rate passes through the threshold value, the energy contained in a mode rises very steeply and the steady-state energy in a mode above threshold is orders of magnitude greater than the energy in the same mode below threshold. Since the laser medium provides gain over a band of frequencies, it may happen that many modes have a gain higher than the loss, and in such a case the laser oscillates in a multi-mode fashion. In Chapter 7 we also briefly discuss various techniques for selecting a single-mode oscillation of the cavity.

The light emitted by ordinary sources of light, like the incandescent lamp, is spread over all directions and is usually over a large range of wavelengths. In contrast, the light from a laser could be highly monochromatic and highly directional. Because of the presence of the optical cavity, only certain frequencies can oscillate in the cavity. In addition, when the laser is oscillating in steady state the losses are exactly compensated by the gain provided by the medium and the wave coming out of the laser can be represented as a nearly continuous wave. The ultimate monochromaticity is determined by the spontaneous emissions occurring inside the cavity because the radiation coming out of the spontaneous emissions is incoherent. The notion of coherence is discussed in Chapter 10 and the expression for the ultimate monochromaticity of the emitted radiation is discussed in Chapter 7. In practice, the monochromaticity is limited due to external factors like temperature fluctuations and mechanical oscillations of the optical cavity. The light coming out of the laser which is oscillating in a single mode is also composed of a well-defined wave front. This comes about because of the effects of propagation and diffraction inside the resonator cavity. This property is also discussed in greater detail in Chapter 10.

In Chapter 11 we briefly discuss some of the important types of lasers. Chapter 12 discusses the very important area of fiber lasers which are now finding widespread applications in many industries. Chapter 13 discusses one of the most important and most widely used lasers, namely semiconductor lasers. In fact semiconductor lasers have revolutionized the consumer application of lasers; they can be found in super markets, in music systems, in printers, etc.

Most lasers work on the principle of population inversion. It is also possible to achieve optical amplification using non-linear optical effects. In Chapter 14 we discuss the concept of parametric amplification using crystals. Since parametric amplifiers do not depend on energy levels of the medium, it is possible to use this process to realize coherent sources over a very broad range of wavelengths. Thus optical parametric oscillators (OPO) are one of the most versatile tunable lasers available in the commercial market.

In Chapter 15–Chapter 19 we discuss some of the important applications of lasers which have come about because of the special properties of lasers. These include spatial frequency filtering and holography, laser-induced fusion, and light wave communications. We also discuss some of the very important applications of lasers in industries and also how lasers are playing a very important role in science. Finally in Part III of the book we reprint the Nobel lectures of Townes, Prochorov, Basov, and Hansch. Townes, Prochorov, and Basov were awarded the 1964 Nobel Prize for physics for their invention of the laser devices. The Nobel lectures of Townes and Prochorov discuss the basic principles of the maser and the laser whereas the Nobel lecture of Basov gives a detailed account of semiconductor lasers. The Nobel lecture of Hansch discusses the very important field of optical clocks. Such clocks are expected to replace atomic clocks in the near future due to their extreme accuracy.

Today lasers span sizes from a few tens of nanometer size to hundreds of kilometers long. The tiniest lasers demonstrated today have a size of only about 44 nm and is referred to as a *SPASER* which stands for *S*urface *P*lasmon *A*mplification by *S*timulated *E*mission of *R*adiation (Ref: Purdue University. "New Nanolaser Key To Future Optical Computers And Technologies." *ScienceDaily* 17 August 2009; 23 January 2010 <http://www.sciencedaily.com /releases/2009/08/090816171003.htm>. The laser emits a wavelength of 530 nm which is much larger than the size of the laser! The longest laser today is the Raman fiber laser (based on stimulated Raman scattering) and has a length of 270 km! (Turitsyn et al. 2009). Such ultralong lasers are expected to find applications in areas such as non-linear science, theory of disordered systems, and wave turbulence. Since loss is a major concern in optical fiber communication systems, such an ultralong laser offers possibilities of having an effectively high-bandwidth lossless fiber optic transmission link.

Lasers can provide us with sources having extreme properties in terms of energy, pulse width, wavelength, etc., and thus help in research in understanding the basic concept of space and matter. Research and development continues unabated to develop lasers with shorter wavelengths, shorter pulses, higher energies etc.

Linac Coherent Light Source is the world's first hard X-ray free-electron laser, located at the SLAC National Accelerator Laboratory in California. Recently the laser produced its first hard X-ray laser pulses of unprecedented energy and ultra-short duration with wavelengths shorter than the size of molecules. Such lasers are expected to enable frontier research into studies on chemical processes and to perhaps understanding ultimately the processes leading to life.

Attosecond (as) is a duration lasting 10^{-18} s, a thousand times shorter than a femtosecond and a million times shorter than a nanosecond. In fact the orbital period of an electron in the ground state of the hydrogen atom is just 152 as. The shortest laser pulses that have been produced are only 80 as long. Attosecond science is still in its infancy and with further development attosecond science should help us understand various molecular processes, electron transition between energy levels, etc.

The world's most powerful laser was recently unveiled in the National Ignition Facility (NIF) at the Lawrence Livermore National Laboratory in California. The NIF has 192 separate laser beams all converging simultaneously on a single target, the size of a pencil eraser. The laser delivers 1.1 MJ of energy into the target; such a high concentration of energy can generate temperatures of more than 100 million degrees and pressures more than 100 billion times earth's atmospheric pressure. These conditions are similar to those in the stars and the cores of giant planets.

The extreme laser infrastructure being designed and realized in France is expected to generate peak powers of more than a petawatt (10^{15} W) with pulse widths lasting a few tens of attoseconds. The expectations are to be able to generate exawatt (10^{18}) lasers. This is expected to make it possible to study phenomena occurring near black holes, to change the refractive index of vacuum, etc. (Gerstner 2007).

Chapter 2
Basic Optics

2.1 Introduction

In this chapter we will discuss the basic concepts associated with polarization, diffraction, and interference of a light wave. The concepts developed in this chapter will be used in the rest of the book. For more details on these basic concepts, the reader may refer to Born and Wolf (1999), Jenkins and White (1981), Ghatak (2009), Ghatak and Thyagarajan (1989), and Tolansky (1955).

2.2 The Wave Equation

All electromagnetic phenomena can be said to follow from Maxwell's equations. For a charge-free homogeneous, isotropic dielectric, Maxwell's equations simplify to

$$\nabla.\mathbf{E} = 0 \tag{2.1}$$

$$\nabla.\mathbf{H} = 0 \tag{2.2}$$

$$\nabla \times \mathbf{E} = -\mu\frac{\partial \mathbf{H}}{\partial t} \tag{2.3}$$

and

$$\nabla \times \mathbf{H} = \varepsilon\frac{\partial \mathbf{E}}{\partial t} \tag{2.4}$$

where ε and μ represent the dielectric permittivity and the magnetic permeability of the medium and $\mathbf{E}$ and $\mathbf{H}$ represent the electric field and magnetic field, respectively. For most dielectrics, the magnetic permeability of the medium is almost equal to that of vacuum, i.e.,

$$\mu = \mu_0 = 4\pi \times 10^{-7} \ \mathrm{N\,C^{-2}\,s^2}$$

If we take the curl of Eq. (2.3), we would obtain

$$\mathrm{curl}\,(\mathrm{curl}\,\mathbf{E}) = -\mu\frac{\partial}{\partial t}\nabla \times \mathbf{H} = -\varepsilon\mu\frac{\partial^2 \mathbf{E}}{\partial t^2} \tag{2.5}$$

K. Thyagarajan, A. Ghatak, *Lasers*, Graduate Texts in Physics,
DOI 10.1007/978-1-4419-6442-7_2, © Springer Science+Business Media, LLC 2010

where we have used Eq. (2.4). Now, the operator $\nabla^2\mathbf{E}$ is *defined* by the following equation:

$$\nabla^2\mathbf{E} \equiv \mathrm{grad}\,(\mathbf{div}\,\mathbf{E}) - \mathrm{curl}\,(\mathrm{curl}\,\mathrm{E}) \tag{2.6}$$

Using Cartesian coordinates, one can easily show that

$$\left(\nabla^2\mathbf{E}\right)_x = \frac{\partial^2 E_x}{\partial x^2} + \frac{\partial^2 E_x}{\partial y^2} + \frac{\partial^2 E_x}{\partial z^2} = \mathrm{div}\,(\mathrm{grad}\,E_x)$$

i.e., a Cartesian component of $\nabla^2\mathbf{E}$ is the div grad of the Cartesian component.[1] Thus, using

$$\nabla \times \nabla \times \mathbf{E} = \nabla(\nabla.\mathbf{E}) - \nabla^2\mathbf{E}$$

we obtain

$$\nabla\,(\nabla.\mathbf{E}) - \nabla^2\,\mathbf{E} = -\varepsilon\mu\frac{\partial^2\mathbf{E}}{\partial t^2} \tag{2.7}$$

or

$$\nabla^2\,\mathbf{E} = \varepsilon\mu\frac{\partial^2\mathbf{E}}{\partial t^2} \tag{2.8}$$

where we have used the equation $\nabla.\mathbf{E} = 0$ [see Eq. (2.1)]. Equation (2.8) is known as the three-dimensional wave equation and each Cartesian component of $\mathbf{E}$ satisfies the scalar wave equation:

$$\nabla^2\psi = \varepsilon\mu\frac{\partial^2\psi}{\partial t^2} \tag{2.9}$$

In a similar manner, one can derive the wave equation satisfied by $\mathbf{H}$

$$\nabla^2\,\mathbf{H} = \varepsilon\mu\frac{\partial^2\mathbf{H}}{\partial t^2} \tag{2.10}$$

For plane waves (propagating in the direction of $\mathbf{k}$), the electric and magnetic fields can be written in the form

$$\mathbf{E} = \mathbf{E}_0\,\exp[i(\omega t - \mathbf{k}.\mathbf{r})] \tag{2.11}$$

and

$$\mathbf{H} = \mathbf{H}_0\,\exp[i(\omega t - \mathbf{k}.\mathbf{r})] \tag{2.12}$$

---

[1] However, $(\mathbf{E})_r \neq \mathrm{div}\,\mathrm{grad}\,E_r$

where $\mathbf{E_0}$ and $\mathbf{H_0}$ are space- and time-independent vectors; but may, in general, be complex. If we substitute Eq. (2.11) in Eq. (2.8), we would readily get

$$\frac{\omega^2}{k^2} = \frac{1}{\varepsilon\mu}$$

where

$$k^2 = k_x^2 + k_y^2 + k_z^2$$

Thus the velocity of propagation (v) of the wave is given by

$$v = \frac{\omega}{k} = \frac{1}{\sqrt{\varepsilon\mu}} \tag{2.13}$$

In free space

$$\varepsilon = \varepsilon_0 = 8.8542 \times 10^{-12}\,\mathrm{C^2\,N^{-1}m^{-2}} \quad \text{and} \quad \mu = \mu_0 = 4\pi \times 10^{-7}\mathrm{N\,C^{-2}\,s^2} \tag{2.14}$$

so that

$$v = c = \frac{1}{\sqrt{\varepsilon_0\mu_0}} = \frac{1}{\sqrt{8.8542 \times 10^{-12} \times 4\pi \times 10^{-7}}} \tag{2.15}$$
$$= 2.99794 \times 10^8 \mathrm{m\ s^{-1}}$$

which is the velocity of light in free space. In a dielectric characterized by the dielectric permittivity ε, the velocity of propagation (v) of the wave will be

$$v = \frac{c}{n} \tag{2.16}$$

where

$$n = \sqrt{\frac{\varepsilon}{\varepsilon_0}} \tag{2.17}$$

is known as the refractive index of the medium. Now, if we substitute the plane wave solution [Eq. (2.11)] in the equation $\nabla.\mathbf{E} = 0$, we would obtain

$$i[k_x E_{0x} + k_y E_{0y} + k_z E_{0z}]\,\exp[i(\omega t - \mathbf{k}.\mathbf{r})] = 0$$

implying

$$\mathbf{k}.\mathbf{E} = 0 \tag{2.18}$$

Similarly the equation $\nabla.\mathbf{H} = 0$ would give us

$$\mathbf{k}.\mathbf{H} = 0 \tag{2.19}$$

Equations (2.18) and (2.19) tell us that $\mathbf{E}$ and $\mathbf{H}$ are at right angles to $\mathbf{k}$; thus the waves are transverse in nature. Further, if we substitute the plane wave solutions

[Eqs. (2.11) and (2.12)] in Eqs. (2.3) and (2.4), we would obtain

$$\mathbf{H} = \frac{\mathbf{k} \times \mathbf{E}}{\omega \mu} \quad \text{and} \quad \mathbf{E} = \frac{\mathbf{H} \times \mathbf{k}}{\omega \varepsilon} \tag{2.20}$$

Thus $\mathbf{E}$, $\mathbf{H}$, and $\mathbf{k}$ are all at right angles to each other. Either of the above equations will give

$$E_0 = \eta H_0 \tag{2.21}$$

where η is known as the intrinsic impedance of the medium given by

$$\eta = \frac{k}{\omega \varepsilon} = \frac{\omega \mu}{k} = \sqrt{\frac{\mu}{\varepsilon}} = \eta_0 \sqrt{\frac{\varepsilon_0}{\varepsilon}} \tag{2.22}$$

and

$$\eta_0 = \sqrt{\frac{\mu_0}{\varepsilon_0}} \approx 377 \ \Omega$$

is known as the impedance of free space. In writing Eq. (2.22) we have assumed $\mu = \mu_0 = 4\pi \times 10^{-7} \ \ \text{N C}^{-2} \text{s}^2$. The (time-averaged) energy density associated with a propagating electromagnetic wave is given by

$$<u> = \frac{1}{2} \varepsilon E_0^2 \tag{2.23}$$

In the SI system, the units of u will be J m^{-3}. In the above equation, E_0 represents the amplitude of the electric field. The intensity I of the beam (which represents the energy crossing an unit area per unit time) will be given by

$$I = <u> v$$

where v represents the velocity of the wave. Thus

$$I = \frac{1}{2} \varepsilon v E_0^2 = \frac{1}{2} \sqrt{\frac{\varepsilon}{\mu_0}} E_0^2 \tag{2.24}$$

Example 2.1 Consider a 5 mW He–Ne laser beam having a beam diameter of 4 mm propagating in air. Thus

$$I = \frac{5 \times 10^{-3}}{\pi \left(2 \times 10^{-3}\right)^2} \approx 400 \ \ \text{J m}^{-2}\text{s}^{-1}$$

Since

$$I = \frac{1}{2} \varepsilon_0 c E_0^2 \Rightarrow E_0 = \sqrt{\frac{2I}{\varepsilon_0 c}}$$

we get

$$E_0 = \sqrt{\frac{2 \times 400}{(8.854 \times 10^{-12}) \times (3 \times 10^8)}} \approx 550 \quad \text{V m}^{-1}$$

2.3 Linearly Polarized Waves

As shown above, associated with a plane electromagnetic wave there is an electric field $\mathbf{E}$ and a magnetic field $\mathbf{H}$ which are at right angles to each other. For a linearly polarized plane electromagnetic wave propagating in the x-direction (in a uniform isotropic medium), the electric and magnetic fields can be written in the form (see Fig. 2.1)

$$E_y = E_0 \cos(\omega t - kx), \; E_z = 0, \; E_x = 0 \tag{2.25}$$

and

$$H_x = 0, \; H_y = 0, \; H_z = H_0 \cos(\omega t - kx) \tag{2.26}$$

Since the longitudinal components E_x and H_x are zero, the wave is said to be a transverse wave. Also, since the electric field oscillates in the y-direction, Eqs. (2.25) and (2.26) describe what is usually referred to as a y-polarized wave. The direction of propagation is along the vector $(\mathbf{E} \times \mathbf{H})$ which in this case is along the x-axis.

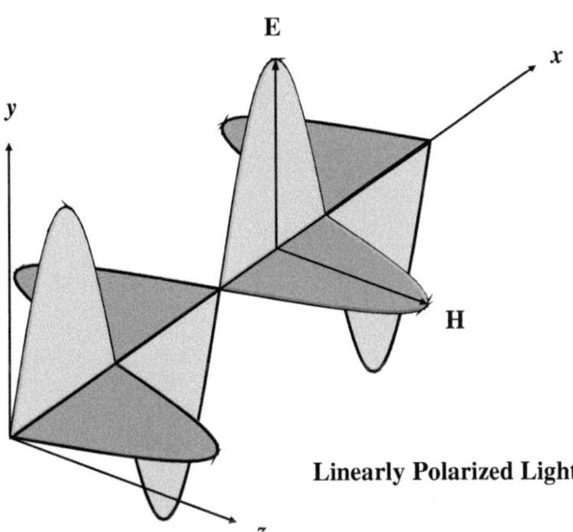

Fig. 2.1 A y-polarized electromagnetic wave propagating in the x-direction

Fig. 2.2 If an ordinary light beam is allowed to fall on a Polaroid, then the emerging beam will be linearly polarized along the pass axis of the Polaroid. If we place another Polaroid P_2, then the intensity of the transmitted light will depend on the relative orientation of P_2 with respect to P_1

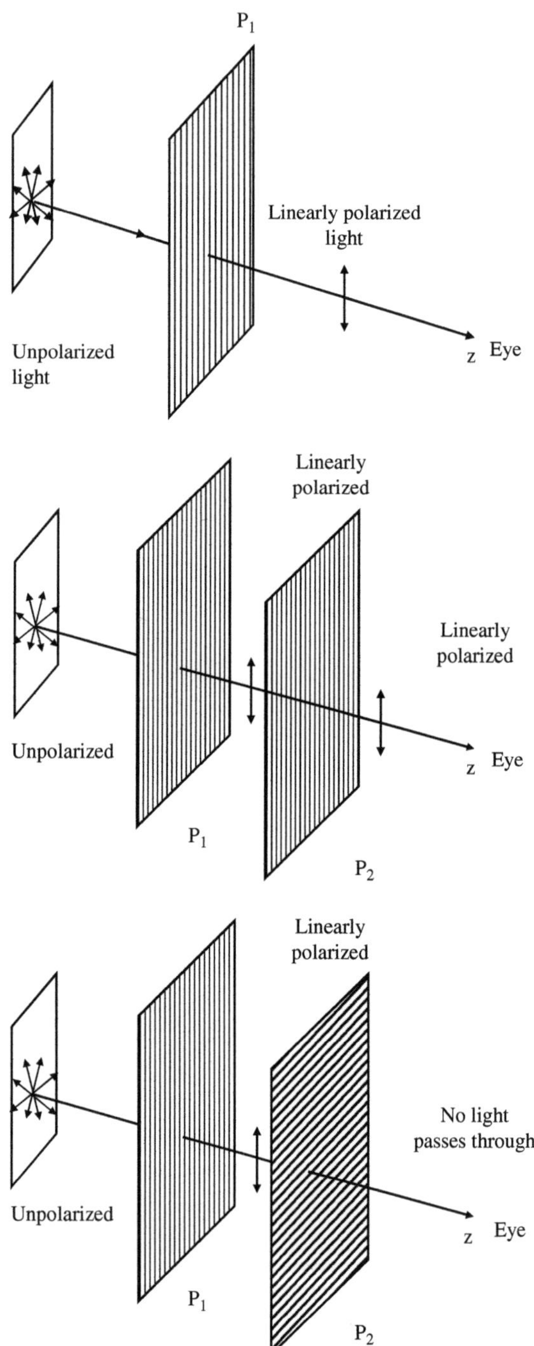

For a z-polarized plane wave (propagating in the $+x$-direction), the corresponding fields would be given by

$$E_x = 0, \ E_y = 0, \quad E_z = E_0 \cos(\omega t - kx), \tag{2.27}$$

and

$$H_x = 0, \ H_y = -H_0 \cos(\omega t - kx) , \ H_z = 0 \tag{2.28}$$

An ordinary light beam, like the one coming from a sodium lamp or from the sun, is unpolarized (or randomly polarized), because its electric vector (on a plane transverse to the direction of propagation) keeps changing its direction in a random manner as shown in Fig. 2.2. If we allow the unpolarized beam to fall on a piece of Polaroid sheet then the beam emerging from the Polaroid will be linearly polarized. In Fig. 2.2 the lines shown on the Polaroid represent what is referred to as the "pass axis" of the Polaroid, i.e., the Polaroid absorbs the electric field perpendicular to its pass axis. Polaroid sheets are extensively used for producing linearly polarized light beams. As an interesting corollary, we may note that if a second Polaroid (whose pass axis is at right angles to the pass axis of the first Polaroid) is placed immediately after the first Polaroid, then no light will come through it; the Polaroids are said to be in a "crossed position" (see Fig. 2.2c).

2.4 Circularly and Elliptically Polarized Waves

We can superpose two plane waves of equal amplitudes, one polarized in the y-direction and the other polarized in the z-direction, with a phase difference of $\pi/2$ between them:

$$\mathbf{E_1} = E_0 \ \hat{\mathbf{y}} \ \cos(\omega t - kx),$$

$$\mathbf{E_2} = E_0 \ \hat{\mathbf{z}} \ \cos\left(\omega t - kx + \frac{\pi}{2}\right), \tag{2.29}$$

The resultant electric field is given by

$$\mathbf{E} = E_0 \ \hat{\mathbf{y}} \ \cos(\omega t - kx) - E_0 \ \hat{\mathbf{z}} \ \sin(\omega t - kx) \tag{2.30}$$

which describes a *left circularly polarized* (usually abbreviated as LCP) wave. At any particular value of x, the tip of the $\mathbf{E}$-vector, with increasing time t, can easily be shown to rotate on the circumference of a circle like a left-handed screw. For example, at $x = 0$ the y and z components of the electric vector are given by

$$E_y = E_0 \ \cos \omega t, \ E_z = -E_0 \ \sin \omega t \tag{2.31}$$

thus the tip of the electric vector rotates on a circle in the anti-clockwise direction (see Fig. 2.3) and therefore it is said to represent an LCP beam. When propagating in air or in any isotropic medium, the state of polarization (SOP) is maintained,

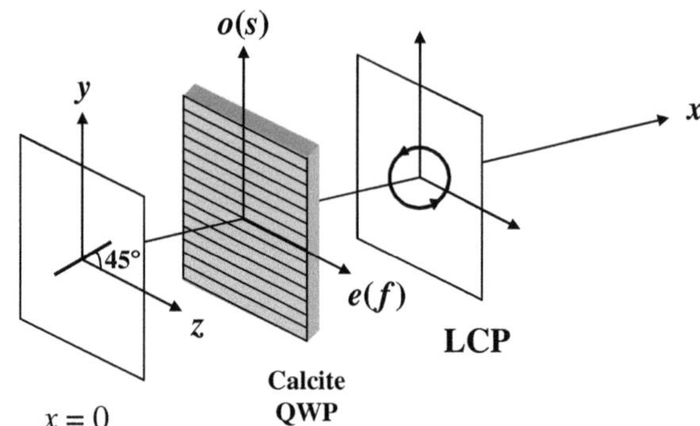

Fig. 2.3 A linearly polarized beam making an angle 45° with the z-axis gets converted to an LCP after propagating through a calcite Quarter Wave Plate (usually abbreviated as QWP); the optic axis in the QWP is along the z-direction as shown by lines parallel to the z-axis

i.e., a linearly polarized beam will remain linearly polarized; similarly, *right circularly polarized* (usually abbreviated as RCP) beam will remain RCP. In general, the superposition of two beams with arbitrary amplitudes and phase

$$E_y = E_0 \cos(\omega t - kx) \text{ and } E_z = E_1 \cos(\omega t - kx + \phi) \tag{2.32}$$

will represent an elliptically polarized beam.

How to obtain a circularly polarized beam? If a linearly polarized beam is passed through a properly oriented quarter wave plate we obtain a circularly polarized beam (see, e.g., Ghatak and Thyagarajan 1989). Crystals such as calcite and quartz are called anisotropic crystals and are characterized by two refractive indices, namely ordinary refractive index n_o and extraordinary refractive index n_e. Inside a crystal-like calcite, there is a preferred direction (known as the optic axis of the crystal); we will assume the crystal to be cut in a way so that the optic axis is parallel to one of the surfaces. In Fig. 2.3 we have assumed the z-axis to be along the optic axis. If the incident beam is y-polarized the beam will propagate as (what is known as) an ordinary wave with velocity (c/n_o). On the other hand, if the incident beam is z-polarized the beam will propagate as (what is known as) an extraordinary wave with velocity (c/n_e). For any other state of polarization of the incident beam, both the extraordinary and the ordinary components will be present. For a crystal-like calcite $n_e < n_o$ and the e-wave will travel faster than the o-wave; this is shown by putting s (slow) and f (fast) inside the parenthesis in Fig. 2.3. Let the electric vector (of amplitude E_0) associated with the incident-polarized beam make an angle ϕ with the z-axis; in Fig. 2.3, ϕ has been shown to be equal to 45°. Such a beam can be assumed to be a superposition of two linearly polarized beams (vibrating in phase), polarized along the y- and z-directions with amplitudes $E_0 \sin \phi$ and $E_0 \cos \phi$, respectively. The y component (whose amplitude is $E_0 \sin \phi$) passes through as an ordinary beam propagating with velocity c/n_o and the z component (whose amplitude is $E_0 \cos \phi$) passes through as an extraordinary beam propagating with velocity c/n_e; thus

$$E_y = E_0 \sin\phi \cos(\omega t - k_o x) = E_0 \sin\phi \cos\left(\omega t - \frac{2\pi}{\lambda_0} n_o x\right) \tag{2.33}$$

and

$$E_z = E_0 \cos \phi \cos (\omega t - k_e x) = E_0 \cos \phi \cos \left(\omega t - \frac{2\pi}{\lambda_0} n_e x \right) \tag{2.34}$$

where λ_0 is the free-space wavelength given by

$$\lambda_0 = \frac{2\pi c}{\omega} \tag{2.35}$$

Since $n_e \neq n_o$, the two beams will propagate with different velocities and, as such, when they come out of the crystal, they will not be in phase. Consequently, the emergent beam (which will be a superposition of these two beams) will be, in general, elliptically polarized. If the thickness of the crystal (denoted by d) is such that the phase difference produced is $\pi/2$, i.e.,

$$\frac{2\pi}{\lambda_0} d (n_o - n_e) = \frac{\pi}{2} \tag{2.36}$$

we have what is known as a quarter wave plate. Obviously, the thickness d of the quarter wave plate will depend on λ_0. For calcite, at $\lambda_0 = 5893$ Å (at 18°C)

$$n_o = 1.65836, \quad n_e = 1.48641$$

and for this wavelength the thickness of the quarter wave plate will be given by

$$d = \frac{5893 \times 10^{-8}}{4 \times 0.17195} \text{ cm} \approx 0.000857 \text{ mm}$$

If we put two identical quarter wave plates one after the other we will have what is known as a half-wave plate and the phase difference introduced will be π. Such a plate is used to change the orientation of an input linearly polarized wave.

2.5 The Diffraction Integral

In order to consider the propagation of an electromagnetic wave in an infinitely extended (isotropic) medium, we start with the scalar wave equation [see Eq. (2.9)]:

$$\nabla^2 \psi = \varepsilon \mu_0 \frac{\partial^2 \psi}{\partial t^2} \tag{2.37}$$

We assume the time dependence of the form $e^{i\omega t}$ and write

$$\psi = U(x, y, z) \, e^{i\omega t} \tag{2.38}$$

to obtain

$$\nabla^2 U + k^2 U = 0 \tag{2.39}$$

where

$$k = \omega\sqrt{\varepsilon\mu_0} = \frac{\omega}{v} \tag{2.40}$$

and U represents one of the Cartesian components of the electric field. The solution of Eq. (2.39) can be written as

$$U(x,y,z) = \int_{-\infty}^{+\infty}\int_{-\infty}^{+\infty} F(k_x, k_y)\, e^{-i(k_x x + k_y y + k_z z)}\, dk_x dk_y \tag{2.41}$$

where

$$k_z = \pm\sqrt{k^2 - k_x^2 - k_y^2} \tag{2.42}$$

For waves making small angles with the z-axis we may write

$$k_z = \sqrt{k^2 - k_x^2 - k_y^2} \;\approx\; k\left[1 - \frac{k_x^2 + k_y^2}{2\,k^2}\right]$$

Thus

$$U(x,y,z) = e^{-ikz}\iint F(k_x,k_y)\exp\left[-i\left(k_x x + k_y y - \frac{k_x^2 + k_y^2}{2\,k}z\right)\right] dk_x dk_y \tag{2.43}$$

and the field distribution on the plane $z = 0$ will be given by

$$U(x,y,z=0) = \iint F(k_x,k_y)\, e^{-i(k_x x + k_y y)}\, dk_x dk_y \tag{2.44}$$

Thus $U(x,y,z=0)$ is the Fourier transform of $F(k_x, k_y)$. The inverse transform will give us

$$F(k_x,k_y) = \frac{1}{(2\pi)^2}\iint U(x',y',0)\; e^{i(k_x x' + k_y y')}\, dx' dy' \tag{2.45}$$

Substituting the above expression for $F(k_x, k_y)$ in Eq. (2.43), we get

$$U(x,y,z) = \frac{e^{-ikz}}{4\pi^2}\iint U(x',y',0)\; I_1 I_2\; dx' dy'$$

where

$$
\begin{aligned}
I_1 &= \int_{-\infty}^{+\infty} \exp\left[ik_x\left(x' - x\right)\right]\, \exp\left[\frac{ik_x^2}{2k}z\right]\,\mathrm{d}k_x \\
&= \sqrt{\frac{i4\pi^2}{\lambda z}}\,\exp\left[-\frac{ik\left(x' - x\right)^2}{2z}\right]
\end{aligned}
\tag{2.46}
$$

and we have used the following integral

$$
\int_{-\infty}^{+\infty} e^{-\alpha x^2 + \beta x}\,\mathrm{d}x = \sqrt{\frac{\pi}{\alpha}}\,\exp\left[\frac{\beta^2}{4\alpha}\right]
\tag{2.47}
$$

Similarly

$$
\begin{aligned}
I_2 &= \int_{-\infty}^{+\infty} \exp\left[ik_y\left(y' - y\right)\right]\, \exp\left[\frac{ik_y^2}{2k}z\right]\,\mathrm{d}k_y \\
&= \sqrt{\frac{i4\pi^2}{\lambda z}}\,\exp\left[-\frac{ik\left(y' - y\right)^2}{2z}\right]
\end{aligned}
\tag{2.48}
$$

Thus

$$
u\left(x, y, z\right) = \frac{i}{\lambda z} e^{-ikz} \iint u\left(x', y', 0\right)\, \exp\left[-\frac{ik}{2z}\left\{\left(x - x'\right)^2 + \left(y - y'\right)^2\right\}\right]\,\mathrm{d}x'\mathrm{d}y'
\tag{2.49}
$$

The above equation (known as the diffraction integral) represents the diffraction pattern in the Fresnel approximation. If we know the field $u(x,y)$ on a plane referred to as $z = 0$, then Eq. (2.49) helps us to calculate the field generated in any plane z. The field changes as it propagates due to diffraction effects.

2.6 Diffraction of a Gaussian Beam

A beam coming out of a laser can be well approximated by a Gaussian distribution of electric field amplitude. We consider a Gaussian beam propagating along the z-direction whose amplitude distribution on the plane $z = 0$ is given by

$$
u(x, y, 0) = A \exp\left[-\frac{x^2 + y^2}{w_0^2}\right]
\tag{2.50}
$$

implying that the phase front is plane at $z = 0$. From the above equation it follows that at a distance w_0 from the z-axis, the amplitude falls by a factor $1/e$ (i.e., the intensity reduces by a factor $1/e^2$). This quantity w_0 is called the *spot size* of the beam. If we substitute Eq. (2.50) in Eq. (2.49) and use Eq. (2.47) to carry out the integration, we would obtain

$$u(x, y, z) \approx \frac{A}{(1 - i\gamma)} \exp\left[-\frac{x^2 + y^2}{w^2(z)}\right] e^{-i\Phi} \tag{2.51}$$

where

$$\gamma = \frac{\lambda z}{\pi w_0^2} \tag{2.52}$$

$$w(z) = w_0 \sqrt{1 + \gamma^2} = w_0 \sqrt{1 + \frac{\lambda^2 z^2}{\pi^2 w_0^4}} \tag{2.53}$$

$$\Phi = kz + \frac{k}{2R(z)}\left(x^2 + y^2\right) \tag{2.54}$$

$$R(z) \equiv z\left(1 + \frac{1}{\gamma^2}\right) = z\left[1 + \frac{\pi^2 w_0^4}{\lambda^2 z^2}\right] \tag{2.55}$$

Thus the intensity distribution varies with z according to the following equation:

$$I(x, y, z) = \frac{I_0}{1 + \gamma^2} \exp\left[-\frac{2\left(x^2 + y^2\right)}{w^2(z)}\right] \tag{2.56}$$

which shows that the transverse intensity distribution remains Gaussian with the beamwidth increasing with z which essentially implies diffraction divergence. As can be seen from Eq. (2.53), for small values of z, the width increases quadratically with z but for values of $z \ggg w_0^2/\lambda$, we obtain

$$w(z) \approx w_0 \frac{\lambda z}{\pi w_0^2} = \frac{\lambda z}{\pi w_0} \tag{2.57}$$

which shows that the width increases linearly with z. This is the Fraunhofer region of diffraction. We define the diffraction angle as

$$\tan \theta = \frac{w(z)}{z} \approx \frac{\lambda}{\pi w_0} \tag{2.58}$$

showing that the rate of increase in the width is proportional to the wavelength and inversely proportional to the initial width of the beam. In order to get some numerical values we assume $\lambda = 0.5 \ \mu$m. Then, for $w_0 = 1$ mm

$$2\theta \approx 0.018° \quad \text{and} \quad w \approx 1.59 \text{ mm} \quad \text{at} \quad z = 10 \text{ m}$$

Similarly, for $w_0 = 0.25$ mm,

$$2\theta \approx 0.073° \quad \text{and} \quad w \approx 6.37 \text{ mm} \quad \text{at} \quad z = 10 \text{ m}$$

Fig. 2.4 A spherical wave diverging from the point O. The *dashed curve* represents a section of the spherical wavefront at a distance R from the source

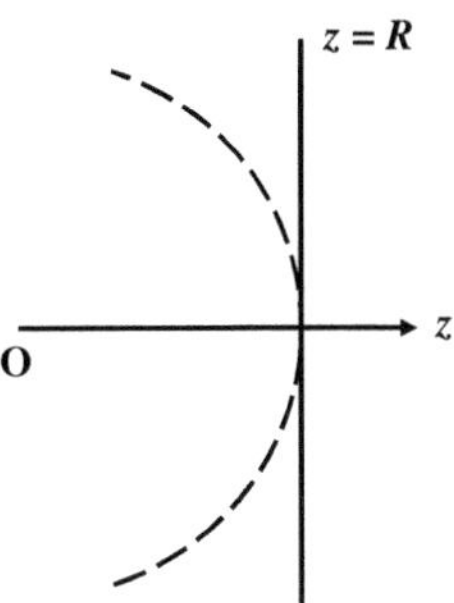

Notice that θ increases with decrease in w_0 (smaller the size of the aperture, greater the diffraction). Further, for a given value of w_0, the diffraction effects decrease with λ. From Eq. (2.51) one can readily show that

$$\int\limits_{-\infty}^{+\infty}\int\limits_{-\infty}^{+\infty} I(x,y,z)\ \mathrm{d}x\mathrm{d}y = \frac{\pi\,w_0^2}{2}\ I_0$$

which is independent of z. This is to be expected, as the total energy crossing the entire x–y plane will not change with z.

Now, for a spherical wave *diverging* from the origin, the field distribution is given by

$$u \sim \frac{1}{r}\,e^{-ikr} \tag{2.59}$$

On the plane $z = R$ (see Fig. 2.4)

$$r = \left[x^2 + y^2 + R^2\right]^{1/2}$$

$$= R\left[1 + \frac{x^2 + y^2}{R^2}\right]^{1/2} \tag{2.60}$$

$$\approx R + \frac{x^2 + y^2}{2R}$$

where we have assumed $|x|$, $|y| \ll R$. Thus on the plane $z = R$, the phase distribution (corresponding to a diverging spherical wave of radius R) would be given by

$$e^{-ikr} \approx e^{-ikR}\ e^{-\frac{ik}{2R}\left(x^2 + y^2\right)} \tag{2.61}$$

From the above equation it follows that a phase variation of the type

$$\exp\left[-i\frac{k}{2R}\left(x^2 + y^2\right)\right] \tag{2.62}$$

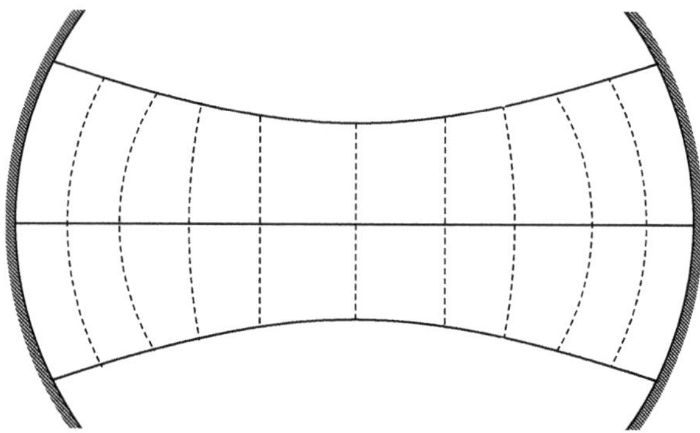

Fig. 2.5 Diffraction divergence of a Gaussian beam whose phase front is plane at $z = 0$. The *dashed curves* represent the phase fronts

(on the x–y plane) represents a *diverging* spherical wave of radius R. If we compare the above expression with Eqs. (2.59) and (2.60) we see that as the Gaussian beam propagates, the phase front curvature changes and we obtain the following approximate expression for the radius of curvature of the phase front at any value z:

$$R(z) \approx z \left(1 + \frac{\pi^2 w_0^4}{\lambda^2 z^2} \right) \tag{2.63}$$

Thus as the beam propagates, the phase front which was plane at $z = 0$ becomes curved. In Fig. 2.5 we have shown a Gaussian beam resonating between two identical spherical mirrors of radius R; the plane $z = 0$, where the phase front is plane and the beam has the minimum spot size, is referred to as the waist of the Gaussian beam. For the beam to resonate, the phase front must have a radius of curvature equal to R on the mirrors. For this to happen we must have

$$R \approx \frac{d}{2} \left(1 + \frac{4\pi^2 w_0^4}{\lambda^2 d^2} \right) \tag{2.64}$$

where d is the distance between the two mirrors. We will discuss more details about the optical resonators in Chapter 7.

It should be mentioned that although in the derivation of Eq. (2.51) we have assumed z to be large, Eq. (2.51) does give the correct field distribution even at $z=0$.

2.7 Intensity Distribution at the Back Focal Plane of a Lens

If a truncated plane wave of diameter $2a$ propagating along the z-axis is incident on a converging lens of focal length f (see Fig. 2.6a), the intensity distribution on the back focal plane is given by (see, e.g., Born and Wolf (1999))

$$I = I_0 \left[\frac{2J_1(v)}{v} \right]^2 \tag{2.65}$$

where

$$v = \frac{2\pi a}{\lambda f} r, \tag{2.66}$$

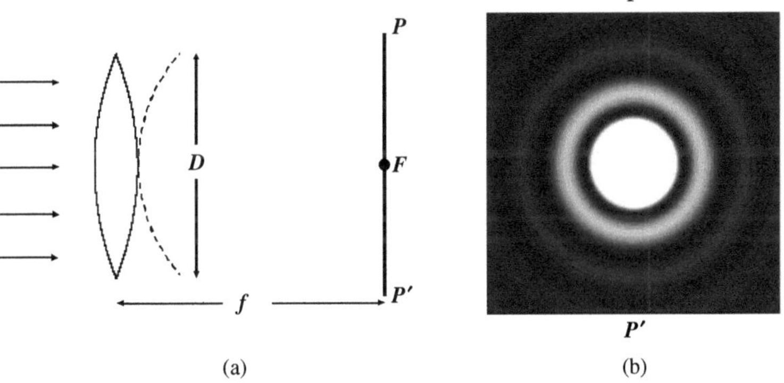

Fig. 2.6 (a) Plane wave falling on a converging lens gets focused at the focus of the lens. (b) The Airy pattern formed at the focus of the lens

I_0 is the intensity at the axial point F and r is the distance from the point F on the focal plane. Equation (2.65) describes the well-known Airy pattern (see Fig. 2.6b). The intensity is zero at the zeroes of the Bessel function $J_1(v)$ and $J_1(v) = 0$ when $v= 3.832, 7.016, 10.174,\ldots$

About 84% of the light energy is contained within the first dark ring and about 7% of light energy is contained in the annular region between the first two dark rings, etc., the first two dark rings occurring at

$$v = 3.832 \quad \text{and} \quad 7.016$$

2.8 Two-Beam Interference

Whenever two waves superpose, one obtains what is known as the interference pattern. In this section, we will consider the interference pattern produced by waves emanating from two point sources. As is well known, a stationary interference pattern is observed when the two interfering waves maintain a constant phase difference. For light waves, due to the very process of emission, one cannot observe a stationary interference pattern between the waves emanating from two independent sources, although interference does take place. Thus one tries to derive the interfering waves from a single wave so that a definite phase relationship is maintained all through.

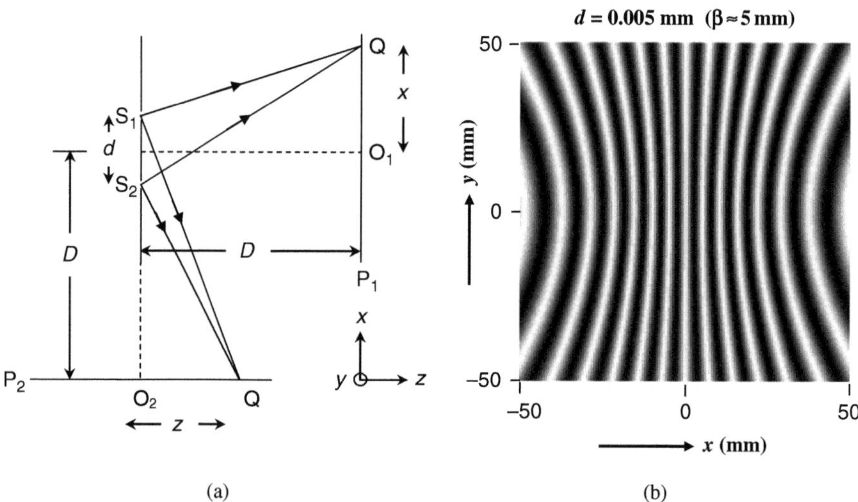

Fig. 2.7 (a) Waves emanating from two point sources interfere to produce interference fringes shown in Fig. 2.7 (b)

Let S_1 and S_2 represent two coherent point sources emitting waves of wavelength λ (see Fig. 2.7a). We wish to determine the interference pattern on the photographic plate P_1; the interference pattern on the photographic plate P_2 is discussed in Problem 2.11. The intensity distribution is given by

$$I = 4I_0 \cos^2 \delta/2 \tag{2.67}$$

where I_0 is the intensity produced by either of the waves independently and

$$\delta = \frac{2\pi}{\lambda} \Delta \tag{2.68}$$

where

$$\Delta = S_1 Q - S_2 Q \tag{2.69}$$

represents the path difference between the two interfering waves. Thus, when

$$\delta = 2n\pi \Rightarrow \Delta = S_1Q - S_2Q = n\lambda, \ n = 0, 1, 2, \ldots .. \text{ (Bright Fringe)} \qquad (2.70)$$

we will have a bright fringe, and when

$$\delta = (2n + 1)\pi \Rightarrow \Delta = S_1Q - S_2Q = \left(n + \frac{1}{2}\right)\lambda, \ n = 0, 1, 2, \ldots .. \text{ (Dark Fringe)}$$
$$(2.71)$$

we will have a dark fringe. Using simple geometry one can show that the locus of the points (on the plane P_1) such that $S_1Q \sim S_2Q = \Delta$ is a hyperbola, given by

$$(d^2 - \Delta^2)x^2 - \Delta^2 y^2 = \Delta^2 \left[D^2 + \frac{1}{4}\left(d^2 - \Delta^2\right) \right] \qquad (2.72)$$

Now,

$$\Delta = 0 \Rightarrow x = 0$$

which represents the central bright fringe. Equation (2.72) can be written in the form (see, e.g., Ghatak (2009))

$$x = \sqrt{\frac{\Delta^2}{d^2 - \Delta^2}\left[y^2 + D^2 + \frac{1}{4}\left(d^2 - \Delta^2\right) \right]^{1/2}} \qquad (2.73)$$

For values of y such that

$$y^2 \ll D^2 \qquad (2.74)$$

the loci are straight lines parallel to the y-axis and one obtains straight line fringes as shown in Fig.2.7b. The corresponding fringe width would be

$$\beta = \frac{\lambda D}{d} \qquad (2.75)$$

Thus for $D = 50$ cm, $d = 0.05$ cm, and $\lambda = 6000$ Å, we get $\beta = 0.06$ cm.

2.9 Multiple Reflections from a Plane Parallel Film

We next consider the incidence of a plane wave on a plate of thickness h (and of refractive index n_2) surrounded by a medium of refractive index n_1 as shown in Fig. 2.8; [the Fabry–Perot interferometer consists of two partially reflecting mirrors (separated by a fixed distance h) placed in air so that $n_1 = n_2 = 1$].

Let A_0 be the (complex) amplitude of the incident wave. The wave will undergo multiple reflections at the two interfaces as shown in Fig. 2.8a. Let r_1 and t_1 represent the amplitude reflection and transmission coefficients when the wave is incident from n_1 toward n_2 and let r_2 and t_2 represent the corresponding coefficients when the wave is incident from n_2 toward n_1. Thus the amplitude of the successive reflected waves will be

$$A_0 r_1, \ A_0 t_1 r_2 t_2 e^{i\delta}, \ A_0 t_1 r_2^3 e^{2i\delta}, \ldots ..$$

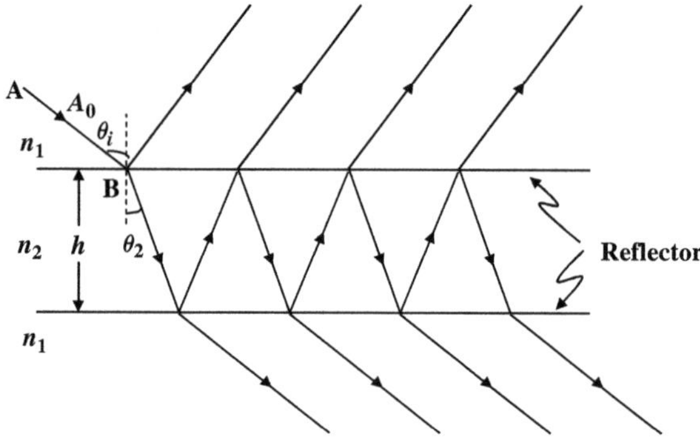

Fig. 2.8 Reflection and transmission of a beam of amplitude A_0 incident at a angle θ_i on a film of refractive index n_2 and thickness h

where

$$\delta = \frac{2\pi}{\lambda_0} \Delta = \frac{4\pi\, n_2\, h\, \cos\theta_2}{\lambda_0} \tag{2.76}$$

represents the phase difference (between two successive waves emanating from the plate) due to the additional path traversed by the beam in the film, and in Eq. (2.76), θ_2 is the angle of refraction *inside* the film (of refractive index n_2), h the film thickness, and λ_0 is the free-space wavelength. Thus the resultant (complex) amplitude of the reflected wave will be

$$A_r = A_0 \left[r_1 + t_1 t_2 r_2\, e^{i\delta} \left(1 + r_2^2\, e^{i\delta} + r_2^4\, e^{2i\delta} + \ldots. \right) \right]$$

$$= A_0 \left[r_1 + \frac{t_1 t_2 r_2\, e^{i\delta}}{1 - r_2^2\, e^{i\delta}} \right] \tag{2.77}$$

Now, if the reflectors are lossless, the reflectivity and the transmittivity at each interface are given by

$$R = r_1^2 = r_2^2$$

$$\tau = t_1 t_2 = 1 - R$$

[We are reserving the symbol T for the transmittivity of the Fabry–Perot etalon]. Thus

$$\frac{A_r}{A_0} = r_1 \left[1 - \frac{(1-R)\, e^{i\delta}}{1 - R\, e^{i\delta}} \right] \tag{2.78}$$

where we have used the fact that $r_2 = -r_1$. Thus the reflectivity of the Fabry–Perot etalon is given by

$$P = \left|\frac{A_r}{A_0}\right|^2 = R \cdot \left|\frac{1 - e^{i\delta}}{1 - R\,e^{i\delta}}\right|^2$$

$$= R \,\frac{(1 - \cos\delta)^2 + \sin^2\delta}{(1 - R\cos\delta)^2 + R^2\sin^2\delta}$$

$$= \frac{4R\,\sin^2\frac{\delta}{2}}{(1-R)^2 + 4R\,\sin^2\frac{\delta}{2}}$$

or

$$P = \frac{F\,\sin^2\frac{\delta}{2}}{1 + F\,\sin^2\frac{\delta}{2}} \tag{2.79}$$

where

$$F = \frac{4R}{(1-R)^2} \tag{2.80}$$

is called the coefficient of Finesse. One can immediately see that when $R \ll 1$, F is small and the reflectivity is proportional to $\sin^2\,\delta/2$. The same intensity distribution is obtained in the two-beam interference pattern; we may mention here that we have obtained $\sin^2\,\delta/2$ instead of $\cos^2\,\delta/2$ because of the additional phase change of π in one of the reflected beams.

Similarly, the amplitude of the successive transmitted waves will be

$$A_0\,t_1\,t_2\,,\ \ A_0\,t_1\,t_2\,r_2^2\,e^{i\delta},\ \ A_0\,t_1\,t_2\,r_2^4\,e^{2i\delta},\ \ \ldots$$

where, without any loss of generality, we have assumed the first transmitted wave to have zero phase. Thus the resultant amplitude of the transmitted wave will be given by

$$A_t = A_0\,t_1\,t_2\left[1 + r_2^2\,e^{i\delta} + r_2^4\,e^{2i\delta} + \ldots\right]$$

$$= A_0\,\frac{t_1\,t_2}{1 - r_2^2\,e^{i\delta}} = A_0\,\frac{1 - R}{1 - R\,e^{i\delta}}$$

Thus the transmittivity T of the film is given by

$$T = \left|\frac{A_t}{A_0}\right|^2 = \frac{(1 - R)^2}{(1 - R\cos\delta)^2 + R^2\sin^2\delta}$$

or

$$T = \frac{1}{1 + F\,\sin^2\frac{\delta}{2}} \tag{2.81}$$

It is immediately seen that the reflectivity and the transmittivity of the Fabry–Perot etalon add up to unity. Further,

$$T = 1$$

when

$$\delta = 2\,m\,\pi \quad , \quad m = 1, 2, 3, \ldots . \tag{2.82}$$

In Fig. 2.9 we have plotted the transmittivity as a function of δ for different values of F. In order to get an estimate of the width of the transmission resources, let

$$T = \frac{1}{2} \quad \text{for} \quad \delta = 2\,m\,\pi \,\pm\, \frac{\Delta \delta}{2}$$

Thus

$$F \sin^2 \frac{\Delta \delta}{4} = 1 \tag{2.83}$$

The quantity $\Delta \delta$ represents the FWHM (full width at half maximum). In almost all cases, $\Delta \delta <<< 1$ and therefore, to a very good approximation, it is given by

$$\Delta \delta \approx \frac{4}{\sqrt{F}} = \frac{2\,(1 - R)}{\sqrt{R}} \tag{2.84}$$

Thus the transmission resonances become sharper as the value of F increases (see Fig. 2.9).

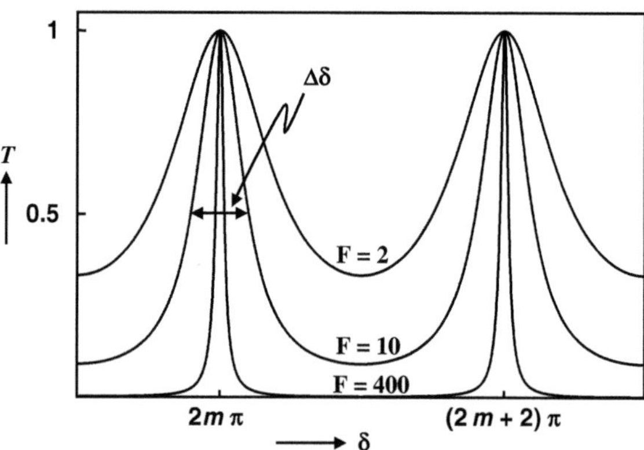

Fig. 2.9 The transmittivity of a Fabry–Perot etalon as a function of d for different values of F; the value of m is usually large. The transmission resonances become sharper as we increase the value of F. The FWHM (Full Width at Half Maximum) is denoted by $\Delta \delta$

2.10 Modes of the Fabry–Perot Cavity

We consider a polychromatic beam incident normally ($\theta_2 = 0$) on a Fabry–Perot cavity with air between the reflecting plates ($n_2 = 1$) – see Fig. 2.8. Equations (2.76) and (2.82) tell us that transmission resonance will occur whenever the incident frequency satisfies the following equation:

$$\nu = \nu_m = m \frac{c}{2h} \tag{2.85}$$

where m is an integer. The above equation represents the different (longitudinal) modes of the (Fabry–Perot) cavity. For $h = 10$ cm, the frequency spacing of two adjacent modes would be given by

$$\delta\nu = \frac{c}{2h} = 1500 \text{ MHz}$$

For an incident beam having a central frequency of

$$\nu = \nu_0 = 6 \times 10^{14} \text{ Hz}$$

and a spectral width[2] of 7000 MHz the output beam will have frequencies

$$\nu_0, \; \nu_0 \pm \delta\nu \text{ and } \nu_0 \pm 2\,\delta\nu$$

as shown in Fig. 2.10. One can readily calculate that the five lines correspond to

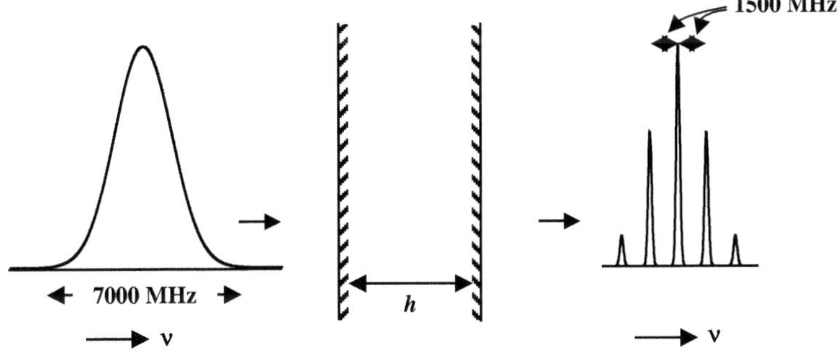

Fig. 2.10 A beam having a spectral width of about 7000 MHz (around $\nu_0 = 6 \times 10^{14}$ Hz) is incident normally on a Fabry–Perot etalon with $h = 10$ cm and $n_2 = 1$. The output has five narrow spectral lines

---

[2]For $\nu_0 = 6 \times 10^{14}$ Hz, $\lambda_0 = 5000$ Å and a spectral width of 7000 MHz would imply $\left| \frac{\Delta\lambda_0}{\lambda_0} \right| = \frac{\Delta\nu}{\nu_0} = \frac{7 \times 10^9}{6 \times 10^{14}} \approx 1.2 \times 10^{-5}$ giving $\Delta\lambda_0 \approx 0.06$ Å. Thus a frequency spectral width of 7000 MHz (around $\nu_0 = 6 \times 10^{14}$ Hz) implies a wavelength spread of only 0.06 Å.

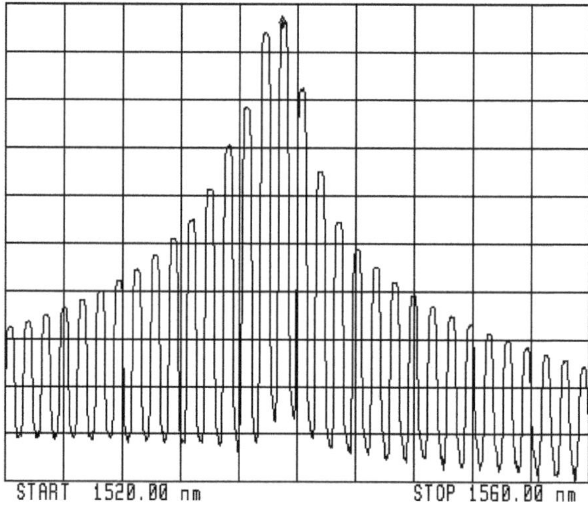

Fig. 2.11 Typical output spectrum of a Fabry–Perot multi longitudinal mode (MLM) laser diode; the wavelength spacing between two modes is about 1.25 nm

$$m = 399998, 399999, 400000, 400001, \text{ and } 400002$$

Figure 2.11 shows a typical output of a multilongitudinal (MLM) laser diode.

Problems

Problem 2.1 The electric field components of a plane electromagnetic wave are

$$E_x = -3E_0 \sin(\omega t - kz); \quad E_y = E_0 \sin(\omega t - kz)$$

Plot the resultant field at various values of time and show that it describes a linearly polarized wave.

Solution The beam will be linearly polarized

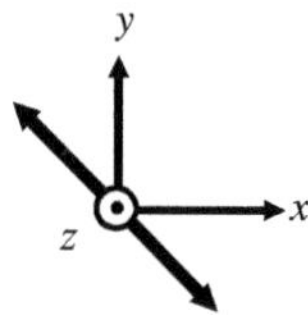

Problem 2.2 The electric field components of a plane electromagnetic wave are

$$E_x = E_0 \sin(\omega t + kz); \quad E_y = E_0 \cos(\omega t + kz)$$

Show that it describes a left circularly polarized wave.

Solution Propagation along the $+z$-direction (coming out of the page). At $z = 0$

$$E_x = E_0 \sin \omega t; \quad E_y = E_0 \cos \omega t$$

$$\Rightarrow E_x^2 + E_y^2 = E_0^2 \quad \Rightarrow \quad \text{Circularly polarized}$$

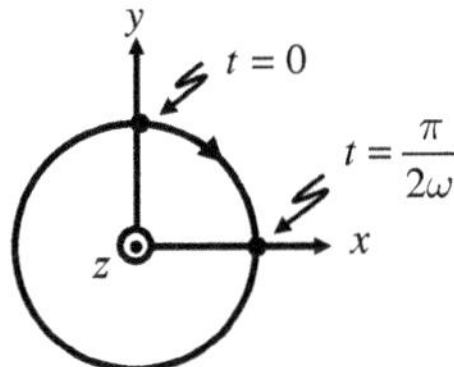

Since propagation is along the $+z$-axis, i.e., coming out of the page, we have an LCP wave.

Problem 2.3 The electric field components of a plane electromagnetic wave are

$$E_x = -2E_0 \cos(\omega t + kz); \quad E_y = E_0 \sin(\omega t + kz)$$

Show that it describes a right elliptically polarized wave.

Problem 2.4 In Fig. 2.3 if we replace the quarter wave plate by a (calcite) half-wave plate, what will be the state of polarization of the output beam?

Problem 2.5 For calcite, at $\lambda_0 = 5893$ Å (at 18°C) $n_0 = 1.65836$, $n_e = 1.48641$. The thickness of the corresponding QWP is 0.000857 mm (see Section 2.4). If in Fig. 2.3 the wavelength of the incident linearly polarized beam is changed to 6328 Å determine the state of polarization of the output beam.

Problem 2.6 A left circularly polarized beam is incident on a calcite half-wave plate. Show that the emergent beam will be right circularly polarized.

Problem 2.7 A 3 mW laser beam ($\lambda_0 \approx 6328$ Å) is incident on the eye. On the retina, it forms a circular spot of radius of about 20 μm. Calculate approximately the intensity on the retina.

Solution Area of the focused spot $A = \pi \left(20 \times 10^{-6}\right)^2 \approx 1.3 \times 10^{-9}$ m^2. On the retina, the intensity will be approximately given by

$$I \approx \frac{P}{A} \approx \frac{3 \times 10^{-3}\,\text{W}}{1.3 \times 10^{-9}\,\text{m}^2} \approx 2.3 \times 10^6\,\text{W/m}^2$$

Problem 2.8 Consider a Gaussian beam propagating along the z-direction whose phase front is plane at $z = 0$ [see Eq. (2.50)]. The spot size of the beam at $z = 0$, w_0 is 0.3 mm. Calculate (a) the spot size and (b) the radius of curvature of the phase front at $z = 60$ cm. Assume $\lambda_0 \approx 6328$ Å.

[Ans : (a) $w(z = 60\,\text{cm}) \approx 0.84$ mm (b) $R(z = 60\,\text{cm}) \approx 93.3$ cm].

Problem 2.9 In continuation of the previous problem, show that for a simple resonator consisting of a plane mirror and a spherical mirror (of radius of curvature 93.3 cm) separated by 60 cm, the spot size of the beam at the plane mirror would be 0.3 mm.

Problem 2.10 Consider a He–Ne laser beam (with $\lambda_0 \approx 6328$ Å) incident on a circular aperture of radius 0.02 cm. Calculate the radii of the first two dark rings of the Airy pattern produced at the focal plane of a convex lens of focal length 20 cm.

Solution The radius of the first dark ring would be [see Eq. (2.66)]

$$r_1 \approx \frac{3.832 \times 6.328 \times 10^{-5} \times 20}{2\pi \times 0.02} \approx 0.039 \text{ cm}$$

Similarly, the radius of the second dark ring is

$$r_2 \approx \frac{7.016 \times 6.328 \times 10^{-5} \times 20}{2\pi \times 0.02} \approx 0.071 \text{ cm}$$

Problem 2.11 Consider two coherent point sources S_1 and S_2 emitting waves of wavelength λ (see Fig. 2.7a). Show that the interference pattern on a plane normal to the line joining S_1 and S_2 will consist of concentric circular fringes.

Problem 2.12 Consider a light beam of all frequencies lying between $\nu = \nu_0 = 5.0 \times 10^{14}$ Hz to $\nu = 5.00002 \times 10^{14}$ Hz incident normally on a Fabry–Perot interferometer (see Fig. 2.10) with $R = 0.95$, $n_0 = 1$, and $d = 25$ cm. Calculate the frequencies (in the above frequency range) and the corresponding mode number which will correspond to transmission resonances.

Solution Transmission resonances occur at
$$\nu = \nu_m = m\frac{c}{2d} = m\frac{3\times10^{10}}{2\times25} = (6 \times 10^8 \, m) \text{ Hz}$$
For $\nu = \nu_0 = 5 \times 10^{14}$ Hz; $m = \frac{5\times10^{14}}{6\times10^8} = 833333.3$
Since m is not an integer the frequency ν_0 does not correspond to a mode.

$$\text{For } m = 833334, \quad \nu = 5.000004 \times 10^{14} \text{ Hz} = \nu_0 + 400 \text{ MHz}$$

$$\text{For } m = 833335, \quad \nu = 5.000010 \times 10^{14} \text{ Hz} = \nu_0 + 1000 \text{ MHz}$$

$$\text{For } m = 833336, \quad \nu = 5.000016 \times 10^{14} \text{ Hz} = \nu_0 + 1600 \text{ MHz}$$

Finally, for $m = 833337$, $\nu = 5.000022 \times 10^{14}$ Hz $= \nu_0 + 2200$ MHz which is beyond the given range.

Chapter 3
Elements of Quantum Mechanics

3.1 Introduction

In this chapter we discuss very briefly the basic principles of quantum mechanics which are used in later chapters. At places, the chapter will appear a bit disconnected; this is inevitable because the subject of quantum mechanics is so vast that it is impossible to present the basic concepts in a coherent fashion in one tiny chapter! Nevertheless, whatever we discuss we will try to do from first principles.

We first give a heuristic derivation of the Schrödinger equation which is followed by its solutions corresponding to some important potential energy functions. We have solved the particle in a box problem and also the harmonic oscillator problem. For the hydrogen atom problem, we just present the results. We have also discussed the physical interpretation of the wave function and the uncertainty principle. Several other "solvable" problems are briefly discussed at the end of the chapter.

3.2 The One-Dimensional Schrödinger Equation

There are many experimental results which show that atomic objects (like electrons, protons, neutrons, α particles) exhibit both wave and particle properties. Indeed the wavelength λ is related to the momentum p through the de Broglie relation

$$\lambda = \frac{h}{p} \tag{3.1}$$

where $h \left(\approx 6.627 \times 10^{-34} \text{J s} \right)$ represents Planck's constant. Thus, we may write

$$p = \hbar k \tag{3.2}$$

where $\hbar = h/2\pi$ and

$$k = \frac{2\pi}{\lambda} \tag{3.3}$$

K. Thyagarajan, A. Ghatak, *Lasers*, Graduate Texts in Physics,
DOI 10.1007/978-1-4419-6442-7_3, © Springer Science+Business Media, LLC 2010

represents the wavenumber. Further, as established by Einstein's explanation of the photoelectric effect, the energy E of the particle is related to the frequency ω by the following equation:

$$E = \hbar\omega \tag{3.4}$$

The simplest type of a wave is a one-dimensional plane wave described by the wave function

$$\Psi(x,t) = A \ \exp\left[i\left(kx - \omega t\right)\right] \tag{3.5}$$

where A is the amplitude of the wave and the propagation is assumed to be in the $+x$ direction. If we now use Eqs. (3.2) and (3.4), we would obtain

$$\Psi = \exp\left[\frac{i}{\hbar}\left(px - Et\right)\right] \tag{3.6}$$

Elementary differentiation will give us

$$i\hbar\frac{\partial\Psi}{\partial t} = E\Psi \tag{3.7}$$

$$-i\hbar\frac{\partial\Psi}{\partial x} = p\Psi \tag{3.8}$$

which suggests that, at least for a free particle, the energy and momentum can be represented by differential operators given by

$$E \rightarrow i\hbar\frac{\partial}{\partial t}, \quad p \rightarrow -i\hbar\frac{\partial}{\partial x} \tag{3.9}$$

Further, if we again differentiate Eq. (3.8), we would obtain

$$-\frac{\hbar^2}{2m}\frac{\partial^2\Psi}{\partial x^2} = \frac{p^2}{2m}\Psi \tag{3.10}$$

For a free particle, the energy and momentum are related by the equation

$$E = \frac{p^2}{2m} \tag{3.11}$$

Thus the right-hand sides of Eqs. (3.7) and (3.10) are equal and we obtain

$$i\hbar\frac{\partial\Psi}{\partial t} = -\frac{\hbar^2}{2m}\frac{\partial^2\Psi}{\partial x^2} \tag{3.12}$$

which is the one-dimensional Schrödinger equation for a free particle. If we use the operator representations of E and p [see Eq. (3.9)], we may write the above equation as

$$E\Psi = \frac{p^2}{2m}\Psi \tag{3.13}$$

We next consider the particle to be in a force field characterized by the potential energy $V(x)$; thus, classically, the total energy of the system is given by

$$E = \frac{p^2}{2m} + V(x) \tag{3.14}$$

If we now assume p and E to be represented by the differential operators, the equation

$$E\Psi = \left[\frac{p^2}{2m} + V(x)\right]\Psi \tag{3.15}$$

would assume the form

$$i\hbar\frac{\partial\Psi}{\partial t} = \left[-\frac{\hbar^2}{2m}\frac{\partial^2}{\partial x^2} + V(x)\right]\Psi \tag{3.16}$$

which represents the one-dimensional time-dependent Schrödinger equation. The above equation can be written in the form

$$i\hbar\frac{\partial\Psi}{\partial t} = H\Psi \tag{3.17}$$

where

$$H = \frac{p^2}{2m} + V(x) = -\frac{\hbar^2}{2m}\frac{\partial^2}{\partial x^2} + V(x) \tag{3.18}$$

is an operator and represents the Hamiltonian of the system. Equations (3.16) and (3.17) represent the one-dimensional time-dependent Schrödinger equation. The above is a very heuristic derivation of the Schrödinger equation and lacks rigor. Strictly speaking Schrödinger equation cannot be derived. To quote Richard Feynman

> Where did we get that equation from ? Nowhere. It is not possible to derive it from anything you know. It came out of the mind of Schrödinger.

Although we have obtained Eq. (3.16) starting from an expression for a plane wave, the Schrödinger equation as described by Eq. (3.16) is more general in the sense that $\psi(\mathbf{r}, t)$ called the wave function contains all information that is knowable about the system. As will be discussed in Section 3.4, $\psi^*(\mathbf{r}, t)\psi(\mathbf{r}, t)\mathrm{d}\tau$ represents the probability of finding the particle in a volume element $\mathrm{d}\tau$. Note also that observables such as momentum energy are represented by operators [see Eq. (3.9)].

When the Hamiltonian, H, is independent of time[1], Eq. (3.16) can be solved by using the method of separation of variables:

$$\Psi(x, t) = \psi(x) T(t) \tag{3.19}$$

Substituting in Eq. (3.16) and dividing by Ψ, we obtain

$$i\hbar \frac{1}{T(t)} \frac{dT}{dt} = \frac{1}{\psi}\left[-\frac{\hbar^2}{2m}\frac{d^2\psi}{dx^2} + V(x)\psi \right] = E \tag{3.20}$$

where E is a constant (and now a number). Thus

$$\frac{dT}{dt} + \frac{i}{\hbar} E T(t) = 0$$

giving

$$T(t) \sim \exp\left(-\frac{i}{\hbar} Et \right) \tag{3.21}$$

Further, Eq. (3.20) gives us

$$-\frac{\hbar^2}{2m}\frac{d^2\psi}{dx^2} + V\psi = E\psi \tag{3.22}$$

or

$$H\psi = E\psi \tag{3.23}$$

which is essentially an eigenvalue equation. For ψ to be "well behaved," the quantity E takes some particular values (see, e.g., Examples 3.1 and 3.2), these are known as the energy eigenvalues and the corresponding forms of ψ are known as eigenfunctions; by "well-behaved" we imply functions which are single valued and square-integrable (i.e., $\int |\psi|^2 d\tau$ should exist).

In Section 3.4 we will interpret the wave function ψ as the probability amplitude; therefore ψ should be single valued and $|\psi(x)|^2 dx$ has to be finite for finite values of dx. Thus

$$\lim_{dx \to 0} |\psi|^2 \, dx = 0$$

In practice this is satisfied by demanding that ψ be finite everywhere. We also have the following theorems:

---

[1] Whenever we are considering bound states of a system (like those of the hydrogen atom or that of the harmonic oscillator) the Hamiltonian is independent of time; however, for problems such as the interaction of an atom with radiation field, the Hamiltonian is not independent of time (see, e.g., Section 4.7).

Theorem 1 *The derivative of the wave function $d\psi/dx$ is always continuous as long as the potential energy $V(x)$ is finite, whether or not it is continuous[2].*

Proof We integrate the Schrödinger equation [Eq. (3.22)] from $x-\varepsilon$ to $x+\varepsilon$ to obtain

$$\int\limits_{x-\varepsilon}^{x+\varepsilon} \frac{d^2\psi}{dx^2} dx = -\frac{2m}{\hbar^2} \int\limits_{x-\varepsilon}^{x+\varepsilon} [E - V(x)]\psi(x)dx$$

or

$$\psi'(x+\varepsilon) - \psi'(x-\varepsilon) = -\frac{2m}{\hbar^2} \int\limits_{x-\varepsilon}^{x+\varepsilon} [E - V(x)]\psi(x)dx$$

Since $V(x)$ is assumed to be finite (it could, however, be discontinuous), the RHS tends to zero as $\varepsilon \to 0$. Thus ψ' is continuous at any value of x. It is obvious that ψ has to be necessarily continuous everywhere. Alternatively one may argue that if $d\psi/dx$ is discontinuous then $d^2\psi/dx^2$ must become infinite; this will be inconsistent with Eq. (3.22) as long as $V(x)$ does not become infinite.

Theorem 2 *If the potential energy function $V(x)$ is infinite anywhere, the proper boundary condition is obtained by assuming $V(x)$ to be finite at that point and carrying out a limiting process making $V(x)$ tend to infinity. Such a limiting process makes the wave function vanish at a point where $V(x)=\infty$.*

Example 3.1 **Particle in a one-dimensional infinitely deep potential well**

We will determine the energy levels and the corresponding eigenfunctions of a particle of mass μ in a one-dimensional infinitely deep potential well characterized by the following potential energy variation (see Fig. 3.1):

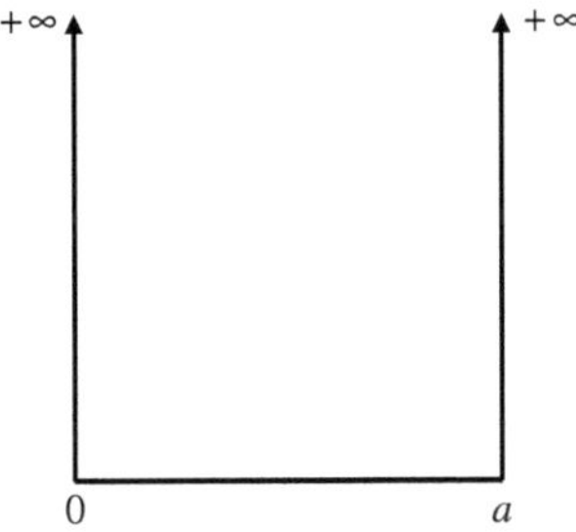

Fig. 3.1 Particle in a one-dimensional box

---

[2]It may be mentioned that in many texts the continuity of ψ and $d\psi/dx$ are taken to be axioms. This is not correct because it follows from the fact that $\psi(x)$ satisfies a second-order differential equation [see Eq. (3.22)]. Indeed, when $V(x)$ becomes infinite, $d\psi/dx$ is *not* continuous.

$$V(x) = 0 \quad \text{for} \quad 0 < x < a$$
$$= \infty \quad \text{for} \quad x < 0 \quad \text{and for } x > a \tag{3.24}$$

For $0 < x < a$, the one-dimensional Schrödinger equation becomes

$$\frac{d^2\psi}{dx^2} + k^2\psi(x) = 0 \tag{3.25}$$

where

$$k^2 = \frac{2\mu E}{\hbar^2} \tag{3.26}$$

The general solution of Eq. (3.25) is

$$\psi(x) = A\sin kx + B\cos kx \tag{3.27}$$

Since the boundary condition at a surface at which there is an infinite potential step is that ψ is zero (see Theorem 2), we must have

$$\psi(x = 0) = \psi(x = a) = 0 \tag{3.28}$$

Using the boundary condition given by the above equation, we get

$$\psi(x = 0) = B = 0 \tag{3.29}$$

and

$$\psi(x = a) = A\sin ka = 0$$

Thus, either $A = 0$ or

$$ka = n\pi, \quad n = 1, 2, \ldots \tag{3.30}$$

The condition $A = 0$ leads to the trivial solution of ψ vanishing everywhere; the same is the case for $n=0$. If we now use Eq. (3.26), the allowed energy values are therefore given by

$$E_n = \frac{n^2\pi^2\hbar^2}{2\mu a^2}, \quad n = 1, 2, 3, \ldots \tag{3.31}$$

The corresponding eigenfunctions are

$$\left.\begin{array}{ll} \psi_n = \sqrt{\frac{2}{a}}\,\sin\left(\frac{n\pi}{a}x\right) & 0 < x < a \\ \quad = 0 & x < 0 \text{ and } x > a \end{array}\right\} \tag{3.32}$$

where the factor $\sqrt{2/a}$ is such that the wave functions form an orthonormal set

$$\int_0^a \psi_m^*(x)\psi_n(x)dx = \delta_{mn} \tag{3.33}$$

and

$$\delta_{kn} = 1 \text{ if } k = n$$
$$= 0 \text{ if } k \neq n \tag{3.34}$$

is known as the Kronecker delta function. It may be noted that whereas $\psi_n(x)$ is continuous everywhere, $d\psi_n(x)/dx$ is discontinuous at $x = 0$ and at $x=a$. This is because of $V(x)$ becoming infinite at $x = 0$ and at $x=a$ (see Theorem 1). Figure 3.2 gives a plot of the first three eigenfunctions and one can see that the eigenfunctions are either symmetric or antisymmetric about the line $x = a/2$; this follows from the fact that $V(x)$ is symmetric about $x = a/2$ (see Problem 3.2).

Fig. 3.2 The energy eigenvalues and eigenfunctions for a particle in an infinitely deep potential well. Notice that the eigenfunctions are either symmetric or antisymmetric about $x = a/2$

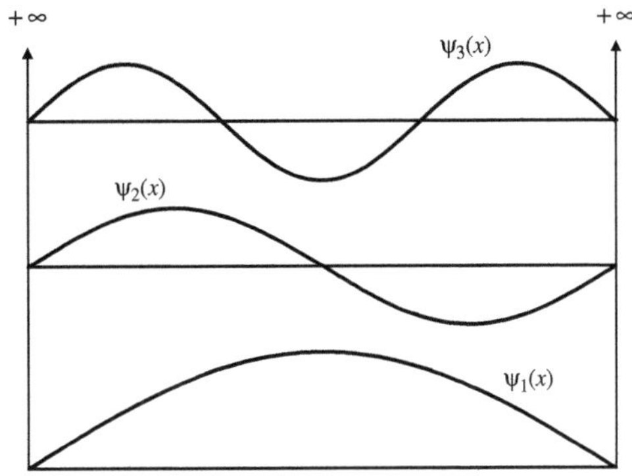

The following points are also to be noted

(i) E cannot be negative because if we assume E to be negative then the boundary conditions at $x=0$ and $x=a$ cannot be simultaneously satisfied.

(ii)   The eigenvalues form a discrete set.

(iii) The eigenfunctions given by Eq. (3.32) form a complete set, i.e., an arbitrary (well-behaved) function $f(x)$ (in the domain $0 < x < a$) can be expanded in terms of the eigenfunctions of H:

$$f(x) = \sum_n c_n \psi_n(x) = \sqrt{\frac{2}{a}} \sum_{n=1,2,...}^{\infty} c_n \sin\left(\frac{n\pi}{a}x\right) \tag{3.35}$$

where c_n are constants which can be determined by multiplying both sides of the above equation by $\psi_m^*(x)$ and integrating from 0 to a to obtain

$$\int_0^a \psi_m^*(x)f(x)\mathrm{d}x = \sum_n c_n \int_0^a \psi_m^*(x)\psi_n(x)\mathrm{d}x = \sum_n c_n \delta_{mn} = c_m \tag{3.36}$$

where we have used the orthonormality condition given by Eq. (3.33).

(iv)   The most general solution of the time-dependent Schrödinger equation

$$i\hbar \frac{\partial \Psi}{\partial t} = H\Psi = -\frac{\hbar^2}{2\mu} \frac{\partial^2 \Psi}{\partial x^2} + V(x)\Psi(x,t) \tag{3.37}$$

with $V(x)$ given by Eq. (3.24) will be

$$\Psi(x,t) = \sum_{n=1,2,...}^{\infty} c_n \psi_n(x) e^{-iE_n t/\hbar} = \sum_{n=1,2,...}^{\infty} c_n \psi_n(x) e^{-in^2\tau} \tag{3.38}$$

where

$$\tau = \frac{t}{t_0}; \quad t_0 = \frac{2\mu L^2}{\pi^2 \hbar} \tag{3.39}$$

Substituting for $\psi_n(x)$ and E_n we get

$$\Psi(x,t) = \sum_{n=1,2,...}^{\infty} c_n \left(\sqrt{\frac{2}{a}} \sin\frac{n\pi x}{a}\right) \exp\left[-i\frac{n^2\pi^2\hbar}{2\mu a^2}t\right] \tag{3.40}$$

Since

$$\Psi(x, 0) = \sum_n c_n \psi_n(x) \tag{3.41}$$

the coefficients c_n can be determined from the initial form of the wave function:

$$c_n = \int_0^a \psi_n^*(x)\Psi(x, 0)dx \tag{3.42}$$

Thus, the recipe for determining $\Psi(x, t)$ is as follows:

If we know $\Psi(x, 0)$ we can determine c_n from Eq. (3.42), we substitute these values in Eq. (3.38), and sum the series to obtain $\Psi(x, t)$.

(v)   We assume $\Psi(x, 0)$ to be normalized:

$$\int_0^a |\Psi(x, 0)|^2 \, dx = 1 \tag{3.43}$$

This would imply

$$1 = \int_0^a \sum_n c_n^* \psi_n^*(x) \sum_m c_m \psi_m(x)dx = \sum_n \sum_n c_n^* c_m \int_0^a \psi_n^*(x)\psi_m(x) \, dx$$
$$= \sum_n \sum_m c_n^* c_m \delta_{mn} = \sum_n |c_n|^2 \tag{3.44}$$

where we have used the orthonormality condition given by Eq. (3.33). Further,

$$\int_0^a |\Psi(x, t)|^2 dx = \sum_n \sum_m c_n^* c_m e^{i(E_n - E_m)t/\hbar} \int_0^a \psi_n^*(x)\psi_m(x) \, dx$$
$$= \sum_n \sum_m c_n^* c_m e^{i(E_n - E_m)t/\hbar} \delta_{mn} = \sum_n |c_n|^2 = 1$$

Thus, if the wave function is normalized at $t=0$, then it will remain normalized at all times. Further, we can interpret Eq. (3.44) by saying that $|c_n|^2$ represents the probability of finding the system in the nth eigenstate which remains the same at all times. Thus there are no transitions. Indeed, whenever the potential energy function is time-independent, we obtain what are known as *stationary states* and there is no transition between states.

(vi)   As a simple example, let us assume that the particle is described by the following wave function (at $t=0$):

$$\Psi(x, 0) = \sqrt{\frac{1}{6}} \, \psi_1(x) + \frac{i}{\sqrt{2}} \, \psi_2(x) + \sqrt{\frac{1}{3}} \, \psi_4(x)$$

Notice that

$$\sum_n |c_n|^2 = 1$$

so that the wave function is normalized. Thus, if we carry out a measurement of energy, the probabilities of obtaining the values E_1, E_2, and E_4 would be 1/6, ½, and 1/3, respectively. How will such a state evolve with time? Well, we just multiply each term by the appropriate time-dependent factor to obtain $\Psi(x, t)$ [see Eq. (3.38)].

$$\Psi(x,t) = \sqrt{\frac{1}{6}}\,\psi_1(x)e^{-it/t_0} + \frac{i}{\sqrt{2}}\psi_2(x)e^{-4it/t_0} + \sqrt{\frac{1}{3}}\,\psi_3(x)e^{-16it/t_0} \tag{3.45}$$

where t_0 is given by Eq. (3.39). Obviously

$$\int_0^a |\Psi(x,t)|^2 dx = 1 \tag{3.46}$$

for all values of t. The quantity

$$P(x,t) = |\Psi(x,t)|^2 \tag{3.47}$$

would represent the time evolution of the probability distribution function. However, at all values of time, the probability of finding the system in a particular state remains the same. Further, the average value of the energy is given by

$$\begin{aligned}
\langle E \rangle &= \frac{1}{6} E_1 + \frac{1}{2} E_2 + \frac{1}{3} E_4 \\
&= \left[\frac{1}{6} + 2 + \frac{16}{3}\right] \frac{\pi^2 \hbar^2}{2\mu a^2} = \frac{15}{2} \frac{\pi^2 \hbar^2}{2\mu a^2}
\end{aligned} \tag{3.48}$$

Thus, if one carries out a large number of measurements (of energy) on identically prepared systems characterized by the same wave function as given by Eq. (3.45), then the average value of the energy would be given by Eq. (3.48).

(vii)   What happens to the wave function if $E \neq E_n$, i.e., if E is not one of the eigenvalues? For such a case the boundary conditions cannot be satisfied and therefore it cannot be an allowed value of energy. For example, if

$$E = \frac{0.81\pi^2 \hbar^2}{2\mu a^2}$$

then the wave function cannot be zero for both $x=0$ and $x=a$.

Example 3.2 **The linear harmonic oscillator**

We next consider the linear harmonic oscillator problem where the potential energy function is given by

$$V(x) = \frac{1}{2} \mu\, \omega^2 x^2 \tag{3.49}$$

and the Schrödinger equation [Eq. (3.22)] can be written in the form

$$\frac{d^2\psi}{d\xi^2} + \left[\Lambda - \xi^2\right]\psi = 0 \tag{3.50}$$

where $\xi = \alpha x$ and we have chosen

$$\alpha = \sqrt{\frac{\mu\omega}{\hbar}} \tag{3.51}$$

so that

$$\Lambda = \frac{2E}{\hbar\omega}$$

For the wave function not to blow up at $x = \pm\infty$ (which represents the boundary condition), Λ must be equal to an odd integer (see Appendix A), i.e.,

$$\Lambda = \frac{2E}{\hbar\omega} = (2m + 1); \quad m = 0, 1, 2, 3, \dots. \tag{3.52}$$

The above equation would give us the following expression for the discrete energy eigenvalues:

$$E = E_m = \left(m + \frac{1}{2}\right)\hbar\omega, \ m = 0, 1, 2, 3, \dots. \tag{3.53}$$

The corresponding normalized wave functions are the Hermite–Gauss functions (see Appendix A):

$$\psi_m(\xi) = N_m H_m(\xi)\exp\left(-\frac{1}{2}\xi^2\right), \ m = 0, 1, 2, 3, \dots. \tag{3.54}$$

where

$$N_m = \left(\frac{\alpha}{\pi^{1/2}2^m m!}\right)^{1/2} \tag{3.55}$$

represents the normalization constant. The first few Hermite polynomials are

$$H_0\left(\xi\right) = 1, \qquad H_1\left(\xi\right) = 2\xi$$

$$H_2\left(\xi\right) = 4\xi^2 - 2, \qquad H_3\left(\xi\right) = 8\xi^3 - 12\xi, \dots \tag{3.56}$$

The wave functions form a complete set of orthonormal functions:

$$\int\limits_{-\infty}^{+\infty} \psi_k^* \psi_n \mathrm{d}x = \delta_{kn} \tag{3.57}$$

The most general solution of the time-dependent Schrödinger equation [Eq. (3.16) with $V(x)$ given by Eq. (3.49)] will be

$$\begin{aligned}
\Psi(x, t) &= \sum_{n=0,1,2,\dots}^{\infty} c_n \psi_n(x) e^{-iE_n t/\hbar} \\
&= \sum_{n=0,1,2,\dots}^{\infty} c_n \psi_n(x) e^{-i\left(n+\frac{1}{2}\right)\omega t}
\end{aligned} \tag{3.58}$$

The values of c_n will be determined by the initial state of the oscillator.

3.3 The Three-Dimensional Schrödinger Equation

The three-dimensional generalization of the Schrödinger equation is quite straight-forward; instead of Eq. (3.29), we have

$$E \rightarrow i\hbar\frac{\partial}{\partial t}$$

$$p_x \rightarrow -i\hbar\frac{\partial}{\partial x}, \quad p_y \rightarrow -i\hbar\frac{\partial}{\partial y}, \quad p_z \rightarrow -i\hbar\frac{\partial}{\partial z} \tag{3.59}$$

Thus the equation [cf. Eq. (3.13)]

$$E\Psi = \left[\frac{1}{2m} \left(p_x^2 + p_y^2 + p_z^2 \right) + V(\mathbf{r}) \right] \Psi \tag{3.60}$$

assumes the form

$$i\hbar \frac{\partial \Psi}{\partial t} = H\Psi \tag{3.61}$$

where

$$H = \frac{p^2}{2m} + V(\mathbf{r}) = -\frac{\hbar^2}{2m}\nabla^2 + V(\mathbf{r}) \tag{3.62}$$

is an operator and represents the Hamiltonian of the system. Equations (3.61) and (3.62) represent the three-dimensional Schrödinger equation. Once again, when the Hamiltonian, H, is independent of time, Eq. (3.61) can be solved by using the method of separation of variables:

$$\Psi(\mathbf{r},t) = \psi(\mathbf{r})\, T(t) \tag{3.63}$$

Substituting in Eq. (3.61) and dividing by Ψ, we obtain

$$i\hbar \frac{1}{T(t)} \frac{dT}{dt} = \frac{1}{\psi}\left[-\frac{\hbar^2}{2m}\nabla^2\psi + V(\mathbf{r})\,\psi \right] = E \tag{3.64}$$

where E is a constant (and now a number). The solution of the time-dependent part is again given by Eq. (3.21). Equation (3.64) gives us

$$-\frac{\hbar^2}{2m}\nabla^2\psi + V\psi = E\psi \tag{3.65}$$

or

$$H\psi = E\psi \tag{3.66}$$

which, once again, is essentially an eigenvalue equation. The solution

$$\Psi_n(\mathbf{r},t) = \psi_n(\mathbf{r})\exp\left(-iE_n t / \hbar\right) \tag{3.67}$$

is said to describe a stationary state; here the subscript n refers to a particular eigenvalue E_n. Thus

$$H\psi_n = E_n\psi_n \tag{3.68}$$

3.4 Physical Interpretation of Ψ and Its Normalization

We rewrite the Schrödinger equation

$$i\hbar\frac{\partial\Psi\left(\mathbf{r},t\right)}{\partial t} = -\frac{\hbar^2}{2m}\nabla^2\Psi + V\left(\mathbf{r}\right)\Psi \tag{3.69}$$

along with its complex conjugate

$$-i\hbar\frac{\partial\Psi^*\left(\mathbf{r},t\right)}{\partial t} = -\frac{\hbar^2}{2m}\nabla^2\Psi^* + V\left(\mathbf{r}\right)\Psi^* \tag{3.70}$$

If we multiply Eq. (3.69) by Ψ^* and Eq. (3.70) by Ψ and subtract, we obtain

$$i\hbar\left(\Psi^*\frac{\partial\Psi}{\partial t} + \Psi\frac{\partial\Psi^*}{\partial t}\right) = -\frac{\hbar^2}{2m}\left(\Psi^*\nabla^2\Psi - \Psi\nabla^2\Psi^*\right) \tag{3.71}$$

Remembering that $\nabla^2 = \partial^2/\partial x^2 + \partial^2/\partial y^2 + \partial^2/\partial z^2$, we may rewrite the above equation in the form

$$\frac{\partial}{\partial t}\left(\Psi^*\Psi\right) + \frac{i\hbar}{2m}\left[\frac{\partial}{\partial x}\left(\Psi\frac{\partial\Psi^*}{\partial x} - \Psi^*\frac{\partial\Psi}{\partial x}\right)\right.$$
$$\left. + \frac{\partial}{\partial y}\left(\Psi\frac{\partial\Psi^*}{\partial y} - \Psi^*\frac{\partial\Psi}{\partial y}\right) + \frac{\partial}{\partial z}\left(\Psi\frac{\partial\Psi^*}{\partial z} - \Psi^*\frac{\partial\Psi}{\partial z}\right)\right] = 0$$

or

$$\frac{\partial\rho}{\partial t} + \nabla\cdot\mathbf{J} = 0 \tag{3.72}$$

where

$$\rho = \Psi^*\Psi \tag{3.73}$$

$$\nabla\cdot\mathbf{J} = \frac{\partial J_x}{\partial x} + \frac{\partial J_y}{\partial y} + \frac{\partial J_z}{\partial z} \tag{3.74}$$

$$J_x = \frac{i\hbar}{2m}\left(\Psi\frac{\partial\Psi^*}{\partial x} - \Psi^*\frac{\partial\Psi}{\partial x}\right) \tag{3.75}$$

and similar expressions for J_y and J_z. Equation (3.72) is the equation of continuity in fluid dynamics and can be physically interpreted by considering a moving gas with ρ representing the number of particles per unit volume and $\mathbf{J}$ representing the current density. Thus, if we normalize Ψ such that

$$\int_{-\infty}^{+\infty}\int_{-\infty}^{+\infty}\int_{-\infty}^{+\infty}\Psi^*\Psi\,\mathrm{d}\tau = 1 \tag{3.76}$$

For all states for which $\int \Psi^* \Psi d\tau$ exists, this normalization is always possible because if Ψ is a solution of Eq. (3.69) then any multiple of Ψ is also a solution and we may always choose the multiplicative constant such that Eq. (3.76) is satisfied. We may associate

$$\rho = \Psi^* \Psi \tag{3.77}$$

with *position probability density* and $\mathbf{J}$ with *probability current density*. This implies that $\Psi^* \Psi d\tau$ represents the probability of finding the particle in the volume element $d\tau$. Further, for an infinitely extended plane wave

$$\Psi = \exp\left[\frac{i}{\hbar}\left(\mathbf{p} \cdot \mathbf{r} - Et\right)\right] \tag{3.78}$$

the current density $\mathbf{J}$ can be easily calculated to give

$$\mathbf{J} = \frac{\mathbf{p}}{m} = \mathbf{v} \tag{3.79}$$

which is just the current for a beam of particles of unit density[3] ($\Psi^* \Psi = 1$) and velocity $\mathbf{v}$.

Example 3.3 **Particle in a three-dimensional box**

For a free particle of mass μ inside a cubical box of side L, the Schrödinger equation is given by

$$\nabla^2 \psi + \frac{2\mu E}{\hbar^2} \psi = 0 \quad \begin{cases} 0 < x < L \\ 0 < y < L \\ 0 < z < L \end{cases} \tag{3.80}$$

with the boundary condition that ψ should vanish everywhere on the surface of the cube. We use the method of separation of variables and write $\psi = X(x)\, Y(y)\, Z(z)$ to obtain

$$\frac{1}{X}\frac{d^2 X}{dx^2} + \frac{1}{Y}\frac{d^2 Y}{dy^2} + \frac{1}{Z}\frac{d^2 Z}{dz^2} = -\frac{2\mu E}{\hbar^2} \tag{3.81}$$

The first term is a function of x alone, the second term of y alone, etc., so that each term has to be set equal to a constant. We write

$$\frac{1}{X}\frac{d^2 X}{dx^2} = -k_x^2 \tag{3.82}$$

and similar equations for $Y(y)$ and $Z(z)$ with

$$k_x^2 + k_y^2 + k_z^2 = \frac{2\mu E}{\hbar^2} \tag{3.83}$$

We have set each term equal to a negative constant; otherwise the boundary conditions cannot be satisfied. The solution of Eq. (3.82) is

---

[3]It may be noted that the plane wave is not normalizable; this is due to the fact that an infinitely extended plane wave corresponds to a constant probability density everywhere.

$$X(x) = A \sin k_x x + B \cos k_x x$$

and since ψ has to vanish on all points on the surface $x=0$ we must have $B=0$. Further, for ψ to vanish on all points on the surface $x=L$, we must have

$$\sin k_x L = 0$$

or

$$k_x = \frac{n_x \pi}{L} \text{ with } n_x = 1, 2, \ldots \tag{3.84}$$

(cf. Example 3.1). Similarly, we would obtain

$$k_y = \frac{n_y \pi}{L}, \ n_y = 1, 2, 3, \ldots$$

and

$$k_z = \frac{n_z \pi}{L}, \ n_z = 1, 2, 3, \ldots$$

Thus using Eq. (3.83) we get the following expression for energy eigenvalues

$$E = \frac{\pi^2 \hbar^2}{2\mu L^2}(n_x^2 + n_y^2 + n_z^2), \ n_x, n_y, n_z = 1, 2, 3, \ldots \tag{3.85}$$

The corresponding normalized wave functions are

$$\psi(x, y, z) = \left(\frac{8}{L^3}\right)^{1/2} \sin\left(\frac{n_x \pi}{L} x\right) \sin\left(\frac{n_y \pi}{L} y\right) \sin\left(\frac{n_z \pi}{L} z\right) \tag{3.86}$$

3.4.1 Density of States

If $g(E) \, dE$ represents the number of states whose energy lies between E and $E+dE$ then $g(E)$ is known as the density of states and it represents a very important quantity in the theory of solids. In order to calculate $g(E)$ we first calculate $N(E)$ which represents the total number of states whose energies are less than E. Obviously

$$N(E) = \int_0^E g(E) dE \tag{3.87}$$

and therefore

$$g(E) = \frac{dN(E)}{dE} \tag{3.88}$$

Now,

$$n_x^2 + n_y^2 + n_z^2 = \frac{2\mu L^2 E}{\pi^2 \hbar^2} = R^2 (\text{say}) \tag{3.89}$$

Thus $N(E)$ will be the number of sets of integers whose sum of square is less than R^2. In the n_x, n_y, and n_z space each point corresponds to a unit volume and if we draw a sphere of radius R then the volume of the positive octant will approximately represent[4] $N(E)$; we have to take the positive octant because n_x, n_y, and n_z take positive values. Thus

$$N(E) = 2 \times \frac{1}{8} \times \frac{4\pi}{3} R^3 = \frac{(2\mu)^{3/2} L^3}{3\pi^2 \hbar^3} E^{3/2} \tag{3.90}$$

where an additional factor of 2 has been introduced as a state can be occupied by two electrons. Using Eq. (3.88) we get

$$g(E) = \frac{(2\mu)^{3/2} V}{2\pi^2 \hbar^3} E^{1/2} \tag{3.91}$$

where $V(=L^3)$ represents the volume of the box. Often it is more convenient to express the density of states in momentum space. Now for a free non-relativistic particle

$$E = \frac{p^2}{2\mu} \tag{3.92}$$

Thus the equation

$$g(p)\mathrm{d}p = g(E)\mathrm{d}E \tag{3.93}$$

would readily give

$$g(p) = \frac{V}{\pi^2 \hbar^3} p^2 \tag{3.94}$$

3.5 Expectation Values of Dynamical Quantities

The interpretation of $|\Psi|^2$ in terms of the position probability density allows us to calculate the expectation value of measurable quantities. For example, the expectation value of the x coordinate is given by

$$\langle x \rangle = \frac{\iiint x \Psi^* \Psi \mathrm{d}\tau}{\iiint \Psi^* \Psi \mathrm{d}\tau} = \iiint \Psi^* (\mathbf{r}, t)\, x \Psi\, (\mathbf{r}, t)\, \mathrm{d}\tau \tag{3.95}$$

where the integration is over the entire space and in the last step we have assumed the wave function to be normalized. Similarly, we may write for $\langle y \rangle$ and $\langle z \rangle$ and also for the expectation value of the potential energy V

$$\langle V \rangle = \iiint \Psi^* (\mathbf{r}, t)\, V\, (\mathbf{r})\, \Psi\, (\mathbf{r}, t)\, \mathrm{d}\tau \tag{3.96}$$

---

[4]If the reader finds it difficult to understand he may first try to make the corresponding two-dimensional calculations in which one is interested in finding the number of sets of integers such that $n_x^2 + n_y^2 < R^2$. If one takes a graph paper then each corner corresponds to a set of integers and each point can be associated with a unit area. Thus the number of sets of integers would be $\pi R^2/4$ where the factor $\frac{1}{4}$ is because of the fact that we are interested only in the positive quadrant.

In order to obtain an expression for the expectation values of quantities like energy and momentum, we multiply the Schrödinger equation [Eq. (3.61)] by Ψ^* and integrate it to obtain

$$\int \Psi^* i\hbar \frac{\partial \Psi}{\partial t} d\tau = \int \Psi^* \left(-\frac{\hbar^2}{2m}\nabla^2\right)\Psi d\tau + \int \Psi^* V\Psi d\tau \qquad (3.97)$$

From now on the single integral sign will be assumed to represent the three-dimensional integral over the entire space. The last term is simply $\langle V \rangle$; further, since

$$\langle E \rangle = \left\langle \frac{p^2}{2m} \right\rangle + \langle V \rangle \qquad (3.98)$$

we may write

$$\langle E \rangle = \int \Psi^* \left(i\hbar \frac{\partial \Psi}{\partial t}\right) d\tau \qquad (3.99)$$

$$\left\langle p_x^2 \right\rangle = \int \Psi^* \left(-\hbar^2 \frac{\partial^2 \Psi}{\partial x^2}\right) d\tau \qquad (3.100)$$

and similar expressions for $\left\langle p_y^2 \right\rangle$ and $\left\langle p_z^2 \right\rangle$. Equations (3.99) and (3.100) suggest that the expectation value of any dynamical quantity O is obtained by operating it on Ψ, premultiplying it by Ψ^*, and then integrating:

$$\langle O \rangle = \int \Psi^* O\Psi \, d\tau \qquad (3.101)$$

In particular

$$\langle p_x \rangle = \int \Psi^* \left(-i\hbar \frac{\partial \Psi}{\partial x}\right) d\tau \qquad (3.102)$$

For the harmonic oscillator wave functions (see Example 3.2) if we use the various properties of the Hermite–Gauss functions, we get

$$\langle x \rangle = \int \Psi_n^* x \Psi_n \, dx = 0 \qquad (3.103)$$

$$\left\langle x^2 \right\rangle = \int \Psi_n^* x^2 \Psi_n \, dx = \frac{\hbar}{m\omega}\left(n + \frac{1}{2}\right) \qquad (3.104)$$

$$\langle p_x \rangle = \int \Psi_n^* \left(-i\hbar \frac{\partial \Psi_n}{\partial x}\right) dx = 0 \qquad (3.105)$$

$$\left\langle p_x^2 \right\rangle = \int \Psi_n^* \left(-\hbar^2 \frac{\partial^2 \Psi_n}{\partial x^2}\right) dx = m\omega\hbar\left(n + \frac{1}{2}\right) \qquad (3.106)$$

We define the uncertainties in the values of position and momentum through the following standard definitions

$$\Delta x = \sqrt{\langle (x - \langle x \rangle)^2 \rangle} = \sqrt{\langle x^2 \rangle - \langle x \rangle^2} \tag{3.107}$$

and

$$\Delta p = \sqrt{\langle (p - \langle p \rangle)^2 \rangle} = \sqrt{\langle p^2 \rangle - \langle p \rangle^2} \tag{3.108}$$

Using the harmonic oscillator wave functions one can show that

$$\Delta x \, \Delta p_x = \left(n + \frac{1}{2} \right) \hbar \tag{3.109}$$

which relates the uncertainties in position and momentum. The minimum uncertainty product occurs for the ground state $(n = 0)$

3.6 The Commutator

The commutator of two operators α and β is defined by the following equation:

$$[\alpha, \beta] = \alpha\beta - \beta\alpha = -[\beta, \alpha] \tag{3.110}$$

Now, the commutator of x and p_x operating on an arbitrary function Ψ is given by

$$\left[x, p_x \right] \Psi = (xp_x - p_x x)\,\Psi = -i\hbar \left[x \frac{\partial \Psi}{\partial x} - \frac{\partial}{\partial x}(x\Psi) \right] = i\hbar\Psi$$

Since Ψ is arbitrary, we obtain

$$\left[x, p_x \right] = xp_x - p_x x = i\hbar \tag{3.111}$$

Similarly

$$\left[y, p_y \right] = \left[z, p_z \right] = i\hbar$$

However,

$$\left[x, p_y \right] = \left[y, p_z \right] = \cdots = 0$$

$$\left[x, y \right] = \left[y, z \right] = \cdots = 0$$

and

$$[p_x, p_y] = [p_y, p_z] = \cdots = 0 \tag{3.112}$$

3.7 Orthogonality of Wave Functions

We shall first prove that all values of E_n [see Eq. (3.68)] are real and that if $E_n \neq E_k$, then the corresponding wave functions are necessarily orthogonal, i.e.,

$$\int \psi_k^* \psi_n \, d\tau = 0 \text{ for } n \neq k \tag{3.113}$$

We start with the Schrödinger equation for the two states:

$$-\frac{\hbar^2}{2m} \nabla^2 \psi_n + V(\mathbf{r}) \, \psi_n = E_n \psi_n \tag{3.114}$$

$$-\frac{\hbar^2}{2m} \nabla^2 \psi_k + V(\mathbf{r}) \, \psi_k = E_k \psi_k \tag{3.115}$$

We multiply Eq. (3.114) by ψ_k^* and the complex conjugate of Eq. (3.115) by ψ_n and subtract:

$$-\frac{\hbar^2}{2m} \left(\psi_k^* \nabla^2 \psi_n - \psi_n \nabla^2 \psi_k^* \right) = \left(E_n - E_k^* \right) \psi_k^* \psi_n \tag{3.116}$$

or

$$-\frac{\hbar^2}{2m} \int \nabla \cdot \left(\psi_k^* \nabla \psi_n - \psi_n \nabla \psi_k^* \right) d\tau = \left(E_n - E_k^* \right) \int \psi_k^* \psi_n d\tau \tag{3.117}$$

Now according to the divergence theorem

$$\int_V \nabla \cdot \mathbf{F} \, d\tau = \int_s \mathbf{F} \cdot d\mathbf{S}$$

where S is the surface bounding the volume V. Thus, the integral on the left-hand side of Eq. (3.117) can be transformed to a surface integral which would vanish if the volume integral is over the entire space; this is because the wave functions vanish at the surface which is at infinity. Thus

$$\left(E_n - E_k^* \right) \int \psi_k^* \psi_n \, d\tau = 0 \tag{3.118}$$

For $n = k$, we must have

$$E_n = E_n^* \tag{3.119}$$

proving that all eigenvalues must be real, and for $E_n \neq E_k$ Eq. (3.113) follows.

If $E_n = E_k$ ($n \neq k$) so that ψ_k and ψ_n are two linearly independent wave functions belonging to the same energy value, then ψ_k and ψ_n are not necessarily orthogonal. An energy level E is said to be degenerate when two or more linearly independent eigenfunctions correspond to it. However, it can easily be shown that any linear combination of the degenerate eigenfunctions (like $C_1\psi_k + C_2\psi_n$) is also a possible eigenfunction belonging to the same eigenvalue:

$$H\psi_k = E_k\psi_k \tag{3.120}$$

$$H\psi_n = E_k\psi_n \tag{3.121}$$

where $H = -\left(\hbar^2 / 2m\right) \nabla^2 + V\left(\mathbf{r}\right)$ [see Eq. (3.114)]. If we multiply Eq. (3.120) by C_1 and Eq. (3.121) by C_2, where C_1 and C_2 are any complex numbers and then add we obtain

$$H\phi = E_k\phi \tag{3.122}$$

where $\phi = C_1\psi_k + C_2\psi_n$. Equation (3.122) tells us that ϕ is also an eigenfunction belonging to the same eigenvalue. Since C_1 and C_2 are arbitrary, it is always possible to construct linearly independent wave functions (belonging to this level) which are mutually orthogonal. Further, one can always multiply an eigenfunction by a suitable constant such that

$$\int \psi_k^* \psi_n \, d\tau = \delta_{kn} \tag{3.123}$$

where δ_{kn} is known as the Kronecker delta function defined through Eq. (3.34)

It may be pointed out that the linear harmonic oscillator states [Eq.3.53)] are nondegenerate; however, for the hydrogen atom problem, the state characterized by the quantum number n is n^2-fold degenerate.

3.8 Spherically Symmetric Potentials

One of the most important problems in quantum mechanics is that of the motion of a particle in a potential which depends only on the magnitude of the distance from a fixed point:

$$V(\mathbf{r}) = V(r) \tag{3.124}$$

Such a potential is referred to as a spherically symmetric potential. Now, in spherical polar coordinates

$$\begin{aligned}
\nabla^2\psi &= \frac{1}{r^2}\frac{\partial}{\partial r}\left(r^2\frac{\partial\psi}{\partial r}\right) + \frac{1}{r^2}\left[\frac{1}{\sin\theta}\frac{\partial}{\partial\theta}\left(\sin\theta\,\frac{\partial\psi}{\partial\theta}\right) + \frac{1}{\sin^2\theta}\frac{\partial^2\psi}{\partial\phi^2}\right] \\
&= \frac{1}{r^2}\frac{\partial}{\partial r}\left(r^2\frac{\partial\psi}{\partial r}\right) - \frac{L^2\psi}{\hbar^2 r^2}
\end{aligned} \tag{3.125}$$

where

$$L^2 = -\hbar^2 \left[\frac{1}{\sin\theta} \frac{\partial}{\partial\theta} \left(\sin\theta \frac{\partial}{\partial\theta} \right) + \frac{1}{\sin^2\theta} \frac{\partial^2}{\partial\phi^2} \right] \tag{3.126}$$

is the operator representation of the square of the angular momentum. Thus the three-dimensional Schrödinger equation

$$\nabla^2\psi + \frac{2\mu}{\hbar^2} [E - V(r)] \psi(r, \theta, \phi) = 0 \tag{3.127}$$

can be written in the form

$$\frac{1}{r^2} \frac{\partial}{\partial r} \left(r^2 \frac{\partial\psi}{\partial r} \right) + \frac{2\mu}{\hbar^2} [E - V(r)] \psi(r, \theta, \phi) = \frac{L^2\psi}{\hbar^2 r^2} \tag{3.128}$$

In order to solve the above equation we use the method of separation of variables and write

$$\psi(r, \theta, \phi) = R(r) Y(\theta, \phi) \tag{3.129}$$

Substituting in Eq. (3.128) we get

$$\frac{Y(\theta,\phi)}{r^2} \frac{d}{dr} \left(r^2 \frac{dR}{dr} \right) + \frac{2\mu}{\hbar^2} [E - V(r)] R(r) Y(\theta, \phi) = \frac{R(r)}{\hbar^2 r^2} L^2 Y(\theta, \phi)$$

Dividing by $R(r)Y(\theta,\phi)/r^2$, we obtain

$$\frac{1}{R(r)} \frac{d}{dr} \left(r^2 \frac{dR}{dr} \right) + \frac{2\mu\, r^2}{\hbar^2} [E - V(r)] = \frac{1}{\hbar^2} \frac{1}{Y(\theta,\phi)} L^2 Y(\theta, \phi) = \lambda \tag{3.130}$$

where we have set the terms equal to a constant λ because the left-hand side of the above equation depends only on r while the other term depends only on θ and ϕ. The above equation gives us the eigenvalue equation

$$L^2 Y(\theta, \phi) = \lambda\hbar^2 Y(\theta, \phi) \tag{3.131}$$

The eigenvalues of L^2 are $l(l+1)\,\hbar^2$, i.e., well-behaved solutions are obtained when

$$\lambda = l(l+1), \quad l = 0, 1, 2, \ldots \tag{3.132}$$

The corresponding eigenfunctions being the spherical harmonics

$$Y_{lm}(\theta, \phi), \quad m = -l, -l+1, \ldots, l-1, l \tag{3.133}$$

Thus Eq. (3.130) can be written in the form

$$\frac{1}{r^2}\frac{d}{dr}\left(r^2\frac{dR}{dr}\right) + \frac{2\mu}{\hbar^2}\left[E - V(r) - \frac{l(l+1)\hbar^2}{2\mu r^2}\right]R(r) = 0 \qquad (3.134)$$

which is known as the radial part of the Schrödinger equation.

3.9 The Two-Body Problem

In this section we will discuss the energy eigenvalues and the corresponding eigenfunctions for the hydrogen-like atom for which the potential energy variation is
given by

$$V(r) = -\frac{Zq^2}{4\pi\varepsilon_0 r} \qquad (3.135)$$

where

$Z = 1$ for the H-atom problem,
$Z = 2$ for the singly ionized He-atom problem (He$^+$),
$Z = 3$ for the doubly ionized Li-atom problem (Li^{++})

where

$$r = |\mathbf{r}_1 - \mathbf{r}_2| \qquad (3.136)$$

represents the magnitude of the distance between the two particles, i.e., between the
electron and the nucleus. In writing Eq. (3.135) we have used the SI system of units
so that

$$q \approx 1.6 \times 10^{-19}\ \text{C}$$

$$\varepsilon_0 \approx 8.854 \times 10^{-12}\quad \text{MKS units}$$

and $V(r)$ is measured in Joules. In this book we will be almost always using the SI
system of units; however, since CGS units are used in many books, we give below
the corresponding expression for $V(r)$ in CGS units:

$$V(r) = -\frac{e^2}{r} \qquad (3.137)$$

where $e \approx 4.8 \times 10^{-10}$ esu represents the electronic charge in CGS units and $V(r)$ is
measured in ergs. We may note that the Coulomb potential described by Eq. (3.135)
depends only on $|\mathbf{r}_1\text{-}\mathbf{r}_2|$, i.e., on the magnitude of the distance between the two
particles. Indeed, for a two-body problem, whenever the potential energy depends
only on the magnitude of the distance between the two particles, the problem can

always be reduced to a one-body problem (describing the internal motion of the atom) along with a uniform translational motion of the centre of mass. Thus the internal motion of the atom is described by the wave function $\psi(\mathbf{r})$ and satisfies the equation

$$\nabla^2 \psi(\mathbf{r}) + \frac{2\mu}{\hbar^2} \left[E - V(r) \right] \psi(\mathbf{r}) = 0 \tag{3.138}$$

where

$$\mathbf{r} = \mathbf{r}_1 - \mathbf{r}_2 \tag{3.139}$$

represents the relative coordinate and

$$\mu = \frac{m_e m_N}{m_e + m_N} \tag{3.140}$$

represents the reduced mass with m_e and m_N represent the mass of the electron and that of the nucleus, respectively. The total energy of the atom is given by

$$E_{\text{total}} = E + E_{\text{cm}} \tag{3.141}$$

where

$$E_{\text{cm}} = \frac{\hbar^2 P^2}{2M} \; ; \; [M = m_1 + m_2] \tag{3.142}$$

represents the uniform translational energy of the center of mass. The different spectroscopic lines emitted by an atom correspond to the transition between different states obtained by solving Eq. (3.138).

3.9.1 The Hydrogen-Like Atom Problem

The radial part of the wave function satisfies the following equation:

$$\frac{1}{r^2} \frac{d}{dr} \left[r^2 \frac{dR}{dr} \right] + \frac{2\mu}{\hbar^2} \left[E + \frac{Zq^2}{4\pi \varepsilon_0 r} - \frac{l(l+1)\hbar^2}{2\mu r^2} \right] R(r) = 0 \tag{3.143}$$

For $R(r)$ to be well behaved at $r=0$ and also as $r \to \infty$, we would obtain the following discrete energy eigenvalues of the problem see Appendix B:

$$E_n = -\frac{|E_1|}{n^2} \tag{3.144}$$

where

$n = 1, 2, 3, \ldots$

represents the total quantum number and

$$|E_1| = \frac{1}{2} \mu Z^2 \alpha^2 c^2 \tag{3.145}$$

represents the magnitude of the ground state energy. Further,

$$\alpha = \frac{q^2}{4\pi\varepsilon_0\hbar c} \approx \frac{1}{137.036} \tag{3.146}$$

represents the fine structure constant and $c (\approx 2.998 \times 10^8 \text{ m/s})$ represents the speed of light in free space. For the hydrogen atom

$$m_N = m_p \approx 1.6726 \times 10^{-27} \text{ kg}$$

giving

$$\mu_H \approx 9.1045 \times 10^{-31} \text{ kg}$$

where we have taken $m_e \approx 9.1094 \times 10^{-31}$ kg. On the other hand, for the deuterium atom

$$m_N = m_D \approx 3.3436 \times 10^{-27} \text{ kg}$$

giving

$$\mu_D \approx 9.1070 \times 10^{-31} \text{ kg}$$

Now, for the $n = n_1 \rightarrow n = n_2$ transition, the wavelength of the emitted radiation is given by

$$\lambda = \frac{hc}{E_{n_1} - E_{n_2}} \tag{3.147}$$

or

$$\lambda = \frac{2h}{\mu Z^2 \alpha^2 c^2}\left[\frac{1}{n_2^2} - \frac{1}{n_1^2}\right]^{-1} \tag{3.148}$$

When $n_2 = 1$, 2, and 3 we have what is known as Lyman series, the Balmer series, and the Paschen series, respectively. For the $n=3 \rightarrow n=2$ transition, the wavelength of the emitted radiation comes out to be

$$6565.2 \text{ Å} \quad \text{and} \quad 6563.4 \text{ Å}$$

for hydrogen and deuterium, respectively. The corresponding wavelength for the $n=4 \rightarrow n=2$ transition is

$$4863.1 \text{ Å} \quad \text{and} \quad 4861.7 \text{ Å}$$

Such a small difference in the wavelength was first observed by Urey in 1932 which led to the discovery of deuterium.

In spectroscopy the energy levels are usually written in wavenumber units which are obtained by dividing by hc:

$$T_n = \frac{E_n}{hc} = -\frac{Z^2}{n^2} R \tag{3.149}$$

where

$$R = \frac{2\pi^2 \mu}{ch^3}\left(\frac{q^2}{4\pi\varepsilon_0}\right)^2 = \frac{\mu c \alpha^2}{2h} \tag{3.150}$$

is known as the Rydberg constant. Values of the Rydberg constant for different hydrogen like atoms are given below:

$$R = 109677.58 \text{ cm}^{-1} \quad \text{(for the hydrogen atom)}$$
$$109707.56 \text{ cm}^{-1} \quad \text{(for the deuterium atom)}$$
$$109722.40 \text{ cm}^{-1} \quad \text{(for the He}^+\text{-atom)}$$
$$109728.90 \text{ cm}^{-1} \quad \text{(for the Li}^{++}\text{ -atom)}$$

The slight difference in the values is because of the difference in the values of the reduced mass μ.

The normalized radial part of the wave function is given by see Appendix B:

$$R_{nl}(\rho) = N\,e^{-\rho/2}\,\rho^l\,{}_1F_1(-n_r,\,2l+2,\,\rho) \tag{3.151}$$

where n_r is known as the radial quantum number and for the ${}_1F_1$ function to be a polynomial, n_r can take only the following values:

$$n_r = 0, 1, 2, 3, \ldots\ldots$$

The total quantum number is given by

$$n = l + 1 + n_r$$

Thus

$$n = 1, 2, 3, \ldots \text{ with } l = 0, 1, 2, \ldots n - 1$$

The normalization constant is given by

$$N = \frac{\gamma^{3/2}}{(2l+1)!}\left\{\frac{(n+l)!}{2n(n-l-1)!}\right\}^{1/2} \tag{3.152}$$

In the above equation

$$\left.\begin{array}{l} \rho = \gamma\,r; \qquad\qquad \gamma = \dfrac{2Z}{na_0} \\[2ex] a_0 = \dfrac{\hbar^2}{\mu\,(q^2/4\pi\varepsilon_0)} \end{array}\right\} \tag{3.153}$$

where a_0 is the Bohr radius. Further

$$_1F_1(a, c, \rho) = 1 + \frac{a}{c}\rho + \frac{a(a+1)}{c(c+1)}\frac{\rho^2}{2!} + \cdots \tag{3.154}$$

represents the confluent hypergeometric function. For given values of n and l, a would be a negative integer or zero and the above function would be a polynomial. For example, for $n=2$, $l=0$, we will have $n_r=1$:

$$N = \gamma^{3/2} \left[\frac{2!}{4 \times 1}\right]^{1/2} = \frac{1}{\sqrt{2}}\left(\frac{Z}{a_0}\right)^{3/2}$$

and

$$_1F_1(-1, 2, \rho) = 1 - \frac{\rho}{2}$$

Thus

$$R_{2\,0}(r) = \frac{1}{\sqrt{2}}\left(\frac{Z}{a_0}\right)^{3/2}\left(1 - \frac{1}{2}\xi\right)e^{-\xi/2} \tag{3.155}$$

where

$$\xi = \frac{r}{a_0} \tag{3.156}$$

Similarly, one can calculate other wave functions. We give below the first few $R_{nl}(r)$

$$R_{10}(r) = 2\left(\frac{Z}{a_0}\right)^{3/2}e^{-\xi} \tag{3.157}$$

$$R_{21}(r) = \frac{1}{2\sqrt{6}}\left(\frac{Z}{a_0}\right)^{3/2}\xi\,e^{-\xi/2} \tag{3.158}$$

$$R_{30}(r) = \frac{2}{3\sqrt{3}}\left(\frac{Z}{a_0}\right)^{3/2}\left(1 - \frac{2}{3}\xi + \frac{2}{27}\xi^2\right)e^{-\xi/3} \tag{3.159a}$$

$$R_{31}(r) = \frac{8}{27\sqrt{6}}\left(\frac{Z}{a_0}\right)^{3/2}\left(\xi - \frac{1}{6}\xi^2\right)e^{-\xi/3} \tag{3.159b}$$

$$R_{32}(r) = \frac{4}{81\sqrt{30}}\left(\frac{Z}{a_0}\right)^{3/2}\xi^2\,e^{-\xi/3} \tag{3.160}$$

The wave functions are normalized so that

$$\int_0^\infty |R_{nl}(r)|^2\ r^2\ dr = 1 \tag{3.161}$$

The complete wave function is given by

$$\psi_{nlm}(r, \theta, \phi) = R_{nl}(r)\ Y_{lm}(\theta, \phi) \tag{3.162}$$

where $Y_{lm}(\theta,\phi)$ are the spherical harmonics tabulated below.

Looking at Eq. (3.162) we see that the energy depends on the total quantum number n. Since for each value of n we have values of l ranging from 0 to $n-1$ and for each value of l, the m values range from $-l$ to $+l$ there are

$$\sum_{l=0}^{n-1} (2l+1) = n^2$$

states ψ_{nlm} belonging to a particular energy. The degeneracy with respect to m is due to spherical symmetry of the potential energy function. But the l-degeneracy is peculiar to the Coulomb field and is, in general, removed for non-Coulomb potentials.

Further,

$$Y_{0,0} = (4\pi)^{-1/2} \tag{3.163}$$

$$Y_{1,1} = -\left(\frac{3}{8\pi}\right)^{1/2} \sin\theta e^{i\phi} \tag{3.164}$$

$$Y_{1,0} = \left(\frac{3}{4\pi}\right)^{1/2} \cos\theta \tag{3.165}$$

$$Y_{1,-1} = \left(\frac{3}{8\pi}\right)^{1/2} \sin\theta e^{-i\phi} \tag{3.166}$$

$$Y_{2,2} = \left(\frac{15}{32\pi}\right)^{1/2} \sin^2\theta e^{2i\phi} \tag{3.167}$$

$$Y_{2,1} = -\left(\frac{15}{8\pi}\right)^{1/2} \sin\theta \cos\theta e^{i\phi} \tag{3.168}$$

$$Y_{2,0} = \left(\frac{5}{16\pi}\right)^{1/2} \left(3\cos^2\theta - 1\right) \tag{3.169}$$

$$Y_{2,-1} = \left(\frac{15}{8\pi}\right)^{1/2} \sin\theta \cos\theta e^{-i\phi} \tag{3.170}$$

$$Y_{2,-2} = \left(\frac{15}{32\pi}\right)^{1/2} \sin^2\theta e^{-2i\phi} \tag{3.171}$$

and so on. The ground state eigenfunction is $\psi_{1,0,0}$ ($n = 1$, $l = 0$, $m = 0$), the first excited state ($n = 2$) is fourfold degenerate $\psi_{2,0,0}$, $\psi_{2,1,-1}$, $\psi_{2,1,0}$, and $\psi_{2,1,1}$. Similarly $n = 3$ state is ninefold degenerate. In general, the states characterized by the quantum number n are n^2-fold degenerate. The wave functions are orthonormal, i.e.,

$$\iiint \psi_{nlm}^* \psi_{n'l'm'} r^2 \, dr \, \sin\theta \, d\theta \, d\phi = \delta_{nn'} \delta_{ll'} \delta_{mm'} \tag{3.172}$$

It may be a worthwhile exercise for the reader to see that the above wave functions satisfy Eq. (3.172) with E given by Eq. (3.144).

Problems

Problem 3.1 Consider a potential energy function given by the following equation (see Fig. 3.3)

$$V(x) = \infty \quad x<0$$
$$= 0 \quad 0<x<a \tag{3.173}$$
$$= V_0 \quad x>a$$

Assume ψ and $\frac{d\psi}{dx}$ to be continuous at $x=a$ and that the wave function vanishes at $x=0$ and as $x \to +\infty$.

(a) Using the above boundary conditions, solve the one-dimensional Schrödinger equation to obtain the following transcendental equations which would determine the discrete values of energy:

$$-\xi \cot \xi = \sqrt{\alpha^2 - \xi^2} \tag{3.174}$$

where

$$\xi = \sqrt{\frac{2\mu E a^2}{\hbar^2}} \quad \text{and} \quad \alpha = \sqrt{\frac{2\mu V_0 a^2}{\hbar^2}} \tag{3.175}$$

(b) Assuming $\alpha = 3\pi$

$$\frac{2\mu V_0 a^2}{\hbar^2} = 9\pi^2$$

show that there will be three bound states (see Fig. 3.3) with

$$\xi = 2.83595, \ 5.64146, \ \text{and} \ 8.33877$$

Fig. 3.3 The first three eigenvalues and eigenfunctions of an isolated well for $\alpha=3\pi$; $\alpha = \sqrt{\frac{2mV_0 a^2}{\hbar^2}}$

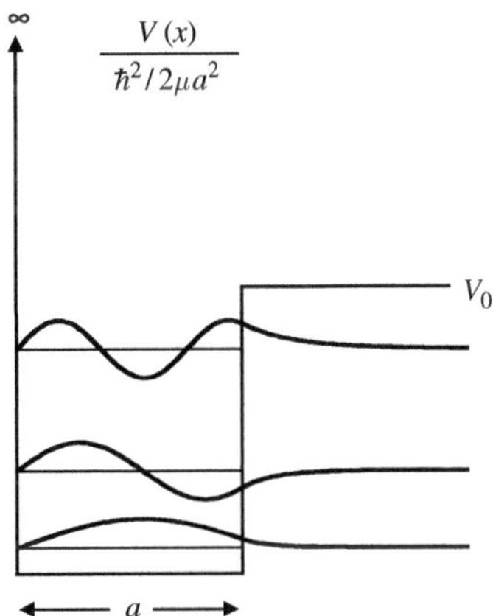

Problem 3.2 Consider a symmetric potential energy function so that $V(-x) = V(x)$. Show that the solutions are either symmetric or antisymmetric functions of x, i.e., either $\psi(-x) = +\psi(x)$ or $\psi(-x) = -\psi(x)$

[Hint: In Eq. (3.22), make the transformation $x \to -x$ and assuming $V(-x) = V(x)$ show that $\psi(-x)$ satisfies the same equation as $\psi(x)$; hence, we must have $\psi(-x)=\lambda\psi(x)$. Make the transformation $x \to -x$ again to prove $\lambda = \pm 1$]

Problem 3.3 Consider a potential energy function given by the following equation

$$V(x) = \begin{cases} 0; & |x| < \frac{d}{2} \\ V_0; & |x| > \frac{d}{2} \end{cases} \tag{3.176}$$

Since the potential energy variation is symmetric about $x=0$, the solutions are either symmetric or antisymmetric functions of x. (see Problem 3.2). For $E < V_0$ solve the Schrödinger equation (in the two regions). Assuming ψ and $\frac{d\psi}{dx}$ continuous at $x=\pm\,d/2$ and that the wave function must vanish as $x \to \pm\infty$ obtain the following transcendental equations which would determine the discrete values of energy

$$\xi \tan \xi = \sqrt{\alpha^2 - \xi^2} \qquad \text{for symmetric states} \tag{3.177}$$

$$-\xi \cot \xi = \sqrt{\alpha^2 - \xi^2} \qquad \text{for antisymmetric states} \tag{3.178}$$

where

$$\xi = \sqrt{\frac{2\mu E d^2}{4\hbar^2}} \quad \text{and} \quad \alpha = \sqrt{\frac{2\mu V_0 d^2}{4\hbar^2}} \tag{3.179}$$

For a given value of α, the solutions of Eqs. (3.177) and (3.178) will give the bound states for the potential well problem given by Eq. (3.176). Obviously, for $\alpha < \pi/2$ we will have only bound state. For given values of V_0, μ and d, as $\hbar \to 0$, the value of α will become large and we will have a continuum of states implying that all energy levels are possible. Thus in the limit of $\hbar \to 0$, we have the results of classical mechanics.

Problem 3.4 Using the results of the previous problem, obtain the energy eigenvalues for a single well corresponding to the following values of various parameters: $\mu = m_e$, $V_0 = 20$ eV; $d = 5$ Å

[**Ans:** $E_1 \approx 1.088$ eV; $E_2 \approx 4.314$ eV, $E_3 \approx 9.527$ eV, $E_4 \approx 16.253$ eV]

Problem 3.5 Consider the three-dimensional harmonic oscillator

$$V = \frac{1}{2}\mu \left(\omega_1^2 x^2 + \omega_2^2 y^2 + \omega_3^2 z^2 \right) \tag{3.180}$$

Use the method of separation of variables to solve the Schrödinger equation (in Cartesian coordinates) and show that the energy eigenvalues are given by

$$E = \left(n_1 + \frac{1}{2} \right) \hbar\omega_1 + \left(n_2 + \frac{1}{2} \right) \hbar\omega_2 + \left(n_3 + \frac{1}{2} \right) \hbar\omega_3 \tag{3.181}$$

$n_1, n_2, n_3 = 0, 1, 2, \ldots$. The corresponding wave functions are products of the Hermite–Gauss functions.

Problem 3.6 Calculate the wavenumbers corresponding to the H$_\alpha$ ($n = 3 \to n = 2$) and the H$_\beta(n = 4 \to n = 2$) lines of the Balmer series for the hydrogen atom. What will be the corresponding wavelengths

[**Ans:** ≈ 6563 Å and 4861 Å]

Problem 3.7 Calculate the wavelengths for the $n = 4 \to n = 3$ transition in the He$^+$ atom

[**Ans:** ≈ 4686 Å]

Problem 3.8 Calculate the wavelengths corresponding to the $n = 2 \to n = 1$; $n = 3 \to n = 1$; $n = 4 \to n = 1$, and $n = 5 \to n = 1$ transitions of the Lyman series of the hydrogen atom

[**Ans:** $\approx$ 1216 Å, 1026 Å, 973 Å, 950 Å]

Problem 3.9 Using the expressions for spherical harmonics, write all wave functions corresponding to the $n = 2$ and $n = 3$ states of the hydrogen atom. Show that they are fourfold and ninefold degenerate. (Actually, if we take into account the spin states, they are 8-fold and 18-fold degenerate)

Problem 3.10 Show that $\int\limits_0^\infty R_{10}(r)\, R_{20}(r)\, r^2\, dr = 0$

Chapter 4
Einstein Coefficients and Light Amplification

4.1 Introduction

In this chapter we discuss interaction of radiation and atoms and obtain the relationship between absorption and emission processes. We show that for light amplification a state of population inversion should be created in the atomic system. We also obtain an expression for the gain coefficient of the system. This is followed by a discussion of two-level, three-level, and four-level systems using the rate equation approach. Finally a discussion of various mechanisms leading to broadening of spectral lines is discussed.

4.2 The Einstein Coefficients

We consider two levels of an atomic system as shown in Fig. 4.1 and let N_1 and N_2 be the number of atoms per unit volume present in the energy levels E_1 and E_2, respectively. The atomic system can interact with electromagnetic radiation in three distinct ways:

(a) An atom in the lower energy level E_1 can absorb the incident radiation at a frequency $\omega = (E_2 - E_1)/\hbar$ and be excited to E_2; this excitation process requires the presence of radiation. The rate at which absorption takes place from level 1 to level 2 will be proportional to the number of atoms present in the level E_1 and also to the energy density of the radiation at the frequency $\omega = (E_2 - E_1)/\hbar$. Thus if $u(\omega)d\omega$ represents the radiation energy per unit volume between ω and $\omega + d\omega$ then we may write the number of atoms undergoing absorptions per unit time per unit volume from level 1 to level 2 as

$$\Gamma_{12} = B_{12}u(\omega)N_1 \tag{4.1}$$

where B_{12} *is a constant of proportionality and depends on the energy levels* E_1 *and* E_2. Notice here that $u(\omega)$ has the units of energy density per frequency interval.

K. Thyagarajan, A. Ghatak, *Lasers*, Graduate Texts in Physics,
DOI 10.1007/978-1-4419-6442-7_4, © Springer Science+Business Media, LLC 2010

Fig. 4.1 Two states of an
atom with energies E_1 and E_2
with corresponding
population densities of N_1
and N_2, respectively

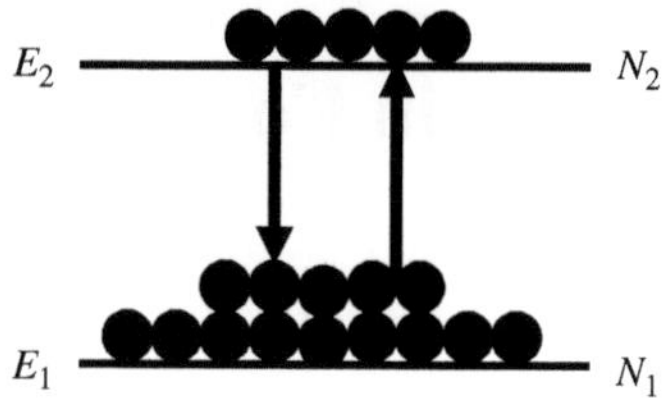

(b) For the reverse process, namely the deexcitation of the atom from E_2 to E_1, Einstein postulated that an atom can make a transition from E_2 to E_1 through two distinct processes, namely *stimulated emission* and *spontaneous emission*. In the case of stimulated emission, the radiation which is incident on the atom stimulates it to emit radiation and the rate of transition to the lower energy level is proportional to the energy density of radiation at the frequency ω. Thus, the number of stimulated emissions per unit time per unit volume will be

$$\Gamma_{21} = B_{21}u(\omega)N_2 \tag{4.2}$$

where B_{21} is the coefficient of proportionality and depends on the energy levels.

(c) An atom which is in the upper energy level E_2 can also make a spontaneous emission; this rate will be proportional to N_2 only and thus we have for the number atoms making spontaneous emissions per unit time per unit volume

$$U_{21} = A_{21}N_2 \tag{4.3}$$

At thermal equilibrium between the atomic system and the radiation field, the number of upward transitions must be equal to the number of downward transitions. Hence, at thermal equilibrium

$$N_1 B_{12}u(\omega) = N_2 A_{21} + N_2 B_{21}u(\omega)$$

or

$$u(\omega) = \frac{A_{21}}{(N_1/N_2)B_{12} - B_{21}} \tag{4.4}$$

Using Boltzmann's law, the ratio of the equilibrium populations of levels 1 and 2 at temperature T is

$$\frac{N_1}{N_2} = e^{(E_2 - E_1)/k_B T} = e^{\hbar\omega/k_B T} \tag{4.5}$$

where $k_B(= 1.38 \times 10^{-23} \mathrm{J/K})$ is the Boltzmann's constant. Hence

$$u(\omega) = \frac{A_{21}}{B_{12}e^{\hbar\omega/k_B T} - B_{21}} \tag{4.6}$$

Now according to Planck's law, the radiation energy density per unit frequency interval is given by (see Appendix F)

$$u(\omega) = \frac{\hbar\omega^3 n_0^3}{\pi^2 c^3} \frac{1}{e^{\hbar\omega/k_B T} - 1} \tag{4.7}$$

where c is the velocity of light in free space and n_0 is the refractive index of the medium.

Comparing Eqs. (4.6) and (4.7), we obtain

$$B_{12} = B_{21} = B \tag{4.8}$$

and

$$\frac{A_{21}}{B_{21}} = \frac{\hbar\omega^3 n_0^3}{\pi^2 c^3} \tag{4.9}$$

Thus the stimulated emission rate per atom is the same as the absorption rate per atom and the ratio of spontaneous to stimulated emission coefficients is given by Eq. (4.9). The coefficients A and B are referred to as the Einstein A and B coefficients.

At thermal equilibrium, the ratio of the number of spontaneous to stimulated emissions is given by

$$R = \frac{A_{21} N_2}{B_{21} N_2 u(\omega)} = e^{\hbar\omega/k_B T} - 1 \tag{4.10}$$

Thus at thermal equilibrium at a temperature T, for frequencies, $\omega \gg k_B T/\hbar$, the number of spontaneous emissions far exceeds the number of stimulated emissions.

Example 4.1 Let us consider an optical source at $T = 1000$ K. At this temperature

$$\frac{k_B T}{\hbar} = \frac{1.38 \times 10^{-23} (\text{J/K}) \times 10^3 (\text{K})}{1.054 \times 10^{-34} (\text{Js})} \approx 1.3 \times 10^{14}\, \text{s}^{-1}$$

Thus for $\omega \gg 1.3 \times 10^{14}\,\text{s}^{-1}$, the radiation would be mostly due to spontaneous emission. For $\lambda \cong$ 500 nm, $\omega \approx 3.8 \times 10^{15}\,\text{s}^{-1}$ and

$$R \approx e^{29.2} \approx 5.0 \times 10^{12}$$

Thus at optical frequencies the emission from a hot body is predominantly due to spontaneous transitions and hence the light from usual light sources is incoherent.

We shall now obtain the relationship between the Einstein A coefficient and the spontaneous lifetime of level 2. Let us assume that an atom in level 2 can make a spontaneous transition only to level 1. Then since the number of atoms making spontaneous transitions per unit time per unit volume is $A_{21} N_2$, we may write the rate of change of population of level 2 with time due to spontaneous emission as

$$\frac{dN_2}{dt} = -A_{21} N_2 \tag{4.11}$$

the solution of which is

$$N_2(t) = N_2(0)e^{-A_{21}t} \tag{4.12}$$

Thus the population of level 2 reduces by 1/e in a time $t_{sp} = 1/A_{21}$ which is called the spontaneous lifetime associated with the transition $2 \rightarrow 1$.

Example 4.2 In the 2P $\rightarrow$ 1S transition in the hydrogen atom, the lifetime of the 2P state for spontaneous emission is given by

$$t_{sp} = \frac{1}{A_{21}} \approx 1.6 \times 10^{-9} s$$

Thus

$$A_{21} \approx 6 \times 10^8 s^{-1}$$

The frequency of the transition is given by

$$\omega \approx 1.55 \times 10^{16} s^{-1} \qquad (\hbar\omega \approx 10.2\, eV)$$

Thus

$$B_{21} = \frac{\pi^2 c^3}{\hbar\omega^3 n_0^3} A_{21} \approx 4.1 \times 10^{20}\, m^3/Js^2$$

where we have assumed $n_0 \approx 1$. (Note the unit for B_{21}.)

Now, if one observes the spectrum of the radiation due to the spontaneous emission from a collection of atoms, one finds that the radiation is not strictly monochromatic but is spread over a certain frequency range. Similarly, if one measures the absorption by a collection of atoms as a function of frequency, one again finds that the atoms are capable of absorbing not just a single frequency but radiation over a band of frequencies. This implies that energy levels have widths and the atoms can interact with radiation over a range of frequencies but the strength of interaction is a function of frequency (see Fig. 4.2). This function that describes the frequency dependence is called the lineshape function and is represented by $g(\omega)$. The function is usually normalized according to

$$\int g(\omega)d\omega = 1 \tag{4.13}$$

Explicit expressions for $g(\omega)$ will be obtained in Section 4.5.

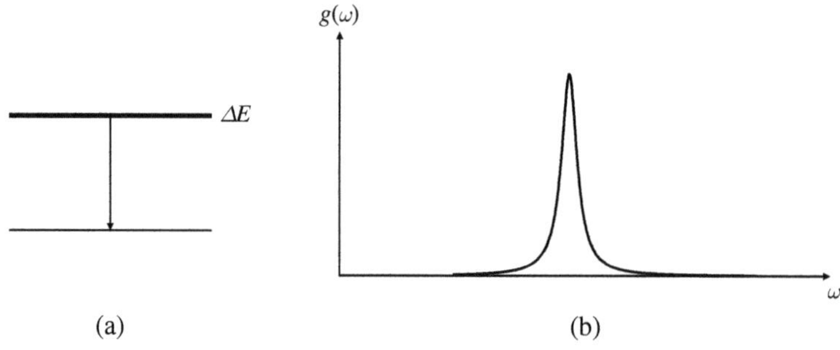

Fig. 4.2 (**a**) Because of the finite lifetime of a state each state has a certain width so that the atom can absorb or emit radiation over a range of frequencies. The corresponding lineshape is shown in (**b**)

From the above we may say that out of the total N_2 and N_1 atoms per unit volume, only $N_2\, g(\omega)d\omega$ and $N_1\, g(\omega)d\omega$ atoms per unit volume will be capable of interacting with radiation of frequency lying between ω and $\omega + d\omega$. Hence the total number of stimulated emissions per unit time per unit volume will now be given by

$$\Gamma_{21} = \int B_{21} u(\omega) N_2 g(\omega) d\omega$$
$$= N_2 \frac{\pi^2 c^3}{\hbar n_0^3 t_{sp}} \int \frac{u(\omega)g(\omega)}{\omega^3} d\omega \tag{4.14}$$

where we have used Eq. (4.9) and $A_{21} = 1/t_{sp}$.

We now consider two specific cases.

(1) If the atoms are interacting with radiation whose spectrum is very broad compared to that of $g(\omega)$ (see Fig. 4.3a), then one may assume that over the region of integration where $g(\omega)$ is appreciable $u(\omega)/\omega^3$ is essentially constant and thus may be taken out of the integral in Eq. (4.14). Using the normalization integral, Eq. (4.14) becomes

$$\Gamma_{21} = N_2 \frac{\pi^2 c^3}{\hbar \omega^3 n_0^3 t_{sp}} u(\omega) \tag{4.15}$$

where ω now represents the transition frequency. Equation (4.15) is consistent with Eq. (4.2) if we use Eq. (4.9) for B_{21}. Thus Eq. (4.15) represents the rate of stimulated emission per unit volume when the atom interacts with broadband radiation.

(2) We now consider the other extreme case in which the atom is interacting with near-monochromatic radiation. If the frequency of the incident radiation is ω', then the $u(\omega)$ curve will be extremely sharply peaked at $\omega = \omega'$ as compared to $g(\omega)$ (see Fig. 4.3b) and thus $g(\omega)/\omega^3$ can be taken out of the integral to obtain

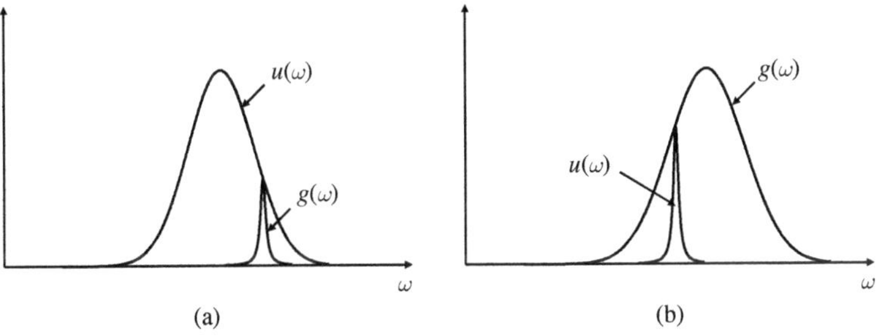

Fig. 4.3 (a) Atoms characterized by the lineshape function $g(\omega)$ interacting with broadband radiation. (b) Atoms interacting with near-monochromatic radiation

$$\Gamma_{21} = N_2 \frac{\pi^2 c^3}{\hbar \omega'^3 n_0^3 t_{sp}} g(\omega') \int u(\omega) d\omega$$

$$(4.16)$$

$$= N_2 \frac{\pi^2 c^3}{\hbar \omega'^3 n_0^3 t_{sp}} g(\omega') u$$

where

$$u = \int u(\omega) d\omega \qquad (4.17)$$

is the energy density of the incident near-monochromatic radiation. It may be noted that u has dimensions of energy per unit volume unlike $u(\omega)$ which has the dimensions of energy per unit volume per unit frequency interval. Thus when the atom described by a lineshape function $g(\omega)$ interacts with near-monochromatic radiation at frequency ω', the stimulated emission rate per unit volume is given by Eq. (4.16).

In a similar manner, the number of stimulated absorptions per unit time per unit volume will be

$$\Gamma_{12} = N_1 \frac{\pi^2 c^3}{\hbar \omega'^3 n_0^3 t_{sp}} g(\omega') u \qquad (4.18)$$

4.2.1 Absorption and Emission Cross Sections

The rates of absorption and stimulated emission can also be characterized in terms of the parameters referred to as absorption and emission cross sections. To do this, we first notice that the energy density u and the intensity I of the propagating electromagnetic wave are related through the following equation (see Section 2.2):

$$u = \frac{I}{c/n_0} = \frac{n_0 I}{c} \qquad (4.19)$$

The number of photons crossing a unit area per unit time also referred to as the photon flux ϕ is related to the intensity I through the following equation:

$$\phi = \frac{I}{\hbar \omega} \qquad (4.20)$$

Thus Eq. (4.18) can be written as

$$\Gamma_{12} = N_1 \frac{\pi^2 c^2}{\omega^2 n_0^2 t_{sp}} g(\omega) \phi \qquad (4.21)$$

$$= \sigma_a N_1 \phi$$

where σ_a represents the absorption cross section (with dimensions of area) for this transition and is given by

$$\sigma_a = \frac{\pi^2 c^2}{\omega^2 n_0^2 t_{sp}} g(\omega) \tag{4.22}$$

Similarly we can define the emission cross section σ_e through the rate Γ_{21}. Since Γ_{12} and Γ_{21} are equal, the absorption and emission cross sections are equal.

Note that the absorption and emission cross sections are functions of frequency and are related to the line broadening function $g(\omega)$ and the lifetime t_{sp}.

The peak emission cross sections for some of the important laser transitions are given in Table 4.1.

Table 4.1 Table giving transition cross section for some important laser lines

Laser transition	Wavelength (nm)	Cross section (m^2)	Lifetime (μs)
He–Ne laser	632.8	5.8×10^{-17}	30×10^{-3}
Argon ion	514.5	2.5×10^{-17}	6×10^{-3}
Nd:YAG	1064	2.8×10^{-23}	230

Example 4.3 Consider the transition in neon atom at the wavelength of 1150 nm. This transition is Doppler broadened with a linewidth of 900 MHz and the upper state spontaneous lifetime is 100 ns. Using Eq. (4.22) we can calculate the peak absorption cross section. If we assume $g(\omega_0) \sim 1/\Delta\omega$, we obtain $\sigma_a \sim 5.8 \times 10^{-16}$ m^2

4.3 Light Amplification

We next consider a collection of atoms and let a near-monochromatic radiation of energy density u at frequency ω' pass through it. We shall now obtain the rate of change of intensity of the radiation as it passes through the medium.

Let us consider two planes P_1 and P_2 of area S situated at z and $z + dz$, z being the direction of propagation of the radiation (see Fig. 4.4). If $I(z)$ and $I(z+dz)$ represent the intensity of the radiation at z and $z + dz$, respectively, then the net amount of energy entering the volume Sdz between P_1 and P_2 will be

$$[I(z) - I(z + dz)]S = [I(z) - I(z) - \frac{dI}{dz}dz]S$$
$$= -\frac{dI}{dz}Sdz \tag{4.23}$$

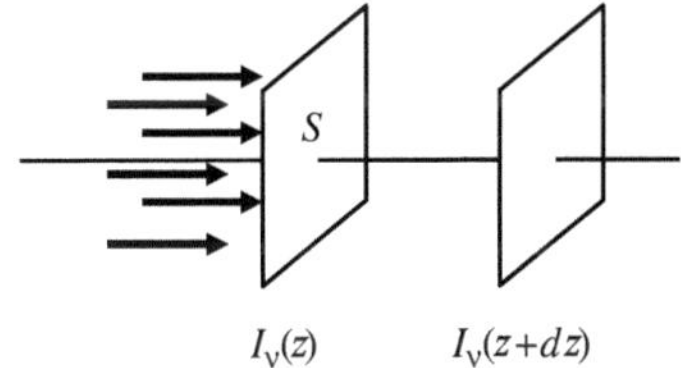

Fig. 4.4 Propagation of radiation at frequency ω' through a medium leading to a change of intensity with propagation

This must be equal to the net energy absorbed by the atoms in the volume Sdz. The energy absorbed by the atoms in going from level 1 to level 2 will be $\Gamma_{12}Sdz\hbar\omega'$ where $\hbar\omega'$ is the energy absorbed when an atom goes from level 1 to level 2. Similarly the energy released through stimulated emissions from level 2 to level 1 will be $\Gamma_{21}Sdz\hbar\omega'$. We shall neglect the energy arising from spontaneous emission since it appears over a broad frequency range and is also emitted in all directions. Thus the fraction of the spontaneous emission which would be at the radiation frequency ω' and which would be traveling along the z-direction will be very small. Thus the net energy absorbed per unit time in the volume Sdz will be

$$
\begin{aligned}
(\Gamma_{12} - \Gamma_{21})\hbar\omega'Sdz &= \frac{\pi^2 c^3}{\hbar\omega'^3 n_0^3}\frac{1}{t_{\rm sp}}ug(\omega')(N_1 - N_2)\hbar\omega'Sdz \\
&= \frac{\pi^2 c^3}{\omega'^2 n_0^3 t_{\rm sp}}ug(\omega')(N_1 - N_2)Sdz
\end{aligned}
\tag{4.24}
$$

Now, the energy density u and the intensity of radiation I are related through Eq. (4.19). Thus using Eqs. (4.23) and (4.24) we obtain

$$
\frac{dI}{dz} = -\alpha I
\tag{4.25}
$$

where

$$
\alpha = \frac{\pi^2 c^2}{\omega^2 n_0^2 t_{\rm sp}}g(\omega)(N_1 - N_2) = -\gamma
\tag{4.26}
$$

and we have removed the prime on ω with the understanding that ω represents the frequency of the incident radiation. Hence if $N_1 > N_2$, α is positive (and γ is negative) and the intensity decreases with z leading to an attenuation of the beam. On the other hand, if $N_2 > N_1$ then α is negative (and γ is positive) the beam is amplified with z. Figure 4.5 shows typical plots of $\alpha(\omega)$ versus ω for $N_1 > N_2$ and $N_2 > N_1$. Obviously the frequency dependence of α will be almost the same as that of the lineshape function $g(\omega)$. The condition $N_2 > N_1$ is called population inversion and it is under this condition that one can obtain optical amplification.

In Eq. (4.26) if $(N_1 - N_2)$ is independent of I, then we have from Eq. (4.25)

$$
I(z) = I(0)e^{-\alpha z}
\tag{4.27}
$$

i.e., an exponential attenuation when $N_1 > N_2$ and an exponential amplification when $N_2 > N_1$. We should mention that such an exponential decrease or increase of intensity is obtained for low intensities; for large intensities saturation sets in and $(N_1 - N_2)$ is no longer independent of I (see Chapter 5).

Example 4.3 We consider a ruby laser (see Chapter 11) with the following characteristics:

$$
n_0 = 1.76,\ t_{sp} = 3 \times 10^{-3}{\rm s},\ \lambda_0 = 6943\mathrm{\AA}
$$

$$
g(\omega_0) \approx 1/\Delta\omega \approx 1.1 \times 10^{-12}{\rm s}
$$

Fig. 4.5 A typical variation of $\alpha(\omega)$ with w for an amplifying medium corresponding to $N_2 > N_1$ (*lower* curve) and for an attenuating medium with $N_2 < N_1$ (*upper* curve)

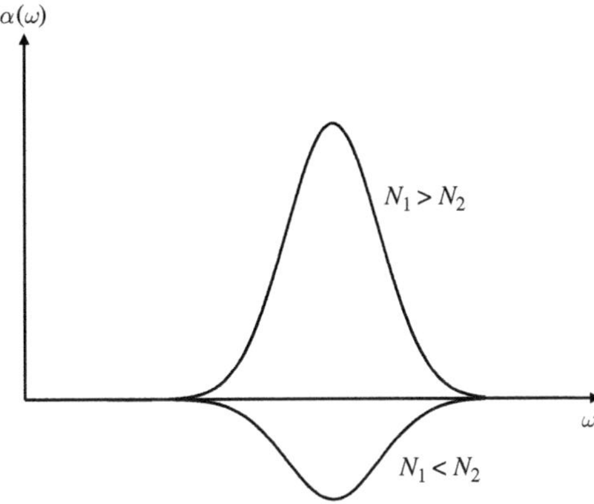

where we have assumed that for the normalized lineshape function[1]

$$g(\omega_0) \approx \frac{1}{\Delta\omega} \tag{4.28}$$

where $\Delta\omega$ represents the full width at half maximum of the lineshape function and ω_0 represents the frequency at the centre of the line. At thermal equilibrium at 300 K,

$$\frac{N_2}{N_1} = e^{-h\nu/k_BT} \approx 10^{-30} \approx 0$$

A typical chromium ion density in a ruby laser is about 1.6×10^{19} cm^{-3} and since at 300 K most atoms are in the ground level, the absorption coefficient at the centre of the line would be

$$\alpha = 1.4 \times 10^{-19}(N_1 - N_2) \approx 1.4 \times 10^{-19} \times 1.6 \times 10^{19}$$
$$\approx 2.2 \, \text{cm}^{-1}$$

If a population inversion density of 5×10^{16} cm^{-3} is generated (which represents a typical value) then the gain coefficient will be

$$-\alpha \approx 1.4 \times 10^{-19} \times 5 \times 10^{16}$$
$$\approx 7 \times 10^{-3} \, \text{cm}^{-1}$$

Example 4.4 As another example we consider the Nd:YAG laser (see Chapter 11) for which

$$n_0 = 1.82, \quad t_{sp} = 0.23 \times 10^{-3}\text{s}, \quad \lambda_0 = 1.06 \, \mu\text{m}$$

$$\Delta\nu = \frac{\Delta\omega}{2\pi} \approx \frac{1}{2\pi g(\omega_0)} \approx 1.95 \times 10^{11} \, \text{Hz}$$

If we want a gain of 1 m^{-1}, the inversion required can be calculated from Eq. (4.26) as

$$(N_2 - N_1) = \frac{4\nu^2 n_0^2 t_{sp}\alpha}{c^2 g(\omega)}$$
$$\approx 3.3 \times 10^{15} \, \text{cm}^{-3}$$

---

[1] We will show in Section 4.5 that $g(\omega_0)\Delta\omega$ equals $(2/\pi)$ and $(4\ln 2/\pi)^{1/2}$ for Lorentzian and Gaussian lineshape functions, respectively.

4.4 The Threshold Condition

In the last section we saw that in order that a medium be capable of amplifying incident radiation, one must create a state of population inversion in the medium. Such a medium will behave as an amplifier for those frequencies which fall within its linewidth. In order to generate radiation, this amplifying medium is placed in an optical resonator which consists of a pair of mirrors facing each other much like in a Fabry–Perot etalon (see Fig. 4.6). Radiation which bounces back and forth between the mirrors is amplified by the amplifying medium and also suffers losses due to the finite reflectivity of the mirrors and other scattering and diffraction losses. If the oscillations have to be sustained in the cavity then the losses must be exactly compensated by the gain. Thus a minimum population inversion density is required to overcome the losses and this is called the threshold population inversion.

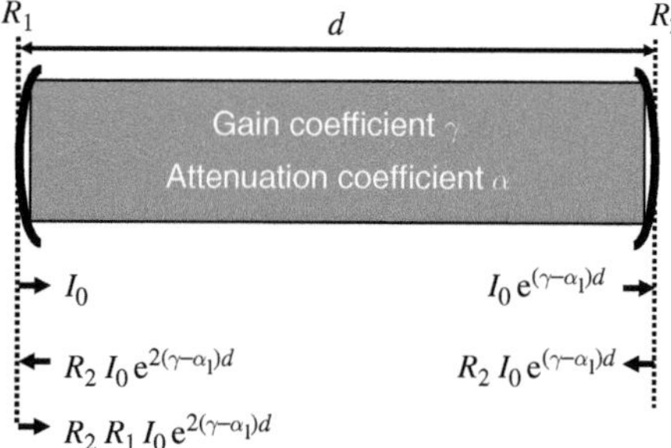

Fig. 4.6 A typical optical resonator consisting of a pair of mirrors facing each other. The active medium is placed inside the cavity

In order to obtain an expression for the threshold population inversion, let d represent the length of the resonator and let R_1 and R_2 represent the reflectivities of the mirrors (see Fig. 4.6). Let α_1 represent the average loss per unit length due to all loss mechanisms (other than the finite reflectivity) such as scattering loss and diffraction loss due to finite mirror sizes. Let us consider a radiation with intensity I_0 leaving mirror M_1. As it propagates through the medium and reaches the second mirror, it is amplified by $e^{\gamma d}$ and also suffers a loss of $e^{-\alpha_l d}$; for an amplifying medium γ is positive and $e^{\gamma d} > 1$. The intensity of the reflected beam at the second mirror will be $I_0 R_2 e^{(\gamma-\alpha_l)d}$. A second passage through the resonator and a reflection at the first mirror leads to an intensity for the radiation after one complete round trip of $I_0 R_1 R_2 e^{2(\gamma-\alpha_l)d}$. Hence for laser oscillation to begin

$$R_1 R_2 e^{2(\gamma-\alpha_l)d} \geq 1 \tag{4.29}$$

the equality sign giving the threshold value for α (i.e., for population inversion). Indeed, when the laser is oscillating in a steady state with a continuous wave oscillation, then the equality sign in Eq. (4.29) must be satisfied. If the inversion is increased then the LHS becomes greater than unity; this implies that the round trip gain is greater than the round trip loss. This would result in an increasing intensity inside the laser till saturation effects take over, which would result in a decrease in the inversion (we shall explicitly show saturation effects in Chapter 5). Thus the gain is brought back to its value at threshold.

Equation (4.29) can be written as

$$\gamma \geq \alpha_1 - \frac{1}{2d}\ln R_1 R_2 \tag{4.30}$$

This RHS of Eq. (4.30) depends on the passive cavity parameters only. This can be related to the passive cavity lifetime t_c which is the time in which energy in the cavity reduces by a factor $1/e$. In the absence of amplification by the medium, the intensity at a point reduces by a factor $R_1 R_2 e^{-2\alpha_1 d} = e^{-(2\alpha_1 d - \ln R_1 R_2)}$ in a time corresponding to one round-trip time. One round-trip time corresponds to $t = 2d(c/n_0) = 2dn_0/c$. Hence if the intensity reduces as e^{-t/t_c}, then in a time $t = 2dn_0/c$, the factor by which the intensity will be reduced is e^{-2dn_0/ct_c}. Thus

$$e^{-(2\alpha_1 d - \ln R_1 R_2)} = e^{-2dn_0/ct_c}$$

or

$$\frac{1}{t_c} = \frac{c}{2dn_0}(2\alpha_1 d - \ln R_1 R_2) \tag{4.31}$$

Using Eqs. (4.26) and (4.31), Eq. (4.30) becomes

$$(N_2 - N_1) \geq \frac{4v^2 n_0^3}{c^3}\frac{t_{sp}}{t_c}\frac{1}{g(\omega)} \tag{4.32}$$

Corresponding to the equality sign, we have the threshold population inversion density required for the oscillation of the laser.

According to Eq. (4.32), in order to have a low threshold value of the population inversion, the following conditions must hold:

(a) The value of t_c should be large, i.e., the cavity losses must be small.
(b) Since $g(\omega)$ is normalized according to Eq. (4.13) the peak value of $g(\omega)$ will be inversely proportional to the width $\Delta\omega$ of the $g(\omega)$ function [see Eq. (4.28)]. Thus smaller widths give larger values of $g(\omega)$ which implies lower threshold values of $(N_2 - N_1)$. Also since the largest $g(\omega)$ appears at the line centre, the resonator mode which lies closest to the line centre will reach threshold first and begin to oscillate.
(c) Smaller values of t_{sp} (i.e., strongly allowed transitions) also lead to smaller values of threshold inversion. At the same time for smaller relaxation times (t_{sp}), larger pumping power will be required to maintain a given population inversion. In general, population inversion is more easily obtained on transitions which have longer relaxation times.
(d) The value of $g(\omega)$ at the centre of the line is inversely proportional to $\Delta\omega$ which, for example, in the case of Doppler broadening is proportional to ω (see Section 4.5). Thus the threshold population inversion increases approximately in proportion to ω^3. Thus it is much easier to obtain laser action at infrared wavelengths than in the ultraviolet region.

Example 4.5 We first consider a ruby laser[2] which has the following typical parameters:

$$\lambda_0 = 6943\text{Å},\ t_{\text{sp}} \approx 3 \times 10^{-3}\text{s},\ n_0 = 1.76,\ d = 5\ \text{cm}$$

$$R_1 = R_2 = 0.9,\ \alpha_1 \approx 0$$

$$g(\omega_0) = \frac{1}{\Delta\omega} = \frac{1}{2\pi\,\Delta\nu} \approx 1.1 \times 10^{-12}\text{s}$$

Thus for the above values

$$t_c \approx 2.8 \times 10^{-9}\text{s}$$

and

$$(N_2 - N_1)_{\text{th}} \approx 1.5 \times 10^{17}\ \text{cm}^{-3}$$

Typical Cr^{+3} ion densities are about 1.6×10^{19} cm^{-3}. Thus the fractional excess population is very small. The above population inversion corresponds to a gain of about $0.02\ \text{cm}^{-1}$ or to 0.09 dB/cm.

Example 4.6 As another example, we consider a He–Ne laser with the following typical characteristics:

$$\lambda_0 = 6328\text{Å},\ t_{\text{sp}} = 10^{-7}\ \text{s},\ n_0 \approx 1,\ d = 20\,\text{cm}$$

$$R_1 = R_2 = 0.98,\ \alpha_1 \approx 0$$

$$\Delta\nu \approx 10^9 \text{Hz}$$

$$g(\omega_0) \approx \frac{1}{2\pi\,\Delta\nu} \approx 0.16 \times 10^{-9}\text{s}$$

for the above values

$$t_c \approx 3.3 \times 10^{-8}\text{s}$$

and

$$(N_2 - N_1)_{\text{th}} \approx 6.24 \times 10^8\ \text{cm}^{-3}$$

4.5 Line Broadening Mechanisms

As we mentioned in Section 4.2 the radiation coming out of a collection of atoms making transitions between two energy levels is never perfectly monochromatic. This line broadening is described in terms of the lineshape function $g(\omega)$ that was introduced in Section 4.2. In this section, we shall discuss some important line broadening mechanisms and obtain the corresponding $g(\omega)$. A study of line broadening is extremely important since it determines the operation characteristics of the laser such as the threshold population inversion and the number of oscillating modes.

The various broadening mechanisms can be broadly classified as homogeneous or inhomogeneous broadening. In the case of homogenous broadening (like natural or collision broadening) the mechanisms act to broaden the response of each atom in an identical fashion, and for such a case the probability of absorption or emission of radiation of a certain frequency is the same for all atoms in the collection. Thus there

---

[2]Ruby laser active medium consists of Cr^{+3}-doped ion Al_2O_3 and is an example of a three level laser. More details regarding the ruby laser are given in Section 10.2.

is nothing which distinguishes one group of atoms from another in the collection. In the case of inhomogeneous broadening, different groups of atoms are distinguished by different frequency responses. Thus, for example, in Doppler broadening groups of atoms having different velocity components are distinguishable and they have different spectral responses. Similarly broadening caused by local inhomogeneities of a crystal lattice acts to shift the central frequency of the response of individual atoms by different amounts, thereby leading to inhomogeneous broadening. In the following, we shall discuss natural, collision, and Doppler broadening.

4.5.1 Natural Broadening

We have seen earlier that an excited atom can emit its energy in the form of spontaneous emission. In order to investigate the spectral distribution of this spontaneous radiation, we recall that the rate of decrease of the number of atoms in level 2 due to transitions from level 2 to level 1 is [see Eq. (4.11)]

$$\frac{dN_2}{dt} = -A_{21}N_2 \tag{4.33}$$

For every transition an energy $\hbar\omega_0 = E_2 - E_1$ is released. Thus the energy emitted per unit time per unit volume will be

$$W(t) = \left|\frac{dN_2}{dt}\right|\hbar\omega_0$$
$$= N_{20}A_{21}\hbar\omega_0 e^{-A_{21}t} \tag{4.34}$$

where we have used Eqs. (4.33) and (4.12). Since Eq. (4.34) describes the variation of the intensity of the spontaneously emitted radiation, we may write the electric field associated with the spontaneous radiation as

$$E(t) = E_0\, e^{i\omega_0 t} e^{-t/2t_{sp}} \tag{4.35}$$

where $t_{sp} = 1/A_{21}$ and we have used the fact that intensity is proportional to the square of the electric field. Thus the electric field associated with spontaneous emission decreases exponentially.

In order to calculate the spectrum associated with the wave described by the Eq. (4.35), we first take the Fourier transform:

$$\tilde{E}(\omega) = \int_{-\infty}^{\infty} E(t)\, e^{-i\omega t} dt$$
$$= E_0 \int_{0}^{\infty} \exp\left[i\,(\omega_0 - \omega)\,t - t/2t_{sp}\right] dt \tag{4.36}$$
$$= E_0 \frac{1}{\frac{1}{2t_{sp}} + i(\omega - \omega_0)}$$

where $t = 0$ is the time at which the atoms start emitting radiation. The power spectrum associated with the radiation will be proportional to $|E_0(\omega)|^2$. Hence we may write the lineshape function associated with the spontaneously emitted radiation as

$$g(\omega) = K \frac{1}{(\omega - \omega_0)^2 + 1/4t_{sp}^2}$$

where K is a constant of proportionality which is determined such that $g(\omega)$ satisfies the normalization condition given by Eq. (4.13). Substituting for $g(\omega)$ in Eq. (4.13) and integrating, one can show that

$$K = \frac{1}{2\pi\, t_{sp}}$$

Thus the normalized lineshape function is

$$g(\omega) = \frac{2t_{sp}}{\pi} \frac{1}{1 + 4(\omega - \omega_0)^2 t_{sp}^2} \tag{4.37}$$

The above functional form is referred to as a Lorentzian and is plotted in Fig. 4.7. The full width at half maximum (FWHM) of the Lorentzian is

$$\Delta\omega_N = \frac{1}{t_{sp}} \tag{4.38}$$

Thus, Eq. (4.37) can also be written as

$$g(\omega) = \frac{2}{\pi\,\Delta\omega_N} \frac{1}{1 + 4(\omega - \omega_0)^2/(\Delta\omega_N)^2} \tag{4.39}$$

A more precise derivation of Eq. (4.39) is given in Appendix G.

Example 4.7 The spontaneous lifetime of the sodium level leading to a D_1 line ($\lambda = 589.1$nm) is 16 ns. Thus the natural linewidth (FWHM) will be

$$\Delta\nu_N = \frac{1}{2\pi\, t_{sp}} \approx 10 \text{ MHz} \tag{4.40}$$

which corresponds to $\Delta\lambda \approx 0.001$ nm.

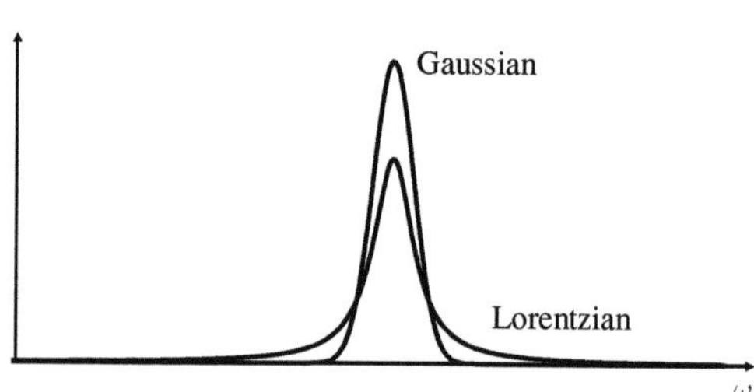

Fig. 4.7 The Lorentzian and Gaussian lineshape functions having the same FWHM

4.5.2 Collision Broadening

In a gas, random collisions occur between the atoms. In such a collision process, the energy levels of the atoms change when the atoms are very close due to their mutual interaction. Let us consider an atom which is emitting radiation and which collides with another atom. When the colliding atoms are far apart, their energy levels are unperturbed and the radiation emitted is purely sinusoidal (if we neglect the decay in the amplitude due to spontaneous emission). As the atoms come close together their energy levels are perturbed and thus the frequency of emission changes during the collision time. After the collision the emission frequency returns to its original value.

If τ_c represents the time between collisions and $\Delta\tau_c$ the collision time then one can obtain order of magnitude expressions as follows:

$$\Delta\tau_c \approx \frac{\text{interatomic distance}}{\text{average thermal velocity}}$$

$$\approx \frac{1\,\text{Å}}{500\,\text{m/s}} \approx 2 \times 10^{-13}\,\text{s}$$

$$\tau_c \approx \frac{\text{mean free path}}{\text{average thermal velocity}} \approx \frac{5 \times 10^{-4}\,\text{m}}{500\,\text{m/s}}$$

$$\approx 10^{-6}\,\text{s}$$

Thus the collision time is very small compared to the time between collisions and hence the collision may be taken to be almost instantaneous. Since the collision time $\Delta\tau_c$ is random, the phase of the wave after the collision is arbitrary with respect to the phase before the collision. Thus each collision may be assumed to lead to random phase changes as shown in Fig. 4.8. The wave shown in Fig. 4.8 is no longer monochromatic and this broadening is referred to as *collision broadening*.

In order to obtain the lineshape function for collision broadening, we note that the field associated with the wave shown in Fig. 4.8 can be represented by

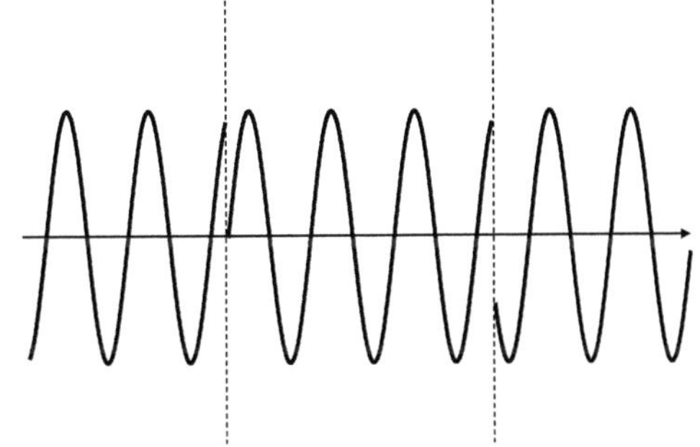

Fig. 4.8 The wave coming out of an atom undergoing random collisions at which there are abrupt phase changes

$$E(t) = E_0\, e^{i(\omega_0 t + \phi)} \tag{4.41}$$

where the phase ϕ remains constant for $t_0 \leq t \leq t_0 + \tau_c$ and at each collision the phase ϕ changes randomly.

Since the wave is sinusoidal between two collisions, the spectrum of such a wave will be given by

$$
\begin{aligned}
\tilde{E}(\omega) &= \frac{1}{2\pi} \int_{t_0}^{t_0 + \tau_c} E_0\, e^{i(\omega_0 t + \phi)}\, e^{-i\omega t}\, dt \\
&= \frac{1}{2\pi} E_0\, e^{i[(\omega_0 - \omega)t_0 + \phi]} \frac{e^{i(\omega_0 - \omega)\tau_c} - 1}{i(\omega_0 - \omega)}
\end{aligned}
\tag{4.42}
$$

The power spectrum of such a wave will be

$$I(\omega) \propto |\tilde{E}(\omega)|^2 = \left(\frac{E_0}{\pi}\right)^2 \frac{\sin^2\left[(\omega - \omega_0)\tau_c/2\right]}{(\omega - \omega_0)^2} \tag{4.43}$$

Now, at any instant, the radiation coming out of the atomic collection would be from atoms with different values of τ_c. In order to obtain the power spectrum we must multiply $I(\omega)$ by the probability $P(\tau_c)d\tau_c$ that the atom suffers a collision in the time interval between τ_c and $\tau_c + d\tau_c$ and integrate over τ_c from 0 to ∞. It can be shown from kinetic theory that (see, e.g., Gopal (1974))

$$P(\tau_c)d\tau_c = \left(\frac{1}{\tau_0}\right) e^{-\tau_c/\tau_0} d\tau_c \tag{4.44}$$

where τ_0 represents the mean time between two collisions. Notice that

$$\int_0^\infty P(\tau_c)d\tau_c = 1, \qquad \int_0^\infty \tau_c P(\tau_c)d\tau_c = \tau_0 \tag{4.45}$$

Hence the lineshape function for collision broadening will be

$$
\begin{aligned}
g(\omega) &\propto \int_0^\infty I(\omega)\, P(\tau_c)d\tau_c \\
&= \left(\frac{E_0}{\pi}\right)^2 \frac{1}{2} \frac{1}{(\omega - \omega_0)^2 + 1/\tau_0^2}
\end{aligned}
$$

which is again a Lorentzian. The normalized lineshape function will thus be

$$g(\omega)d\omega = \frac{\tau_0}{\pi} \frac{1}{1 + (\omega - \omega_0)^2 \tau_0^2} d\omega \tag{4.46}$$

and the FWHM will be

$$\Delta\omega_c = 2/\tau_0 \tag{4.47}$$

Thus a mean collision time of $\sim 10^{-6}$ s corresponds to a $\Delta\nu$ of about 0.3 MHz.

The mean time between collisions depends on the mean free path and the average speed of the atoms in the gas which in turn would depend on the pressure and temperature of the gas as well as the mass of the atom. An approximate expression for the average collision time is

$$\tau_0 = \frac{1}{8\pi} \left(\frac{2}{3}\right)^{1/2} \frac{(Mk_B T)^{1/2}}{pa^2}$$

where M is the atomic mass, a is the radius of the atom (assumed to be a hard sphere), and p is the pressure of the gas.

Example 4.8 In a He–Ne laser the pressure of gas is typically 0.5 torr. (Torr is a unit of pressure and 1 Torr = 1 mm of Hg). If we assume $a \sim 0.1$ nm, $T = 300$ K, $M = 20 \times 1.67 \times 10^{-27}$ kg , we obtain $\tau_0 \sim 580$ ns.

Problem 4.1 In the presence of both natural and collision broadening, in addition to the sudden phase changes at every collision, there will also be an exponential decay of the field as represented by Eq. (4.35). Show that in such a case, the FWHM is given by

$$\Delta\omega = \frac{1}{t_{sp}} + \frac{2}{t_0} \tag{4.48}$$

4.5.3 Doppler Broadening

In a gas, atoms move randomly and when a moving atom interacts with electromagnetic radiation, the apparent frequency of the wave is different form that seen from a stationary atom; this is called the Doppler effect and the broadening caused by this is termed Doppler broadening.

In order to obtain $g(\omega)$ for Doppler broadening, we consider radiation of frequency ω passing through a collection of atoms which have a resonant frequency ω_0 and which move randomly (we neglect natural and collision broadening in this discussion). In order that an atom may interact with the incident radiation, it is necessary that the apparent frequency seen by the atom in its frame of reference be ω_0. If the radiation is assumed to propagate along the z-direction, then the apparent frequency seen by the atom having a z-component of velocity v_z will be

$$\tilde{\omega} = \omega\left(1 - \frac{v_z}{c}\right) \tag{4.49}$$

Hence for a strong interaction, the frequency of the incident radiation must be such that $\tilde{\omega} = \omega_0$. Thus

$$\omega = \omega_0\left(1 - \frac{v_z}{c}\right)^{-1} \approx \omega_0\left(1 + \frac{v_z}{c}\right) \tag{4.50}$$

where we have assumed $v_z << c$. Thus the effect of the motion is to change the resonant frequency of the atom.

In order to obtain the $g(\omega)$ due to Doppler broadening, we note that the probability that an atom has a z component of velocity lying between v_z and $v_z + dv_z$ is given by the Maxwell distribution

$$P(v_z)\,dv_z = \left(\frac{M}{2\pi k_B T}\right)^{\frac{1}{2}} \exp\left(-\frac{Mv_z^2}{2k_B T}\right) dv_z \tag{4.51}$$

where M is the mass of the atom and T the absolute temperature of the gas. Hence the probability $g(\omega)d\omega$ that the transition frequency lies between ω and $\omega + d\omega$ is equal to the probability that the z component of the velocity of the atom lies between v_z and $v_z + dv_z$ where

$$v_z = \frac{(\omega - \omega_0)}{\omega_0}c$$

Thus

$$g(\omega)d\omega = \frac{c}{\omega_0}\left(\frac{M}{2\pi k_B T}\right)^{\frac{1}{2}} \exp\left[-\frac{Mc^2}{2k_B T}\frac{(\omega - \omega_0)^2}{\omega_0^2}\right]d\omega \tag{4.52}$$

which corresponds to a Gaussian distribution. The lineshape function is peaked at ω_0, and the FWHM is given by

$$\Delta\omega_D = 2\omega_0\left(\frac{2k_B T}{Mc^2}\ln 2\right)^{\frac{1}{2}} \tag{4.53}$$

In terms of $\Delta\omega_D$ Eq. (4.52) can be written as

$$g(\omega)d(\omega) = \frac{2}{\Delta\omega_D}\left(\frac{\ln 2}{\pi}\right)^{\frac{1}{2}} \exp\left[-4\ln 2\frac{(\omega - \omega_0)^2}{(\Delta\omega_D)^2}\right]d\omega \tag{4.54}$$

Figure 4.7 shows a comparative plot of a Lorentzian and a Gaussian line having the same FWHM. It can be seen that the peak value of the Gaussian is more and that the Lorentzian has a much longer tail. As an example, for the D_1 line of sodium $\lambda \cong$ 589.1 nm at $T = 500$ K, $\Delta v_D = 1.7 \times 10^9$ Hz which corresponds to $\Delta\lambda_D \approx 0.02$Å. For neon atoms corresponding to $\lambda = 6328$Å (the red line of the He–Ne laser) at 300 K, we have $\Delta v_D \approx 1600$ MHz where we have used $M_{Ne} \approx 20 \times 1.67 \times 10^{-27}$ kg. For the vibrational transition of the carbon dioxide molecule leading to the 10.6 μm radiation, at $T = 300$ K, we have

$$\Delta v_D \approx 5.6 \times 10^7 \text{Hz} \Rightarrow \Delta\lambda_D \approx 0.19\text{Å}$$

where we have used $M_{CO_2} \approx 44 \times 1.67 \times 10^{-27}$kg

In all the above discussions we have considered a single broadening mechanisms at a time. In general, all broadening mechanisms will be present simultaneously and the resultant lineshape function has to be evaluated by performing a convolution of the different lineshape functions.

Problem 4.2 Obtain the lineshape function in the presence of both natural and Doppler broadening

Solution From Maxwell's velocity distribution, the fraction of atoms with their center frequency lying between ω' and $\omega' + d\omega'$ is given by

$$f(\omega')d\omega' = \left(\frac{M}{2\pi k_B T}\right)^{\frac{1}{2}} \frac{c}{\omega_0} \exp\left[-\frac{Mc^2}{2k_B T} \frac{(\omega' - \omega_0)^2}{\omega_0^2}\right] d\omega' \tag{4.55}$$

These atoms are characterized by a naturally broadened lineshape function described by

$$h(\omega - \omega') = \frac{2t_{sp}}{\pi} \frac{1}{1 + (\omega - \omega')^2 4t_{sp}^2} \tag{4.56}$$

Thus the resultant lineshape function will be given by

$$g(\omega) = \int f(\omega')h(\omega - \omega')d\omega' \tag{4.57}$$

which is nothing but the convolution of $f(\omega)$ with $h(\omega)$

Example 4.9 Neodymium doped in YAG and in glass are two very important lasers. The host YAG is crystalline while glass is amorphous. Thus the broadening in YAG host is expected to be much smaller than in glass host. In fact the linewidth at 300 K for Nd:YAG is about 120 GHz while that for Nd:glass is about 5400 GHz.

4.6 Saturation Behavior of Homogeneously and Inhomogeneously Broadened Transitions

In Section 4.5 we discussed the various line broadening mechanisms belonging to both homogeneous and inhomogeneous broadenings. In this section, we briefly discuss the difference in saturation behavior between the two kinds of broadenings.

Let us first consider a homogeneously broadened laser medium placed inside a resonator and let us assume that there is a resonator mode coinciding exactly with the center of the line. Initially as the pumping rate is below threshold, the gain in the resonator is less than the losses and the laser does not oscillate. As the pumping rate is increased, first to reach threshold is the mode at the center as it has the minimum threshold. We have seen earlier that when the laser is oscillating in steady state, the gain is exactly equal to the loss at the oscillating frequency. Thus at steady state even when the pumping power is increased beyond threshold, the gain at the oscillating frequency does not increase beyond the threshold value; this is because of the fact that the losses remain constant. In fact, increasing the pumping power will be accompanied by an increase in the power in the mode which in turn would be accompanied by a stronger saturation of the laser transition, thus reducing the gain at the oscillation frequency again to the value at threshold. It may be mentioned that the gain could exceed the threshold value on a transient basis but not under steady state operation.

Now in a homogenously broadened transition all the atoms have identical line-shapes peaked at the same frequency. Thus all atoms interact with the same oscillating mode and the increase in pumping power cannot increase the gain at other frequencies and thus the laser will oscillate only in a single longitudinal mode (see Fig. 4.9). This observation has been verified experimentally on some homogeneously broadened transitions such as Nd:YAG laser. The fact that a laser with homogeneously broadened transition can oscillate in many modes is due to spatial hole burning. This can be understood from the fact that each mode is a standing wave pattern between the resonator mirrors. Thus there are regions of high population inversion (at the nodes of the field where the field amplitude is very small) and regions of saturated population inversion (at the antinodes of the field where the field has maximum value). If one considers another mode which has (at least over some portions) antinodes at the nodes corresponding to the central oscillating mode, then this mode can draw energy from the atoms and, if the loss can be compensated by gain, this mode can also oscillate.

In contrast to the case of homogeneous broadening, if the laser medium is inhomogeneouly broadened then a given mode at a central frequency can interact with only a group of atoms whose response curve contains the mode frequency (see Fig. 4.10). Thus if the pumping is increased beyond threshold, the gain at the oscillating frequency remains fixed but the gain at other frequencies can go on increasing (see Fig. 4.10). Thus, in an inhomogeneously broadened line one can have multimode oscillation and as one can see from Fig. 4.10. Each oscillating mode "burns holes in the frequency space" of the gain profile. These general conclusions regarding homogeneously and inhomogeneoulsy broadened lines have been verified experimentally.

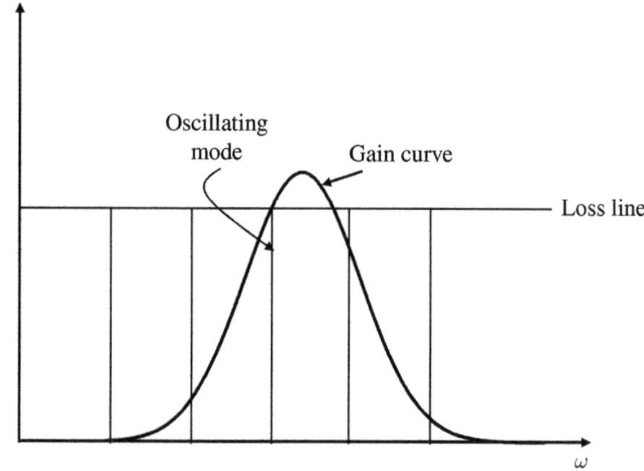

Fig. 4.9 In a homogeneously broadened transition, gain can compensate loss at only one oscillating mode leading to single longitudinal mode operation

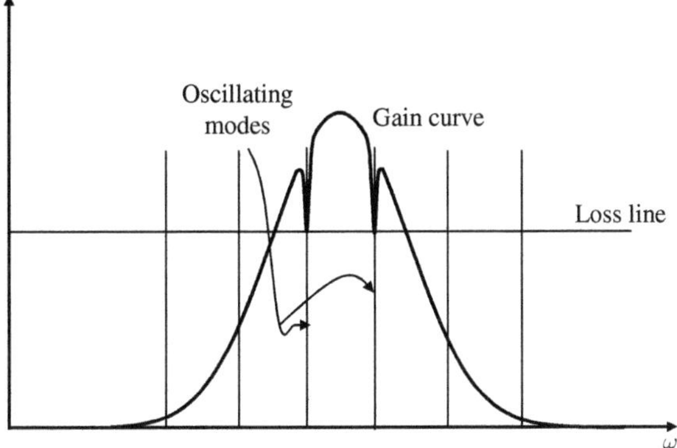

Fig. 4.10 As the pumping is increased beyond threshold, under steady-state operation the gain at the various oscillating frequencies cannot increase beyond the threshold value but the gain at other frequencies may be much above the threshold value. The various frequencies are said to burn holes in the gain curve

Various techniques for single longitudinal mode oscillation of inhomogeneously broadened lasers are discussed in Chapter 7.

Let us now consider an inhomogeneously broadened laser medium and let us assume that only a single mode exists within the entire gain profile. Let us also assume to begin with that the frequency of the mode does not coincide with the line center and that we slowly change the frequency of the mode so that it passes through the center of the profile to the other side of the peak in the gain profile. In order to determine the variation of the power output as the frequency is scanned through the line center, we observe that a mode of the laser is actually made up of two traveling waves traveling along opposite directions along the resonator axis. Thus when the mode frequency does not coincide with the line center, the wave travelling from left to right in the resonator will interact with those atoms whose z-directed velocities are near to [see Eq. (4.49)]:

$$v_z = \frac{\omega - \omega_{21}}{\omega_{21}} c \tag{4.58}$$

while the wave moving from right to left would interact with those atoms whose z-directed velocity would be

$$v_z = -\frac{\omega - \omega_{21}}{\omega_{21}} c \tag{4.59}$$

Thus there are two groups of atoms with equal and opposite z-directed velocities which are strongly interacting with the mode. As the frequency of the mode is tuned to the center these groups of atoms change with the frequency, and at the line center, the mode can interact only with the groups of atoms having a zero value of z-directed velocity. Thus the power output must decrease slightly when the mode frequency is tuned through the line center. In fact, this has been observed experimentally and is referred to as the Lamb dip – the presence of a Lamb dip in a He–Ne laser was shown by McFarlane, Bennet, and Lamb (1963).

4.7 Quantum Theory for the Evaluation of the Transition Rates and Einstein Coefficients

For the calculation of transition rates we consider the atom to be in the presence of an oscillating electric field given by

$$\mathbf{E}(t) = \hat{\mathbf{e}}\,E_0\,\cos\omega t \qquad (4.60)$$

which is switched on at $t = 0$; $\hat{\mathbf{e}}$ represents the unit vector along the direction of the electric field. The frequency ω is assumed to be very close to the resonant frequency $\left[(E_2 - E_1)/\hbar\right]$ corresponding to the transition from state 1 to 2 (see Fig. 4.1). We will show that the presence of the higher excited states can be neglected because of the corresponding transition frequencies are far away from ω. In the presence of the electric field, the time-dependent Schrödinger equation becomes

$$i\hbar\frac{\partial\Psi}{\partial t} = \left(H_0 + H'\right)\Psi \qquad (4.61)$$

where

$$H' = -e\mathbf{E}.\mathbf{r} = -eE_0(\hat{\mathbf{e}}.\mathbf{r})\cos\omega t \qquad (4.62)$$

represents the interaction energy of the electron with the electric field and H_0 (which is independent of time) represents the Hamiltonian of the atom; $e\ (< 0)$ represents the charge of the electron.[3] Since H_0 is independent of time, the solution of the Schrödinger equation

$$i\hbar\frac{\partial\Psi}{\partial t} = H_0\Psi \qquad (4.63)$$

is of the form

$$\Psi = \sum \psi_n(\mathbf{r})e^{-iE_nt/\hbar} \qquad (4.64)$$

where $\psi_n(\mathbf{r})$ and E_n are the eigenfunctions and eigenvalues of H_0:

$$H_0\psi_n(\mathbf{r}) = E_n\psi_n(\mathbf{r}) \qquad (4.65)$$

The functions $\psi_n(\mathbf{r})$ are known as the atomic wave functions and satisfy the orthonormality condition

$$\int \psi_n^*(\mathbf{r})\psi_m(\mathbf{r})d\tau = \delta_{mn} = \begin{cases} 0 & \text{if } n \neq m \\ 1 & \text{if } n = m \end{cases} \qquad (4.66)$$

---

[3]We are considering here a single electron atom with r representing the position of the electron with respect to the nucleus. Thus the electric dipole moment of the atom is given by $p = e\,r$ because the direction of the dipole moment is from negative to the positive charge. The interaction energy of a dipole placed in an electric field E is $-\vec{p}.E$ is which leads to Eq. (4.62).

The solution of Eq. (4.63) can be written as a linear combination of the atomic wave functions:

$$\Psi(\mathbf{r}, t) = \sum_n C_n(t)\psi_n(\mathbf{r})e^{-i\omega_n t} \tag{4.67}$$

where

$$\omega_n = \frac{E_n}{\hbar} \tag{4.68}$$

and the coefficients are now time dependent to account for transitions among the various energy levels due to the perturbation. Substituting from Eq. (4.67) in Eq. (4.63) we obtain

$$i\hbar \sum_n \left(\frac{dC_n}{dt} - i\omega_n C_n\right)e^{-i\omega_n t}\psi_n(\mathbf{r}) = \sum_n E_n C_n(t)\psi_n(\mathbf{r})e^{-i\omega_n t}$$
$$- eE_0\left(\hat{\mathbf{e}}.\mathbf{r}\right)\sum_n C_n(t)\psi_n(\mathbf{r})e^{-i\omega_n t}\cos\omega t$$

where we have used Eq. (4.65). It is immediately seen that the second term on the left-hand side exactly cancels with the first term on the right-hand side. If we multiply by ψ_m^* and integrate we would get

$$i\hbar\frac{dC_m}{dt} = \frac{1}{2}E_0\sum_n D_{mn}C_n(t)\left(e^{i(\omega_{mn}+\omega)t} + e^{i(\omega_{mn}-\omega)t}\right) \tag{4.69}$$

where use has been made of the orthogonality relation [Eq. (4.66)] and

$$\omega_{mn} = \omega_m - \omega_n = \frac{E_m - E_n}{\hbar} \tag{4.70}$$

$$D_{mn} = \hat{\mathbf{e}}.\mathbf{P}_{mn} \tag{4.71}$$

$$\mathbf{P}_{mn} = -e\int \psi_m^*(\mathbf{r})\mathbf{r}\psi_n(\mathbf{r})d\tau = |e|\int \psi_m^*(\mathbf{r})\mathbf{r}\psi_n(\mathbf{r})d\tau \tag{4.72}$$

We wish to solve Eq. (4.69) subject to the boundary condition

$$\begin{aligned} C_k(t=0) &= 1 \\ C_n(t=0) &= 0 \qquad \text{for } n \neq k \end{aligned} \tag{4.73}$$

i.e., at $t = 0$, the atom is assumed to be in the state characterized by the wave function ψ_k. Equation (4.69) represents an infinite set of coupled equations, and as a first approximation, one may replace $C_n(t)$ by $C_n(0)$ on the right-hand side of Eq. (4.69). Thus

$$i\hbar\frac{dC_m}{dt} = \frac{1}{2}E_0 D_{mk}\left(e^{i(\omega_{mk}+\omega)t} + e^{i(\omega_{mk}-\omega)t}\right) \tag{4.74}$$

Integrating, one obtains

$$C_m(t) - C_m(0) \approx -\frac{E_0}{2\hbar}\mathbf{D}_{mk}\left[\frac{e^{i(\omega_{mk}+\omega)t} - 1}{(\omega_{mk} + \omega)} + \frac{e^{i(\omega_{mk}-\omega)t} - 1}{(\omega_{mk} - \omega)}\right] \quad (4.75)$$

or, for $m \neq k$

$$C_m(t) \approx -i\frac{E_0}{\hbar}\mathbf{D}_{mk}\left[e^{i(\omega_{mk}+\omega)t/2}\frac{\sin(\omega_{mk}+\omega)\,t/2}{(\omega_{mk} + \omega)} + e^{i(\omega_{mk}-\omega)t/2}\frac{\sin(\omega_{mk}-\omega)\,t/2}{(\omega_{mk} - \omega)}\right] \quad (4.76)$$

It can be easily seen that for large values of t, the function

$$\frac{\sin(\omega_{mk} - \omega)\,t/2}{(\omega_{mk} - \omega)}$$

is very sharply peaked around $\omega \approx \omega_{mk}$ and negligible everywhere else (see Fig. 4.11). Thus for states for which ω_{mk} is significantly different from ω, $C_m(t)$ would be negligible and transitions between such states will not be stimulated by the incident field. This justifies our earlier statement that the presence of only those excited states be considered which are close to the resonance frequency.

In an emission process, $\omega_k > \omega_m$ and hence ω_{mk} is negative; thus it is the first term on the right-hand side of Eq. (4.76) which contributes. On the other hand, in an absorption process, $\omega_{mk} > 0$ and the second term in Eq. (4.76) contributes.

We consider absorption of radiation and assume that at $t = 0$ the atom is in state 1, the corresponding wave function being $\psi_1(\vec{r})$. We also assume ω to be close to $\omega_{21}\left[= (E_2 - E_1)/\hbar\right]$ – see Fig. 4.1. The probability for the transition to occur to state 2 is given by

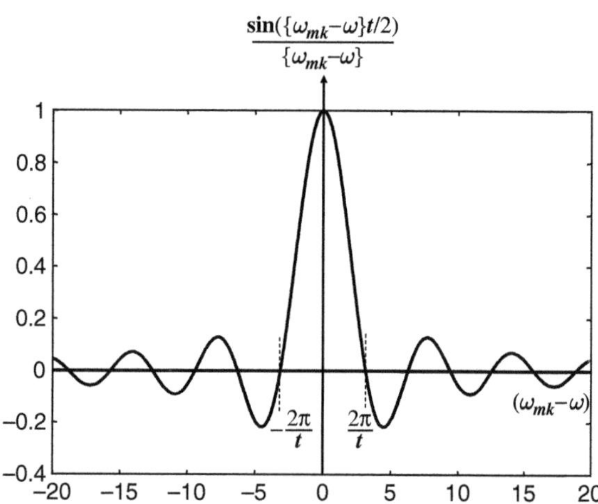

Fig. 4.11 For large values of t, the function $\frac{\sin(\omega_{mk}-\omega)t/2}{(\omega_{mk}-\omega)}$ is very sharply peaked around $\omega = \omega_{mk}$

$$|C_2(t)|^2 = \frac{1}{4} \frac{D_{21}^2 E_0^2}{\hbar^2} \left\{ \frac{\sin\left[(\omega_{21} - \omega)/2\right]t}{(\omega_{21} - \omega)/2} \right\}^2 \tag{4.77}$$

The above expression represents the probability for stimulated absorption of radiation. In deriving Eq. (4.77) we have assumed that $|C_2(t)|^2 \ll 1$; thus the result is accurate when

$$\frac{D_{21}^2 E_0^2 t^2}{\hbar^2} \ll 1 \quad \text{or} \quad \frac{\left(\frac{D_{21}^2 E_0^2}{\hbar^2}\right)}{(\omega_{21} - \omega)^2} \ll 1 \tag{4.78}$$

A more exact result for a two-state system will be discussed in Section 4.8.

We next assume that the quantity $(\omega_{21} - \omega)$ has a range of values either on account of the field having a continuous spectrum or the atom is capable of interaction with radiation having a range of frequencies.

4.7.1 Interaction with Radiation Having a Broad Spectrum

We first consider the field having a continuous spectrum characterized by $u(\omega)$ which is defined such that $u(\omega)\, d\omega$ represents the energy associated with the field per unit volume within the frequency interval ω and $\omega + d\omega$. Since the average energy density associated with an electromagnetic wave is $(1/2)\varepsilon_0 E_0^2$ where ε_0 is the permittivity of free space, we replace E_0^2 in Eq. (4.77) by $(2/\varepsilon_0)\, u(\omega) d\omega$ and integrate over all frequencies, which gives us the following expression for the transition probability:

$$\Gamma_{12} = \frac{1}{2\varepsilon_0} \frac{D_{21}^2}{\hbar^2} \int u(\omega) \left\{ \frac{\sin\left[(\omega_{21} - \omega)/2\right]t}{(\omega_{21} - \omega)/2} \right\}^2 d\omega \tag{4.79}$$

Assuming that $u(\omega)$ varies very slowly in comparison to the quantity inside the square brackets, we replace $u(\omega)$ by its value at $\omega = \omega_{21}$ and take it out of the integral to obtain

$$\Gamma_{12} \approx \frac{1}{2\varepsilon_0} \frac{D_{21}^2}{\hbar^2} u(\omega_{21}) \left(\int \frac{\sin^2 \xi}{\xi^2} d\xi \right) 2t$$

$$= \frac{\pi}{\varepsilon_0} \frac{D_{21}^2}{\hbar^2} u(\omega_{21}) t \tag{4.80}$$

where $\xi = \frac{\omega_{21} - \omega}{2} t$. The above expression shows that[4] the probability of transition is proportional to time; thus the probability per unit time (which we denote by w_{12})

---

[4]It may be noted that Eq. (4.80) predicts an indefinite increase in the transition probability with time; however, the first-order perturbation theory itself breaks down when Γ_{21} is not appreciably less than unity. Thus Eq. (4.80) gives correct results as long as $\Gamma_{21} \ll 1$.

would be given by

$$w_{12} \approx \frac{\pi}{\varepsilon_0} \frac{D_{21}^2}{\hbar^2} u(\omega_{21}) \tag{4.81}$$

Now (omitting the subscripts) we have

$$\mathrm{D} = \hat{\mathbf{e}}.\mathbf{P} = P \cos\theta \tag{4.82}$$

where θ is the angle that $\hat{\mathbf{e}}$ (i.e., the electric field) makes with the dipole moment vector $\mathbf{P}$. Assuming that the dipole moment vector is randomly oriented, the average value of D^2 is given by

$$\overline{\mathrm{D}}^2 = P^2 \left\langle \cos^2\theta \right\rangle = \frac{1}{3} P^2 \tag{4.83}$$

where use has been made of the following relation:

$$\left\langle \cos^2\theta \right\rangle = \frac{1}{4\pi} \int\limits_0^{2\pi} \int\limits_0^{\pi} \cos^2\theta \, \sin\theta \, \mathrm{d}\theta \, \mathrm{d}\phi = \frac{1}{3} \tag{4.84}$$

Thus

$$w_{12} = \frac{\pi}{3\varepsilon_0} \frac{P^2}{\hbar^2} u(\omega_{21}) \tag{4.85}$$

If there are N_1 atoms per unit volume in state 1 then the total number of absorptions per unit time per unit volume would be $N_1 w_{12}$, which would be equal to

$$N_1 \frac{\pi}{3\varepsilon_0} \frac{P^2}{\hbar^2} u(\omega_{21}) \tag{4.86}$$

Comparing Eqs. (4.86) and (4.1), we obtain

$$B_{12} = \frac{\pi}{3\varepsilon_0} \frac{P^2}{\hbar^2} = \frac{4\pi^2}{3\hbar^2} \left(\frac{e^2}{4\pi\varepsilon_0} \right) \left| \int \psi_2^* \mathbf{r} \psi_1 \mathrm{d}\tau \right|^2 \tag{4.87}$$

The corresponding expression for stimulated emission is obtained by starting with the first term on the right-hand side of Eq. (4.76) and proceeding in a similar fashion. The final expression is identical to Eq. (4.87) except for an interchange of indices 1 and 2.

Using Eq. (4.87) we get the following expression for the A coefficient

$$A = \frac{4}{3} \left(\frac{e^2}{4\pi\varepsilon_0} \frac{1}{\hbar c} \right) \frac{\omega^3}{c^2} \left| \int \psi_2^* \mathbf{r} \psi_1 \mathrm{d}\tau \right|^2 \tag{4.88}$$

It may be of interest to note that

$$\left(\frac{e^2}{4\pi\varepsilon_0}\frac{1}{hc}\right) \approx \frac{1}{137} \tag{4.89}$$

Using this value, we obtain

$$A = \frac{4}{3}\frac{1}{137}\frac{\omega^3}{c^2}\left|\int \psi_2^* \mathbf{r}\psi_1 d\tau\right|^2 \tag{4.90}$$

As an example we calculate the A coefficient for the 2P$\rightarrow$ 1S transition in the hydrogen atom, i.e., the transition from the $(n=2, l=1, m=0)$ state to the $(n=1, l=0, m=0)$ state. For these states (see, e.g., Ghatak and Lokanathan (2004))

$$\psi_1 = \frac{1}{(4\pi)^{1/2}}\frac{2}{a_0^{3/2}}\exp\left(-\frac{r}{a_0}\right) \tag{4.91}$$

and

$$\psi_2 = \frac{1}{(2a_0)^{3/2}}\frac{r}{a_0\sqrt{3}}\exp\left(-\frac{r}{2a_0}\right)\left[\left(\frac{3}{4\pi}\right)^{1/2}\cos\theta\right] \tag{4.92}$$

where $a_0 = \left(\hbar^2/m\right)\left(4\pi\varepsilon_0/e^2\right) \approx 0.5 \times 10^{-10}$m. In order to evaluate the matrix element, we write

$$\begin{aligned}
x &= r\sin\theta\cos\phi \\
y &= r\sin\theta\sin\phi \\
z &= r\cos\theta
\end{aligned} \tag{4.93}$$

Now,

$$\int \psi_1^* x\psi_2 d\tau = \frac{1}{4\pi\sqrt{2}}\frac{1}{a_0^4}\left(\int_0^\infty r^2 dr\, e^{-3r/2a_0}r^2\right) \times \left(\int_0^\pi \cos\theta\,\sin^2\theta\,d\theta\right)$$
$$\times \left(\int_0^{2\pi}\cos\phi\,d\phi\right) = 0$$

because the integral over ϕ vanishes. Similarly

$$\int \psi_1^* y\psi_2 d\tau = 0 \tag{4.94}$$

The only non-vanishing integral is

$$\int \psi_1^* z \psi_2 d\tau = \frac{1}{4\pi\sqrt{2}} \frac{1}{a_0^4} \left(\int_0^\infty r^2 dr e^{-3r/2a_0} r^2 \right)$$

$$\times \left(\int_0^\pi \cos^2\theta \, \sin\theta \, d\theta \right) \times \left(\int_0^{2\pi} d\phi \right) = 4\sqrt{2} \left(\frac{2}{3} \right)^5 a_0$$

Thus[5]

$$\left| \int \psi_1^* \mathbf{r} \psi_2 d\tau \right|^2 = 2^5 \left(\frac{2}{3} \right)^{10} a_0^2 \tag{4.95}$$

Further for the 2P$\rightarrow$ 1S transition

$$\omega = \frac{1}{\hbar} \frac{3}{8a_0} \left(\frac{e^2}{4\pi\varepsilon_0} \right) = \frac{3c}{8a_0} \left(\frac{e^2}{4\pi\varepsilon_0} \frac{1}{\hbar c} \right) \approx \frac{3 \times 3 \times 10^8}{8 \times 0.51 \times 10^{-10}} \frac{1}{137} \approx 1.5 \times 10^{16} \mathrm{s}^{-1} \tag{4.96}$$

Substituting in Eq. (4.90), we obtain

$$A \approx \frac{4}{3} \frac{1}{137} \frac{(1.5 \times 10^{16})^3}{(3 \times 10^8)^2} 2^5 \left(\frac{2}{3} \right)^{10} \left(0.5 \times 10^{-10} \right)^2 \tag{4.97}$$

$$= 6 \times 10^8 \mathrm{s}^{-1}$$

The mean lifetime of the state, τ, is the inverse of A giving

$$\tau \approx 1.6 \times 10^{-9} s$$

Thus the lifetime of the hydrogen atom in the upper level corresponding to the 2P$\rightarrow$1S transition is about 1.6 ns. Transitions having such small lifetimes are referred to as strongly allowed transitions.

In contrast, the levels used in laser transitions are such that the upper laser level has a very long lifetime ($\sim 10^{-3}$–10^{-6} s). A level having such a long lifetime is referred to as a metastable level, and such transitions come under the class of weakly allowed or nearly forbidden transitions. The strength of an atomic transition is usually expressed in terms of the f-value defined by the following equation:

$$f_{21} = \frac{2}{3} \frac{m\omega_{21}}{\hbar} |D_{21}|^2 \tag{4.98}$$

---

[5] It can be shown that $\left| \int \psi_1^* \mathbf{r} \psi_2 d\tau \right|^2$ has the same value for transition from anyone of the states $(n = 2, l = 1, m = 0)$ or $(n = 2, l = 1, m = -1)$ or $(n = 2, l = 1, m = -1)$ to $(n = 2, l = 0, m = 0)$ state. However, the matrix element for the transition from $(n = 2, l = 0, m = 0)$ state to the $(n = 1, l = 0, m = 0)$ state is zero. This implies that the corresponding dipole transition is forbidden.

For strongly allowed transitions, f is of the order of unity, for example, for the $2P \to 1S$ transition in the hydrogen atom, $f = 0.416$. On the other hand, for the transitions from the upper laser level, $f \sim 10^{-3}-10^{-6}$.

4.7.2 Interaction of a Near-Monochromatic Wave with an Atom Having a Broad Frequency Response

We next consider a nearly monochromatic field interacting with atoms characterized by the lineshape function $g(\omega)$. For such a case the probability for the atom being in the upper state would be given by

$$
\begin{aligned}
\Gamma_{12} &= \frac{1}{4}\frac{D_{21}^2 E_0^2}{\hbar^2} \int g(\omega') \left\{ \frac{\sin\left[(\omega'-\omega)/2\right]t}{(\omega'-\omega)/2} \right\}^2 d\omega' \\
&= \frac{1}{4}\frac{D_{21}^2 E_0^2}{\hbar^2} g(\omega) 2\pi t \\
&= \frac{\pi P^2}{3\hbar^2\varepsilon_0} g(\omega)u_\omega t
\end{aligned}
\tag{4.99}
$$

where in the last step we have replaced D_{21}^2 and E_0^2 by $\frac{1}{3}P^2$ and $2u_\omega/\varepsilon_0$, respectively. Since

$$
B_{12} = B_{21} = \frac{\pi}{3\varepsilon_0}\frac{P^2}{\hbar^2} = \frac{\pi^2 c^3}{\hbar\omega^3 t_{sp}}
\tag{4.100}
$$

we obtain the following expression for the transition rate (per unit time) per unit volume:

$$
W_{12} = N_1 \frac{\pi^2 c^3}{\hbar\omega^3 t_{sp}} u_\omega g(\omega)
\tag{4.101}
$$

which is consistent with Eq. (4.18).

4.8 More Accurate Solution for the Two-Level System

A more accurate solution of the time-dependent Schrödinger equation can be obtained if we assume that the atom can exist in only two possible states characterized by $\psi_1(\mathbf{r})$ and $\psi_2(\mathbf{r})$. Thus Eq. (4.67) gets replaced by

$$
\Psi(\mathbf{r},t) = C_1(t)\psi_1(\mathbf{r})e^{-i\omega_1 t} + C_2(t)\psi_2(\mathbf{r})e^{-i\omega_2 t}
\tag{4.102}
$$

If we substitute from Eq. (4.102) into Eq. (4.61), multiply by ψ_1^* and integrate, we would get [cf. Eq. (4.74)]

$$
i\hbar\frac{dC_1}{dt} = \frac{1}{2}E_0 D_{12} C_2(t)\left(e^{-i(\omega'-\omega)t} + e^{-i(\omega'+\omega)t}\right)
\tag{4.103}
$$

Similarly

$$i\hbar\frac{dC_2}{dt} = \frac{1}{2}E_0 D_{12} C_1(t)\left(e^{i(\omega'+\omega)t} + e^{i(\omega'-\omega)t}\right) \tag{4.104}$$

where use has been made of the fact that

$$\int \psi_1^* \mathbf{r}\psi_1 d\tau = \int \psi_2^* \mathbf{r}\psi_2 d\tau = 0$$

and

$$\omega' = \frac{(E_2 - E_1)}{\hbar} = \omega_{21} = -\omega_{12} \tag{4.105}$$

In the rotating wave approximation, considering absorption we neglect the terms $e^{-i(\omega'+\omega)t}$ and $e^{i(\omega'+\omega)t}$ in Eqs. (4.103) and (4.104) and obtain

$$\frac{dC_1}{dt} = -\frac{i}{2\hbar}E_0 D_{12} C_2(t) e^{i(\omega-\omega')t} \tag{4.106}$$

$$\frac{dC_2}{dt} = -\frac{i}{2\hbar}E_0 D_{12} C_1(t) e^{-i(\omega-\omega')t} \tag{4.107}$$

If we assume a solution of the form

$$C_1(t) = e^{i\Omega t} \tag{4.108}$$

then from Eq. (4.107),

$$C_2(t) = -\frac{2\hbar\Omega}{E_0 D_{12}} e^{i(\Omega-\omega+\omega')t} \tag{4.109}$$

Substituting in Eq. (4.106), we get

$$-i\frac{2\hbar\Omega}{E_0 D_{12}}\left(\Omega - \omega + \omega'\right) = -\frac{i}{2\hbar}E_0 D_{21}$$

or

$$\Omega\left(\Omega - \omega + \omega'\right) - \frac{\Omega_0^2}{4} = 0 \tag{4.110}$$

where

$$\Omega_0^2 = \frac{E_0^2 D_{12} D_{21}}{\hbar^2} = \frac{E_0^2 D^2}{\hbar^2} \tag{4.111}$$

and

$$D = D_{12} = D_{21} \tag{4.112}$$

Equation (4.110) gives

$$\Omega_{1,2} = \frac{1}{2}\left\{-(\omega' - \omega) \pm \left[(\omega' - \omega)^2 + \Omega_0^2\right]^{1/2}\right\} \tag{4.113}$$

Thus the general solution will be

$$C_1(t) = A_1 e^{i\Omega_1 t} + A_2 e^{i\Omega_2 t} \tag{4.114}$$

$$C_2(t) = -\frac{2}{\Omega_0} e^{i(\omega' - \omega)t}\left(A_1\Omega_1 e^{i\Omega_1 t} + A_2\Omega_2 e^{i\Omega_2 t}\right) \tag{4.115}$$

If we now assume that the atom is initially in the ground state, i.e.,

$$C_1(0) = 1, \qquad C_2(0) = 0 \tag{4.116}$$

then

$$A_1 = -\frac{\Omega_2}{\Omega_1}A_2 \tag{4.117}$$

and

$$1 = A_1 + A_2 = A_2\frac{\Omega_1 - \Omega_2}{\Omega_1} = \left[(\omega' - \omega)^2 + \Omega_0^2\right]^{1/2}\frac{A_2}{\Omega_1} \tag{4.118}$$

or

$$A_2 = \frac{\Omega_1}{\Omega'} \tag{4.118}$$

where

$$\Omega' = \left[(\omega' - \omega)^2 + \Omega_0^2\right]^{1/2} \tag{4.119}$$

On substitution we finally obtain

$$C_2(t) = -i\frac{\Omega_0}{\Omega'} e^{i(\omega' - \omega)t/2} \sin\left(\frac{\Omega' t}{2}\right) \tag{4.120}$$

Thus the transition probability for absorption is given by

$$|C_2(t)|^2 = \left(\frac{\sin\left(\Omega' t/2\right)}{\Omega'/2}\right)^2 \left(\frac{\Omega_0}{2}\right)^2 \tag{4.121}$$

Fig. 4.12 Variation of the transition probability with time for a two-level system for different frequencies of the electromagnetic field. The curves correspond to the function $DE_0/\hbar = 0.1\omega'$. The *solid line* corresponds to Eq. (4.121) and the *dotted curve* corresponds to an accurate numerical computation (Reprinted with permission from Salzman (1917). © 1971 American Institute of Physics)

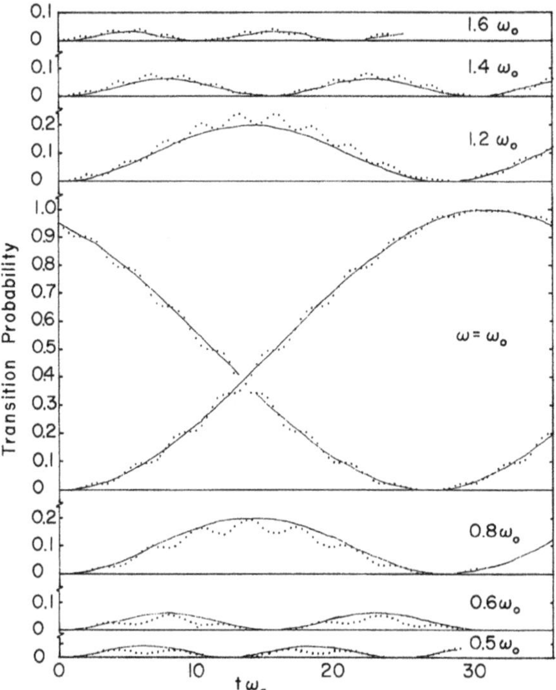

which has been plotted in Fig. 4.12. Also shown in the figure are the results of the exact numerical calculations without resorting to the rotating wave approximation. At resonance $\omega = \omega'$ and one obtains

$$|C_2(t)|^2 = \sin^2\left(\frac{\Omega_0 t}{2}\right) \tag{4.122}$$

which shows that the system flip flops between states 1 and 2. A comparison of Eqs. (4.122) and (4.77) shows that the perturbation theory result is valid if

$$\left(\frac{D_{21}E_0}{\hbar}t\right)^2 \ll 1 \quad \text{or} \quad \left(\frac{D_{21}E_0}{\hbar}\right)^2 \frac{1}{(\omega'-\omega)^2} \ll 1 \tag{4.123}$$

It may be of interest to note that the solutions obtained in this section are exact when $\omega = 0$ (i.e., a constant electric field) and if D_{21} is replaced by $2D_{21}$ in the solution given by Eq. (4.121). This follows from the fact that for $\omega = 0$, the exact equations [Eqs. (4.103) and (4.104)] are the same as Eqs. (4.106) and (4.107) with D_{21} replaced by $2D_{21}$.

Problems

Problem 4.3 Consider the two-level system shown in Fig. 4.1 with $E_1 = -13.6\,\text{eV}$ and $E_2 = -3.4\,\text{eV}$. Assume $A_{21} \approx 6 \times 10^8\,\text{s}^{-1}$. (a) What is the frequency of light emitted due to transitions from E_2 and E_1? Assuming the emission to have only natural broadening, what is the FWHM of the emission? What is the population ratio N_2/N_1 at $T = 300\,\text{K}$?

$$[\text{Answer}: (a) v \approx 2.5 \times 10^{15}\,\text{Hz}, \ \Delta v = A_{21}/2\pi \cong 10^8\,\text{Hz}, \ N_2/N_1 \approx e^{-394}$$

Problem 4.4 Given that the gain coefficient in a Doppler-broadened line is

$$\alpha(v) = \alpha(v_0) \exp\left[-4\ln 2(v - v_0)^2/(\Delta v_0)^2\right]$$

where v_0 is the centre frequency and Δv_0 is the FWHM and that the gain coefficient at the line centre is twice the loss averaged per unit length, calculate the bandwidth over which oscillation can take place. [Answer: Δv_0].

Problem 4.5 Consider an atomic system as shown below:

$$3 \text{———} \quad E_3 = 3\,\text{eV}$$
$$2 \text{———} \quad E_2 = 1\,\text{eV}$$
$$1 \text{———} \quad E_1 = 0\,\text{eV}$$

The A coefficient of the various transitions are given by

$$A_{32} = 7 \times 10^7\,\text{s}^{-1}, \quad A_{31} = 10^7\,\text{s}^{-1}, \quad A_{21} = 10^8\,\text{s}^{-1}$$

(a) What is the spontaneous lifetime of level 3?

(b) If the steady-state population of level 3 is 10^{15} atoms/cm^3, what is the power emitted spontaneously in the $3 \to 2$ transition? [Answer: (a) $t_{\text{sp}} = 1.2 \times 10^{-8}\,\text{s}$ (b) 2.2×10^{10} W/m^3]

Problem 4.6 Consider the transition in neon that emits 632.8 nm in the He–Ne laser and assume a temperature of 300 K. For a collision time of 500 ns, and a lifetime of 30 ns, obtain the broadening due to collisions, lifetime, and Doppler and show that the Doppler broadening is the dominant mechanism.

Problem 4.7 Consider an atomic system under thermal equilibrium at $T = 1000$ K. The number of absorptions per unit time corresponding to a wavelength of 1 μm is found to be $10^{22}\,\text{s}^{-1}$. What would be the number of stimulated emissions per unit time between the two energy levels? [Ans: $10^{22}\,5^{-1}$]

Problem 4.8 Consider a laser with plane mirrors having reflectivities of 0.9 each and of length 50 cm filled with the gain medium. Neglecting scattering and other cavity losses, estimate the threshold gain coefficient (in m^{-1}) required to start laser oscillation. [Ans: 0.21 m^{-1}]

Problem 4.9 An atomic transition has a linewidth of $\Delta v = 10^8$ Hz. Estimate the approximate value of $g(\omega)$ at the center of the line. [Ans: $\sim 1.6 \times 10^{-9}\,5^{-1}$]

Problem 4.10 There is a 10% loss per round trip in a ruby laser resonator having a 10 cm long ruby crystal as the active medium. Calculate the cavity lifetime, assuming that the mirrors are coated on the ends of the ruby crystal. Given: Refractive index of ruby at the laser wavelength is 1.78 [Ans: 11.3 ns]

Problem 4.11 In a ruby crystal, a population inversion density of $(N_2 - N_1) = 5 \times 10^{17} cm^{-3}$ is generated by pumping. Assuming $g(\nu_0) = 5 \times 10^{-12}$s, $t_{sp} = 3 \times 10^{-3}$s, wavelength of 694.3 nm and a refractive index of 1.78, obtain the gain coefficient $\gamma(\nu_0)$. By what factor will a beam get amplified if it passes through 5 cm of such a crystal? [Ans: $5 \times 10^{-2} cm^{-1}$, 1.28]

Problem 4.12 An optical amplifier of length 10 cm amplifies an input power of 1 to 1.1 W. Calculate the gain coefficient in m^{-1}. [Ans: $0.95\ m^{-1}$]

Problem 4.13 Doppler broadening leads to a linewidth given by

$$\Delta\nu_D = 2\nu_0 \sqrt{\frac{2k_B T}{Mc^2} \ln 2}$$

Estimate the broadening for the 632.8 nm transition of Ne (used in the He–Ne laser) assuming $T = 300$ K and atomic mass of Ne to be 20. What would be the corresponding linewidth of the 10.6 μm transition of the CO_2 molecule? [Ans: 1.6×10^9 Hz, 6×10^7 Hz]

Problem 4.14 In a typical He–Ne laser the threshold population inversion density is $10^9 cm^{-3}$. What is the value of the population inversion density when the laser is oscillating in steady state with an output power of 2 mW?

Problem 4.15 Given that the gain coefficient in a Doppler-broadened line is

$$\gamma(\nu) = \gamma_0 \exp\left[-\frac{4\ln 2(\nu - \nu_0)^2}{(\Delta\nu_0)^2} \right]$$

and that the gain coefficient at the center of the line is four times the loss averaged per unit length, obtain the bandwidth over which oscillation will take place. [Ans: $\sqrt{2}\Delta\nu_0$]

Problem 4.16 A laser resonator 1 m long is filled with a medium having a gain coefficient of 0.02 m^{-1}. If one of the mirrors is 100 % reflecting, what should be the minimum reflectivity of the other mirror so that the laser may oscillate? [Ans: $\sim 96\%$]

Chapter 5
Laser Rate Equations

5.1 Introduction

In Chapter 4 we studied the interaction of radiation with matter and found that under the action of radiation of proper frequencies, the atomic populations of various energy levels change. In this chapter, we will be studying the rate equations which govern the rate at which populations of various energy levels change under the action of the pump and in the presence of laser radiation. The rate equations approach provides a convenient means of studying the time dependence of the atomic populations of various levels in the presence of radiation at frequencies corresponding to the different transitions of the atom. It also gives the steady-state population difference between the actual levels involved in the laser transition and allows one to study whether an inversion of population is achievable in a transition and, if so, what would be the minimum pumping rate required to maintain a steady population inversion between two levels, the gain that such a medium would provide at and near the transition frequency, and the phase shift effects that such a medium would introduce are discussed in detail in Chapter 6. Thus Chapter 6 discusses the behavior of a system having two levels when there is a population inversion between the two levels, and this chapter deals with the means of obtaining an inversion between two levels of an atomic system by making use of other energy levels. The rate equations can also be solved to obtain the transient behavior of the laser, which gives rise to phenomena like Q-switching and spiking.

The atomic rate equations along with the rate equation for the photon number in the cavity form a set of coupled nonlinear equations. These equations can be solved under the steady-state regime and one can study the evolution of the photon number as one passes through the threshold pumping region.

In Section 5.2 we discuss a two-level system and show that it is not possible to achieve population inversion in steady state in a two-level system. Sections 5.3 and 5.4 discuss three-level and four-level laser systems and obtain the dependence of inversion on the pump power. In Section 5.5 we obtain the variation of laser power around threshold showing the sudden increase in the output power as a function of pumping. This is a very characteristic behavior of a laser. Finally in Section 5.6 we discuss the optimum output coupling for maximizing the output power of a laser.

K. Thyagarajan, A. Ghatak, *Lasers*, Graduate Texts in Physics,
DOI 10.1007/978-1-4419-6442-7_5, © Springer Science+Business Media, LLC 2010

5.2 The Two-Level System

We first consider a two-level system consisting of energy levels E_1 and E_2 with N_1 and N_2 atoms per unit volume, respectively [see (Fig. 5.1)]. Let radiation at frequency ω with energy density u be incident on the system. The number of atoms per unit volume which absorbs the radiation and is excited to the upper level will be [see Eq. (4.18)]

$$\Gamma_{12} = \frac{\pi^2 c^3}{\hbar \omega^3 t_{sp} n_0^3} u g(\omega) N_1 = W_{12} N_1 \tag{5.1}$$

where

$$W_{12} = \frac{\pi^2 c^3}{\hbar \omega^3 t_{sp} n_0^3} u g(\omega) \tag{5.2}$$

The number of atoms undergoing stimulated emissions from E_2 to E_1 per unit volume per unit time will be [see Eqs. (4.16) and (4.18)]

$$\Gamma_{21} = W_{21} N_2 = W_{12} N_2 \tag{5.3}$$

where we have used the fact that the absorption probability is the same as the stimulated emission probability. In addition to the above two transitions, atoms in the level E_2 would also undergo spontaneous transitions from E_2 to E_1. If A_{21} and S_{21} represent the radiative and non-radiative transition[1] rates from E_2 to E_1, then the number of atoms undergoing spontaneous transitions per unit time per unit volume from E_2 to E_1 will be $T_{21} N_2$ where

$$T_{21} = A_{21} + S_{21} \tag{5.4}$$

Thus we may write the rate of change of population of energy levels E_2 and E_1 as

$$\frac{dN_2}{dt} = W_{12}(N_1 - N_2) - T_{21} N_2 \tag{5.5}$$

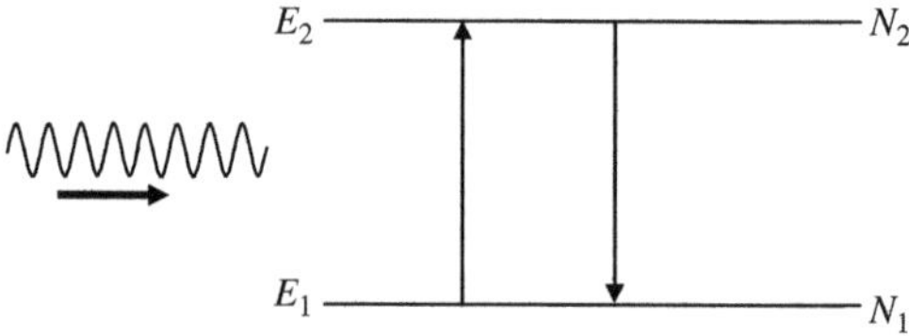

Fig. 5.1 A two-level system

---

[1]In a non-radiative transitions when the atom de-excites, the energy is transferred to the translational, vibrational or rotational energies of the surrounding atoms or molecules.

$$\frac{dN_1}{dt} = -W_{12}(N_1 - N_2) + T_{21}N_2 \tag{5.6}$$

As can be seen from Eqs. (5.5) and (5.6)

$$\frac{d}{dt}(N_1 + N_2) = 0$$
$$\Rightarrow N_1 + N_2 = \text{a constant} = N \quad (\text{say}) \tag{5.7}$$

which is nothing but the fact that the total number of atoms N per unit volume is constant. At steady state

$$\frac{dN_1}{dt} = 0 = \frac{dN_2}{dt} \tag{5.8}$$

which gives us

$$\frac{N_2}{N_1} = \frac{W_{12}}{W_{12} + T_{21}} \tag{5.9}$$

Since both W_{12} and T_{21} are positive quantities, Eq. (5.9) shows us that we can never obtain a steady-state population inversion by optical pumping between just two levels.

Let us now have a look at the population difference between the two levels. From Eq. (5.9) we have

$$\frac{N_2 - N_1}{N_2 + N_1} = -\frac{T_{21}}{2W_{12} + T_{21}}$$

or if we write $\Delta N = N_2 - N_1$, we have

$$\frac{\Delta N}{N} = -\frac{1}{1 + 2W_{12}/T_{21}} \tag{5.10}$$

In order to put Eq. (5.10) in a slightly different form, we first assume that the transition from 2 to 1 is mostly radiative, i.e., $A_{21} \gg S_{21}$ and $T_{21} \approx A_{21}$. We also introduce a lineshape function $\tilde{g}(\omega)$ which is normalized to have unit value at $\omega = \omega_0$, the center of the line, i.e.,

$$\tilde{g}(\omega) = \frac{g(\omega)}{g(\omega_0)} \tag{5.11}$$

Since $g(\omega) \leq g(\omega_0)$ for all ω, we have $0 < \tilde{g}(\omega) < 1$. Substituting the value of W_{12} in terms of u from Eq. (5.2) and observing that $u = n_0 I/c$, where I is the intensity of the incident radiation at ω, we have

$$\frac{W_{12}}{T_{21}} = \frac{\pi^2 c^3}{\hbar \omega^3 t_{sp} n_0^3} I \frac{n_0}{c} \tilde{g}(\omega) g(\omega_0) \frac{1}{A_{21}}$$

$$= \frac{\pi^2 c^3}{\hbar \omega^3 n_0^2} g(\omega_0) \tilde{g}(\omega) I \tag{5.12}$$

where we have used the fact that $A_{21}t_{sp} = 1$. Hence Eq. (5.10) becomes

$$\frac{\Delta N}{N} = -\frac{1}{1 + (I/I_s)\tilde{g}(\omega)} \qquad (5.13)$$

where

$$I_s \equiv \frac{\hbar\omega^3 n_0^2}{2\pi^2 c^2 \, g(\omega_0)} \qquad (5.14)$$

is called the saturation intensity. In order to see what I_s represents let us consider a monochromatic wave at frequency ω_0 interacting with a two-level system. Since $\tilde{g}(\omega_0) = 1$, we see from Eq. (5.13) that for $I \ll I_s$, the density of population difference between the two levels ΔN is almost independent of the intensity of the incident radiation. On the other hand for I comparable to I_s, ΔN becomes a function of I and indeed for $I = I_s$, the value of ΔN is half the value at low incident intensities.

We showed in Section 4.3 that the loss/gain coefficient for a population difference $\Delta N = N_2 - N_1$ between two levels is given by [see Eq. (4.26)]

$$\alpha = -\frac{\pi^2 c^2}{\omega^2 t_{sp} n_0^2} g(\omega)\Delta N$$

$$= \frac{\alpha_0}{1 + (I/I_s)\tilde{g}(\omega)} \qquad (5.15)$$

where

$$\alpha_0 = \frac{\pi^2 c^2}{\omega^2 t_{sp} n_0^2} g(\omega)N \qquad (5.16)$$

corresponds to the small signal loss, i.e., the loss coefficient when $I \ll I_s$. We can see from Problem 5.1 that with α given by Eq. (5.15), the loss is exponential for $I \ll I_s$ while it becomes linear for $I \gg I_s$. Thus we see that the attenuation caused by a medium decreases as the incident intensity increases to values comparable to the saturation intensity. Organic dyes having reasonably low values of $I_s (\sim 5 \text{ MW/cm}^2)$ are used as saturable absorbers in mode locking and Q-switching of lasers (see Section 7.7.1).

Problem 5.1 Using Eq. (5.15) in Eq. (4.25) obtain the variation of I with z.
[Answer:

$$\ln\frac{I}{I_0} + \frac{\tilde{g}(\omega)}{I_s}(I - I_0) = -\alpha_0 z$$

where I_0 is the intensity at $z = 0$.]

5.3 The Three-Level Laser System

In the last section we saw that one cannot create a steady-state population inversion between two levels just by using pumping between these levels. Thus in order to produce a steady-state population inversion, one makes use of either a three-level or a four-level system. In this section we shall discuss a three-level system.

We consider a three-level system consisting of energy levels E_1, E_2, and E_3 all of which are assumed to be nondegenerate. Let N_1, N_2, and N_3 represent the population densities of the three levels [see (Fig. 5.2)]. The pump is assumed to lift atoms from level 1 to level 3 from which they decay rapidly to level 2 through some nonradiative process. Thus the pump effectively transfers atoms from the ground level 1 to the excited level 2 which is now the upper laser level; the lower laser level being the ground state 1. If the relaxation from level 3 to level 2 is very fast, then the atoms will relax down to level 2 rather than to level 1. Since the upper level 3 is not a laser level, it can be a broad level (or a group of broad levels) so that a broadband light source may be efficiently used as a pump source (see, e.g., the ruby laser discussed in Chapter 11).

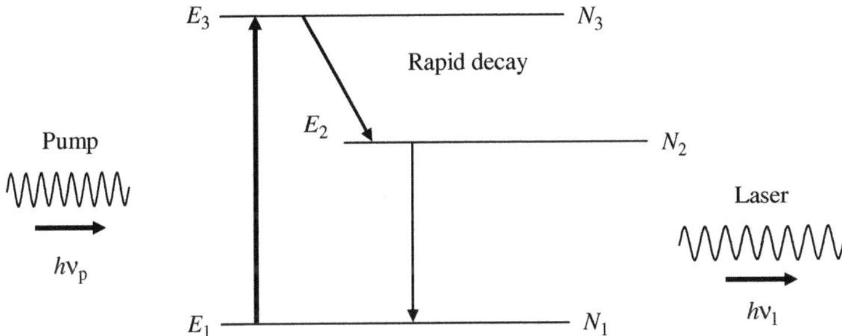

Fig. 5.2 A three-level system. The pump excites the atoms from level E_1 to level E_3 from where the atoms undergo a fast decay to level E_2. The laser action takes place between levels E_2 and E_1

If we assume that transitions take place only between these three levels then we may write

$$N = N_1 + N_2 + N_3 \tag{5.17}$$

where N represents the total number of atoms per unit volume.

We may now write the rate equations describing the rate of change of N_1, N_2 and N_3. For example, the rate of change of N_3 may be written as

$$\frac{dN_3}{dt} = W_p(N_1 - N_3) - T_{32}N_3 \tag{5.18}$$

where W_p is the rate of pumping per atom from level 1 to level 3 which depends on the pump intensity. The first term in Eq. (5.18) represents stimulated transitions between levels 1 and 3 and $T_{32}N_3$ represents the spontaneous transition from level 3 to level 2:

$$T_{32} = A_{32} + S_{32} \tag{5.19}$$

A_{32} and S_{32} correspond, respectively, to the radiative and nonradiative transition rates between levels 3 and 2. In writing Eq. (5.18) we have neglected $T_{31}N_3$ which corresponds to spontaneous transitions between levels 3 and 1 since most atoms raised to level 3 are assumed to make transitions to level 2 rather than to level 1.

In a similar manner, we may write

$$\frac{dN_2}{dt} = W_1(N_1 - N_2) + N_3 T_{32} - N_2 T_{21} \tag{5.20}$$

and

$$\frac{dN_1}{dt} = W_{\mathrm{p}}(N_3 - N_1) + W_1(N_2 - N_1) + N_2 T_{21} \tag{5.21}$$

where

$$W_1 = \frac{\pi^2 c^2}{\hbar \omega^3 n_0^2} A_{21} g(\omega) I_1 \tag{5.22}$$

represents the stimulated transition rate per atom between levels 1 and 2, I_1 is the intensity of the radiation in the $2 \rightarrow 1$ transition and $g(\omega)$ represents the lineshape function describing the transitions between levels 1 and 2. Further,

$$T_{21} = A_{21} + S_{21} \tag{5.23}$$

with A_{21} and S_{21} representing the radiative and nonradiative relaxation rates between levels 1 and 2. For efficient laser action since the transition must be mostly radiative, we shall assume $A_{21} \gg S_{21}$.

At steady state we must have

$$\frac{dN_1}{dt} = 0 = \frac{dN_2}{dt} = \frac{dN_3}{dt} \tag{5.24}$$

From Eq. (5.18) we obtain

$$N_3 = \frac{W_{\mathrm{p}}}{W_{\mathrm{p}} + T_{32}} N_1 \tag{5.25}$$

Using Eqs. (5.20), (5.21), and (5.25) we get

$$N_2 = \frac{W_1(T_{32} + W_{\mathrm{p}}) + W_{\mathrm{p}} T_{32}}{(W_{\mathrm{p}} + T_{32})(W_1 + T_{21})} N_1 \tag{5.26}$$

Thus from Eqs. (5.17), (5.25), and (5.26) we get

$$\frac{N_2 - N_1}{N} = \frac{[W_p(T_{32} - T_{21}) - T_{32}T_{21}]}{[3W_pW_1 + 2W_pT_{21} + 2T_{32}W_1 + T_{32}W_p + T_{32}T_{21}]} \tag{5.27}$$

From the above equation, one may see that in order to obtain population inversion between levels 2 and 1, i.e., for $(N_2 - N_1)$ to be positive, a necessary (but not sufficient) condition is that

$$T_{32} > T_{21} \tag{5.28}$$

Since the lifetimes of levels 3 and 2 are inversely proportional to the relaxation rates, according to Eq. (5.28), the lifetime of level 3 must be smaller than that of level 2 for attainment of population inversion between levels 1 and 2. If this condition is satisfied then according to Eq. (5.27), there is a minimum pumping rate required to achieve population inversion which is given by

$$W_{pt} = \frac{T_{32}T_{21}}{T_{32} - T_{21}} \tag{5.29}$$

If $T_{32} \gg T_{21}$,

$$W_{pt} \approx T_{21} \tag{5.30}$$

and under the same approximation, Eq. (5.27) becomes

$$\frac{N_2 - N_1}{N} = \frac{(W_p - T_{21})/(W_p + T_{21})}{\left[1 + \frac{3W_p + 2T_{32}}{T_{32}(W_p + T_{21})}W_1\right]} \tag{5.31}$$

Below the threshold for laser oscillation, W_1 is very small and hence we may write

$$\frac{N_2 - N_1}{N} = \frac{(W_p - T_{21})}{(W_p + T_{21})} \tag{5.32}$$

Thus when W_1 is small, i.e., when the intensity of the radiation corresponding to the laser transition is small [see Eq. (5.22)], then the population inversion is independent of I_1 and there is an exponential amplification of the beam. As the laser starts oscillating, W_1 becomes large and from Eq. (5.31) we see that this reduces the inversion $N_2 - N_1$ which in turn reduces the amplification. When the laser oscillates under steady-state conditions, the intensity of the radiation at the laser transition increases to such a value that the value of $N_2 - N_1$ is the same as the threshold value.

Recalling Eq. (5.31), we see that for a population inversion $N_2 - N_1$, the gain coefficient of the laser medium is

$$\gamma = \frac{\pi^2 c^2}{\omega^2 t_{sp} n_0^2} g(\omega)(N_2 - N_1)$$

$$= \frac{\gamma_0}{1 + \frac{3W_p + 2T_{32}}{T_{32}(W_p + T_{21})}W_1} \tag{5.33}$$

where

$$\gamma_0 = \frac{\pi^2 c^2}{\omega^2 t_{sp} n_0^2} g(\omega) N \frac{W_p - T_{21}}{W_p + T_{21}}$$ (5.34)

is the small signal gain coefficient. If we now carry out a similar analysis to that in Section 5.2, we may write

$$\gamma = \frac{\gamma_0}{1 + (I/I_s)\tilde{g}(\omega)}$$ (5.35)

where

$$\tilde{g}(\omega) = g(\omega)/g(\omega_0)$$

$$I_s = \frac{\hbar\omega^3 n_0^2}{\pi^2 c^2 A_{21} g(\omega_0)} \frac{T_{32}(W_p + T_{21})}{(3W_p + 2T_{32})}$$ (5.36)

I_s being the saturation intensity [see the discussion following Eq. (5.16)].

If T_{32} is very large then there will be very few atoms residing in level 3. Consequently, we may write

$$N = N_1 + N_2 + N_3 \approx N_1 + N_2$$ (5.37)

Substituting in Eq. (5.32), we get

$$\frac{N_2 - N_1}{N_2 + N_1} = \frac{W_p - T_{21}}{W_p + T_{21}}$$

or

$$W_p N_1 = T_{21} N_2$$ (5.38)

The left-hand side of the above equation represents the number of atoms being lifted (by the pump) per unit volume per unit time from level 1 to level 2 via level 3 and the right-hand side corresponds to the spontaneous emission rate per unit volume from level 2 to level 1. These rates must be equal under steady-state conditions for $W_1 \approx 0$, i.e., below the threshold.

We shall now estimate the threshold pumping power required to start laser oscillation. In order to do this, we first observe that the threshold inversion required is usually very small compared to N (i.e., $N_2 - N_1 \ll N$ – see the example of the ruby laser discussed in Chapter 11). Thus from Eq. (5.38), we see that the threshold value of W_p required to start laser oscillation is also approximately equal to T_{21} Now the number of atoms being pumped per unit time per unit volume from level 1 to level 3 is $W_p N_1$. If ν_p represents the average pump frequency corresponding to excitation to E_3 from E_1, then the power required per unit volume will be

$$P = W_p N_1 h\nu_p$$ (5.39)

Thus the threshold pump power for laser oscillation is given by

$$P_t = T_{21} N_1 h\nu_p$$ (5.40)

Since $N_2 - N_1 \ll N$ and $N_3 \approx 0$, $N_1 \approx N_2 \approx N/2$. Also assuming the transition from level 2 to level 1 to be mainly radiative (i.e., $A_{21} \gg S_{21}$), we have

$$P_t \approx Nh\nu_p/2t_{sp} \tag{5.41}$$

where we have used $A_{21} = 1/t_{sp}$.

As an example, we consider the ruby laser for which we have the following values of the various parameters:

$$N \approx 1.6 \times 10^{19}\,\text{cm}^{-3} \qquad t_{sp} \approx 3 \times 10^{-3}\,\text{s} \qquad \nu_p \approx 6.25 \times 10^{14}\,\text{Hz} \tag{5.42}$$

Substitution in Eq. (5.41) gives us

$$P_t \approx 1100\,\text{W/cm}^3 \tag{5.43}$$

If we assume that the efficiency of the pumping source to be 25% and also that only 25% of the pump light is absorbed on passage through the ruby rod, then the electrical threshold power comes out to be about 18 kW/cm^3 of the active medium. This is consistent with the threshold powers obtained experimentally.

Under pulsed operation if we assume that the pumping pulse is much shorter than the lifetime of level 2, then the atoms excited to the upper laser level do not appreciably decay during the duration of the pulse and the threshold pump energy would be

$$U_{pt} = \frac{N}{2}h\nu_p$$

per unit volume of the active medium. For the case of ruby laser, with the above efficiencies of pumping and absorption, one obtains

$$U_{pt} \approx 54\,\text{J/cm}^3$$

It may be noted here that even though ruby laser is a three-level laser system, because of various other factors mentioned below it does operate with not too large a pumping power. Thus, for example, the absorption band of ruby crystal is very well matched to the emission spectrum of available pump lamps so that the pumping efficiency is quite high. Also most of the atoms pumped to level 3 drop down to level 2 which has a very long lifetime which is nearly radiative. In addition the line width of laser transition is also very narrow.

5.4 The Four-Level Laser System

In the last section we found that since the lower laser was the ground level, one has to lift more than 50% of the atoms in the ground level in order to obtain population inversion. This problem can be overcome by using another level of the atomic system and having the lower laser level also as an excited level. The four-level laser

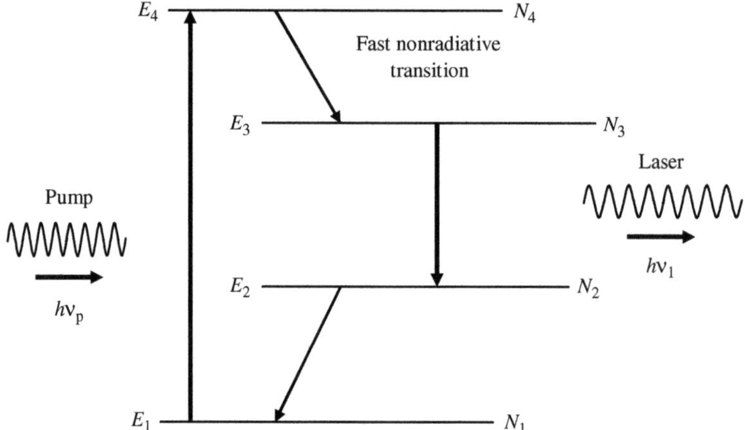

Fig. 5.3 A four-level system; the pump lifts atoms from level E_1 to level E_4 from where they decay rapidly to level E_3 and laser emission takes place between levels E_3 and E_2. Atoms drop down from level E_2 to level E_1

system is shown in Fig 5.3. Level 1 is the ground level and levels 2, 3, and 4 are excited levels of the system. Atoms from level 1 are pumped to level 4 from where they make a fast nonradiative relaxation to level 3. Level 3 which corresponds to the upper laser level is usually a metastable level having a long lifetime. The transition from level 3 to level 2 forms the laser transition. In order that atoms do not accumulate in level 2 and hence destroy the population inversion between levels 3 and 2, level 2 must have a very small lifetime so that atoms from level 2 are quickly removed to level 1 ready for pumping to level 4. If the relaxation rate of atoms from level 2 to level 1 is faster than the rate of arrival of atoms to level 2 then one can obtain population inversion between levels 3 and 2 even for very small pump powers. Level 4 can be a collection of a large number of levels or a broad level. In such a case an optical pump source emitting over a broad range of frequencies can be used to pump atoms from level 1 to level 4 effectively. In addition, level 2 is required to be sufficiently above the ground level so that, at ordinary temperatures, level 2 is almost unpopulated. The population of level 2 can also be reduced by lowering the temperature of the system.

We shall now write the rate equations corresponding to the populations of the four levels. Let N_1, N_2, N_3, and N_4 be the population densities of levels 1, 2, 3, and 4, respectively. The rate of change of N_4 can be written as

$$\frac{dN_4}{dt} = W_p(N_1 - N_4) - T_{43}N_4 \qquad (5.44)$$

where, as before, W_pN_1 is the number of atoms being pumped per unit time per unit volume, W_pN_4 is the stimulated emission rate per unit volume,

$$T_{43} = A_{43} + S_{43} \qquad (5.45)$$

is the relaxation rate from level 4 to level 3 and is the sum of the radiative (A_{43}) and nonradiative (S_{43}) rates. In writing Eq. (5.44) we have neglected (T_{42}) and (T_{41}) in comparison to (T_{43}), i.e., we have assumed that the atoms in level 4 relax to level 3 rather than to levels 2 and 1.

Similarly, the rate equation for level 3 may be written as

$$\frac{dN_3}{dt} = W_1(N_2 - N_3) + T_{43}N_4 - T_{32}N_3 \tag{5.46}$$

where

$$W_1 = \frac{\pi^2 c^2}{\hbar \omega^3 n_0^2} A_{32} g_1(\omega) I_1 \tag{5.47}$$

represents the stimulated transition rate per atom between levels 3 and 2 and the subscript 1 stands for laser transition; $g_1(\omega)$ is the lineshape function describing the $3 \leftrightarrow 2$ transition and I_1 is the intensity of the radiation at the frequency $\omega = (E_3 - E_2)/\hbar$. Also

$$T_{32} = A_{32} + S_{32} \tag{5.48}$$

is the net spontaneous relaxation rate from level 3 to level 2 and consists of the radiative (A_{32}) and the nonradiative (S_{32}) contributions. Again we have neglected any spontaneous transition from level 3 to level 1. In a similar manner, we can write

$$\frac{dN_2}{dt} = -W_1(N_2 - N_3) + T_{32}N_3 - T_{21}N_2 \tag{5.49}$$

$$\frac{dN_1}{dt} = -W_p(N_1 - N_4) + T_{21}N_2 \tag{5.50}$$

where

$$T_{21} = A_{21} + S_{21} \tag{5.51}$$

is the spontaneous relaxation rate from $2 \rightarrow 1$.

Under steady-state conditions

$$\frac{dN_1}{dt} = \frac{dN_2}{dt} = \frac{dN_3}{dt} = \frac{dN_4}{dt} = 0 \tag{5.52}$$

We will thus get four simultaneous equations in N_1, N_2, N_3, and N_4 and in addition we have

$$N = N_1 + N_2 + N_3 + N_4 \tag{5.53}$$

for the total number of atoms per unit volume in the system.

From Eq. (5.44) we obtain, setting $dN_4/dt = 0$

$$\frac{N_4}{N_1} = \frac{W_p}{(W_p + T_{43})} \tag{5.54}$$

If the relaxation from level 4 to level 3 is very rapid then $T_{43} \gg W_p$ and hence $N_4 \ll N_1$. Using this approximation in the remaining three equations we can obtain for the population difference,

$$\frac{N_3 - N_2}{N} \approx \frac{W_p(T_{21} - T_{32})}{W_p(T_{21} + T_{32}) + T_{32}T_{21} + W_l(2W_p + T_{21})} \tag{5.55}$$

Thus in order to be able to obtain population inversion between levels 3 and 2, we must have

$$T_{21} > T_{32} \tag{5.56}$$

i.e., the spontaneous rate of deexcitation of level 2 to level 1 must be larger than the spontaneous rate of deexcitation of level 3 to level 2.

If we now assume $T_{21} \gg T_{32}$, then from Eq. (5.55) we obtain

$$\frac{N_3 - N_2}{N} \approx \frac{W_p}{W_p + T_{32}} \frac{1}{1 + W_l(T_{21} + 2W_p)/T_{21}(W_p + T_{32})} \tag{5.57}$$

From the above equation we see that even for very small pump rates one can obtain population inversion between levels 3 and 2. This is contrary to what we found in a three-level system, where there was a minimum pump rate, W_{pt}, required to achieve inversion. The first factor in Eq. (5.57) which is independent of W_l [i.e., independent of the intensity of radiation corresponding to the laser transition – see Eq. (5.47)] – gives the small signal gain coefficient whereas the second factor in Eq. (5.57) gives the saturation behavior.

Just below threshold for laser oscillation, $W_l \approx 0$, and hence from Eq. (5.57) we obtain

$$\frac{\Delta N}{N} \approx \frac{W_p}{(W_p + T_{32})} \tag{5.58}$$

where $\Delta N = N_3 - N_2$ is the population inversion density. We shall now consider two examples of four-level systems.

Example 5.1 The Nd:YAG laser corresponds to a four-level laser system (see Chapter 11). For such a laser, typical values of various parameters are

$$\lambda_0 = 1.06 \, \mu m \, (v = 2.83 \times 10^{14} \, Hz), \quad \Delta v = 1.95 \times 10^{11} \, Hz,$$
$$t_{sp} = 2.3 \times 10^{-4} s, N = 6 \times 10^{19} \, cm^{-3}, \quad n_0 = 1.82 \tag{5.59}$$

If we consider a resonator cavity of length 7 cm and $R_1 = 1.00$, $R_2 = 0.90$, neglecting other loss factors (i.e., $\alpha_l = 0$)

$$t_c = -\frac{2n_0 d}{c \ln R_1 R_2} \approx 8 \times 10^{-9} s$$

We now use Eq. (4.32) to estimate the population inversion density to start laser oscillation corresponding to the center of the laser transition:

$$(\Delta N)_t = \frac{4v^2 n_0^3}{c^3} \frac{1}{g(\omega)} \frac{t_{sp}}{t_c}$$

$$= \frac{4v^2 n_0^3}{c^3} \pi^2 \Delta v \frac{t_{sp}}{t_c}$$

(5.60)

where for a homogenous transition (see Section 4.5)

$$g(\omega_0) = 2/\pi \Delta\omega = 1/\pi^2 \Delta v$$

(5.61)

Thus substituting various values, we obtain

$$(\Delta N)_t \approx 4 \times 10^{15} \ cm^{-3}$$

(5.62)

Since $(\Delta N)_t \ll N$, we may assume in Eq. (5.58) $T_{32} \gg W_p$ and hence we obtain for the threshold pumping rate required to start laser oscillation

$$W_{pt} \approx \frac{(\Delta N)_t}{N} T_{32} \approx \frac{(\Delta N)_t}{N} \frac{1}{t_{sp}}$$

$$= \frac{4 \times 10^{15}}{6 \times 10^{19}} \times \frac{1}{2.3 \times 10^{-4}} \approx 0.3\,s^{-1}$$

At this pumping rate the number of atoms being pumped from level 1 to level 4 is $W_{pt}N_1$ and since N_2, N_3 and N_4 are all very small compared to N_1, we have $N_1 \approx N$. For every atom lifted from level 1 to level 4 an energy hv_p has to be given to the atom where v_p is the average pump frequency corresponding to the $1 \rightarrow 4$ transition. Assuming $v_p \approx 4 \times 10^{14}$ Hz we obtain for the threshold pump power required per unit volume of the laser medium

$$P_{th} = W_{pt}N_1 hv_p \approx W_{pt}Nhv_p$$

$$= 0.3 \times 6 \times 10^{19} \times 6.6 \times 10^{-34} \times 4 \times 10^{14}$$

$$\approx 4.8 \ W/cm^3$$

which is about three orders of magnitude smaller than that obtained for ruby.

Example 5.2 As a second example of a four-level laser system, we consider the He–Ne laser (see Chapter 11). We use the following data:

$$\lambda_0 = 0.6328 \times 10^{-4} cm (v = 4.74 \times 10^{14} Hz),$$

$$t_{sp} = 10^{-7} s, \quad \Delta v = 10^9 \ Hz, \quad n_0 \approx 1$$

(5.63)

If we consider the resonator to be of length 10 cm and having mirrors of reflectivities $R_1 = R_2 = 0.98$, then assuming the absence of other loss mechanisms ($\alpha_l = 0$),

$$t_c = -2n_0 d/c \ln R_1 R_2$$

$$\approx 1.6 \times 10^{-8} s$$

(5.64)

For an inhomogeneously broadened transition (see Section 4.5)

$$g(\omega_0) = \frac{2}{\Delta\omega}\left(\frac{\ln 2}{\pi}\right)^{\frac{1}{2}}$$

$$\approx 1.5 \times 10^{-10}s \tag{5.65}$$

Thus the threshold population inversion required is

$$(\Delta N)_t \approx 1.4 \times 10^9 \mathrm{cm}^{-3} \tag{5.66}$$

Hence the threshold pump power required to start laser oscillation is

$$P_{th} = W_{pt}N_1(E_4 - E_1)$$

$$\approx \frac{(\Delta N)_t}{t_{sp}}h v_p \tag{5.67}$$

where again we assume $(\Delta N)_t \ll N$ and $T_{32} \approx A_{32} = 1/t_{sp}$. Assuming $v_p \approx 5 \times 10^{15}$ Hz, we obtain

$$P_{th} = \frac{1.4 \times 10^9 \times 6.6 \times 10^{-34} \times 5 \times 10^{15}}{10^{-7}}$$

$$\approx 50\,\mathrm{mW/cm}^3 \tag{5.68}$$

which again is very small compared to the threshold powers required for ruby laser.

5.5 Variation of Laser Power Around Threshold

In the earlier sections we considered the three-level and four-level laser systems and obtained conditions for the attainment of population inversion. In this section we shall discuss the variation of the power in the laser transition as the pumping rate passes through threshold.

We consider the two levels involved in the laser transition in a four-level laser[2] and assume that the lower laser level has a very fast relaxation rate to lower levels so that it is essentially unpopulated. We will assume that only one mode has sufficient gain to oscillate and that the line is homogenously broadened so that the same induced rate applies to all atoms (see Section 4.5). Let R represent the number of atoms that are being pumped into the upper level per unit time per unit volume [see

---

[2]A similar analysis can also be performed for a three-level laser system but the general conclusions of this simple analysis remain valid.

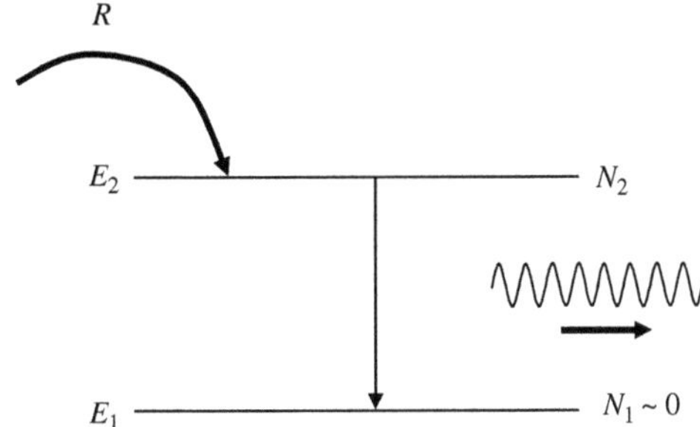

Fig. 5.4 The upper level is pumped at a rate R per unit volume and the lower level is assumed to be unpopulated due to rapid relaxation to other lower levels

(Fig. 5.4)]. If the population density of the upper level is N_2, then the number of atoms undergoing stimulated emissions from level 2 to level 1 per unit time will be [see Eq. (4.16)]

$$F_{21} = \Gamma_{21}V = \frac{\pi^2 c^3}{\hbar \omega^3 n_0^3} A_{21} u g(\omega) N_2 V \tag{5.69}$$

where u is the density of radiation at the oscillating mode frequency ω, V represents the volume of the active medium, and n_0 is the refractive index of the medium.

Instead of working with the energy density u, we introduce the number of photons n in the oscillating cavity mode. Since each photon carries an energy $\hbar \omega$, the number of photons n in the cavity mode will be given by

$$n = uV/\hbar \omega \tag{5.70}$$

Thus

$$F_{21} = \frac{\pi^2 c^3}{\omega^2 n_0^3} A_{21} g(\omega) N_2 n = K n N_2 \tag{5.71}$$

where

$$K \equiv (\pi^2 c^3 / \omega^2 n_0^3) A_{21} g(\omega) \tag{5.72}$$

The spontaneous relaxation rate from level 2 to level 1 in the whole volume will be $T_{21} N_2 V$ where

$$T_{21} = A_{21} + S_{21} \tag{5.73}$$

is the total relaxation rate consisting of the radiative (A_{21}) and the nonradiative (S_{21}) components. Hence we have for the net rate of the change of population of level 2

$$\frac{d}{dt}(N_2 V) = -K n N_2 - T_{21} N_2 V + RV$$

or

$$\frac{dN_2}{dt} = -\frac{K n N_2}{V} - T_{21} N_2 + R \tag{5.74}$$

In order to write a rate equation describing the variation of photon number n in the oscillating mode in the cavity, we note that n change due to

a) All stimulated emissions caused by the n photons existing in the cavity mode which results in a rate of increase of n of KnN_2 since every stimulated emission from level 2 to level 1 caused by radiation in that mode will result in the addition of a photon in that mode. There is no absorption since we have assumed the lower level to be unpopulated.

b) In order to estimate the increase in the number of photons in the cavity mode due to spontaneous emission, we must note that not all spontaneous emission occurring from the $2 \rightarrow 1$ transition will contribute to a photon in the oscillating mode. As we will show in Section 7.2 for an optical resonator which has dimensions which are large compared to the wavelength of light, there are an extremely large number of modes ($\sim 10^8$) that have their frequencies within the atomic linewidth. Thus when an atom deexcites from level 2 to level 1 by spontaneous emission it may appear in any one of these modes. Since we are only interested in the number of photons in the oscillating cavity mode, we must first obtain the rate of spontaneous emission into a mode of oscillation of the cavity. In order to obtain this we recall from Section 4.2 that the number of spontaneous emissions occurring between ω and $\omega + d\omega$ will be

$$G_{21}d\omega = A_{21}N_2 g(\omega)d\omega V \tag{5.75}$$

We shall show in Appendix E that the number of oscillating modes lying in a frequency interval between ω and $\omega + d\omega$ is

$$N(\omega)d\omega = n_0^3 \frac{\omega^2}{\pi^2 c^3} V\, d\omega \tag{5.76}$$

where n_0 is the refractive index of the medium. Thus the spontaneous emission rate per mode of oscillation at frequency ω is

$$S_{21} = \frac{G_{21}d\omega}{N(\omega)d\omega} = \frac{\pi^2 c^3}{n_0^3 \omega^2} g(\omega)A_{21}N_2 \tag{5.77}$$

$$= KN_2$$

i.e., the rate of spontaneous emission into a particular cavity mode is the same as the rate of stimulated emission into the same mode when there is just one photon in that mode. This result can indeed be obtained by rigorous quantum mechanical derivation (see Chapter 9).

c) The photons in the cavity mode are also lost due to the finite cavity lifetime. Since the energy in the cavity reduces with time as e^{-t/t_c} (see Section 4.4) the rate of decrease of photon number in the cavity will also be n/t_c.

Thus we can write for the total rate of change of n

$$\frac{dn}{dt} = KnN_2 + KN_2 - \frac{n}{t_c} \tag{5.78}$$

Eqs. (5.74) and (5.78) represent the pair of coupled rate equations describing the variation of N_2 and n with time.

Under steady-state conditions both time derivatives are zero. Thus we obtain from Eq. (5.78),

$$N_2 = \frac{n}{n+1} \frac{1}{Kt_c} \tag{5.79}$$

The above equation implies that under steady-state conditions $N_2 \leq 1/Kt_c$. When the laser is oscillating under steady-state conditions $n \gg 1$ and $N_2 \approx 1/Kt_c$. If we substitute the value of K from Eq. (5.72) we find that (for $n \gg 1$)

$$N_2 \approx \frac{\omega^2 n_0^3}{\pi^2 c^3} \frac{t_{sp}}{t_c} \frac{1}{g(\omega)} \tag{5.80}$$

which is nothing but the threshold population inversion density required for laser oscillation (cf. Eq. (4.32)). Thus Eq. (5.79) implies that when the laser oscillates under steady-state conditions, the population inversion density is almost equal to and can never exceed the threshold value. This is also obvious since if the inversion density exceeds the threshold value, the gain in the cavity will exceed the loss and thus the laser power will start increasing. This increase will continue till saturation effects take over and reduce N_2 to the threshold value.

Substituting from Eq. (5.79) into Eq. (5.74) and putting $dN_2/dt = 0$, we get

$$\frac{K}{VT_{21}} n^2 + n \left(1 - \frac{R}{R_t}\right) - \frac{R}{R_t} = 0 \tag{5.81}$$

where

$$R_t = \frac{T_{21}}{Kt_c} \tag{5.82}$$

The solution of the above equation which gives a positive value of n is

$$n = \frac{VT_{21}}{2K} \left\{ \left(\frac{R}{R_t} - 1\right) + \left[\left(1 - \frac{R}{R_t}\right)^2 + \frac{4K}{VT_{21}} \frac{R}{R_t}\right]^{\frac{1}{2}} \right\} \tag{5.83}$$

The above equation gives the photon number in the cavity under steady-state conditions for a pump rate R.

For a typical laser system, for example an Nd:glass laser (see Chapter 11),

$$V \approx 10\,\text{cm}^3, \qquad n_0 \approx 1.5$$

$$\lambda \approx 1.06\,\mu\text{m}, \qquad \Delta\upsilon \approx 3 \times 10^{12}\,\text{Hz}$$

so that

$$\frac{K}{VT_{21}} = \frac{c^3}{8v^2 n_0^3} \frac{1}{V\pi \Delta v} \approx 1.3 \times 10^{-13} \tag{5.84}$$

where we have used $T_{21} \approx A_{21}$. For such small values of K/VT_{21}, unless R/R_t is extremely close to unity, we can make a binomial expansion in Eq. (5.83) to get

$$\begin{aligned} n &\approx \frac{R/R_t}{1-R/R_t} &&\text{for} &&\frac{R}{R_t} < 1 - \Delta \\ n &\approx \frac{VT_{21}}{K}\left(\frac{R}{R_t} - 1\right) &&\text{for} &&\frac{R}{R_t} > 1 + \Delta \end{aligned} \tag{5.85}$$

where $\Delta \gg (2K/VT_{21})^{\frac{1}{2}}$. Further

$$n \approx \left(\frac{VT_{21}}{K}\right)^{\frac{1}{2}} \qquad \text{for} \qquad \frac{R}{R_t} = 1 \tag{5.86}$$

Figure 5.5 shows a typical variation of n with R/R_t. As is evident $n \approx 1$ for $R < R_t$ and approaches 10^{12} for $R > R_t$. Thus R_t as given by Eq. (5.82) gives the threshold pump rate for laser oscillation.

Fig. 5.5 Variation of photon number n in the cavity mode as a function of pumping rate R; R_t corresponds to the threshold pumping rate. Note the steep rise in the photon number as one crosses the threshold for laser oscillation

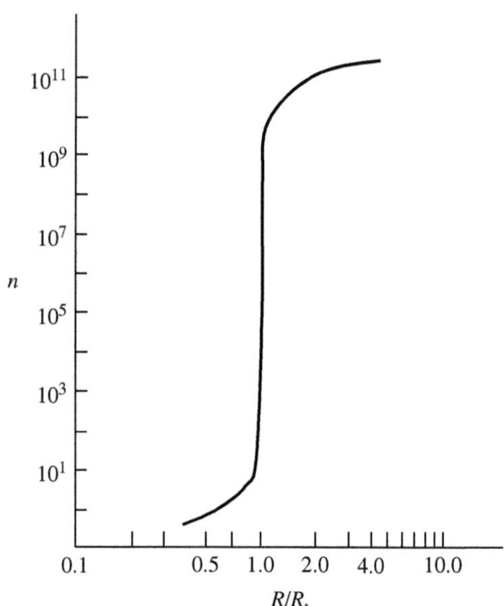

Problem 5.2 Show that the threshold pump rate R_t given by Eq. (5.82) is consistent with that obtained in Section 4.4.

From the above analysis it follows that when the pumping rate is below threshold ($R < R_t$) then the number of photons in the cavity mode is very small (~ 1). As one approaches the threshold, the number of photons in the preferred cavity mode (having higher gain and lower cavity losses) increases at a tremendous rate and as one passes the threshold, the number of photons in the oscillating cavity mode becomes extremely large. At the same time the number of photons in other cavity modes which are below threshold remains orders of magnitude smaller.

In addition to the sudden increase in the number of photons in the cavity mode and hence laser output power, the output also changes from an incoherent to a coherent emission. The output becomes an almost pure sinusoidal wave with a well-defined wave front, apart from small amplitude and phase fluctuations caused by the ever-present spontaneous emission.[3] It is this spontaneous emission which determines the ultimate linewidth of the laser.

If the only mechanism in the cavity is that arising from output coupling due to the finite reflectivity of one of the mirrors, then the output laser power will be

$$P_{out} = \frac{nhv}{t_c} \tag{5.87}$$

where n/t_c is the number of photons escaping from the cavity per unit time and hv is the energy of each photon. Taking K/VT_{21} as given by Eq. (5.84) and $t_c \approx 10^{-8}$s, for $R/R_t = 2$ we obtain

$$P_{out} = 144\,\text{W}$$

Example 5.3 It is interesting to compare the number of photons per cavity mode in an oscillating laser and in a black body at a temperature T. The number of photons/mode in a black body is (see Appendix D)

$$n = \frac{1}{e^{\hbar\omega/k_B T} - 1} \tag{5.88}$$

Hence for $\lambda = 1.06\,\mu$m, $\quad T = 1000$ K, we obtain

$$n \approx \frac{1}{e^{13.5} - 1} \approx 1.4 \times 10^{-6}$$

which is orders of magnitude smaller than in an oscillating laser [see (Fig 5.5)].

From Eq. (5.85) we may write for the change in number of photons dn for a change dR in the pump rate as

$$\frac{dn}{dR} = \frac{VT_{21}}{K}\frac{1}{R_t} = Vt_c$$

or

$$VdR = \frac{dn}{t_c} \tag{5.89}$$

where we have used Eq. (5.82). The LHS of Eq. (5.89) represents the additional number of atoms that are being pumped per unit time into the upper laser level and

---

[3]In an actual laser system, the ultimate purity of the output beam is restricted due to mechanical vibrations of the laser, mirrors, temperature fluctuations, etc.

the RHS represents the additional number of photons that is being lost from the cavity. Thus above threshold all the increase in pump rate goes toward the increase in the laser power.

Example 5.4 Let us consider an Nd:glass laser (see Chapter 11) with the parameters given in page 119 and having

$$d = 10 \, \text{cm}$$

$$R_1 = 0.95, \qquad R_2 = 1.00$$

For these values of the parameters, using Eq. (4.31) we have

$$t_c \approx -\frac{2n_0 d}{c \ln R_1 R_2} \approx 1.96 \times 10^{-8} \text{s}$$

and

$$\frac{VT_{21}}{K} \approx \frac{V}{Kt_{sp}} = \frac{4v^2 Vn_0^3}{c^3 g(\omega)} \tag{5.90}$$

Thus for $R/R_t = 2$, i.e., for a pumping rate twice the threshold value (see Eq. (5.85))

$$n = VT_{21}/K \approx 7.7 \times 10^{12}$$

Hence the energy inside the cavity is

$$E = nhv$$
$$\approx 1.4 \times 10^{-6} \text{J} \tag{5.91}$$

If the only loss mechanism is the finite reflectivity of one of the mirrors, then the output power will be

$$P_{\text{out}} = \frac{nhv}{t_c} \approx 74 \, \text{W}$$

Problem 5.3 In the above example, if it is required that there be 1 W of power from the mirror at the left and 73 W of power from the right mirror, what should the reflectivities of the two mirrors be? Assume the absence of all other loss mechanisms in the cavity.

[Answer: $R_1 = 0.9993, \quad R_2 = 0.9507$]

Example 5.5 In this example, we will obtain the relationship between the output power of the laser and the energy present inside the cavity by considering radiation to be making to and fro oscillations in the cavity. Figure 5.6a shows the cavity of length l bounded by mirrors of reflectivities 1 and R and filled by a medium characterized by the gain coefficient α. Let us for simplicity assume absence of all other loss mechanisms. Figure 5.6b shows schematically the variation of intensity along the length of the resonator when the laser oscillates under steady-state conditions. For such a case, the intensity after one round trip I_4 must be equal to the intensity at the same point at the start of the round trip. Hence

$$R \, e^{2\alpha l} = 1 \tag{5.92}$$

Also, recalling the definition of cavity lifetime (see Eq. (4.31) with $\alpha_1 = 0$), we have

$$t_c = -\frac{2l}{c} \ln R = \frac{1}{\alpha c} \tag{5.93}$$

Now let us consider a plane P inside the resonator. Let the distance of the plane from mirror M_1 be x. Thus if I_1 is the intensity of the beam at mirror M_1, then assuming exponential amplification, the intensity of the beam going from left to right at P is

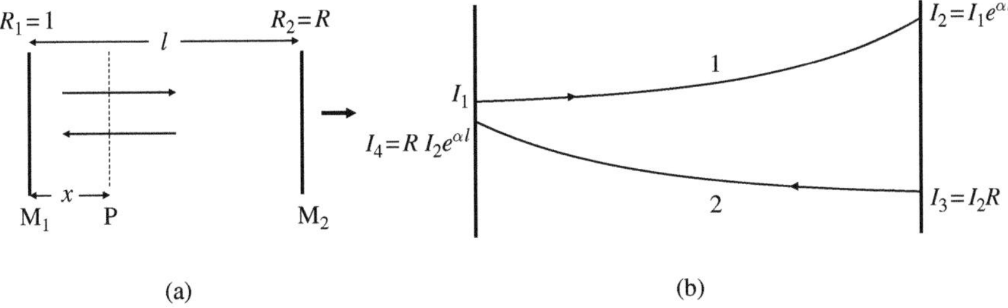

Fig. 5.6 (**a**) A resonator of length l bound by mirrors of reflectivities 1 and R and filled by a medium of gain coefficient α. (**b**) Curves 1 and 2 represent the qualitative variation of intensity associated with the waves propagating in the forward and backward directions within the cavity. The sudden drop in intensity from I_2 to I_3 is due to the finite reflectivity of the mirror M_2

$$I_+ = I_1 e^{\alpha x} \tag{5.94}$$

Similarly, the intensity of the beam going from right to left at P is

$$\begin{aligned} I_- &= I_1 e^{\alpha l} R e^{\alpha(l-x)} = I_1 R e^{2\alpha l} e^{-\alpha x} \\ &= I_1 e^{-\alpha x} \end{aligned} \tag{5.95}$$

Hence the energy density at x is

$$u(x) = \frac{I_+ + I_-}{c} = \frac{I_1}{c}\left(e^{\alpha x} + e^{-\alpha x}\right) \tag{5.96}$$

If A is the area of cross section, then the total energy in the cavity is

$$\begin{aligned} W &= \int\int u\,dA\,dx = A\int_0^l u\,dx \\ &= \frac{AI_1}{\alpha c} e^{\alpha l}(1-R) \\ &= AI_1 t_c e^{\alpha l}(1-R) \end{aligned} \tag{5.97}$$

where we have used Eqs. (5.92) and (5.93) and have assumed, for the sake of simplicity, uniform intensity distribution in the transverse direction. Now the power emerging from mirror M_2 is

$$\begin{aligned} P_{\text{out}} &= I_2 A(1-R) \\ &= I_1 A e^{\alpha l}(1-R) \\ &= W/t_c \end{aligned} \tag{5.98}$$

which is consistent with Eq. (5.87)

5.6 Optimum Output Coupling

In the last section we obtained the steady-state energy inside the resonator cavity as a function of the pump rate. In order to get an output laser beam, one of the mirrors is made partially transparent so that a part of the energy is coupled out. In this section we shall obtain the optimum reflectivity of the mirror so as to have a maximum output power.

The fact that an optimum output coupling exists can be understood as follows. If one has an almost zero output coupling (i.e., if both mirrors are almost 100% reflecting) then even though the laser may be oscillating, the output power will be almost zero. As one starts to increase the output coupling, the energy inside the cavity will start to decrease since the cavity loss is being increased but, since one is taking out a larger fraction of power the output power starts increasing. The output power will start decreasing again if the reflectivity of the mirror is continuously reduced since if it is made too small, then for that pumping rate, the losses will exceed the gain and the laser will stop oscillating. Thus for a given pumping rate, there must be an optimum output coupling which gives the maximum output power.

In Section 4.4 we showed that the cavity lifetime of a passive resonator is

$$\frac{1}{t_c} = \frac{c}{2dn_0}(2\alpha_1 d - \ln R_1 R_2)$$
$$= \frac{1}{t_i} + \frac{1}{t_e}$$

(5.99)

where

$$\frac{1}{t_i} = \frac{c\alpha_1}{n_0}, \quad \frac{1}{t_e} = -\frac{c}{2dn_0}\ln R_1 R_2$$

(5.100)

t_i accounts for all loss mechanisms except for the output coupling due to the finite mirror reflectivities and t_e for the loss due to output coupling only. Thus, the number photons escaping the cavity due to finite mirror reflectivity will be n/t_e and hence the output power will be

$$P_{\text{out}} = \frac{nh\nu}{t_e}$$
$$= \frac{h\nu}{t_e} \frac{VT_{21}}{K} \left[\frac{RK}{T_{21}} \left(\frac{1}{t_i} + \frac{1}{t_e}\right)^{-1} - 1 \right]$$

(5.101)

where we have used Eqs. (5.82), (5.85), and (5.99). The optimum output power will correspond to the value of t_e satisfying $\partial P_{\text{out}}/\partial t_e = 0$ which gives

$$\frac{1}{t_e} = \left(\frac{RK}{T_{21}t_i}\right)^{\frac{1}{2}} - \frac{1}{t_i}$$

(5.102)

Using Eqs. (5.99) and (5.82), the above equation can be simplified to

$$\frac{1}{t_e} = \frac{1}{t_i}\left(\frac{R}{R_t} - 1\right)$$

(5.103)

Substituting for t_e from Eq. (5.103) in Eq. (5.101) we obtain the maximum output power as

$$P_{\text{max}} = h\nu RV \left[1 - \left(\frac{T_{21}}{KRt_i} \right)^{1/2} \right]^2 \tag{5.104}$$

It is interesting to note that the optimum t_e and hence the optimum reflectivity is a function of the pump rate R.

Even though the output power passes through a maximum as the transmittivity $T = (1-R)$ of the mirror is increased, the energy inside the cavity monotonically reduces from a maximum value as T is increased. This may be seen from the fact that the energy in the cavity is

$$E = nh\nu = \frac{VT_{21}}{K} \left(\frac{KRt_c}{T_{21}} - 1 \right) h\nu \tag{5.105}$$

Thus as T is increased, t_c reduces and hence E reduces monotonically finally becoming zero when

$$t_c = \frac{T_{21}}{KR} \tag{5.106}$$

beyond which the losses become more than the gain.

Problems

Problem 5.4 Using Eq. (5.103) calculate the optimum reflectivity of one of the mirrors of the resonator (assuming the other mirror to have 100% reflectivity) for $R = 2R_t$. Assume the length of the resonator to be 100 cm, $n_0 = 1$ and the intrinsic loss per unit length to be $3 \times 10^{-5} \text{cm}^{-1}$.

Problem 5.5 Consider an atomic system as shown below:

```
3————— E3 = 3 eV
2————— E2 = 1 eV
1————— E1 = 0 eV
```

The A coefficient of the various transitions are given by

$$A_{32} = 7 \times 10^7 \text{s}^{-1}, \ A_{31} = 10^7 \text{s}^{-1}, \qquad A_{21} = 10^8 \text{s}^{-1}$$

(a) Show that this system cannot be used for continuous wave laser oscillation between levels 2 and 1.
(b) Suppose at $t = 0, N_0$ atoms are lifted to level 3 by some external mechanism describe the change of populations in levels 1, 2, and 3.

Problem 5.6 Using Eq. (5.103) calculate the optimum reflectivity of one of the mirrors of the resonator (assuming the other mirror to be 100% reflecting) for $R = 2R_t$. Assume the length of the resonator to be 50 cm, $n_0 = 1$ and the intrinsic loss per unit length to be $3 \times 10^{-4} \mathrm{m}^{-1}$. If the power output at the optimum coupling is 10 mW, what is the corresponding energy inside the cavity?

[Answer: $R \approx 0.9997$, energy $\cong 1.1 \times 10^{-7}$ J]

Problem 5.7 Consider a laser with plane mirrors having reflectivities of 0.9 each and of length 50 cm filled with the gain medium. Neglecting scattering and other cavity losses, estimate the threshold gain coefficient (in m^{-1}) required to start laser oscillation.

Problem 5.8 The cavity of a 6328 Å He–Ne laser is 1 m long and has mirror of reflectivities 100 and 98%; the internal cavity losses are negligible. (a) If the steady-state power output is 10 mW, what is the energy stored in the resonator? (b) What is the linewidth of the above passive cavity? (c) If the oscillating linewidth is 1500 MHz, how many longitudinal modes would oscillate?

Problem 5.9 Consider a two-level system shown below:

$$\text{————} \quad E_2 = 2\,eV$$
$$A_{21} = 10^7 \ \mathrm{s}^{-1}$$
$$\text{————} \quad E_1 = 0$$

a) What is the frequency of light emitted due to transitions from E_2 to E_1?
b) Assuming the emission to have only natural broadening, what is the FWHM of the emission?
c) What is the population ratio N_2/N_1 at 300 K?
d) An atomic system containing N_0 atoms/cm^3 of the above atoms is radiated by a beam of intensity I_0 at the line center. Write down the rate equation and obtain the population difference between E_2 and E_1 under steady-state condition. Calculate the incident intensity required to produce a population ratio $N_1 = 2\,N_2$.

Problem 5.10 The active medium of a three-level atomic system is characterized by the following spontaneous emission rates: $A_{21} = 10^8 \mathrm{s}^{-1}, A_{31} = 10^6 \mathrm{s}^{-1}, A_{32} = 10^5 \mathrm{s}^{-1}$. (Neglect non-radiative transitions.) Can we use the atomic system to realize a laser? (YES or NO). Justify your answer.

Problem 5.11 Consider a three-level laser system with lasing between levels E_2 and E_1. The level E_2 has a lift time of 1 μs. Assuming the transition $E_3 \rightarrow E_2$ to be very rapid, estimate the number of atoms that needs to be pumped per unit time per unit volume from level E_1 to reach threshold for achieving population inversion. Given that the total population density of the atoms is $10^{19} \mathrm{cm}^{-3}$.

Chapter 6
Semiclassical Theory of the Laser

6.1 Introduction

The present chapter deals with the semiclasscial theory of the laser as developed by Lamb (1964). In this analysis, we will treat the electromagnetic field classically with the help of Maxwell's equations and the atom will be treated using quantum mechanics. We will consider a collection of two-level atoms placed inside an optical resonator. The electromagnetic field of the cavity mode produces a macroscopic polarization of the medium. This macroscopic polarization is calculated using quantum mechanics. The polarization then acts as a source for the electromagnetic field in the cavity. Since this field must be self-consistent with the field already assumed, one gets, using this condition, the amplitude and frequencies of oscillation. We will obtain explicit expressions for the real and imaginary parts of the electric susceptibility of the medium. The real part is responsible for additional phase shifts due to the medium and leads to the phenomenon of mode pulling. On the other hand, the imaginary part of the susceptibility is responsible for loss or gain due to the medium. Under normal conditions, the population of the upper level is less than that of the lower level and the medium adds to the losses of the cavity. In the presence of population inversion, the medium becomes an amplifying medium; however, a minimum population inversion is necessary to sustain oscillations in the cavity. We will show that in the first-order theory, the electric field in the cavity can grow indefinitely, but using a third-order theory we would show that the field would indeed saturate rather than growing indefinitely.

Since the analysis is semiclassical in nature, the effects of spontaneous emission do not appear. Thus, the analysis does not give the ultimate linewidth of the laser oscillator which is caused by spontaneous emissions.

6.2 Cavity Modes

We consider a laser cavity with plane mirrors at $z = 0$ and $z = L$ (see Fig. 6.1). The electromagnetic radiation inside the cavity can be described by Maxwell's equations, which in the MKS system of units are

K. Thyagarajan, A. Ghatak, *Lasers*, Graduate Texts in Physics,
DOI 10.1007/978-1-4419-6442-7_6, © Springer Science+Business Media, LLC 2010

Fig. 6.1 A plane parallel resonator bounded by a pair of plane mirrors facing each other. The active medium is placed inside the resonator

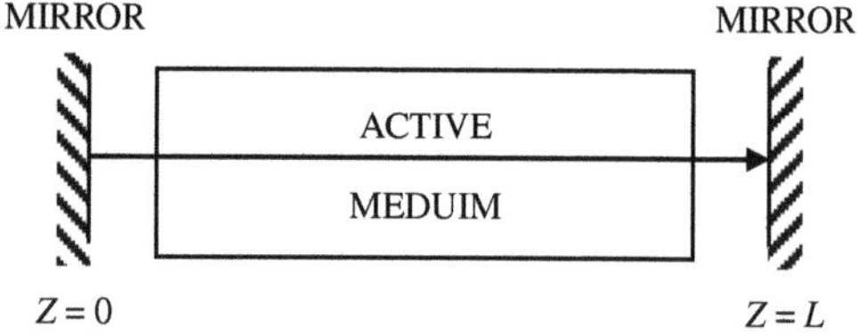

$$\nabla \times \mathbf{E} = -\frac{\partial \mathbf{B}}{\partial t} \tag{6.1}$$

$$\nabla \times \mathbf{H} = \mathbf{J}_f + \frac{\partial \mathbf{D}}{\partial t} \tag{6.2}$$

$$\nabla \cdot \mathbf{D} = \rho_f \tag{6.3}$$

$$\nabla \cdot \mathbf{B} = 0 \tag{6.4}$$

where ρ_f represents the free charge density and $\mathbf{J}_f$ the free current density; $\mathbf{E}, \mathbf{D}, \mathbf{B},$ and $\mathbf{H}$ represent the electric field, electric displacement, magnetic induction, and magnetic field, respectively. Inside the cavity we may assume

$$\rho_f = 0 \tag{6.5}$$

$$\mathbf{B} = \mu_0 \mathbf{H} \tag{6.6}$$

$$\mathbf{D} = \varepsilon_0 \mathbf{E} + \mathbf{P} \tag{6.7}$$

$$\mathbf{J}_f = \sigma \mathbf{E} \tag{6.8}$$

where $\mathbf{P}$ is the polarization, σ the conductivity, and ε_0 and μ_0 are the dielectric permittivity and magnetic permeability of free space. It will be seen that the conductivity term leads to the medium being lossy which implies attenuation of the field; we will assume that other losses like those due to diffraction and finite transmission at the mirrors are taken into account in σ. Now,

$$\nabla \times (\nabla \times \mathbf{E}) = -\mu_0 \frac{\partial}{\partial t} (\nabla \times \mathbf{H}) = -\mu_0 \frac{\partial \mathbf{J}}{\partial t} - \mu_0 \frac{\partial^2 \mathbf{D}}{\partial t^2} \tag{6.9}$$

or

$$\nabla \times (\nabla \times \mathbf{E}) + \mu_0 \sigma \frac{\partial \mathbf{E}}{\partial t} + \varepsilon_0 \mu_0 \frac{\partial^2 \mathbf{E}}{\partial t^2} = -\mu_0 \frac{\partial^2 \mathbf{P}}{\partial t^2} \tag{6.10}$$

If we assume the losses to be small and the medium to be dilute, we may neglect the second term on the left-hand side and the term on the right-hand side of the above equation to approximately obtain

$$\nabla \times (\nabla \times \mathbf{E}) + \varepsilon_0 \mu_0 \frac{\partial^2 \mathbf{E}}{\partial t^2} = 0 \tag{6.11}$$

Further since **P** is small, Eq. (6.3) gives

$$0 = \nabla \cdot \mathbf{D} \approx \varepsilon_0 \nabla \cdot \mathbf{E} \tag{6.12}$$

Thus

$$\nabla \times \nabla \times \mathbf{E} = -\nabla^2 \mathbf{E} + \nabla \left(\nabla \cdot \mathbf{E} \right) \approx -\nabla^2 \mathbf{E} \tag{6.13}$$

or

$$\nabla \times \nabla \times \mathbf{E} \approx -\nabla^2 \mathbf{E} = -\frac{\partial^2 \mathbf{E}}{\partial z^2} \tag{6.14}$$

where, in writing the last equation, we have neglected the x and y derivatives; this is justified when intensity variations in the directions transverse to the laser axis is small in distances $\sim \lambda$, which is indeed the case (see Chapter 7). Thus Eq. (6.11) becomes

$$-\frac{\partial^2 \mathbf{E}}{\partial z^2} + \frac{1}{c^2} \frac{\partial^2 \mathbf{E}}{\partial t^2} = 0 \tag{6.15}$$

where $c = (\varepsilon_0 \mu_0)^{-1/2}$ represents the speed of light in free space. If we further assume a specific polarization of the beam, Eq. (6.15) becomes a scalar equation:

$$\frac{\partial^2 E}{\partial z^2} = \frac{1}{c^2} \frac{\partial^2 E}{\partial t^2} \tag{6.16}$$

which we solve by the method of separation of variables:

$$E\left(z,t\right) = Z(z)T(t) \tag{6.17}$$

to obtain

$$\frac{1}{Z} \frac{d^2 Z}{dz^2} = \frac{1}{c^2} \frac{1}{T} \frac{d^2 T}{dt^2} = -K^2 \;(\text{say}) \tag{6.18}$$

Thus

$$Z\left(z\right) = A \sin\left(Kz + \theta\right) \tag{6.19}$$

where the quantity K corresponds to the wave number. At the cavity ends (i.e., at $z = 0$ and $z = L$), the field [and hence $Z(z)$] will vanish, giving

$$\theta = 0$$

and

$$K = \frac{n\pi}{L}, \quad n = 1, 2, 3, \ldots \tag{6.20}$$

We designate different values of K by K_n, $(n = 1, 2, 3, \ldots)$. The corresponding time dependence will be of the form

$$\cos \Omega_n t$$

where

$$\Omega_n = K_n c = \frac{n \pi c}{L} \tag{6.21}$$

Thus the complete solution of Eq. (6.16) would be given by

$$E(z, t) = \sum_n A_n \cos(\Omega_n t) \sin(K_n z) \tag{6.22}$$

If we next include the term describing the loses, we would have (instead of Eq. 6.16)

$$\frac{\partial^2 E}{\partial z^2} - \mu_0 \sigma \frac{\partial E}{\partial t} = \frac{1}{c^2} \frac{\partial^2 E}{\partial t^2} \tag{6.23}$$

We assume the same spatial dependence ($\sim \sin K_n z$) and the time dependence to be of the form $e^{i \Lambda_n t}$ to obtain

$$\Lambda_n^2 - \mu_0 \sigma\, c^2 i \Lambda_n - \Omega_n^2 = 0 \tag{6.24}$$

or

$$\Lambda_n = \frac{1}{2} \left[i \mu_0 \sigma c^2 \pm \left(-\mu_0^2 \sigma^2 c^4 + 4\Omega_n^2 \right)^{1/2} \right] \tag{6.25}$$
$$\approx \pm \Omega_n + i \sigma / 2\varepsilon_0$$

Thus the time dependence is of the form

$$\exp\left(-\frac{\sigma}{2\varepsilon_0} t \right) e^{\pm i \Omega_n t} \tag{6.26}$$

the first term describing the attenuation of the beam. In the expression derived above, the attenuation coefficient does not depend on the mode number n; however, in general, there is a dependence on the mode number which we explicitly indicate by writing the time-dependent factor as[1]

$$\exp\left(-\frac{\Omega_n}{2Q_n} t \right) e^{\pm i \Omega_n t} \tag{6.27}$$

where

----

[1] Because of the losses, the field in the cavity decays with time as $\exp\left(-\Omega_n t/2Q_n\right)$ and hence the energy decays as $\exp\left(-\Omega_n t/Q_n\right)$. Thus, the energy decays to $1/e$ of the value at $t = 0$ in a time $t_c = Q_n/\Omega_n$ which is referred to as the cavity lifetime (see also Section 7.4).

$$Q_n = \frac{\varepsilon_0}{\sigma} \Omega_n \tag{6.28}$$

represents the quality factor (see Section 7.4). Thus the solution of Eq. (6.23) would be

$$E(z,t) = \sum_n A_n \exp\left(-\frac{\Omega_n}{2Q_n}t\right) \cos(\Omega_n t) \sin(K_n z) \tag{6.29}$$

Finally, we try to solve the equation which includes the term involving the polarization:

$$\frac{\partial^2 E}{\partial z^2} - \mu_0 \sigma \frac{\partial E}{\partial t} - \frac{1}{c^2}\frac{\partial^2 E}{\partial t^2} = \mu_0 \frac{\partial^2 P}{\partial t^2} \tag{6.30}$$

[cf. Eqs. (6.16) and (6.23)]. We assume E to be given by

$$E = \frac{1}{2}\sum_n \{E_n(t) \exp[-i(\omega_n t + \phi_n(t))] + c.c.\} \sin K_n z \tag{6.31}$$

where $c.c.$ stands for the complex conjugate (so that E is necessarily real), $E_n(t)$ and $\phi_n(t)$ are real slowly varying amplitude and phase coefficients, and ω_n is the frequency of oscillation of the mode which may, in general, be slightly different from Ω_n. We assume P to be of the form

$$P = \frac{1}{2}\sum_n \{P_n(t,z) \exp[-i(\omega_n t + \phi_n(t))] + c.c.\} \tag{6.32}$$

where $P_n(t,z)$ may be complex but is a slowly varying component of the polarization. On substitution of E and P in Eq. (6.30), we get[2] (after multiplying by c^2)

$$\Omega_n^2 E_n - i\left(\frac{\sigma}{\varepsilon_0}\right)\omega_n E_n - 2i\omega_n \dot{E}_n - \left(\omega_n + \dot{\phi}_n\right)^2 E_n = \frac{\omega_n^2}{\varepsilon_0}p_n(t) \tag{6.33}$$

$$p_n(t) = \frac{2}{L}\int_0^L P_n(t,z) \sin K_n z \, dz$$

where we have neglected small terms involving $\ddot{E}_n$, $\ddot{\phi}_n$, $\ddot{P}_n$, $\dot{E}_n\dot{\phi}_n$, $\sigma\dot{E}_n$, $\sigma\dot{\phi}_n$, $\dot{\phi}_n\dot{p}_n$, and $\dot{p}_n$ which are all of second order. Now, since ω_n will be very close to Ω_n, we may write

$$\Omega_n^2 - \left(\omega_n + \dot{\phi}_n\right)^2 \approx 2\omega_n\left(\Omega_n - \omega_n - \dot{\phi}_n\right) \tag{6.34}$$

---

[2] Actually we have equated each Fourier component; this follows immediately by multiplying Eq. (6.32) by $\sin K_m z$ and integrating from 0 to L.

Thus, equating real and imaginary parts of both sides of Eq. (6.33), we get

$$\left(\omega_n + \dot{\phi}_n - \Omega_n\right) E_n\left(t\right) = -\frac{1}{2}\frac{\omega_n}{\varepsilon_0}\,\mathrm{Re}\,\left(p_n\left(t\right)\right) \tag{6.35}$$

$$\dot{E}_n\left(t\right) + \frac{1}{2}\frac{\omega_n}{Q_n'}\,E_n\left(t\right) = -\frac{\omega_n}{2\varepsilon_0}\,\mathrm{Im}\,\left(p_n\left(t\right)\right) \tag{6.36}$$

where

$$Q_n' = \frac{\varepsilon_0\omega_n}{\sigma} \tag{6.37}$$

When $p_n = 0$, $\omega_n = \Omega_n$ and $E_n\left(t\right)$ will decrease exponentially with time – consistent with our earlier findings. In general, if we define the susceptibility χ through the equation

$$p_n\left(t\right) = \varepsilon_0\chi_n E_n\left(t\right) = \varepsilon_0\left(\chi_n' + i\chi_n''\right) E_n\left(t\right) \tag{6.38}$$

where χ_n' and χ_n'' represent, respectively, the real and imaginary parts of χ_n, then

$$\omega_n + \dot{\phi}_n = \Omega_n - \frac{1}{2}\omega_n\chi_n' \tag{6.39}$$

and

$$\dot{E}_n = -\frac{1}{2}\frac{\omega_n}{Q_n'}E_n - \frac{1}{2}\omega_n\chi_n''E_n\left(t\right) \tag{6.40}$$

The first term on the right-hand side of Eq. (6.40) represents cavity losses and the second term represents the effect of the medium filling the cavity. It can be easily seen that if χ_n'' is positive, then the cavity medium adds to the losses. On the other hand if χ_n'' is negative, the second term leads to gain.[3] If

$$-\chi_n'' = \frac{1}{Q_n'} \tag{6.41}$$

the losses are just compensated by the gain and Eq. (6.41) is referred to as the threshold condition. If $-\chi_n'' > 1/Q_n'$, there would be a buildup of oscillation.

From Eq. (6.39), one may note that if we neglect the term $\dot{\phi}_n$ the oscillation frequency differs from the passive cavity frequency by $-\frac{1}{2}\omega_n\chi_n'$, which is known as the pulling term. In order to physically understand the gain and pulling effects due to the cavity medium, we consider a plane wave propagating through the cavity medium. If χ_n represents the electric susceptibility of the medium for the wave, then the permittivity ε of the medium would be

---

[3]We will show in Section 6.3 that χ_n'' is negative for a medium with a population inversion.

$$\varepsilon = \varepsilon_0 + \varepsilon_0 \chi_n = \varepsilon_0 \left(1 + \chi_n\right) \tag{6.42}$$

This implies that the complex refractive index of the cavity medium is

$$\left(\frac{\varepsilon}{\varepsilon_0}\right)^{1/2} = (1 + \chi_n)^{1/2} \approx \left(1 + \frac{1}{2}\chi_n\right) = 1 + \frac{1}{2}\chi_n' + \frac{i}{2}\chi_n'' \tag{6.43}$$

The propagation constant of the plane wave in such a medium would be

$$\beta = \frac{\omega}{c}\left(\frac{\varepsilon}{\varepsilon_0}\right)^{1/2} = \frac{\omega}{c}\left(1 + \frac{1}{2}\chi_n'\right) + \frac{1}{2}i\frac{\omega}{c}\chi_n''$$

$$= \alpha + i\delta \tag{6.44}$$

where

$$\alpha = \frac{\omega}{c}\left(1 + \frac{1}{2}\chi_n'\right) ; \qquad \delta = \frac{1}{2}\frac{\omega}{c}\chi_n'' \tag{6.45}$$

Thus, a plane wave propagating along the z-direction would have a z dependence of the form

$$e^{i\beta z} = e^{i\alpha z}e^{-\delta z} \tag{6.46}$$

In the absence of the component due to the laser transition $\chi_n' = \chi_n'' = 0$ and the plane wave propagating through the medium undergoes a phase shift per unit length of ω/c. The presence of the laser transition contributes both to the phase change and to the loss or amplification of the beam. Thus if χ_n'' is positive, then δ is positive and the beam is attenuated as it propagates along the z-direction. On the other hand if χ_n'' is negative, then the beam is amplified as it propagates through the medium. As the response of the medium is stimulated by the field, the applied field and the stimulated response are phase coherent.

In addition to the losses or amplification caused by the cavity medium, there is also a phase shift caused by the real part of the susceptibility χ_n'. We will show in the next section that χ_n' is zero exactly at resonance, i.e., if the frequency of the oscillating mode is at the center of the atomic line and it has opposite signs on either side of the line center. This additional phase shift causes the frequencies of oscillation of the optical cavity filled with the laser medium to be different from the frequencies of oscillation of the passive cavity (i.e., the cavity in the absence of the laser medium). The actual oscillation frequencies are slightly pulled toward the center of the atomic line and hence the phenomenon is referred to as mode pulling.

6.3 Polarization of the Cavity Medium

In the last section, we obtained equations describing the cavity field and the oscillation frequency of the cavity in terms of the polarization associated with the cavity medium. In the present section, we consider a collection of two-level atoms and obtain an explicit expression for the macroscopic polarization (and hence the electric susceptibility) of the cavity medium in terms of the atomic populations in the two levels of the system. The time-dependent Schrödinger equation is given by.

$$i\hbar\frac{\partial\Psi}{\partial t} = H\Psi \tag{6.47}$$

where H is the Hamiltonian and Ψ represents the time-dependent wave function of the atomic system.

Let H_0 represent the Hamiltonian of the atom and let $\psi_1(\mathbf{r})\,e^{-i\omega_1 t}$ and $\psi_2(\mathbf{r})\,e^{-i\omega_2 t}$ be the normalized wave functions associated with the lower level 1 and the upper level 2, respectively, of the atom. Then

$$H_0\psi_1(\mathbf{r}) = E_1\psi_1(\mathbf{r}) \tag{6.48a}$$

$$H_0\psi_2(\mathbf{r}) = E_2\psi_2(\mathbf{r}) \tag{6.48b}$$

where $E_1 = \hbar\omega_1$ and $E_2 = \hbar\omega_2$ are the energies of the lower and the upper levels, respectively. The interaction of the atom with the electromagnetic field is described by

$$H' = -e\mathbf{E}.\mathbf{r} \tag{6.49}$$

which is assumed to be a perturbation on the Hamiltonian H_0; here E represents the electric field associated with the radiation. In the presence of such an interaction we write the wave function as

$$\Psi(\mathbf{r},t) = C_1(t)\,\psi_1(\mathbf{r}) + C_2(t)\,\psi_2(\mathbf{r}) \tag{6.50}$$

where $C_1(t)$ and $C_2(t)$ are time-dependent factors. The physical significance of $C_1(t)$ and $C_2(t)$ is that $|C_1(t)|^2$ and $|C_2(t)|^2$ represent, respectively, the probability of finding the atom in the lower state ψ_1 and in the upper state ψ_2 at time t. Also, since we are considering a collection of N_v atoms per unit volume and each atom has a probability $|C_1(t)|^2$ of being found in the level ψ_1 at time t, the mean number of atoms per unit volume in the lower level 1, namely N_1 would be

$$N_1 = N_v\,|C_1(t)|^2 \tag{6.51a}$$

Similarly, the mean number of atoms per unit volume in the level 2, N_2, would be

$$N_2 = N_v\,|C_2(t)|^2 \tag{6.51b}$$

Using Eqs. (6.48) and (6.50), Eq. (6.47) gives

$$i\hbar \sum_{n=1,2} \dot{C}_n \psi_n = \sum_{n=1,2} \left(\hbar \omega_n + H' \right) C_n \psi_n \tag{6.52}$$

Multiplying both sides by ψ_1^* and integrating over spatial coordinates, one obtains

$$i\hbar \dot{C}_1 = E_1 C_1 + H'_{11} C_1 + H'_{12} C_2 \tag{6.53}$$

where

$$H'_{mn} = \int \psi_m^* H' \psi_n d\tau \tag{6.54}$$

Similarly by multiplying Eq. (6.52) by ψ_2^* and integrating one obtains

$$i\hbar \dot{C}_2 = E_2 C_2 + H'_{22} C_2 + H'_{21} C_1 \tag{6.55}$$

But

$$H'_{11} = -e.E. \int \psi_1^* (\mathbf{r})\, \mathbf{r} \psi_1 (\mathbf{r})\, d\tau = 0 \tag{6.56}$$

because $\mathbf{r}$ is an odd function. Similarly

$$H'_{22} = 0 \quad \text{and} \quad H'_{12} = H'^*_{21} \tag{6.57}$$

Thus

$$\dot{C}_1 (t) = \frac{1}{i\hbar} \left[E_1 C_1 (t) + H'_{12} C_2 (t) \right] \tag{6.58a}$$

and

$$\dot{C}_2 (t) = \frac{1}{i\hbar} \left[E_2 C_2 (t) + H'_{21} C_1 (t) \right] \tag{6.58b}$$

In deriving the above equations, we have not considered any damping mechanism. We wish to do so now by introducing phenomenological damping factors. Even though we are considering only two levels of the atomic system, the phenomenological damping factors take account of not only spontaneous transitions from the two levels but also, for example, collisions, etc. This we do by rewriting Eqs. (6.58a) and (6.58b) as

$$\dot{C}_1 (t) = -\frac{i}{\hbar} E_1 C_1 - \frac{1}{2} \gamma_1 C_1 (t) - \frac{i}{\hbar} H'_{12} C_2 (t) \tag{6.59a}$$

$$\dot{C}_2 (t) = -\frac{i}{\hbar} E_2 C_2 - \frac{1}{2} \gamma_2 C_2 (t) - \frac{i}{\hbar} H'_{21} C_1 (t) \tag{6.59b}$$

where γ_1 and γ_2 represent damping factors for levels 1 and 2, respectively. In order to see more physically, we find that in the absence of any interaction when $H'_{12} = 0 = H'_{21}$ the solutions of Eqs. (6.59a) and (6.59b) would be

$$C_1(t) = \text{const} \times \exp\left(-i\frac{E_1}{\hbar}t\right) e^{-(\gamma_1/2)t} \tag{6.60a}$$

$$C_2(t) = \text{const} \times \exp\left(-i\frac{E_2}{\hbar}t\right) e^{-(\gamma_2/2)t} \tag{6.60b}$$

Hence the probability of finding the atom in levels 1 and 2 (which are, respectively, proportional to $|C_1|^2$ and $|C_2|^2$) decays as $e^{-\gamma_1 t}$ and $e^{-\gamma_2 t}$, respectively. Thus, the lifetimes of levels 1 and 2 are $1/\gamma_1$ and $1/\gamma_2$, respectively.

We now define the following quantities[4]:

$$\rho_{11} = C_1^* C_1, \qquad \rho_{12} = C_1 C_2^* \tag{6.61}$$

$$\rho_{21} = C_1^* C_2, \qquad \rho_{22} = C_2^* C_2 \tag{6.62}$$

Notice that ρ_{11} and ρ_{22} are nothing but the probabilities of finding the system in states 1 and 2, respectively. Since we know the time dependence of C_1 and C_2 we can write down the time variation of the quantities ρ_{11}, etc. Thus,

$$\dot{\rho}_{11} = \dot{C}_1 C_1^* + C_1 \dot{C}_1^*$$

$$= -\left(i\frac{E_1}{\hbar} + \frac{\gamma_1}{2}\right) C_1 C_1^* - \frac{i}{\hbar}H'_{12}C_2 C_1^*$$

$$- \left(-i\frac{E_1}{\hbar} + \frac{\gamma_1}{2}\right) C_1^* C_1 + \frac{i}{\hbar}H'^*_{12}C_2^* C_1$$

or

$$\dot{\rho}_{11} = -\gamma_1 \rho_{11} + \left(\frac{i}{\hbar}H'_{21}\rho_{12} + c.c.\right) \tag{6.63}$$

where $c.c.$ represents the complex conjugate. Similarly

$$\dot{\rho}_{22} = -\gamma_2 \rho_{22} - \left(\frac{i}{\hbar}H'_{21}\rho_{12} + c.c.\right) \tag{6.64}$$

$$\dot{\rho}_{21} = -(i\omega_{21} + \gamma_{12})\rho_{21} + \frac{i}{\hbar}H'_{21}(\rho_{22} - \rho_{11}) \tag{6.65}$$

where

$$\omega_{21} = \frac{E_2 - E_1}{\hbar} \quad \text{and} \quad \gamma_{12} = \frac{\gamma_1 + \gamma_2}{2} \tag{6.66}$$

---

[4]The four quantities ρ_{11}, ρ_{12}, ρ_{21}, and ρ_{22} form the elements of what is known as the density matrix ρ.

Now,

$$H'_{21} = -e\mathbf{E}.\int \psi_2^* \mathbf{r}\psi_1 d\tau \tag{6.67}$$

and if we consider single-mode operation with $\mathbf{E}$ along with the x-axis then

$$H'_{21} = -eE_n \int \psi_2^* x\psi_1 d\tau = -E_n P \tag{6.68}$$

where

$$P = e \int \psi_2^* x\psi_1 d\tau \tag{6.69}$$

and

$$\begin{aligned}E_n &= \frac{1}{2}E_n\,(t)\exp\left\{-i\left[\omega_n t + \phi_n\,(t)\right]\right\}\sin K_n z + c.c. \\ &= E_n\,(t)\cos\left[\omega_n t + \phi_n\,(t)\right]\sin K_n z\end{aligned} \tag{6.70}$$

Further when the system is in the state $\Psi\,(t)$, the average dipole moment is given by

$$\begin{aligned}P_a &= e\int \Psi^* x\Psi\,d\tau \\ &= e\int \left(C_1^*\psi_1^* + C_2^*\psi_2^*\right) x\left(C_1\psi_1 + C_2\psi_2\right) d\tau \\ &= P\left(\rho_{21} + \rho_{12}\right)\end{aligned} \tag{6.71}$$

where we have used the relation

$$\int \psi_1^* x\psi_2\,d\tau = \int \psi_2^* x\psi_1\,d\tau \tag{6.72}$$

which can always be made to satisfy by appropriate choice of phase factors. Thus, in order to calculate P_a we must know ρ_{12} and its complex conjugate ρ_{21}. We first present the first-order theory which will be followed by the more rigorous third-order theory.

6.3.1 First-Order Theory

In the first-order theory, we assume $(\rho_{22} - \rho_{11})$ to be dependent only on z and to be independent of time[5]:

$$\rho_{22} - \rho_{11} = N\,(z) \tag{6.73}$$

Referring to Eqs. (6.51a) and (6.51b), we see that $N_v\,(\rho_{22} - \rho_{11})$ represents the difference (per unit volume) in the population of the upper and lower states. Thus,

---

[5] This will be justified in Section 6.3.2

Eq. (6.65) becomes

$$\dot{\rho}_{21} = -\left(i\omega_{21} + \gamma_{12}\right)\rho_{21} - \frac{i}{2\hbar}PE_n\left(t\right)\left(e^{-i(\omega_n t + \phi_n)} + c.c.\right)\sin K_{nz}\,N\left(z\right) \quad (6.74)$$

where we have used Eqs. (6.68) and (6.70). Now, if we neglect the second term on the right-hand side of the above equation, the solution would be of the form

$$\rho_{21}\left(t\right) = \rho_{21}^{(0)}\exp\left[-\left(i\omega_{21} + \gamma_{12}\right)t\right] \quad (6.75)$$

We next assume the solution of Eq. (6.74) to be of the above form with $\rho_{21}^{(0)}$ now depending on time. On substitution in Eq. (6.74), we obtain

$$\dot{\rho}_{21}^{(0)} = -P\frac{i}{2\hbar}E_n\left(t\right)N\left(z\right)\sin K_{nz}\left\{\exp\left[-i\left(\omega_n - \omega_{21} + i\gamma_{12}\right)t - i\phi_n\right]\right.$$
$$\left. + \exp\left[i\left(\omega_n + \omega_{21} - i\gamma_{12}\right)t - i\phi_n\right]\right\} \quad (6.76)$$

We neglect the time dependence of E_n and ϕ_n and integrate the above equation to get

$$\rho_{21}^{(0)} \approx \frac{P}{2\hbar}N\left(z\right)E_n\left(t\right)\sin K_{nz}\left\{\frac{\exp\left[-i\left(\omega_n - \omega_{21} + i\gamma_{12}\right)t - i\phi_n\right]}{\omega_n - \omega_{21} + i\gamma_{12}}\right.$$
$$\left. - \frac{\exp\left[i\left(\omega_n + \omega_{21} - i\gamma_{12}\right)t - i\phi_n\right]}{\omega_n + \omega_{21} + i\gamma_{12}}\right\} \quad (6.77)$$

We neglect the second exponential term in the curly brackets in the above equation as it has very rapid variations and we obtain

$$\rho_{21}^{(0)} \approx P\frac{1}{2\hbar}N\left(z\right)E_n\left(t\right)\sin K_{nz}\,e^{\gamma_{12}t}\frac{\exp\left\{-i\left[\left(\omega_n - \omega_{21}\right)t + \phi_n\left(t\right)\right]\right\}}{\left(\omega_n - \omega_{21}\right) + i\gamma_{12}} \quad (6.78)$$

Thus,[6]:

$$\rho_{21} \approx P\frac{1}{2\hbar}\frac{N\left(z\right)\sin K_{nz}\,E_n\left(t\right)}{\left(\omega_n - \omega_{21}\right) + i\gamma_{12}}\exp\left[-i\left(\omega_n t + \phi_n\right)\right]$$

or

$$\rho_{21} \approx P\frac{1}{2\hbar}\frac{N\left(z\right)\sin K_{nz}\,E_n\left(t\right)}{\Gamma_n}e^{-i\theta_n}\exp\left[-i\left(\omega_n t + \phi_n\right)\right] \quad (6.79)$$

---

[6]The constant of integration in Eq. (6.77) would have led to an exponentially decaying term in Eq. (6.79).

where

$$\cos\theta_n = \frac{\omega_n - \omega_{21}}{\Gamma_n}, \quad \sin\theta_n = \frac{\gamma_{12}}{\Gamma_n} \tag{6.80}$$

$$\Gamma_n \equiv \left[(\omega_n - \omega_{21})^2 + \gamma_{12}^2\right]^{1/2} \tag{6.81}$$

Thus, from Eq. (6.71) we get for the average dipole moment per atom

$$P_a = P(\rho_{21} + \rho_{12}) = P(\rho_{21} + c.c.)$$

If we assume that there are N_v atoms per unit volume in the cavity, then the macroscopic polarization would be given by

$$P = N_v P_a$$

$$= P^2 \frac{N_v}{2\hbar} \left[\frac{N(z)\sin K_n z}{\Gamma_n} E_n(t)\, e^{-i\theta_n} e^{-i(\omega_n t + \phi_n)} + c.c. \right] \tag{6.82}$$

where we have used Eq. (6.79). Comparing with Eq. (6.78), we get

$$P_n(t, z) = \frac{N_v\, P^2\, E_n(t)\, N(z)\sin K_n z}{\hbar \Gamma_n} e^{-i\theta_n} \tag{6.83}$$

and

$$p_n(t) = \frac{2}{L}\int_0^L P_n(t, z)\,\sin K_n z\, dz$$

$$= P^2 \frac{E_n(t)}{\hbar \Gamma_n} e^{-i\theta_n} \overline{N} N_v \tag{6.84}$$

where

$$\overline{N} = \frac{2}{L}\int_0^L N(z)\sin^2 K_n z\, dz \approx \frac{1}{L}\int_0^L N(z)\, dz \tag{6.85}$$

In writing the last step, we have assumed that $N(z)$ (which represents the population inversion density) varies slowly in an optical wavelength. Comparing Eq. (6.84) with Eq. (6.38), we get

$$\chi_n' = P^2 \frac{\overline{N} N_v}{\hbar \varepsilon_0} \frac{\omega_n - \omega_{21}}{(\omega_n - \omega_{21})^2 + \gamma_{12}^2} \tag{6.86}$$

and

$$\chi_n'' = -P^2 \frac{\overline{N} N_v}{\varepsilon_0 \hbar} \frac{\gamma_{12}}{(\omega_n - \omega_{21})^2 + \gamma_{12}^2} \tag{6.87}$$

The above two equations represent the variation of the real and imaginary parts of the susceptibility with the mode frequency ω_n.

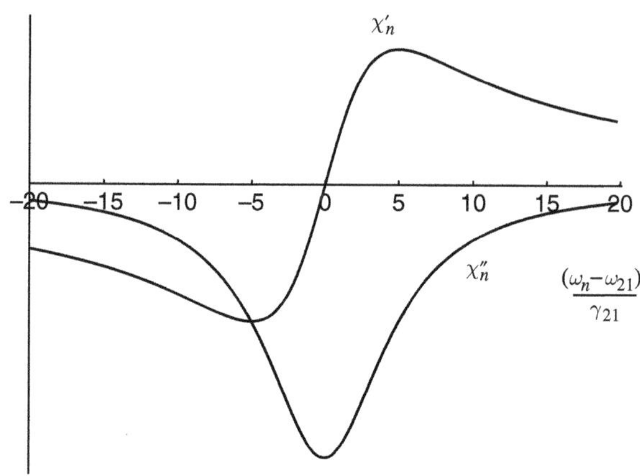

Fig. 6.2 Variation of χ_n' and χ_n'', which represent, respectively, the real and imaginary parts of the electric susceptibility of the medium, as a function of $\omega_n - \omega_{21}$; χ_n' is peaked at $\omega_n = \omega_{21}$ and thus maximum gain appears at $\omega_n = \omega_{21}$

In Fig. 6.2 we have plotted the variation of χ_n' and χ_n'' with ω_n for a medium with a population inversion (i.e., $\overline{N} > 0$).

Substituting for χ_n'' in Eq. (6.40), we get

$$\dot{E}_n(t) = \left[-\frac{\omega_n}{2Q_n'} + \frac{\omega_n P^2 \overline{N} N_v}{2\varepsilon_0 \hbar} \frac{\gamma_{12}}{(\omega_n - \omega_{21})^2 + \gamma_{12}^2} \right] E_n(t) \qquad (6.88)$$

Thus, for the amplitude to grow with time, the quantity inside the square brackets should be positive or

$$P^2 \frac{\overline{N} N_v}{\varepsilon_0 \hbar} \frac{\gamma_{12}}{(\omega_n - \omega_{21})^2 + \gamma_{12}^2} > \frac{1}{Q_n'} \qquad (6.89)$$

When the two sides of the above inequality are equal, then the losses are exactly compensated by the gain due to the cavity medium and this corresponds to the threshold condition.

The sign of the second term in the square brackets in Eq. (6.88) depends on the sign of $\overline{N}$. It may be recalled that $\overline{N}$ is proportional to the population difference between the upper and the lower states[7]. Thus if $\overline{N}$ is negative, i.e., if there are more atoms in the lower level than in the upper level, then the second term contributes an additional loss. On the other hand, if there is a population inversion between the levels 1 and 2 then $\overline{N}$ is positive and the medium acts as an amplifying medium. In order that the mode may oscillate, the losses have to be compensated by the gain and this leads to the threshold condition, for which we must have

---

[7] In fact $\overline{N} N_v$ represents the population inversion density in the cavity medium, i.e., it is equal to $(N_2 - N_1)$ of Chapter 5 [see discussion after Eq. (6.73)].

$$\overline{N}_t N_v = (N_2 - N_1)_t = \frac{\varepsilon_0 \hbar}{P^2 \gamma_{12}} \frac{(\omega_n - \omega_{21})^2 + \gamma_{12}^2}{Q'_n}$$
$$= \frac{\varepsilon_0 \hbar}{\pi g(\omega) P^2 Q'_n} \tag{6.90}$$

where the subscript t implies the threshold value and $g(\omega)$ represents the normalized lineshape function.

$$g(\omega) = \frac{\gamma_{12}}{\pi} \frac{1}{(\omega_n - \omega_{21})^2 + \gamma_{12}^2} \tag{6.91}$$

which is identical to Eq. (4.37) with $2t_{sp} = 1/\gamma_{12}$. Further

$$P^2 = e^2 \left[\int \psi_2^* x \psi_1 d\tau \right]^2 = \frac{e^2}{3} \left| \int \psi_2' \mathbf{r} \psi_1 d\tau \right|^2$$
$$= \frac{\pi \varepsilon_0 \hbar c^3}{\omega^3} A = \frac{\pi \varepsilon_0 \hbar c^3}{\omega^3} \frac{1}{t_{sp}} \tag{6.92}$$

where t_{sp} is the spontaneous relaxation time of level 2. Substituting for P^2 in Eq. (6.90), we get an expression for $(N_2 - N_1)_t$ identical to Eq. (4.32) for the case of natural broadening.

The minimum value of threshold inversion would correspond to $\omega_n = \omega_{21}$ (i.e., at resonance) giving[8]

$$\overline{N}_{tm} = \frac{\varepsilon_0 \hbar \gamma_{12}}{P^2 Q'_n N_v} \tag{6.93}$$

Next, we substitute for χ'_n from Eq. (6.86) in Eq. (6.39) to obtain

$$\omega_n - \Omega_n = \frac{\omega_n}{2 \hbar \varepsilon_0} P^2 \overline{N} N_v \frac{\omega_{21} - \omega_n}{(\omega_n - \omega_{21})^2 + \gamma_{12}^2} \tag{6.94}$$

where we have neglected the term $\dot{\phi}_n$, in Eq. (6.39). Thus, in the presence of the active medium, the oscillations do not occur at the passive cavity resonances but are shifted because of the presence of the χ'_n term. In general, this shift is small and one can obtain the approximate oscillation frequencies as

$$\omega_n \approx \Omega_n + \frac{\Omega_n P^2}{2 \hbar \varepsilon_0} \overline{N} N_v \frac{\omega_{21} - \Omega_n}{(\Omega_n - \omega_{21})^2 + \gamma_{12}^2} \tag{6.95}$$

---

[8]Notice that for $\omega_n \neq \omega_{21}$, i.e., for a mode shifted away from resonance, the value of N_t increases with increase in the value of $|\omega_n - \omega_{21}|$.

If Ω_n coincides exactly with the resonance frequency ω_{21} then $\omega_n = \Omega_n$ and in such a case the frequency of oscillation in the active resonator is the same as in the passive case. If $\Omega_n < \omega_{21}$, then for an inverted medium $\omega_n > \Omega_n$. Similarly for $\Omega_n > \omega_{21}, \omega_n < \Omega_n$. Thus, in the presence of the active medium, the oscillation frequencies are pulled toward the line center.

At threshold, we substitute for $\overline{N}_t$ from Eq. (6.90) to obtain

$$\omega_n - \Omega_n \approx \frac{\omega_n}{2Q'_n \gamma_{12}} (\omega_{21} - \omega_n) \tag{6.96}$$

where we have assumed $\omega_n \approx \omega_{21}$. We define a parameter

$$S = \frac{\omega_n / 2Q'_n}{\gamma_{12}} \tag{6.97}$$

which is known as the stabilizing factor[9], so that

$$\omega_n - \Omega_n \approx S (\omega_{21} - \omega_n)$$

or

$$\omega_n \approx \frac{\Omega_n + S\omega_{12}}{1 + S} \tag{6.98}$$

For a gas laser $S \sim 0.01 - 0.1$ so that the oscillation frequency lies very close to the normal mode frequency of the passive cavity mode.

6.3.2 Higher Order Theory

We have shown earlier that if the laser operates above threshold [see Eq. (6.89)], the power will grow exponentially with time [see Eq. (6.88)]. This unlimited growth is due to the assumption that the population difference remains constant with time [see Eq. (6.73)]. However, as the power increases, the population of the upper level would decrease (because of increase in stimulated emission), and hence in an actual laser, the power level would saturate at a certain value. We will show this explicitly in this section[10].

Similar to the rate equations discussed in Chapter 5, we start with the equations describing the population of the two levels:

$$\dot{\rho}_{11} = \lambda_1 - \gamma_1 \rho_{11} + \left(\frac{i}{\hbar} H'_{21} \rho_{12} + c.c. \right) \tag{6.99a}$$

$$\dot{\rho}_{22} = \lambda_2 - \gamma_2 \rho_{22} - \left(\frac{i}{\hbar} H'_{21} \rho_{12} + c.c. \right) \tag{6.99b}$$

---

[9]It represents the ratio of the cavity bandwidth to the natural linewidth.

[10]See also Section 5.5, where we showed that on a steady-state basis the inversion can never exceed the threshold value.

Fig. 6.3 λ_1 and λ_2 represent the rates of pumping of the lower and upper levels, respectively, and γ_1 and γ_2 represent their decay constants

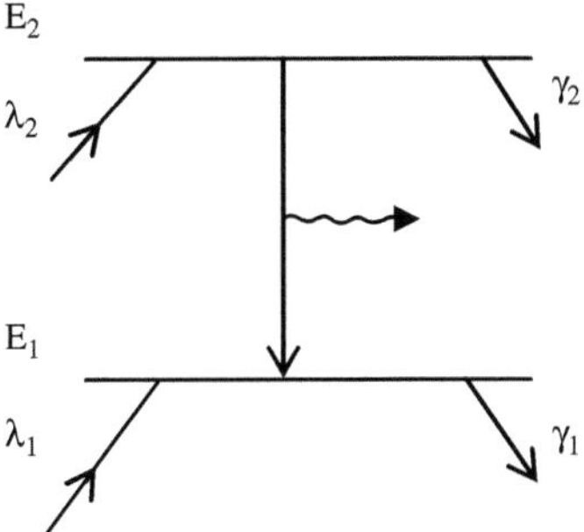

Just as in Section 5.4, the quantities λ_1 and λ_2 represent constant rates of pumping of atoms into levels 1 and 2, respectively (see Fig. 6.3). In order to solve the above equations, we substitute for ρ_{12} from the first-order solution obtained in the previous section. Thus

$$\frac{i}{\hbar}H'_{21}\rho_{12} = \frac{i}{\hbar}\left[-E_n P \cos(\omega_n t + \phi_n) \sin K_n z\right]\frac{P}{2\hbar}N(z)\frac{\sin K_n z}{\Gamma_n}E_n(t)\, e^{i\theta_n}e^{i(\omega_n t + \phi_n)}$$

$$= -\frac{i}{2\hbar^2}P^2\frac{E_n^2(t)}{\Gamma_n}(\rho_{22} - \rho_{11})\sin^2 K_n z \cos(\omega_n t + \phi_n)\, e^{i(\omega_n t + \phi_n + \theta_n)}$$

Hence

$$\frac{i}{\hbar}H'_{21}\rho_{12} + c.c. = \frac{P^2}{2\hbar^2}\frac{E_n^2(t)}{\Gamma_n}\sin^2 K_n z\,(\rho_{22} - \rho_{11})\,G \qquad (6.100)$$

where

$$G = -\cos(\omega_n t + \phi_n)\left[ie^{i(\omega_n t + \phi_n + \theta_n)} - ie^{-i(\omega_n t + \phi_n + \theta_n)}\right]$$

$$= 2\cos(\omega_n t + \phi_n)\sin(\omega_n t + \phi_n + \theta_n)$$

$$= 2\cos^2(\omega_n t + \phi_n)\sin\theta_n + \sin 2(\omega_n t + \phi_n)\cos\theta_n$$

$$\approx \sin\theta_n = \frac{\gamma_{12}}{\Gamma_n} \qquad (6.101)$$

where we have replaced G by its time average value. Substituting in Eq. (6.99a), we get

$$\dot{\rho}_{11} = \lambda_1 - \gamma_1\rho_{11} + R(\rho_{22} - \rho_{11}) \qquad (6.102)$$

Similarly

$$\dot{\rho}_{22} = \lambda_2 - \gamma_2\rho_{22} - R(\rho_{22} - \rho_{11}) \qquad (6.103)$$

where

$$R = \frac{\gamma_{12}}{2\hbar^2} P^2 E_n^2 \frac{\sin^2 K_n z}{\Gamma_n^2} \tag{6.104}$$

At steady state we must have $\dot{\rho}_{11} = \dot{\rho}_{22} = 0$, and

$$\rho_{11} - \frac{R}{\gamma_1}(\rho_{22} - \rho_{11}) = \frac{\lambda_1}{\gamma_1} \tag{6.105a}$$

$$\rho_{22} + \frac{R}{\gamma_2}(\rho_{22} - \rho_{11}) = \frac{\lambda_2}{\gamma_2} \tag{6.105b}$$

or

$$\rho_{22} - \rho_{11} = \frac{N(z)}{1 + R/R_s} \tag{6.106}$$

where

$$N(z) = \frac{\lambda_2}{\gamma_2} - \frac{\lambda_1}{\gamma_1} \tag{6.107}$$

and

$$R_s = \left(\frac{1}{\gamma_1} + \frac{1}{\gamma_2}\right)^{-1} = \frac{\gamma_1 \gamma_2}{2\gamma_{12}} \tag{6.108}$$

It follows from Eq. (6.106) that the population inversion depends on the field value also. In the absence of the field, $R = 0$ and the population difference density is simply $N(z) N_v$; however, as the field strength E_n (and hence R) increases, the population difference decreases. Since $R(z)$ has a sinusoidal dependence on z [see Eq. (6.104)], the population difference also varies with z. Whenever $K_n z$ is an odd multiple of $\pi/2$ [i.e., wherever the field has a maximum amplitude – see Eq. (6.70)], the population difference has a minimum value, which is often referred to as hole burning in the population difference and the holes have a spacing of half of a wavelength.

If instead of Eq. (6.73), we now use Eq. (6.106) for $\rho_{22} - \rho_{11}$, Eq. (6.79) would be replaced by

$$\rho_{21} \approx \frac{P}{2\hbar} \frac{E_n \sin K_n z}{\Gamma_n} \frac{N(z)}{1 + R/R_s} e^{-i\theta_n} e^{-i(\omega_n t + \phi_n)} \tag{6.109}$$

Thus [cf. Eq. (6.84)]

$$p_n(t) = \frac{P^2 E_n(t) e^{-i\theta_n}}{\hbar \Gamma_n} N_v \left[\frac{2}{L} \int_0^L \frac{N(z)}{1 + R/R_s} \sin^2 K_n z \, dz\right] \tag{6.110}$$

We next assume E_n (and hence R) to be small enough so that

$$\left(1 + \frac{R}{R_s}\right)^{-1} \approx 1 - \frac{R}{R_s} \tag{6.111}$$

Substituting this in Eq. (6.110) and carrying out a term-by-term integration, we obtain

$$\frac{2}{L} \int_0^L \frac{N(z)}{1 + R/R_s} \sin^2 K_n z\, dz \approx \frac{2}{L} \int_0^L N(z) \sin^2 K_n z\, dz$$

$$- \frac{P^2 E_n^2 \gamma_{12}^2}{\hbar^2 \gamma_1 \gamma_2} \frac{1}{\Gamma_n^2} \left[\frac{2}{L} \int_0^L N(z) \sin^4 K_n z\, dz \right] \tag{6.112}$$

$$\approx \overline{N} \left[1 - \frac{3}{4} \frac{P^2 E_n^2}{\hbar^2 \gamma_1 \gamma_2} \frac{\gamma_{12}^2}{\Gamma_n^2} \right]$$

where we have used Eq. (6.85) and the relation

$$\frac{2}{L} \int_0^L N(z) \sin^4 K_n z\, dz = \frac{2}{L} \left[\int_0^L N(z)\, dz \right] \left\langle \sin^4 K_n z \right\rangle \approx \frac{3}{4} \overline{N} \tag{6.113}$$

Thus,

$$p_n(t) \approx \frac{P^2 E_n}{\hbar \Gamma_n} e^{-i\theta_n} \overline{N} N_v \left(1 - \frac{3}{4} \frac{P^2 E_n^2}{\hbar^2 \gamma_1 \gamma_2} \frac{\gamma_{12}^2}{\Gamma_n} \right) \tag{6.114}$$

or

$$p_n(t) \approx \frac{P^2 E_n}{\hbar \Gamma_n} e^{-i\theta_n} \overline{N} N_v \left(1 + \frac{3}{4} \frac{P^2 E_n^2}{\hbar^2 \gamma_1 \gamma_2} \frac{\gamma_{12}^2}{\Gamma_n^2} \right)^{-1} \tag{6.115}$$

where in the last step we have assumed the two terms inside the square brackets in Eq. (6.114) to be the first two terms of a geometric series. This way the gain saturates as the electric field increases indefinitely. Eq. (6.115) may be compared with Eq. (6.84); hence instead of Eqs. (6.86) and (6.87), we get

$$\chi_n' \approx \frac{P^2 \overline{N} N_v}{\hbar \varepsilon_0 \Gamma_n} \cos \theta_n \left(1 + \frac{3}{4} \frac{P^2 E_n^2}{\hbar^2 \gamma_1 \gamma_2} \frac{\gamma_{12}^2}{\Gamma_n^2} \right)^{-1} \tag{6.116}$$

$$\chi_n'' \approx - \frac{P^2 \overline{N} N_v}{\hbar \varepsilon_0 \Gamma_n} \sin \theta_n \left(1 + \frac{3}{4} \frac{P^2 E_n^2}{\hbar^2 \gamma_1 \gamma_2} \frac{\gamma_{12}^2}{\Gamma_n^2} \right)^{-1} \tag{6.117}$$

Substituting this value of χ_n'' in Eq. (6.40), we get at steady state,

$$0 = \dot{E}_n = \left[-\frac{1}{2}\frac{\omega_n}{Q_n'} + \frac{1}{2}\omega_n \frac{P^2 \overline{N} N_v}{\hbar \varepsilon_0 \Gamma_n} \sin \theta_n \left(1 + \frac{3}{4} \frac{P^2 E_n^2}{\hbar^2 \gamma_1 \gamma_2} \frac{\gamma_{12}^2}{\Gamma_n^2} \right)^{-1} \right] E_n \quad (6.118)$$

which after simplification gives

$$E_n^2 = \frac{4\hbar^2}{3P^2}\gamma_1\gamma_2 \left[\frac{\overline{N}}{\overline{N}_{tm}} - 1 - \frac{(\omega_n - \omega_{21})^2}{\gamma_{12}^2} \right] \quad (6.119)$$

where $\overline{N}_{tm}$ is given by Eq. (6.93). The above equation gives the dependence of the saturation value of the intensity as a function of the detuning $(\omega_n - \omega_{21})$. At resonance $\omega_n = \omega_{21}$ and we get

$$E_n^2 = \frac{4\hbar^2\gamma_1\gamma_2}{3P^2} \left(\frac{\overline{N}}{\overline{N}_m} - 1 \right) \quad (6.120)$$

It is clear from the above equation that the intensity of the field inside the cavity increases linearly with the pumping rate above threshold.

It should be pointed out that in the above equation, $\overline{N}_{tm}$ is proportional to the pumping rate at threshold [see Eq. (6.107)] and happens to be equal to the inversion density at the threshold [see Eq. (6.73)]. On the other hand, $\overline{N}$ is proportional to the pumping rate corresponding to the actual laser operation which is greater than $\overline{N}_t$. Thus if we assume $\gamma_1 \ggg \gamma_2$, then $\overline{N}N_v$ is nothing but R/T_{21} of Section 5.5. Using this value of $\overline{N}$ one obtains

$$E_n^2 \simeq \frac{4}{3}\frac{\hbar^2\gamma_1\gamma_2}{P^2} \left(\frac{R}{R_t} - 1 \right) \quad (6.121)$$

where we have used the relation

$$R_t = N_{2t}T_{21} \approx N_v \overline{N}_t \gamma_2 \quad (6.122)$$

Further, in order to relate the photon number of the cavity to E_n^2, we note that the energy density of the field in the cavity is given by $\frac{1}{2}\varepsilon_0 E_n^2$ and the total energy in the cavity of volume V would be $\frac{1}{2}\varepsilon_0 V E_n^2$. If the frequency of the cavity mode is ω_n, the number of photons in the cavity mode would be

$$n = \frac{1}{2}\varepsilon_0 \frac{E_n^2 V}{\hbar\omega_n} \quad (6.123)$$

Substituting this in Eq. (6.121) and using the fact that K defined in Eq. (5.72) is identical to[11]

$$K = \frac{P^2 \omega_n}{\hbar \varepsilon_0 \gamma_{12}} \tag{6.124}$$

we obtain

$$n = \frac{4}{3} V \left(\frac{\gamma_1 \gamma_2}{2\gamma_{12}} \right) \frac{1}{K} \left(\frac{R}{R_t} - 1 \right) \tag{6.125}$$

For $\gamma_1 \gg \gamma_2$ (i.e., the lower level has a very short lifetime as compared to the upper state)

$$\frac{\gamma_1 \gamma_2}{2\gamma_{12}} = \frac{\gamma_1 \gamma_2}{\gamma_1 + \gamma_{12}} \approx \gamma_2 \tag{6.126}$$

Thus Eq. (6.125) becomes

$$n = \frac{4}{3} V \frac{\gamma_2}{K} \left(\frac{R}{R_t} - 1 \right) \tag{6.127}$$

which is the same as Eq. (5.85) obtained in the last chapter (γ_2 correspond to T_{21} of Section 5.5), apart from the factor 4/3 which has appeared because of the consideration of the spatial dependence of the modal field in this chapter.

---

[11]Use has been made of Eq. (6.92)

Chapter 7
Optical Resonators

7.1 Introduction

In Chapter 4 we discussed briefly the optical resonator, which consists of a pair of mirrors facing each other in between which is placed the active laser medium which provides for optical amplification. As we discussed, the mirrors provide optical feedback and the system then acts as an oscillator generating light rather than just amplifying. In this chapter we give a more detailed account of optical resonators. In Section 7.2 we will discuss the modes of a rectangular cavity and show that there exist an extremely large number of modes of oscillation under the linewidth of the active medium in a closed cavity of practical dimensions (which are large compared to the wavelength of light). Section 7.3 discusses the important concept of the quality factor of an optical resonator. In this section we obtain the linewidth corresponding to the passive cavity in terms of the parameters of the resonator. We also introduce the concept of cavity lifetime. In Section 7.4 we discuss the ultimate linewidth of the laser oscillator – this is, as discussed earlier in Chapter 6, determined by spontaneous emissions occurring in the cavity. In practice the observed linewidth is much larger than the ultimate linewidth and is determined by mechanical stability, temperature fluctuations, etc. Section 7.5 discusses some techniques for selecting a single transverse and longitudinal mode in a laser oscillator. In Sections 7.6 and 7.7 we discuss the techniques for producing short intense pulses of light using Q-switching and mode locking. Using the mode locking techniques one can obtain ultrashort pulses of very high peak power which find widespread applications.

In Section 7.8 we give a scalar wave analysis of the modes of a symmetrical confocal resonator which consists of a pair of concave mirrors of equal radii of curvatures and separated by a distance equal to the radius of curvature. We will show that in such a structure, the lowest order transverse mode has a Gaussian field distribution across its wave front. Most practical lasers are made to oscillate in this mode. In Section 7.9 we give the results for the beamwidth and the field distributions corresponding to a general spherical resonator.

K. Thyagarajan, A. Ghatak, *Lasers*, Graduate Texts in Physics,
DOI 10.1007/978-1-4419-6442-7_7, © Springer Science+Business Media, LLC 2010

7.2 Modes of a Rectangular Cavity and the Open Planar Resonator

Consider a rectangular cavity of dimensions $2a \times 2b \times d$ as shown in Fig. 7.1. Starting from Maxwell's equations [see Eqs. (2.1), (2.2), (2.3), and (2.4)] one can show that the electric and magnetic fields satisfy a wave equation of the form given by

$$\nabla^2 \mathbf{E} - \frac{n_0^2}{c^2} \frac{\partial^2 \mathbf{E}}{\partial t^2} = 0 \tag{7.1}$$

where c represents the velocity of light in free space and n_0 represents the refractive index of the medium filling the rectangular cavity. Equation (7.1) has been derived in Chapter 2.

If the walls of the rectangular cavity are assumed to be perfectly conducting then the tangential component of the electric field must vanish at the walls. Thus if $\hat{n}$ represents the unit vector along the normal to the wall then we must have

$$\mathbf{E} \times \hat{n} = 0 \tag{7.2}$$

on the walls of the cavity.

Let us consider a Cartesian component (say x component) of the electric vector; this will also satisfy the wave equation, which in the Cartesian system of coordinates will be given by

$$\frac{\partial^2 \mathbf{E}_x}{\partial x^2} + \frac{\partial^2 \mathbf{E}_x}{\partial y^2} + \frac{\partial^2 \mathbf{E}_x}{\partial z^2} = \frac{n_0^2}{c^2} \frac{\partial^2 \mathbf{E}_x}{\partial t^2} \tag{7.3}$$

In order to solve Eq. (7.3) we use the method of separation of variables and write

$$\mathbf{E}_x = X(x)Y(y)Z(z)T(t) \tag{7.4}$$

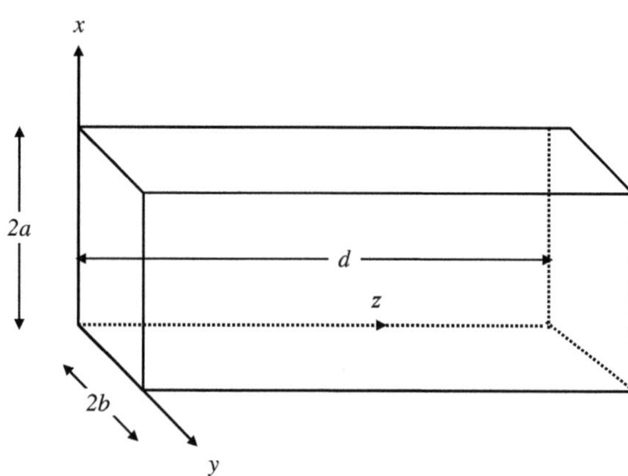

Fig. 7.1 A rectangular cavity of dimensions $2a \times 2b \times d$

Substituting this in Eq. (7.3) and dividing by E_x We obtain

$$\frac{1}{X}\frac{\partial^2 X}{\partial x^2} + \frac{1}{Y}\frac{\partial^2 Y}{\partial y^2} + \frac{1}{Z}\frac{\partial^2 Z}{\partial z^2} = \frac{n_0^2}{c^2 T}\frac{\partial^2 T}{\partial t^2} \tag{7.5}$$

Thus the variables have indeed separated out and we may write

$$\frac{1}{X}\frac{\partial^2 X}{\partial x^2} = -k_x^2 \tag{7.6}$$

$$\frac{1}{Y}\frac{\partial^2 Y}{\partial y^2} = -k_y^2 \tag{7.7}$$

$$\frac{1}{Z}\frac{\partial^2 Z}{\partial z^2} = -k_z^2 \tag{7.8}$$

and

$$\frac{n_0^2}{c^2 T}\frac{\partial^2 T}{\partial t^2} = -k^2 \tag{7.9}$$

where

$$k^2 = k_x^2 + k_y^2 + k_z^2 \tag{7.10}$$

Equation (7.9) tells us that the time dependence is of the form

$$T(t) = Ae^{-i\omega t} \tag{7.11}$$

where $\omega = c\,k/n_0$ represents the angular frequency of the wave and A is a constant. It should be mentioned that we could equally well have chosen the time dependence to be of the form $e^{i\omega t}$. Since E_x is a tangential component on the planes $y = 0$, $y = 2b$, $z = 0$, and $z = d$, it has to vanish on these planes and the solution of Eqs. (7.7) and (7.8) would be $\sin k_y y$ and $\sin k_z z$, respectively, with

$$k_y = \frac{n\pi}{2b}, \qquad k_z = \frac{q\pi}{d}, \qquad n,q = 0,1,2,3,... \tag{7.12}$$

where we have intentionally included the value 0, which in this case would lead to the trivial solution of E_x vanishing everywhere (The above solutions are very similar to the ones discussed in Example 3.3). In a similar manner, the x and z dependences of E_y would be $\sin k_x x$ and $\sin k_z z$, respectively, with

$$k_x = \frac{m\pi}{2a}, \qquad m = 0,1,2,3,\ldots \tag{7.13}$$

and k_z given by Eq. (7.12). Finally the x and y dependences of E_z would be $\sin k_x x$ and $\sin k_y y$ respectively.

Now, because of the above forms of the x dependence of E_y and E_z, $\partial E_y / \partial y$, and $\partial E_z / \partial z$ would vanish on the surfaces $x = 0$ and $x = 2\,a$. Thus on the planes $x = 0$ and $x = 2a$, the equation $\nabla.\mathbf{E} = 0$ leads to $\partial E_x / \partial x = 0$. Hence the x dependence of E_x will be of the form $\cos k_x x$ with k_x given by Eq. (7.13). Notice that the case $m = 0$ now corresponds to a nontrivial solution.

In a similar manner, one may obtain the solutions for E_y and E_z. The complete solution (apart from the time dependence) would therefore be given by

$$
\begin{aligned}
E_x &= E_{0x} \cos k_x x \sin k_y y \sin k_z z \\
E_y &= E_{0y} \sin k_x x \cos k_y y \sin k_z z \\
E_z &= E_{0z} \sin k_x x \sin k_y y \cos k_z z
\end{aligned}
\tag{7.14}
$$

where E_{0x}, E_{0y}, and E_{0z} are constants. The use of Maxwell's equation $\nabla.\mathbf{E} = 0$, immediately gives

$$
\vec{E}_0.\vec{k} = 0
\tag{7.15}
$$

where $\vec{k} = \hat{x} k_x + \hat{y} k_y + \hat{z} k_z$. Since the coefficients E_{0x}, E_{0y} and E_{0z} have to satisfy Eq. (7.15) it follows that for a given mode, i.e., for a given set of values of m, n, and q only two of the components of E_0 can be chosen independently. Thus a given mode can have two independent states of polarization.

Note that when one of the quantities m, n, or q is zero, then there is only one possible polarization state associated with the mode. Thus if we consider the use with $m = 0$, $n \neq 0$, $q \neq 0$, then $E_x = E_{0x} \sin k_y y \sin k_z z$, $E_y = 0$, $E_z = 0$. Thus the only possible case is with the electric vector oriented along the x-direction.

Using Eqs (7.10), (7.12), and (7.13), we obtain

$$
\begin{aligned}
\omega^2 &= \frac{c^2 k^2}{n_0^2} = \frac{c^2}{n_0^2} \left(k_x^2 + k_y^2 + k_z^2 \right) \\
&= \frac{c^2 \pi^2}{n_0^2} \left(\frac{m^2}{4a^2} + \frac{n^2}{4b^2} + \frac{q^2}{d^2} \right)
\end{aligned}
$$

or

$$
\omega = \frac{c\pi}{n_0} \left(\frac{m^2}{4a^2} + \frac{n^2}{4b^2} + \frac{q^2}{d^2} \right)^{1/2}
\tag{7.16}
$$

which gives us the allowed frequencies of oscillation of the field in the cavity. Field configurations given by Eq. (7.14) represent standing wave patterns in the cavity and are called modes of oscillation of the cavity. These are similar to the acoustic modes of vibration of an acoustic cavity (like in a musical instrument such as a guitar and veena) and represent the only possible frequencies that can exist within the cavity.

Example 7.1 As a specific example we consider a mode with

$$
m = 0, n = 1, \text{ and } q = 1
$$

Thus $k_x = 0, k_y = \pi / 2b, k_z = \pi / d$ and using Eq. (7.14), we have

$$E_x = E_{0x} \sin k_y y \sin k_z z = E_{0x} \sin \left(\frac{\pi}{2b}y\right) \sin \left(\frac{\pi}{d}z\right)$$

$$E_y = 0$$

$$E_z = 0$$

Using the time dependence of the form $e^{-i\omega t}$ and expanding the sine functions into exponentials, we may write

$$E = \frac{1}{(2i)^2}\hat{x}\left(e^{-i(\omega t - k_y y - k_z z)} + e^{-i(\omega t - k_y y + k_z z)} + e^{-i(\omega t + k_y y - k_z z)} + e^{-i(\omega t + k_y y + k_z z)}\right) \qquad (7.17)$$

Thus the total field inside the cavity has been broken up into four propagating plane waves; in Eq. (7.17) the first term on the right-hand side represents a wave propagating along the $(+y, +z)$ direction, the second along $(+y, -z)$ direction, the third along $(-y, +z)$ direction, and the fourth along $(-y, -z)$ direction, respectively. These four plane waves interfere at every point inside the cavity to produce a standing wave pattern. However, since k_y and k_z take discrete values, the plane waves which constitute the mode make discrete angles with the axes.

Example 7.2 If we take a cavity with $a = b = 1$ cm and $d = 20$ cm and consider the mode with $m = 0, n = 1, q = 10^6$
then

$$k_x = 0, k_y = \pi / 2\,\text{cm}^{-1}, k_z = 10^6 \pi / 20\,\text{cm}^{-1}$$

implying

$$k \approx 10^6 \pi / 20\,\text{cm}^{-1} \text{ and } \nu = ck/2\pi = 7.5 \times 10^{14}\,\text{Hz}$$

which lies in the optical region. For such a case

$$\theta_y = \cos^{-1}\left(\frac{k_y}{k}\right) \approx 89.9994°$$

$$\theta_z = \cos^{-1}\left(\frac{k_z}{k}\right) \approx 0.0006°$$

and $\theta_x = 0$ because of which $\theta_y + \theta_z = 90°$. It may be noted that the component waves are propagating almost along the z-axis. In general,

$$\cos^2 \theta_x + \cos^2 \theta_y + \cos^2 \theta_z = 1$$

Further for $m \neq 0$, $n \neq 0$, $q \neq 0$ the cavity mode can be thought of as a standing wave pattern formed by eight plane waves with components of $\vec{k}$ given by $(\pm k_x, \pm k_y, \pm k_z)$.

Example 7.3 Let us now consider a few hundred nanometer-sized rectangular cavity (also referred to as a microcavity) filled with free space. Let $2a = 2b = d = 500$ nm. We now calculate the wavelengths ($\lambda = c/\nu$) of oscillation corresponding to some of the lower order modes which can be obtained from Eq. (7.16) as

m	n	q	λ (nm)
1	0	0	1000
0	1	0	1000
0	0	1	1000
1	1	0	707.1
1	0	1	707.1
0	1	1	707.1
1	1	1	577.4
2	0	0	500

Note that since the cavity dimensions are of the order of optical wavelength, in the optical wavelength region, the wavelengths of oscillation of the modes are well separated. Also the cavity cannot support any mode at wavelengths longer than 1000 nm. If we place an atom in such a cavity and if the atom has energy levels separated by energy difference corresponding to a wavelength of say 800 nm with emission spectral width of about 10 nm, then since there are no possible modes in the cavity corresponding to this wavelength region, the atom would be inhibited from emitting radiation. Thus it is possible to inhibit spontaneous emission from atoms and increase the lifetime of the level. Microcavities of dimensions comparable to optical wavelength are now being extensively investigated for various applications including suppressing spontaneous emission or for enhancing spontaneous emission, for lowering threshold for laser oscillation, etc. (see, e.g., Vahala (2003) and Gerard (2003)).

If we had chosen even one of the dimensions to be much larger then the mode spacing would be much smaller. As an example if we assume $2a = 2b = 500$ nm and $d = 10{,}000$ nm, then the wavelength corresponding to various low-order modes would be

m	n	q	λ (nm)
1	0	0	1000
0	1	0	1000
1	1	0	707.1
1	0	1	998.8
0	1	1	998.8
1	1	1	706.7
2	0	0	500
0	0	21	952.3
0	0	22	909.1
0	0	23	869.5
0	0	24	833.3

It can be noted that since the value of d is large compared to wavelength around 900 nm, the mode spacing is small.

Using Eq. (7.16) we can show (see Appendix E) that the number of modes per unit volume in a frequency interval from ν to $\nu + d\nu$ will be given by

$$p(\nu)d\nu = \frac{8\pi\, n_0^3}{c^3}\nu^2 d\nu \tag{7.18}$$

where n_0 represents the refractive index of the medium filling the cavity. For a typical atomic system $d\nu \sim 3 \times 10^9$ Hz at $\nu = 3 \times 10^{14}$ Hz and the number of modes per unit volume would be (assuming $n_0 = 1$)

$$p(\nu)d\nu = \frac{8\pi\, n_0^3}{c^3}\nu^2 d\nu = \frac{8 \times \pi \times 1 \times \left(3 \times 10^{14}\right)^2}{\left(3 \times 10^8\right)^3} \times 3 \times 10^9 \approx 2 \times 10^8 \mathrm{cm}^{-3}$$

Thus for cavities having typical volumes of 10 cm^3, the number of possible oscillating modes within the linewidth will be 2×10^9 which is very large. To achieve a very small number of possible oscillating modes within the linewidth of the atomic transition, the volume of the cavity has to be made very small. Thus to achieve a single mode of oscillation within the linewidth the volume of the cavity should be of the order of 5×10^{-9} cm^3. This corresponds to a cube of linear dimension of the order of 17 μm. Optical microcavities having such small dimensions can be fabricated using various techniques and are finding applications for studying strong interactions between atoms and radiation (cavity quantum electrodynamics), for inhibiting spontaneous emission, or for enhancing spontaneous emission and as filters for optical fiber communication systems. For a nice review, readers are referred to Vahala (2003).

In the case of conventional lasers the volume of the cavity is large and thus the number of oscillating modes within the linewidth of the atomic transition is very large. Thus all these oscillating modes can draw energy from the atomic system and the resulting emission would be far from monochromatic. In order to have very few oscillating modes within the cavity, if the dimensions of the cavity are chosen to be of the order of the wavelength, then the volume of the atomic system available for lasing becomes quite small and the power would be quite small.

The problem of the extremely large number of oscillating modes can be overcome by using *open cavities* (as against closed cavities) which consist of a pair of plane or curved mirrors facing each other. As we have seen earlier, a mode can be considered to be a standing wave pattern formed between plane waves propagating within the cavity with $\vec{k}$ given by $(\pm k_x, \pm k_y, \pm k_z)$. Thus the angles made by the component plane waves with the x-, y- and z-directions will, respectively, be, $\cos^{-1}(m\lambda / 2a)$, and $\cos^{-1}(n\lambda / 2b)$, $\cos^{-1}(q\lambda / d)$. Since in open resonators, the side walls of the cavity have been removed, those modes which are propagating almost along the z-direction (i.e., with large value of q and small values of m and n) will have a loss which is much smaller than the loss of modes which make large angles with the z-axis (i.e., modes with large values of m and n). Thus on removing the side walls of the cavity, only modes having small values of m and n ($\sim 0, 1, 2..,$) will have a small loss, and thus as the amplifying medium placed inside the cavity is pumped, only these modes will be able to oscillate. Modes with larger values of m and n will have a large loss and thus will be unable to oscillate.

It should be noted here that since the resonator cavity is now open, *all* modes would be lossy. Thus even the modes that have plane wave components traveling almost along the z-direction will suffer losses. Since m and n specify the field patterns along the transverse directions x and y and q that along the longitudinal direction z, modes having different values of (m, n) are referred to as various transverse modes while modes differing in q-values are referred to as various longitudinal modes.

The oscillation frequencies of the various modes of the closed cavity are given by Eq. (7.16). In order to obtain an approximate value for the oscillation frequencies of the modes of an open cavity, we may again use Eq. (7.16) with the condition m, $n \ll q$. Thus making a binomial expansion in Eq. (7.16) we obtain

$$\nu_{mnq} = \frac{c}{2n_0} \left(\frac{q}{d} + \left(\frac{m^2}{a^2} + \frac{n^2}{b^2} \right) \frac{d}{8q} \right)^{1/2} \tag{7.19}$$

The difference in frequency between two adjacent modes having same values of m and n and differing in q value by unity would be very nearly given by

$$\Delta v_q \approx \frac{c}{2 n_0 d} \qquad (7.20)$$

which corresponds to the longitudinal mode spacing. In addition if we completely neglect the terms containing m and n in Eq. (7.19) we will obtain

$$v_q \approx q \frac{c}{2 n_0 d} \qquad (7.21)$$

The above equation is similar to the frequencies of oscillation of a stretched string of length d.

Example 7.4 For a typical laser resonator $d \sim 100$ cm and assuming free space filling the cavity, the longitudinal mode spacing comes out to be ~ 150 MHz which corresponds to a wavelength spacing of approximately 0.18 pm $(= 0.18 \times 10^{-12}$ m) at a wavelength of 600 nm.

Problem 7.1 Show that the separation between two adjacent transverse modes is much smaller than Δv_q.

Solution The frequency separation between two modes differing in m values by unity would be

$$\Delta v_m \approx \frac{c}{2 n_0} \frac{d}{8 a^2 q} \left[m^2 - (m-1)^2 \right] \approx \Delta v_q \frac{\lambda d}{8 a^2} \left(m - \frac{1}{2} \right)$$

where we have used $q \approx 2d / \lambda$ [see Eq. (7.21)]. For typical values of $\lambda = 600$ nm, $d = 100$ cm, $a = 1$ cm, $\frac{\lambda d}{8 a^2} = 7.5 \times 10^{-4}$. Thus for $m \sim 1$, $\Delta v_m << \Delta v_q$.

It is of interest to mention that an open resonator consisting of two plane mirrors facing each other is, in principle, the same as a Fabry–Perot interferometer or an etalon (see Section 2.9). The essential difference in respect of the geometrical dimensions is that in a Fabry–Perot interferometer the spacing between the mirrors is very small compared to the transverse dimensions of the mirrors while in an optical resonator the converse is true. In addition, in the former case the radiation is incident from outside while in the latter the radiation is generated within the cavity.

Earlier we showed that the modes in closed cavities are essentially superpositions of propagating plane waves. Because of diffraction effects, plane waves cannot represent the modes in open cavities. Indeed if we start with a plane wave traveling parallel to the axis from one of the mirrors, it will undergo diffraction as it reaches the second mirror and since the mirror is of finite transverse dimension the energy in the diffracted wave that lies outside the mirror would be lost. The wave reflected from the mirror will again undergo diffraction losses when it is reflected from the first mirror. Fox and Li (1961) performed numerical calculations of such a planar resonator. The analysis consisted of assuming a certain field distribution at one of the mirrors of the resonator and calculating the Fresnel diffracted field at the second mirror. The field reflected at the second mirror is used to calculate back the field distribution at the first mirror. It was shown that after many traversals between the mirrors, the field distribution settles down to a steady pattern, i.e., it does not change between successive reflections but only the amplitude of the field decays exponentially in time due to diffraction losses. Such a field distribution represents a normal mode of the resonator and by changing the initial field distribution on the first mirror other modes can also be obtained.

A pair of curved mirrors instead of plane mirrors can also form an optical cavity. Depending on the curvature of the mirrors and the spacing between them, the resonator so formed can be stable or unstable. In stable resonators, the field distribution can keep bouncing back and forth between the mirrors without much loss due to the finite size of the mirrors. On the other hand, in unstable resonators, the field escapes from the sides of the mirrors and is not well confined to the cavity. Thus the diffraction losses in resonators formed by curved mirrors can be much smaller. In fact if the mirrors are sufficiently large in the transverse dimensions, the diffraction losses can be made almost negligible.

7.3 Spherical Mirror Resonators

An open resonator with plane mirrors would have significant diffraction losses on account of the finite transverse size of the mirrors. If focusing action is provided in the cavity then the diffraction losses can be much reduced; this can be achieved by replacing plane mirrors by spherical mirrors.

In spherical mirror resonators, the resonator is formed by a pair of spherical mirrors or a plane mirror and a concave mirror. Figure 7.2 shows various spherical mirror resonators. These include the symmetric confocal resonator which consists of a pair of identical concave mirrors each having a radius of curvature R and the separation between the mirrors is R so that the foci of the two mirrors coincide at the center of the resonator. In a symmetric concentric resonator, identical concave mirrors of radii of curvature R are separated by a distance $2R$ so that the centers of curvatures of the mirrors coincide. In general one can form a spherical mirror resonator with plane, concave, or convex mirrors. Depending on the curvatures of the

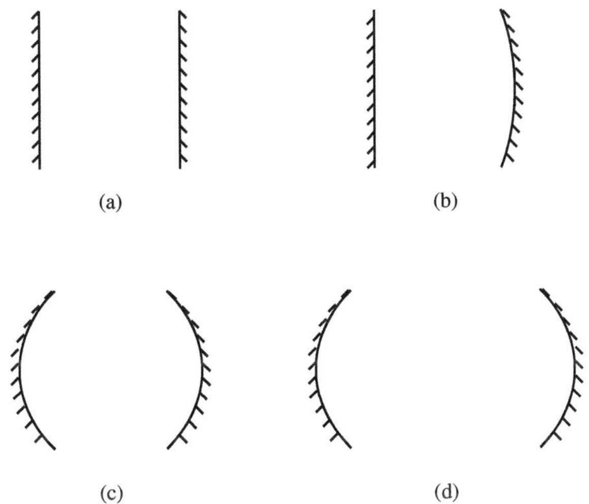

Fig. 7.2 Different spherical mirror resonators

mirrors and the separation between the mirrors, the resonator is stable or unstable. In the language of geometrical optics, in stable resonators, a family of light rays may keep bouncing back and forth between the mirrors of the cavity indefinitely without ever escaping from the cavity. On the other hand in unstable resonator system, there are no ray families that can bounce back and forth without escaping from the cavity; the ray diverges away from the axis after every pass and thus escapes from the resonator after a few traversals.

In Section 7.9 we will give a detailed scalar analysis of spherical mirror resonators and will show that the modes of such a stable resonator system are given by Hermite–Gauss functions:

$$E_{mn}(x, y) = E_0 H_m \left(\frac{\sqrt{2}\,x}{w_0} \right) H_n \left(\frac{\sqrt{2}\,y}{w_0} \right) e^{-(x^2+y^2)/w_0^2} \tag{7.22}$$

where m and n represent the transverse mode numbers, $H_m\left(\sqrt{2}\,x/w_0\right)$ and $H_n\left(\sqrt{2}\,y/w_0\right)$ represent Hermite polynomials (see Chapter 3), and w_0 is the characteristic mode width which depends on the wavelength of operation, the resonator characteristics such as the radii of the mirrors, and the distance between them. Figure 7.3 shows the intensity patterns of some of the lower order modes of a stable resonator cavity formed by spherical mirrors.

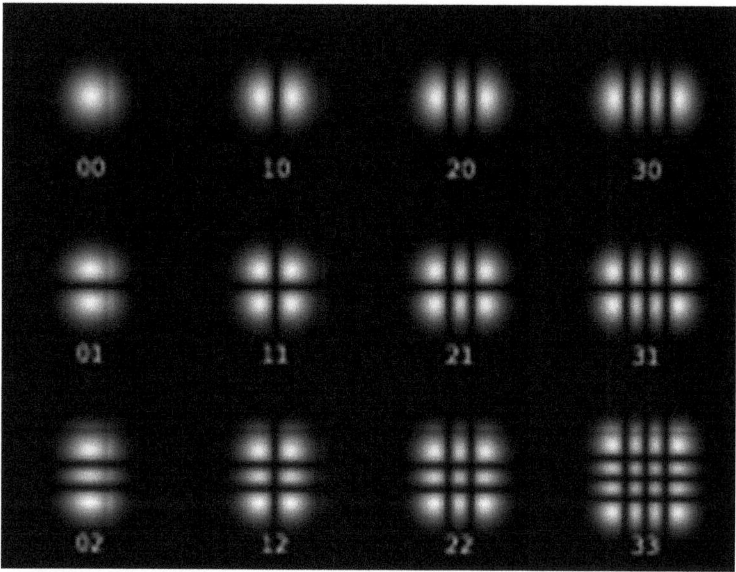

Fig. 7.3 Photograph showing some of the lower order resonator modes (www.absolute astromy.com/topics/Transverse_mode)

The lowest order mode of such a resonator system has a Gaussian amplitude distribution across its wave front and is given by

$$E_{00}(x, y) = E_0 e^{-(x^2+y^2)/w_0^2} \tag{7.23}$$

Higher order modes are characterized by a larger transverse dimension.

7.4 The Quality Factor

The quality factor (Q factor) is a dimensionless parameter that characterizes the energy dissipation in a resonant system by comparing the time constant of decay of energy to the oscillation period of the system. The smaller the loss, the slower would be the decay rate compared to the oscillating period and larger would be the Q factor. If we consider a closed optical cavity made up of perfectly conducting walls which do not have any loss and if the cavity is filled with free space, then in principle energy fed into the cavity will never die and thus would correspond to an infinite quality factor. Since in practice the walls of the cavity would have some loss and also the medium within the cavity would have some loss the Q factor of the cavity would not be infinite. On the other hand, open cavities by definition are lossy even if the two mirrors at the ends of the cavity are lossless with the free space filling the cavity; diffraction loss would still occur.

In an actual resonator, the mirrors would not be having 100% reflectivity; the medium filling the cavity would have some losses.

We define the Q factor of the cavity by the following equation:

$$Q = \omega_0 \frac{\text{energy stored in the mode}}{\text{energy lost per unit time}} \tag{7.24}$$

Here ω_0 is the oscillation frequency of the mode. If $W(t)$ represents the energy in the mode at time t, then from Eq. (7.24) we obtain

$$Q = \omega_0 \frac{W(t)}{-dW/dt}$$

or

$$\frac{dW}{dt} = -\left(\frac{\omega_0}{Q}\right) W(t)$$

whose solution is

$$W(t) = W(0)e^{-\omega_0 t/Q} \tag{7.25}$$

Thus if t_c represents the cavity lifetime, i.e., the time in which the energy in the mode decreases by a factor 1/e, then,

$$t_c = \frac{Q}{\omega_0} = \frac{Q}{2\pi \nu_0} \tag{7.26}$$

We can write for the electric field associated with the mode as

$$E(t) = E_0 e^{i\omega_0 t} e^{-\omega_0 t / 2Q} \tag{7.27}$$

The frequency spectrum of this wave train can be obtained in a manner similar to that used in Section 4.5.1 to obtain the spontaneous emission spectrum and it comes out to be

$$\left|\tilde{E}(\nu)\right|^2 = \frac{E_0^2}{4\pi^2} \frac{1}{(\nu - \nu_0)^2 + \frac{\nu_0^2}{4Q^2}} \tag{7.28}$$

which represents a Lorentzian (see Fig. 4.7). The FWHM of the spectrum is

$$\Delta \nu_P = \frac{\nu_0}{Q} \tag{7.29}$$

Thus the linewidth of the passive mode depends inversely on the quality factor. The higher the quality factor (i.e., longer the cavity lifetime) the smaller will be the FWHM.

In order to calculate the Q of a passive resonator, we first find the energy left in the cavity after one complete cycle of oscillation and then use Eq. (7.25) to obtain an explicit expression for Q. Let W_0 be the total energy contained within the cavity at $t = 0$. One complete cycle of oscillation in the cavity corresponds to a pair of reflections from the mirrors M_1 and M_2 (with power reflection coefficients R_1 and R_2, respectively) and two traversals through the medium filling the cavity, which is assumed to have a net power attenuation coefficient α_l per unit length. Thus the energy remaining within the cavity after one complete cycle would be

$$W_0 R_1 R_2 e^{-2\alpha_l d} = W_0 e^{-2\alpha_l d + \ln R_1 R_2} \tag{7.30}$$

where d is the length of the cavity. Also one complete cycle corresponds to a time interval of

$$t_r = \frac{2n_0 d}{c}$$

where n_0 is the refractive index of the medium filling the cavity. Thus from Eq. (7.25) we obtain for the energy inside the cavity

$$W(t_r) = W_0 \exp\left(-\frac{2\pi \nu_0}{Q} \frac{2n_0 d}{c}\right) \tag{7.31}$$

From Eqs. (7.30) and (7.31) we obtain

$$2\alpha_l d - \ln R_1 R_2 = \frac{4\pi \nu_0 n_0 d}{cQ}$$

or

$$Q = \frac{4\pi \nu_0 n_0 d}{c} \frac{1}{2\alpha_l d - \ln R_1 R_2} \tag{7.32}$$

We can also obtain an expression for the cavity lifetime t_c in terms of the fractional loss per round trip. The initial energy W_0 becomes $W_0 \exp(-\kappa)$ after one round trip; here

$$\kappa = 2\alpha_l d - \ln R_1 R_2$$

Thus the fractional loss per round trip would be

$$x = \frac{W_0 - W_0 e^{-\kappa}}{W_0} = 1 - e^{-\kappa}$$

or

$$\kappa = \ln\left(\frac{1}{1-x}\right)$$

Hence from Eqs. (7.26) and (7.32) we have

$$t_c = \frac{2n_0 d}{c \ln (1/(1-x))} = \frac{2n_0 d}{c (2\alpha_l d - \ln R_1 R_2)} \tag{7.33}$$

From Eqs. (7.29) and (7.32) we may write

$$\Delta \nu_P = \frac{c}{4\pi n_0 d} (2\alpha_l d - \ln R_1 R_2) \tag{7.34}$$

where the subscript P stands for passive cavity.

Example 7.5 Let us consider a typical cavity of a He–Ne laser with the following specifications: $d = 20$ cm, $n_0 = 1$, $R_1 = 1$, $R_2 = 0.98$, $\alpha_l \sim 0$
   For such a cavity

$$\Delta \nu_P \approx 2.4 \text{ MHz}$$

For the same cavity, the frequency separation between adjacent longitudinal modes is

$$\delta \nu \approx \frac{c}{2d} = 750 \text{ MHz}$$

   Thus the spectral width of each mode is much smaller than the separation between adjacent modes.

Example 7.6 As another example we consider a GaAs semiconductor laser (see Chapter 13) with the following values of various parameters: $d = 500$ μm, $n_0 = 3.5$, $R_1 = R_2 = 0.3$, $\alpha_l \sim 0$
   For such a cavity we obtain

$$\Delta \nu_P \approx 3.3 \times 10^{10} \text{ Hz}$$

7.5 The Ultimate Linewidth of a Laser

One of the most important properties of a laser is its ability to produce light of high spectral purity or high temporal coherence. The finite spectral width of a laser operating continuously in a single mode is caused by two primary mechanisms. One is the external factors, which tend to perturb the cavity, for example, temperature fluctuations, vibrations. randomly alter the oscillation frequency which results in a finite spectral width. The second more fundamental mechanism which determines the ultimate spectral width of the laser is that due to the ever present random spontaneous emissions in the cavity. Since spontaneous emission is completely incoherent with respect to the existing energy in the cavity mode, it leads to a finite spectral width of the laser. In this section, we shall give a heuristic derivation of the ultimate linewidth of a laser (Gordon, Zeiger and Townes 1955, Maitland and Dunn 1969). In order to obtain a value for the ultimate laser linewidth, we assume that the radiation arising out of spontaneous emission represents a loss as far as the coherent energy is concerned. This loss will then lead to a finite linewidth of the laser. We recall from Section 5.5 that the number of spontaneous emissions per unit time into a mode of the cavity is given by KN_2 [where K is defined by Eq. (5.72)] and N_2 represents the number of atoms per unit volume in the upper laser level. We are assuming that $N_1 \sim 0$. When the laser oscillates in steady state, then we know from Eq. (5.79) that $N_2 \sim 1/Kt_c$ where we are assuming that $n \gg 1$ and t_c is the passive cavity lifetime. Thus above threshold the number of spontaneous emissions per unit time would be $KN_2 = 1/t_c$. Hence the energy appearing per unit time in a mode due to spontaneous emission will be $h\nu_0/t_c$ where ν_0 is the oscillation frequency of the mode.

Now the total energy contained in the mode is $nh\nu_0$ and since the output power P_{out} is given by $nh\nu_0/t_c$, the energy contained in the mode is $P_{out}\, t_c$.

We now use Eq (7.29) and denote the linewidth of the oscillating laser caused by spontaneous emission by $\delta\nu_{sp}$ to obtain

$$Q = \frac{\nu_0}{\delta\nu_{sp}} = 2\pi\nu_0 \frac{P_{out} t_c}{h\nu_0 / t_c} = \frac{2\pi P_{out} t_c^2}{h} \tag{7.35}$$

Now if $\Delta\nu_P$ is the passive cavity linewidth then $t_c = 1/2\pi\Delta\nu_P$ and thus from Eq. (7.35) we obtain

$$\delta\nu_{sp} = \frac{2\pi\,(\Delta\nu_P)^2\, h\nu_0}{P_{out}} \tag{7.36}$$

The above equation gives the ultimate linewidth of an oscillating laser and is similar to the one given by Schawlow and Townes (1958). It is interesting to note that $\delta\nu_{sp}$ depends inversely on the output power P_{out}. This is physically due to the fact that for a given mirror reflectivity, an increase in P_{out} corresponds to an increase in the energy in the mode inside the cavity which in turn implies a greater dominance of stimulated emission over spontaneous emission. Figure 7.4 shows a typical measured variation of the linewidth of a GaAs semiconductor laser which shows the linear increase in the linewidth with inverse optical power.

The above derivation is rather heuristic: an analysis based on the random-phase additions due to spontaneous emission is given by Jacobs (1979) which gives a result half that predicted by Eq. (7.36).

Example 7.7 Let us first consider a He–Ne laser given in Example 7.5 and we assume that it oscillates with an output power of 1 mW at a wavelength of 632.8 nm. The spontaneous emission linewidth of the laser will be

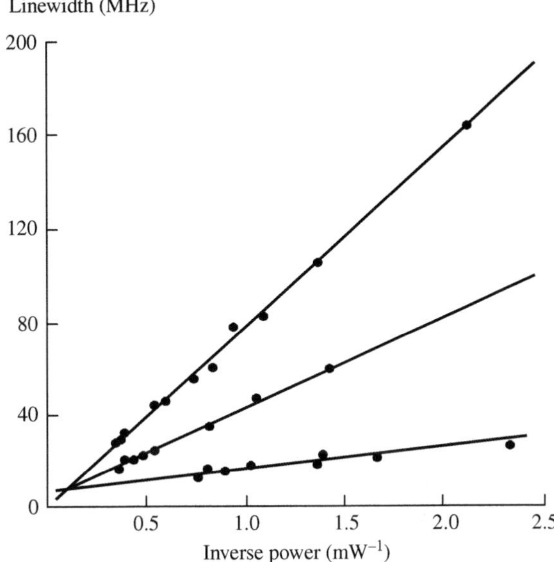

Fig. 7.4 Experimentally measured variation of laser linewidth in a single-frequency GaAs semiconductor laser as a function of inverse optical power (Reprinted with permission from A. Mooradian, Laser linewidth, Physics Today, May 1985, p. 43, © 1985, American Institute of Physics.)

$$\delta \nu_{sp} \sim 0.01\,\text{Hz}$$

which is extremely small. To emphasize how small these widths are, let us try to estimate the precision with which the length of the cavity has to be controlled in order that the oscillation frequency changes by 0.01 Hz. We know that the approximate oscillation frequency of a mode is $\nu = qc/2d$. Thus the change in frequency $\Delta \nu$ caused by a change in length Δd is

$$\delta \nu = \left(\frac{\nu}{d}\right) \Delta d$$

Using $d = 20$ cm, $\lambda_0 = 632.8$ nm, and $\delta \nu = 0.01$ Hz, we obtain

$$\Delta d \sim 4 \times 10^{-18}$$

which corresponds to a stability of less than nuclear dimensions.

Example 7.8 As another example consider a GaAs semiconductor laser operating at a wavelength of 850 nm with an output power of 1 mW with cavity dimensions as given in Example 7.6. For such a laser

$$\delta \nu_{sp} \approx 1.6\,\text{MHz}$$

which is much larger than that of a He–Ne laser. This is primarily due to the very large value of $\Delta \nu_P$ in the case of semiconductor lasers. Since $\Delta \nu_P$ can be reduced by increasing the length of the cavity, one can use an external mirror for feedback and thus reduce the linewidth. For a more detailed discussion of semiconductor laser linewidth, readers are referred to Mooradian (1985).

7.6 Mode Selection

Since conventional laser resonators have dimensions that are large compared to the optical wavelength there are, in general, a large number of modes which fall within the gain bandwidth of the active medium and which can oscillate in the laser. Hence the output may consist of various transverse and longitudinal modes leading to a greater divergence of the output beam as well as containing a number of oscillating frequencies. In fact the power per unit frequency and solid angle interval is maximum when the laser oscillates in a single mode which is the fundamental mode. In order to obtain highly directional and spectrally pure output, various techniques have been developed both for transverse and for longitudinal mode selection in lasers. In this section we will describe some techniques that are used to select a single transverse mode (well-defined transverse field pattern with minimal divergence) and a single longitudinal mode (small spectral width) of oscillation.

7.6.1 Transverse Mode Selection

As we mentioned in Section 7.3 different transverse modes are characterized by different transverse field distributions. The lowest order transverse mode has a Gaussian amplitude distribution across the transverse plane. It is this mode that one usually prefers to work with as it does not have any abrupt phase changes across the wave front (as the higher order transverse modes have) and has also a monotonically decreasing amplitude away from the axis. This leads to the fact that it can be focused to regions of the order of wavelength of light, producing enormous intensities.

Since the fundamental Gaussian mode has the narrowest transverse dimension, an aperture placed inside the resonator cavity can preferentially introduce higher losses for higher order modes. Thus if a circular aperture is introduced into the laser cavity such that the loss suffered by all higher order modes is greater than the gain while the loss suffered by the fundamental mode is lower that its gain, then the laser would oscillate only in the fundamental mode (see Fig. 7.5). It is interesting to note that specific higher order transverse modes can also be selected at a time by choosing complex apertures which introduce high loss for all modes except

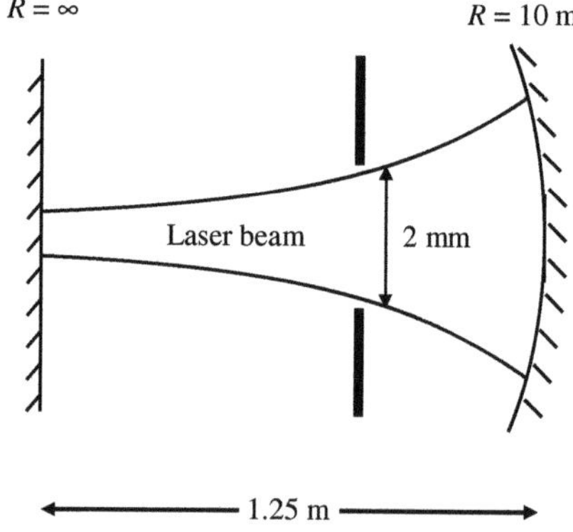

Fig. 7.5 A typical configuration for achieving single transverse mode oscillation. The aperture introduces differential losses between the fundamental mode and the higher order modes

the required mode or by profiling the reflectivity of one of the mirrors to suit the required mode. Thus a wire placed normal to the axis would select the second-order TE_{01} mode.

Problem 7.2 Consider a Gaussian mode and the first-order Hermite–Gauss mode with $w_0 = 1$ mm. Assume that both the beams pass through a circular aperture of radius a placed with its center coinciding with the axis. Obtain the fraction of light passing through the aperture for the Gaussian and the first-order Hermite–Gauss mode. For what value of a would there be maximum discrimination between the two modes?

7.6.2 Longitudinal Mode Selection

We have seen in Section 7.2 that the various longitudinal modes corresponding to a transverse mode are approximately separated by a frequency interval of $c/2n_0d$. As an example if we consider a 50 cm long laser cavity with $n_0 = 1$, then the longitudinal mode spacing would be 300 MHz. If the gain bandwidth of the laser is 1500 MHz, then in this case even if the laser is oscillating in a single transverse mode it would still oscillate in about five longitudinal modes. Thus the output would consist of five different frequencies separated by 300 MHz. This would result in a much reduced coherence length of the laser (see Problem 7.3). Thus in applications such as holography and interferometry where a long coherence length is required or where a well-defined frequency is required (e.g., in spectroscopy) one would require the laser to oscillate in a single longitudinal mode in addition to its single transverse mode oscillation.

Referring to Fig. 7.6 we can have a simple method of obtaining single longitudinal mode oscillation by reducing the cavity length to a value such that the intermode

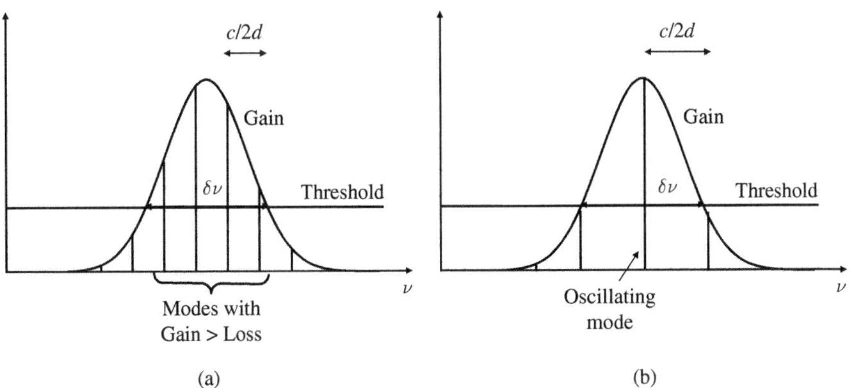

Fig. 7.6 (a) The longitudinal mode spacing of a resonator of length d is $c/2d$. Different modes having gain more than loss would oscillate simultaneously. (b) If the resonator length is reduced, the mode spacing can become less than the gain bandwidth and if there is a mode at the line center, then it would result in single longitudinal mode oscillation of the laser

spacing is larger than the spectral width over which gain exceeds loss in the cavity. Thus if this bandwidth is $\Delta \nu_g$ then for single longitudinal mode laser oscillation the cavity length must be such that

$$\frac{c}{2n_0 d} > \Delta \nu_g$$

For a He–Ne laser $\Delta \nu_g \sim 1500$ MHz and for single longitudinal mode oscillation one must have $d < 10$ cm. We should note here that if one can ensure that a resonant mode exists at the center of the gain profile, then single mode oscillation can be obtained even with a cavity length of $c / n_0 \Delta \nu_g$.

One of the major drawbacks with the above method is that since the volume of the active medium gets very much reduced due to the restriction on the length of the cavity, the output power is small. In addition, in solid-state lasers where the gain bandwidth is large, the above technique becomes impractical. Hence other techniques have been developed which can lead to single longitudinal mode oscillation without any restriction in the length of the cavity and hence capable of high powers.

It is important to understand that even if the laser oscillates in a single longitudinal mode (single-frequency output), there could be a temporal drift in the frequency of the output. In many applications it is important to have single-frequency lasers in which the frequency of the laser should not deviate beyond a desired range. To achieve this the frequency of oscillation of the laser can be locked by using feedback mechanisms in which the frequency of the output of the laser is monitored continuously and any change in the frequency of the laser is fed back to the cavity as an error signal which is then used to control the mirror positions of the cavity to keep the frequency stable. There are many techniques to monitor the frequency of the laser output; this includes using very accurate wavelength meters capable of giving frequency accuracy in the range of 2 MHz.

Oscillation of a laser in a given resonant mode can be achieved by introducing frequency selective elements such as Fabry–Perot etalons (see Chapter 2) into the laser cavity. The element should be so chosen that it introduces losses at all but the desired frequency so that the losses of the unwanted frequencies are larger than gain resulting in a single-frequency oscillation. Figure 7.7 shows a tilted Fabry–Perot etalon placed inside the resonator. The etalon consists of a pair of highly reflecting parallel surfaces which has a transmission versus frequency variation as shown in Fig. 7.8. As discussed in Chapter 2, such an etalon has transmission peaks centered at frequencies given by

$$\nu_p = p \frac{c}{2nt \cos \theta} \tag{7.37}$$

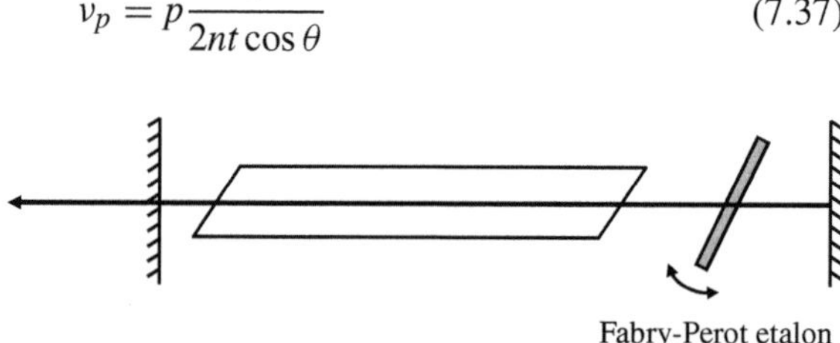

Fig. 7.7 A laser resonator with a Fabry–Perot etalon placed inside the cavity

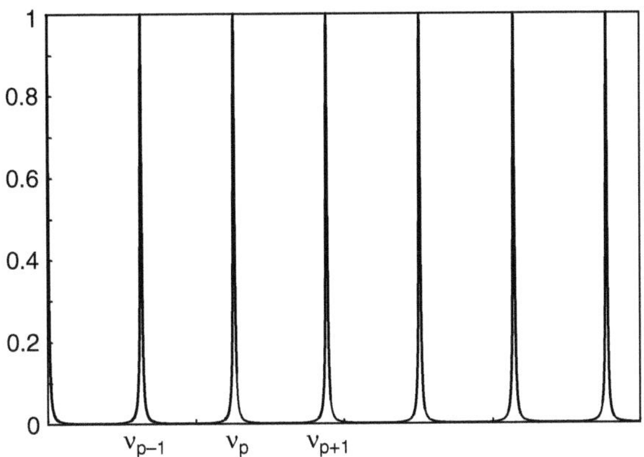

Fig. 7.8 Transmittance versus frequency of a Fabry–Perot etalon. The higher the reflectivity the sharper are the resonances

where t is the thickness of the etalon and n is the refractive index of the medium between the reflecting plates and θ is the angle made by the wave inside the etalon. The width of each peak depends on the reflectivity of the surfaces, the higher the reflectivity, the shaper are the resonance peaks. The frequency separation between two adjacent peaks of transmission is

$$\Delta \nu = \frac{c}{2nt \cos \theta} \tag{7.38}$$

which is also referred to as free spectral range (FSR).

If the etalon is so chosen that its free spectral range is greater than the spectral width of the gain profile, then the Fabry–Perot etalon can be tilted inside the resonator so that one of the longitudinal modes of the resonator cavity coincides with the peak transmittance of the etalon (see Fig. 7.9) and other modes are reflected away from the cavity. If the finesse of the etalon is high enough so as to introduce sufficiently high losses for the modes adjacent to the mode selected, then one can have single longitudinal mode oscillation (see Fig. 7.9).

Example 7.9 Consider an argon ion laser for which the FWHM of the gain profile is about 8 GHz. Thus for near normal incidence ($\theta \sim 0$) the free spectral range of the etalon must be greater than about 10 GHz. Thus

$$\frac{c}{2nt} > 10^{10} \, \text{Hz}$$

Taking fused quartz as the medium of the etalon, we have $n \sim 1.462$ (at $\lambda \sim 510$ nm) and thus $t \leq 1$ cm

Another very important method used to obtain single-frequency oscillation is to replace one of the mirrors of the resonator with a Fox–Smith interferometer as shown in Fig. 7.10. Waves incident on the beam splitter BS from M_1 will suffer multiple reflections as follows:

Reflection 1: $M_1 \rightarrow BS \rightarrow M_2 \rightarrow BS \rightarrow M_1$

Reflection 2: $M_1 \rightarrow BS \rightarrow M_2 \rightarrow BS \rightarrow M_3 \rightarrow BS \rightarrow M_2 \rightarrow BS \rightarrow M_1$, etc.

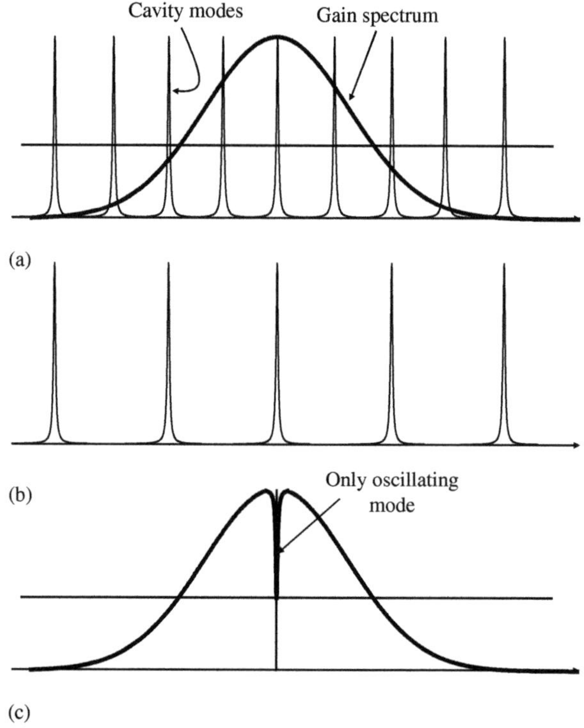

Fig. 7.9 The figure shows how by inserting a Fabry–Perot etalon in the laser cavity one can achieve single longitudinal mode oscillation

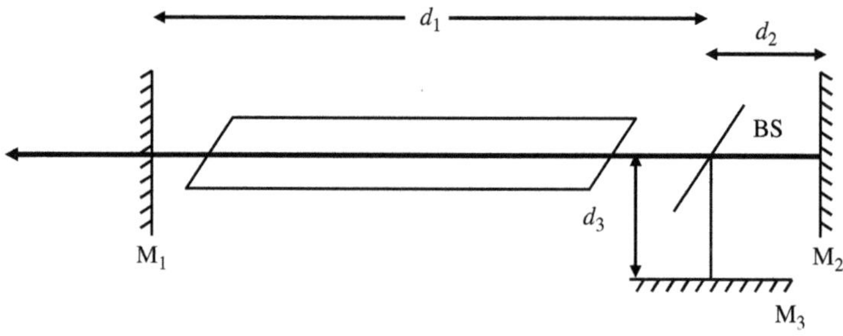

Fig. 7.10 The Fox–Smith interferometer arrangement for selection of a single longitudinal mode

Thus the structure still behaves much like a Fabry–Perot etalon if the beam splitter BS has a high reflectivity.[1] For constructive interference among waves reflected toward M_1 from the interferometer, the path difference between two consecutively reflected waves must be $m\lambda$, i.e.,

---

[1] It is interesting to note that if mirror M_3 is put above BS (in Fig. 7.10), then it would correspond to a Michelson interferometer arrangement and the transmittivity would not be sharply peaked.

$$(2d_2 + 2d_3 + 2d_2 - 2d_2) = m\lambda$$

or

$$\nu = m\frac{c}{2\,(d_2 + d_3)}, \qquad m = \text{any integer} \qquad (7.39)$$

Thus the frequencies separated by $\Delta\nu = c/2\,(d_2 + d_3)$ will have a low loss. Hence if $\Delta\nu$ is greater than the bandwidth of oscillation of the laser, then one can achieve single mode oscillation. Since the frequencies of the resonator formed by mirrors M_1 and M_2 are

$$\nu = q\frac{c}{2\,(d_1 + d_2)}, \qquad q = \text{any integer} \qquad (7.40)$$

for the oscillation of a mode one must have

$$\frac{m}{(d_2 + d_3)} = \frac{q}{(d_1 + d_2)} \qquad (7.41)$$

which can be adjusted by varying d_3 by placing the mirror M_3 on a piezoelectric movement.

Example 7.10 If one wishes to choose a particular oscillating mode out of the possible resonator modes which are separated by 300 MHz, what is the approximate change in d_3 required to change oscillation from one mode to another? Assume $\lambda_0 = 500$ nm, $d_2 + d_3 = 5$ cm.
   Solution Differentiating Eq. (7.39) we have

$$\delta\nu = \frac{mc}{2\,(d_2 + d_3)^2}\,\delta d_3 = \frac{\nu}{(d_2 + d_3)}\,\delta d_3$$

which gives us $\delta d_3 = 25$ nm.

Problem 7.3 Consider a laser which is oscillating simultaneously at two adjacent frequencies ν_1 and ν_2. If this laser is used in an interference experiment, what is the minimum path difference between the interfering beams for which the interference pattern disappears?

Solution The interference pattern disappears when the interference maxima produced by one wavelength fall on the interference minima produced by the other wavelength. This will happen when

$$l = m\frac{c}{\nu_1} = \left(m + \frac{1}{2}\right)\frac{c}{\nu_2}, \qquad m = 0, 1, 2...$$

Eliminating m from the two equations we get

$$l = \frac{c}{2\,(\nu_2 - \nu_1)} = d$$

where d is the length of the laser resonator. Thus the laser can be considered coherent only for path differences of $l = d$, the length of the laser. On the other hand, if the laser was oscillating in a single mode the coherence length would have been much larger and would be determined by the frequency width of the oscillating mode only.

7.7 Pulsed Operation of Lasers

In many applications of lasers, one wishes to have a pulsed laser source. In principle it is possible to generate pulses of light from a continuously operating laser, but it would be even more efficient if the laser itself could be made to emit pulses of light. In this case the energy contained in the population inversion would be much more efficiently utilized. There are two standard techniques for the pulsed operation of a laser; these are Q-switching and mode locking. Q-switching is used to generate pulses of high energy but nominal pulse widths in the nanosecond regime. On the other hand mode locking produces ultrashort pulses with smaller energy content. We shall see that using mode locking it is possible to produce laser pulses in the femtosecond regime.

7.7.1 Q-Switching

Imagine a laser cavity within which we have placed a shutter which can be opened and closed at will (see Fig. 7.11). Let us assume that the shutter is closed (i.e., does not transmit) and we start to pump the amplifying medium. Since the shutter is closed, there is no feedback from the mirror and the laser beam does not build up. Since the pump is taking the atoms from the ground state and disposing them into the excited state and there is no stimulated emission the population inversion keeps on building up. This value could be much higher than the threshold inversion required for the same laser in the absence of the shutter. When the inversion is built to a reasonably high value, if we now open the shutter, then the spontaneous emission is now able to reflect from the mirror and pass back and forth through the amplifying medium. Since the population inversion has been built up to a large value, the gain provided by the medium in one round trip will be much more than the loss in one round trip and as such the power of the laser beam would grow very quickly with every passage. The growing laser beam consumes the population inversion, which then decreases rapidly resulting in the decrease of power of the laser beam. Thus when the shutter is suddenly opened, a huge light pulse gets generated and this technique is referred to as Q-switching. High losses imply low Q while low losses

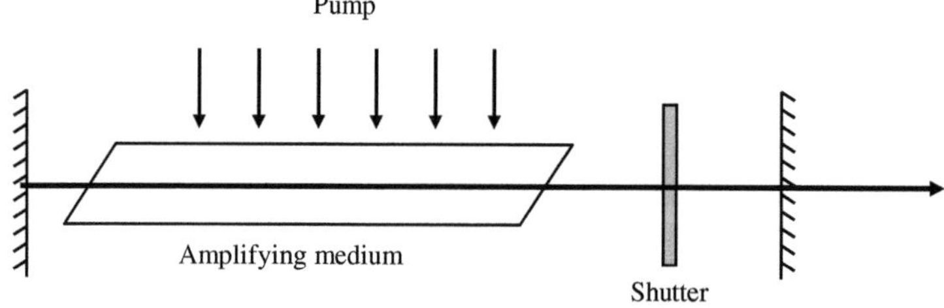

Fig. 7.11 A laser resonator with a shutter placed in front of one of the mirrors to achieve Q-switching

imply high Q. Thus when the shutter is kept closed and suddenly opened, the Q of the cavity is suddenly increased from a very small value to a large value and hence the name Q-switching. For generating another pulse the medium would again need to be pumped while the shutter is kept closed and the process repeated again.

Figure 7.12 shows schematically the time variation of the cavity loss, cavity Q, population inversion, and the output power. As shown in the figure an intense pulse is generated with the peak intensity appearing when the population inversion in the cavity is equal to the threshold value. Figure 7.13 shows a Q-switched pulse emitted from a neodymium–YAG laser. The energy per pulse is 850 mJ and the pulse width is about 6 ns. This corresponds to a peak power of about 140 MW. The pulse repetition frequency is 10 Hz, i.e., the laser emits 10 pulses per second. Using this phenomenon it is possible to generate extremely high power pulses for use in various applications such as cutting, drilling, or in nuclear fusion experiments.

We will now write down rate equations corresponding to Q-switching and obtain the most important parameters such as peak power, total energy, and duration of the pulse. We shall consider only one mode of the laser resonator and shall examine

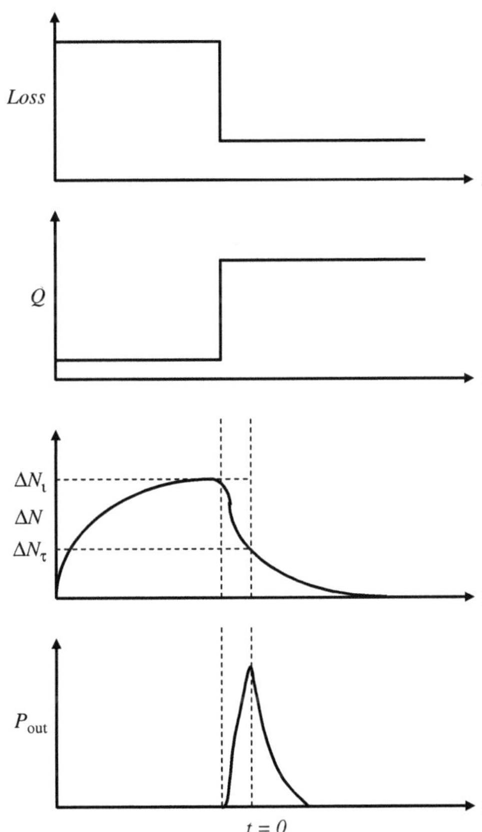

Fig. 7.12 Schematic of the variation of loss, Q value, population inversion, and the laser output power with time

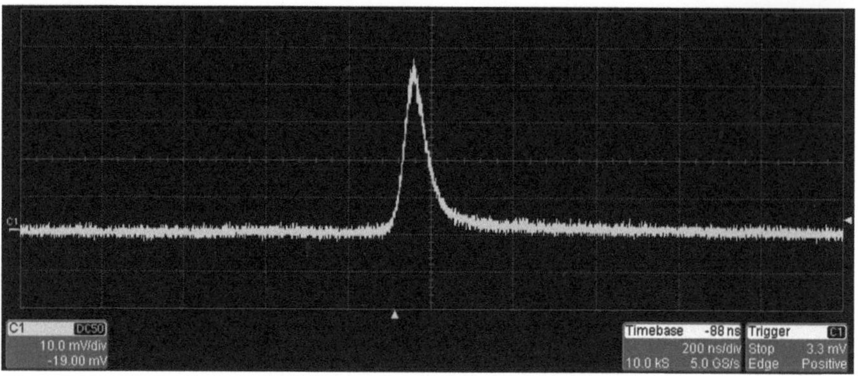

Fig. 7.13 An acousto optically Q-switched output from an Nd:YAG laser. The average power of the pulse train is 15 W and the repetition frequency is 2 kHz with the pulse duration of 97 ns. [Figure provided by Brahmanand Upadhyaya, RRCAT, Indore]

the specific case of a three-level laser system such as that of ruby. In Section 5.5 while writing the rate equations for the population N_2 and the photon number n, we assumed the lower laser level to be essentially unpopulated. If this is not the case then instead of Eq. (5.74) we will have

$$\frac{\mathrm{d}(N_2 V)}{\mathrm{d}t} = -KnN_2 + KnN_1 - T_{21}N_2 V + RV \tag{7.42}$$

where the second term on the right-hand side is the contribution due to absorption by N_1 atoms per unit volume in the lower level. Since the Q-switched pulse is of a very short duration, we will neglect the effect of the pump and spontaneous emission during the generation of the Q-switched pulse. It must, at the same time, be noted that for the start of the laser oscillation, spontaneous emission is essential. Thus we get from Eq. (7.42)

$$\frac{\mathrm{d}N_2'}{\mathrm{d}t} = -\left(\frac{Kn}{V}\right)\Delta N' \tag{7.43}$$

where

$$\Delta N' = (N_2 - N_1)V, \qquad N_2' = N_2 V \tag{7.44}$$

and V is the volume of the amplifying medium. Similarly one can also obtain or the rate of change of population of the lower level

$$\frac{\mathrm{d}N_1'}{\mathrm{d}t} = \left(\frac{Kn}{V}\right)\Delta N' \tag{7.45}$$

where $N_1' = N_1 V$. Subtracting Eq. (7.45) from Eq. (7.43) we get

$$\frac{d(\Delta N')}{dt} = -2\left(\frac{Kn}{V}\right)\Delta N' \qquad (7.46)$$

We can also write the equation for the rate of change of the photon number n in the cavity mode in analogy to Eq. (5.78) as

$$\begin{aligned}
\frac{dn}{dt} &= Kn\,(N_2 - N_1) - \frac{n}{t_c} + KN_2 \\
&\approx \left(\frac{Kn}{V}\right)\Delta N' - \frac{n}{t_c}
\end{aligned} \qquad (7.47)$$

where we have again neglected the spontaneous emission term KN_2. From Eq. (7.47) we see that the threshold population inversion is

$$(\Delta N')_t = \frac{V}{Kt_c} \qquad (7.48)$$

when the gain represented by the first term on the right-hand side becomes equal to the loss represented by the second term [see Eq. (7.47)]. Replacing V/K in Eqs. (7.46) and (7.47) by $(\Delta N')_t t_c$ and writing

$$\tau = \frac{t}{t_c} \qquad (7.49)$$

we obtain

$$\frac{d(\Delta N')}{d\tau} = -2n\frac{\Delta N'}{(\Delta N')_t} \qquad (7.50)$$

and

$$\frac{dn}{d\tau} = n\left[\frac{(\Delta N')}{(\Delta N')_t} - 1\right] \qquad (7.51)$$

Equations (7.51) and (7.50) give us the variation of the photon number n and the population inversion $\Delta N'$ in the cavity as a function of time. As can be seen the equations are nonlinear and solutions to the above set of equations can be obtained numerically by starting from an initial condition

$$\left(\Delta N'\right)(\tau = 0) = \left(\Delta N'\right)_i \quad \text{and} \quad n(\tau = 0) = n_i \tag{7.52}$$

where the subscript i stands for initial values. Here n_i represents the initial small number of photons excited in the cavity mode through spontaneous emission. This spontaneous emission is necessary to trigger laser oscillation.

From Eq. (7.51) we see that since the system is initially pumped to an inversion $\Delta N' > \left(\Delta N'\right)_t$, $dn/d\tau$ is positive; thus the number of photons in the cavity increases with time. The maximum number of photons in the cavity appear when $dn/d\tau = 0$, i.e., when $\Delta N' = \left(\Delta N'\right)_t$. At such an instant n is very large and from Eq. (7.50) we see that $\Delta N'$ will further reduce below $\left(\Delta N'\right)_t$ and thus will result in a decrease in n.

Although the time-dependent solution of Eqs. (7.50) and (7.51) requires numerical computation, we can analytically obtain the variation of n with $\Delta N'$ and from this we can draw some general conclusions regarding the peak power, the total energy in the pulse, and the approximate pulse duration. Indeed, dividing Eq. (7.51) by (7.50) we obtain

$$\frac{dn}{d(\Delta N')} = \frac{1}{2}\left[\frac{\left(\Delta N'\right)_t}{\left(\Delta N'\right)} - 1\right]$$

Integrating we get

$$n - n_i = \frac{1}{2}\left\{\Delta N'_t \ln\left[\frac{\Delta N'}{\left(\Delta N'\right)_i}\right] + \left[\left(\Delta N'\right)_i - \Delta N'\right]\right\} \tag{7.53}$$

7.7.1.1 Peak Power

Assuming the only loss mechanism to be output coupling and recalling our discussion in Section 7.5 we have for the instantaneous power output

$$P_{\text{out}} = \frac{nh\nu}{t_c} \tag{7.54}$$

Thus the peak power output will correspond to maximum n which occurs when $\Delta N' = \left(\Delta N'\right)_t$. Thus

$$\begin{aligned} P_{\max} &= \frac{n_{\max}h\nu}{t_c} \\ &= \frac{h\nu}{2t_c}\left[\left(\Delta N'\right)_t \ln\left(\frac{\left(\Delta N'\right)_t}{\left(\Delta N'\right)_i}\right) + \left(\left(\Delta N'\right)_t - \left(\Delta N'\right)_i\right)\right] \end{aligned} \tag{7.55}$$

where we have neglected n_i (the small number of initial spontaneously emitted photons in the cavity). This shows that the peak power is inversely proportional to cavity lifetime.

7.7.1.2 Total Energy

In order to calculate the total energy in the Q-switched pulse we return to Eq. (7.51) and substitute for $(\Delta N') / (\Delta N')_t$ from Eq (7.50) to get

$$\frac{dn}{d\tau} = -\frac{1}{2}\frac{\mathrm{d}\left(\Delta N'\right)}{\mathrm{d}\tau} - n$$

Integrating the above equation from $t = 0$ to ∞ we get

$$n_f - n_i = \frac{1}{2}\left[\left(\Delta N'\right)_i - \left(\Delta N'\right)_f\right] - \int_0^\infty n\mathrm{d}\tau$$

or

$$\int_0^\infty n\mathrm{d}\tau = \frac{1}{2}\left[\left(\Delta N'\right)_i - \left(\Delta N'\right)_f\right] - (n_f - n_i) \tag{7.56}$$

where the subscript f denotes final values. Since n_i and n_f are very small in comparison to the total integrated number of photons we may neglect them and obtain

$$\int_0^\infty n\mathrm{d}\tau \approx \frac{1}{2}\left[\left(\Delta N'\right)_i - \left(\Delta N'\right)_f\right]$$

Thus the total energy of the Q-switched pulse is

$$\begin{aligned}
E &= \int_0^\infty P_{\text{out}}\mathrm{d}t \\
&= h\nu \int_0^\infty n\mathrm{d}\tau \\
&= \frac{1}{2}\left[\left(\Delta N'\right)_i - \left(\Delta N'\right)_f\right]h\nu
\end{aligned} \tag{7.57}$$

The above expression could also have been derived through physical arguments as follows: for every additional photon appearing in the cavity mode there is an atom making a transition from the upper level to the lower level and for every atom making this transition the population inversion reduces by 2. Thus if the population inversion changes from $(\Delta N')_i$ to $(\Delta N')_f$, the number of photons emitted must be $\frac{1}{2}\left[\left(\Delta N'\right)_i - \left(\Delta N'\right)_f\right]$ and Eq. (7.57) follows immediately.

7.7.1.3 Pulse Duration

An approximate estimate for the duration of the Q-switched pulse can be obtained by dividing the total energy by the peak power. Thus

$$t_{\mathrm{d}} = \frac{E}{P_{\max}} = \frac{(\Delta N')_i - (\Delta N')_{\mathrm{f}}}{\left[(\Delta N')_{\mathrm{t}} \ln\left(\frac{(\Delta N')_{\mathrm{t}}}{(\Delta N')_i}\right) + ((\Delta N')_{\mathrm{t}} - (\Delta N')_i)\right]} t_{\mathrm{c}} \qquad (7.58)$$

In the above formulas we still have the unknown quantity $(\Delta N')_{\mathrm{f}}$ the final inversion. In order to obtain this, we may use Eq. (7.53) for $t \to \infty$. Since the final number of photons in the cavity is small, we have

$$\left((\Delta N')_i - (\Delta N')_f\right) = (\Delta N')_t \ln\left(\frac{(\Delta N')_i}{(\Delta N')_f}\right) \qquad (7.59)$$

from which we can obtain $(\Delta N')_f$ for a given set of $(\Delta N')_i$ and $(\Delta N')_t$

Example 7.11 We consider the Q-switching of a ruby laser with the following characteristics:

Length of ruby rod $= 10$ cm
Area of cross section $= 1$ cm^2
Resonator length $= 10$ cm
Mirror reflectivities $= 1$ and 0.7
Cr^{3+} population density $= 1.58 \times 10^{19}$ cm^{-3}
$\lambda_0 = 694.3$ nm
$n_0 = 1.76$
$t_{\mathrm{sp}} = 3 \times 10^{-3}$ s
$g(\omega_0) = 1.1 \times 10^{-12}$ s

The above parameters yield a cavity lifetime of 3.3×10^{-9} s and the required threshold population density of 1.25×10^{17} cm^{-3}. Thus

$$(\Delta N')_t = 1.25 \times 10^{18}$$

Choosing

$$(\Delta N')_i = 4(\Delta N')_t = 5 \times 10^{18}$$

we get

$$P_{\max} = 8.7 \times 10^7 \text{ W}$$

Solving Eq. (7.59) we obtain $(\Delta N')_f \approx 0.02 (\Delta N')_i$ Thus

$$E \approx 0.7\text{J}$$

and

$$t_d \approx 8 \text{ ns}$$

7.7.2 *Techniques for Q-Switching*

As discussed, for Q-switching the feedback to the amplifying medium must be initially inhibited and when the inversion is well past the threshold inversion, the optical feedback must be restored very rapidly. In order to perform this, various devices are available which include mechanical movements of the mirror or shutters which can be electronically controlled. The mechanical device may simply rotate one of the mirrors about an axis perpendicular to the laser axis which would restore the Q of the resonator once every rotation. Since the rotation speed cannot be made very large (typical rotation rates are 24,000 revolutions per second) the switching of the Q from a low to a high value does not take place instantaneously and this leads to multiple pulsing.

In comparison to mechanical rotation, electronically controlled shutters employing the electro optic or acousto optic effect can be extremely rapid. A schematic arrangement of Q-switching using the electro optic effect is shown in Fig. 7.14 The electro optic effect is the change in the birefringence of a material on application of an external electric field (see, e.g., Yariv (1977), Ghatak and Thyagarajan (1989)). Thus the electro optic modulator (EOM) shown in Fig. 7.14 could be a crystal which is such that in the absence of any applied electric field the crystal does not introduce any phase difference between two orthogonally polarized components traveling along the laser axis. On the other hand, if a voltage V_0 is applied across the crystal, then the crystal introduces a phase difference of $\pi/2$ between the orthogonal components, i.e., it behaves like a quarter wave plate. If we now consider the polarizer–modulator–mirror system, when there is no applied voltage, the

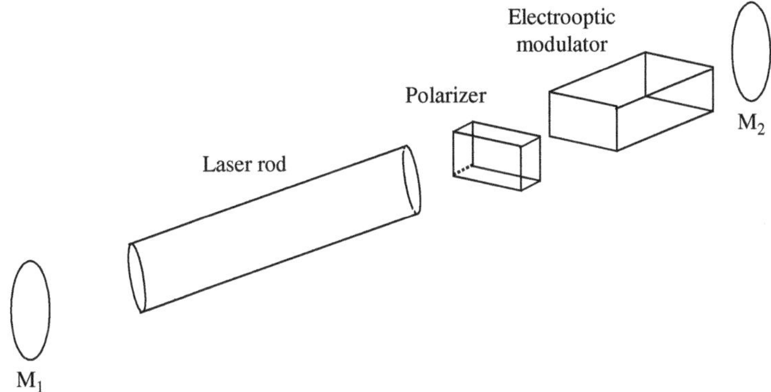

Fig. 7.14 A typical arrangement for achieving Q-switching using an electro optic modulator placed within the cavity

state of polarization (SOP) of the light incident on the polarizer after reflection by the mirror is along the pass axis of the polarizer and thus corresponds to a high Q state. When a voltage V_0 is applied the linearly polarized light on passage through the EOM becomes circularly polarized (say right circularly polarized). Reflection from the mirror converts this to left circularly polarized light and passage through the EOM the wave becomes linearly polarized but now polarized perpendicular to the pass axis of the polarizer. The polarizer does not allow this to pass through and this leads to essentially no feedback which corresponds to the low Q state. Hence Q-switching can be accomplished by first applying a voltage across the crystal and removing it at the instant of highest inversion in the cavity. Some important electro optic crystals used for Q-switching include potassium dihydrogen phosphate (KDP) and lithium niobate ($LiNbO_3$).

An acousto optic Q-switch is based on the acousto optic effect. In the acousto optic effect a propagating acoustic wave in a medium creates a periodic refractive index modulation due to the periodic strain in the medium, and this periodic refractive index modulation leads to diffraction of a light wave interacting with it (see, e.g., Ghatak and Thyagarajan (1989), Yariv (1977)). The medium in the presence of the acoustic wave behaves like a phase grating. Thus if an acousto optic cell is placed inside the resonator, it can be used to deflect the light beam out of the cavity, thus leading to a low Q value. The Q can be switched to high value by switching off the acoustic wave.

Q-switching can also be obtained by using a saturable absorber inside the laser cavity. In a saturable absorber (which essentially consists of an organic dye dissolved in an appropriate solvent) the absorption coefficient of the medium reduces with an increase in the incident intensity. This reduction in the absorption is caused by the saturation of a transition (see Chapter 5). In order to understand how a saturable absorber can be made to Q-switch, consider a laser resonator with the amplifying medium and the saturable absorber placed inside the cavity as shown in Fig. 7.15. As the amplifying medium is pumped, the intensity level inside the cavity is initially low since the saturable absorber does not allow any feedback from the mirror M_2. As the pumping increases, the intensity level inside the cavity increases which, in turn, starts to bleach the saturable absorber. This leads to an increase in feedback which gives rise to an increased intensity and so on. Thus the energy stored inside the medium is released in the form of a giant pulse leading to Q-switching. If the relaxation time of the absorber is short compared to the cavity transit time then as we will discuss in Section 7.7.3, the saturable absorber would simultaneously mode lock and Q-switch the laser.

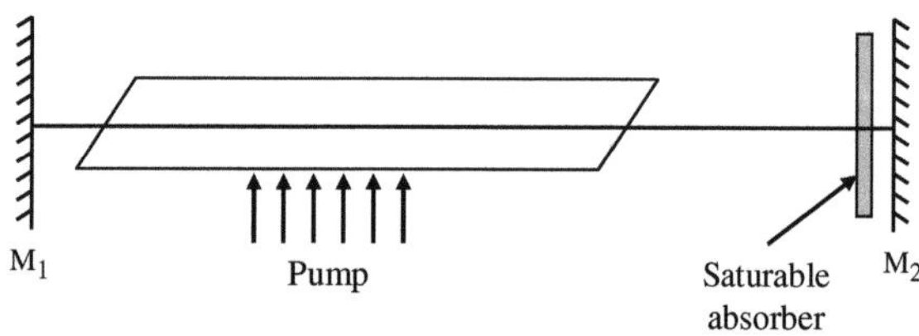

Fig. 7.15 A laser resonator with a saturable absorber placed inside the cavity for passive Q-switching

7.7.3 Mode Locking

Q-switching produces very high energy pulses but the pulse durations are typically in the nanosecond regime. In order to produce ultrashort pulses of durations in picoseconds or shorter, the technique most commonly used is mode locking. In order to understand mode locking let us first consider the formation of beats when two closely lying sound waves interfere with each other. In this case we hear beats due to the fact that the two sound waves (each of constant intensity) being of slightly different frequency will get into and out of phase periodically (see Fig. 7.16). When they are in phase then the two waves add constructively to produce a larger intensity. When they are out of phase, then they will destructively interfere to produce no sound. Hence in such a case we hear a waxing and waning of sound waves and call them as beats.

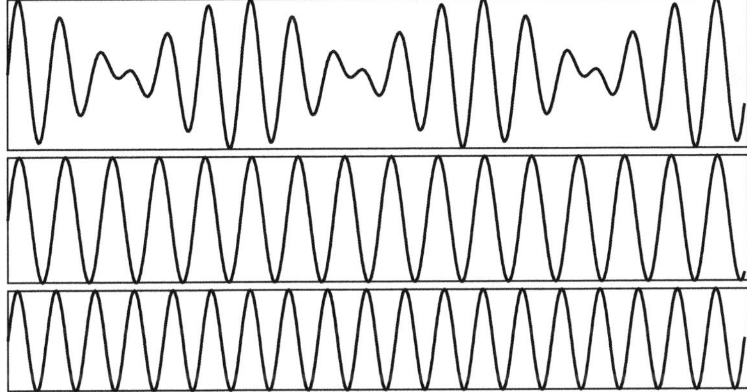

Fig. 7.16 The top curve is obtained by adding the lower two sinusoidal variations and corresponds to beats as observed when two sound waves at closely lying frequencies interfere with each other

Mode locking is very similar to beating except that instead of just two waves now we are dealing with a large number of closely lying frequencies of light. Thus we expect beating between the waves; of course this beating will be in terms of intensity of light rather than intensity of sound. In order to understand mode locking, we first consider a laser oscillating in many frequencies simultaneously. Usually these waves at different frequencies are not correlated and oscillate almost independently of each other, i.e., there is no fixed-phase relationship between the different frequencies. In this case the output consists of a sum of these waves with no correlation among them. When this happens the output is almost the sum of the intensities of each individual mode and we get an output beam having random fluctuations in intensity. In Fig. 7.17 we have plotted the output intensity variation with time obtained as a sum of eight different equally spaced frequencies but with random phases. It can be seen that the output intensity varies randomly with time resembling noise.

Now, if we can lock the phases of each of the oscillating modes, for example bring them all in phase at any time and maintain this phase relationship, then just

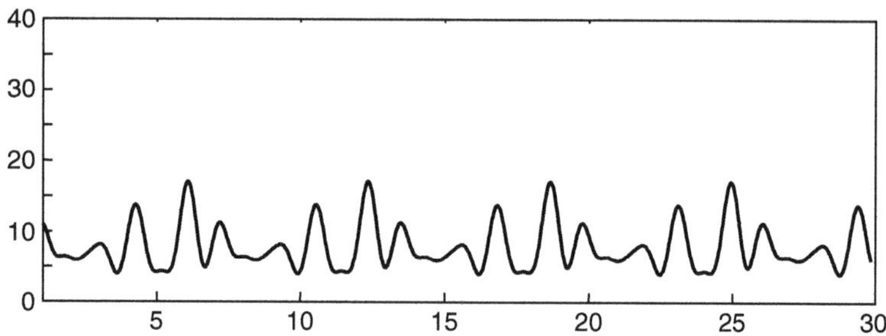

Fig. 7.17 The intensity variation obtained by adding eight equally spaced frequencies with random phases. The intensity variation is noise like

like in the case of beats, once in a while the waves will have their crests and troughs coinciding to give a very large output and at other times the crests and troughs will not be overlapping and thus giving a much smaller intensity (see Fig. 7.18). In such a case the output from the laser would be a repetitive series of pulses of light and such a pulse train is called mode-locked pulse train and this phenomenon is called mode locking. Figure 7.19 shows the output intensity variation with time corresponding to the same set of frequencies as used to plot Fig.7.17, but now the different waves have the same initial phase. In this case the output intensity consists of a periodic series of pulses with intensity levels much higher than obtained with random phases. The peak intensity in this case is higher than the average intensity in the earlier case by the number of modes beating with each other. Also the pulse width is inversely proportional to the number of frequencies.

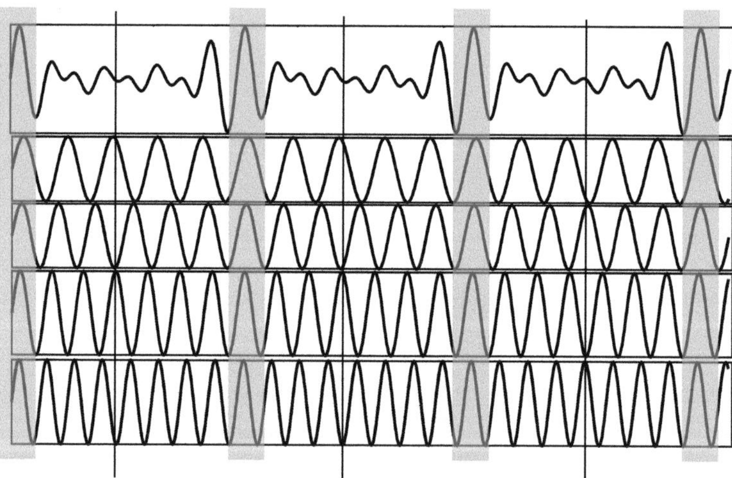

Fig. 7.18 Figure showing interference between four closely lying and equally spaced frequencies and which are in phase at the beginning and retain a constant phase relationship. Note that the waves add constructively periodically

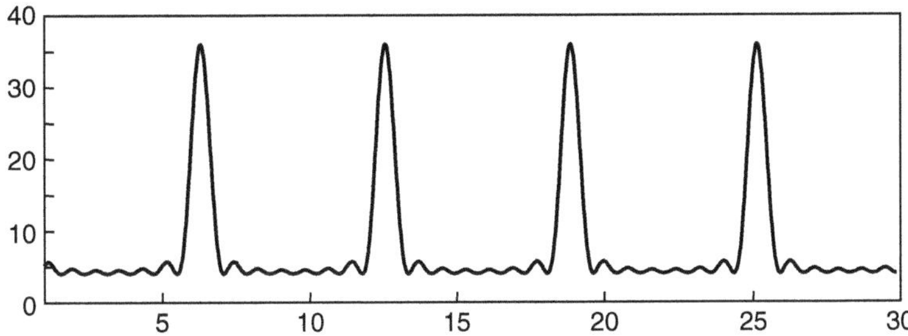

Fig. 7.19 The output intensity variation for the same situation as in Fig. 7.17 but now the waves are having the same phase at a given instant of time and they maintain their phase relationship. Note that the resultant intensity peaks periodically

If the number of waves interfering becomes very large, e.g., a hundred or so, then the peak intensity can be very high and the pulse widths can be very small. Figure 7.20 shows the output mode-locked pulse train coming out of a titanium sapphire laser. Such mode-locked pulse train can be very short in duration (in picoseconds) and have a number of applications.

Mode locking is very similar to the case of diffraction of light from a grating. In this case the constructive interference among the waves diffracted from different slits appears at specific angles and at other angular positions, the waves almost cancel each other. The angular width of any of the diffracted order depends on the number of slits in the grating similar to the temporal width of the mode-locked pulse train depending on the number of modes that are locked in phase.

In order to understand the concept of mode locking, we consider a laser formed by a pair of mirrors separated by a distance d. If the bandwidth over which gain exceeds losses in the cavity is Δv (see Fig.7.21), then, since the intermodal spacing is $c/2n_0d$, the laser will oscillate simultaneously in a large number of frequencies. The number of oscillating modes will be approximately (assuming that the laser is oscillating only in the fundamental transverse mode)

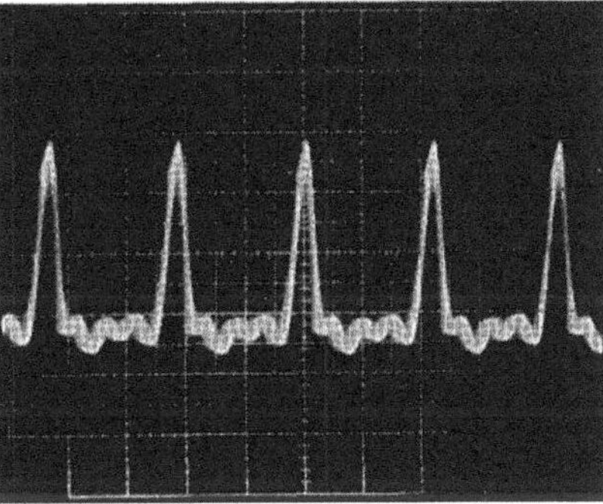

Fig. 7.20 Mode-locked pulse train from a titanium sapphire laser. Each division corresponds to 2 ns (Adapted from French et al. (1990) © 1990 OSA)

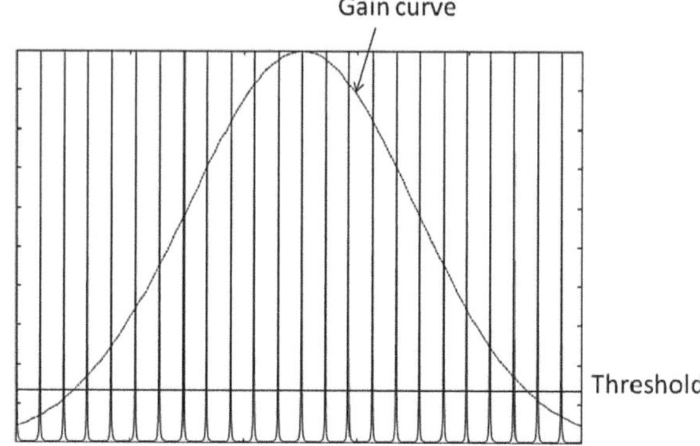

Fig. 7.21 Gain profile of an active medium centered at the frequency ν_0 and of width $\Delta\nu$

$$N + 1 \approx 1 + \text{ integer closest to but less than } \frac{\Delta\nu}{c/2n_0 d} \qquad (7.60)$$

For example if we consider a ruby laser with an oscillating bandwidth of 6×10^{10} Hz, with a cavity length of about 50 cm, the number of oscillating modes will be ~200. This large number of modes oscillates independently of each other and their relative phases are, in general, randomly distributed over the range $-\pi$ to $+\pi$. In order to obtain the output intensity of the laser when it oscillates in such a condition, we note that the total electric field of the laser output will be given by a superposition of the various modes of the laser. Thus

$$E(t) = \sum_{n=-N/2}^{N/2} A_n \exp\left(2\pi i \nu_n t + i\phi_n\right) \qquad (7.61)$$

where A_n and ϕ_n represent the amplitude and phase of the nth mode whose frequency is given by

$$\nu_n = \nu_0 + n\delta\nu, \qquad n = -\frac{N}{2}, -\left(\frac{N}{2}+1\right)\ldots\frac{N}{2} \qquad (7.62)$$

with $\delta\nu = c/2n_0 d$, the intermode spacing, and ν_0 represents the frequency of the mode at the line center.

In general the various modes represented by different values of ν oscillate with different amplitudes A_n and also different phases ϕ_n. The intensity at the output of the laser will be given by

$$I = K\,|E(t)|^2 = K\left|\sum_{n=-N/2}^{N/2} A_n \exp\left(2\pi i \nu_n t + i\phi_n\right)\right|^2$$

$$=K\left|\sum_{n=-N/2}^{N/2} A_n \exp\left(2\pi i\, n\delta\nu\, t + i\phi_n\right)\right|^2 \qquad (7.63)$$

where K represents a constant of proportionality and we have used Eq. (7.62) for ν_n. Equation (7.63) can be rewritten as

$$I = K \sum_n |A_n|^2 + K \sum_{n \neq m} \sum_m A_n A_m^* \exp[2\pi i(n-m)\delta v\, t + i(\phi_n - \phi_m)] \qquad (7.64)$$

From the above equation the following observations can be made:

(a) Since the phases ϕ_n are randomly distributed in the range $-\pi$ to $+\pi$ for the various modes, if the number of modes is sufficiently large, the second term on the right-hand side in Eq. (7.64) will have a very small value. Thus the intensity at the output would have an average value equal to the first term which is nothing but the sum of the intensities of various modes. In this case the output is an incoherent sum of the intensities of the various modes.

(b) Although the output intensity has an average value of the sum of the individual mode intensities, it is fluctuating with time due to the second term in Eq. (7.64). It is obvious from Eq. (7.64) that if t is replaced by $t + q/\delta v$ where q is an integer, then the intensity value repeats itself. Thus the output intensity fluctuation repeats itself every time interval of $1/\delta v = 2n_0 d/c$ which is nothing but the round-trip transit time in the resonator (see Fig. 7.17).

(c) It also follows from Eq. (7.64) that within this periodic repetition in intensity, the intensity fluctuates. The time interval of this intensity fluctuation (which is caused by the beating between the two extreme modes) will be

$$t_f \approx \left[\left(v_0 + \frac{1}{2}N\delta v \right) - \left(v_0 - \frac{1}{2}N\delta v \right) \right]^{-1} \approx \frac{1}{\Delta v} \qquad (7.65)$$

i.e., the inverse of the oscillation bandwidth of the laser medium.

When the laser is oscillating below threshold the various modes are largely uncorrelated due to the absence of correlation among the various spontaneously emitting sources. The fluctuations become much less on passing about threshold but the different modes remain essentially uncorrelated and the output intensity fluctuates with time.

Let us now consider the case in which the modes are locked in phase such that $\phi_n = \phi_0$, i.e., they are all in phase at some arbitrary instant of time $t = 0$. For such a case we have from Eq. (7.63)

$$I = K \left| \sum_{n=-N/2}^{N/2} A_n \exp(2\pi i v_n t) \right|^2 \qquad (7.66)$$

If we also assume that all the modes have the same amplitude, i.e, $A_n = A_0$, then we have

$$I = I_0 \left| \sum_{n=-N/2}^{N/2} \exp(2\pi i v_n t) \right|^2 \qquad (7.67)$$

where $I_0 = KA_0^2$ is the intensity of each mode. The sum in Eq. (7.67) can be easily evaluated and we obtain

$$I = I_0 \left\{ \frac{\sin\left[\pi(N+1)\delta\nu\, t\right]}{\sin\left[\pi\delta\nu\, t\right]} \right\}^2 \tag{7.68}$$

The variation of intensity with time as given by Eq. (7.68) is shown in Fig. 7.22. It is interesting to note that Eq. (7.68) represents a variation in time similar to that exhibited by a diffraction grating in terms of angle. Indeed the two situations are very similar, in the case of diffraction grating, the principal maxima are caused due to the constructive interference among waves diffracted by different slits and in the present case it is the interference in the temporal domain among modes of various frequencies.

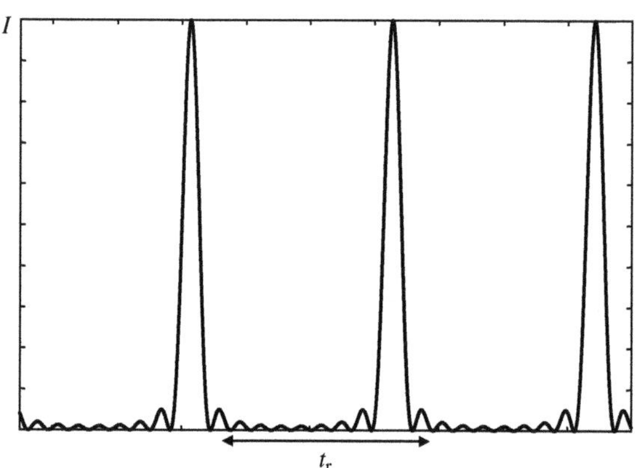

Fig. 7.22 Time variation of the output intensity of a mode-locked laser

From Eq. (7.68) we can conclude the following:

(a) The output of a mode-locked laser will be in the form of a series of pulses and the pulses are separated by a duration

$$t_r = \frac{1}{\delta\nu} = \frac{2n_0 d}{c} \tag{7.69}$$

i.e., the cavity round-trip time. Thus the mode-locked condition can also be viewed as a condition in which a pulse of light is bouncing back and forth inside the cavity and every time it hits the mirror, a certain fraction is transmitted as the output pulse.

(b) From Eq. (7.68) it follows that the intensity falls off very rapidly around every peak (for large N) and the time interval between the zeroes of intensity on either side of the peak is

$$\Delta t \sim \frac{2}{(N+1)\delta\nu} = \frac{2}{\Delta\nu}$$

We may define the pulse duration as approximately (like FWHM)

$$t_D \sim \frac{1}{\Delta\nu} \tag{7.70}$$

i.e., the inverse of the oscillating bandwidth of the laser. Thus the larger the oscillating bandwidth the smaller would be the pulse width. Table 7.1 shows the bandwidth and expected and observed pulse duration for some typical laser systems. As can be seen one can obtain pulses of duration as short as a few hundred femtoseconds. Such ultrashort pulses are finding wide applications in various scientific investigations and industrial applications.

Table 7.1. Some typical laser systems and their mode locked pulse widths

Laser	$\lambda\,(\mu m)$	$\Delta\nu\,(GHz)$	$t_D \sim (\Delta\nu)^{-1}(ps)$	Observed pulse width (ps)
He-Ne	0.6328	1.5	670	600
Argon ion	0.488	7	150	250
Nd: YAG	1.06	12	83	76
Ruby	0.6943	60	17	12
Nd: Glass	1.06	3000	0.33	0.3
Dye	0.6	10000	0.1	0.1

(c) The peak intensity of each mode-locked pulse is given by

$$I = (N + 1)^2 I_0 \tag{7.71}$$

which is $(N+1)$ times the average intensity when the modes are not locked. Thus for typical solid-state lasers which can oscillate simultaneously in 10^3–10^4 modes, the peak power due to mode locking can be very large.

Figure 7.23 shows the output from a mode-locked He–Ne laser operating at 632.8 nm. Notice the regular train of pulses separated by the round-trip time of the cavity. Figure 7.23b shows an expanded view of one of the pulses, the pulse width is about 330 ps.

7.7.3.1 Techniques for Mode Locking

As we have seen above, mode locking essentially requires that the various longitudinal modes be coupled to each other. In practice this can be achieved either by modulating the loss or optical path length of the cavity externally (active mode locking) or by placing saturable absorbers inside the laser cavity (passive mode locking). In Chapter 12 we also discuss the concept of Kerr lens mode locking.

In order to understand how a periodic loss modulation inside the resonator cavity can lead to mode locking, we consider a laser resonator having a loss modulator inside the cavity with the modulation frequency equal to the intermode frequency spacing $\delta\nu$. Consider one of the modes at a frequency ν_q. Since the loss of the cavity is being modulated at a frequency $\delta\nu$, the amplitude of the mode will also be modulated at the same frequency $\delta\nu$ and thus the resultant field in the mode may be written as

Fig. 7.23 Output train of a mode-locked He–Ne laser (Reprinted with permission from A.G. Fox, S.E. Schwarz, and P.W. Smith, use of neon as a nonlinear absorber for mode locking a He-Ne laser, Appl. Phys. Lett. 12 (1969), 371, © 1969 American Institute of Physics.)

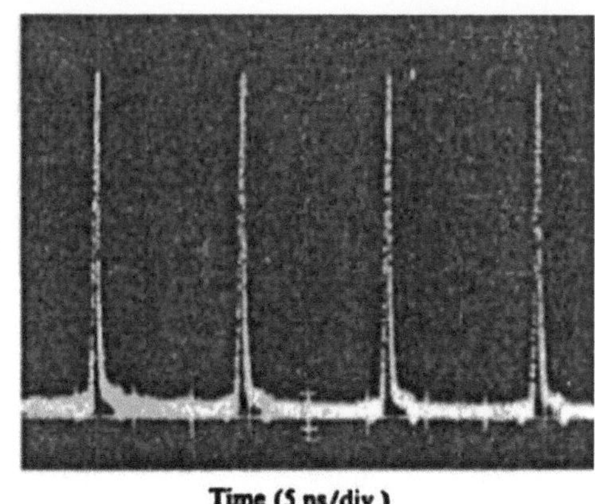

$$(A + B\cos 2\pi\,\delta\nu t)\cos 2\pi\,\nu_q t = A\cos 2\pi\,\nu_q t + \frac{1}{2}B\cos[2\pi(\nu_q + \delta\nu)t]$$

$$+ \frac{1}{2}B\cos[2\pi(\nu_q - \delta\nu)t]$$

Thus the amplitude-modulated mode at a frequency ν_q generates two waves at frequencies $\nu_q + \delta\nu$ and $\nu_q - \delta\nu$. Since $\delta\nu$ is the intermode spacing, these new frequencies correspond to the two modes lying on either side of ν_q. The oscillating field at the frequencies $(\nu_q \pm \delta\nu) = \nu_{q\pm1}$ forces the modes corresponding to these frequencies to oscillate such that a perfect phase relationship now exists between the three modes. Since the amplitudes of these new modes are also modulated at the frequency $\delta\nu$, they generate new side bands at $(\nu_{q+1} + \delta\nu) = \nu_{q+2}$ and $(\nu_{q-1} - \delta\nu) = \nu_{q-2}$. Thus all modes are forced to oscillate with a definite phase relationship and this leads to mode locking.

The above phenomenon of mode locking can also be understood in the time domain by noticing that the intermode frequency spacing $\delta v = c/2n_0 d$ corresponds to the time for the light wave to go through one round trip through the cavity. Hence considering the fluctuating intensity present inside the cavity (see Fig. 7.17), we observe that since the loss modulation has a period equal to a round-trip time, the portion of the fluctuating intensity incident on the loss modulator at a given value of loss would after every round trip be incident at the same loss value. Thus the portion incident at the highest loss instant will suffer the highest loss at every round trip. Similarly, the portion incident at the instant of lowest loss will suffer the lowest loss at every round trip. This will result in the buildup of narrow pulses of light which pass through the loss modulator at the instant of lowest loss. The pulse width must be approximately the inverse of the gain bandwidth since wider pulses would experience higher losses in the modulator and narrower pulses (which would have a spectrum broader than the gain bandwidth) would have lower gain. Thus the above process leads to mode locking.

The loss modulator inside the cavity could be an electro optic modulator or an acousto optic modulator. The electro optic modulator changes the state of polarization of the propagating light beam and the output state of polarization depends on the voltage applied across the modulator crystal. Thus the SOP of the light passing through the modulator, reflected by the mirror and returning to the polarizer, can be changed by the applied voltage and consequently the feedback provided by the polarizer–modulator–mirror system. This leads to a loss modulation. Some of the electro optic materials used include potassium dihydrogen phosphate (KDP), lithium niobate ($LiNbO_3$).

Acousto optic modulators can also be used for mode locking. In this case the acousto optic modulator is used to diffract the propagating light beam out of the cavity (see Fig. 7.24) and thus can be used to modulate the loss of the cavity. Thus standing acoustic waves at frequency Ω produce a loss modulation at the frequency 2Ω and if this equals the intermode spacing, then this would result in mode locking.

The above techniques in which an external signal is used to mode lock the laser are referred to as active mode locking. One can also obtain mode locking using a saturable absorber inside the laser cavity. This technique does not require an external signal to mode lock and is referred to as passive mode locking. As described in Section 7.7.2 in a saturable absorber the absorption coefficient decreases with an increase in the incident light intensity. Thus, the material becomes more and more transparent as the intensity of the incident light increases. In order to understand how a saturable absorber can mode lock a laser, consider a laser cavity with a cell containing the saturable absorber placed adjacent to one of the resonator mirrors (see Fig. 7.15). Initially the saturable absorber does not transmit fully and the intensity inside the resonator has a noise-like structure. The intensity peaks arising from this fluctuation bleach the saturable absorber more than the average intensity values. Thus the intensity peaks suffer less loss than the other intensity values and are amplified more rapidly as compared to the average intensity. If the saturable absorber has a rapid relaxation time (i.e., the excited atoms relax back rapidly to the ground state)

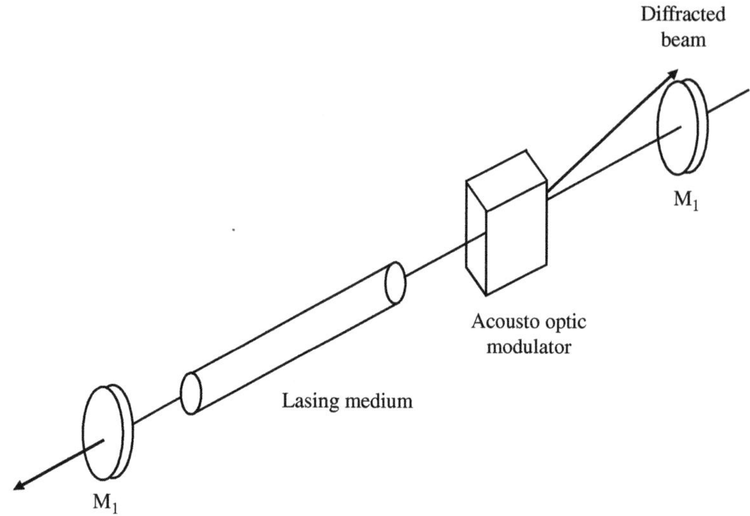

Fig. 7.24 A typical configuration for mode locking of lasers using an acousto optic modulator placed within the cavity

so that it can follow the fast oscillations in the intensity in the cavity, one will obtain mode locking. Since the transit time of the pulse through the resonator is $2n_0d/c$, the mode-locked pulse train will appear at a frequency of $c/2n_0d$. Table 7.2 gives some organic dyes used to mode lock ruby and Nd laser systems; I_s and τ_D represent the saturation intensity and relaxation time, respectively.

Table 7.2. Some organic dyes used to mode lock laser systems

	DDI*	Cryptocyanine	Eastman No. 9740	Eastman No. 9860
Laser	Rudy	Rudy	Nd:	Nd:
I_s (W/m^2)	$\sim 2 \times 10^7$	$\sim 5 \times 10^6$	$\sim 4 \times 10^7$	$\sim 5.6 \times 10^7$
τ_p(ps)	~ 14	~ 22	~ 8.3	~ 9.3
Solvents	Methanol	Nitrobenzene		1,2-
	Ethanol	Acetone		Dichloroethane
		Ethanol		Chlorobenzene
		Methanol		

DDI* stands for 1,1'-diethyl-2,2'-dicarbocyanine iodide. Table adapted from Koechner (1976)

7.8 Modes of Confocal Resonator System

In this section, we shall obtain the modes of a symmetric confocal resonator system which consists of a pair of mirrors of equal radii of curvatures separated by a distance equal to the radius of curvature – see Fig. 7.25. Since the resonator system

Fig. 7.25 A symmetric
confocal resonator system

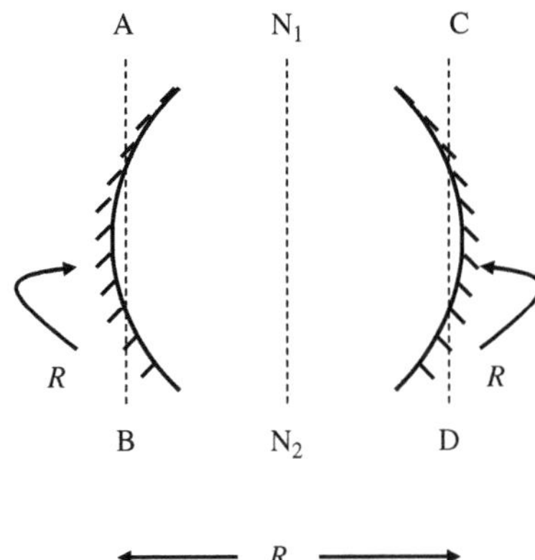

is symmetric about the midplane N_1N_2 the modes of the resonator can be obtained
by simply requiring that the field distribution across the plane AB (say) after com-
pleting half a round trip (i.e., after traversing a distance R and getting reflected from
mirror M_2) must repeat itself on the plane CD. Such a condition would give us the
transverse modes of the resonator. The oscillation frequencies can also be obtained
by requiring that the phase shift suffered by the wave in half a round trip must be
equal to an integral multiple of π.

In order to obtain the modes, we recall that if the field in a plane $z = 0$ is given
by $f(x,y,0)$, then the field on a plane z is given by (see Chapter 2)

$$f(x, y, z) = \frac{i}{\lambda z} e^{-ikz} \int \int f(x', y', 0) \exp\left[-i\frac{k}{2z}\left\{(x - x')^2 + (y - y')^2\right\}\right] dx'dy'$$

$$(7.72)$$

We also need to know the effect of a mirror on the field distribution as it gets
reflected from a mirror of radius of curvature R. In order to calculate this we note
that a spherical wave emanating from an axial point situated at a distance u from the
mirror, after getting reflected from the mirror, becomes a spherical wave converging
to a point at a distance v from the mirror (see Fig. 7.26); u and v are related through
the mirror equation

$$\frac{1}{u} + \frac{1}{v} = \frac{2}{R} \qquad\qquad (7.73)$$

Thus the mirror converts the incident diverging spherical wave of radius u
to a converging spherical wave of radius v. The phase variation produced on
the plane AB by the diverging spherical wave would be given by e^{-ikr}, where
$r = \sqrt{x^2 + y^2 + z^2}$; x and y being the transverse coordinates on the plane
AB. If we assume $x, y \ll z$ (which is the paraxial approximation), then we

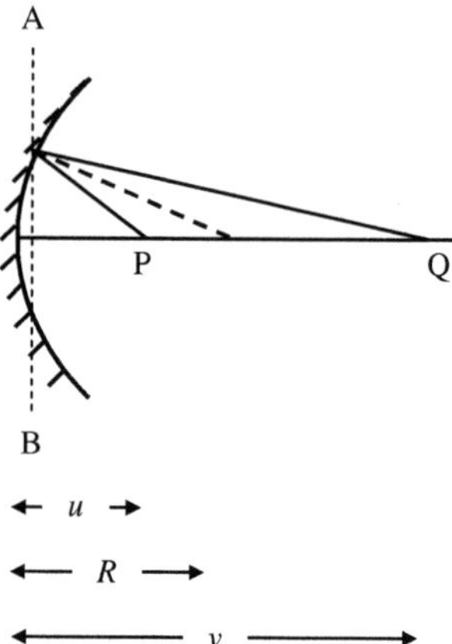

Fig. 7.26 Diverging spherical waves are converted to converging spherical waves by a concave mirror

may write

$$r = u \left(1 + \frac{x^2 + y^2}{u^2}\right)^{1/2} \approx u + \frac{x^2 + y^2}{2u} \tag{7.74}$$

Thus the phase distribution on the plane AB would be

$$\exp\left(-\frac{ik}{2u}(x^2 + y^2)\right)$$

where we have omitted the constant phase term $\exp(-iku)$. Similarly the phase distribution produced on the plane AB by the spherical wave converging to the point Q at a distance v from the mirror would be

$$\exp\left(\frac{ik}{2v}(x^2 + y^2)\right)$$

where we have again omitted the constant phase term $\exp(+ikv)$. Hence if p_m represents the factor which when multiplied to the incident-phase distribution gives the emergent phase distribution, then we have

$$p_m \exp\left(-\frac{ik}{2u}(x^2 + y^2)\right) = \exp\left(\frac{ik}{2v}(x^2 + y^2)\right)$$

or

$$p_m = \exp\left(\frac{ik}{2f}(x^2 + y^2)\right) \tag{7.75}$$

where $f = R/2$ represents the focal length of the mirror. Equation (7.75) represents the effect of the mirror on the incident field distribution.

A field distribution $f(x,y)$ would be a transverse mode of the resonator if it reproduces itself after traversing from plane AB to plane CD (see Fig. 7.25). Thus if $f(x,y)$ represents the field distribution on the plane AB, then the field distribution on the plane CD (after half a round trip) would be given by

$$g(x, y) = \frac{i}{\lambda R} e^{-ikR} \int \int f(x',y') \exp\left[-i\frac{k}{2R}\left\{(x-x')^2 + (y-y')^2\right\}\right] dx'dy'$$

$$\exp\left[i\frac{k}{2R}(x^2+y^2)\right]$$

$$(7.76)$$

where the integration is performed over the surface represented by AB. The field distribution $f(x,y)$ would be a mode of the resonator if

$$g(x, y) = \sigma f(x, y) \tag{7.77}$$

where σ is some complex constant. The losses suffered by the field would be governed by the magnitude of σ, and the phase shift suffered by the wave (which determines the oscillation frequencies of the resonator) would be determined by the phase of σ. Using Eq. (7.77), Eq. (7.76) may be written as

$$\sigma f(x, y) = \frac{i}{\lambda R} e^{-ikR} \int \int f(x',y') \exp\left[-i\frac{k}{2R}\left\{(x-x')^2 + (y-y')^2\right\}\right] dx'dy'$$

$$\exp\left[i\frac{k}{R}(x^2+y^2)\right]$$

$$(7.78)$$

where we have used the fact that $f = R/2$.

In order to solve Eq. (7.78) we define a function $u(x,y)$ through the following relation:

$$u(x, y) = f(x, y) \exp\left[-i\frac{k}{2R}(x^2+y^2)\right] \tag{7.79}$$

We also introduce a set of dimensionless variables

$$\xi = \left(\frac{k}{R}\right)^{1/2} x = \left(\frac{2\pi}{\lambda R}\right)^{1/2} x \tag{7.80}$$

$$\eta = \left(\frac{k}{R}\right)^{1/2} y = \left(\frac{2\pi}{\lambda R}\right)^{1/2} y \tag{7.81}$$

Using Eqs. (7.79), (7.80), and (7.81), Eq. (7.78) becomes

$$\sigma u(\xi, \eta) = \frac{i}{2\pi} e^{-ikR} \int \int u(\xi', \eta') e^{i(\xi\xi'+\eta\eta')} d\xi'd\eta' \tag{7.82}$$

In order to simplify the analysis, we assume the mirrors to be rectangular with dimensions $2a \times 2b$. In such a case we may write

$$\sigma u(\xi, \eta) = \frac{i}{2\pi} e^{-ikR} \int_{-\xi_0}^{\xi_0} \int_{-\eta_0}^{\eta_0} u(\xi', \eta') e^{i(\xi\xi' + \eta\eta')} \, d\xi' d\eta' \tag{7.83}$$

where

$$\xi_0 = \left(\frac{k}{R}\right)^{1/2} a; \qquad \eta_0 = \left(\frac{k}{R}\right)^{1/2} b \tag{7.84}$$

In order to solve Eq. (7.83) we try the separation of variable technique and write

$$\sigma = \kappa \tau \tag{7.85}$$

$$u(\xi, \eta) = p(\xi) q(\eta) \tag{7.86}$$

On substituting Eq. (7.86) in Eq. (7.83) we find that the variables indeed separate out and we obtain

$$\kappa p(\xi) = \sqrt{\frac{i}{2\pi}} e^{-ikR/2} \int_{-\xi_0}^{\xi_0} p(\xi') e^{i\xi\xi'} \, d\xi' \tag{7.87}$$

and

$$\tau q(\eta) = \sqrt{\frac{i}{2\pi}} e^{-ikR/2} \int_{-\eta_0}^{\eta_0} q(\eta') e^{i\eta\eta'} \, d\eta' \tag{7.88}$$

The integrals appearing in Eqs. (7.87) and (7.88) are referred to as finite Fourier transforms; they reduce to the usual Fourier transforms in the limit $\xi_0 - > \infty$ and $\eta_0 - > \infty$. It has been shown by Slepian and Pollack (1961) that the solutions of Eqs. (7.87) and (7.88) are prolate spheroidal functions. We will only consider the case when $\xi_0 \gg 1$ and $\eta_0 \gg 1$, i.e, resonators having large Fresnel numbers. For such a case, we may extend the limits of integration in Eq. (7.87) from $-\infty$ to $+\infty$. Thus Eq. (7.87) becomes

$$A p(\xi) = \int_{-\infty}^{\infty} p(\xi') e^{i\xi\xi'} \, d\xi' \tag{7.89}$$

where

$$A = \kappa \sqrt{\frac{2\pi}{i}} e^{+ikR/2} \tag{7.90}$$

An identical equation is satisfied by $q(\eta)$. Equation (7.89) requires that (apart from some constant factors) $p(\xi)$ be its own Fourier transform.

In order to solve Eq. (7.89) for $p(\xi)$, we differentiate Eq. (7.89) twice with respect to ξ and obtain

$$A \frac{d^2 p}{d\xi^2} = -\int_{-\infty}^{\infty} p(\xi') \xi'^2 e^{i\xi\xi'} \, d\xi' \tag{7.91}$$

We now consider the integral

$$I = \int_{-\infty}^{\infty} \frac{d^2 p}{d\xi'^2} e^{i\xi\xi'} d\xi' \tag{7.92}$$

For the mode, we assume $p(\xi)$ and its derivative to vanish at infinity and integrate Eq. (7.92) twice by parts and obtain

$$\int_{-\infty}^{\infty} \frac{d^2 p}{d\xi'^2} e^{i\xi\xi'} d\xi' = -\xi^2 \int_{-\infty}^{\infty} p(\xi') e^{i\xi\xi'} d\xi'$$
$$= A\xi^2 p(\xi) \tag{7.93}$$

Combining Eqs. (7.91) and (7.93) we get

$$A\left[\frac{d^2 p}{d\xi^2} - \xi^2 p(\xi)\right] = \int_{-\infty}^{\infty} \left(\frac{d^2 p}{d\xi'^2} - \xi'^2 p(\xi')\right) e^{i\xi\xi'} d\xi' \tag{7.94}$$

Comparing Eq. (7.89) with Eq. (7.94) we note that both $p(\xi)$ and $\left[d^2 p/d\xi^2 - \xi^2 p(\xi)\right]$ satisfy the same equation and as such one must have

$$\frac{d^2 p}{d\xi^2} - \xi^2 p(\xi) = -Kp(\xi) \tag{7.95}$$

where K is a constant. We may rewrite Eq. (7.95) as

$$\frac{d^2 p}{d\xi^2} + (K - \xi^2)p(\xi) = 0 \tag{7.96}$$

The solutions of Eq. (7.96) with the condition that $p(\xi)$ vanish at large values of ξ are the Hermite–Gauss functions (see Chapter 3):

$$p_m(\xi) = N_m H_m(\xi) e^{-\xi^2/2} \tag{7.97}$$

where N_m is the normalization constant, $H_m(\xi)$ represents the mth order Hermite polynomial of argument ξ. A few lower order Hermite polynomials are

$$H_0(\xi) = 1, \qquad H_1(\xi) = 2\xi, \qquad H_2(\xi) = 4\xi^2 - 2... \tag{7.98}$$

Thus the complete solution of Eq. (7.78) may be written as

$$f(x,y) = CH_m(\xi)H_n(\eta) e^{-(\xi^2+\eta^2)/2} e^{i(\xi^2+\eta^2)/2} \tag{7.99}$$

where C is some constant; here m and n represent the transverse mode numbers and determine the transverse field distributions of the mode. The Hermite–Gauss functions satisfy the equation

$$i^m H_m(\xi) e^{-\xi^2/2} = \frac{1}{\sqrt{2\pi}} \int\limits_{-\infty}^{\infty} \mathrm{d}\xi' H_m(\xi') e^{-\xi'^2/2} e^{i\xi\xi'} \qquad (7.100)$$

If we use the fact that $p(\xi)$ in Eq. (7.87) is a Hermite–Gauss function, then using Eq. (7.100) one readily obtains

$$\kappa = i^m \sqrt{i} e^{-ikR/2} = \exp\left\{-i\left[\frac{kR}{2} - \left(m + \frac{1}{2}\right)\frac{\pi}{2}\right]\right\} \qquad (7.101)$$

Similarly

$$\tau = i^n \sqrt{i} e^{-ikR/2} = \exp\left\{-i\left[\frac{kR}{2} - \left(n + \frac{1}{2}\right)\frac{\pi}{2}\right]\right\} \qquad (7.102)$$

Thus from Eq. (7.85) we have

$$\sigma = \kappa\tau = \exp\left\{-i\left[kR - (m + n + 1)\frac{\pi}{2}\right]\right\} \qquad (7.103)$$

Observe that $|\sigma| = 1$ implying the absence of any losses. This can be attributed to the fact that, in our analysis we have essentially assumed the mirrors to be of extremely large transverse dimensions.

Since the phase of σ represents the phase shift suffered by the wave in half a round trip, one must have

$$kR - (m + n + 1)\frac{\pi}{2} = q\pi, \qquad q = 1, 2, 3... \qquad (7.104)$$

where q refers to the longitudinal mode number. Using $k = 2\pi v/c$, we obtain the frequencies of oscillation of the cavity as

$$v_{mnq} = (2q + m + n + 1)\frac{c}{4R} \qquad (7.105)$$

Observe that all modes having the same value of $(2q + m + n)$ would have the same oscillation frequency and hence would be degenerate. The frequency separation between two modes having the same value of m and n but adjacent values of q is

$$\Delta v_q = \frac{c}{2R} \qquad (7.106)$$

The frequency separation between two transverse modes corresponding to the same value of q is

$$\Delta v_m = \Delta v_n = \frac{c}{4R} \qquad (7.107)$$

which is half that between two consecutive longitudinal modes.

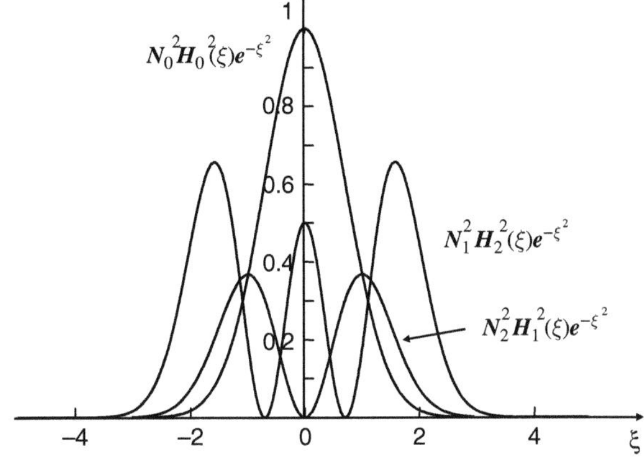

Fig. 7.27 Transverse intensity distributions corresponding to the three lowest order Hermite–Gauss modes

In Fig. 7.27 we depict the transverse intensity distribution corresponding to the mode amplitude distributions given by Eq. (7.99). Figure 7.3 shows the photographs of the intensity distribution corresponding to different transverse modes of the resonator. Observe that higher order modes extend more in the transverse dimension and hence would have higher diffraction losses.

The field given by Eq. (7.99) is the field distribution in a transverse plane passing through the pole of the mirror. The field distribution at any other plane can be derived by using the diffraction formula given in Eq. (7.72). Thus the field distribution midway between the mirrors would be

$$
\begin{aligned}
f_M(x,y) &= \frac{iC}{\pi} e^{-ikR/2} \int\int H_m(\xi')H_n(\eta')\exp\left[-\frac{1}{2}\xi'^2(1-i)-\frac{1}{2}\eta'^2(1-i)\right] \\
&\quad \times \exp\left[-i\left(\xi^2+\xi'^2-2\xi\xi'\right)-i\left(\eta^2+\eta'^2-2\eta\eta'\right)\right]\mathrm{d}\xi'\mathrm{d}\eta' \\
&= \frac{iC}{\pi} e^{-ikR/2}e^{-i(\xi^2+\eta^2)}\int H_m(\xi')\exp\left[-\frac{1}{2}\xi'^2(1+i)+2i\xi\xi'\right]\mathrm{d}\xi' \\
&\quad \times \int H_n(\eta')\exp\left[-\frac{1}{2}\eta'^2(1+i)+2i\eta\eta'\right]\mathrm{d}\eta'
\end{aligned}
$$

$$(7.108)$$

Using the integral

$$
\int H_n(\alpha x)e^{-x^2+2xy}\mathrm{d}x = \sqrt{\pi}\,e^{y^2}\left(1-\alpha^2\right)^{n/2} H_n\left[\frac{\alpha y}{\left(1-\alpha^2\right)^{n/2}}\right] \tag{7.109}
$$

we obtain

$$
f_M(x,y) = C\exp\left[-i\left(\frac{kR}{2}-\frac{\pi}{2}\right)\right]\exp\left[-\left(\xi^2+\eta^2\right)\right]H_m(\sqrt{2}\xi)H_n(\sqrt{2}\eta) \tag{7.110}
$$

which shows that the phase along the transverse plane midway between the mirrors is constant. Thus the phase fronts are plane midway between the mirrors. From Eq. (7.99) it can also be seen that the phase front of the modal field distribution has a radius of curvature R which is equal to the radius of curvature of the resonator mirror.

7.9 Modes of a General Spherical Resonator

In Section 7.8 we showed that the transverse modes of a symmetric confocal resonator are Hermite–Gauss functions. In fact Hermite–Gauss functions describe the transverse modes of stable resonators formed using spherical mirrors. In this section we consider a general spherical resonator and obtain the characteristics of the fundamental Gaussian mode and also obtain the stability condition.

We consider a general spherical resonator consisting of two mirrors of radii of curvatures R_1 and R_2 separated by a distance d (see Fig. 7.28). The radius of curvature is assumed to be positive if the mirror is concave toward the resonator and negative if it is convex toward the resonator. We will now show that such a resonator is stable or unstable depending on the values of R_1, R_2, and d and if the resonator is stable, then the fundamental transverse mode of such a resonator is a Gaussian.

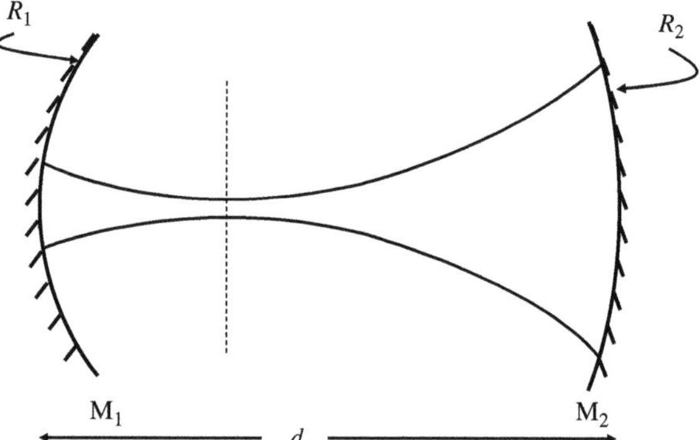

Fig. 7.28 A general spherical resonator

In Chapter 2 we had shown that a Gaussian beam propagating along the z-direction and whose amplitude distribution on the plane $z = 0$ is given by

$$u(x, y, 0) = a \exp\left[-\frac{x^2 + y^2}{w_0^2}\right] \tag{7.111}$$

has the following electric field distribution at a plane z (see Eq. 2.51):

$$u(x, y, z) \approx \frac{a}{(1 - i\gamma)} \exp\left[-\frac{x^2 + y^2}{w^2(z)}\right] e^{i\Phi} \tag{7.112}$$

where

$$\gamma = \frac{\lambda z}{\pi w_0^2} \tag{7.113}$$

$$w(z) = w_0 \left[1 + \gamma^2\right]^{1/2} = w_0 \left[1 + \frac{\lambda^2 z^2}{\pi^2 w_0^4}\right]^{1/2} \tag{7.114}$$

$$\Phi = kz + \frac{k}{2R(z)}\left(x^2 + y^2\right) \tag{7.115}$$

$$R(z) \equiv z\left(1 + \frac{1}{\gamma^2}\right) = z\left[1 + \frac{\pi^2 w_0^4}{\lambda^2 z^2}\right] \tag{7.116}$$

Let the poles of the mirrors M_1 and M_2 be at $z = z_1 = -d_1$ and at $z = z_2 = +d_2$, respectively. We are assuming the origin somewhere between the mirrors so that both d_1 and d_2 are positive quantities. Thus the distance between the two mirrors is given by

$$d = d_1 + d_2 \tag{7.117}$$

Now, for the Gaussian beam to resonate between the two mirrors, the radii of the phase front (at the mirrors) should be equal to the radii of curvatures of the mirrors:

$$-R_1 = -d_1 - \frac{\alpha}{d_1} \quad \text{and} \quad R_2 = d_2 + \frac{\alpha}{d_2} \tag{7.118}$$

where $\alpha = \pi^2 w_0^4/\lambda^2$. In such a case, the Gaussian beam would be normally incident on the mirrors and hence will retrace its path to the other mirror where it is normally incident. Thus such a Gaussian beam can resonate in the resonator and would form a mode of the resonator.

With the sign convention mentioned earlier, for the type of mirrors shown in Fig. 7.28, both R_1 and R_2 are positive. Thus

$$\alpha = d_1(R_1 - d_1) = d_2(R_2 - d_2) \tag{7.119}$$

If we use the relation $d_2 = d - d_1$, we would readily get

$$d_1 = \frac{(R_2 - d)d}{R_1 + R_2 - 2d} \quad \text{and} \quad d_2 = \frac{(R_1 - d)d}{R_1 + R_2 - 2d} \tag{7.120}$$

We define

$$g_1 = 1 - \frac{d}{R_1} \quad \text{and} \quad g_2 = 1 - \frac{d}{R_2} \tag{7.121}$$

From the above equations we may write $R_1 = \frac{d}{1-g_1}$ and $R_2 = \frac{d}{1-g_2}$ and we obtain

$$d_1 = \frac{g_2 (1 - g_1) d}{g_1 + g_2 - 2g_1 g_2} \quad \text{and} \quad d_2 = \frac{g_1 (1 - g_2) d}{g_1 + g_2 - 2g_1 g_2} \tag{7.122}$$

Thus (see Eq. 7.119)

$$\alpha = d_1 (R_1 - d_1)$$
$$= \frac{g_1 g_2 d^2 (1 - g_1 g_2)}{(g_1 + g_2 - 2g_1 g_2)^2} \tag{7.123}$$

Since, $\alpha = \frac{\pi^2 w_0^4}{\lambda^2}$ we get for the spot size at the waist

$$w_0^2 = \frac{\lambda d}{\pi (g_1 + g_2 - 2g_1 g_2)} \sqrt{g_1 g_2 (1 - g_1 g_2)} \tag{7.124}$$

For w_0 to be real we must have $0 \le g_1 g_2 \le 1$, or

$$0 \le \left(1 - \frac{d}{R_1}\right)\left(1 - \frac{d}{R_2}\right) \le 1 \tag{7.125}$$

where R_1 and R_2 are the radii of curvatures of the mirrors. The above equation represents the stability condition for a resonator consisting of two spherical mirrors. Figure 7.2 shows different resonator configurations. Figure 7.29 shows the stability diagram and the shaded region correspond to stable resonator configurations.

The spot sizes of the Gaussian beam at the two mirrors are given by

$$w^2 (z_1) = \frac{\lambda d}{\pi} \sqrt{\frac{g_2}{g_1 (1 - g_1 g_2)}} \tag{7.126}$$

and

$$w^2 (z_2) = \frac{\lambda d}{\pi} \sqrt{\frac{g_1}{g_2 (1 - g_1 g_2)}} \tag{7.127}$$

Since most of the energy in a Gaussian beam is contained within a radius of about twice the beamwidth, if the transverse dimensions of the mirrors are large compared to the spot sizes at the mirrors, then most of the energy is reflected back and the loss due to diffraction spill over from the edges of the mirrors is small. It can be easily

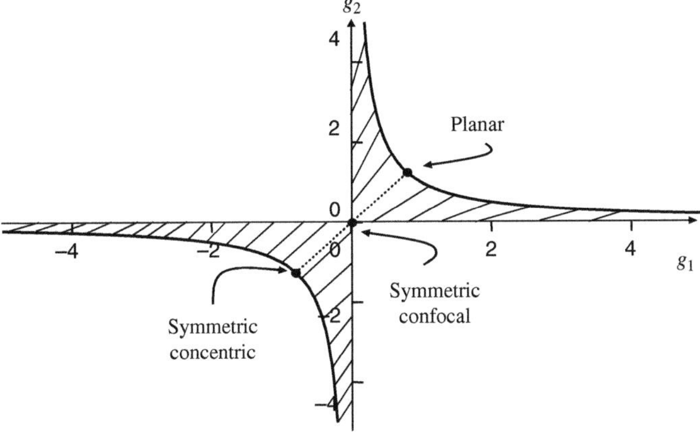

Fig. 7.29 The stability diagram of spherical mirror resonators. The shaded region corresponds to stable resonators

seen from Eqs. (7.126) and (7.127) that when $g_1g_2 \to 0$ or $g_1g_2 \to 1$, $w(z_1)$ or $w(z_2)$ both become very large and our analysis would not remain valid.

Problems

Problem 7.4 A Fabry–Perot (F.P.) etalon made of glass of refractive index 1.5 and having faces of reflectivities of 0.9 each is used inside a laser resonator cavity having a gain medium with a bandwidth $\Delta \nu \sim 10$ GHz. What thickness would you choose so that the F.P. etalon, placed perpendicular to the cavity, can lead to single longitudinal mode operation of the laser?

Problem 7.5 I wish to make a resonator in which one of the mirrors is a convex mirror of radius of curvature 1 m. If the length of the resonator is to be 1 m, what type of mirror (plane, convex or concave) and what radius of curvature will you choose so that the resonator is stable?

Problem 7.6 A gas laser of length 20 cm oscillates simultaneously in two adjacent longitudinal modes around a wavelength of 800 nm. Calculate the wavelength spacing between the modes and estimate the coherence length of the laser.

Problem 7.7 Consider a symmetric spherical resonator consisting of two concave mirrors of radii of curvature 1 m and separated by 20 cm operating at 1 μm. What will be the angle of divergence of the laser beam (oscillating in the fundamental mode) emanating from such a laser?

Problem 7.8 Consider a resonator shown below:

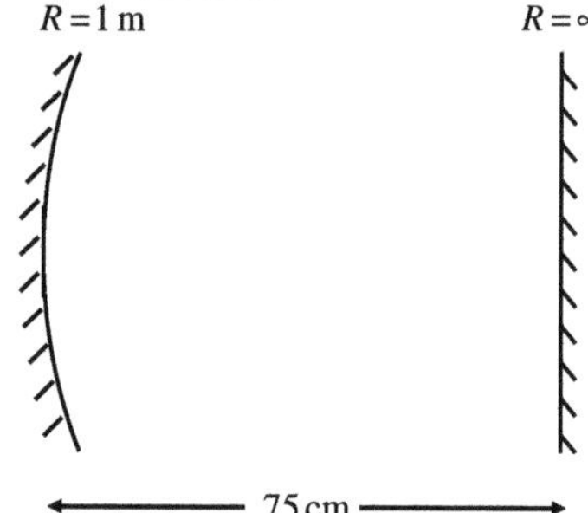

Given that; $\lambda_0 = 1\ \mu m$

(a) Obtain the transverse intensity distributions of the fundamental mode at the position of the two mirrors

(b) If both the mirrors are partially reflecting which beam (one coming from the right and the other coming from the left) would have a larger diffraction divergence and why?

(c) If the same resonator is used for oscillation at 1.5 μm by what approximate factor would the diffraction divergence of the beam at 1.5 μm increase or decrease compared to 1 μm?

(d) For what range of mirror separation will the above resonator be stable.

Problem 7.9 The length of a laser resonator oscillating in a single longitudinal mode varies randomly by 20 nm from the equilibrium position. Assuming $d = 10$ cm and $\lambda_0 = 500$ nm, calculate the corresponding variation in the frequency of oscillation of the laser.

Problem 7.10 A He–Ne laser with a gain tube of length 10 cm and a mirror separation of 40 cm oscillates over a bandwidth of 1.5 GHz. Estimate the shortest pulse that can be generated by mode locking such a laser. What is the duration of each pulse and the pulse repetition frequency.

Problem 7.11 The gain coefficient (in m^{-1}) of a laser medium with a center wavelength of 500 nm depends on frequency through the following equation:

$$\gamma\,(\upsilon) = \gamma\,(\upsilon_0)\exp\left[-4\left\{\frac{\upsilon - \upsilon_0}{\Delta\upsilon}\right\}^2\right]$$

where υ_0 is the center frequency, $\gamma(\upsilon_0) = 1\ m^{-1}$ and $\Delta\upsilon = 3$ GHz. The length of the laser cavity is 1 m and the mirror reflectivities are 99% each. Obtain the number of longitudinal modes that will oscillate in the laser. Neglect all other losses in the cavity.

Problem 7.12 Shown below is the output power from a mode-locked laser as a function of time:

(a) What is the length of the laser resonator? (Assume the refractive index of the medium within the cavity to be unity).

(b) What is the approximate number of oscillating modes?

(c) What would be the average output power if the same laser operates without mode locking?

(d) What should be the frequency of the loss modulation for mode locking?

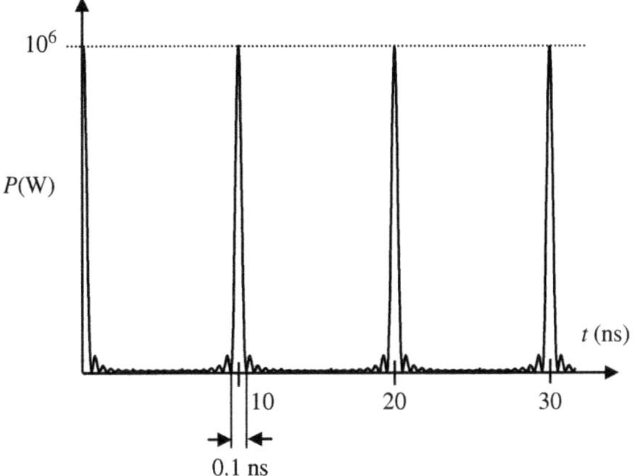

Problem 7.13 The cavity of a 6328 Å He–Ne laser is 1 m long and has mirror of reflectivities 100 and 98%; the internal cavity losses are negligible (a) if the steady-state power output is 10 mW, what is the energy stored in the resonator (b) What is the linewidth of the above passive cavity. (c) If the oscillating band width is 1500 MHz, how many longitudinal mode would oscillate?

Problem 7.14 Limiting apertures are used to suppress higher order transverse mode oscillation. Consider a Gaussian beam of waist size $w_0 = 0.5$ mm and a total power of 1 mW. An aperture of radius 1 mm is introduced at the position of the waist. Calculate the power which goes through the aperture.

Problem 7.15 What is the approximate angular divergence of the output beam from a He–Ne laser (operating at 6328 Å) having a 1 m long confocal cavity?

Problem 7.16 The Gaussian beam coming out of a 100 cm long symmetric spherical resonator oscillating at a wavelength of 1 μm has an angular divergence of 0.06° of arc. Calculate the radius of curvature of the mirrors.

Problem 7.17 Consider a He–Ne laser with Doppler broadened linewidth of 1700 MHz. What should be the length of the resonator cavity so that only a single longitudinal mode would oscillate?

Problem 7.18 The cavity of a 6328 Å He–Ne laser is 1 m long and has mirror of reflections 98 and 100%; the internal cavity losses being negligible. If the steady-state power output is 35 mW, what is the steady state photon number in the cavity?

Problem 7.19 Consider a confocal resonator of length 1 m used for a He–Ne laser at $\lambda = 6328$ Å. Calculate the spot size at the resonator center and at the mirrors. If the total power in the laser beam is 1 mW, calculate the intensity of the beam on the axis as it comes out of the laser.

Problem 7.20 Consider a laser of resonator length 100 cm and oscillating at a frequency of 3×10^{14} Hz. Calculate the variation in the cavity length that will lead to a frequency fluctuation of 1 kHz

Problem 7.21 A stable resonator is to be made using a convex mirror of radius of curvature 1 m and a concave mirror of radius of curvature 0.5 m. What is the condition on the separation so that the resonator is stable?

Problem 7.22 Consider a resonator consisting of a convex mirror of radius of curvature 1 m and a concave mirror of radius of curvature 0.5 m. Take a separation between the mirrors so that the resonator is stable. Obtain the position of the waist of the Gaussian mode. $\lambda = 1\ \mu m$.

Problem 7.23 Consider a laser with a resonator consisting of mirrors of reflectivity 100 and 98%, oscillating in a single longitudinal mode; the length and width of the resonator are 20 and 1 cm, respectively. There are no intrinsic losses in the laser resonator. The laser is oscillating in steady state with a frequency ω_0, which corresponds to the peak of the lineshape function $g(\omega)$. A monochromatic wave at a frequency $[\omega_0 + (\Delta\omega/2)]$, where $\Delta\omega$ is the FWHM of the lineshape function, passes perpendicular through the laser (see figure). Will the wave (at frequency $[\omega_0 + (\Delta\omega/2)]$) get amplified or attenuated ? Obtain the corresponding amplification/attenuation factor for the intensity of this wave, in going across the laser.

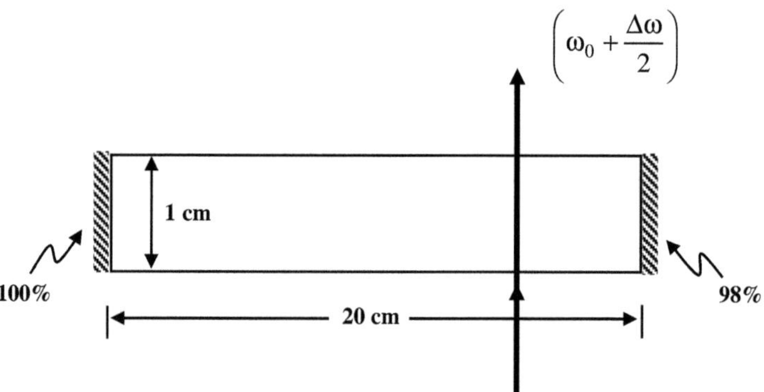

Problem 7.24 Consider a resonator made of a concave mirror of radius of curvature 20 cm, and silvered thin plano-convex lens, as shown in the figure. The focal length of the unsilvered plano-convex lens is 20 cm. For what values of d will the resonator be stable?

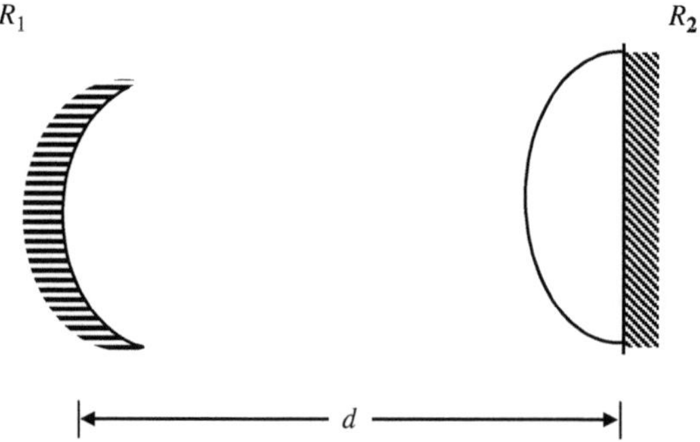

Problem 7.25 The cavity of a 6328 Å laser is 1 m long and has mirror of reflectivities 100 and 98%, with negligible internal cavity losses (a) what is the cavity lifetime? (b) If the output of the laser is 50 mW, calculate the energy inside the cavity. (c) What is the value of the gain coefficient required to reach threshold for laser oscillation?

Problem 7.26 A Fabry–Perot interferometer with mirrors of reflectivity 90% and separated by a distance h is kept inclined at 45° with the axis of the laser within the cavity of a He–Ne laser. If the gain bandwidth of the laser is 1 GHz, what is the condition on h so that only one longitudinal mode can oscillate? Are there any other conditions that need to be satisfied so that the laser oscillates?

Problem 7.27 There are 10^{11} photons in the cavity of an Ar-ion laser oscillating in steady state at the wavelength of 514 nm. If the laser resonator is formed by two plane mirrors of reflectivities 100 and 90%, separated by a distance of 50 cm, calculate the output power and the energy inside the cavity.[Ans : $P_O = 1.22\,W, E = 38nJ$].

Problem 7.28 State whether resonators made with the following mirror pairs are stable or not?

(a) $R_1 = \infty, R_2 = 20$ cm, $d = 25$ cm

(b) $R_1 = 20$ cm, $R_2 = 20$ cm, $d = 40$ cm

(c) $R_1 = 20$ cm, $R_2 = -20$ cm, $d = 15$ cm.

Problem 7.29 When a laser oscillates, the emerging transverse field distribution is characterized by a Hermite–Gaussian distribution:

$$E_{mn}\,(x,y,z=0) = E_0 H_m\left(\frac{\sqrt{2}x}{w_0}\right) e^{-\frac{x^2}{w_0^2}} H_n\left(\frac{\sqrt{2}x}{w_0}\right) e^{-\frac{y^2}{w_0^2}}$$

These correspond to the various modes of oscillation of the laser. The fundamental mode corresponds to $m = 0, n = 0$ and has a Gaussian field distribution. Obtain the Fraunhofer diffraction pattern of such a field distribution.

Problem 7.30 In a ruby crystal, a population inversion density of $(N_2 - N_1) = 5 \times 10^{17}$ cm^{-3} is generated by pumping. Assuming $g(v_0) = 5 \times 10^{-12}$ s, $t_{sp} = 3 \times 10^{-3}$ s, wavelength of 694.3 nm and a refractive index of 1.78, obtain the gain coefficient $\gamma(v_0)$. By what factor will a beam get amplified if it passes through 5 cm of such a crystal?
    [Ans: 5×10^{-2} cm^{-1}, 1.28]

Problem 7.31 The longitudinal mode spectrum of a passive resonator consisting of plane mirrors having one mirror of reflectivity 98% is shown below.

(a) What is the length of the cavity? (1.5 m)

(b) What is the reflectivity of the second mirror? (Assume $n_0 = 1$ and absence of any internal losses in the cavity) (95.8%).

(c) If the cavity is filled with an amplifying medium having a single pass gain G, what is the threshold value of G to start laser oscillation? (1.03).

(d) If for a given pumping rate, the output from the mirror having reflectivity 98% is 1 mW, what is the output from the other mirror? (2.12 mW).

Problem 7.32 The beam coming out of a laser resonator 50 cm long is seen to have an angular divergence of 1.1 min of arc and it is seen that the beam coming out of one of the mirrors is converging and reaches a minimum diameter at a distance of 10 cm from the mirror. Assuming the wavelength to be 1000 nm

(a)   Calculate the radii of curvatures of the mirrors.

(b)   Calculate the beam size at the two mirrors.

(c)   For what range of distance between the mirrors would the resonator be stable?

Problem 7.33 A He–Ne laser of length 20 cm oscillates in two longitudinal modes. If the output of the laser is incident on a photodetector (a device which converts light into electrical current) whose output current is proportional to the incident intensity, what will be the time variation of the output current ?

Problem 7.34 Consider a resonator consisting of a plane mirror and a concave mirror of radius of curvature R (see figure). Assume $\lambda = 1\ \mu$m, $R = 100$ cm, and the distance between the two mirrors to be 50 cm. Calculate the spot size of the Gaussian beam.

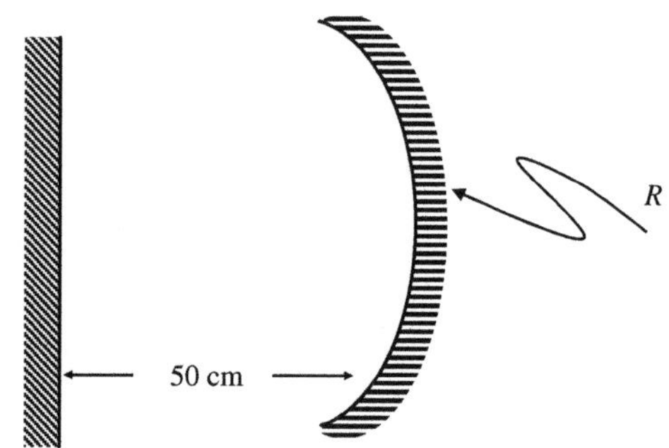

Solution

$$R = d\left[1 + \frac{\pi^2 w_0^4}{\lambda^2 d^2}\right] \Rightarrow w_0 = \sqrt{\frac{\lambda d}{\pi}}\left[\frac{R}{d} - 1\right]^{1/4} \approx 4 \times 10^{-4}\,\text{m} = 0.4\,\text{mm}$$

Problem 7.35 Show that a phase variation of the type $\exp\left[-i\frac{k}{2R}\left(x^2 + y^2\right)\right]$ (on the x–y plane) represents a diverging spherical wave of radius R.

Solution For a spherical wave diverging from the origin, the field distribution is given by

$$u \sim \frac{1}{r}e^{-ikr}$$

Now, on the plane $z = R$

$$r = \left[x^2 + y^2 + R^2\right]^{1/2}$$
$$= R\left[1 + \frac{x^2 + y^2}{R^2}\right]^{1/2}$$
$$\approx R + \frac{x^2 + y^2}{2R}$$

where we have assumed $|x|$, $|y| \ll R$. Thus on the plane $z = R$, the phase distribution (corresponding to a spherical wave of radius R) would be given by

$$e^{ikr} \approx e^{-ikR}e^{-\frac{ik}{2R}\left(x^2+y^2\right)}$$

Thus, a phase variation of the type $\exp\left[-i\frac{k}{2R}\left(x^2+y^2\right)\right]$ (on the x–y plane) represents a diverging spherical wave of radius R. Similarly, a phase variation of the type $\exp\left[+i\frac{k}{2R}\left(x^2+y^2\right)\right]$ (on the x–y plane) represents a converging spherical wave of radius R.

Problem 7.36 The output of a semiconductor laser can be approximately described by a Gaussian function with two different widths along the transverse (w_T) and lateral (w_L) directions as

$$\psi(x,y) = A\,\exp\left(-\frac{x^2}{w_L^2} - \frac{y^2}{w_T^2}\right)$$

where x and y represent axes parallel and perpendicular to the junction plane. Typically $w_T \approx 0.5$ μm and $w_L = 2$ μm. Discuss the far field of this beam (see figure below).

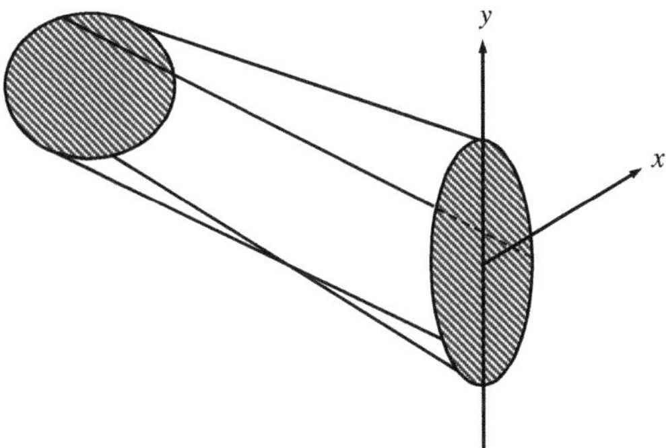

Solution

$$u(x,y,z) = \frac{a}{\sqrt{(1+i\gamma_T)\,(1+i\gamma_L)}}\,\exp\left[-\frac{x^2}{w_1^2} - \frac{y^2}{w_2^2}\right]e^{i\Phi}$$

Thus

$$I(x,y,z) = \frac{I_0}{\sqrt{\left(1+\gamma_T^2\right)\left(1+\gamma_L^2\right)}}\,\exp\left[-\frac{2x^2}{w_1^2(z)} - \frac{2y^2}{w_2^2(z)}\right]$$

where

$$w_1^2(z) = w_T\left(1+\gamma_T^2\right)^{1/2} = w_T\left[1 + \frac{\lambda^2 z^2}{\pi^2 w_T^4}\right]^{1/2} \approx \frac{\lambda z}{\pi w_T} \quad \text{(for large } z\text{)}$$

and

$$w_2^2(z) = w_L\left(1+\gamma_L^2\right)^{1/2} = w_L\left[1 + \frac{\lambda^2 z^2}{\pi^2 w_L^4}\right]^{1/2} \approx \frac{\lambda z}{\pi w_L} \quad \text{(for large } z\text{)}$$

Problem 7.37 The output of a He–Ne laser ($\lambda = 6328$ Å) can be assumed to be Gaussian with plane phase front. For $w_0 = 1$ mm and $w_0 = 0.2$ mm, calculate the beam diameter at $z = 20$ m.

Solution The beam diameter is given by

$$2w = 2w_0 \left[1 + \gamma^2\right]^{1/2} = 2w_0 \left[1 + \frac{\lambda^2 z^2}{\pi^2 w_0^4}\right]^{1/2}, \ \gamma = \frac{\lambda z}{\pi w_0^2}$$

For $w_0 = 0.1$ cm, $\lambda = 6.328 \times 10^{-5}$ cm, $z = 2000$ cm

$$\frac{\lambda^2 z^2}{\pi^2 w_0^4} \approx 16.23 \Rightarrow 2w \approx 0.83 \, \text{cm}$$

For $w_0 = 0.2$ mm $= 0.02$ cm and with same values of λ and z

$$\frac{\lambda^2 z^2}{\pi^2 w_0^4} \approx 10143 \Rightarrow 2w \approx 4.0 \, \text{cm}$$

The above results show that the divergence increases as w_0 becomes smaller.

Chapter 8
Vector Spaces and Linear Operators: Dirac Notation

8.1 Introduction

In this chapter we will introduce Dirac's bra and ket notation and also discuss the representation of observables by linear operators. By imposing the commutation relations, we will solve the linear harmonic oscillator problem which will be used in the next chapter to study the quantized states of the radiation field. Here, we will discuss only those aspects of the bra and ket algebra which will be used later on. For a thorough account, the reader is referred to the classic treatise by Dirac (1958a).

8.2 The Bra and Ket Notation

A state of a system can be represented by a certain type of vector, which we call a ket vector and represent by the symbol $| \rangle$.[1] In order to distinguish the ket vectors corresponding to different states, we insert a label; thus, the ket vector (or simply the ket) corresponding to the state A is described by the symbol $|A\rangle$. The kets form a linear vector space implying that if we have two states described by the kets $|A\rangle$ and $|B\rangle$, then the linear combination

$$C_1 |A\rangle + C_2 |B\rangle \tag{8.1}$$

is also a vector in the same space; here C_1 and C_2 are arbitrary complex numbers. The state $|P\rangle = C_1 |A\rangle$ represents the same state $|A\rangle$; thus, if a ket is superposed on itself, it corresponds to the same state. Further, to quote Dirac (p.16)

> ...each state of a dynamical system at a particular time corresponds to a ket vector, the correspondence being such that if a state results from the superposition of certain other states, its corresponding ket vector is expressible linearly in terms of the corresponding ket vectors of the other states, and conversely.

The conjugate imaginary of a ket vector $|A\rangle$ is denoted by $\langle A|$ and is called a bra vector (or simply a bra). The scalar product of $|A\rangle$ and $\langle B|$ is denoted by $\langle B|A\rangle$,

---

[1] The analysis will be based on the book by Dirac (1958a).

K. Thyagarajan, A. Ghatak, *Lasers*, Graduate Texts in Physics,
DOI 10.1007/978-1-4419-6442-7_8, © Springer Science+Business Media, LLC 2010

which is a complex number. Further

$$\langle B \,|A \rangle = \overline{\langle A \,|B \rangle} \tag{8.2}$$

where $\overline{\langle B \,|A \rangle}$ denotes the complex conjugate of the scalar product. If we put $|B\rangle = |A\rangle$, we obtain from Eq. (8.2) that $\langle A \,|A\rangle$ should be a real number. We further impose that

$$\langle A \,|A \rangle \geq 0 \tag{8.3}$$

the equality sign holds if and only if $|A\rangle = 0$, i.e., $|A\rangle$ is a null vector.

The bra corresponding to $C\,|A\rangle$ is denoted by $C^*\,\langle A|$ where C is a complex number and C^* its complex conjugate. A ket $|A\rangle$ is said to be normalized if

$$\langle A \,|A \rangle = 1 \tag{8.4}$$

and the two kets are said to be orthogonal to each other if

$$\langle A \,|B \rangle = 0 \tag{8.5}$$

We may mention here the relationship between the Schrödinger wave functions developed in Chapter 3 to the bra and kets developed in this section. If $|\Psi\rangle$ and $|\Phi\rangle$ represent the kets corresponding to the states described by the wave functions $\psi\,(\mathbf{r})$ and $\varphi\,(\mathbf{r})$, respectively, then

$$\langle \Phi \,|\Psi \rangle = \int \varphi^*\,(\mathbf{r})\,\psi\,(\mathbf{r})\,\mathrm{d}\tau = \overline{\langle \Psi \,|\Phi \rangle} \tag{8.6}$$

8.3 Linear Operators

Let $\hat{\alpha}$ be a linear operator which, acting on $|A\rangle$, produces $|B\rangle$:

$$\hat{\alpha}\,|A\rangle = |B\rangle \tag{8.7}$$

The hat on $\hat{\alpha}$ represents the fact that $\hat{\alpha}$ is an operator. If $|B\rangle = 0$ for all possible $|A\rangle$, then $\hat{\alpha} = 0$, i.e., $\hat{\alpha}$ is a null operator. The operator is said to be linear if

$$\hat{\alpha}\,(C_1\,|A\rangle + C_2\,|B\rangle) = C_1\hat{\alpha}\,|A\rangle + C_2\hat{\alpha}\,|B\rangle \tag{8.8}$$

where C_1 and C_2 are any complex numbers and $|A\rangle$ and $|B\rangle$ are arbitrary kets. Two linear operators $\hat{\alpha}$ and $\hat{\beta}$ are said to be equal if

$$\langle A|\,\hat{\alpha}\,|A\rangle = \langle A|\,\hat{\beta}\,|A\rangle \tag{8.9}$$

for any $|A\rangle$. The addition and multiplication of two linear operators $\hat{\alpha}$ and $\hat{\beta}$ are defined through the equations

$$\left.\begin{array}{l}\left(\hat{\alpha}+\hat{\beta}\right)|A\rangle = \hat{\alpha}\,|A\rangle + \hat{\beta}\,|A\rangle \\[2mm] \left\{\hat{\alpha}\hat{\beta}\right\}|A\rangle = \hat{\alpha}\left\{\hat{\beta}\,|A\rangle\right\}\end{array}\right\} \quad \text{for any } |A\rangle \qquad (8.10)$$

An operator $\hat{\alpha}$ acting on the bra $\langle A|$ results in $\langle A|\,\hat{\alpha}$ and is defined through the equation

$$\left\{\langle A|\,\hat{\alpha}\right\}|B\rangle = \langle A|\left\{\hat{\alpha}\,|B\rangle\right\} \text{ for any } |B\rangle \qquad (8.11)$$

The adjoint of the operator $\hat{\alpha}$ is denoted by $\hat{\alpha}^{\dagger}$ and is defined through the following equation:

$$\langle A|\,\hat{\alpha}^{\dagger}\,|B\rangle = \overline{\langle B|\,\hat{\alpha}\,|A\rangle} \qquad (8.12)$$

where $\overline{\langle B|\,\hat{\alpha}\,|A\rangle}$ is the complex conjugate of the number $\langle B|\,\hat{\alpha}\,|A\rangle$. Now

$$\begin{aligned}\langle A|\,\hat{\alpha}^{\dagger\dagger}\,|B\rangle &= \langle A|\,\hat{\beta}^{\dagger}\,|B\rangle \;;\left(\hat{\beta}\equiv\hat{\alpha}^{\dagger}\right)\\[1mm] &= \overline{\langle B|\,\hat{\beta}\,|A\rangle}\\[1mm] &= \overline{\langle B|\,\hat{\alpha}^{\dagger}\,|A\rangle}\\[1mm] &= \overline{\overline{\langle A|\,\hat{\alpha}\,|B\rangle}}\\[1mm] &= \langle A|\,\hat{\alpha}\,|B\rangle\end{aligned} \qquad (8.13)$$

because the complex conjugate of the complex conjugate of a number is the original number itself. Since the above equation holds for arbitrary $|A\rangle$ and $|B\rangle$, we must have

$$\hat{\alpha}^{\dagger\dagger} = \hat{\alpha} \qquad (8.14)$$

implying that the adjoint of the adjoint of an operator is the original operator itself. If $\hat{\alpha}^{\dagger} = \hat{\alpha}$, then $\hat{\alpha}^{\dagger}$ is said to be a *real* or a *Hermitian* operator.

Let $\hat{\alpha}\,|A\rangle = |P\rangle$, then

$$\begin{aligned}\langle A|\,\hat{\alpha}^{\dagger}\,|B\rangle &= \overline{\langle B|\,\hat{\alpha}\,|A\rangle} = \overline{\langle B\,|P\rangle}\\[1mm] &= \langle P\,|B\rangle \text{ [using Eq.(8.2)]}\end{aligned} \qquad (8.15)$$

Since the above equation is valid for arbitrary $|B\rangle$ we have

$$\langle P| = \langle A|\,\hat{\alpha}^{\dagger} = \text{conjugate of } \hat{\alpha}\,|A\rangle \qquad (8.16)$$

We next consider two linear operators $\hat{\alpha}$ and $\hat{\beta}$ whose adjoints are denoted by $\hat{\alpha}^{\dagger}$ and $\hat{\beta}^{\dagger}$, respectively. Let

$$|P\rangle = \hat{\alpha}\hat{\beta}\,|A\rangle \qquad (8.17)$$

then

$$\langle P| = \langle A|\,(\hat{\alpha}\hat{\beta})^{\dagger} \tag{8.18}$$

Further, if $|Q\rangle = \hat{\beta}\,|A\rangle$, then $|P\rangle = \hat{\alpha}\,|Q\rangle$ and

$$\langle P| = \langle Q|\,\hat{\alpha}^{\dagger} = \langle A|\,\hat{\beta}^{\dagger}\hat{\alpha}^{\dagger} \tag{8.19}$$

Thus

$$(\hat{\alpha}\hat{\beta})^{\dagger} = \hat{\beta}^{\dagger}\hat{\alpha}^{\dagger} \tag{8.20}$$

and, in general,

$$(\hat{\alpha}\hat{\beta}\hat{\gamma}\,\ldots\ldots)^{\dagger} = \cdots\hat{\gamma}^{\dagger}\hat{\beta}^{\dagger}\hat{\alpha}^{\dagger} \tag{8.21}$$

It may be mentioned that the quantity

$$|P\rangle\,\langle Q| \tag{8.22}$$

is a linear operator, because when it acts on the ket $|A\rangle$ it produces the ket

$$C\,|P\rangle \tag{8.23}$$

where $C = \langle Q\,|A\rangle$ is a complex number. However, operations like $|P\rangle\,|Q\rangle$ or $\langle Q|\,\langle P|$ or $\alpha\,\langle Q|$ (where α is a linear operator) are meaningless.

The bras and kets follow the same axiomatic rules as row and column matrices, and linear operators follow the axiomatic rules of a square matrix. Thus the multiplication of a row matrix by a column matrix (cf. $\langle B\,|\,A\rangle$) gives a number, and multiplication of a column matrix by a row matrix (cf. $|A\rangle\,\langle B|$) gives a square matrix. The multiplication of a square matrix by a column matrix (cf. $\hat{\alpha}\,|P\rangle$) gives another column matrix, whereas multiplications of a square matrix by a row matrix (cf. $\hat{\alpha}\,\langle P|$) or of two column matrices (cf. $|P\rangle\,|Q\rangle$) are meaningless.

8.4 The Eigenvalue Equation

The equation

$$\hat{\alpha}\,|P\rangle = a\,|P\rangle \tag{8.24}$$

where a is a complex number, defines an eigenvalue equation; $|P\rangle$ is said to be an eigenket of the operator $\hat{\alpha}$ belonging to the eigenvalue a. We premultiply Eq. (8.24) by $\langle P|$ to obtain

$$\langle P|\,\hat{\alpha}\,|P\rangle = a\,\langle P\,|P\rangle \tag{8.25}$$

If we assume $\hat{\alpha}$ to be a real operator, then by taking the complex conjugate of both sides we get

$$a^* \langle P|P \rangle = \overline{\langle P|\,\hat{\alpha}\,|P \rangle} = \langle P|\,\hat{\alpha}^\dagger\,|P \rangle = \langle P|\,\hat{\alpha}\,|P \rangle = a \langle P|P \rangle \qquad (8.26)$$

where we have used Eq. (8.14) and the fact that $\hat{\alpha}^\dagger = \hat{\alpha}$. Since $\langle P|P \rangle$ is always positive (unless $|P \rangle$ is a null ket – which corresponds to the trivial solution) we obtain that a is a real number. Thus, all eigenvalues of a real linear operator are real.

We next show that eigenkets (of a real linear operator) belonging to different eigenvalues are necessarily orthogonal. The conjugate of Eq. (8.24) gives us

$$\langle P|\,\hat{\alpha} = a^* \langle P| = a \langle P| \qquad (8.27)$$

because a is just a real number. Now, if $|Q \rangle$ is another eigenket of $\hat{\alpha}$ belonging to the eigenvalue b then

$$\hat{\alpha}\,|Q \rangle = b\,|Q \rangle \qquad (8.28)$$

Thus, postmultiplying Eq. (8.27) by $|Q \rangle$ and premultiplying Eq. (8.28) by $\langle P|$ we get

$$\langle P|\,\hat{\alpha}\,|Q \rangle = a \langle P|Q \rangle = b \langle P|Q \rangle \qquad (8.29)$$

Since $b \neq a$, we get

$$\langle P|Q \rangle = 0 \qquad (8.30)$$

implying that two eigenkets of a real linear operator (belonging to different eigenvalues) are orthogonal. Further, if $|P_1 \rangle$ and $|P_2 \rangle$ are eigenkets of $\hat{\alpha}$ belonging to the same eigenvalue a, i.e.,

$$\hat{\alpha}\,|P_1 \rangle = a\,|P_1 \rangle \quad \text{and} \quad \hat{\alpha}\,|P_2 \rangle = a\,|P_2 \rangle$$

then

$$\hat{\alpha}\,(C_1\,|P_1 \rangle + C_2\,|P_2 \rangle) = a\,(C_1\,|P_1 \rangle + C_2\,|P_2 \rangle)$$

implying that the ket $C_1\,|P_1 \rangle + C_2\,|P_2 \rangle$ is also an eigenket belonging to the *same* eigenvalue a. Consequently, we can always choose appropriate linear combinations so that the eigenkets form an orthonormal set.

8.5 Observables

We assume that all measurable quantities (like position, momentum, energy) can be represented by linear operators. Further, if one makes a precise measurement of such a quantity, one gets one of the eigenvalues of the corresponding linear operator.

The operator corresponding to the observable must be real (Hermitian) because the result of measurement of any observable must be a real number; however, not every linear operator need to correspond to an observable. Finally, if a system is in a state described by $|P\rangle$, then the expectation value of an observable $\hat{\alpha}$ is given by

$$\langle\hat{\alpha}\rangle = \langle P|\,\hat{\alpha}\,|P\rangle \tag{8.31}$$

The above equation may be compared with Eq. (3.101). And, if $|P\rangle$ is such that

$$|P\rangle = C_1\,|\alpha_1\rangle + C_2\,|\alpha_2\rangle \tag{8.32}$$

where $|\alpha_1\rangle$ and $|\alpha_2\rangle$ are the normalized eigenkets of $\hat{\alpha}$ belonging to the eigenvalues α_1 and α_2, respectively, then a measurement of an observable represented by the operator $\hat{\alpha}$ on $|P\rangle$ would lead to either α_1 or α_2 with probabilities $|C_1|^2$ and $|C_2|^2$, respectively; here we have assumed $|P\rangle$ to be normalized (i.e., $\langle P|P\rangle = 1$) implying

$$|C_1|^2 + |C_2|^2 = 1 \tag{8.33}$$

In general, if the system is in any state then a measurement of $\hat{\alpha}$ will definitely lead to one of its eigenvalues. Thus any ket $|P\rangle$ can be expressed as a linear combination of the eigenkets of the observable:

$$|P\rangle = \sum_n C_n\,|\alpha_n\rangle \tag{8.34}$$

implying that the eigenkets of an observable form a complete set.

8.6 The Harmonic Oscillator Problem

In Section 3.2 we solved the Schrödinger equation for the linear harmonic oscillator problem. In this section we will use Dirac algebra to solve the harmonic oscillator problem for which the Hamiltonian is given by

$$\hat{H} = \frac{\hat{p}^2}{2m} + \frac{1}{2}m\omega^2\hat{x}^2 \tag{8.35}$$

Since $\hat{H}, \hat{p}$, and $\hat{x}$ are observables,

$$\hat{H} = \hat{H}^\dagger, \quad \hat{p} = \hat{p}^\dagger, \quad \hat{x} = \hat{x}^\dagger \tag{8.36}$$

Further, the operators $\hat{x}$ and $\hat{p}$ will be assumed to satisfy the commutation relation (see Section 3.6):

$$[\hat{x},\hat{p}] = \hat{x}\hat{p} - \hat{p}\hat{x} = i\hbar \tag{8.37}$$

Our objective is to solve the eigenvalue equation

$$\hat{H}\,|H'\rangle = H'\,|H'\rangle \tag{8.38}$$

where $|H'\rangle$ is the eigenket of the operator $\hat{H}$ belonging to the eigenvalue H'.
   It is convenient to introduce the dimensionless complex operator

$$\hat{a} = \frac{1}{(2m\hbar\omega)^{1/2}}\left(m\omega\hat{x} + i\hat{p}\right) \tag{8.39}$$

The adjoint of $\hat{a}$ would be given by

$$\hat{a}^{\dagger} = \frac{1}{(2m\hbar\omega)^{1/2}}\left(m\omega\hat{x} - i\hat{p}\right) \tag{8.40}$$

where we have used Eq. (8.36). In terms of the above operators

$$\begin{aligned}
\hbar\omega\hat{a}\hat{a}^{\dagger} &= \frac{1}{2m}\left(m\omega\hat{x} + i\hat{p}\right)\left(m\omega\hat{x} - i\hat{p}\right) \\
&= \frac{1}{2m}\left[m\omega^2\hat{x}^2 + \hat{p}^2 - im\omega\left(\hat{x}\hat{p} - \hat{p}\hat{x}\right)\right] \\
&= \hat{H} + \frac{1}{2}\hbar\omega
\end{aligned} \tag{8.41}$$

where we have used Eqs. (8.35) and (8.37). Similarly

$$\hbar\omega\hat{a}^{\dagger}\hat{a} = \hat{H} - \frac{1}{2}\hbar\omega \tag{8.42}$$

Thus

$$\hat{H} = \frac{1}{2}\hbar\omega\left(\hat{a}^{\dagger}\hat{a} + \hat{a}\hat{a}^{\dagger}\right) \tag{8.43}$$

and

$$\hat{a}\hat{a}^{\dagger} - \hat{a}^{\dagger}\hat{a} = \left[\hat{a}, \hat{a}^{\dagger}\right] = 1 \tag{8.44}$$

From Eq. (8.41)

$$\hbar\omega\hat{a}\hat{a}^{\dagger}\hat{a} = \hat{H}\hat{a} + \frac{1}{2}\hbar\omega\hat{a} \tag{8.45}$$

and from Eq. (8.42)

$$\hbar\omega\hat{a}\hat{a}^{\dagger}\hat{a} = \hat{a}\hat{H} - \frac{1}{2}\hbar\omega\hat{a} \tag{8.46}$$

Thus

$$\hat{a}\hat{H} - \hat{H}\hat{a} = \left[\hat{a}, \hat{H}\right] = \hbar\omega\hat{a} \tag{8.47}$$

Similarly

$$\hat{a}^{\dagger}\hat{H} - \hat{H}\hat{a}^{\dagger} = \left[\hat{a}^{\dagger}, \hat{H}\right] = -\hbar\omega\hat{a}^{\dagger} \tag{8.48}$$

Let

$$|P\rangle = \hat{a}\,|H'\rangle$$

where $|H'\rangle$ is an eigenket of H belonging to the eigenvalue H' [see Eq. (8.38)]. Then

$$\begin{aligned}
\hbar\omega\,\langle P|P\rangle &= \hbar\omega\,\langle H'|\,\hat{a}^{\dagger}\hat{a}\,|H'\rangle \\
&= \langle H'|\,\hat{H} - \frac{1}{2}\hbar\omega\,|H'\rangle \qquad \text{[using Eq. (8.42)]} \\
&= \left(H' - \frac{1}{2}\hbar\omega\right)\langle H'\,|H'\rangle \quad \text{[using Eq. (8.38)]}
\end{aligned}$$

But $\langle P|P\rangle$ and $\langle H'\,|H'\rangle$ are positive numbers [see Eq. (8.3)] and therefore

$$H' \geq \frac{1}{2}\hbar\omega \tag{8.49}$$

and $H' = \frac{1}{2}\hbar\omega$ if and only if $|P\rangle = \hat{a}\,|H'\rangle = 0$ (conversely, $\hat{a}\,|H'\rangle$ is also a null ket only when $H' = \frac{1}{2}\hbar\omega$). That H' should be positive follows from Eq. (8.35) and the fact that the expectation values of $\hat{x}^2$ and $\hat{p}^2$ should be positive or zero for any state of the system.

Next, let us consider the operator $\hat{H}\hat{a}$ operating on $|H'\rangle$:

$$\begin{aligned}
\hat{H}\hat{a}\,|H'\rangle &= \left(\hat{a}\hat{H} - \hbar\omega\hat{a}\right)|H'\rangle \\
&= \left(\hat{a}H' - \hbar\omega\hat{a}\right)|H'\rangle \\
&= \left(H' - \hbar\omega\right)\hat{a}\,|H'\rangle
\end{aligned} \tag{8.50}$$

Thus, if $|H'\rangle$ is an eigenket of $\hat{H}$ then $\hat{a}\,|H'\rangle$ is also an eigenket of $\hat{H}$ belonging to the eigenvalue $H'-\hbar\omega$ provided, of course, $\hat{a}\,|H'\rangle$ is not a null ket, which will occur only if $H' = \frac{1}{2}\hbar\omega$. Thus if $H' \neq \frac{1}{2}\hbar\omega$, $H' - \hbar\omega$ is also an eigenvalue (provided $\hat{a}\,|H'\rangle \neq 0$). Similarly if $H'-\hbar\omega \neq \frac{1}{2}\hbar\omega$ then $H'-2\hbar\omega$ is also an eigenvalue of $\hat{H}$. We can thus say that $H' - \hbar\omega$, $H' - 2\hbar\omega, \ldots$ are also eigenvalues provided $\hat{a}\,|H'\rangle, \hat{a}\hat{a}\,|H'\rangle, \ldots$ are not null kets. This, however, cannot go on indefinitely because it will then contradict Eq. (8.49). Further, it can terminate only at $H' = \frac{1}{2}\hbar\omega$ because then $\hat{a}\left|\frac{1}{2}\hbar\omega\right\rangle = 0$.

Now, using Eq. (8.48)

$$\hat{H}\hat{a}^{\dagger}\,|H'\rangle = \left(\hat{a}^{\dagger}\hat{H} + \hbar\omega\hat{a}\right)\,|H'\rangle$$

$$= \left(\hat{a}^{\dagger}H' + \hbar\omega\hat{a}^{\dagger}\right)\,|H'\rangle \tag{8.51}$$

$$= \left(H' + \hbar\omega\right)\hat{a}^{\dagger}\,|H'\rangle$$

implying that $\left(H' + \hbar\omega\right)$ is another eigenvalue of $\hat{H}$, with $\hat{a}^{\dagger}\,|H'\rangle$ as the eigenket belonging to it, unless $\hat{a}^{\dagger}\,|H'\rangle = 0$. However, $\hat{a}^{\dagger}\,|H'\rangle$ can never be equal to zero, since it would lead to

$$0 = \hbar\omega\hat{a}\hat{a}^{\dagger}\,|H'\rangle = \left(\hat{H} + \frac{1}{2}\hbar\omega\right)\,|H'\rangle \qquad \text{[using Eq. (8.41)]}$$

$$= \left(H' + \frac{1}{2}\hbar\omega\right)\,|H'\rangle$$

giving $H' = -\frac{1}{2}\hbar\omega$, which contradicts Eq. (8.49). Thus if H' is an eigenvalue, then $H' + \hbar\omega$ is always another eigenvalue of $\hat{H}$, and so are

$$H' + 2\hbar\omega, \quad H' + 3\hbar\omega, \ldots$$

Hence the eigenvalues of the Hamiltonian for the linear harmonic oscillator problem are

$$\frac{1}{2}\hbar\omega, \quad \frac{3}{2}\hbar\omega, \quad \frac{5}{2}\hbar\omega, \quad \frac{7}{2}\hbar\omega, \ldots \tag{8.52}$$

extending to infinity, which is the same as obtained in Section 3.2.

We now relabel the eigenfunctions with the index n; thus $|n\rangle$ denotes the eigenfunction corresponding to the eigenvalues $\left(n + \frac{1}{2}\right)\hbar\omega$:

$$\hat{H}\,|n\rangle = \left(n + \frac{1}{2}\right)\hbar\omega\,|n\rangle; \quad n = 0, 1, 2, 3, \ldots \tag{8.53}$$

The eigenkets corresponding to different eigenvalues will necessarily be orthogonal. We assume that the states $|n\rangle$ are normalized, so that

$$\langle m\,|n\rangle = \delta_{mn} \tag{8.54}$$

Further, since $|0\rangle$ corresponds to $H' = \frac{1}{2}\hbar\omega$, we must have

$$\hat{a}\,|0\rangle = 0 \tag{8.55}$$

Now, for $n = 1, 2, 3, \ldots$, $\hat{a}\,|n\rangle$ is an eigenket of $\hat{H}$ belonging to the eigenvalue $n - \frac{1}{2}\hbar\omega$ [see Eq. (8.50)]; therefore $\hat{a}\,|n\rangle$ must be a multiple of $|n-1\rangle$:

$$\hat{a}\,|n\rangle = \alpha_n\,|n-1\rangle \tag{8.56}$$

In order to determine α_n we calculate the square of the length of $\hat{a}\,|n\rangle$:

$$\langle n|\,\hat{a}^\dagger \hat{a}\,|n\rangle = |\alpha_n|^2\,\langle n-1\,|\,n-1\rangle = |\alpha_n|^2$$

But

$$\hbar\omega\,\langle n|\,\hat{a}^\dagger\hat{a}\,|n\rangle = \langle n|\left(\hat{H} - \frac{1}{2}\hbar\omega\right)|n\rangle$$

$$= \langle n|\left(n + \frac{1}{2}\right)\hbar\omega - \frac{1}{2}\hbar\omega\,|n\rangle$$

$$= n\hbar\omega\,\langle n\,|\,n\rangle = n\hbar\omega$$

Thus

$$|\alpha_n|^2 = n$$

and therefore

$$\hat{a}\,|n\rangle = \sqrt{n}\,|n-1\rangle \tag{8.57}$$

Similarly

$$\hat{a}^\dagger\,|n\rangle = \sqrt{n+1}\,|n+1\rangle \tag{8.58}$$

Thus, if $|0\rangle$ denotes the ground-state eigenket, then

$$|1\rangle = \frac{\hat{a}^\dagger}{\sqrt{1}}\,|0\rangle\,, \quad |2\rangle = \frac{\hat{a}^\dagger\hat{a}^\dagger}{\sqrt{1.2}}\,|0\rangle = \frac{\left(\hat{a}^\dagger\right)^2}{\sqrt{2!}}\,|0\rangle\,, \quad |2\rangle = \frac{\left(\hat{a}^\dagger\right)^3}{\sqrt{3!}}\,|0\rangle\,,\ldots$$

and, in general,

$$|n\rangle = \frac{\left(\hat{a}^\dagger\right)^n}{\sqrt{n!}}\,|0\rangle \tag{8.59}$$

We can think of an excited state $|n\rangle$ of the oscillator as containing n quanta of energy $\hbar\omega$ in addition to the zero-point energy of $\frac{1}{2}\hbar\omega$. The operator $\hat{a}^\dagger$, according to Eq. (8.58), creates a quantum of energy; and, therefore, $\hat{a}^\dagger$ is called the *creation* operator. Similarly, the operator $\hat{a}$, according to Eq. (8.57), annihilates a quantum of energy, and therefore, $\hat{a}$ is called the *annihilation* or *destruction* operator.

8.6.1 The Number Operator

Consider the operator

$$\hat{N}_{op} = \hat{a}^{\dagger}\hat{a} \tag{8.60}$$

Using Eq. (8.42) we may write

$$\hat{H} = \left(\hat{N}_{op} + \frac{1}{2}\right)\hbar\omega \tag{8.61}$$

Since

$$\hat{H}\,|n\rangle = \left(n + \frac{1}{2}\right)\hbar\omega\,|n\rangle \tag{8.62}$$

we have

$$\hbar\omega\left(\hat{N}_{op} + \frac{1}{2}\right)|n\rangle = \left(n + \frac{1}{2}\right)\hbar\omega\,|n\rangle$$

or

$$\hat{N}_{op}\,|n\rangle = n\,|n\rangle \tag{8.63}$$

Thus $|n\rangle$ are also the eigenkets of $\hat{N}_{op}$, the corresponding eigenvalue being n, and since n takes the values 0, 1, 2,..., the operator $\hat{N}_{op}$ is called the number operator. Obviously

$$\langle m|\,\hat{N}_{op}\,|n\rangle = n\delta_{mn} \tag{8.64}$$

8.6.2 The Uncertainty Product

The quantities Δx and Δp, which represent the uncertainties in $\hat{x}$ and $\hat{p}$, are defined through the equations

$$\Delta x = \left(\langle\hat{x}^2\rangle - \langle\hat{x}\rangle^2\right)^{1/2} \tag{8.65}$$

$$\Delta p = \left(\langle\hat{p}^2\rangle - \langle\hat{p}\rangle^2\right)^{1/2} \tag{8.66}$$

where $\langle\hat{x}^2\rangle$ represents the expectation value of $\hat{x}^2$, etc. We will calculate the expectation values of $\hat{x}$, $\hat{x}^2$, etc., when the harmonic oscillator is in the state $|n\rangle$. Now

$$\langle n|\,\hat{x}\,|n\rangle = \left(\frac{\hbar}{2m\omega}\right)^{1/2}\langle n|\,(\hat{a} + \hat{a}^{\dagger})\,|n\rangle \qquad \text{[using Eqs. (8.39) and (8.40)]}$$

$$= \left(\frac{\hbar}{2m\omega}\right)^{1/2}\left[\sqrt{n}\,\langle n\,|n-1\rangle + \sqrt{n+1}\,\langle n\,|n+1\rangle\right] \tag{8.67}$$

$$\qquad \text{[using Eqs. (8.57) and (8.58)]}$$

$$= 0 \quad \text{[using Eqs. (8.54)]}$$

and

$$\langle n| \hat{x}^2 |n\rangle = \left(\langle n| \hat{a}\hat{a} |n\rangle + \langle n| \hat{a}\hat{a}^\dagger |n\rangle + \langle n| \hat{a}^\dagger\hat{a} |n\rangle + \langle n| \hat{a}^\dagger\hat{a}^\dagger |n\rangle \right)$$

$$= \frac{\hbar}{2m\omega} \left[0 + (n+1) + n + 0 \right]$$

$$= \frac{\hbar}{m\omega} \left(n + \frac{1}{2} \right)$$

Thus

$$\Delta x = \sqrt{ \frac{\hbar}{m\omega} \left(n + \frac{1}{2} \right) } \tag{8.68}$$

Similarly

$$\langle n| \hat{p} |n\rangle = 0 \tag{8.69}$$

and

$$\Delta p = \sqrt{ \langle n| p^2 |n\rangle } = \sqrt{ m\omega\hbar \left(n + \frac{1}{2} \right) } \tag{8.70}$$

Thus

$$\Delta x \Delta p = \left(n + \frac{1}{2} \right) \hbar \tag{8.71}$$

The minimum uncertainty product $\left(= \frac{1}{2}\hbar \right)$ occurs for the ground state ($n = 0$). The result given by Eq. (8.71) is consistent with the uncertainty principle.

8.6.3 The Coherent States

Consider the eigenvalue equation

$$\hat{a} |\alpha\rangle = \alpha |\alpha\rangle \tag{8.72}$$

where $\hat{a}$ is the annihilation operator defined through Eq. (8.39). The eigenkets defined by Eq. (8.72) are known as the coherent states.[2] In this section we will study some of the properties of the coherent states; these properties will be used in Chapter 9.

---

[2]In Section 9.4 we will show that when a laser is operated much beyond the threshold, it generates a coherent-state excitation of the cavity mode. It is left as an exercise for the reader to show that the operator $\hat{a}^\dagger$ cannot have any eigenkets and similarly $\hat{a}$ cannot have any eigenbras.

Since the eigenkets of an observable form a complete set, we expand $|\alpha\rangle$ in terms of the kets $|n\rangle$:

$$|\alpha\rangle = \sum_{n=0,1\ldots} C_n |n\rangle \tag{8.73}$$

Now

$$\hat{a} |\alpha\rangle = \sum C_n \hat{a} |n\rangle = \sum_{n=1}^{\infty} C_n \sqrt{n} \, |n-1\rangle \tag{8.74}$$

Also

$$\hat{a} |\alpha\rangle = \alpha |\alpha\rangle = \alpha \sum C_n |n\rangle \tag{8.75}$$

Thus

$$\alpha \left(C_0 |0\rangle + C_1 |1\rangle + \cdots \right) = C_1 |0\rangle + C_2 \sqrt{2} \, |1\rangle + C_3 \sqrt{3} \, |2\rangle + \cdots$$

or

$$C_1 = \alpha C_0, \quad C_2 = \frac{\alpha C_1}{\sqrt{2}} = \frac{\alpha^2}{\sqrt{2}} C_0$$

$$C_3 = \alpha \frac{C_2}{\sqrt{3}} = \frac{\alpha^3}{\sqrt{3!}} C_0, \ldots \tag{8.76}$$

In general,

$$C_n = \frac{\alpha^n}{\sqrt{n!}} C_0 \tag{8.77}$$

Thus

$$|\alpha\rangle = C_0 \sum_n \frac{\alpha^n}{\sqrt{n!}} |n\rangle \tag{8.78}$$

If we normalize $|\alpha\rangle$, we would get

$$1 = \langle \alpha | \alpha \rangle = |C_0|^2 \sum_n \sum_m \frac{\alpha^n \alpha^{*m}}{\sqrt{n!}\sqrt{m!}} \delta_{nm}$$

$$= |C_0|^2 \sum_n \frac{\left(|\alpha|^2\right)^n}{n!} = |C_0|^2 \exp\left(|\alpha|^2\right)$$

or

$$C_0 = \exp\left(-\frac{1}{2} |\alpha|^2\right) \tag{8.79}$$

within an arbitrary phase factor. Substituting in Eq. (8.78) we obtain

$$|\alpha\rangle = \exp\left(-\frac{1}{2} |\alpha|^2\right) \sum_n \frac{\alpha^n}{\sqrt{n!}} |n\rangle \tag{8.80}$$

Notice that there is no restriction on the value of α, i.e., α can take *any* complex value. Further, if $|\beta\rangle$ is another eigenket of $\hat{a}$ belonging to the eigenvalue β, then

$$
\begin{aligned}
|\langle\alpha\,|\beta\rangle|^2 &= \left|\exp\left(-\frac{1}{2}|\alpha|^2\right)\exp\left(-\frac{1}{2}|\beta|^2\right)\sum_n\sum_m\frac{\alpha^{*n}\beta^m}{\sqrt{n!m!}}\,\langle n\,|m\rangle\right|^2 \\
&= \exp\left(-|\alpha|^2-|\beta|^2\right)\left|\sum_n\frac{(\alpha^*\beta)^n}{n!}\right|^2 \\
&= \exp\left(-|\alpha|^2-|\beta|^2+\alpha^*\beta+\alpha\beta^*\right)=\exp\left(-|\alpha-\beta|^2\right)
\end{aligned}
\tag{8.81}
$$

Thus the eigenkets are not orthogonal (this is because $\hat{a}$ is not a real operator); they, however, become approximately orthogonal for large values of $|\alpha-\beta|^2$. Further, the kets $|\alpha\rangle$ can be shown to satisfy the following relations:

$$
|\langle n\,|\alpha\rangle|^2 = \frac{1}{n!}\left|\alpha^2\right|^n\exp\left(-|\alpha|^2\right)
\tag{8.82}
$$

$$
\Delta x\Delta p = \frac{1}{2}\hbar
\tag{8.83}
$$

where

$$
\Delta x \equiv \sqrt{\langle\alpha|\,\hat{x}^2\,|\alpha\rangle-\langle\alpha|\,\hat{x}\,|\alpha\rangle^2}
\tag{8.84}
$$

and

$$
\Delta p \equiv \sqrt{\langle\alpha|\,\hat{p}^2\,|\alpha\rangle-\langle\alpha|\,\hat{p}\,|\alpha\rangle^2}
\tag{8.85}
$$

Thus the uncertainty product $\Delta x\,\Delta p$ has the minimum value $\left(=\frac{1}{2}\hbar\right)$ for *all* coherent states. Indeed it can be shown that if we solve the eigenvalue equation

$$
\hat{a}\psi = \frac{1}{(2m\hbar\omega)^{1/2}}\left(m\omega\hat{x}+i\hat{p}\right)\psi = \alpha\psi
\tag{8.86}
$$

by replacing $\hat{p}$ by $-i\hbar d/dx$ [see Eq. (3.9)] then the eigenfunctions would be displaced Gaussian functions and the uncertainty product is a minimum for the Gaussian function (see, e.g., Ghatak and Lokanathan (2004)). It may be pointed out that the ground state ($n=0$) wave function for the harmonic oscillator problem is also Gaussian [see Eq. (3.54) with $H_0(\xi)=1$]; the corresponding uncertainty product is a minimum [see Eq. (8.71) with $n=0$].

8.7 Time Development of States

Let us consider a system described by the *time-independent* Hamiltonian $\hat{H}$. Let $|n\rangle$ represent the eigenkets of $\hat{H}$ belonging to the eigenvalue E_n:

$$\hat{H}\,|n\rangle = E_n\,|n\rangle \tag{8.87}$$

Now, since the eigenkets form a complete set, we may express an arbitrary ket $|\psi\rangle$ as a linear combination of $|n\rangle$:

$$|\psi\rangle = \sum_n C_n\,|n\rangle \tag{8.88}$$

The constants C_n can be obtained by premultiplying the above equation by

$$\langle m\,|\psi\rangle = \sum C_n\,\langle m\,|n\rangle = \sum C_n \delta_{mn} = C_m \tag{8.89}$$

where we have used the orthonormality condition satisfied by the eigenkets. Thus

$$|\psi\rangle = \sum |n\rangle\,\langle n\,|\psi\rangle \tag{8.90}$$

which may be rewritten as

$$|\psi\rangle = \left\{ \sum_n |n\rangle\,\langle n| \right\}\,|\psi\rangle \tag{8.91}$$

Since the above equation is valid for an arbitrary ket, the quantity inside the curly brackets must be the unit operator:

$$\sum_n |n\rangle\,\langle n| = 1 \tag{8.92}$$

Equation (8.92) is often referred to as the completeness condition.

Next, we are interested in finding out how a system (described by the time-independent Hamiltonian $\hat{H}$) will evolve with time, if the state of the system, at $t = 0$, is described by the ket $|\psi\,(0)\rangle$. Now, the ket $|\psi\,(t)\rangle$ (which describes the evolution of the system with time) would satisfy the time-dependent Schrödinger equation [Eq. (3.61)]:

$$i\hbar\frac{\partial}{\partial t}\,|\Psi\,(t)\rangle = \hat{H}\,|\Psi\,(t)\rangle \tag{8.93}$$

Since the Hamiltonian is independent of time, we can "integrate" the above equation to obtain

$$|\Psi\,(t)\rangle = e^{-i\hat{H}t/\hbar}\,|\Psi\,(0)\rangle \tag{8.94}$$

where the exponential of an operator is *defined* through the power series:

$$e^{\hat{O}} = 1 + \hat{O} + \frac{1}{2!}\hat{O}\hat{O} + \frac{1}{3!}\hat{O}\hat{O}\hat{O} + \cdots \tag{8.95}$$

That Eq. (8.94) is the solution of Eq. (8.93) can be immediately seen by direct substitution. Using Eq. (8.91), we obtain

$$|\Psi(t)\rangle = e^{-i\hat{H}t/\hbar}\sum_n |n\rangle\,\langle n\,|\Psi(0)\rangle \tag{8.96}$$

Replacing $e^{-i\hat{H}t/\hbar}$ by the power series and using Eq. (8.87), we get

$$|\Psi(t)\rangle = \sum_n e^{-iE_n t/\hbar}\,|n\rangle\,\langle n\,|\Psi(0)\rangle \tag{8.97}$$

We may use the above equation to study the time development of coherent states. Thus,

$$|\Psi(0)\rangle = |\alpha\rangle = \exp\left(-\frac{1}{2}|\alpha|^2\right)\sum_n \frac{\alpha^n}{\sqrt{n!}}\,|n\rangle \tag{8.98}$$

[see Eq. (8.80)]. Thus

$$\langle n\,|\Psi(0)\rangle = \langle n\,|\alpha\rangle = \exp\left(-\frac{1}{2}|\alpha|^2\right)\frac{\alpha^n}{\sqrt{n!}} \tag{8.99}$$

Hence

$$|\Psi(t)\rangle = \exp\left(-\frac{1}{2}|\alpha|^2\right)\sum_n\left[\frac{\alpha^n}{\sqrt{n!}}\right]\exp\left[-i\left(n+\frac{1}{2}\right)\omega t\right]|n\rangle \tag{8.100}$$

where we have used the fact that

$$E_n = \left(n+\frac{1}{2}\right)\hbar\omega \tag{8.101}$$

8.8 The Density Operator

Let $|0\rangle$, $|1\rangle$, $|2\rangle$, ... form a complete set of orthonormal kets, i.e.,

$$\langle n\,|m\rangle = \delta_{mn} \tag{8.102}$$

and

$$\sum_n |n\rangle\,\langle n| = 1 \tag{8.103}$$

[see Eqs. (8.54) and (8.92)]. An arbitrary ket can be expanded in terms of $|n\rangle$:

$$|P\rangle = \sum_n C_n\,|n\rangle, \quad C_n = \langle n\,|P\rangle \tag{8.104}$$

A state can be characterized by the density operator $\hat{\rho}$ defined by the following equation:

$$\hat{\rho} = |P\rangle \langle P| \tag{8.105}$$

The trace[3] of an operator is defined to be equal to the sum of the diagonal matrix elements for any complete set of states; thus

$$\mathrm{Tr}\hat{O} = \sum_n \langle n| \hat{O} |n\rangle \tag{8.106}$$

where $\hat{O}$ is an arbitrary operator and

$$O_{nm} \equiv \langle n| \hat{O} |m\rangle \tag{8.107}$$

is known as the (nm)th matrix element of the operator $\hat{O}$ with respect to kets $|0\rangle, |1\rangle, \ldots$ being the basis states; the $n = m$ terms represent the diagonal elements. Now

$$\mathrm{Tr}\,(|P\rangle \langle Q|) = \sum_n \langle n|P\rangle \langle Q|n\rangle$$

$$= \sum_n \langle Q|n\rangle \langle n|P\rangle$$

[because $\langle n|P\rangle$ and $\langle Q|n\rangle$ are complex numbers]

$$= \langle Q| \left\{ \sum_n |n\rangle \langle n| \right\} |P\rangle \tag{8.108}$$

or

$$\mathrm{Tr}\,(|P\rangle \langle Q|) = \langle Q|P\rangle \tag{8.109}$$

where, in the last step, we have used Eq. (8.103). Thus

$$\mathrm{Tr}\,\rho = \mathrm{Tr}\,|P\rangle \langle P| = \langle P|P\rangle = 1 \tag{8.110}$$

where we have assumed $|P\rangle$ to be normalized. Further, the expectation value of the operator $\hat{O}$ (when the system is in the state $|P\rangle$) is given by

$$\left\langle \hat{O} \right\rangle = \langle P| \hat{O} |P\rangle = \sum_n \langle P|\hat{O}|n\rangle \langle n|P\rangle \quad \text{[using Eq. (8.103)]}$$

$$= \sum_n \langle n|P\rangle \langle P|\hat{O}|n\rangle \quad \text{[because } \langle n|P\rangle \text{ is a number]}$$

$$= \sum_n \langle n| \hat{\rho} \hat{O} |n\rangle \quad \text{[using Eq. (8.105)]}$$

---

[3] Abbreviated as Tr.

or

$$\left\langle \hat{O} \right\rangle = \mathrm{Tr}\left(\hat{\rho}\,\hat{O} \right) \tag{8.111}$$

Also

$$\rho_{nm} = \langle n \,|P \rangle \, \langle P\,|n \rangle = |C_n|^2 \tag{8.112}$$

implying that the diagonal matrix elements of the density operator represent the probabilities of finding the system in the basis states.

Perhaps the most important application of the density operator is in the field of statistical mechanics where we consider a large number of identical systems, each system having a certain probability of being in a certain state. If w_ψ represents the probability of finding the system in the state characterized by $|\psi\rangle$, then the corresponding density operator is given by

$$\hat{\rho} = \sum_\Psi w_\Psi \, |\Psi\rangle \, \langle\Psi| \tag{8.113}$$

where the summation is carried over all possible states of the system; the density operator contains all the information about the ensemble. Since

$$\sum_\Psi w_\Psi = 1 \tag{8.114}$$

we obtain

$$\begin{aligned}
\mathrm{Tr}\,\hat{\rho} &= \sum_\Psi w_\Psi \,\mathrm{Tr}\, |\Psi\rangle \, \langle\Psi| \\
&= \sum_\Psi w_\Psi \, \langle\Psi\,|\,\Psi\rangle \\
&= \sum_\Psi w_\Psi = 1
\end{aligned} \tag{8.115}$$

Equation of Motion of the Density Operator

We calculate the time dependence of the density operator by differentiating Eq. (8.113) with respect to time:

$$i\hbar\frac{\mathrm{d}\hat{\rho}}{\mathrm{d}t} = \sum_\Psi w_\Psi \left\{ i\hbar\frac{\mathrm{d}\,|\Psi\rangle}{\mathrm{d}t} \, \langle\Psi| + |\Psi\rangle \left[i\hbar\frac{\mathrm{d}}{\mathrm{d}t} \, \langle\Psi| \right] \right\} \tag{8.116}$$

But

$$i\hbar\frac{\mathrm{d}}{\mathrm{d}t} \, |\Psi\rangle = \hat{H} \, |\Psi\rangle \tag{8.117}$$

and taking its conjugate imaginary

$$-i\hbar\frac{\mathrm{d}}{\mathrm{d}t} \, \langle\Psi| = \langle\Psi|\,\hat{H} \tag{8.118}$$

Thus

$$i\hbar\frac{d\hat{\rho}}{dt} = \sum_{\Psi} w_{\Psi} \left[\hat{H} \, |\Psi\rangle \, \langle\Psi| - |\Psi\rangle \, \langle\Psi| \hat{H} \right]$$

$$= \hat{H}\hat{\rho} - \hat{\rho}\hat{H} \tag{8.119}$$

or

$$i\hbar\frac{d\hat{\rho}}{dt} = -\left[\hat{\rho},\hat{H}\right] \tag{8.120}$$

8.9 The Schrödinger and Heisenberg Pictures

While solving the linear harmonic oscillator problem in Section 8.6, we had assumed the observables x, p, and H to be real operators and independent of time. This is the so-called Schrödinger picture, and the time development of the ket describing the quantum mechanical system is obtained by solving the time-dependent Schrödinger equation:

$$i\hbar\frac{\partial}{\partial t} |\Psi(t)\rangle = \hat{H} |\Psi(t)\rangle \tag{8.121}$$

If the Hamiltonian is independent of time then we can "integrate" the above equation to obtain (see Section 8.7)

$$|\Psi(t)\rangle = e^{-i\hat{H}t/\hbar} |\Psi(0)\rangle \tag{8.122}$$

where

$$e^{-i\hat{H}t/\hbar} \equiv 1 - \left(\frac{i\hat{H}t}{\hbar}\right) + \frac{1}{2!}\left(\frac{i\hat{H}t}{\hbar}\right)\left(\frac{i\hat{H}t}{\hbar}\right) + \cdots \tag{8.123}$$

Now, the expectation value of an observable characterized by the operator $\hat{O}$ is given by

$$\left\langle\hat{O}\right\rangle = \langle\Psi(t)| \, \hat{O} \, |\Psi(t)\rangle \tag{8.124}$$

Next, let $|n\rangle$ represent the eigenkets of the Hamiltonian $\hat{H}$ belonging to the eigenvalue E_n, i.e.,

$$\hat{H}|n\rangle = E_n |n\rangle \tag{8.125}$$

then [see Eq. (8.97)]

$$|\Psi(t)\rangle = \sum e^{-i\hat{H}t/\hbar} |n\rangle \, [\langle n|\Psi(0)\rangle] \tag{8.126}$$

Thus, in the Schrödinger picture, we may visualize the basis vectors (here $|n\rangle$) as a fixed set of vectors and $|\psi(t)\rangle$ (describing the system) as moving.

Now, if we substitute for $|\psi(t)\rangle$ from Eq. (8.126) in Eq. (8.124), we would get

$$
\begin{aligned}
\left\langle \hat{O} \right\rangle &= \langle \Psi(0)| \, e^{+i\hat{H}t/\hbar}\hat{O}\,e^{-i\hat{H}t/\hbar} |\Psi(0)\rangle \\
&= \langle \Psi(0)| \, \hat{O}_H(t) \, |\Psi(0)\rangle
\end{aligned}
\tag{8.127}
$$

where the operator $\hat{O}_H(t)$ is defined by the following equation:

$$
\hat{O}_H(t) = e^{i\hat{H}t/\hbar}\hat{O}\,e^{-i\hat{H}t/\hbar}
\tag{8.128}
$$

Equations (8.124) and (8.127) tell us that the expectation values remain the same if we endow the operator $\hat{O}_H(t)$ with the entire time dependence but assume that the kets are time independent. This is known as the Heisenberg picture (and hence the subscript H) in which operators representing the observables change with time but the ket describing the state of the system is time independent. From Eq. (8.128), we have

$$
\frac{d\hat{O}_H}{dt} = e^{+i\hat{H}t/\hbar}\frac{\partial \hat{O}}{\partial t}\,e^{-i\hat{H}t/\hbar} + \frac{i}{\hbar}e^{i\hat{H}t/\hbar}\left[\hat{H}\hat{O}-\hat{O}\hat{H}\right]e^{-i\hat{H}t/\hbar}
\tag{8.129}
$$

where the first term on the right-hand side allows for any explicit time dependence of the operator. If there is no such explicit time dependence, we may write

$$
i\hbar\,\frac{d\hat{O}_H(t)}{dt} = \left[e^{i\hat{H}t/\hbar}\hat{O}e^{-i\hat{H}t/\hbar}\right]\hat{H} - \hat{H}\left[e^{i\hat{H}t/\hbar}\hat{O}e^{-i\hat{H}t/\hbar}\right]
\tag{8.130}
$$

$$
= \hat{O}_H(t)\hat{H} - \hat{H}\hat{O}_H(t)
\tag{8.131}
$$

or

$$
i\hbar\,\frac{d\hat{O}_H(t)}{dt} = \left[\hat{O}_H(t),\,\hat{H}\right]
\tag{8.132}
$$

The above equation is known as the Heisenberg equation of motion and gives the time dependence of an operator in the Heisenberg picture.

If the Hamiltonian is assumed to be independent of time in the Schrödinger representation, then it is also independent of time in the Heisenberg representation:

$$
\hat{H}_H(t) = e^{i\hat{H}t/\hbar}\hat{H}\,e^{-i\hat{H}t/\hbar} = e^{i\hat{H}t/\hbar}\,e^{-i\hat{H}t/\hbar}\hat{H} = \hat{H}
\tag{8.133}
$$

It should be mentioned that if $\hat{H}$ had an explicit time dependence, the analysis would have been much more involved (see, e.g., Baym (1969), Chapter 5).

We next consider the operators $\hat{\alpha}$ and $\hat{\beta}$, which, in the Schrödinger representation, satisfy the commutation relation

$$\left[\hat{\alpha},\hat{\beta}\right] = i\hat{\gamma} \tag{8.134}$$

or

$$\hat{\alpha}\hat{\beta} - \hat{\beta}\hat{\alpha} = i\hat{\gamma} \tag{8.135}$$

If we multiply on the left by $e^{i\hat{H}t/\hbar}$ and on the right by $e^{-i\hat{H}t/\hbar}$, we obtain

$$e^{i\hat{H}t/\hbar}\hat{\alpha}e^{-i\hat{H}t/\hbar}e^{i\hat{H}t/\hbar}\hat{\beta}e^{-i\hat{H}t/\hbar} - e^{i\hat{H}t/\hbar}\hat{\beta}e^{-i\hat{H}t/\hbar}e^{i\hat{H}t/\hbar}\hat{\alpha}e^{-i\hat{H}t/\hbar} = ie^{i\hat{H}t/\hbar}\hat{\gamma}e^{-i\hat{H}t/\hbar} \tag{8.136}$$

where we have inserted $e^{-i\hat{H}t/\hbar}e^{i\hat{H}t/\hbar}\,(=1)$ between $\hat{\alpha}$ and $\hat{\beta}$. Using Eq. (8.128), we get

$$\left[\hat{\alpha}_H(t),\,\hat{\beta}_H(t)\right] = i\hat{\gamma}_H(t) \tag{8.137}$$

which shows the physical equivalence of Heisenberg and Schrödinger pictures.

As an illustration we consider the harmonic oscillator problem. However, before we do so, we note that

$$\left[\hat{x},\hat{p}^n\right] = i\hbar n\hat{p}^{n-1} = i\hbar\frac{\partial}{\partial\hat{p}}\hat{p}^n \tag{8.138}$$

and

$$\left[\hat{p},\hat{x}^n\right] = -i\hbar n\hat{x}^{n-1} = -i\hbar\frac{\partial}{\partial\hat{x}}\hat{x}^n \tag{8.139}$$

Thus if $P(\hat{p})$ and $X(\hat{x})$ can be expanded in a power series in p and x, respectively, we will have

$$\left[\hat{x},P(\hat{p})\right] = i\hbar\frac{\partial\hat{P}}{\partial\hat{p}} \tag{8.140}$$

and

$$\left[\hat{p},X(\hat{x})\right] = -i\hbar\frac{\partial\hat{X}}{\partial\hat{x}} \tag{8.141}$$

Now for the harmonic oscillator problem

$$\hat{H} = \frac{\hat{p}^2}{2m} + \frac{1}{2}m\omega^2\hat{x}^2 = \hat{H}_H = \frac{\hat{p}_H^2(t)}{2m} + \frac{1}{2}m\omega^2\hat{x}_H^2(t) \tag{8.142}$$

where we have used Eq. (8.133). Thus, using Eq. (8.132),

$$\frac{d\hat{x}_H\,(t)}{dt} = \frac{1}{i\hbar}\left[\hat{x}_H\,(t),\hat{H}_H\right] = \frac{\partial\hat{H}_H}{\partial\hat{p}_H} = \frac{1}{m}\hat{p}_H\,(t) \tag{8.143}$$

and

$$\frac{d\hat{p}_H\,(t)}{dt} = \frac{1}{i\hbar}\left[\hat{p}_H\,(t),\hat{H}_H\right] = \frac{\partial\hat{H}_H}{\partial\hat{x}_H} = -m\omega^2\hat{x}_H\,(t) \tag{8.144}$$

These are the Hamilton equations of motion (see, e.g., Goldstein (1950)). The solutions of the above equations are

$$\hat{x}_H\,(t) = \hat{x}\,\cos\,\omega t + \frac{1}{m\omega}\hat{p}\,\sin\,\omega t \tag{8.145}$$

$$\hat{p}_H\,(t) = -m\omega\hat{x}\,\sin\,\omega t + \hat{p}\,\cos\,\omega t \tag{8.146}$$

where $\hat{x} = \hat{x}_H\,(t=0)$ and $\hat{p} = \hat{p}_H\,(t=0)$ represent the operators in the Schrödinger representation. Further,

$$\hat{a}_H\,(t) = \frac{1}{(2m\hbar\omega)^{1/2}}\left[m\omega\hat{x}_H\,(t) + i\hat{p}_H\,(t)\right] \qquad \text{[see Eq. (8.39)]}$$

$$= \hat{a}e^{-i\omega t} \tag{8.147}$$

where $\hat{a} \equiv \hat{a}_H\,(t=0)$. Similarly

$$\hat{a}_H^\dagger\,(t) = \hat{a}^\dagger e^{+i\omega t} \tag{8.148}$$

where

$$\hat{a}^\dagger \equiv \hat{a}_H^\dagger\,(t=0) \tag{8.149}$$

These relations will be used in the next chapter.

Problems

Problem 8.1 If $|\delta\rangle$ is an eigenket of an operator $\hat{g}$ with an eigenvalue δ, then $\langle\delta|$ is an eigenbra of which operator and what is the corresponding eigenvalue?

Problem 8.2 Consider a harmonic oscillator state given by the following superposition state:

$$|\psi\rangle = \alpha\,|0\rangle + \beta\,|1\rangle$$

Normalize $|\psi\rangle$ and obtain the relationship between α and β.

Problem 8.3 Using the commutation relations between $\hat{a}$ and $\hat{a}^\dagger$ obtain the value of the commutator $\left[\hat{a}, \hat{a}^{\dagger 2}\right]$.

Problem 8.4 A harmonic oscillator is in a superposition of number states as given by

$$|\psi\rangle = \frac{1}{2} |2\rangle + \frac{\sqrt{3}}{2} |3\rangle$$

Obtain the expectation value of energy of the harmonic oscillator.

Problem 8.5 Consider a harmonic oscillator in a number state $|n\rangle$ with $n = 10$. (a) what is the uncertainty in the energy of the oscillator? (b) What are the variances of the position and momentum of the oscillator in this state? (c) Is this a minimum uncertainty state?

Chapter 9
Quantum Theory of Interaction of Radiation Field with Matter

9.1 Introduction

In this chapter we show that the electromagnetic field can be considered as an infinite set of harmonic oscillators, each corresponding to a particular value of the frequency, wave vector, and a particular state of polarization. Comparing with the quantum mechanical treatment of harmonic oscillators, we replace the generalized coordinates and generalized momenta by operators. By imposing the commutation relations between the canonical variables, it is shown that the energy of each oscillator can increase or decrease by integral multiples of a certain quantum of energy; this quantum of energy is known as the *photon*. Having quantized the field, we show that the state which corresponds to a given number of photons (also referred to as the number state) for a particular mode does *not* correspond to the classical plane wave. Indeed, we show that the eigenstates of the annihilation operator (which are known as the *coherent states*) resemble the classical plane wave for large intensities.

In Section 6.3 we considered the interaction of the radiation field with matter using the semiclassical theory; i.e., we used a quantum mechanical description of the atom and a classical description of the electromagnetic field. In Section 9.6 we use the quantum mechanical description of the radiation field to study its interaction with an atom and thereby obtain explicit expressions for the Einstein A and B coefficients, which are shown to be identical to the one obtained in Chapter 6. We may mention here that the fully quantum mechanical theory automatically leads to spontaneous emissions which in the semiclassical theory had to be introduced in an ad hoc manner. Further, the theory also shows that the natural lineshape function is Lorentzian; this is explicitly shown in Appendix G.

In Section 9.7 we show that it is difficult to give a quantum mechanical description of the phase of the electromagnetic field.

9.2 Quantization of the Electromagnetic Field

We start with Maxwell's equations in free space

$$\nabla \times \mathbf{H} = \frac{\partial \mathbf{D}}{\partial t} = \varepsilon_0 \frac{\partial \mathbf{E}}{\partial t} \tag{9.1}$$

K. Thyagarajan, A. Ghatak, *Lasers*, Graduate Texts in Physics,
DOI 10.1007/978-1-4419-6442-7_9, © Springer Science+Business Media, LLC 2010

$$\nabla \times \mathbf{E} = -\frac{\partial \mathbf{B}}{\partial t} = -\mu_0 \frac{\partial \mathbf{H}}{\partial t} \tag{9.2}$$

$$\nabla \cdot \mathbf{E} = 0 \tag{9.3}$$

$$\nabla \cdot \mathbf{B} = 0 \tag{9.4}$$

where we have assumed the absence of free currents and free charges; $\mathbf{E}, \mathbf{D}, \mathbf{B}$ and $\mathbf{H}$ represent the electric field, electric displacement, magnetic field, and magnetic $\mathbf{H}$ vector, respectively; ε_0 and μ_0 represent the permittivity and magnetic permeability of free space. From Eq. (9.4), it follows that $\mathbf{B}$ can be expressed as the curl of a vector:

$$\mathbf{B} = \nabla \times \mathbf{A} = \mu_0 \mathbf{H} \tag{9.5}$$

where $\mathbf{A}$ is called the vector potential. Thus Eqs. (9.2) and (9.5) give us

$$\nabla \times \left(\mathbf{E} + \frac{\partial \mathbf{A}}{\partial t} \right) = 0 \tag{9.6}$$

Hence we may set

$$\mathbf{E} + \frac{\partial \mathbf{A}}{\partial t} = -\nabla \phi \tag{9.7}$$

or,

$$\mathbf{E} = -\nabla \phi - \frac{\partial \mathbf{A}}{\partial t} \tag{9.8}$$

where ϕ is known as the scalar potential. Substituting Eq. (9.8) into Eq. (9.3), we get

$$\nabla^2 \phi + \frac{\partial}{\partial t} (\nabla \cdot \mathbf{A}) = 0 \tag{9.9}$$

Now, $\mathbf{B}$ is left unchanged if the gradient of any scalar quantity is added to $\mathbf{A}$:

$$\mathbf{A} \to \mathbf{A}' = \mathbf{A} + \nabla \chi \tag{9.10}$$

where χ is any scalar function.[1] We may choose χ such that

$$\nabla \cdot \mathbf{A} = 0 \tag{9.11}$$

---

[1] This is because $\nabla \times (\nabla \chi) = 0$ for arbitrary χ.

This is known as the Coulomb gauge. The scalar potential then satisfies the equation $\nabla^2 \phi = 0$ and we may assume $\phi = 0$ Thus, we finally obtain

$$\mathbf{B} = \mu_0 \mathbf{H} = \nabla \times \mathbf{A} \tag{9.12}$$

and

$$\mathbf{E} = -\frac{\partial \mathbf{A}}{\partial t} \tag{9.13}$$

Substituting for $\mathbf{H}$ and $\mathbf{E}$ in Eq. (9.1), we get

$$\nabla \times (\nabla \times \mathbf{A}) = -\varepsilon_0 \mu_0 \frac{\partial^2 \mathbf{A}}{\partial t^2} \tag{9.14}$$

If we now use the identity

$$\nabla \times (\nabla \times \mathbf{A}) = \nabla (\nabla \cdot \mathbf{A}) - \nabla^2 \mathbf{A} = -\nabla^2 \mathbf{A} \tag{9.15}$$

(because $\nabla \cdot \mathbf{A} = 0$), we finally obtain

$$\nabla^2 \mathbf{A} = \frac{1}{c^2} \frac{\partial^2 \mathbf{A}}{\partial t^2} \tag{9.16}$$

where $c[= (\varepsilon_0 \mu_0)^{-1/2}]$ represents the speed of light in free space. Equation (9.16) represents the wave equation. In order to solve the wave equation, we use the method of separation of variables:

$$\mathbf{A}(\mathbf{r}, t) = \mathbf{A}(\mathbf{r}) q(t) \tag{9.17}$$

Thus

$$q(t) \nabla^2 \mathbf{A}(\mathbf{r}) = \mathbf{A}(\mathbf{r}) \frac{1}{c^2} \frac{d^2 q}{dt^2} \tag{9.18}$$

We next consider a Cartesian component (say the x component) of $\mathbf{A}(\mathbf{r})$ which we denote by $\mathbf{A}_x(\mathbf{r})$; thus

$$\frac{c^2}{\mathbf{A}_x(\mathbf{r})} \nabla^2 \mathbf{A}_x(\mathbf{r}) = \frac{1}{q(t)} \frac{d^2 q}{dt^2} = -\omega^2 \quad (say) \tag{9.19}$$

Thus

$$q(t) \sim e^{-i\omega t}, \; e^{+i\omega t} \tag{9.20}$$

and

$$\nabla^2 \mathbf{A}_x(\mathbf{r}) + k^2 \mathbf{A}_x(\mathbf{r}) = 0 \tag{9.21}$$

where

$$k^2 \equiv \frac{\omega^2}{c^2} \tag{9.22}$$

The solutions of Eq. (9.21) are plane waves; and similarly if we consider the y and z components, we obtain

$$\mathbf{A}\,(\mathbf{r}) \sim \hat{\mathbf{e}}\, e^{i\mathbf{k}\cdot\mathbf{r}}, \quad \hat{\mathbf{e}}\, e^{-i\mathbf{k}\cdot\mathbf{r}} \tag{9.23}$$

where $\mathbf{k} = \hat{\mathbf{x}}k_x + \hat{\mathbf{y}}k_y + \hat{\mathbf{z}}k_z$ and $\mathbf{k} \cdot \mathbf{k} = k^2$ and $\hat{\mathbf{e}}$ is the unit vector along $\mathbf{A}$. The condition $\nabla \cdot \mathbf{A} = 0$ gives us

$$\mathbf{k} \cdot \hat{\mathbf{e}} = 0 \tag{9.24}$$

implying that $\hat{e}$ is at right angles to the direction of propagation k; this is nothing but the transverse character of the wave, the vector $\hat{e}$ denotes the polarization of the wave. For a given $\mathbf{k}$ there are two independent states of polarization.

In free space all values of ω are allowed and the total electromagnetic field is, in general, an integral over all possible frequencies and direction of propagation. In order to simplify the analysis, it is convenient to impose boundary conditions over a finite volume so that the frequency spectrum becomes discrete and the total electromagnetic field then can be written as a sum (rather than an integral) of various terms. It is then possible to let the volume of the region considered tend to infinity to go over to fields in free space.

One can impose two types of boundary conditions; one in which the cavity is bounded by perfectly conducting walls and the other in which we apply periodic boundary conditions. The former leads to standing wave solutions, while the latter to propagating wave solutions. Here we consider a cube of side L and use periodic boundary conditions, i.e., assume that the fields on the sides facing each other to be equal. Thus we will have

$$\mathbf{A}\,(x = 0, y, z) = \mathbf{A}\,(x = L, y, z), \quad \text{etc.} \tag{9.25}$$

giving

$$e^{ik_x L} = 1 = e^{ik_y L} = e^{ik_z L} \tag{9.26}$$

Thus

$$\left.\begin{aligned}
k_x &= \frac{2\pi \nu_x}{L} \\[4pt]
k_y &= \frac{2\pi \nu_y}{L} \\[4pt]
k_z &= \frac{2\pi \nu_z}{L}
\end{aligned}\right\} \quad \nu_x, \nu_y, \nu_z = 0,\, \pm 1,\, \pm 2, \ldots \tag{9.27}$$

The complete solution of Eq. (9.16) is therefore given by

$$\mathbf{A}(\mathbf{r}, t) = \sum_{\lambda=-\infty}^{\infty} \left[q_\lambda(t)\, \mathbf{A}_\lambda(\mathbf{r}) + q_\lambda^*(t)\, \mathbf{A}_\lambda^*(\mathbf{r}) \right] \tag{9.28}$$

where

$$\mathbf{A}_\lambda(\mathbf{r}) = \hat{\mathbf{e}}_\lambda e^{i k_\lambda \cdot \mathbf{r}}\,; \qquad q_\lambda(t) = |q_\lambda|\, e^{-i\omega_\lambda t} \tag{9.29}$$

and the subscript λ signifies the various modes of the field [see (Eq. 9.27)] including the two states of polarization. Thus, a particular value of λ corresponds to a particular set of values of v_x, v_y, v_z (which implies a particular frequency) and a particular direction $\hat{\mathbf{e}}$. In Eq. (9.28), the second term on the right-hand side is the complex conjugate of the first term, making $\mathbf{A}$ necessarily real. We also assume that

$$k_{-\lambda} = -k_\lambda \text{ and } \omega_{-\lambda} = \omega_\lambda \tag{9.30}$$

Thus negative values of λ in Eq. (9.28) correspond to plane waves traveling in opposite direction to the waves with positive values of λ.

Because of the allowed values of $\mathbf{k}_\lambda$ [see Eq. (9.27)], we readily obtain

$$\iiint_V \mathbf{A}_\lambda \cdot \mathbf{A}_\mu^* \, d\tau = \iiint_V \mathbf{A}_\lambda \cdot \mathbf{A}_{-\mu} d\tau = V\delta_{\lambda,\mu} \tag{9.31}$$

where the integration is over the entire volume of the cavity.

Using Eq. (9.28), we obtain the following expressions for the electric and magnetic fields:

$$\mathbf{E} = -\frac{\partial \mathbf{A}}{\partial t} = \sum_\lambda \mathbf{E}_\lambda \tag{9.32}$$

$$\mathbf{H} = \frac{1}{\mu_0} \nabla \times \mathbf{A} = \sum_\lambda \mathbf{H}_\lambda \tag{9.33}$$

where

$$\mathbf{E}_\lambda = i\omega_\lambda \left[q_\lambda(t)\, \mathbf{A}_\lambda(\mathbf{r}) - q_\lambda^*(t)\, \mathbf{A}_\lambda^*(\mathbf{r}) \right] \tag{9.34}$$

and

$$\mathbf{H}_\lambda = \frac{i}{\mu_0}\, \mathbf{k}_\lambda \times \left(q_\lambda \mathbf{A}_\lambda - q_\lambda^* \mathbf{A}_\lambda^* \right) \tag{9.35}$$

The total energy of the radiation field is given by

$$\mathbf{H} = \frac{1}{2} \int \left(\varepsilon_0 \mathbf{E} \cdot \mathbf{E} + \mu_0 \mathbf{H} \cdot \mathbf{H} \right) d\tau \tag{9.36}$$

Now

$$\frac{1}{2}\varepsilon_0 \int_v \mathbf{E} \cdot \mathbf{E}\, d\tau = -\frac{1}{2}\varepsilon_0 \int_v \sum_\lambda \sum_\mu \omega_\lambda \omega_\mu \left[q_\lambda q_\mu \int \mathbf{A}_\lambda \cdot \mathbf{A}_\mu d\tau - q_\lambda q_\mu^* \int \mathbf{A}_\lambda \cdot \mathbf{A}_\mu^* d\tau \right.$$

$$\left. - q_\lambda^* q_\mu \int \mathbf{A}_\lambda{}^* \cdot \mathbf{A}_\mu \, d\tau + q_\lambda^* q_\mu^* \int \mathbf{A}_\lambda{}^* \cdot \mathbf{A}_\mu^* d\tau \right]$$

$$= -\frac{1}{2}\varepsilon_0 V \sum_\lambda \sum_\mu \omega_\lambda \omega_\mu \left[q_\lambda\, q_\mu \delta_{\lambda,-\mu} - q_\lambda q_\mu^* \delta_{\lambda,\mu} \right.$$

$$\left. - q_\lambda^* q_\mu \delta_{\lambda,\mu} + q_\lambda^* q_\mu^* \delta_{\lambda,-\mu} \right]$$

$$- \frac{1}{2}\varepsilon_0 V \sum_\lambda \omega_\lambda^2 \left[q_\lambda q_{-\lambda} + q_\lambda^* q_{-\lambda}^* - 2 q_\lambda q_\lambda^* \right]$$

$$\tag{9.37}$$

Similarly one can evaluate $\int \mathbf{H} \cdot \mathbf{H}\, d\tau$. The final result is

$$\frac{1}{2}\mu_0 \int \mathbf{H} \cdot \mathbf{H}\, d\tau = +\frac{1}{2}\varepsilon_0 V \sum_\lambda \omega_\lambda^2 \left[q_\lambda q_{-\lambda} + q_\lambda^* q_{-\lambda}^* + 2 q_\lambda q_\lambda^* \right] \tag{9.38}$$

where use has to be made of the vector identity

$$(\mathbf{a} \times \mathbf{b}) \cdot (\mathbf{c} \times \mathbf{d}) = (\mathbf{a} \cdot \mathbf{c})\,(\mathbf{b} \cdot \mathbf{d}) - (\mathbf{b} \cdot \mathbf{c})\,(\mathbf{a} \cdot \mathbf{d}) \tag{9.39}$$

and the relation

$$k_\lambda^2 = \frac{\omega_\lambda^2}{c^2} = \varepsilon_0 \mu_0 \omega_\lambda^2 \tag{9.40}$$

Thus

$$\mathbf{H} = 2\varepsilon_0 V \sum_\lambda \omega_\lambda^2 q_\lambda\,(t)\, q_\lambda^*\,(t) \tag{9.41}$$

We next introduce the dimensionless variables $Q_\lambda\,(t)$ and $P_\lambda\,(t)$ defined through the equations

$$Q_\lambda\,(t) \equiv (\varepsilon_0 V)^{1/2} \left[q_\lambda\,(t) + q_\lambda^*\,(t) \right] \tag{9.42}$$

and

$$P_\lambda\,(t) \equiv \frac{1}{i} \left(\varepsilon_0 V \omega_\lambda^2 \right)^{1/2} \left[q_\lambda\,(t) - q_\lambda^*\,(t) \right] \tag{9.43}$$

Thus

$$q_\lambda(t) = \left(4\varepsilon_0 V\omega_\lambda^2\right)^{-1/2}\left[\omega_\lambda Q_\lambda(t) + iP_\lambda(t)\right] \tag{9.44}$$

$$q_\lambda^*(t) = \left(4\varepsilon_0 V\omega_\lambda^2\right)^{-1/2}\left[\omega_\lambda Q_\lambda(t) - iP_\lambda(t)\right] \tag{9.45}$$

and

$$\mathbf{H} = \sum_\lambda \mathbf{H}_\lambda \tag{9.46}$$

where[2]

$$\mathbf{H}_\lambda = \frac{1}{2}\left[P_\lambda^2 + \omega_\lambda^2 Q_\lambda^2\right] \tag{9.47}$$

The Hamiltonian given by Eq. (9.47) is identical to that of the linear harmonic oscillator [see Eq. (8.35) with mass $m = 1$] which suggests that the electromagnetic field can be regarded as an infinite set of harmonic oscillators with each mode (i.e., for each value of $\mathbf{k}_\lambda$ and to a particular direction of polarization) being associated with a harmonic oscillator. Just to remind, the analysis carried out till now is completely classical.

In order to quantize the electromagnetic field we use the same approach as in Section 8.6; we consider Q_λ and P_λ to be Hermitian operators $\hat{Q}_\lambda$ and $\hat{P}_\lambda$ satisfying the following commutation relations [cf. Eqs. (8.37) and (8.137)]:

$$\left[\hat{Q}_\lambda(t),\hat{P}_\lambda(t)\right] \equiv \hat{Q}_\lambda(t)\hat{P}_\lambda(t) - \hat{P}_\lambda(t)\hat{Q}_\lambda(t) = i\hbar \tag{9.48}$$

$$\left[\hat{Q}_\lambda(t),\hat{P}_{\lambda'}(t)\right] = 0, \quad \lambda \neq \lambda' \tag{9.49}$$

$$\left[\hat{Q}_\lambda(t),\hat{Q}_\mu(t)\right] = 0 = \left[\hat{P}_\lambda(t),\hat{P}_\mu(t)\right] \tag{9.50}$$

---

[2]Notice that

$$\frac{\partial \mathbf{H}_\lambda}{\partial Q_\lambda} = \omega_\lambda^2 Q_\lambda = \omega_\lambda^2 (\varepsilon_0 V)^{1/2}(q_\lambda + q_\lambda^*) = i\left(\varepsilon_0 V\omega_\lambda^2\right)^{1/2}(\dot{q}_\lambda - \dot{q}_\lambda^*) = -\dot{P}_\lambda$$

Similarly

$$\frac{\partial \mathbf{H}_\lambda}{\partial P_\lambda} = \dot{Q}_\lambda$$

which are nothing but Hamilton's equations of motion (see, e.g., Goldstein (1950)). Thus Q_λ and P_λ are the canonical coordinates.

where all the operators are in the Heisenberg representation (see Section 8.9). We next introduce the dimensionless variables

$$\hat{a}_\lambda\,(t) = (2\hbar\omega_\lambda)^{-1/2}\left[\omega_\lambda\hat{Q}_\lambda\,(t) + i\hat{P}_\lambda\,(t)\right] \tag{9.51}$$

$$\hat{a}_\lambda^\dagger\,(t) = (2\hbar\omega_\lambda)^{-1/2}\left[\omega_\lambda\hat{Q}_\lambda\,(t) - i\hat{P}_\lambda\,(t)\right] \tag{9.52}$$

Since $\left[\omega_\lambda\hat{Q}_\lambda\,(t) + i\hat{P}_\lambda\,(t)\right]$ is proportional to $q_\lambda\,(t)$ [see Eq. (9.44)] which has a time dependence of the form $e^{-i\omega_\lambda t}$, we may write [cf. Eq. (8.147)]

$$\hat{a}_\lambda\,(t) = \hat{a}_\lambda e^{-i\omega_\lambda t} \tag{9.53}$$

Similarly

$$\hat{a}_\lambda^\dagger\,(t) = \hat{a}_\lambda^\dagger e^{+i\omega_\lambda t} \tag{9.54}$$

where

$$\hat{a}_\lambda \equiv \hat{a}_\lambda(0) \quad \text{and} \quad \hat{a}_\lambda^\dagger \equiv \hat{a}_\lambda^\dagger(0) \tag{9.55}$$

Solving Eqs. (9.51) and (9.52) for $\hat{Q}_\lambda\,(t)$ and $\hat{P}_\lambda\,(t)$ we obtain

$$\hat{Q}_\lambda\,(t) = \left(\frac{\hbar}{2\omega_\lambda}\right)^{1/2}\left[\hat{a}_\lambda^\dagger\,(t) + \hat{a}_\lambda\,(t)\right] \tag{9.56}$$

$$\hat{P}_\lambda\,(t) = i\left(\frac{\hbar\omega_\lambda}{2}\right)^{1/2}\left[\hat{a}_\lambda^\dagger\,(t) - \hat{a}_\lambda\,(t)\right] \tag{9.57}$$

Substituting the above expressions for $\hat{Q}_\lambda\,(t)$ and $\hat{P}_\lambda\,(t)$ in Eq. (9.47), we obtain

$$\hat{H} = \sum_\lambda \hat{H}_\lambda = \sum_\lambda \frac{1}{2}\hbar\omega_\lambda\left[\hat{a}_\lambda^\dagger\,(t)\,\hat{a}_\lambda\,(t) + \hat{a}_\lambda\,(t)\,\hat{a}_\lambda^\dagger\,(t)\right]$$

$$= \sum_\lambda \frac{1}{2}\hbar\omega_\lambda\left[\hat{a}_\lambda^\dagger\hat{a}_\lambda + \hat{a}_\lambda\hat{a}_\lambda^\dagger\right] \tag{9.58}$$

Using the commutation relation between $\hat{a}_\lambda$ and $\hat{a}_\lambda^\dagger$, we can write Eq. (9.58) as

$$\hat{H} = \sum_\lambda \hbar\omega_\lambda\left[\hat{a}_\lambda^\dagger\hat{a}_\lambda + \frac{1}{2}\right]$$

which is the quantum mechanical Hamiltonian operator of the total electromagnetic field. If we now carry out an analysis similar to that followed in Section 8.6, we obtain

$$\left(n_\lambda + \frac{1}{2}\right)\hbar\omega_\lambda, \quad n_\lambda = 0, 1, 2,\ldots \tag{9.59}$$

as the eigenvlaues of $\hat{H}_\lambda$ and

$$\sum_\lambda \left(n_\lambda + \frac{1}{2}\right)\hbar\omega_\lambda \tag{9.60}$$

as the eigenvalues of the total Hamiltonian $\hat{H}\left(=\sum_\lambda \hat{H}_\lambda\right)$ Thus, quantum mechanically, we can visualize the radiation field as consisting of an infinite number of simple harmonic oscillators; the energy of each oscillator can increase or decrease only by integral multiples of $\hbar\omega_\lambda$. If we consider $\hbar\omega_\lambda$, as the energy of a *photon*, then we can say that each oscillator can have energy corresponding to n_λ photons or that the λth mode is occupied by n_λ photons.

Note from Eq. (9.60) that even in a state in which none of the modes are occupied, i.e., $n_\lambda = 0$ for all λ, each mode still possesses an energy of $\hbar\omega_\lambda/2$ also referred to as zero-point energy. Since there are infinite number of radiation modes, this implies that the zero-point energy is infinite. This infinite value of zero-point energy is a major unresolved issue in the quantization of electromagnetic waves.

The eigenkets of the total Hamiltonian would be

$$|n_1\rangle |n_2\rangle \cdots |n_\lambda\rangle \cdots = |n_1, n_2, \ldots n_\lambda, \ldots\rangle \tag{9.61}$$

where n_λ represents the number of photons in the mode characterized by λ. Thus

$$\hat{H}|n_1, n_2, \ldots n_\lambda, \ldots\rangle = \left[\sum_\lambda \left(n_\lambda + \frac{1}{2}\right)\hbar\omega_\lambda\right]|n_1, n_2, \ldots n_\lambda, \ldots\rangle \tag{9.62}$$

The multimode vacuum state is a state in which none of the modes is occupied and is represented by

$$|\psi\rangle \cdots = |0_1, 0_2, 0_3 \ldots .0_i, \ldots\rangle \tag{9.63}$$

Further

$$\hat{a}_\lambda |n_1, n_2, \ldots n_\lambda, \ldots\rangle = (n_\lambda)^{1/2} |n_1, n_2, \ldots n_\lambda - 1, \ldots\rangle \tag{9.64}$$

$$\hat{a}_\lambda^\dagger |n_1, n_2, \ldots n_\lambda, \ldots\rangle = (n_\lambda + 1)^{1/2} |n_1, n_2, \ldots n_\lambda + 1, \ldots\rangle \tag{9.65}$$

[cf. Eqs. (8.57) and (8.58)]; and

$$\langle n'_1, n'_2, \ldots, n'_\lambda, \ldots \mid n_1, n_2, \ldots, n_\lambda, \ldots \rangle$$

$$= \delta_{n_1 n'_1} \, \delta_{n_2 n'_2} \cdots \delta_{n_\lambda n'_\lambda} \cdots \qquad (9.66)$$

Finally, the state of the radiation field need not be an eigenstate of $\hat{H}$, it could be a superposition of the eigenstates like that given by the following equation:

$$|\Psi\rangle = \sum_{n_1,n_2,\ldots} C_{n_1,n_2,\ldots n_\lambda,\ldots} \, |n_1, n_2, \ldots, n_\lambda, \ldots\rangle \qquad (9.67)$$

Physically $\left|C_{n_1,n_2,\ldots}\right|^2$ would represent the probability of finding n_1 photons in the first mode, n_2 in the second mode, etc.

9.3 The Eigenkets of the Hamiltonian

In this section we study the properties of the radiation field when it is in one of the eigenkets of the Hamiltonian. In order to do so we have to first express the electric field in terms of the operators $\hat{a}_\lambda$ and $\hat{a}_\lambda^\dagger$. Now, using Eqs. (9.44) and (9.51), we get for a mode λ,

$$\hat{q}_\lambda(t) = \left(\frac{\hbar}{2\varepsilon_0 V \omega_\lambda}\right)^{1/2} \hat{a}_\lambda(t) \qquad (9.68)$$

which is now to be considered as an operator. Using Eq. (9.34), we may write

$$\hat{E}_\lambda(t) = i\left(\frac{\hbar\omega_\lambda}{2\varepsilon_0 V}\right)^{1/2} \left[\hat{a}_\lambda(t)\, e^{i\mathbf{k}_\lambda \cdot \mathbf{r}} - \hat{a}_\lambda^\dagger(t)\, e^{-i\mathbf{k}_\lambda \cdot \mathbf{r}}\right] \hat{\mathbf{e}}_\lambda \qquad (9.69)$$

where all the operators $\hat{E}_\lambda(t)$, $\hat{a}_\lambda(t)$ and $\hat{a}_\lambda^\dagger(t)$ are in the Heisenberg representation (see Section 8.9). In the Schrödinger representation, we will have

$$\hat{E}_\lambda = i\left(\frac{\hbar\omega_\lambda}{2\varepsilon_0 V}\right)^{1/2} \left(\hat{a}_\lambda e^{i\mathbf{k}_\lambda \cdot \mathbf{r}} - \hat{a}_\lambda^\dagger e^{-i\mathbf{k}_\lambda \cdot \mathbf{r}}\right) \hat{\mathbf{e}}_\lambda \qquad (9.70)$$

which will be independent of time.

We consider the state of the radiation field for which there are precisely n_λ photons in the state λ Operating with the number operator of the λth state we have

$$\hat{a}_\lambda^\dagger \hat{a}_\lambda \, |n_1, n_2, \ldots n_\lambda, \ldots\rangle = n_\lambda \, |n_1, n_2, \ldots n_\lambda, \ldots\rangle \qquad (9.71)$$

which implies that the state is an eigenstate of the number operator with eigenvalue n_λ.

The expectation value of the electric field operator $\hat{E}_\lambda$ in this state would be given by

$$\langle n_1, n_2, \ldots n_\lambda, \ldots | \hat{E}_\lambda | n_1, n_2, \ldots n_\lambda, \ldots \rangle$$

$$= \langle n_1 \mid n_1 \rangle \langle n_2 \mid n_2 \rangle \cdots \langle n_\lambda | \hat{E}_\lambda | n_\lambda \rangle \cdots = 0 \tag{9.72}$$

because

$$\langle n_\lambda | \hat{a}_\lambda | n_\lambda \rangle = 0 = \langle n_\lambda | \hat{a}_\lambda^\dagger | n_\lambda \rangle \tag{9.73}$$

Similarly the expectation value of $\hat{E}_\lambda^2$ would be

$$\langle n_1, n_2, \ldots n_\lambda, \ldots | \hat{E}_\lambda^2 | n_1, n_2, \ldots n_\lambda, \ldots \rangle$$

$$= \langle n_1 \mid n_1 \rangle \langle n_2 \mid n_2 \rangle \cdots \langle n_\lambda | \hat{E}_\lambda^2 | n_\lambda \rangle \cdots$$

$$= -\frac{\hbar \omega_\lambda}{2 \varepsilon_0 V} \langle n_\lambda | \left(\hat{a}_\lambda e^{i\mathbf{k}_\lambda \cdot \mathbf{r}} - \hat{a}_\lambda^\dagger e^{i\mathbf{k}_\lambda \cdot \mathbf{r}} \right) \left(\hat{a}_\lambda e^{i\mathbf{k}_\lambda \cdot \mathbf{r}} - \hat{a}_\lambda^\dagger e^{-i\mathbf{k}_\lambda \cdot \mathbf{r}} \right) | n_\lambda \rangle \tag{9.74}$$

$$= \left(\frac{\hbar \omega_\lambda}{\varepsilon_0 V} \right) \left(n_\lambda + \frac{1}{2} \right)$$

where use has been made of relations like (see Section 8.6)

$$\langle n_\lambda | \hat{a}_\lambda \hat{a}_\lambda^\dagger | n_\lambda \rangle = (n_\lambda + 1)^{1/2} \langle n_\lambda | \hat{a}_\lambda | n_\lambda + 1 \rangle = (n_\lambda + 1) \tag{9.75}$$

$$\langle n_\lambda | \hat{a}_\lambda^\dagger \hat{a}_\lambda | n_\lambda \rangle = (n_\lambda)^{1/2} \langle n_\lambda | \hat{a}_\lambda^\dagger | n_\lambda - 1 \rangle = n_\lambda \tag{9.76}$$

$$\langle n_\lambda | \hat{a}_\lambda \hat{a}_\lambda | n_\lambda \rangle = 0 \tag{9.77}$$

$$\langle n_\lambda | \hat{a}_\lambda^\dagger \hat{a}_\lambda^\dagger | n_\lambda \rangle = 0 \tag{9.78}$$

The uncertainty $\Delta \mathbf{E}_\lambda$ in the electric field $\mathbf{E}_\lambda$ of the λth mode can be defined through the variance

$$(\Delta \mathbf{E}_\lambda)^2 = \left\langle \hat{E}_\lambda^2 \right\rangle - \left\langle \hat{E}_\lambda \right\rangle^2$$

$$= \frac{\hbar \omega_\lambda}{\varepsilon_0 V} \left(n_\lambda + \frac{1}{2} \right) \tag{9.79}$$

Equation (9.72) tells us that the *expectation value* of the electric field in the state $|n\rangle \, (= |n_1, n_2, \ldots, n_\lambda, \ldots\rangle)$ is zero. Since the average of sine waves with random phases is zero, we may *loosely* say that in the state $|n\rangle$ the phase of the electric field is completely uncertain.[3] Thus if we take an ensemble of identical states and make measurements of the phase of the field we will obtain random values spanning all values giving an average field value of zero. Some authors tend to explain this by resorting to the uncertainty principle

$$\Delta \mathbf{E}\, \Delta t \geq \hbar \tag{9.80}$$

where $\Delta \mathbf{E}$ is the uncertainty in the energy of the radiation field and Δt is related to the uncertainty in the phase angle through the relation

$$\Delta \phi = \omega \cdot \Delta t \tag{9.81}$$

Since $\mathbf{E} = \left(n + \frac{1}{2}\right) \hbar \omega$, $\Delta \mathbf{E} = \hbar \omega \, \Delta n$ and we obtain

$$\Delta n \, \Delta \phi \geq 1 \tag{9.82}$$

If the number of photons is exactly known, then $\Delta n = 0$ and consequently there is no knowledge of the phase. However, such arguments are not rigorously correct because it is not possible to give a precise definition of $\Delta \phi$ (see Section 9.6). Nevertheless, we can say that the states described by $|n\rangle$ do not correspond to the classical electromagnetic wave with a certain phase.

Returning to Eq. (9.74), we notice that the states have an amplitude $\left(\hbar \omega_\lambda / \varepsilon_0 V\right)^{1/2} \left(n_\lambda + \frac{1}{2}\right)^{1/2}$, for the mode λ, which is directly related to the number of photons.

A state with all $n_\lambda = 0$ for all λ, i.e., the state

$$|0\rangle = |0_1, 0_2, \ldots 0_\lambda, \ldots\rangle \tag{9.83}$$

is referred to as the vacuum state. In this state the variance in the electric field will be [see Eq. (9.79)]

$$\left(\Delta \mathbf{E}_\lambda^2\right) = \frac{\hbar \omega_\lambda}{2 \varepsilon_0 V} \tag{9.84}$$

Thus even when none of the modes are occupied, i.e., in vacuum, the variance of the electric field is finite. This is referred to as vacuum fluctuations and is responsible for many effects such as spontaneous emission, parametric down conversion, etc.

We can also introduce two quadrature operators defined as follows:

$$\hat{X}_1 = \frac{\hat{a} + \hat{a}^\dagger}{2}; \qquad \hat{X}_2 = \frac{\hat{a} - \hat{a}^\dagger}{2i} \tag{9.85}$$

---

[3] We say it loosely because it is not possible to define a phase operator which is real (see Section 9.7).

Note that the operators $\hat{a}$ and $\hat{a}^{\dagger}$ are not Hermitian operators, while $\hat{X}_1$ and $\hat{X}_2$ are Hermitian operators. Here we have dropped the subscript λ to keep the notation simple. We can express the electric field operator $\hat{E}$ [see Eq. (9.69)] in terms of the quadrature operators as follows:

$$\hat{E}(t) = \left(\frac{2\hbar\omega}{\varepsilon_0 V}\right)^{1/2} \left[\hat{X}_1 \sin(\omega t - \mathbf{k}.\mathbf{r}) - \hat{X}_2 \cos(\omega t - \mathbf{k}.\mathbf{r})\right] \hat{\mathbf{e}}_\lambda \qquad (9.86)$$

Since the operators $\hat{X}_1$ and $\hat{X}_2$ are factors multiplying sine and cosine terms, they are called quadrature operators. It is easy to show that the commutator

$$\left[\hat{X}_1, \hat{X}_2\right] = \frac{i}{2} \qquad (9.87)$$

For the number state we can show that the expectation values of $\hat{X}_1$ and $\hat{X}_2$ are zero and the variances are given by

$$(\Delta X_1)^2 = \frac{1}{2}\left(n + \frac{1}{2}\right) \qquad (9.88)$$

$$(\Delta X_2)^2 = \frac{1}{2}\left(n + \frac{1}{2}\right) \qquad (9.89)$$

and the product of the uncertainties in the two quadratures is given by

$$(\Delta X_1)(\Delta X_2) = \frac{1}{2}\left(n + \frac{1}{2}\right) \qquad (9.90)$$

The minimum value of the uncertainty product is $\tfrac{1}{4}$ and is for the vacuum state with $n = 0$. A state with the uncertainty product of $\tfrac{1}{4}$ is referred to as minimum uncertainty state (MUS). States with higher occupation numbers have larger uncertainty product. Also note that the uncertainties in both quadratures are equal.

Equation (9.90) implies that it is not possible to simultaneously measure precisely both the quadratures of the electromagnetic field. At the same time there are states in which the product of the uncertainties in the quadratures is $\tfrac{1}{4}$ but the uncertainty in either one of the quadrature is below the value of $\tfrac{1}{2}$. Such states are referred to as squeezed states (see Section 9.5).

Example 9.1 Let us consider a state described by the following ket

$$|\psi\rangle = \frac{1}{\sqrt{2}}\left(|1\rangle_i + |1\rangle_j\right) \qquad (9.91)$$

where $|1\rangle_i$ represents a number ket with the ith mode of the radiation field occupied by a single photon and all other modes being unoccupied. Similarly $|1\rangle_j$ represents a number ket with the jth mode of the radiation field occupied by a single photon and all other modes being unoccupied.

Now the total number operator for the radiation field is given by

$$\hat{N} = \sum_{l=0}^{\infty} \hat{a}_l^{\dagger}\hat{a}_l \qquad (9.92)$$

Thus

$$\hat{N}\,|\psi\rangle = \sum_{l=0}^{\infty} \hat{a}_l^\dagger \hat{a}_l\,|\psi\rangle = \frac{1}{\sqrt{2}} \sum_{l=0}^{\infty} \hat{a}_l^\dagger \hat{a}_l \left(|1\rangle_i + |1\rangle_i\right)$$
$$= \frac{1}{\sqrt{2}} \left(1\,|1\rangle_i + 1\,|1\rangle_j\right) \tag{9.93}$$
$$= 1\,|\psi\rangle$$

where we have used the fact that

$$\sum_{l=0}^{\infty} \hat{a}_l^\dagger \hat{a}_l\,|1\rangle_i = \left(\hat{a}_1^\dagger \hat{a}_1 + \hat{a}_2^\dagger \hat{a}_2 + \ldots \hat{a}_i^\dagger \hat{a}_i + \ldots\right)|1\rangle_i$$
$$= (0 + 0 + \ldots.1 + 0\ldots.)\,|1\rangle_i \tag{9.94}$$
$$= |1\rangle_i$$

and similarly for $|1\rangle_j$

Thus the state $|\psi\rangle$ is occupied by a single photon and is in a superposition state of occupying the ith mode and jth mode with equal probabilities of $\tfrac{1}{2}$.

Example 9.2 We consider another state in which only one mode is occupied. Let the state be

$$|\psi\rangle = \frac{1}{\sqrt{2}} \left(|0\rangle + |10\rangle\right)$$

i.e., the state is a superposition of two eigenstates, one a vacuum state $|0\rangle$ and the other a number state with $n = 10$. For simplicity we have omitted the subscript identifying the mode. Now the probability of detecting no photons is

$$|\langle 0 \mid \psi\rangle|^2 = \frac{1}{2} \tag{9.96}$$

Similarly the probability of detecting 10 photons is also $\tfrac{1}{2}$. The probability of detecting any other number of photons is zero. You can show that $|\psi\rangle$ is not an eigenket of the number operator. What would be the expectation value of the number of photons in this state?

Example 9.3 Let us consider two modes represented by subscripts 1 and 2 propagating along two different directions. For each direction of propagation we can have two independent states of polarization referred to as horizontal and vertical. We shall represent the ket corresponding to the occupation of mode 1 by a single horizontally polarized photon by $|\mathbf{H}\rangle_1$. Similarly let us represent the ket corresponding to the occupation of mode 1 by a single vertically polarized photon by $|V\rangle_1$. Similarly we shall have for mode 2 occupied by a single horizontally polarized photon or a single vertically polarized photon the following kets: $|\mathbf{H}\rangle_2$ and $|V\rangle_2$

Now let us consider the following ket:

$$|\psi\rangle = \frac{1}{\sqrt{2}} \left(|\mathbf{H}\rangle_1\,|V\rangle_2 - |V\rangle_1\,|\mathbf{H}\rangle_2\right) \tag{9.97}$$

First note that the ket cannot be written as a product of kets belonging to mode 1 and mode 2. Such a state is referred to as an *entangled* state.

To calculate the number of photons in this state we first see that the number operator that we must use to operate is given by

$$\hat{N} = \hat{a}_{1\mathbf{H}}^\dagger \hat{a}_{1\mathbf{H}} + \hat{a}_{1V}^\dagger \hat{a}_{1V} + \hat{a}_{2\mathbf{H}}^\dagger \hat{a}_{2\mathbf{H}} + \hat{a}_{2V}^\dagger \hat{a}_{2V} \tag{9.98}$$

since these are the only occupied modes. Now

$$\hat{N}\,|\psi\rangle = \left(\hat{a}_{1H}^{\dagger}\hat{a}_{1H} + \hat{a}_{1V}^{\dagger}\hat{a}_{1V} + \hat{a}_{2H}^{\dagger}\hat{a}_{2H} + \hat{a}_{2V}^{\dagger}\hat{a}_{2V}\right)\frac{1}{\sqrt{2}}\,(|\mathbf{H}\rangle_1\,|V\rangle_2 - |V\rangle_1\,|\mathbf{H}\rangle_2)$$

$$= \frac{1}{\sqrt{2}}\,(1\,|\mathbf{H}\rangle_1\,|V\rangle_2 + 0 + 0 + 1\,|\mathbf{H}\rangle_1\,|V\rangle_2 - 0 - 1\,|V\rangle_1\,|\mathbf{H}\rangle_2 - 1\,|V\rangle_1\,|\mathbf{H}\rangle_2 - 0) \qquad (9.99)$$

$$= 2\,\frac{1}{\sqrt{2}}\,(|\mathbf{H}\rangle_1\,|V\rangle_2 - |V\rangle_1\,|\mathbf{H}\rangle_2) = 2\,|\psi\rangle$$

which implies that the state is an eigenstate of the number operator and is occupied by 2 photons.

Now, in this state the polarization of the photon occupying mode 1 or mode 2 is undefined. For example if we pass mode 1 through a polarizer which has its pass axis oriented in the horizontal direction, then the probability of its getting transmitted through the polarizer is $\tfrac{1}{2}$. This is so because the probability that the photon in mode 1 is horizontal and the photon in mode 2 is also horizontal is zero since

$$|(_1\langle\mathbf{H}|\,_2\langle\mathbf{H}|)\,|\psi\rangle|^2 = \frac{1}{2}\,\big|\big[_1\langle\mathbf{H}\,|\,\mathbf{H}\rangle_1\,_2\langle\mathbf{H}\,|\,V\rangle_2 - {}_1\langle\mathbf{H}\,|\,V\rangle_1\,_2\langle\mathbf{H}\,|\,\mathbf{H}\rangle_2\big]\big|^2 = 0 \qquad (9.100)$$

Similarly the probability that the photon in mode 1 is horizontally polarized and the photon in mode 2 is vertically polarized is given by

$$|(_1\langle\mathbf{H}|\,_2\langle V|)\,|\psi\rangle|^2 = \frac{1}{2}\,\big|\big[_1\langle\mathbf{H}\,|\,\mathbf{H}\rangle_1\,_2\langle V\,|\,V\rangle_2 - {}_1\langle\mathbf{H}\,|\,V\rangle_1\,_2\langle V\,|\,\mathbf{H}\rangle_2\big]\big|^2 = \frac{1}{2} \qquad (9.101)$$

Thus the total probability that photon in mode 1 is horizontally polarized and the photon in mode 2 is either horizontally or vertically polarized is $\tfrac{1}{2}$. Similarly it can be shown that the probability that photon in mode 1 is vertically polarized and the photon in mode 2 is also vertically polarized is zero and the probability of photon in mode 1 being vertically polarized and the photon in mode 2 is horizontally polarized is $\tfrac{1}{2}$.

Now let us assume that we pass the photon through a polarizer with pass axis which is oriented in the horizontal direction. Now from the earlier discussions the probability of the photon passing through the polarizer is $\tfrac{1}{2}$. If the photon in mode 1 is detected after the polarizer then it implies that the photon in mode 1 is projected into the horizontal state of polarization. From the state vector it can be seen that the photon in mode 2 must be automatically projected into a vertical state of polarization. There is no more any uncertainty in the state of polarization of the photon in mode 2. Before the measurement on mode 1, the state of polarization of mode 2 was uncertain. However, measuring the state of polarization of mode 1 forces the photon in mode 2 to have a definite polarization state. This happens irrespective of the distance between the experimental arrangements detecting photons in mode 1 and 2. Similarly if the photon in mode 1 is not detected, then automatically the photon in mode 2 gets projected into a horizontal state of polarization. This mysterious correlation between the two photons is a characteristic of entangled states and is finding wide applications in the branch of quantum information science including quantum cryptography, quantum teleportation, and quantum computing. The process of spontaneous parametric down conversion discussed in Chapter 14 leads to such entangled states of photons.

9.4 The Coherent States

We next consider the radiation field to be in one of the coherent states which are the eigenkets of the operator $\hat{a}_\lambda$ (see Section 8.6.3). We will show that when the radiation field is in a coherent state, the field has properties very similar to that of a classical electromagnetic wave with a certain phase and amplitude. However, before we do so, we would like to discuss some of the properties of the coherent state.

The coherent states satisfy the equation

$$\hat{a}_\lambda \, |\alpha_\lambda\rangle = \alpha_\lambda \, |\alpha_\lambda\rangle \tag{9.102}$$

where α_λ, which represents the eigenvalue of $\hat{a}_\lambda$, can be an arbitrary complex number. In Section 8.6, we showed that

$$|\alpha_\lambda\rangle = \exp\left(-\frac{1}{2}\,|\alpha_\lambda|^2\right) \sum_{n_\lambda=0,1,\dots} \frac{\alpha_\lambda^{n_\lambda}}{(n_\lambda!)^{1/2}} \, |n_\lambda\rangle \tag{9.103}$$

For convenience we drop the subscript λ so that the above equation becomes

$$|\alpha\rangle = \exp\left(-\frac{1}{2}\,|\alpha|^2\right) \sum_{n=0,1,2,\dots} \frac{\alpha^n}{(n!)^{1/2}} \, |n\rangle \tag{9.104}$$

Some of the important properties of $|\alpha\rangle$ are discussed below:

(i) The expectation value of the number operator $\hat{N}_{op}\,(=\hat{a}^\dagger\hat{a})$ is given by

$$\langle\alpha|\,\hat{N}_{\mathrm{op}}\,|\alpha\rangle = e^{-|\alpha|^2} \sum_m \frac{\alpha^{*m}}{(m!)^{1/2}} \langle m| \sum_n \frac{\alpha^n}{(n!)^{1/2}} n \, |n\rangle$$

$$= e^{-|\alpha|^2} \sum_m \sum_n \frac{\alpha^{*m}\alpha^n}{(n!m!)^{1/2}} n \, \delta_{mn} \qquad \text{[using Eq. (8.102)]}$$

$$= e^{-|\alpha|^2} \sum_{n=1,2,\dots} \frac{n\,|\alpha|^{2n}}{n!}$$

or

$$\langle\alpha|\,\hat{N}_{\mathrm{op}}\,|\alpha\rangle = e^{-|\alpha|^2}\,|\alpha|^2 \sum_{n=0,1,\dots} \frac{|\alpha|^{2n}}{n!} = |\alpha|^2 = N \tag{9.105}$$

Thus the average number of photons (which we will denote by N) in the state $|\alpha\rangle$ is $|\alpha|^2$, and we may write

$$|\alpha\rangle = e^{-N/2} \sum_n \frac{\alpha^n}{n!} \, |n\rangle \tag{9.106}$$

(ii) From Eq. (9.106) it readily follows that the probability of finding n photons in a coherent state is given by

$$|\langle n \mid \alpha\rangle|^2 = \frac{1}{n!}\,|\alpha|^{2n} \exp\left(-|\alpha|^2\right) = \frac{N^n e^{-N}}{n!} \tag{9.107}$$

which is a Poisson distribution about the mean $|\alpha|^2$ [see also Section (18.9)].

(iii) In Section 8.9, we showed that if the field is in the coherent state at $t = 0$, then at a later time t the state will be given by

$$|\Psi(t)\rangle = e^{-N/2} \sum_n \frac{\alpha^n}{(n!)^{1/2}} |n\rangle \, e^{-i(n+1/2)\omega t} \tag{9.108}$$

It is easy to see that

$$|\Psi(0)\rangle = e^{-N/2} \sum_n \frac{\alpha^n}{(n!)^{1/2}} |n\rangle = |\alpha\rangle \tag{9.109}$$

Further

$$\langle \Psi(t)| \hat{N}_{\text{op}} |\Psi(t)\rangle = N \quad \text{(independent of time)} \tag{9.110}$$

and[4]

$$\langle \Psi(t)| \hat{a} |\Psi(t)\rangle = e^{-N} \sum_m \sum_n \frac{\alpha^{*m}\alpha^n}{(m!n!)^{1/2}} e^{i(m-n)\omega t} (n+1)^{1/2} \langle m \mid n+1 \rangle$$

$$= e^{i\omega t} e^{-N} \sum \frac{\alpha^* |\alpha|^{2n}}{n!} = \alpha^* e^{i\omega t} \tag{9.111}$$

Similarly (or, taking the complex conjugate of the above equation)

$$\langle \Psi(t)| \hat{a}^\dagger |\Psi(t)\rangle = \alpha e^{i\omega t} \tag{9.112}$$

We now consider the radiation field to be in the coherent state and calculate the expectation value of $\hat{E}$ and $\hat{E}^2$

$$\begin{aligned}
\langle \Psi(t)| \hat{E} |\Psi(t)\rangle &= i \left(\frac{\hbar\omega}{2\varepsilon_0 V} \right)^{1/2} \Big[\langle \Psi(t)| \hat{a} |\Psi(t)\rangle \, e^{i\mathbf{k}\cdot\mathbf{r}} \\
&\quad - \langle \Psi(t)| \hat{a}^\dagger |\Psi(t)\rangle \, e^{-i\mathbf{k}\cdot\mathbf{r}} \Big] \hat{\mathbf{e}} \\
&= i \left(\frac{\hbar\omega}{2\varepsilon_0 V} \right)^{1/2} \Big[\alpha e^{i(\mathbf{k}\cdot\mathbf{r}-\omega t)} - \alpha^* e^{-i(\mathbf{k}\cdot\mathbf{r}-\omega t)} \Big] \hat{\mathbf{e}} \\
&= \left(\frac{2\hbar\omega}{\varepsilon_0 V} \right)^{1/2} |\alpha| \sin(\omega t - \mathbf{k}\cdot\mathbf{r} + \varphi) \, \hat{\mathbf{e}}
\end{aligned} \tag{9.113}$$

where

$$\alpha = |\alpha| \, e^{i\varphi} \tag{9.114}$$

---

[4]In the Heisenberg representation, the expectation value of $\hat{a}$ would have been $\langle \Psi(0)| \hat{a}(t) |\Psi(0)\rangle = \langle \alpha| \hat{a} e^{i\omega t} |\alpha\rangle = a^* e^{i\omega t}$, which is the same as expressed by Eq. (9.111).

which resembles a classical sinusoidal electromagnetic wave. Thus the coherent state can be interpreted to represent a harmonic wave with phase φ. In a similar manner, we can calculate the expectation value of $\hat{E}^2$. The result is

$$\langle \Psi(t)| \hat{E} \cdot \hat{E} |\Psi(t)\rangle = \frac{\hbar\omega}{2\varepsilon_0 V} \left[1 + 4\,|\alpha|^2 \sin^2\left(\omega t - \mathbf{k}\cdot\mathbf{r} + \varphi\right)\right] \qquad (9.115)$$

Finally, the variance in $\hat{E}$ would be given by [cf. Eq. (9.79)]

$$(\Delta\mathbf{E})^2 = \langle \Psi(t)| \hat{E}\cdot\hat{E}|\Psi(t)\rangle - \langle\Psi(t)|\hat{E}|\Psi(t)\rangle^2$$

$$= \left(\frac{\hbar\omega}{2\varepsilon_0 V}\right) \qquad (9.116)$$

which is the same as was found for the vacuum state [see Eq. (9.84)]. Thus the coherent state has an expectation value of electric field resembling a classical electromagnetic wave and has a noise which is equal to the noise of vacuum field. Notice that the uncertainty $\Delta\mathbf{E}$ is independent of the amplitude $|\alpha|$; thus, the greater the intensity of the beam (i.e., larger is the expectation value of $\hat{E}\cdot\hat{E}$), the greater will be the proximity of the radiation field (corresponding to the coherent state) to the classical plane wave.

Earlier in this section, we had evaluated $\langle\alpha|\hat{N}_{\rm op}|\alpha\rangle$ – see Eq. (9.105); in a similar manner, we can calculate $\langle\alpha|\hat{N}_{\rm op}^2|\alpha\rangle$ from which we obtain[5]

$$\Delta N = \left[\langle\alpha|\hat{N}_{\rm op}^2|\alpha\rangle - \langle\alpha|\hat{N}_{\rm op}|\alpha\rangle^2\right]^{1/2} = N^{1/2} \qquad (9.117)$$

or

$$\frac{\Delta N}{N} = \frac{1}{N^{1/2}} \qquad (9.118)$$

implying that the fractional uncertainty in the average number of photons goes to zero with increase in intensity.

9.5 Squeezed States of Light

Another very important class of quantum states are the squeezed states. In order to understand squeezed states, we consider the following operator:

$$\hat{b} = \mu\,\hat{a} + \nu\,\hat{a}^\dagger \qquad (9.119)$$

---

[5]It is of interest to point out that even in nuclear counting, the uncertainty in the actual count is $N^{1/2}$ (see, e.g., Bleuler and Goldsmith (1952)).

where $\hat{a}$ and $\hat{a}^\dagger$ are the annihilation and creation operators for an electromagnetic mode and μ and ν are complex coefficients. We are again omitting the subscript for brevity. From Eq. (9.119) we have

$$\hat{b}^\dagger = \mu^* \hat{a}^\dagger + \nu^* \hat{a} \tag{9.120}$$

We also assume that μ and ν are related through

$$|\mu|^2 - |\nu|^2 = 1 \tag{9.121}$$

Squeezed states $|\beta\rangle$ are defined as the eigenstates of the operator $\hat{b}$ with eigenvalue β:
$$\hat{b}\,|\beta\rangle = \beta\,|\beta\rangle \tag{9.122}$$

Now the commutator between $\hat{b}$ and $\hat{b}^\dagger$ is given by

$$\begin{aligned}
\left[\hat{b}, \hat{b}^\dagger\right] = \hat{b}\hat{b}^\dagger - \hat{b}^\dagger \hat{b} &= \left(\mu\,\hat{a} + \nu\,\hat{a}^\dagger\right)\left(\mu^*\,\hat{a}^\dagger + \nu^*\,\hat{a}\right) \\
&\quad - \left(\mu^*\,\hat{a}^\dagger + \nu^*\,\hat{a}\right)\left(\mu\,\hat{a} + \nu\,\hat{a}^\dagger\right) = 1
\end{aligned} \tag{9.123}$$

where we have used the commutation relations between $\hat{a}$ and $\hat{a}^\dagger$. From Eqs. (9.119) and (9.120) we obtain

$$\hat{a} = \mu^*\hat{b} - \nu\,\hat{b}^\dagger \tag{9.124}$$

$$\hat{a}^\dagger = \mu\,\hat{b}^\dagger - \nu^*\,\hat{b} \tag{9.125}$$

The states $|\beta\rangle$ are referred to as squeezed states.

In order to understand the nature of these quantum states we assume for simplicity that μ, ν, and β are real. Now the time dependence of the operators $\hat{a}(t)$ and $\hat{a}^\dagger(t)$ are given by

$$\hat{a}(t) = \hat{a}(0)e^{-i\omega t} \tag{9.126}$$

$$\hat{a}^\dagger(t) = \hat{a}^\dagger(0)e^{i\omega t} \tag{9.127}$$

Thus we have

$$\hat{b}(0) = \mu\,\hat{a}(0) + \nu\,\hat{a}^\dagger(0) = \mu\,\hat{a}_0 + \nu\,\hat{a}_0^\dagger \tag{9.127}$$

$$\hat{b}(t) = \mu\,\hat{a}_0 e^{-i\omega t} + \nu\,\hat{a}_0^\dagger e^{i\omega t} \tag{9.128}$$

$$\hat{b}^\dagger(t) = \mu\,\hat{a}_0^\dagger e^{i\omega t} + \nu\hat{a}_0 e^{-i\omega t} \tag{9.129}$$

We now calculate the expectation value of the electric field in the state $|\beta\rangle$. Referring to Eq. (9.70), we note that the electric field operator is given by

$$\hat{E} = i\left(\frac{\hbar\omega}{2\varepsilon_0 V}\right)^{1/2}\left(\hat{a}e^{i(kz-\omega t)} - \hat{a}^\dagger e^{-i(kz-\omega t)}\right) \tag{9.130}$$

where we consider a mode propagating in the z-direction. Now

$$\langle\beta|\,\hat{a}\,|\beta\rangle = \langle\beta|\,\mu\,\hat{b} - \nu\,\hat{b}^\dagger\,|\beta\rangle = (\mu - \nu)\,\beta \tag{9.131}$$

Similarly

$$\langle\beta|\,\hat{a}^\dagger\,|\beta\rangle = \langle\beta|\,\mu\,\hat{b}^\dagger - \nu\,\hat{b}\,|\beta\rangle = (\mu - \nu)\,\beta \tag{9.132}$$

Hence

$$\begin{aligned}
\langle\hat{E}\rangle &= \langle\beta|\,\hat{E}\,|\beta\rangle = i\left(\frac{\hbar\omega}{2\varepsilon_0 V}\right)^{1/2}\langle\beta|\left(\hat{a}e^{i(\mathbf{k}\cdot\mathbf{r}-\omega t)} - \hat{a}^\dagger e^{-i(\mathbf{k}\cdot\mathbf{r}-\omega t)}\right)|\beta\rangle\\
&= i\left(\frac{\hbar\omega}{2\varepsilon_0 V}\right)^{1/2}(\mu - \nu)\,\beta 2i\,\sin(\omega t - kz) \\
&= \left(\frac{2\hbar\omega}{\varepsilon_0 V}\right)^{1/2}(\mu - \nu)\,\beta\,\sin(\omega t - kz)
\end{aligned} \tag{9.133}$$

The expectation value of $\hat{E}^2$ is given by

$$\begin{aligned}
\langle\hat{E}^2\rangle &= \langle\beta|\,\hat{E}\hat{E}\,|\beta\rangle = -i\frac{\hbar\omega}{2\varepsilon_0 V}[\langle\beta|\,\hat{a}\hat{a}\,|\beta\rangle\,e^{-2i(\omega t - kz)} - \langle\beta|\,\hat{a}\hat{a}^\dagger\,|\beta\rangle \\
&\quad - \langle\beta|\,\hat{a}^\dagger\hat{a}\,|\beta\rangle + \langle\beta|\,\hat{a}^\dagger\hat{a}^\dagger\,|\beta\rangle\,e^{2i(\omega t - kz)}]
\end{aligned} \tag{9.134}$$

Now

$$\begin{aligned}
\langle\beta|\,\hat{a}\hat{a}\,|\beta\rangle &= \langle\beta|\,(\mu\hat{b} - \nu\hat{b}^\dagger)(\mu\hat{b} - \nu\hat{b}^\dagger)\,|\beta\rangle \\
&= \mu^2\langle\beta|\,\hat{b}^2\,|\beta\rangle - \mu\nu\langle\beta|\,\hat{b}\hat{b}^\dagger\,|\beta\rangle - \mu\nu\langle\beta|\,\hat{b}^\dagger\hat{b}\,|\beta\rangle + \nu^2\langle\beta|\,\hat{b}^{\dagger 2}\,|\beta\rangle \\
&= \mu^2\beta^2 - \mu\nu - \mu\nu\beta^2 - \mu\nu\beta^2 + \nu^2\beta^2 \\
&= \beta^2(\mu - \nu)^2 - \mu\nu
\end{aligned} \tag{9.135}$$

Similarly

$$\langle\beta|\,\hat{a}^\dagger\hat{a}^\dagger\,|\beta\rangle = \beta^2(\mu - \nu)^2 - \mu\nu \tag{9.136}$$

$$\langle\beta|\,\hat{a}^\dagger\hat{a}\,|\beta\rangle = \beta^2(\mu - \nu)^2 + \nu^2 \tag{9.137}$$

$$\langle\beta|\,\hat{a}\hat{a}^\dagger\,|\beta\rangle = \beta^2(\mu - \nu)^2 + \nu^2 + 1 \tag{9.138}$$

Substituting the values in Eq. (9.134) and simplifying we obtain

$$\left\langle \hat{E}^2 \right\rangle = \frac{\hbar\omega}{\varepsilon_0 V}\left[-\left\{ \beta^2 (\mu - v)^2 - \mu v \right\} \cos 2(\omega t - kz) + \beta^2 (\mu - v)^2 + v^2 + \frac{1}{2} \right]$$
$$(9.139)$$

Thus the variance in the electric field is given by

$$\left\langle \Delta \hat{E} \right\rangle^2 = \left\langle \mathbf{E}^2 \right\rangle - \langle \mathbf{E} \rangle^2$$
$$= \frac{\hbar\omega}{2\varepsilon_0 V}\left[\mu^2 + v^2 + 2\mu v \, \cos 2(\omega t - kz) \right]$$
$$(9.140)$$

From Eq. (9.140) we notice that the variance in the electric field in the squeezed state oscillates between $\frac{\hbar\omega}{2\varepsilon_0 V} (\mu + v)^2$ and $\frac{\hbar\omega}{2\varepsilon_0 V} (\mu - v)^2$ at a frequency 2ω. Comparing this with a coherent state for which the variance is a constant and equals $\frac{\hbar\omega}{2\varepsilon_0 V}$ (which can be obtained by putting $v = 0$ in which case the squeezed state becomes a coherent state) we find that the variance can go below that in the vacuum state at certain times. Note that μ and v are related through Eq. (9.121).

From the commutation relations it is possible to evaluate the expectation values and variances of the quadrature operators. We give here the results:

$$\left\langle \hat{X}_1 \right\rangle = \langle \beta | \hat{X}_1 | \beta \rangle = (\mu - v) \, \beta \, \cos \omega t \qquad (9.141)$$

$$\left\langle \hat{X}_2 \right\rangle = \langle \beta | \hat{X}_2 | \beta \rangle = -(\mu - v) \, \beta \, \sin \omega t \qquad (9.142)$$

$$\left\langle \hat{X}_1^2 \right\rangle = \frac{1}{4} (\mu - v)^2 + (\mu - v)^2 \beta^2 \cos^2 \omega t + \mu v \sin^2 \omega t \qquad (9.143)$$

$$\left\langle \hat{X}_2^2 \right\rangle = \frac{1}{4} (\mu - v)^2 + (\mu - v)^2 \beta^2 \sin^2 \omega t + \mu v \cos^2 \omega t \qquad (9.144)$$

Hence

$$(\Delta X_1)^2 = \frac{1}{4} (\mu - v)^2 + \mu v \sin^2 \omega t \qquad (9.145)$$

$$(\Delta X_2)^2 = \frac{1}{4} (\mu - v)^2 + \mu v \cos^2 \omega t \qquad (9.146)$$

As can be seen from the above equations, for example, for positive values of μ and v, the uncertainty in the quadrature $\hat{X}_1$ varies from $\frac{1}{2} (\mu - v)$ at $t = 0, \pi/\omega$ to $\frac{1}{2} (\mu + v)$ at $\pi/2\omega, 3\pi/2\omega$, etc. The uncertainty oscillates periodically with time and attains value below that for vacuum at periodic instants of time. Of course at those times the noise in the other quadrature is more than for vacuum and the product of their uncertainties will always be equal to or more than $\frac{1}{4}$.

This implies that the uncertainty in the quadrature $\hat{X}_1$ (which is the coefficient of the sine term) goes below that of vacuum state at periodic intervals. Such squeezed states of light find many applications such as in communication, gravitational wave detection.

9.6 Transition Rates

The Hamiltonian of a system consisting of an atom in a radiation field can be written as

$$\hat{H} = \hat{H}_0 + \hat{H}'$$

$$= \hat{H}_a + \hat{H}_r + \hat{H}' \tag{9.147}$$

where $\hat{H}_a$ represents the Hamiltonian of the atom, $\hat{H}_r$ the Hamiltonian corresponding to the pure radiation field [see Eq. (9.58)], and H' represents the interaction between the atom and the radiation field. We will consider $H_0 \ (= H_a + H_r)$ as the unperturbed Hamiltonian and H' will be considered as a perturbation which will be assumed to be of the form [see Eqs. (9.32) and (9.70)]

$$\hat{H}' = -e\mathbf{E} \cdot \mathbf{r}$$

$$= -e\sum_\lambda \mathbf{E}_\lambda \cdot \mathbf{r} \tag{9.148}$$

Now, the eigenvalue equations for $\hat{H}_a$ and $\hat{H}_r$ are

$$\hat{H}_a \, |\psi_i\rangle = \mathbf{E}_i \, |\psi_i\rangle \tag{9.149}$$

and

$$\hat{H}_r \, |n_1, n_2, \ldots n_\lambda, \ldots\rangle = \left[\sum_\lambda \left(n_\lambda + \frac{1}{2}\right) \hbar\omega_\lambda\right] |n_1, n_2, \ldots n_\lambda, \ldots\rangle \tag{9.150}$$

where $|\psi_i\rangle$ and $\mathbf{E}_i$ represent, respectively, the eigenkets and energy eigenvalues of the isolated atom and $|n_1, n_2, \ldots, n_\lambda, \ldots\rangle$ represents the eigenket of the pure radiation field with $\sum_\lambda \left(n_\lambda + \frac{1}{2}\right) \hbar\omega_\lambda$ representing the corresponding eigenvalue [see Eq. (9.62)]. Thus the eigenvalue equation for $\mathbf{H}_0$ will be

$$\hat{H}_0 \, |u_n\rangle = W_n \, |u_n\rangle \tag{9.151}$$

where

$$W_n = \mathbf{E}_n + \sum_\lambda \left(n_\lambda + \frac{1}{2}\right) \hbar\omega_\lambda \tag{9.152}$$

and

$$|u_n\rangle = |i\rangle \, |n_1, n_2, \ldots, n_\lambda, \ldots\rangle = |i; \, n_1, n_2, \ldots, n_\lambda, \ldots\rangle \tag{9.153}$$

represents the ket corresponding to the atom being in state $|i\rangle$ and the radiation being in the state $|n_1, n_2, \ldots, n_\lambda, \ldots\rangle$.

Now the Schrödinger equation for the system consisting of the atom and the radiation field is

$$i\hbar\frac{\partial}{\partial t} \, |\Psi\rangle = \left(\hat{H}_0 + \hat{H}'\right) |\Psi\rangle \tag{9.154}$$

As in Section 3.3, the solution of the above equation can be written as a linear combination of the eigenkets of $\hat{H}_0$ [cf. Eq. (3.38)]:

$$|\Psi\rangle = \sum_n C_n(t)\, e^{-iW_n t/\hbar}\, |u_n\rangle \tag{9.155}$$

Substituting in Eq. (9.154), we obtain

$$i\hbar \sum_n \left[\frac{dC_n}{dt} - \frac{iW_n}{\hbar} C_n \right] e^{-iW_n t/\hbar} |u_n\rangle$$
$$= \sum_n C_n(t)\, W_n e^{-iW_n t/\hbar} |u_n\rangle + H' \sum_n C_n(t)\, e^{-iW_n t/\hbar} |u_n\rangle \tag{9.156}$$

where we have used Eq. (9.151). It is immediately seen that the second term on the left-hand side cancels exactly with the first term on the right-hand side. If we now multiply by $\langle u_m|$ on the left, we would get

$$i\hbar \frac{dC_m}{dt} = \sum_n \langle u_m|\hat{H}'|u_n\rangle\, e^{i(W_m - W_n)t/\hbar} C_n(t) \tag{9.157}$$

Now, using Eq. (9.69) for $\mathbf{E}_\lambda$, we obtain for H':

$$\hat{H}' = -ie \sum_\lambda \left(\frac{\hbar\omega_\lambda}{2\varepsilon_0 V} \right)^{1/2} \left[\hat{a}_\lambda e^{i\mathbf{k}_\lambda \cdot \mathbf{r}} - \hat{a}_\lambda^\dagger e^{-i\mathbf{k}_\lambda \cdot \mathbf{r}} \right] \hat{\mathbf{e}}_\lambda \cdot r \tag{9.158}$$

Now, because of the appearance of $\hat{a}_\lambda$ and $\hat{a}_\lambda^\dagger$ in the expression for $\hat{H}'$, the various terms in $\langle u_m|\hat{H}'|u_n\rangle$ will be non-zero only if the number of photons in $|u_m\rangle$ differs by unity from the number of photons in $|u_n\rangle$. If we write out completely the right-hand side of Eq. (9.157) it will lead to a coupled set of an infinite number of equations which would be impossible to solve. We employ the perturbation theory and consider the absorption of one photon (of energy $\hbar\omega_i$) from the ith mode. Further, if we assume the frequency ω_i to be very close to the resonant frequency corresponding to the transition from the atomic state $|a\rangle$ to $|b\rangle$, then Eq. (9.157) reduces to the following two coupled equations:

$$i\hbar \frac{dC_1}{dt} = H'_{12} e^{i(W_1 - W_2)t/\hbar} C_2(t) \tag{9.159}$$

$$i\hbar \frac{dC_2}{dt} = H'_{21} e^{-i(W_1 - W_2)t/\hbar} C_1(t) \tag{9.160}$$

where

$$|1\rangle = |a; n_1, n_2, \ldots, n_\lambda, \ldots, n_i, \ldots\rangle \tag{9.161}$$

and

$$|2\rangle = |b; n_1, n_2, \ldots, n_\lambda, \ldots, n_i - 1, \ldots\rangle \qquad (9.162)$$

represent the initial and final states of the system. Obviously, because of relations like Eqs. (9.73),

$$\langle 1| \hat{H}' |1\rangle = H'_{11} = 0 = H'_{22}$$

Further,

$$W_1 = \mathbf{E}_a + \sum_{\substack{\lambda \\ \lambda \neq i}} \left(n_\lambda + \frac{1}{2} \right) \hbar \omega_\lambda + \left(n_i + \frac{1}{2} \right) \hbar \omega_i \qquad (9.163)$$

and

$$W_2 = \mathbf{E}_b + \sum_{\substack{\lambda \\ \lambda \neq i}} \left(n_\lambda + \frac{1}{2} \right) \hbar \omega_\lambda + \left(n_i - 1 + \frac{1}{2} \right) \hbar \omega_i \qquad (9.164)$$

Thus,

$$W_1 - W_2 = (\mathbf{E}_a - \mathbf{E}_b) + \hbar \omega_i \qquad (9.165)$$

Now,

$$
\begin{aligned}
H_{21}^{\prime *} = H'_{12} &= \sum_\lambda \langle a; n_1, n_2, \ldots, n_i, \ldots | \, (-ie) \left(\frac{\hbar \omega_\lambda}{2\varepsilon_0 V} \right)^{1/2} \\
&\quad \times \left[\hat{a}_\lambda e^{i\mathbf{k}_\lambda \cdot \mathbf{r}} - \hat{a}_\lambda^\dagger e^{-i\mathbf{k}_\lambda \cdot \mathbf{r}} \right] \hat{\mathbf{e}}_\lambda \cdot \mathbf{r} \, |b; n_1, n_2, \ldots, n_i - 1, \ldots\rangle \\
&= \langle a; n_1, n_2, \ldots, n_i, \ldots | \left[+ie \left(\frac{\hbar \omega_i}{2\varepsilon_0 V} \right)^{1/2} e^{-i\mathbf{k}_\lambda \cdot \mathbf{r}} \hat{\mathbf{e}}_i \cdot \mathbf{r} \, (n_i)^{1/2} \right] \\
&\quad \times |b; n_1, n_2, \ldots, n_i, \ldots\rangle \\
&= ie \left(\frac{\hbar \omega_i}{2\varepsilon_0 V} \right)^{1/2} (n_i)^{1/2} \langle a| \, e^{-i\mathbf{k}_\lambda \cdot \mathbf{r}} \mathbf{r} \, |b\rangle \cdot \hat{\mathbf{e}}_i \qquad (9.166)
\end{aligned}
$$

where we have used Eqs. (9.64), (9.65), and (9.66). In calculating the above matrix element between the atomic states a and b we note that the atomic wave functions are almost zero for $r \geq 10^{-8}$ cm. On the other hand, since in the optical region, $|\mathbf{k}| \, (= 2\pi/\lambda) \approx 10^5$ cm^{-1}, in the domain of integration, in the evaluation of the matrix element $\mathbf{k}_\lambda \cdot \mathbf{r} << 1$. Thus we may replace $e^{-i\mathbf{k}_\lambda \cdot \mathbf{r}}$ by unity; this is known as the dipole approximation, and we obtain

$$H'_{12} = -i \left(\frac{\hbar \omega_i}{2\varepsilon_0 V} \right)^{1/2} (n_i)^{1/2} D_{ab} \qquad (9.167)$$

where D_{ab} is defined through Eq. (4.71). We next try to solve Eqs. (9.159) and (9.160) by using a method similar to that employed in Section 4.7. We assume that at $t = 0$ the system is in the state represented by $|1\rangle$, i.e.,

$$C_1(0) = 1, \quad C_2(0) = 0 \tag{9.168}$$

On working out the solution one obtains

$$|C_2(t)|^2 = \left(\frac{\tilde{\Omega}_0}{2}\right)^2 \left[\frac{\sin\left(\tilde{\Omega}'t/2\right)}{\tilde{\Omega}'/2}\right]^2 \tag{9.169}$$

where

$$\tilde{\Omega}_0^2 = \frac{2\omega_i n_i D_{ab}^2}{\hbar \varepsilon_0 V} \tag{9.170}$$

$$\tilde{\Omega}' = \left[\left(\frac{\mathbf{E}_b - \mathbf{E}_a}{\hbar} - \omega_i\right)^2 + \tilde{\Omega}_0^2\right]^{1/2} \tag{9.171}$$

For $\left(\tilde{\Omega}_0^2 t^2/\hbar^2\right) << 1$, we obtain

$$|C_2(t)|^2 \approx \frac{\omega n}{2\hbar \varepsilon_0 V} D_{ab}^2 \left(\frac{\sin\left\{\left[(\mathbf{E}_b - \mathbf{E}_a)/\hbar - \omega\right](t/2)\right\}}{\frac{1}{2}\left[(\mathbf{E}_b - \mathbf{E}_a)/\hbar - \omega\right]}\right)^2 \tag{9.172}$$

where we have dropped the subscript i. Equation (9.172) is the same as Eq. (4.77) provided we replace $\mathbf{E}_0^2$ by $2\hbar\omega n/\varepsilon_0 V$ (see Section 4.7.1). Similarly, if we consider the emission process, we would obtain

$$|C_2(t)|^2 = \frac{\omega(n+1)}{2\hbar \varepsilon_0 V} D_{ab}^2 \left(\frac{\sin\left\{\left[(\mathbf{E}_b - \mathbf{E}_a)/\hbar - \omega\right](t/2)\right\}}{\frac{1}{2}\left[(\mathbf{E}_b - \mathbf{E}_a)/\hbar - \omega\right]}\right)^2 \tag{9.173}$$

where the states $|1\rangle$ and $|2\rangle$ are now given by

$$|1\rangle = |b; n_1, n_2, \ldots, n_i, \ldots\rangle \tag{9.174}$$

$$|2\rangle = |a; n_1, n_2, \ldots, n_i + 1, \ldots\rangle \tag{9.175}$$

Notice the presence of the term[6] $(n+1)$ in Eq. (9.173). This implies that even if the number of photons were zero originally, the emission probability is finite. The term proportional to n in Eq. (9.173) gives the probability for induced or stimulated

---

[6]The appearance of the term $(n+1)$ is because of the relation

$$\hat{a}^\dagger |n\rangle = (n+1)^{1/2} |n+1\rangle$$

emission since the rate at which it occurs if proportional to the intensity of the applied radiation. On the other hand, the second term, which is independent of n, gives the spontaneous emission rate into the mode. Observe that the spontaneous emission probability into a particular mode is exactly the same as the stimulated emission probability caused by a single photon into the same mode, a fact which we have used in Section 5.5.

We next calculate the probability per unit time for spontaneous emission of radiation. If we consider the emission to be in the solid angle $d\Omega$ then the number of modes for which the photon frequency lies between ω and $\omega + d\omega$ is (see Appendix E)

$$N(\omega)\, d\omega\, d\Omega = \frac{V\omega^2\, d\omega}{8\pi^3 c^3} d\Omega \tag{9.176}$$

Thus, the total probability of emission in the solid angle $d\Omega$ would be given by

$$\Gamma = \frac{|D_{ab}|^2}{2\hbar\varepsilon_0 V} \int \left\{ \frac{\sin\left[(\omega_{ba} - \omega)\,(t/2)\right]}{(\omega_{ba} - \omega)/2} \right\}^2 \omega \frac{V}{8\pi^3 c^3}\omega^2\, d\omega\, d\Omega$$

$$\approx \frac{|D_{ab}|^2}{16\pi^3\hbar\varepsilon_0 c^3} d\Omega\, \omega_{ba}^3 \int \left\{ \frac{\sin\left[(\omega_{ba} - \omega)\,(t/2)\right]}{(\omega_{ba} - \omega)/2} \right\}^2 d\omega \tag{9.177}$$

where use has been made of the fact that the quantity inside the curly brackets is a sharply peaked function around $\omega = \omega_{ba}$. Carrying out the integration, using the fact that

$$\int_{-\infty}^{+\infty} \frac{\sin^2 x}{x^2}\, dx = \pi \tag{9.178}$$

we obtain

$$\Gamma \approx \frac{1}{2\pi} \left(\frac{e^2}{4\pi\varepsilon_0\hbar c} \right) \frac{\omega^3}{c^2} \left| \langle b|\, \mathbf{r}\,|a\rangle \cdot \hat{\mathbf{e}} \right|^2 t\, d\Omega \tag{9.179}$$

Thus the transition rate is given by

$$w_{sp} = \frac{1}{2\pi} \left[\frac{e^2}{4\pi\varepsilon_0\hbar c} \right] \frac{\omega^3}{c^2} \left| \langle b|\, \mathbf{r}\,|a\rangle \cdot \hat{\mathbf{e}} \right|^2 d\Omega \tag{9.180}$$

In order to calculate the total probability per unit time for the spontaneous emission to occur (the inverse of which will give the spontaneous lifetime of the state), we must sum over the two independent states of polarization and integrate over the solid angle. Assuming the direction of $\mathbf{k}$ to be along the z-axis, we may choose $\mathbf{E}$ to be along the x or y axes. Thus, if we sum

$$\left| \langle b|\, \mathbf{r}\,|a\rangle \cdot \hat{\mathbf{e}} \right|^2$$

over the two independent states of polarization, we obtain

$$\left| \langle b|\, \mathbf{r}\,|a\rangle \cdot \hat{\mathbf{x}} \right|^2 + \left| \langle b|\, \mathbf{r}\,|a\rangle \cdot \hat{\mathbf{y}} \right|^2 = P_x^2 + P_y^2 = P^2 \sin^2\theta \tag{9.181}$$

where $\mathbf{P} \equiv \langle b| \mathbf{r} |a\rangle$ and θ is the angle that $\mathbf{P}$ makes with the z-axis. Thus in order to obtain the Einstein A coefficient (which represents the total probability per unit time for the spontaneous emission to occur), in Eq. (9.180), we replace $|\langle b| \mathbf{r} |a\rangle \cdot \hat{\mathbf{e}}|^2$ by $|\langle b| \mathbf{r} |a\rangle|^2 \sin^2 \theta$ and integrate over the solid angle $d\Omega$ to obtain

$$
\begin{aligned}
A &= \frac{1}{2\pi} \left[\frac{e^2}{4\pi \varepsilon_0 \hbar c} \right] \frac{\omega^3}{c^2} |\langle b| \mathbf{r} |a\rangle|^2 \int_0^\pi \int_0^{2\pi} \sin^2 \theta \, \sin \theta \, d\theta \, d\varphi \\
&= \frac{4}{3} \left[\frac{e^2}{4\pi \varepsilon_0 \hbar c} \right] \frac{\omega^3}{c^2} |\langle b| \mathbf{r} |a\rangle|^2
\end{aligned}
\tag{9.182}
$$

which is identical to Eq. (4.88).

9.7 The Phase Operator[7]

Classically, for the Hamiltonian given by

$$
\mathbf{H} = \frac{p^2}{2m} + \frac{1}{2} m\omega^2 x^2
\tag{9.183}
$$

the Hamilton equations of motion (see, e.g., Goldstein (1950))

$$
\dot{x} = \frac{\partial \mathbf{H}}{\partial p} \quad \text{and} - \dot{p} = \frac{\partial \mathbf{H}}{\partial x}
\tag{9.184}
$$

gives us

$$
\dot{x} = \frac{p}{m} \quad \text{and} - \dot{p} = m\omega^2 x^2
\tag{9.185}
$$

Thus

$$
\ddot{x} = \frac{1}{m} \dot{p} = -\omega^2 x
\tag{9.186}
$$

or

$$
\ddot{x} + \omega^2 x = 0
\tag{9.187}
$$

giving

$$
x = Ae^{+i\phi} + Ae^{-i\phi}, \quad \phi = \omega t
\tag{9.188}
$$

and

$$
p = m\dot{x} = im\omega \left(Ae^{i\phi} - Ae^{-i\phi} \right)
\tag{9.189}
$$

---

[7] A major part of the discussion in this section is based on a paper by Susskind and Glogower (1964).

where A has been assumed to be real. In general, we should have written $\phi = \omega t + \alpha$, but with proper choice of the time origin, we can always choose $\alpha = 0$. In quantum mechanics

$$\hat{x} = \left(\frac{2\hbar}{m\omega}\right)^{1/2}\left(\hat{a}^{\dagger} + \hat{a}\right) \tag{9.190}$$

and

$$\hat{p} = im\omega\left(\frac{2\hbar}{m\omega}\right)^{1/2}\left(\hat{a}^{\dagger} - \hat{a}\right) \tag{9.191}$$

[see Eqs. (8.39) and (8.40)]. If we compare Eqs. (9.190) and (9.191) with Eqs. (9.188) and (9.189), we are tempted to define the phase operator ϕ through the equations

$$\hat{a}^{\dagger} = \hat{R}e^{i\hat{\phi}} \quad \text{and} \quad \hat{a} = e^{-i\hat{\phi}}\hat{R} \tag{9.192}$$

where $\hat{R}$ and $\hat{\phi}$ are assumed to be Hermitian operators. This is indeed what has been done by Dirac (1958b) and Heitler (1954); however, this definition leads to inconsistent results as will be shown later. Now, the number operator is given by

$$\hat{N}_{\text{op}} = \hat{a}^{\dagger}\hat{a} = \hat{R}e^{i\hat{\phi}}e^{-i\hat{\phi}}\hat{R} = \hat{R}^2 \tag{9.193}$$

Thus

$$\hat{R} = \hat{N}_{\text{op}}^{1/2} \tag{9.194}$$

where the square root operator is defined through the relation

$$\hat{N}_{\text{op}} = \hat{N}_{\text{op}}^{1/2}\hat{N}_{\text{op}}^{1/2} \tag{9.195}$$

Now

$$\left[\hat{a}, \hat{a}^{\dagger}\right] = \hat{a}\hat{a}^{\dagger} - \hat{a}^{\dagger}\hat{a} = 1 \quad \text{[see Eq. (8.44)]} \tag{9.196}$$

Thus

$$e^{-i\hat{\phi}}\hat{R}\,\hat{R}e^{i\hat{\phi}} - \hat{R}e^{i\hat{\phi}}e^{-i\hat{\phi}}\hat{R} = 1 \tag{9.197}$$

or, premultiplying by $e^{i\hat{\phi}}$, we get

$$\hat{N}_{\text{op}}e^{i\hat{\phi}} - e^{i\hat{\phi}}\hat{N}_{\text{op}} = e^{i\hat{\phi}} \tag{9.198}$$

Hence

$$\left[\hat{N}_{\text{op}}, e^{i\hat{\phi}}\right] = e^{i\hat{\phi}} \tag{9.199}$$

The above equation is satisfied if $\hat{\phi}$ and $\hat{N}_{op}$ satisfy the commutation relation[8]

$$\left[\hat{N}_{op}, \hat{\phi}\right] = \hat{N}_{op}\hat{\phi} - \hat{\phi}\hat{N}_{op} = -i \tag{9.200}$$

The above equation suggests the uncertainty relation[9]

$$\Delta N \, \Delta\phi \geq 1 \tag{9.201}$$

From the above equation it follows that if the number of light quanta of a wave are given, the phase of this wave is entirely undetermined and vice versa (Heitler 1954). However, the above definition of the phase operator (and hence the uncertainty relation) is not correct because the definition leads to inconsistent results. For example, let us consider the matrix element:

$$\langle m| \left[\hat{N}_{op}, \hat{\phi}\right] |n\rangle = -i \langle m \mid n\rangle = -i\delta_{m,n} \tag{9.202}$$

But

$$\langle m| \left[\hat{N}_{op}, \hat{\phi}\right] |n\rangle = \langle m| \hat{N}_{op}\hat{\phi} - \hat{\phi}\hat{N}_{op} |n\rangle = (m - n) \langle m| \phi |n\rangle \tag{9.203}$$

Thus

$$(m - n) \langle m| \phi |n\rangle = -i\delta_{m,n} \tag{9.204}$$

which is certainly an impossibility. One can also show that $e^{-i\hat{\phi}}e^{+i\hat{\phi}}$ is not a unit operator. Thus the definition of a Hermitian $\hat{\phi}$ through Eq. (9.192) leads

---

[8] This can be shown by noting that repeated application of Eq. (9.200) gives

$$N_{op}\phi^m - \phi^m N_{op} = -im\phi^{m-1}$$

Now

$$[N_{op}, e^{i\phi}] = \sum_m \frac{i^m}{m!} [N_{op}, e^m] = \sum \frac{i^{m-1}\phi^{m-1}}{(m-1)!} = e^{i\phi}$$

[9] Equation (8.152) may be compared with the equation [see Eq. (7.37)]

$$[x, p_x] = xp_x - p_x x = i\hbar$$

from which one can derive the uncertainty relation

$$\Delta x \, \Delta p_x \geq \hbar$$

(see any text on quantum mechanics, e.g., Powell and Craseman (1961)).

to inconsistent results. Susskind and Glogower (1964) define the phase operator through the equations

$$\hat{P}_{\exp -} \equiv \hat{a} \left(\hat{N}_{\mathrm{op}} + 1 \right)^{-1/2} \tag{9.205}$$

and

$$\hat{P}_{\exp +} \equiv \left(\hat{N}_{\mathrm{op}} + 1 \right)^{-1/2} \hat{a}^{\dagger} \tag{9.206}$$

The subscripts to $\hat{P}$ imply that a certain limiting sense $\hat{P}_{\exp -}$ and $\hat{P}_{\exp +}$ behave as $e^{-i\hat{\phi}}$ and $e^{+i\hat{\phi}}$, respectively. These operators are used to define trigonometric functions of phase:

$$\hat{P}_{\cos} = \frac{1}{2} \left(\hat{P}_{\exp +} + \hat{P}_{\exp -} \right) \tag{9.207}$$

$$\hat{P}_{\sin} = \frac{1}{2i} \left(\hat{P}_{\exp +} - \hat{P}_{\exp -} \right) \tag{9.208}$$

Both $\hat{P}_{\cos}$ and $\hat{P}_{\sin}$ are Hermitian operators; however, they do not commute. Indeed, it is the non-commuting nature of $\hat{P}_{\cos}$ and $\hat{P}_{\sin}$ which makes $e^{+i\hat{\phi}}$ not unitary.

Although we leave the discussion on the phase operator rather abruptly here, what we have shown is that the phase operator for an oscillator cannot exist. We refer the reader to the works of Susskind and Glogower (1964) and of Loudon (1973), where they show that the operators $\hat{P}_{\cos}$ and $\hat{P}_{\sin}$ are observables and that in a certain limiting sense they do become the classical functions of phase. Indeed these operators can be used to define uncertainty relations.

9.8 Photons Incident on a Beam Splitter

A beam splitter is a device which is used in many optical experiments and from an incident beam gives rise to a transmitted beam and a reflected beam (see Fig. 9.1). In this section we consider a lossless beam splitter and assume that it has a transmittivity of 0.5 and a reflectivity of 0.5. It can be shown that for a symmetric beam splitter covered by the same medium on both sides, the phases of amplitude reflectivity and transmittivity differ by $\pi/2$. If the input arms of the beam splitter are denoted by the subscripts 1 and 2 and the two other arms are denoted by subscripts 3 and 4 (see Fig. 9.1), the electric field of the waves emerging in arms 3 and 4 are related to the electric fields of the waves incident in arms 1 and 2 on the beam splitter by the following equations:

$$E_3 = \frac{i}{\sqrt{2}} E_1 + \frac{1}{\sqrt{2}} E_2$$
$$E_4 = \frac{1}{\sqrt{2}} E_1 + \frac{i}{\sqrt{2}} E_2 \tag{9.209}$$

For a quantum mechanical description of the beam splitter we replace the classical electric fields by annihilation operators corresponding to the electromagnetic fields in each arm. Thus the annihilation operators of the electromagnetic waves in

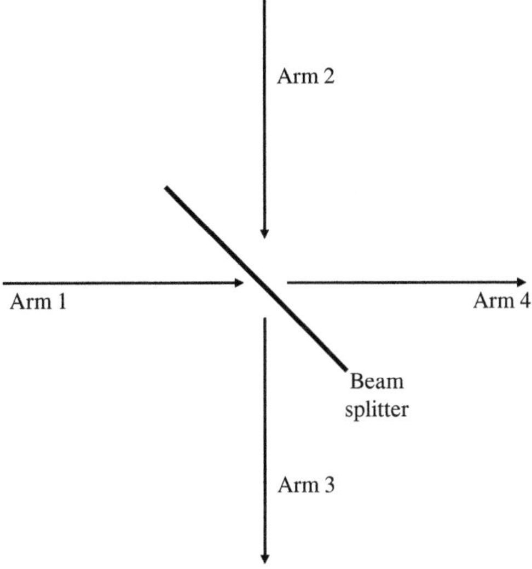

Fig. 9.1 A beam splitter has two input arms and two output arms

arms 3 and 4 are related to the annihilation operators of the electromagnetic waves in arms 1 and 2 by the relations:

$$\hat{a}_3 = \frac{i}{\sqrt{2}}\hat{a}_1 + \frac{1}{\sqrt{2}}\hat{a}_2$$

$$\hat{a}_4 = \frac{1}{\sqrt{2}}\hat{a}_1 + \frac{i}{\sqrt{2}}\hat{a}_2 \tag{9.210}$$

The above equations can be inverted to obtain

$$\hat{a}_1 = -\frac{i}{\sqrt{2}}\hat{a}_3 + \frac{1}{\sqrt{2}}\hat{a}_4$$

$$\hat{a}_2 = \frac{1}{\sqrt{2}}\hat{a}_3 - \frac{i}{\sqrt{2}}\hat{a}_4 \tag{9.211}$$

Since vacuum state is the lowest state of an electromagnetic field, vacuum states in arms 1 and 2 will lead to output vacuum states in arms 3 and 4. Hence we can say that the beam splitter converts vacuum states in arms 1 and 2 to vacuum states in arms 3 and 4:

$$|0\rangle_1 |0\rangle_2 - > |0\rangle_3 |0\rangle_4 \tag{9.212}$$

9.8.1 Single-Photon Incident on a Beam Splitter

Let us consider the incidence of a single photon on arm 1 of the beam splitter and vacuum state in arm 2. The input state is then given by

$$|1\rangle_1 |0\rangle_2 = \hat{a}_1^{\dagger} |0\rangle_1 |0\rangle_2 \tag{9.213}$$

The effect of the beam splitter is taken into account using Eq. (9.211) and we get

$$|1\rangle_1 |0\rangle_2 = \hat{a}_1^\dagger |0\rangle_1 |0\rangle_2 - > \left(\frac{i}{\sqrt{2}} \hat{a}_3^\dagger + \frac{1}{\sqrt{2}} \hat{a}_4^\dagger \right) |0\rangle_3 |0\rangle_4$$

$$= \frac{i}{\sqrt{2}} |1\rangle_3 |0\rangle_4 + \frac{1}{\sqrt{2}} |0\rangle_3 |1\rangle_4 \tag{9.214}$$

The output state is a superposition state of finding either 1 photon in arm 3 and no photon in arm 4 or finding one photon in arm 4 and no photon in arm 3. Both these have equal probabilities of $\frac{1}{2}$. Thus the incident photon goes into a superposition state.

Example 9.4 Consider now a situation in which one photon each is simultaneously incident in each of the input arms. Thus the incident state becomes

$$|1\rangle_1 |1\rangle_2 = \hat{a}_1^\dagger \hat{a}_2^\dagger |0\rangle_1 |0\rangle_2 \tag{9.215}$$

Using Eq. (9.211) we can introduce the effect of the beam splitter through the following transformation:

$$|1\rangle_1 |1\rangle_2 = \hat{a}_1^\dagger \hat{a}_2^\dagger |0\rangle_1 |0\rangle_2 - > \left(\frac{i}{\sqrt{2}} \hat{a}_3^\dagger + \frac{1}{\sqrt{2}} \hat{a}_4^\dagger \right) \left(\frac{1}{\sqrt{2}} \hat{a}_3^\dagger + \frac{i}{\sqrt{2}} \hat{a}_4^\dagger \right) |0\rangle_3 |0\rangle_4 \tag{9.216}$$

$$= \frac{i}{2} \left(|2\rangle_3 |0\rangle_4 + |0\rangle_3 |2\rangle_4 \right)$$

This implies that both photons exit from the same port with equal probabilities. The probability of one photon exiting each of the arms is zero. Thus detectors placed at the two output ports of the beam splitter will never register simultaneous detection events. An experimental demonstration of this was first carried out by Hong, Ou, and Mandel (1987)). In the experiment as the delay of arrival of the photon on one of the arms of the beam is changed, the rate of coincidence counting would show a dip when the two photons arrive simultaneously at the beam splitter. This is referred to as Hong–Ou–Mandel dip.

Mach Zehnder Interferometer

Let us now consider the incidence of a photon on one of the arms of a Mach Zehnder interferometer as shown in Fig. 9.2. Let us assume that in one of the arms, there is a phase shifter which shifts the phase by ϕ. Now

Input to the interferometer:

$$|1\rangle_1 |0\rangle_2 = \hat{a}_1^\dagger |0\rangle_1 |0\rangle_2 \tag{9.217}$$

After beam splitter BS_1 the state becomes:

$$\left(\frac{i}{\sqrt{2}} \hat{a}_3^\dagger + \frac{1}{\sqrt{2}} \hat{a}_4^\dagger \right) |0\rangle_3 |0\rangle_4 = \frac{i}{\sqrt{2}} |1\rangle_3 |0\rangle_4 + \frac{1}{\sqrt{2}} |0\rangle_3 |1\rangle_4 \tag{9.218}$$

Let us assume that mirrors M_1 and M_2 introduce a phase shift of $\pi/2$ each. Thus the state of the field just after the two mirrors would be

$$- \frac{1}{\sqrt{2}} |1\rangle_3 |0\rangle_4 + \frac{i}{\sqrt{2}} |0\rangle_3 |1\rangle_4$$

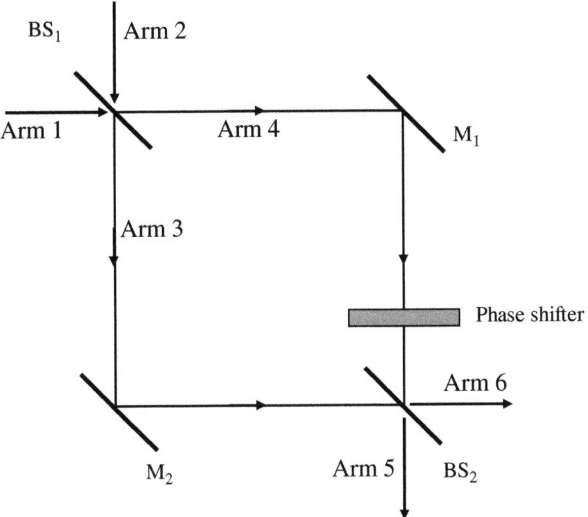

Fig. 9.2 A Mach Zehnder interferometer set up with a single-photon incident in arm 1

After the phase shifter the state would become

$$-\frac{1}{\sqrt{2}}\,|1\rangle_3\,|0\rangle_4 + \frac{i}{\sqrt{2}}\,|0\rangle_3\,|1\rangle_4\,e^{i\phi} \tag{9.219}$$

since the phase shift takes place only along path 4 of the interferometer.

In order to take account of the effect of beam splitter BS$_2$ we write the state incident on beam splitter 2 as

$$-\frac{1}{\sqrt{2}}\hat{a}_3^\dagger\,|0\rangle_3\,|0\rangle_4 + \frac{i}{\sqrt{2}}\hat{a}_4^\dagger\,|0\rangle_3\,|0\rangle_4\,e^{i\phi}$$

Since BS$_2$ will convert $|0\rangle_3\,|0\rangle_4$ to $|0\rangle_5\,|0\rangle_6$, we write for the state after BS$_2$ as

$$-\frac{1}{\sqrt{2}}\left(\frac{i}{\sqrt{2}}\hat{a}_5^\dagger + \frac{1}{\sqrt{2}}\hat{a}_6^\dagger\right)|0\rangle_5\,|0\rangle_6 + \frac{i}{\sqrt{2}}\left(\frac{1}{\sqrt{2}}\hat{a}_5^\dagger + \frac{i}{\sqrt{2}}\hat{a}_6^\dagger\right)|0\rangle_5\,|0\rangle_6\,e^{i\phi}$$

which on simplification gives us

$$-\frac{i}{2}\left(1 - e^{i\phi}\right)|1\rangle_5\,|0\rangle_6 - \frac{1}{2}\left(1 + e^{i\phi}\right)|0\rangle_5\,|1\rangle_6 \tag{9.220}$$

Thus the probability of finding a photon in arm 5 will be

$$\left|-\frac{i}{2}\left(1 - e^{i\phi}\right)\right|^2 = \frac{1}{2}(1 - \cos\phi) \tag{9.221}$$

Similarly the probability of finding a photon in arm 6 would be

$$\left| -\frac{1}{2}\left(1 + e^{i\phi}\right) \right|^2 = \frac{1}{2}(1 + \cos\phi) \tag{9.222}$$

This clearly shows interference effects between the two possible paths of photons to the output of the interferometer. Note that it is a single photon which is interfering with itself; actually the probability amplitudes of the two indistinguishable paths are interfering to produce the interference effects in the probability. If we choose $\phi = 0$, then the probability of detecting a photon in arm 5 is 0; the photon will always exit from the arm 6. Thus in this setup it is incorrect to state that the photon chooses one of the arms when it arrives on BS_1; it chooses both the paths of the interferometer!

9.8.2 Moving Mirror in One Arm

Let us now consider a slight modification of the above interferometer. We now assume that mirror M_2 is not fixed but can move. We assume that the mirror is so sensitive that even if one photon hits it, it will vibrate. Let us denote the two states of the mirror namely stationary and vibrating mirror by the following two orthogonal states:

$$|\psi\rangle_S \text{ and } |\psi\rangle_V \tag{9.223}$$

Now let us again assume the incidence of a single photon in arm 1 or the interferometer and assume $\phi = 0$. The analysis remains the same until the state passes through the mirrors 1 and 2. Thus to repeat, the state after BS_1 is

$$\frac{i}{\sqrt{2}}\,|1\rangle_3\,|0\rangle_4 + \frac{1}{\sqrt{2}}\,|0\rangle_3\,|1\rangle_4$$

After mirrors M_1 and M_2 the state becomes

$$\frac{i}{\sqrt{2}}\,|1\rangle_3\,|0\rangle_4\,|\psi\rangle_S + \frac{1}{\sqrt{2}}\,|0\rangle_3\,|1\rangle_4\,|\psi\rangle_V \tag{9.224}$$

where we have used the fact that for the portion corresponding to the photon being found in path 3 the mirror will remain stationary and for the path in which the photon may be found in arm 4 the mirror would vibrate.

Continuing as before the state after BS$_2$ would be

$$
\begin{aligned}
|\chi\rangle &= -\frac{1}{\sqrt{2}}\left(\frac{i}{\sqrt{2}}\hat{a}_5^\dagger + \frac{1}{\sqrt{2}}\hat{a}_6^\dagger\right)|0\rangle_5\,|0\rangle_6\,|\psi\rangle_S \\
&\quad + \frac{i}{\sqrt{2}}\left(\frac{1}{\sqrt{2}}\hat{a}_5^\dagger + \frac{i}{\sqrt{2}}\hat{a}_6^\dagger\right)|0\rangle_5\,|0\rangle_6\,|\psi\rangle_V \\
&= \left(-\frac{i}{2}|1\rangle_5\,|0\rangle_6\,|\psi\rangle_S - \frac{1}{2}|0\rangle_5\,|1\rangle_6\,|\psi\rangle_S\right) \\
&\quad + \left(\frac{i}{2}|1\rangle_5\,|0\rangle_6\,|\psi\rangle_V - \frac{1}{2}|0\rangle_5\,|1\rangle_6\,|\psi\rangle_V\right) \\
&= -\frac{i}{2}|1\rangle_5\,|0\rangle_6\left(|\psi\rangle_S - |\psi\rangle_V\right) - \frac{1}{2}|0\rangle_5\,|1\rangle_6\left(|\psi\rangle_S + |\psi\rangle_V\right)
\end{aligned}
\tag{9.225}
$$

Thus the expectation value of the photon number is arm 5 would be

$$
\begin{aligned}
\left|\langle\chi|\hat{a}_5^\dagger\hat{a}_5|\chi\rangle\right|^2 &= \frac{1}{4}\,|_5\langle1|1\rangle_5\;_6\langle0|0\rangle_6 \\
\left(_S\langle\psi|\psi\rangle_S + _V\langle\psi|\psi\rangle_V - _S\langle\psi|\psi\rangle_V - _V\langle\psi|\psi\rangle_S\right)\Big|^2 &= \frac{1}{2}
\end{aligned}
\tag{9.226}
$$

Notice that now the probability of detecting a photon on arm 5 is $\tfrac{1}{2}$. By having a vibrating mirror in one of the arms, we have a device in the interferometer which can tell us about the path of the photon in the interferometer and immediately the interference vanishes from the output.

In fact if we had an absorbing object in the arm 4 of the interferometer, then the probability of detecting a photon in arm 5 is still half (show this result). Thus appearance of a photon in arm 5 in a balanced interferometer wherein no photon is supposed to have exited arm 5 signifies the presence of an object in arm 4. Also since the photon has exited arm 5, it has not been absorbed by the absorber. Thus the photon is able to detect the presence of an object in arm 4 without interacting with the object. Such effects are referred to as interaction free measurements and are very interesting from both a philosophical perspective and for practical applications.

Problems

Problem 9.1 Given $\vec{S} = \vec{E} \times \vec{H}$ show that

$$\left\langle n\left|\vec{S}\right| n\right\rangle = \frac{c^2}{V}(n + \frac{1}{2})\hbar\vec{k}$$

The energy flow for a radiation mode is proportional to the photon momentum $\hbar\vec{k}$

Problem 9.2 Prove that the creation operator $\hat{a}^\dagger$ has no normalizable eigenstates.

Problem 9.3 Consider two coherent states $|\alpha\rangle$ and $|\beta\rangle$. Show that they are not orthonormal, i.e., show that $\langle \beta \mid \alpha \rangle \neq 0$.

Problem 9.4 The vector potential for a single-mode plane electromagnetic wave propagating along the z-direction and polarized along x is given by

$$A = q\, e^{-i(\omega t - kz)} + c.c.$$

Obtain the corresponding expressions for the electric and magnetic fields.

Problem 9.5 If $|\delta\rangle$ is an eigenket of an operator $\hat{g}$ with an eigenvalue δ, then $\langle \delta|$ is an eigenbra of which operator and what is the corresponding eigenvalue?

Problem 9.6 A single-mode electromagnetic wave is in a state given by

$$|\psi\rangle = \frac{1}{2}\,|2\rangle + \frac{\sqrt{3}}{2}\,|3\rangle$$

Obtain the expectation value of energy.

Problem 9.7 Consider a single-mode quantum state given by

$$|\psi\rangle = \alpha\,|0\rangle + \beta\,|1\rangle$$

a)   Normalize $|\psi\rangle$ and obtain the relationship between α and β.
b)   Obtain the values of α and β for which the quantum state will exhibit squeezing in the quadrature represented by $\hat{X}_1$?

Problem 9.8 Show that the expectation value of photon number in a coherent vacuum state is zero while that in a squeezed vacuum state is not zero. For positive real μ and v for the squeezed state, plot the uncertainty area in the X_1 -X_2 plane at $t = 0$, $t = \pi/2\omega$ where ω is the frequency of the electromagnetic wave.

Problem 9.9 If $|\beta\rangle$ is an eigenstate of the operator $\hat{b} = \sqrt{2}\hat{a} - \hat{a}^\dagger$ with eigenvalue β, obtain the variances in $\hat{X}_1$ and $\hat{X}_2$ and plot the corresponding uncertainty area graphically in a figure. Assume β to be real.

Problem 9.10 Consider a harmonic oscillator in a state descried by

$$|\psi\rangle = \frac{1}{2}\,|n\rangle + \frac{\sqrt{3}}{2}\,|n+1\rangle$$

Obtain the expectation value of the harmonic oscillator energy in this state.

Problem 9.11 A squeezed state $|\beta\rangle$ is the eigenstate of the operator $\hat{b}$ defined by

$$\hat{b} = \mu\hat{a} + v\widehat{a}^{\,\dagger}$$

with $|\mu|^2 - |\nu|^2 = 1$. Obtain the expectation value of $\widehat{X}_1 = \frac{1}{2}(\widehat{a} + \widehat{a}^{\dagger})$ for this state. Assume μ, ν, and β to be real quantities.

Problem 9.12 Consider a coherent state $|\alpha\rangle$. The value of α is such that the probability of detecting 2 photons is 1% of the probability of detecting one photon. What is the probability of detecting no photons? Assume α to be real.

Problem 9.13 Consider a state of an electromagnetic wave described by $|\psi\rangle = A\left(|n\rangle_i + |n\rangle_j\right)$ where the number state $|n\rangle_i$ corresponds to having n photons in i^{th} mode with all other modes unoccupied. Similarly for the state $|n\rangle_j$ corresponding to the jth mode. Obtain the number of photons in this state. A is a constant.

Problem 9.14 Consider a number state $|n\rangle$ with $n = 10$.

a)   Obtain the value of uncertainty in the photon number in this state.

b)   Obtain the variances of the two quadrature operators.

c)   Is this a minimum uncertainty state?

Problem 9.15 Show that the expectation value of photon number in a coherent vacuum state (defined by $\widehat{a}\,|0\rangle_c = 0$) is zero while that in a squeezed vacuum state is not zero where squeezed vacuum is defined by $\widehat{b}\,|0\rangle_s = 0$ with $\widehat{b} = \mu\,\widehat{a} + \nu\,\widehat{a}^{\dagger}$ (assume μ and ν to be real). Obtain the expectation value of photon number in the squeezed vacuum state with $\widehat{b} = 2\widehat{a} - \sqrt{3}\widehat{a}^{\dagger}$

Problem 9.16 Consider a superposition of two coherent states $|\alpha\rangle$ and $-|\alpha\rangle$:

$$|\psi\rangle = N(|\alpha\rangle + |-\alpha\rangle)$$

a)   Normalize $|\psi\rangle$.

b)   Show that the state $|\psi\rangle = N(|\alpha\rangle + |-\alpha\rangle)$ is an eigenstate of the operator $\widehat{a}^2$.

Show that the probability of finding an odd number of photons in this state is zero.

Chapter 10
Properties of Lasers

10.1 Introduction

So far we have discussed the physics behind laser operation. Basically the light from both a laser and any ordinary source of light is electromagnetic in nature, but laser light can be extremely monochromatic, highly directional, and very intense. Apart from these laser light also differs from light produced by thermal emission in the basic quantum properties.

In this chapter we shall look at the different properties of laser light and in Part II of the book we shall discuss some of the most interesting applications of lasers.

10.2 Laser Beam Characteristics

Light from the laser arises primarily from stimulated emission and the resonator cavity within which the amplifying medium is kept leads to the following special properties:

Directionality
Spectral purity
High power
Extremely short pulse durations

Table 10.1 gives the achievable laser characteristics; these special properties lead to many applications of lasers.

Directionality: Light from a source of light such as a torchlight diverges significantly as it propagates (see Fig. 10.1). But the beam coming from a laser is in the form of a pencil of rays and seems to propagate without any divergence. The laser beam also diverges but by a much smaller magnitude. The wave nature of light imparts an intrinsic divergence to the beam due to the phenomenon of diffraction (see Chapter 2). Thus unlike a torchlight where the divergence is due to the finite size of the filament, the divergence of the laser beam is limited by diffraction depending on the laser types and can be less than 10^{-5} radians (~ 2 s of arc). This extremely small divergence leads to the many application of the laser in surveying, remote sensing, lidar, etc.

K. Thyagarajan, A. Ghatak, *Lasers*, Graduate Texts in Physics,
DOI 10.1007/978-1-4419-6442-7_10, © Springer Science+Business Media, LLC 2010

Table 10.1 Some of the special properties possessed by laser beams from different types of lasers

• Directionality	(Divergence $\sim 10^{-7}$ rad)
• Spectral purity	($\Delta\lambda \sim 10^{-9}$ μm)
• High power	($P \sim 10^{18}$ W/cm^2)
• Ultra short pulses	($\Delta t \sim 10^{-15}$ s)
• High electric fields	($E \sim 10^{12}$ V/m)
• Small focused areas	($\sim 10^{-12}$ m^2)

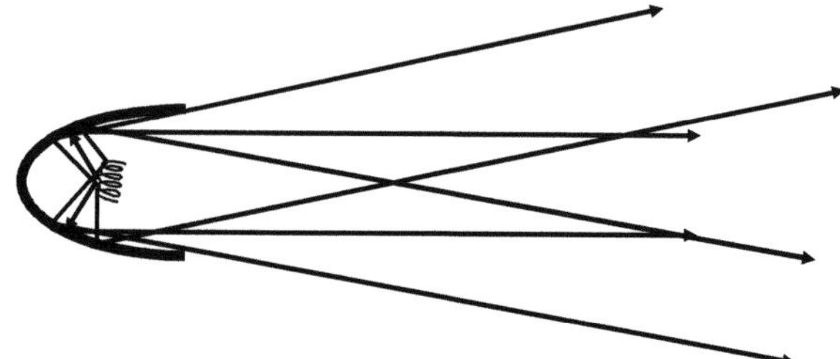

Fig. 10.1 Light from a torch has a divergence primarily due to the fact that light emanating from different points on the filament propagates along different directions after reflection from the parabolic mirror

As an example consider a tiny filament lamp placed at the focus of a convex lens as shown in Fig. 10.2. The filament can be considered to be made up of a number of point sources, and thus the light emanating from different points on the filament will travel along different directions after passing through the lens and the exiting beam will diverge. If the linear dimension of the filament is about 2 mm and if the focal length of the convex lens is 10 cm then the angular divergence of the beam (due to the finite size of the filament) is approximately 1° (= 0.02 radians). This divergence could be reduced provided we reduce the dimension of the filament, but then the amount of light will also get correspondingly reduced.

Compared to the filament, the divergence of a laser beam is primarily due to diffraction. For most laser beams, the *spot size* (the radius of the cross section of the laser beam) of the beam is about a few millimeters. As discussed in Chapter 2, if the laser beam has a free space wavelength of λ_0 and a spot size w_0, then the divergence angle of the beam is given by

$$\theta \approx \frac{\lambda_0}{\pi\, w_0} \tag{10.1}$$

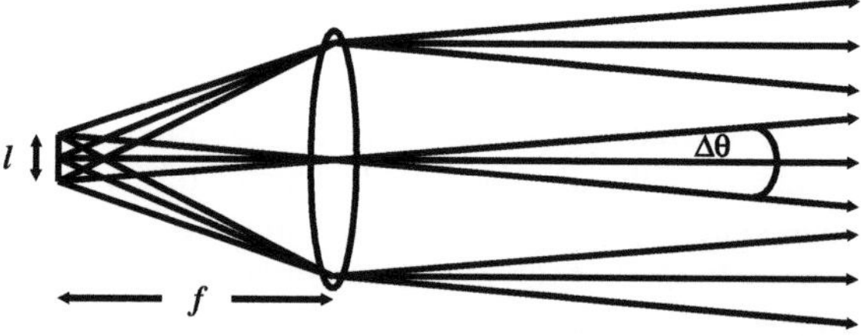

Fig. 10.2 Light from the filament of a bulb placed at the focus of a convex lens diverges after passing through it

For a typical spot size of 1 mm and a wavelength of 0.6 μm, the divergence angle is given approximately by 0.01°. A beam is said to be *diffraction limited* if it diverges only due to diffraction and usually laser beams are diffraction limited. We may mention that the laser beam from a laser diode has a significant divergence due to the small spot size of the beam. At the same time, unlike the case of the torch, the divergence of the beam can be reduced by simply using a lens in front of the laser diode.

Tight Focusing: Because of highly directional properties of the laser beams, they can be focused to very small areas of a few $(\mu m)^2$. The limits to focusing are again determined by diffraction effects. Smaller the wavelength, smaller the size of the focused spot. This property leads to applications in surgery, material processing, compact discs, etc.

When a convex lens images a point object, the size of the image point is directly proportional to the wavelength of the light wave and also to the ratio of focal length to the diameter. The ratio of focal length to the diameter of a lens is also called the *f*-number. This parameter is used in specifying the quality of camera lenses. Thus an *f*/2 lens implies that the focal length to diameter ratio is 2. If the focal length of this camera lens is 50 mm then its diameter is 25 mm. Smaller the *f*-number for a given focal length larger is the diameter of the lens. Smaller the wavelength, smaller the spot size, and similarly smaller the *f*-number, smaller the image size. Since an object can be considered to be made up of points, if we consider the imaging by the convex lens, the resolution provided by the lens will depend on the *f*-number and the wavelength. For a given wavelength, for better resolution we must have a smaller *f*-number. Smaller the *f*-number of a camera, better will be the resolution of the camera.

Thus when a laser beam is allowed to fall on a convex lens then the radius of the focused spot is directly proportional to the wavelength and to the *f*-number, provided the laser beam fills the entire area of the lens (see Fig. 10.3). If we take a lens having an *f*-number of 2 (i.e., focal length is twice the lens diameter), then for a laser wavelength of 600 nm the radius of the focused spot will be about 1.5 μm. Thus the area of such a focused spot would be about 7 μm^2. If the laser beam has a power of 1 MW ($= 10^6$ W) then the intensity at the focused spot would be approximately 14 TW/cm^2. Such intensities of light lead to electric fields of 10^9 V/m. Such high

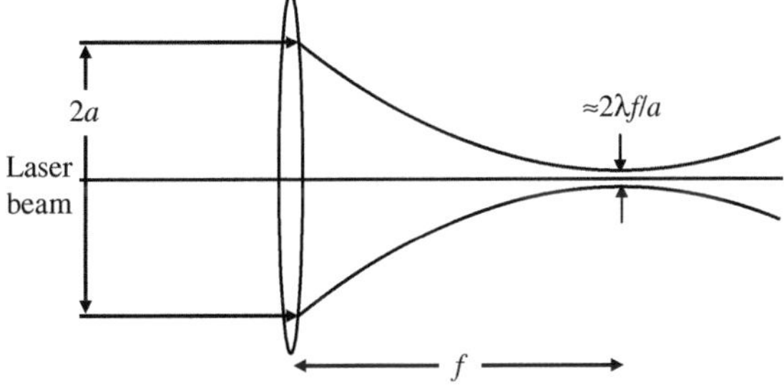

Fig. 10.3 If a truncated plane wave (of diameter 2*a*) is incident on a lens without any aberration of focal length *f*, then the wave emerging from the lens will get focused to spot of radius $\approx \lambda\, f/a$

Fig. 10.4 Focusing of a 3 MW peak power-pulsed ruby laser beam. At the focus, the electric field strengths are of the order of a billion Volts per meter which results in the creation of a spark in the air. (Photograph courtesy Dr. R. W. Terhune)

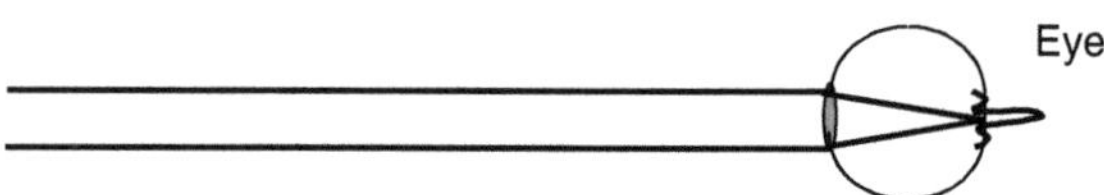

Fig. 10.5 When a laser beam falls on the eye, then it gets focused to a very small diffraction-limited size producing very high intensities even for small powers as 2 mW

electric fields can create a spark in air – see Fig. 10.4. This shows that laser beams (because of their high directionality) can be focused to extremely small regions producing very high intensities and electric fields.

When the electric fields are very high inside a medium, the light beam can change the properties of the medium. Such effects are termed non-linear effects; some effects of this non-linearity are discussed in Chapters 14 and 18.

We may mention here that a low-power (≈ 2 mW) diffraction-limited laser beam incident on the eye gets focused to a very small spot (see Fig. 10.5) and can produce an intensity of about 100 W/cm^2 on the retina – this could indeed damage the retina. On the other hand, when we look at a 20 W bulb at a distance of about 5 m from the eye, the eye produces an image of the bulb on the retina and this would produce an intensity of only about 10 W/m^2 on the retina of the eye (see Fig. 10.6). Thus, whereas it is quite safe to look at a 20 W bulb, it is very dangerous to look directly into a 2 mW laser beam. Indeed, because a laser beam can be focused to very narrow areas, it has found important applications in areas like eye surgery and laser cutting.

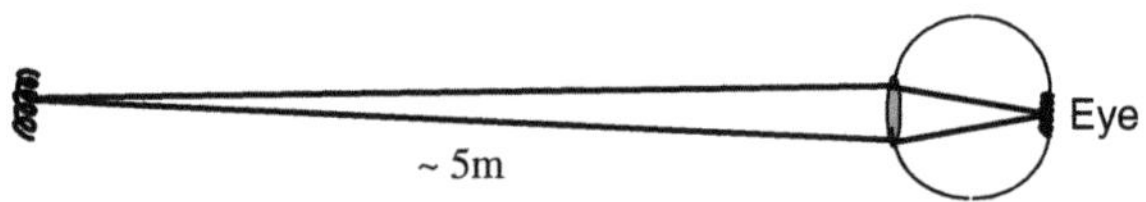

Fig. 10.6 Looking at a bulb produces an image of the bulb on the retina and even 20 W bulb does not produce very high intensities

It may be of interest to mention that, if we are directly looking at the sun, the power density in the image formed is about 30 kW/m^2. This follows from the fact that on the earth, about 1.35 kW of solar energy is incident (normally) on an area of 1 m^2. Thus the energy entering the eye is about 4 mW. Since the sun subtends about 0.5° on the earth, the radius of the image of the sun (on the retina) is about 2×10^{-4} m. Therefore if we are directly looking at the sun the power density in the image formed is about 30 kW/m^2. *Thus, never look into the sun; the retina will be damaged not only because of high intensities but also because of large ultraviolet content of the sunlight.*

A very interesting application of the extreme directionality of laser beams is in the realization of artificial stars in the sky. At a height of about 95 km above the surface of earth there is a layer containing sodium atoms. If a laser beam at a wavelength of 589 nm is sent up, then the sodium atoms absorb the radiation, get excited to a higher energy level, and then emit spontaneously when they get de-excited. Some of this radiation is traveling toward the earth and resembles a star. The position of this artificial star or guide star can be adjusted by changing the direction of the laser beam. Typically pulsed lasers emitting a power of about 20 W and pulse widths of 100 ns are used to create the star. One of the interesting applications of this guide star is in the correction of images formed by telescopes on the earth. Since the light coming from objects outside the earth has to pass via the turbulent atmosphere, the image of any extra terrestrial object will not be stable. By looking at the image of this artificial star it is possible to determine the correction to the optical system required in real time for canceling the effects of turbulence. Figure 10.7 shows the images of the application of laser guide star in imaging the dense star cluster at the center of the Milky Way.

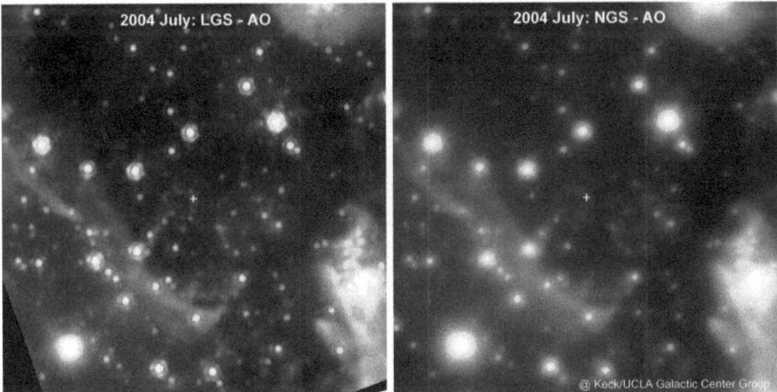

Fig. 10.7 Images of the dense star cluster at the center of the Milky Way Galaxy in infrared light at 3.6 μm wavelength. *Left*: image using laser guide star adaptive optics. *Right*: best natural guide star image. The laser guide star image has a total integration time of 8 min, while the natural guide star image has 150 min. The plus sign marks the position of the central million-solar-mass black hole, Sgr A*, in both images. (Credit: UCLA Galactic Center Group and W.M. Keck Observatory Laser Guide Star Team; photograph provided by Prof Claire Max, University of California, USA)

Problem 10.1 Consider a confocal laser resonator made of mirrors of radii of curvature 1 m each. Assuming that the laser is oscillating in the fundamental Gaussian mode, obtain the divergence of the exiting laser beam if the free space wavelength of the laser is 500 nm. [Ans: $w_0 \sim 300$ μm, 90 s of arc].

Problem 10.2 Referring to Problem 10.1 if the separation between the mirrors is kept 1 m while the mirror radii of curvatures are increased to 5 m, obtain the corresponding divergence. [Ans: $w_0 \sim 5$ mm, 6 s of arc].

Spectral Purity: Laser beams can have an extremely small spectral width, of the order of 10^{-6} Å. Compare this with a typical source such as a sodium lamp which has a spectral width of about 0.1 Å. The stimulated emission process coupled with the optical resonator within which the amplifying medium is placed is responsible for the very small spectral widths. In general a laser may oscillate in a number of frequencies simultaneously unless special techniques are adopted (see Chapter 7). This includes using Fabry–Perot filters within the laser cavity to allow only one frequency to oscillate. Even in a laser oscillating in a single frequency, there could be random but small variations in the frequency of oscillation due to temperature fluctuations and vibrations of the mirrors of the cavity. Stabilization of frequency is achieved using various techniques, for example, the laser is coupled to another very stable cavity and the emission from the laser gets locked to this stable cavity. Frequency-stabilized lasers with frequency stability of better than 10^{-8} (i.e.,

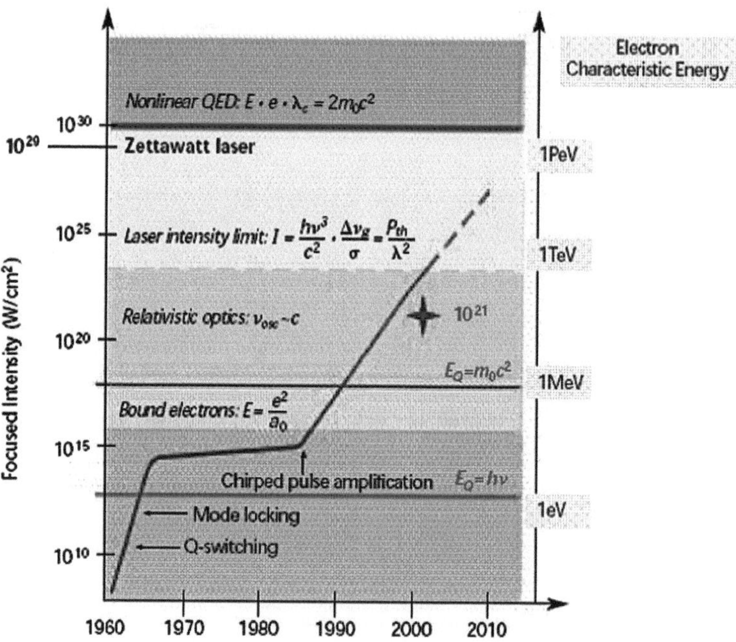

Fig. 10.8 Increase in achievable laser intensity with the year. The increase has a large slope around 1960 due to the invention of the laser and then again after 1985. (Adapted from Mourou and Yanovsky (2004) © 2004 OSA)

fractional frequency shift of less than 10 parts per billion) are commercially available. Because of high spectral purity, lasers find applications in holography, optical communications, spectroscopy, etc.

High Power: Lasers can generate extremely high powers, and since they can also be focused to very small areas, it is possible to generate extremely high-intensity values. Figure 10.8 shows how the intensity achievable using laser beams has increased every year. At intensities such as 10^{21} W/m^2, the electric fields are so high that electrons can get accelerated to relativistic velocities (velocities approaching that of light) leading to very interesting effects. Apart from scientific investigations of extreme conditions, continuous wave lasers having power levels $\sim 10^5$ W and pulsed lasers having a total energy ~ 50000 J have applications in welding, cutting, laser fusion, star wars, etc.

10.3 Coherence Properties of Laser Light

In this section we will introduce the concepts of temporal and spatial coherence since these play a significant role in their applications.

10.3.1 Temporal Coherence

In order to understand the concept of temporal coherence, we consider a Michelson interferometer arrangement as shown in Fig. 10.9. S represents an extended near monochromatic source, G represents a beam splitter, and M_1 and M_2 are two plane mirrors. The mirror M_2 is fixed while the mirror M_1 can be moved either toward

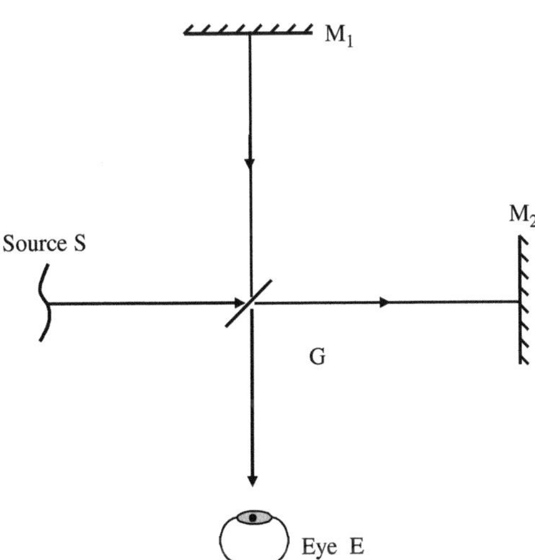

Fig. 10.9 A Michelson interferometer setup

or away from G. Light from the source S is incident on G and is divided into two equal portions; one part travels toward M_1 and is reflected back and the other part is reflected back from M_2. The two reflected waves interfere and produce interference fringes which are visible from E. When the mirrors M_1 and M_2 are nearly equidistant from G, i.e, when the two waves traversing the two different paths take the same amount of time, then it is observed that the contrast of the interference fringes formed is good. If now the mirror M_1 is slowly moved away from G, then it is seen that for ordinary extended source of light (like a sodium lamp), the contrast in the fringes goes on decreasing and when the difference between the distances from G to M_1 and M_2 is about a few millimeters to a few centimeters, the fringes are no longer visible. This decrease in contrast of the fringes can be explained as follows: The source S is emitting small wave trains of an average duration τ_c (say) and there is no phase relationship between different wave trains (see Section 4.5). This is in contrast to an infinitely long pure sinusoidal wave train, which is also referred to as a monochromatic wave. When the difference in time taken by the wave trains to travel the paths G to M_1 and back and G to M_2 and back is much less than the average duration τ_c, then the interference is produced between two wave trains each one being derived from the same wave train. Hence even though different wave trains emanating from the source S do not have definite phase relationship, since one is superimposing two wave trains derived from the same wave train, fringes of good contrast will be seen. On the other hand, if the difference in the time taken to traverse the paths to M_1 and back and to M_2 and back is much more than τ_c, then one is superimposing two wave trains which are derived from two different wave trains, and since there is no definite phase relationship between two wave trains emanating from S, interference fringes will not be observed. Hence as the mirror M_1 is moved, the contrast in the fringes becomes poorer and poorer and for large separations no fringes would be seen. The time τ_c is referred to as the coherence time and the length of the wave train $c\tau_c$ is referred to as the longitudinal coherence length. It may be mentioned that there is no definite distance at which the interference pattern disappears; as the distance increases, the contrast in the fringes becomes gradually poorer and eventually the fringes disappear.

As an example, for the neon 632.8 nm line from a discharge lamp, the interference fringes would vanish if the path difference between the two mirrors is about a few centimeters. Thus for this source, $\tau_c \sim 100$ ps. On the other hand, for the red cadmium line at 643.8 nm, the coherence length is about 30 cm, which gives $\tau_c \sim 1$ ns.

The decrease in contrast of the fringes can also be interpreted as being due to the fact that the source S is not emitting a single frequency but emits over a band of frequencies (see Section 4.5). When the path difference is zero or very small, the different wavelength components produce fringe patterns superimposed on one another and the fringe contrast is good. On the other hand, when the path difference is increased, different wavelength components produce fringe patterns which are slightly displaced with respect to one another and the fringe contract becomes poor. Thus the non-monochromaticity of the light source can equally well be interpreted as the reason for poor fringe visibility for large optical path differences.

The equivalence of the above two approaches can also be seen using Fourier analysis (see Section 4.5). One can indeed show that a wave having a coherence time $\sim \tau_c$ is essentially a superposition of harmonic waves having frequencies in the range $\nu_0 - \Delta\nu/2 \leq \nu \leq \nu_0 + \Delta\nu/2$ where

$$\Delta\nu \sim \frac{1}{2\pi\tau_c} \qquad (10.2)$$

Thus the longer the coherence time, the smaller the frequency width. For ordinary sources $\tau_c \sim 100$ ps and

$$\Delta\nu \sim 10^{10}\,\mathrm{Hz}$$

For $\lambda = 600$ nm, $\nu = 5 \times 10^{14}$ Hz and

$$\frac{\Delta\nu}{\nu_0} \sim 0.00002$$

The quantity $\Delta\nu/\nu_0$ represents the monochromaticity and one can see that even for ordinary light sources it is quite small. In Section 4.5 we have discussed some of the mechanisms leading to the broadening of spectral lines emitted by atoms.

In a laser, in contrast to an ordinary source of light, the optical resonant cavity is excited in different longitudinal modes of the cavity which are specified by discrete frequencies of oscillation. In an optical resonator without the amplifying medium, the finite loss of the resonator leads to an exponentially decaying output amplitude which leads to a finite linewidth of the output. On the other hand, in an actual laser oscillating in steady state, the loss is exactly compensated by the gain provided by the laser medium, and when the laser is oscillating in a single mode, the output is essentially a pure sinusoidal wave. Superposed on this are the random emissions arising out of spontaneous emission and it is this spontaneous emission which limits the ultimate monochromaticity of the laser (see Section 7.5).

In contrast to $\Delta\nu \sim 10^{10}$ Hz, for an ordinary source of light, for a well-controlled laser one can obtain $\Delta\nu \sim 500$ Hz, which gives $\tau_c \sim 2$ ms. The corresponding coherence length is about 600 km. Such long coherence lengths imply that the laser could be used for performing interference experiments with very large path differences.

For a laser oscillating in many modes, the monochromaticity depends obviously on the number of oscillating modes (see Problem 7.3). Also for a pulsed laser, the minimum linewidth is limited by the duration of the pulse. Thus for a 1 ps pulse, the coherence time is 1 ps and the spectral width would be about 10^{12} Hz.

10.3.2 Spatial Coherence

In order to understand the concept of spatial coherence, we consider the Young's double-hole experiment as shown in Fig. 10.10. S represents a source placed in front of a screen with two holes S_1 and S_2 and the interference pattern between the

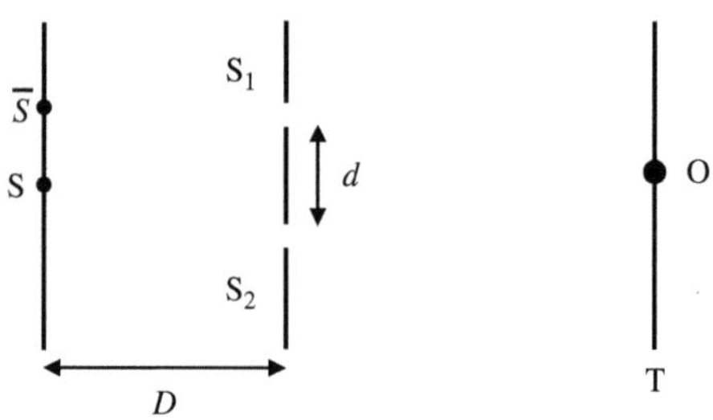

Fig. 10.10 Young's double-slit experimental arrangement

waves emanating from S_1 and S_2 is observed on screen T. We restrict ourselves to the region near O for which the optical path lengths S_1O and S_2O are equal. If S represents a point source then it illuminates the pinholes S_1 and S_2 with spherical waves. Since the holes S_1 and S_2 are being illuminated coherently, the interference fringes formed near O will be of good contrast. Consider now another point source $\bar{S}$ placed near S and assume that the waves from S and $\bar{S}$ have no phase relationship. In such a case the interference pattern observed on the screen T will be a superposition of the *intensity* distributions of the interference patterns formed due to S and $\bar{S}$. If $\bar{S}$ is moved slowly away from S, the contrast in the interference pattern on T will become poorer because of the fact that the interference pattern produced by $\bar{S}$ is slightly shifted in relation to that produced by S. For a particular separation, the interference maximum produced by S falls on the interference minimum produced by $\bar{S}$ and the minimum produced by S falls on the maximum produced by $\bar{S}$. For such a position the interference fringe pattern on the screen T is washed away.

In order to obtain an approximate expression for the separation $S\bar{S}$ for disappearance of fringes, we assume that S and O are equidistant from S_1 and S_2. If the position of $\bar{S}$ is such that the path difference between $\bar{S}S_2$ and $\bar{S}S_1$ is $\lambda/2$ (where λ is the wavelength of light used), then the source $\bar{S}$ produces an interference minimum at O and the two fringe patterns would be out of step. If we assume $S\bar{S} = l$, $S_1S_2 = d$, and the distance between S and the plane of the pinholes is D, we obtain

$$\bar{S}S_2 = \left[D^2 + \left(\frac{d}{2} + l \right)^2 \right]^{1/2} \approx D + \frac{1}{2D}\left(\frac{d}{2} + l \right)^2 \qquad (10.3)$$

$$\bar{S}S_1 = \left[D^2 + \left(\frac{d}{2} - l \right)^2 \right]^{1/2} \approx D + \frac{1}{2D}\left(\frac{d}{2} - l \right)^2 \qquad (10.4)$$

where we have assumed that $D \gg d, l$. Thus for disappearance of fringes,

$$\bar{S}S_2 - \bar{S}S_1 = \frac{\lambda}{2} \approx \frac{ld}{D}$$

or

$$l \approx \frac{\lambda D}{2d} \qquad (10.5)$$

For an extended source made up of independent point sources, one may say that
good interference fringes will be observed as long as

$$l \ll \frac{\lambda D}{d} \tag{10.6}$$

Equivalently for a given source of width l, interference fringes of good contrast will
be formed by interference of light from two point sources S_1 and S_2 separated by a
distance d as long as

$$d \ll \frac{\lambda D}{l}$$

Since l/d is the angle (say θ) subtended by the source at the slits above equation can
also be written as

$$d \ll \frac{\lambda}{\theta} \tag{10.7}$$

The distance l_w $(=\lambda/\theta)$ is referred to as the lateral coherence width. It can be seen
from Eq. (10.7) that l_w depends inversely on θ.

Example 10.1 The angle subtended by sun on the earth is 32 s of arc which is approximately 0.01 radians.
Thus assuming a wavelength of 500 nm, the lateral coherence width of the sun would be 50 μm. Thus
if we have a pair of pinholes separated by a distance much less than 50 μm, and illuminated by the sun,
interference pattern of good contrast will be obtained on the screen.

Using ordinary extended sources, one must pass the light through a pinhole in
order to produce a spatially coherent light beam. In contrast, the laser beam is highly
spatially coherent. For example, Fig. 10.11 shows the interference pattern obtained
by Nelson and Collins (1961) by placing a pair of slits of width 7.5 μm separated
by a distance of 54 μm on the end of a ruby rod of a ruby laser. The interference
pattern agrees with the theoretical prediction to within 20%. To show that the spatial

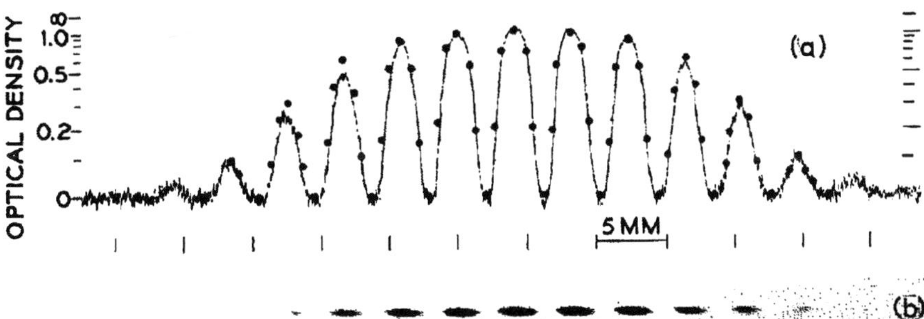

Fig. 10.11 Interference fringes observed by placing a pair of slits in front of a ruby laser show-
ing the spatial coherence of the laser beam (Reprinted with permission from D.F. Nelson and
R.J. Collins, Spatial coherence in the output of a maser, J. Appl. Phys. *32* (1961) 739. © 1961
American Institute of Physics)

coherence is indeed due to laser action, they showed that below threshold no inter-ference pattern was observed; only a uniform darkening of the photographic plate was obtained.

Problem 10.3 A 1550 nm semiconductor laser emits an elliptical Gaussian beam (i.e., the spot sizes in the x- and y-directions are not equal) with minimum spot sizes of 100 and 10 μm in two orthogonal directions at the output facet of the laser. At what distance from the facet does the beam become circular and what is the spot size at this point?

Problem 10.4 Consider a symmetric spherical resonator consisting of two concave mirrors of radii of curvature 1 m and separated by 20 cm operating at 1 mm. What will be the angle of divergence of the laser beam emanating from such a laser?

Problem 10.5 Consider an optical resonator shown below:

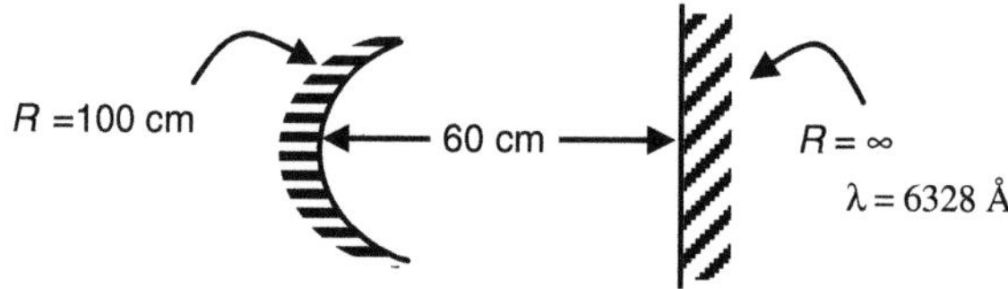

(a)   Calculate the intensity distribution of the fundamental Gaussian mode at the plane mirror and at the spherical mirror

(b)   If both mirrors are partially reflecting, which of the beams coming out (one from the plane mirror and the other from the spherical mirror) would have a larger diffraction divergence?

(c)   If the same resonator is used for oscillation at 1.5 μm by what approximate factor would the diffraction divergence of the beam at 1.5 μm increase or decrease compared to 1 μm?

Problem 10.6 What is the approximate angular divergence of the output beam from a He–Ne laser (operating at 6328 Å) having a 1 m long confocal cavity?

Problem 10.7 A parallel laser beam with a diameter of 2 mm and a power of 10 W falls on a convex lens of diameter 25 mm and focal length 10 mm. If the wavelength of the laser beam is 500 nm, estimate the intensity at the focused spot?

Problem 10.8 Estimate the intensity levels produced on the retina if someone accidentally happens to look straight into a laser beam of 10 mW power and a diameter of 2 mm. Assume the pupil diameter to be 5 mm and the wavelength of light as 600 nm.

Problem 10.9 In continuation, estimate the intensity levels produced on the retina if someone acciden-tally looks at the sun. Assume the pupil diameter to be 2 mm and the wavelength of light as 600 nm. Also calculate the intensity levels produced on the retina when someone looks straight at a 60 W bulb. Take reasonable values of various parameters.

Problem 10.10 An interference experiment is to be conducted using a He–Ne laser. We have two lasers one oscillating in a single longitudinal mode with a linewidth of 10 MHz and the other with two modes with linewidths of 10 MHz and separated by a frequency of 600 MHz. Estimate the minimum path difference for which the interference pattern will disappear when either of the lasers is used.

Problem 10.11 A He–Ne laser of length 20 cm oscillates in two longitudinal modes. If the output of the laser is incident on a photodetector (a device that converts light into electrical current) whose output current is proportional to the incident intensity, what will be the time variation of the output current?

Problem 10.12 Consider a laser beam of circular cross section of diameter 10 cm and of wavelength 800 nm pointed toward the moon which is at a distance of 3.76×10^5 km. What will be the approximate diameter of the laser spot on the moon? Neglect effects due to atmospheric turbulence, etc. [Ans: ~ 6 km].

Problem 10.13 Consider a lens having a focal length of 5 mm and an f-number of 2. Obtain the area of the focused spot using a laser at 650 nm and a laser at 400 nm.

Problem 10.14 Consider two sources, one having a spectral width of 0.1 nm and the other with a spectral width of 0.001 nm. Obtain the coherence length of the two sources.

Problem 10.15 Consider a laser emitting pulses of duration 100 fs. What would be the approximate spectral width of the pulse. What will be the physical length of the optical pulse in free space?

Problem 10.16 The breakdown electric field of air is 30 V/μm. To what intensity level does this correspond?

Chapter 11
Some Laser Systems

11.1 Introduction

In this chapter we shall discuss some specific laser systems and their important operating characteristics. The systems that we shall consider are some of the more important lasers that are in widespread use today for different applications. The lasers considered are

(a) solid-state lasers: ruby, Nd:YAG, Nd:glass;
(b) gas lasers: He–Ne, argon ion, and CO_2;
(c) liquid lasers: dyes;
(d) excimer lasers;

In Chapters 12–14 we shall discuss in detail fiber lasers, semiconductor lasers and coherent sources based on non-linear optical effect, namely parametric oscillators.

11.2 Ruby Lasers

The first laser to be operated successfully was the ruby laser which was fabricated by Maiman in 1960. Ruby, which is the lasing medium, consists of a matrix of aluminum oxide in which some of the aluminum ions are replaced by chromium ions. It is the energy levels of the chromium ions which take part in the lasing action. Typical concentrations of chromium ions are ~0.05% by weight. The energy level diagram of the chromium ion is shown in Fig. 11.1. As is evident from figure this a three-level laser.[1] The pumping of the chromium ions is performed with the help of flash lamp (e.g., a xenon or krypton flashlamp) and the chromium ions in the ground state absorb radiation around wavelengths of 5500 Å and 4000 Å and are excited to the levels marked E_1 and E_2. The chromium ions excited to these levels relax rapidly through a non-radiative transition (in a time $\sim 10^{-8}$–10^{-9}s) to the level

---

[1] The level M actually consists of a pair of levels corresponding to wavelengths of 6943 and 6929 Å. However, laser action takes place only on the 6943 Å line because of higher inversion.

K. Thyagarajan, A. Ghatak, *Lasers*, Graduate Texts in Physics,
DOI 10.1007/978-1-4419-6442-7_11, © Springer Science+Business Media, LLC 2010

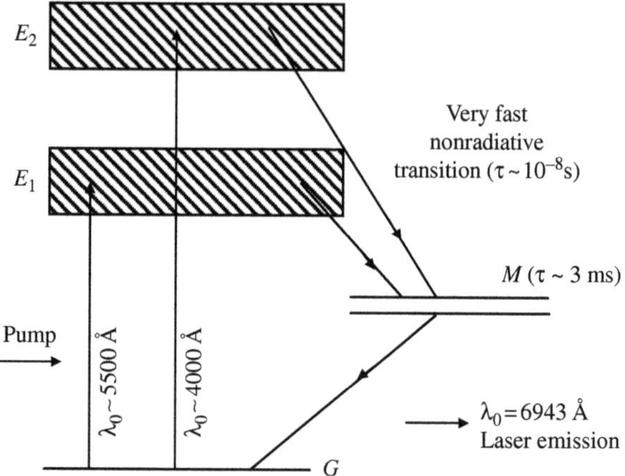

Fig. 11.1 The energy levels of the chromium ions in the ruby laser

marked M which is the upper laser level. The level M is a metastable level with a lifetime of ~ 3 ms. Laser emission occurs between level M and the ground state G at an output wavelength of $\lambda_0 = 6943$ Å.

The flashlamp operation of the laser leads to a pulsed output of the laser. As soon as the flashlamp stops operating the population of the upper level is depleted very rapidly and lasing action stops till the arrival of the next flash. Even during the short period of a few tens of microseconds in which the laser is oscillating, the output is a highly irregular function of time with the intensity having random amplitude fluctuations of varying duration as shown in Fig. 11.2. This is called laser spiking, the formation of which can be understood as follows: when the pump is turned on, the intensity of light at the laser transition is small and hence the pump builds up the inversion rapidly. Although under steady-state conditions the inversion cannot exceed the threshold inversion, on a transient basis it can go beyond the threshold value due to the absence of sufficient laser radiation in the cavity which causes stimulated emission. Thus the inversion goes beyond threshold when the radiation density in the cavity builds up rapidly. Since the inversion is greater than threshold, the radiation density goes beyond the steady-state value which in turn depletes the upper level population and reduces the inversion below threshold. This leads to an interruption of laser oscillation till the pump can again create an inversion beyond threshold. This cycle repeats itself to produce the characteristic spiking in lasers.

Figure 11.3 shows a typical setup of a flashlamp pumped pulsed ruby laser. The helical flashlamp is surrounded by a cylindrical reflector to direct the pump light onto the ruby rod efficiently. The ruby rod length is typically 2–20 cm with diameters of 0.1–2 cm. As we have seen in Section 5.3, typical input electrical energies required are in the range of 10–20 kJ. In addition to the helical flashlamp pumping scheme shown in Fig. 11.3, one may use other pumping schemes such as that shown in Fig. 11.4 in which the pump lamp and the laser rod are placed along the foci of

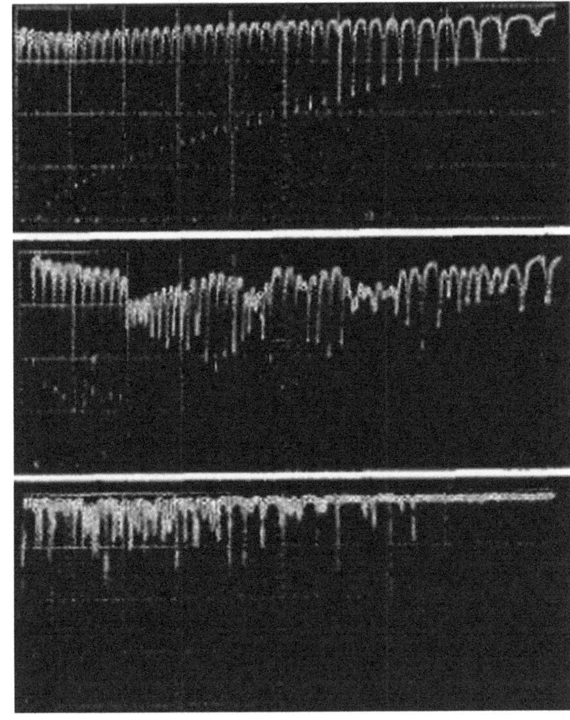

Fig. 11.2 Temporal output power variations of a ruby laser beam leading to what is referred to as laser spiking. The three figures show regular spiking, partially regular spiking and irregular spiking. (Adapted with permission from Sacchi and Svelto (1965) © 1965 IEEE)

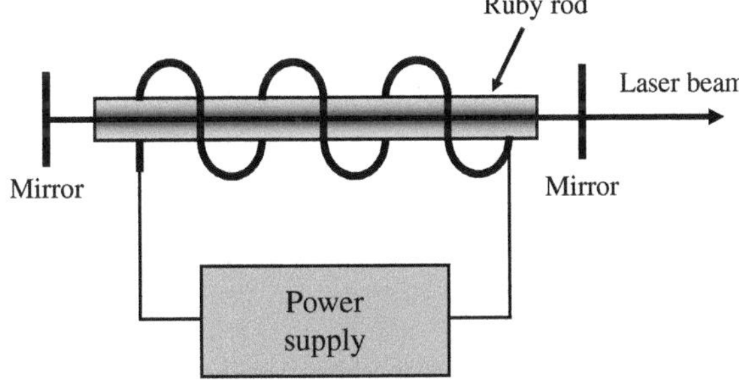

Fig. 11.3 A typical setup of a flashlamp pumped-pulsed ruby laser. The flashlamp is covered by a cylindrical reflector for efficient coupling of the pump light to the ruby rod

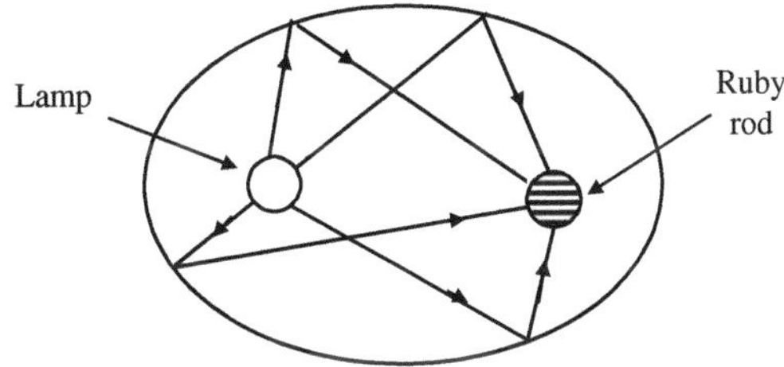

Fig. 11.4 Elliptical pump cavity in which the lamp and the ruby rod are placed along the foci of the elliptical cylindrical reflector

an elliptical cylindrical reflector. It is well known that the elliptical reflector focuses the light emerging from one focus into the other focus of the ellipse, thus leading to an efficient focusing of pump light on the laser rod.

In spite of the fact that the ruby laser is a three-level laser, it still is one of the important practical lasers. The absorption bands of ruby are very well matched with the emission spectra of practically available flashlamps so that an efficient use of the pump can be made. It also has a favorable combination of a long lifetime and a narrow linewidth. The ruby laser is also attractive from an application point of view since its output lies in the visible region where photographic emulsions and photodetectors are much more sensitive than they are in the infrared region. Ruby lasers find applications in pulsed holography, in laser ranging, etc.

11.3 Neodymium-Based Lasers

The Nd:YAG laser (YAG stands for yttrium aluminum garnet which is $Y_3Al_5O_{12}$) and the Nd:glass laser are two very important solid-state laser systems in which the energy levels of the neodymium ion take part in laser emission. They both correspond to a four-level laser. Using neodymium ions in a YAG or glass host has specific advantages and applications.

(a) Since glass has an amorphous structure the fluorescent linewidth of emission is very large leading to a high value of the laser threshold. On the other hand YAG is a crystalline material and the corresponding linewidth is much smaller which implies much over thresholds for laser oscillation.

(b) The fact that the linewidth in the case of the glass host is much larger than in the case of the YAG host can be made use of in the production of ultrashort pulses using mode locking since as discussed in Section 7.7.3, the pulsewidth obtainable by mode locking is the inverse of the oscillating linewidth.

(c) The larger linewidth in glass leads to a smaller amplification coefficient and thus the capability of storing a larger amount of energy before the occurrence of saturation. This is especially important in obtaining very high-energy pulses using Q-switching.

(d) Other advantages of the glass host are the excellent optical quality and excellent uniformity of doping that can be obtained and also the range of glasses with different properties that can be used for solving specific design problems.

(e) As compared to YAG, glass has a much lower thermal conductivity which may lead to induced birefringence and optical distortion.

From the above discussion we can see that for continuous or very high pulse repetition rate operation the Nd:YAG laser will be preferred over Nd:glass. On the other hand for high energy-pulsed operation, Nd:glass lasers maybe preferred. In the following we discuss some specific characteristics of Nd:YAG and Nd:glass laser systems.

11.3.1 Nd:YAG Laser

The Nd:YAG laser is a four-level laser and the energy level diagram of the neodymium ion is shown in Fig. 11.5. The laser emission occurs at $\lambda_0 \approx 1.06\,\mu m$. Since the energy difference between the lower laser level and the ground level is ~ 0.26 eV, the ratio of its population to that of the ground state at room temperature ($T= 300$ K) is $e^{-\Delta E}/k_B T \approx e^{-9} << 1$. Thus the lower laser level is almost unpopulated and hence inversion is easy to achieve. The main pump bands for excitation of the neodymium ions are in the 0.81 and 0.75 μm wavelength regions and pumping is done using arc lamps (e.g., the Krypton arc lamp). Typical neodymium ion concentrations used are $\sim 1.38 \times 10^{20}$ cm^{-3}. The spontaneous lifetime corresponding to the laser transition is 550 μs and the emission line corresponds to homogeneous broadening and has a width $\Delta v \sim 1.2 \times 10^{11}$ Hz which corresponds to $\Delta \lambda \sim 4.5$Å. We have shown in Section 5.4 that the Nd:YAG laser has a much lower threshold of oscillation than a ruby laser.

With the availability of high-power compact and efficient semiconductor lasers, efficient pumping of Nd ions to upper laser level can be accomplished using laser diodes. This leads to very compact diode pumped Nd-based lasers. Diode laser pumping is simpler than lamp pumping and also produces much less heat in the laser medium leading to increased overall efficiency. Since the laser diode output is narrow band unlike a normal lamp, the output at 808 nm can be efficiently used for pumping. Typical output powers of 150 W are commercially available. In fact an intracavity second-harmonic generator can efficiently convert the laser wavelength to 532 nm (the second harmonic of 1064 nm of Nd:YAG) leading to very efficient green lasers.

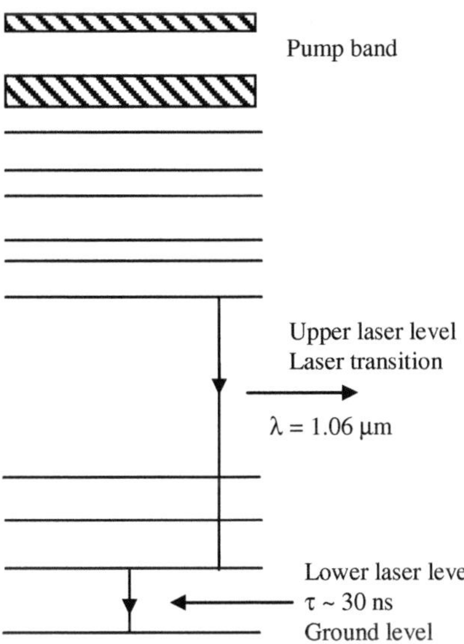

Fig. 11.5 The energy levels of neodymium ion in the Nd:YAG laser

Nd:YAG lasers find many applications in range finders, illuminators with Q-switched operation giving about 10–50 pulses per second with output energies in the range of 100 mJ per pulse, and pulse width ~ 10 ns. They also find applications in resistor trimming, scribing, micromachining operations as well as welding, hole drilling, etc.

11.3.2 Nd:Glass Laser

The Nd:glass laser is again a four-level laser system with a laser emission around 1.06 μm. Typical neodymium ion concentrations are $\sim 2.8 \times 10^{20}$ cm^{-3} and various silicate and phosphate glasses are used as the host material. Since glass has an amorphous structure different neodymium ions situated at different sites have slightly different surroundings. This leads to an inhomogeneous broadening and the resultant linewidth is $\Delta v \sim 7.5 \times 10^{12}$ Hz which corresponds to $\Delta \lambda \sim 260$ Å. This width is much larger than in Nd:YAG lasers and consequently the threshold pump powers are also much higher. The spontaneous lifetime of the laser transition is ~ 300 μs.

Nd-doped fiber lasers are also efficient. Chapter 12 discusses fiber lasers and primarily erbium-doped fiber lasers. But similar analysis can also be carried out for Nd-doped fiber lasers.

Nd:glass lasers are more suitable for high energy-pulsed operation such as in laser fusion where the requirement is of subnanosecond pulses with an energy content of several kilojoules (i.e., peak powers of several tens of terawatts). Other applications are in welding or drilling operations requiring high pulse energies.

Table 11.1 gives a comparison of some important characteristics of ruby, Nd:YAG, and Nd:glass laser systems.

Table 11.1 Comparison of ruby, Nd:YAG, and Nd:glas laser systems

Laser	Ruby	Nd:YAG	Nd:glass
Wavelength (Å)	6943	10,641	10,623
Spontaneous lifetime (μs)	3000	240	300
Active ion concentration (cm^{-3})	1.58×10^{19}	1.38×10^{20}	2.83×10^{20}
Linewidth (GHz)	330	120	7500
(Å)	5.5	4.0	260
Population inversion density for 1% gain/cm (cm^{-3})	4×10^7 $+ 7.6 \times 10^{18}$	1.1×10^{16}	3.3×10^{17}
Index of refraction (n) (at laser λ)	$n_\mathrm{o} = 1.763$ $n_\mathrm{e} = 1.755$	1.82	1.55
Major pump	4040	5800	5800
bands (Å)	5540	7500	7500
		8100	8100

Table adapted from Koechner (1976).

11.4 Titanium Sapphire Laser

Titanium sapphire (Ti:sapphire) laser is one of the most important solid-state lasers since it is a continuously tunable laser from about 650 to 1100 nm and due to its large gain bandwidth can produce mode-locked pulses in the tens of femtosecond regime. Titanium sapphire laser consists of titanium-doped sapphire (Al_2O_3), in which the energy levels of titanium take part in laser action. The titanium sapphire laser is pumped by another laser, usually argon ion laser emitting in the wavelength region of 514 nm; other lasers such as frequency-doubled Nd:YAG laser emitting 532 nm wavelength is also used for pumping the laser. Figure 11.6 shows the absorption and emission spectra of titanium sapphire. The broad absorption and emission spectra are evident. The broad emission spectrum leads to a broad gain spectrum which in turn allows for ultrashort pulse generation using mode locking techniques.

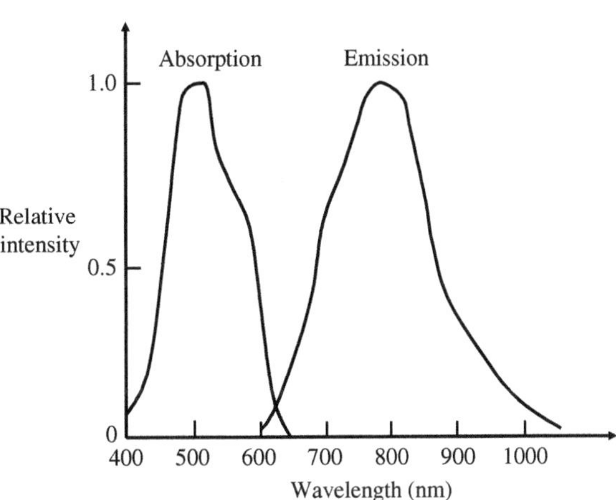

Fig. 11.6 Absorption and emission spectrum of titanium sapphire laser medium [Adapted from http://www.olympusmicro.com/ primer/techniques/fluorescence/ multiphoton/images/ tisapspectra.jpg]

11.5 The He–Ne Laser

The first gas laser to be operated successfully was the He–Ne laser. As we discussed earlier in solid-state lasers, the pumping is usually done using a flashlamp or a continuous high-power lamp. Such a technique is efficient if the lasing system has broad absorption bands. In gas lasers since the atoms are characterized by sharp energy levels as compared to those in solids, one generally uses an electrical discharge to pump the atoms.

The He–Ne laser consists of a long and narrow discharge tube (diameter ~ 2– 8 mm and length 10–100 cm) which is filled with helium and neon with typical pressures of 1 torr[2] and 0.1 torr. The actual lasing atoms are the neon atoms and as we shall discuss helium is used for a selective pumping of the upper laser level

---

[2]Torr is a unit of pressure and 1 torr = 1 mm Hg.

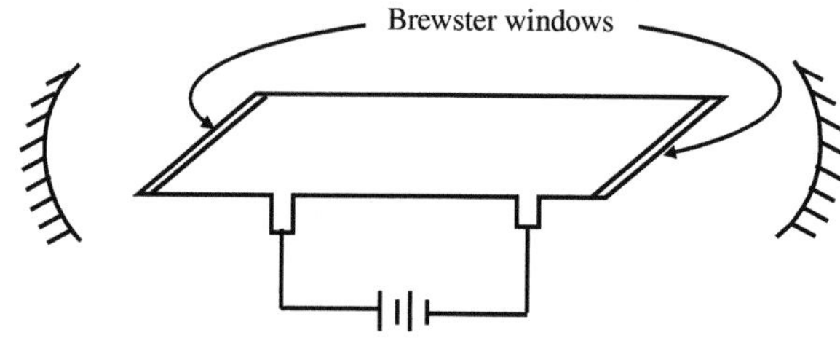

Fig. 11.7 A typical He–Ne laser with external mirrors. The ends of the discharge tube are fitted with Brewster windows

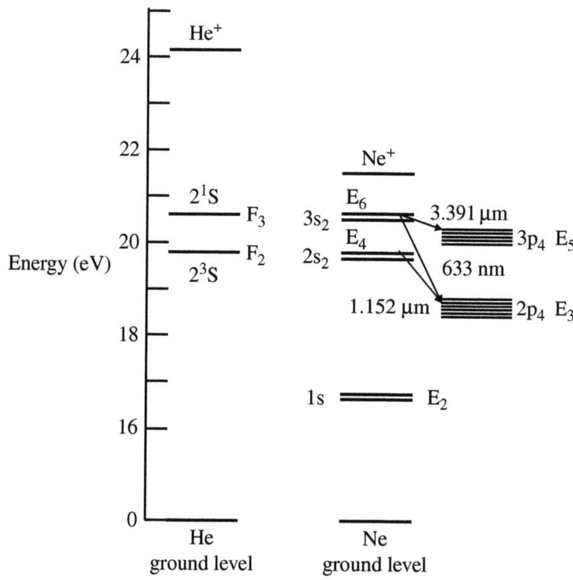

Fig. 11.8 Low lying energy levels of helium and neon taking part in the He–Ne laser

of neon. The laser resonator may consist of either internal or external mirrors (see Fig. 11.7). Figure 11.8 shows the energy levels of helium and neon. When an electrical discharge is passed through the gas, the electrons which are accelerated down the tube collide with helium and neon atoms and excite them to higher energy levels. The helium atoms tend to accumulate at levels F_2 and F_3 due to their long lifetimes of $\sim 10^{-4}$ and 5×10^{-6} s, respectively. Since the levels E_4 and E_6 of neon atoms have almost the same energy as F_2 and F_3, excited helium atoms colliding with neon atoms in the ground state can excite the neon atoms to E_4 and E_6. Since the pressure of helium is ten times that of neon, the levels E_4 and E_6 of neon are selectively populated as compared to other levels of neon.

Transition between E_6 and E_3 produces the very popular 6328 Å line of the He–Ne laser. Neon atoms de-excite through spontaneous emission from E_3 to E_2 (lifetime $\sim 10^{-8}$ s). Since this time is shorter than the lifetime of level E_6 (which is $\sim 10^{-7}$ s) one can achieve steady-state population inversion between E_6 and E_3.

Level E_2 is metastable and thus tends to collect atoms. The atoms from this level relax back to the ground level mainly through collisions with the walls of the tube. Since E_2 is metastable it is possible for the atoms in this level to absorb the spontaneously emitted radiation in the $E_3 \rightarrow E_2$ transition to be re-excited to E_3. This will have the effect of reducing the inversion. It is for this reason that the gain in this laser transition is found to increase with decreasing tube diameter.

The other two important wavelengths from the He–Ne laser are 1.15 and 3.39 μm, which correspond to the $E_4 \rightarrow E_3$ and $E_6 \rightarrow E_5$ transitions. It is interesting to observe that both 3.39 μm and 6328 Å transitions share the same upper laser level. Now since the 3.39 μm transition corresponds to a much lower frequency than the 6328 Å line, the Doppler broadening is much smaller at 3.39 μm and also since gain depends inversely on ν^2 (see Eq. (4.26)), the gain at 3.39 μm is much higher than at 6328 Å. Thus due to the very large gain, oscillations will normally tend to occur at 3.39 μm rather than at 6328 Å. Once the laser starts to oscillate at 3.39 μm, further build up of population in E_6 is not possible. The laser can be made to oscillate at 6328 Å by either using optical elements in the path which strongly absorb the 3.39 μm wavelength or increasing the linewidth through the Zeeman effect by applying an inhomogeneous magnetic field across the tube.

If the resonator mirrors are placed outside the discharge tube then reflections from the ends of the discharge tube can be avoided by placing the windows at the Brewster angle (see Fig. 11.7). In such a case the beam polarized in the plane of incidence suffers no reflection at the windows while the perpendicular polarization suffers reflection losses. This leads to a polarized output of the laser.

11.6 The Argon Ion Laser

In an argon ion laser, one uses the energy levels of the ionized argon atom and the laser emits various discrete lines in the 3500–5200 Å wavelength region. Figure 11.9

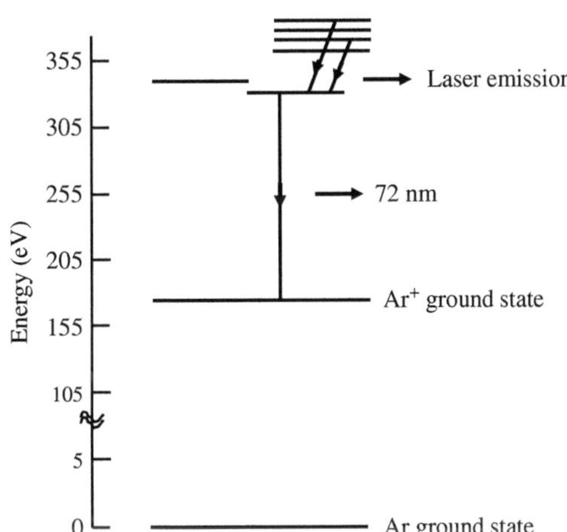

Fig. 11.9 Some of the levels taking part in the laser transition corresponding to the argon ion laser

shows some of the energy levels taking part in the laser transition. The argon atoms have to be first ionized and then excited to the higher energy levels of the ion. Because of the large energies involved in this, the argon ion laser discharge is very intense; typical values being 40 A at 165 V. A particular wavelength out of the many possible lines is chosen by placing a dispersive prism inside the cavity close to one of the mirrors. Rotation of the prism–mirror system provides feedback only at the wavelength which is incident normally on the mirror. Typical output power in a continuous wave argon ion laser is 3–5 W. Some of the important emission wavelengths include 5145 Å, 4965 Å, 4880 Å, 4765 Å, and 4579 Å.

11.7 The CO_2 Laser

The lasers discussed above use transitions among the various excited electronic states of an atom or an ion. In a CO_2 laser one uses the transitions occurring between different vibrational states of the carbon dioxide molecule. Figure 11.10 shows the carbon dioxide molecule consisting of a central carbon atom with two oxygen atoms attached one on either side. Such a molecule can vibrate in the three independent modes of vibration shown in Fig. 11.10.

These correspond to the symmetric stretch, the bending, and the asymmetric stretch modes. Each of these modes is characterized by a definite frequency of vibration. According to basic quantum mechanics these vibrational degrees of freedom are quantized, i.e., when a molecule vibrates in any of the modes it can have only a discrete set of energies. Thus if we call ν_1 the frequency corresponding to the symmetric stretch mode then the molecule can have energies of only

$$E_1 = \left(m + \frac{1}{2} \right) h\nu_1, \qquad m = 0, 1, 2, \ldots \tag{11.1}$$

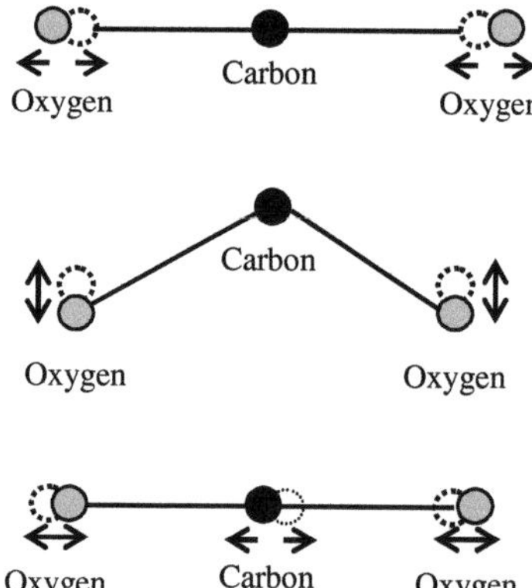

Fig. 11.10 The three independent modes of vibration of the carbon dioxide molecule

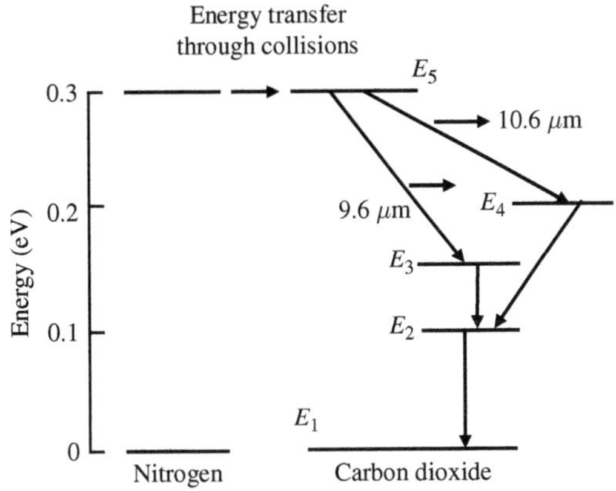

Fig. 11.11 The low lying vibrational levels of nitrogen and carbon dioxide molecules. Energy transfer from excited nitrogen molecules to carbon dioxide molecules results in the excitation of carbon dioxide molecules. Important lasing transitions occur at 9.6 and 10.6 μm

when it vibrates in the symmetric stretch mode. Thus the degree of excitation is characterized by the integer m when the carbon dioxide molecule vibrates in the symmetric stretch mode. In general, since the carbon dioxide molecule can vibrate in a combination of the three modes the state of vibration can be described by three integers (mnq); the three integers correspond, respectively, to the degree of excitation in the symmetric stretch, bending, and asymmetric stretch modes, respectively. Figure 11.11 shows the various vibrational energy levels taking part in the laser transition.

The laser transition at 10.6 μm occurs between the (001) and (100) levels of carbon dioxide. The excitation of the carbon dioxide molecules to the long-lived level (001) occurs both through collisional transfer from nearly resonant excited nitrogen molecules and also from the cascading down of carbon dioxide molecules from higher energy levels.

The CO$_2$ laser possesses an extremely high efficiency of $\sim$30%. This is because of efficient pumping to the (001) level and also because all the energy levels involved are close to the ground level. Thus the atomic quantum efficiency which is the ratio of the energy difference corresponding to the laser transition to the energy difference of the pump transition, i.e.,

$$\eta = \frac{E_5 - E_4}{E_5 - E_1}$$

is quite high ($\sim$45%). Thus a large portion of the input power can be converted into useful laser power.

Output powers of several watts to several kilowatts can be obtained from CO$_2$ lasers. High-power CO$_2$ lasers find applications in materials processing, welding, hole drilling, cutting, etc., because of their very high output power. In addition, the atmospheric attenuation is low at 10.6 μm which leads to some applications of CO$_2$ lasers in open air communications.

11.8 Dye Lasers

One of the most widely used tunable lasers in the visible region is the organic dye laser. The dyes used in the lasers are organic substances which are dissolved in solvents such as water, ethyl alcohol, methanol, and ethylene glycol. These dyes exhibit strong and broad absorption and fluorescent spectra and because of this they can be made tunable. By choosing different dyes one can obtain tenability from 3000 Å to 1.2 μm.

The levels taking part in the absorption and lasing correspond to the various vibrational sublevels of different electronic states of the dye molecule. Figure 11.12 shows a typical energy level diagram of a dye in which S_0 is the ground state, S_1 is the first excited singlet state, and T_1, T_2 are the excited triplet states of the dye molecule. Each state consists of a large number of closely spaced vibrational and rotational sublevels. Because of strong interaction with the solvent, the closely spaced sublevels are collision broadened to such an extent that they almost form a continuum.

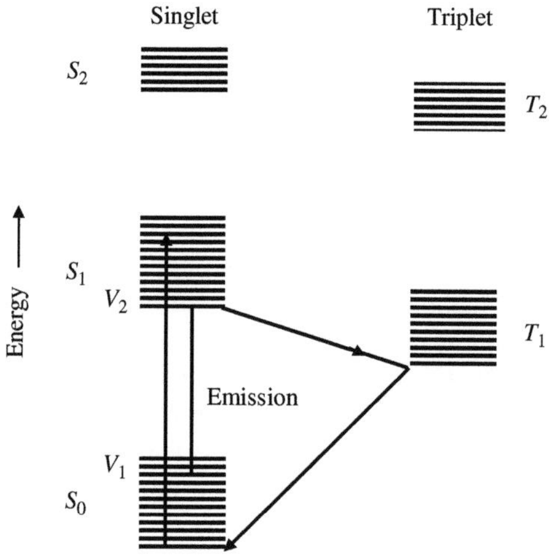

Fig. 11.12 Typical energy level diagram of a dye molecule

When dye molecules in the solvent are irradiated by visible or ultraviolet radiation then the molecules are excited to the various sublevels of the state S_1. Due to collisions with the solvent molecules, the molecules excited to higher vibrational and rotational states of S_1 relax very quickly (in times $\sim 10^{-11}$–10^{-12} s) to the lowest level V_2 of the state S_1. Molecules from this level emit spontaneously and de-excite to the different sublevels of S_0. Thus the fluorescent spectrum is found to be red shifted against the absorption spectrum.

Problems

Problem 11.1 Consider mode locking an Nd:YAG laser and an Nd:glass laser. Compare the minimum pulse widths obtainable using these two lasers. If the length of the resonators in both cases is 20 cm, what typical pulse widths are possible by mode locking these two lasers?

Problem 11.2 Estimate the Doppler-broadened linewidth of a carbon dioxide laser system assuming a temperature of 300 K.

Problem 11.3 Consider a He–Ne laser having a cavity length of 30 cm and oscillating over a bandwidth of 1500 MHz. What will be the coherence length of the laser?

Problem 11.4 Compare the Doppler-broadened linewidth of the 632.8 and 3.39 mm emission lines of He–Ne laser. Assume a temperature of 300 K.

Chapter 12
Doped Fiber Amplifiers and Lasers

12.1 Introduction

The most common solid-state lasers are Nd:YAG laser and Ti:sapphire laser and are used extensively in various laboratories for R&D and also in many applications. These lasers usually require laboratory-like environments and have a reasonably high power consumption requiring maintenance. In this context, optical fiber lasers in which the gain medium is in the form of an optical fiber are revolutionizing the applications of solid-state lasers. Some of the most attractive features of fiber lasers are the direct pumping using semiconductor lasers, high gain achievable with broad bandwidths, and a laser beam with excellent beam quality. Using components developed specifically for the telecommunication application of fiber optics, fiber lasers have seen an explosive growth in terms of high output powers, ultrashort pulses, and extensive wavelength region. With developments in large mode area optical fibers, photonic crystal fibers, etc., the field is continuing to grow. Figure 12.1 shows the growth of fiber lasers during the last 15 years with a steep rise after 2002.

There are many important fiber lasers such as erbium-doped fiber laser (for operation at 1550 nm) and ytterbium-doped fiber laser (for operation at 1060 nm). They are either three-level or four-level lasers and can be analyzed using the standard rate equations. Compared to standard lasers, in the case of fiber lasers, the amplifying medium is quite long and the signal and pump powers vary significantly over the length of the fiber. Hence the analysis of fiber laser is somewhat different from standard lasers discussed in Chapter 5. As an example of a fiber laser, in Section 12.2, we will consider an erbium-doped fiber laser in detail and obtain its characteristics. Similar analysis can be performed for other fiber laser systems.

12.2 The Fiber Laser

In a fiber laser, the active medium providing optical amplification is an optical fiber with its core doped with suitable dopants (more details on optical fibers and their characteristics can be found in Chapter 16). When such a fiber is pumped by a suitable light source (usually another laser), population inversion between two of the energy levels is achieved, thus providing optical amplification. As discussed in

K. Thyagarajan, A. Ghatak, *Lasers*, Graduate Texts in Physics,
DOI 10.1007/978-1-4419-6442-7_12, © Springer Science+Business Media, LLC 2010

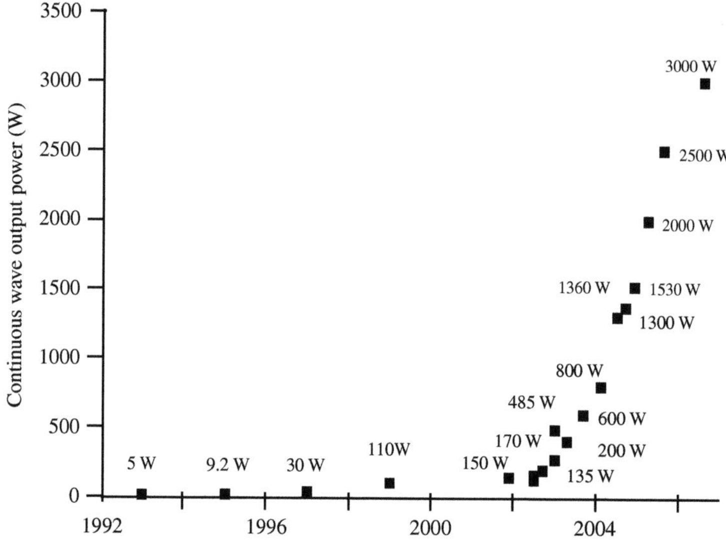

Fig. 12.1 Growth of fiber lasers since 1993 showing a steep rise after 2002. (Adapted with permission from Limpert et al. (2007) © 2007 IEEE)

Chapter 4, by providing an optical feedback using mirrors at the end of the amplifying fiber, it is possible to achieve laser oscillation leading to a fiber laser. As early as 1961, Elias Snitzer wrapped a flashlamp around a glass fiber (having a 300-μm core doped with Nd^{3+} ions clad in a lower index glass) and when suitable feedback was applied, the first fiber laser was born. Thus, the fiber laser was fabricated within a year of the demonstration of the first ever laser by Theodore Maiman.

Figure 12.2 shows a schematic of a fiber laser. As an example, let us consider an erbium-doped fiber laser. Light from a pump laser emitting at 980 nm is coupled into a short length of erbium-doped fiber using a wavelength division multiplexing (WDM) coupler. The WDM coupler is a device which combines two different wavelengths from two different fibers into a single output fiber and acts like a dichroic beam splitter. It can also be used to split light waves at two different wavelengths propagating in the same fiber into two different fibers. The WDM coupler at the

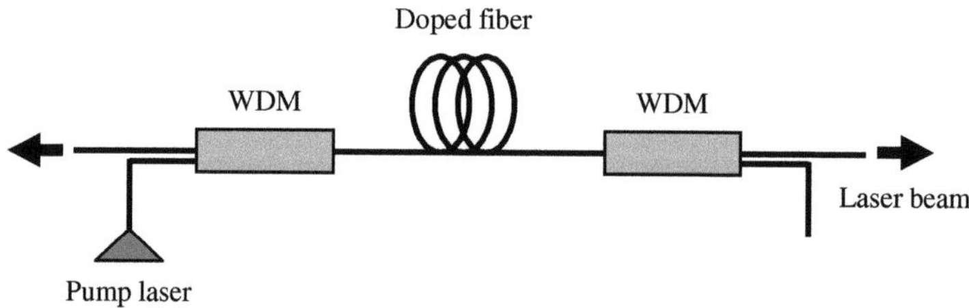

Fig. 12.2 A schematic of an erbium-doped fiber laser

output removes any unused pump laser power. To provide optical feedback, the two ends of the signal ports of the WDM coupler are cut properly and coated to have high reflectivity or have fiber Bragg gratings as reflectors. Since the gains provided by erbium ions are very large, even a small reflectivity is sufficient to satisfy the condition for laser oscillation, namely compensation of loss by the gain provided by population inversion. This would result in an output laser beam from both coupler ends. The wavelength of emission is usually determined by the wavelength satisfying the maximum gain and minimum loss; this is around 1530 nm for erbium-doped fibers. Figure 12.3 shows the output from an erbium-doped fiber laser as the pumping in increased. Just before starting to lase, the pump power is insufficient to overcome the losses in the cavity and thus the output is only amplified spontaneous emission (lower curve in Fig. 12.3). As we increase the pump power, the erbium-doped fiber starts to lase usually emitting multiple wavelengths seen as spikes in the figure.

Fig. 12.3 Lower curve corresponds to the spectrum at the output of the fiber laser below the threshold and the upper curve represents the spectrum above the threshold for laser oscillation

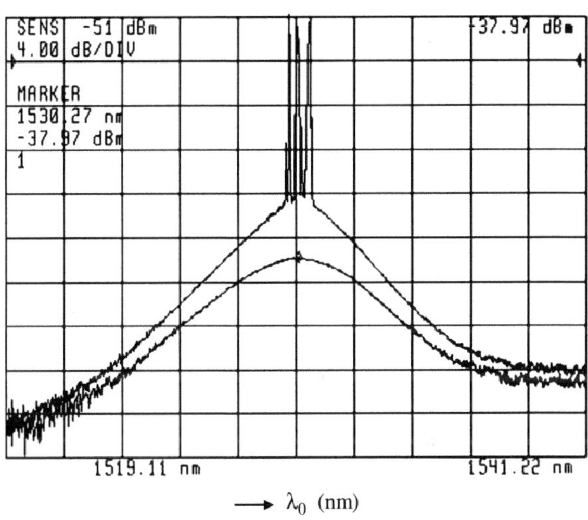

In case it is required to have the laser oscillate at a specific wavelength within the gain bandwidth of the erbium ion, then this can be achieved by using a fiber Bragg grating (FBG) at one end of the laser. Fiber Bragg gratings are periodic variations of refractive index within the core of an optical fiber. The periodic variations cause a specific wavelength to be reflected back along the fiber; the wavelength of peak reflection depends on the period of the grating and the optical fiber. With sufficient length and refractive index modulation, reflectivities of more than 99% can be achieved (see, e.g., Ghatak and Thyagarajan (1998)).

By placing an FBG at one end of the fiber laser, only the wavelength where the FBG reflects strongly is fed back into the fiber cavity. This wavelength would thus suffer much lower loss compared to other wavelengths. This would ensure that the fiber laser oscillates at the frequency as determined by the FBG. The lasing wavelength can be tuned by tuning the peak reflection wavelength of the FBG.

Fiber lasers possess many interesting advantages vis-a-vis other laser systems. In particular, since the laser beam is confined to a very small cross-sectional area within the core of the fiber, large pump intensities can be achieved even with small pump powers and thus leading to lower pump power thresholds. Since both the pump and the laser beam are propagating within the fiber, they overlap very well and this also adds to increased efficiency of the laser and efficiencies of 80% are possible. Since the fiber guides the pump beam, one can use very long-length cavities without bothering about the divergence of the pump laser beam. Since the ratio of surface area to volume of fiber laser is very large, it does not suffer from thermal problems and heat dissipation is much easier. The output beam is of very good quality since it emerges as the fundamental mode of the fiber. Also since the components in the laser are made up of fibers which are all spliced, there are no mechanical perturbation problems such as in bulk lasers with separate mirrors.

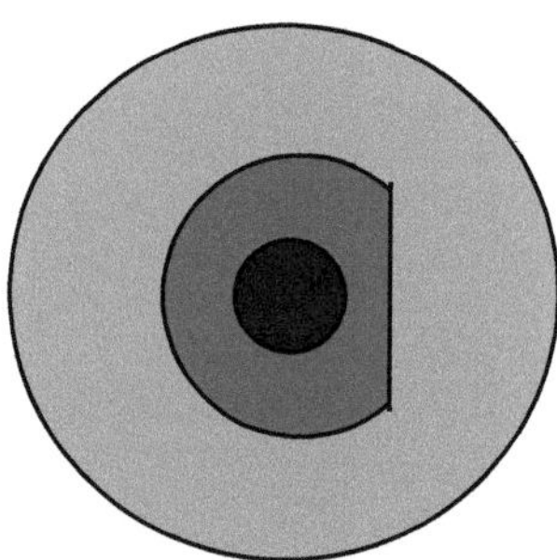

Fig. 12.4 D-shaped fiber cladding of a doped fiber for efficient pump utilization

Using conventional fibers with doped single-mode core and a cladding, the laser power is restricted to about 1 W. To achieve higher output powers, fiber lasers use double-clad fibers as the amplifying medium (see Fig. 12.4). In this fiber, the central core guides the laser wavelength and is single moded at this wavelength. The inner cladding is surrounded by an outer cladding and this region acts as a multi-moded guide for the pump wavelength. The radius of the inner cladding is large and so is the refractive index difference between the inner cladding and the outer cladding. This ensures that power from large area diode lasers can be launched into the fiber efficiently. At the same time, since the laser wavelength is propagating as a fundamental mode in the inner core, the laser output would be single moded. The pump power propagating in the inner cladding propagates in the form of different rays (or modes). If the cladding is circular in cross section, then it is possible that some of the rays (skew rays) propagating in the cladding would never have an opportunity to cross the core and this portion of the pump would never be used in creating inversion and thus leading to reduced conversion efficiencies. In order that all the rays corresponding to the pump power propagating in the inner cladding of the fiber cross the core, the inner cladding is made non-circular (see Fig. 12.4). This type of design can lead to very much increased pump conversion efficiencies.

12.3 Basic Equations for Amplification in Erbium-Doped Fiber

Figure 12.5 shows the absorption spectrum of an erbium-doped fiber. Several absorption peaks are apparent; these correspond to different pairs of energy levels of erbium ion. The most important absorption peak correspond to 980 nm and a broad absorption band around 1550 nm. Figure 12.6 shows a schematic of the energy levels involved in the absorption process.

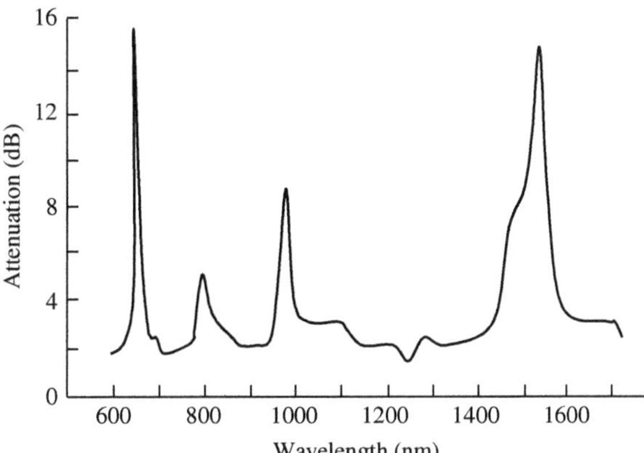

Fig. 12.5 Typical absorption spectrum of an erbium-doped fiber

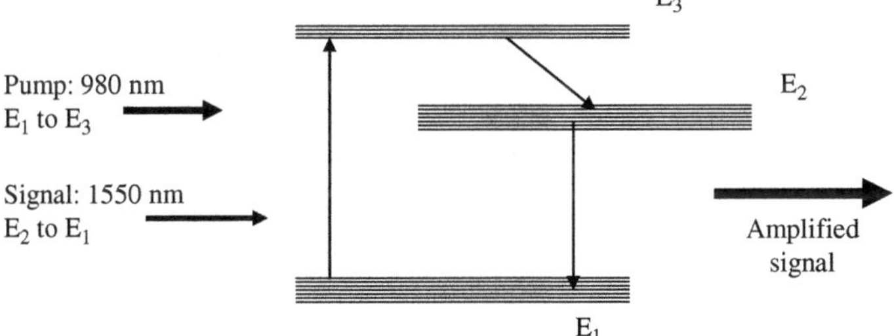

Fig. 12.6 Three lowest lying energy levels of erbium ions in silica matrix. Each level consists of many sublevels

We will now derive expressions describing an erbium-doped fiber laser (EDFL). We will consider the EDFL to be a three-level laser and assume that ions pumped into level E_3 by a 980-nm pump laser jump rapidly to level E_2, which is the upper laser level (see Fig. 12.6), and assume that ions pumped from level E_1 to level E_3 by a 980-nm pump laser jump rapidly to level E_2, the upper laser level.

Let N_1, N_2, and N_t represent the erbium ion density (number of erbium ions per unit volume) in energy levels E_1 and E_2 and the total erbium ion density, respectively. Since we are assuming that ions relax rapidly from level E_3, we have $N_3 \sim 0$ and

$$N_t(r) = N_1(r, z) + N_2(r, z) \tag{12.1}$$

Usually the entire fiber core of radius a is uniformly doped with erbium ions. Hence

$$
\begin{aligned}
N_t(r) &= N_t; && 0 < r < a \\
&= 0; && r > a
\end{aligned}
\tag{12.2}
$$

By pumping appropriately it is possible to generate population inversion between levels E_2 and E_1. In such a case, light at a frequency corresponding to $(E_2 - E_1)/h$ can get amplified as it propagates through the fiber. Such a device behaves as an optical amplifier. Erbium-doped fiber amplifiers (EDFAs) based on erbium-doped fibers are extremely important components in today's long-distance fiber optical communication systems. In such an amplifier, the pump is usually a 980-nm laser diode and the amplification is provided over a large band of wavelengths around 1550 nm. We shall discuss the properties of EDFA in Section 12.5; here we analyze an erbium-doped fiber laser.

Unlike an EDFA, in which the signal propagates only along one direction, in the case of a laser, the signal will form a standing wave in the laser cavity. Hence there would be signal beams propagating along both the +z- and the −z-directions. We will assume the pump to be travelling along the +z-direction (see Fig. 12.7).

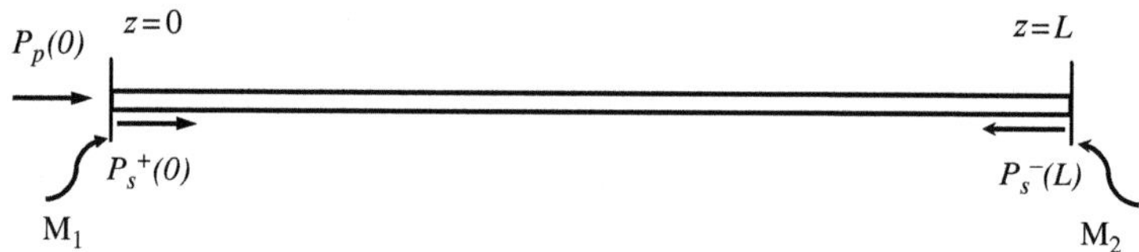

Fig. 12.7 Pump power travels along the fiber in the +z-direction, while the power at the lasing wavelength travels in the +z- and −z-directions. M_1 and M_2 represent the two mirrors at the ends of the fiber cavity; they could be just cleaved facets of the fiber

Recalling the discussion on rate equations in Chapter 5, we can write the rate equation describing the time rate of change of erbium ion population density N_2 in level E_2 as

$$\frac{dN_2}{dt} = -\frac{N_2}{\tau_{sp}} + \frac{\sigma_p I_p}{h\nu_p} N_1 - \frac{\sigma_{se} I_s^+}{h\nu_s} N_2 + \frac{\sigma_{sa} I_s^+}{h\nu_s} N_1 - \frac{\sigma_{se} I_s^-}{h\nu_s} N_2 + \frac{\sigma_{sa} I_s^-}{h\nu_s} N_1 \tag{12.3}$$

On the right-hand side of Eq. (12.3)

- the first term represents spontaneous emission per unit time per unit volume with τ_{sp} representing the spontaneous lifetime of the level E_2;
- the second term represents induced absorption per unit time per unit volume due to the pump with I_p representing the pump intensity and σ_p the absorption cross section at the pump frequency ν_p (see Section 4.2.1 for the definition of absorption and emission cross sections);

- the third term represents stimulated emission per unit time per unit volume induced by the signal of intensity I_s^+ travelling along the $+z$-direction and σ_{se} represents the emission cross section at the signal frequency ν_s;
- the fourth term represents induced absorption per unit time per unit volume due to the signal frequency ν_s induced by signal of intensity I_s^+ travelling along the $+z$-direction;
- the fifth and sixth terms represent stimulated emission and absorption per unit time per unit volume induced by the signal of intensity I_s^- travelling along the $-z$-direction.

Unlike conventional lasers, the pump and signal waves travel as modes in the fiber lasers. These modes are characterized by specific intensity distributions along the transverse cross section of the fiber. Due to the guidance mechanism, unlike conventional lasers where the laser light within the cavity will diffract, there would be no diffraction of the waves in fiber lasers. In view of this we should describe the propagation of pump and signal along the fiber length in terms of powers rather than in terms of intensities. In order to do so, we introduce two new functions $f_p(r)$ and $f_s(r)$ at the pump and signal frequencies as

$$I_p(r, z) = P_p(z) f_p(r) \tag{12.4}$$

and

$$I_s^{\pm}(r, z) = P_s^{\pm}(z) f_s(r) \tag{12.5}$$

Here $P_p(z)$ and $P_s^{\pm}(z)$ represent the powers carried by the pump propagating along the $+z$-direction and the signal propagating along the $+z$- and $-z$-directions, respectively, and $f_p(r)$ and $f_s(r)$ represent the transverse dependence of the modal intensity patterns at the pump and signal frequencies. Usually the pump and signal beams travel as the fundamental modes of the fiber and thus we assume that the intensity distributions at the pump and the signal are dependent only on the cylindrical radial coordinate r and are independent of the azimuthal coordinate φ. By integrating Eqs. (12.4) and (12.5) along the entire transverse cross section, we note that the functions $f_p(r)$ and $f_s(r)$ satisfy the following normalization conditions:

$$2\pi \int_0^\infty f_p(r) r \, dr = 1 \tag{12.6}$$

and

$$2\pi \int_0^\infty f_s(r) r \, dr = 1 \tag{12.7}$$

Since the population of a level depends on the intensity of the interacting light wave and the intensities at the pump and signal frequencies depend on the coordinate r, in general the populations N_1 and N_2 also depend on r. In order to simplify the analysis we shall neglect this dependence and assume that N_1 and N_2 are independent of r and depend only on the longitudinal coordinate z.

Integrating Eq. (12.3) over the transverse cross section and using Eqs. (12.4) and (12.5) we obtain

$$\frac{dN_2(z)}{dt}A = -\frac{N_2}{\tau_{sp}}A + \frac{\sigma_p P_p}{h\nu_p}N_1 2\pi \int_0^a f_p(r)r\,dr$$

$$- \frac{(\sigma_{se}N_2 - \sigma_{sa}N_1)}{h\nu_s}\left(P_s^+ + P_s^-\right)2\pi \int_0^a f_s(r)r\,dr \tag{12.8}$$

where $A \ (= \pi a^2)$ represents the area of cross section of the doped region of the fiber. We now define

$$\Gamma_p = 2\pi \int_0^a f_p(r)r\,dr; \qquad \Gamma_s = 2\pi \int_0^a f_s(r)r\,dr \tag{12.9}$$

which represent the fractional powers inside the core at the pump and signal wavelengths, respectively. We also define the normalized population densities

$$\tilde{N}_1 = \frac{N_1}{N_t}; \qquad \tilde{N}_2 = \frac{N_2}{N_t} \tag{12.10}$$

Using these definitions, Eq. (12.8) becomes

$$\frac{d\tilde{N}_2}{dt} = -\frac{\tilde{N}_2}{\tau_{sp}} + \frac{\sigma_p \Gamma_p}{h\nu_p A}P_p\tilde{N}_1 - \frac{(\sigma_{se}\tilde{N}_2 - \sigma_{sa}\tilde{N}_1)\Gamma_s}{h\nu_s A}(P_s^+ + P_s^-) \tag{12.11}$$

We now write for the equations describing the evolution of the pump and signal powers along the fiber. In the case of pump, there are only transitions from level E_1 to E_3; thus the variation of pump intensity I_p along z would be given as

$$\frac{dI_p}{dz} = -\sigma_p N_1 I_p \tag{12.12}$$

Now

$$\frac{dP_p}{dz} = \frac{d}{dz}\int I_p(r,z)r\,dr\,d\varphi = -2\pi\sigma_p N_1 P_p \int_0^a f_p(r)r\,dr = -\sigma_p \Gamma_p P_p N_1 \tag{12.13}$$

Similarly the variation of signal intensity along $+z$ is given as

$$\frac{dI_s^+}{dz} = \sigma_{se}I_s^+(r,z)N_2 - \sigma_{sa}I_s^+(r,z)N_1 \tag{12.14}$$

The first term corresponds to stimulated emission and the second to stimulated absorption. The contribution from spontaneous emission is neglected here since spontaneous emission occurs over a large spectral bandwidth and in all directions. The fraction of the spontaneous emission coupling into the forward direction in the mode and at the signal frequency is negligible. Of course, spontaneous emission is necessary to start the laser oscillation.

Thus

$$\frac{dP_s^+}{dz} = \frac{d}{dz} \int I_s^+(r,z) r\, dr\, d\varphi = (\sigma_{se}N_2 - \sigma_{sa}N_1)P_s^+\Gamma_s \qquad (12.15)$$

Similarly for the signal propagating along the $-z$-direction

$$\frac{dP_s^-}{dz} = -(\sigma_{se}N_2 - \sigma_{sa}N_1)P_s^-\Gamma_s \qquad (12.16)$$

We can combine Eqs. (12.15) and (12.16) as

$$\frac{dP_s^\pm}{dz} = \pm(\sigma_{se}N_2 - \sigma_{sa}N_1)P_s^\pm\Gamma_s \qquad (12.17)$$

We now define powers in terms of photon flux; thus the photon flux (i.e., the total number of photons crossing per unit time across a plane perpendicular to the fiber axis) at pump and signal wavelengths is denoted by n_p and $n_s^\pm$, respectively, and is given as

$$n_p = \frac{P_p}{h\nu_p}; \qquad n_s^\pm = \frac{P_s^\pm}{h\nu_s} \qquad (12.18)$$

Using these definitions Eqs. (12.13) and (12.17) can be written as

$$\frac{dn_p}{dz} = -\sigma_p N_t \Gamma_p \tilde{N}_1 n_p \qquad (12.19)$$

$$\frac{dn_s^\pm}{dz} = \pm(\sigma_{se}\tilde{N}_2 - \sigma_{sa}\tilde{N}_1)N_t\Gamma_s n_s^\pm$$
$$= \pm[(\sigma_{se} + \sigma_{sa})\tilde{N}_2 - \sigma_{sa}]N_t\Gamma_s n_s^\pm \qquad (12.20)$$

where we have used the fact that $\tilde{N}_1 + \tilde{N}_2 = 1$. Using Eqs. (12.18)–(12.20) in Eq. (12.11), we obtain

$$\frac{d\tilde{N}_2}{dt} = -\frac{\tilde{N}_2}{\tau_{sp}} - \frac{1}{N_t A}\frac{dn_p}{dz} - \frac{1}{N_t A}\left(\frac{dn_s^+}{dz} - \frac{dn_s^-}{dz}\right) \qquad (12.21)$$

Equations (12.19)–(12.21) describe the evolution of population and pump and signal powers along the length of the doped fiber. These equations can be solved to obtain various parameters of the erbium-doped fiber amplifier and laser.

12.3.1 Gaussian Approximation

The actual transverse intensity patterns $f_p(r)$ and $f_s(r)$ of a step-index single-mode fiber are described in terms of Bessel functions. If the fiber has some other refractive index variation in the transverse cross section, it is difficult to obtain an analytical expression for the actual intensity pattern. However, since the fundamental mode has a bell-shaped distribution, a very good approximation for the transverse intensity variation is the Gaussian approximation. Under this approximation, we assume the transverse intensity pattern to be given by the following variation:

$$f(r) = \frac{1}{\pi \Omega^2} e^{-r^2/\Omega^2} \tag{12.22}$$

where the quantity Ω also referred to as the spot size is determined by the fiber parameters such as core radius, numerical aperture, and wavelength. An empirical expression for Ω is given as (see Marcuse (1978))

$$\Omega = \frac{a}{\sqrt{2}} \left(0.65 + \frac{1.619}{V^{1.5}} + \frac{2.879}{V^6} \right) \tag{12.23}$$

where

$$V = ak_0 \left(n_1^2 - n_2^2 \right)^{1/2} \tag{12.24}$$

is the normalized V parameter of the fiber described in terms of the core radius a, the free space wavelength λ_0 ($= 2\pi/k_0$), and the numerical aperture of the fiber $\left(n_1^2 - n_2^2 \right)^{1/2}$. Since V value depends on the wavelength, the value of Ω at pump wavelength and various signal wavelengths is different. The multiplicative factor in Eq. (12.22) is to ensure that $f(r)$ satisfies the normalization condition [Eqs. (12.6) and (12.7)]

Example 12.1 When the pump power is small, then most of the erbium ions would be found in the ground state and in such a case, $\tilde{N}_1 \approx 1$. Equation (12.19) can then be integrated to give $n_p(z) = n_p(0)e^{-\sigma_p N_t \Gamma_p z}$. This implies that the pump gets absorbed as it propagates along the fiber and the corresponding absorption coefficient is given as $\alpha_p = \sigma_p N_t \Gamma_p$. Note that the absorption coefficient depends on the absorption cross section, the erbium ion population density, and the overlap factor Γ_p. In a similar fashion, by putting $\tilde{N}_2 \approx 0$ in Eq. (12.20) we obtain the signal variation along z to be given as $n_s^+(z) = n_s^+(0)e^{-\sigma_{sa} N_t \Gamma_s z}$ giving an absorption coefficient at the signal wavelength of $\alpha_s = \sigma_{sa} N_t \Gamma_s$. Greater the confinement of the modes, closer the Γ_p and Γ_s to unity and larger the absorption coefficient.

Example 12.2 Consider step-index fiber with a core radius of 1.5 μm and an NA of 0.24. The V values at the pump wavelength of 980 nm and at the signal wavelength of 1530 nm are 2.31 and 1.48, respectively. Using Eq. (12.23) we can get the corresponding spot sizes as 1.2 and 1.93 μm, respectively. Note that the spot size at the pump wavelength is smaller than that at the signal wavelength. This is due to greater penetration of the field with increase in wavelength. In fact higher signal wavelengths would have even larger spot sizes.

Example 12.3 Under the Gaussian approximation, we have for the pump and signal intensity patterns

$$f_p(r) = \frac{1}{\pi \Omega_p^2} e^{-r^2/\Omega_p^2}$$

and

$$f_s(r) = \frac{1}{\pi \Omega_s^2} e^{-r^2/\Omega_s^2}$$

respectively, where Ω_p and Ω_s are the spot sizes at the pump and signal wavelengths, respectively. Using these expressions we can immediately obtain the following:

$$\Gamma_p = 1 - e^{-a^2/\Omega_p^2}; \tag{12.25}$$
$$\Gamma_s = 1 - e^{-a^2/\Omega_s^2}; \tag{12.26}$$

For a doped fiber with a core radius of 1.5 μm and a numerical aperture of 0.24, from Eq. (12.23) we have

$$\Omega_p \approx 1.2 \, \mu m, \quad \Omega_s \approx 1.93 \, \mu m$$

and $\Gamma_p \approx 0.79$, $\Gamma_s \approx 0.45$. Here we have neglected the wavelength dependence of the numerical aperture of the fiber.

Example 12.4 For a pump power of 100 mW and a value of $\Omega_p = 1.35$ μm, the peak intensity which occurs along the fiber axis is given as

$$I_p = \frac{P_p}{\pi \Omega_p^2} \approx 1.75 \times 10^{10} \, W/m^2$$

12.3.2 Gaussian Envelope Approximation

Instead of using Eq. (12.23) a more accurate expression can be obtained under what is referred to as Gaussian envelope approximation. In this approximation, Ω is given as

$$\Omega = a \, J_0(U) \frac{V \, K_1(W)}{U \, K_0(W)} \tag{12.27}$$

where

$$V = k_0 a \sqrt{\left(n_1^2 - n_2^2\right)} \tag{12.28}$$

$$U = a \sqrt{\left(k_0^2 n_1^2 - \beta^2\right)} \tag{12.29}$$

$$W = a \sqrt{\left(\beta^2 - k_0^2 n_2^2\right)} \tag{12.30}$$

β is the propagation constant of the fundamental mode and J_0, K_0, and K_1 are Bessel functions. For a step-index fiber, W can be approximated as

$$W = 1.1428 \, V - 0.996 \tag{12.31}$$

which is accurate in the range $1.5 < V < 2.5$. Table 12.1 gives the values of Ω as predicted by Eq. (12.31). These values can be used to estimate the Gaussian envelope parameters for a given erbium-doped fiber. Corresponding to pump and signal wavelengths, one can obtain the values of V, U, and W and hence Ω.

Table 12.1 Gaussian spot
size of a single-mode fiber for
different V values

V	Ω/a
1.2	1.4709
1.25	1.3792
1.3	1.3032
1.35	1.2394
1.4	1.1851
1.45	1.1384
1.5	1.0979
1.55	1.0624
1.6	1.031
1.65	1.0032
1.7	0.97824
1.75	0.95582
1.8	0.93554
1.85	0.91712
1.9	0.90031
1.95	0.8849
2	0.87074
2.05	0.85766

12.3.3 Solutions Under Steady State

Under steady state, the time derivative is zero and hence from Eq. (12.21), we have

$$\tilde{N}_2 = -\frac{\tau_{\mathrm{sp}}}{N_{\mathrm{t}}A}\left(\frac{dn_{\mathrm{p}}}{dz} + \frac{dn_{\mathrm{s}}^+}{dz} - \frac{dn_{\mathrm{s}}^-}{dz}\right) \tag{12.32}$$

Substituting in Eq. (12.20) we have

$$\frac{dn_{\mathrm{s}}^{\pm}}{n_{\mathrm{s}}^{\pm}} = \pm N_{\mathrm{t}}\Gamma_{\mathrm{s}}\left[-\frac{(\sigma_{\mathrm{se}} + \sigma_{\mathrm{sa}})\tau_{\mathrm{sp}}}{N_{\mathrm{t}}A}\left(\frac{dn_{\mathrm{p}}}{dz} + \frac{dn_{\mathrm{s}}^+}{dz} - \frac{dn_{\mathrm{s}}^-}{dz}\right) - \sigma_{\mathrm{sa}}\right]dz$$

$$= \mp\left[N_{\mathrm{t}}\Gamma_{\mathrm{s}}\sigma_{\mathrm{sa}} + \frac{\Gamma_{\mathrm{s}}\tau_{\mathrm{sp}}(\sigma_{\mathrm{se}} + \sigma_{\mathrm{sa}})}{A}\left(\frac{dn_{\mathrm{p}}}{dz} + \frac{dn_{\mathrm{s}}^+}{dz} - \frac{dn_{\mathrm{s}}^-}{dz}\right)\right]dz$$

We define

$$\alpha_{\mathrm{s}} = \Gamma_{\mathrm{s}}N_{\mathrm{t}}\sigma_{\mathrm{sa}} \tag{12.33}$$

$$n_{\mathrm{ss}} = \frac{A}{\Gamma_{\mathrm{s}}\tau_{\mathrm{sp}}(\sigma_{\mathrm{se}} + \sigma_{\mathrm{sa}})} \tag{12.34}$$

where α_{s} is the signal absorption coefficient (see Example 12.1) and n_{ss} is referred to as the intrinsic signal saturation photon number ($h\nu_{\mathrm{s}}n_{\mathrm{ss}}$ is the intrinsic signal saturation power). Thus

$$\frac{dn_s^{\pm}}{n_s^{\pm}} = \mp\left[\alpha_s + \frac{1}{n_{ss}}\left(\frac{dn_p}{dz} + \frac{dn_s^+}{dz} - \frac{dn_s^-}{dz}\right)\right]dz \tag{12.35}$$

Similarly from Eq. (12.19) we have

$$\begin{aligned}
\frac{dn_p}{n_p} &= -\sigma_p N_t \Gamma_p (1 - \tilde{N}_2)dz \\
&= -\left[\alpha_p + \frac{1}{n_{ps}}\left(\frac{dn_p}{dz} + \frac{dn_s^+}{dz} - \frac{dn_s^-}{dz}\right)\right]dz
\end{aligned} \tag{12.36}$$

where

$$\alpha_p = \sigma_p N_t \Gamma_p \tag{12.37}$$

and

$$n_{ps} = \frac{A}{\Gamma_p \tau_{sp} \sigma_p} \tag{12.38}$$

represent the pump absorption coefficient and the intrinsic pump saturation photon number, respectively ($h\nu_p n_{ps}$ is the intrinsic pump saturation power).

Equation (12.35) can be integrated from $z = 0$ to $z = L$ to get the forward-propagating signal photon number n_s^+:

$$\ln\left(\frac{n_s^+(L)}{n_s^+(0)}\right) = -\alpha_s L + \frac{1}{n_{ss}}\left(n_{pa} + n_{sa}^+ + n_{sa}^-\right) \tag{12.39}$$

where

$$\begin{aligned}
n_{pa} &= n_p(0) - n_p(L); \\
n_{sa}^+ &= n_s^+(0) - n_s^+(L); \\
n_{sa}^- &= n_s^-(L) - n_s^-(0)
\end{aligned} \tag{12.40}$$

represent the number of photons at pump and signal wavelengths propagating in the forward and backward directions, respectively, that have been absorbed by the doped fiber (negative value of the quantity would imply net emission rather than absorption). Note that for signal propagating in the $-z$-direction, input photon flux is $n_s^-(L)$ and output photon flux is $n_s^-(0)$. Thus

$$n_s^+(L) = n_s^+(0)e^{-\alpha_s L}e^{(n_{pa}+n_{sa}^++n_{sa}^-)/n_{ss}} \tag{12.41}$$

In a similar fashion we obtain

$$n_s^-(0) = n_s^-(L)e^{-\alpha_s L}e^{(n_{pa}+n_{sa}^++n_{sa}^-)/n_{ss}} \tag{12.42}$$

For the pump photon flux, we have from Eq. (12.36)

$$n_{\mathrm{p}}(L) = n_{\mathrm{p}}(0)e^{-\alpha_{\mathrm{p}}L}e^{\left(n_{\mathrm{pa}}+n_{\mathrm{sa}}^{+}+n_{\mathrm{sa}}^{-}\right)/n_{\mathrm{ps}}} \qquad (12.43)$$

The above equations can be used to study erbium-doped fiber amplifiers as well as erbium-doped fiber lasers.

Example 12.5 Let the fiber described in Example 12.2 have an erbium concentration of 5.4×10^{24} m^{-3}. The absorption cross section at the pump and signal wavelengths are $\sigma_{\mathrm{pa}} = 2.7 \times 10^{-25}$ m^{2} and $\sigma_{\mathrm{sa}} = 8.2 \times 10^{-25}$ m^{-2}. The absorption coefficients of the fiber at the pump and signal wavelengths are given by Eqs. (12.37) and (12.33) and would be 1.15 and 1.99 m^{-1}, respectively. These can be written in units of decibel per meter by multiplying the quantity in m^{-1} by 4.34. Thus the fiber would have 5 and 8.64 dB/m absorption at 980 and 1532 nm, respectively.

12.4 Fiber Lasers

We will now use the equations derived above to analyze the characteristics of erbium-doped fiber lasers. In a fiber laser, the pump creates population inversion in the doped fiber and a pair of reflectors on either end of the doped fiber provide for optical feedback. If the cavity losses are compensated by the gain provided by the doped fiber, then lasing begins. Thus unlike an optical amplifier, there is no signal input into the doped fiber; signal light is generated by spontaneous emissions from the excited-level erbium ions.

For steady-state lasing we require that the signal photon flux be the same after one round-trip, i.e., the losses suffered by the signal photon flux in one round-trip be compensated exactly by the gain from the inversion.

Let R_1 and R_2 represent the reflectivities of the two mirrors (see Fig. 12.7). Let $n_{\mathrm{s}}^{+}(0)$ represent the photon flux propagating to the right at $z = 0$, which is the left end of the cavity. The photon flux at $z = L$ incident on the mirror M_2 would then be [using Eq. (12.41)]

$$n_{\mathrm{s}}^{+}(L) = n_{\mathrm{s}}^{+}(0)e^{-\alpha_{\mathrm{s}}L}e^{\left(n_{\mathrm{pa}}+n_{\mathrm{sa}}^{+}+n_{\mathrm{sa}}^{-}\right)/n_{\mathrm{ss}}} \qquad (12.44)$$

A fraction of these photons get reflected back by the mirror M_2 with reflectivity R_2 into the fiber giving for the photon flux propagating to the left at $z = L$:

$$n_{\mathrm{s}}^{-}(L) = R_2 n_{\mathrm{s}}^{+}(L) \qquad (12.45)$$

In propagating from $z = L$ to $z = 0$, the photon flux becomes [using Eq. (12.42)]

$$n_{\mathrm{s}}^{-}(0) = n_{\mathrm{s}}^{-}(L)e^{-\alpha_{\mathrm{s}}L}e^{\left(n_{\mathrm{pa}}+n_{\mathrm{sa}}^{+}+n_{\mathrm{sa}}^{-}\right)/n_{\mathrm{ss}}} \qquad (12.46)$$

A fraction of these photons are reflected by mirror M_1 with reflectivity R_1 into the fiber giving for the photon flux propagating to the right at $z = 0$ as

$$n_{\mathrm{s}1}^{+}(0) = R_1 n_{\mathrm{s}}^{-}(0) \qquad (12.47)$$

For laser oscillation we impose the condition that the signal photon flux after one round-trip be the same as the starting flux. Hence

$$n_{s1}^{+}(0) = n_{s}^{+}(0) \tag{12.48}$$

which using Eqs. (12.44)–(12.47) gives

$$R_1 R_2 \, e^{-2\alpha_s L} e^{2(n_{pa}+n_{sa}^{+}+n_{sa}^{-})/n_{ss}} = 1$$

or

$$\frac{(n_{pa} + n_{sa}^{+} + n_{sa}^{-})}{n_{ss}} = \alpha_s L - \frac{1}{2}\ln(R_1 R_2) \tag{12.49}$$

We shall now discuss some general features from the set of equations obtained above.

12.4.1 Minimum Required Doped Fiber Length

Since the pump power exiting the fiber must be less than the pump power entering the fiber, n_{pa} must be a positive quantity. Now

$$\begin{aligned} n_{pa} &= n_p(0) - n_p(L) \\ &= n_p(0)\left[1 - e^{-\alpha_p L}e^{(\alpha_s L - \frac{1}{2}\ln R_1 R_2)n_{ss}/n_{ps}}\right] \end{aligned} \tag{12.50}$$

where we have used Eqs. (12.43) and (12.49). For n_{pa} to be positive, we must have

$$e^{-\alpha_p L}e^{(\alpha_s L - \frac{1}{2}\ln R_1 R_2)n_{ss}/n_{ps}} < 1$$

or

$$-\alpha_p L + \frac{n_{ss}}{n_{ps}}\left(\alpha_s L - \frac{1}{2}\ln(R_1 R_2)\right) < 0$$

which gives us a condition on the minimum length L_m of the doped fiber for laser oscillation as

$$L_m = \frac{n_{ss}}{2n_{ps}}\ln\left(\frac{1}{R_1 R_2}\right)\left(\alpha_p - \frac{n_{ss}}{n_{ps}}\alpha_s\right)^{-1}$$

Using the expressions for n_{ss}, n_{ps}, α_s, and α_p, the above equation simplifies to

$$L_m = \frac{1}{2\sigma_{se}N_t\Gamma_s}\ln\left(\frac{1}{R_1 R_2}\right) \tag{12.51}$$

The minimum length requirement implies that for a given doped fiber and a cavity with specific mirror reflectivities, no amount of pump power can lead to laser oscillation if the doped fiber length is less than L_m given by Eq. (12.51). This is due to the fact that for $L < L_m$, there are not enough erbium ions in the laser cavity to provide sufficient gain to overcome the losses in the cavity.

We also note from Eq. (12.51) that cavities with larger mirror reflectivities would require shorter fiber lengths for lasing. Also since largest value of σ_{se} occurs at about 1532 nm, lasing at this wavelength would require the shortest length.

Example 12.6 As an example we consider an erbium-doped fiber with a doping concentration of $N_t = 5.4 \times 10^{24}$ m^{-3} and $\Gamma_s = 0.54$. At the wavelength of 1532 nm, $\sigma_{se} = 8.2 \times 10^{-25}$ m^2. If the doped fiber ends are bare, then they would have an approximate reflectivity of 0.04. Using these values we obtain the minimum length for the fiber to lase to be approximately 1.34 m.

Example 12.7 For the same values as in Example 12.6, if the mirror reflectivities are 0.9 each, then the minimum length becomes 4.4 cm.

Problem 12.1 Consider a fiber laser with a doped fiber length of 1 m with mirror reflectivities of 0.9 each and an input pump power of 100 mW. Assuming the absorption coefficient at the pump and signal wavelengths to be 1.4 and 0.9 m^{-1}, obtain the amount of pump absorbed by the fiber when the laser is oscillating, given $n_{ss} = 2.35 \times 10^{15}$ s^{-1} and $n_{ps} = 4.2 \times 10^{15}$ s^{-1}.

Solution We can use Eq. (12.50) to obtain the absorbed pump power:

$$P_{pa} = P_p(0) \left[1 - e^{-\alpha_p L} e^{(\alpha_s L - \frac{1}{2}\ln R_1 R_2)n_{ss}/n_{ps}} \right] \approx 57\,\text{mW}$$

If in this problem, the reflectivity is reduced to 0.5 for each mirror, then the pump power absorbed would be about 40 mW.

12.4.2 Threshold

We can estimate the threshold pump power required to start laser oscillation. We first note that at threshold since the laser power is still small, we can put $n_{sa}^+ \approx n_{sa}^- \approx 0$. Thus from Eq. (12.49) we get

$$n_{pa} \approx \left(\alpha_s L - \frac{1}{2}\ln R_1 R_2 \right) n_{ss} \tag{12.52}$$

Using this value in Eq. (12.50) we get

$$n_{p,th} = n_{p,th}(0) = \frac{n_{ss}\left(\alpha_s L - \frac{1}{2}\ln R_1 R_2\right)}{\left[1 - e^{-\alpha_p L} e^{(\alpha_s L - \frac{1}{2}\ln R_1 R_2)n_{ss}/n_{ps}} \right]} \tag{12.53}$$

where $n_{p,th}(0)$ is the input pump power at threshold. Thus the threshold pump power would be

$$P_{p,th} = h\nu_p n_{p,th} \tag{12.54}$$

If most of the incident pump power is absorbed by the doped fiber, then $n_p(L) \ll n_p(0)$ and $n_{pa} \approx n_p(0)$. In such a case, we obtain a simplified expression for $P_{p,th}$:

$$P_{p,th} = h\nu_p n_p(0) = \frac{h\nu_p A}{\Gamma_s \tau_{sp}(\sigma_{se} + \sigma_{sa})}\left(\alpha_s L - \frac{1}{2}\ln R_1 R_2\right) \tag{12.55}$$

where we have substituted the expression for n_{ss} from Eq. (12.34).

Example 12.8 Using Eq. (12.55) we can estimate the threshold pump power required to start laser oscillation in an erbium-doped fiber cavity. Taking typical parameter values of $A = 7.1 \; \mu m^2$ (core radius of 1.5 μm), $\tau_{sp} = 12 \times 10^{-3}$ s, $\sigma_{se} = 8.2 \times 10^{-25}$ m^2, $\sigma_{sa} = 7.8 \times 10^{-25}$ m^2, $R_1 = R_2 = 0.9$, $L = 10$ m, $\alpha_s = 2.4$ m^{-1}, the threshold pump power comes out to be 3.3 mW.

12.4.3 Laser Output Power

We will now obtain an expression for the output power of the laser in terms of various parameters of the fiber, cavity, and pump power.

The laser power exiting the fiber laser from mirror M_2 is

$$P_{laser} = n_s^+(L)(1 - R_2)h\nu_s \tag{12.56}$$

which using Eqs. (12.44) and (12.49) gives

$$P_{laser} = n_s^+(0)(1 - R_2)h\nu_s \, e^{-\frac{1}{2}\ln R_1 R_2} = \frac{(1 - R_2)}{\sqrt{R_1 R_2}} n_s^+(0)h\nu_s \tag{12.57}$$

We now need to obtain an expression for $n_s^+(0)$. Now

$$n_{sa}^+ = n_s^+(0) - n_s^+(L) = n_s^+(0)\left(1 - \frac{1}{\sqrt{R_1 R_2}}\right)$$

where we have used Eqs. (12.44) and (12.49). Similarly

$$n_{sa}^- = n_s^-(L) - n_s^-(0) = n_s^-(0)\left(\sqrt{R_1 R_2} - 1\right) = n_s^+(0)\frac{\left(\sqrt{R_1 R_2} - 1\right)}{R_1}$$

where we have used Eqs. (12.47) and (12.49). Thus

$$n_{sa}^+ + n_{sa}^- = n_s^+(0)\left[\left(1 - \frac{1}{\sqrt{R_1 R_2}}\right) + \left(\frac{\sqrt{R_1 R_2} - 1}{R_1}\right)\right] \tag{12.58}$$

The pump power absorbed by the doped fiber is given by Eq. (12.50). Adding Eqs. (12.58) and (12.50) and using Eq. (12.49), we obtain

$$n_s^+(0) = \left[\left(\frac{1}{\sqrt{R_1 R_2}} - 1\right) + \frac{(1 - \sqrt{R_1 R_2})}{R_1}\right]^{-1}$$

$$\left[n_p(0)\left\{1 - e^{-\alpha_p L}e^{(\alpha_s L - \frac{1}{2}\ln R_1 R_2)n_{ss}/n_{ps}}\right\} - n_{ss}\left\{\alpha_s L - \frac{1}{2}\ln(R_1 R_2)\right\}\right]$$

(12.59)

Using this value of $n_s^+(0)$ in Eq. (12.57) we can obtain the output power from the laser.

Problem 12.2 Consider a fiber laser cavity with $R_1 = R_2 = 0.9$, $L = 2$ m, and input pump power $= 20$ mW. Assume $n_{ss} = 2.35 \times 10^{15}$ s^{-1} and $n_{ps} = 4.2 \times 10^{15}$ s^{-1}, $\alpha_s = 0.9$ m^{-1}, and $\alpha_p = 1.4$ m^{-1}. Obtain the steady-state laser power exiting from the end of the fiber. [Ans: 0.8 mW.]

Problem 12.3 If in the above problem we take a length of 10 m for the doped fiber length, keeping all other parameters the same, calculate the output laser power. [Ans. 5 mW.]

Figure 12.8 shows the variation of output laser power as a function of the input pump power for the set of fiber parameters given in Table 12.2. For low input pump powers, there is no output. After reaching threshold, the output laser power monotonically increases with the input pump power. The figure corresponds to a slope efficiency (the rate of increase of laser power with increase in pump power) of 31.5%. The slope efficiency depends on the fiber parameters as well as the cavity parameters. Note that the maximum efficiency would correspond to a situation when every pump photon is converted into a lasing photon. In such a case the maximum efficiency would be $0.98/1.532 = 0.64$. If the cavity has mirrors of equal reflectivity on both sides, then half the output photons would be coming from each direction and the maximum efficiency for one of the outputs would be about 32%.

Figure 12.9 shows a measured variation of laser power with the input pump power. As can be seen, the threshold for this laser is less than about 10 mW and

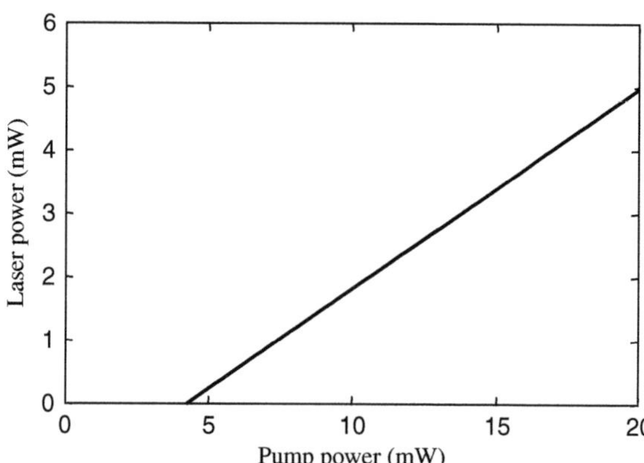

Fig. 12.8 Typical simulated variation of laser power with input pump power

Table 12.2 Erbium-doped fiber parameters used in the simulations

Core diameter	3.5 μm
NA	0.23
Doping density	$5.4 \times 10^{24} \mathrm{m}^{-3}$
Pump wavelength	980 nm
Signal wavelength	1550 nm
α_s	0.874 m^{-1}
α_p	1.14 m^{-1}

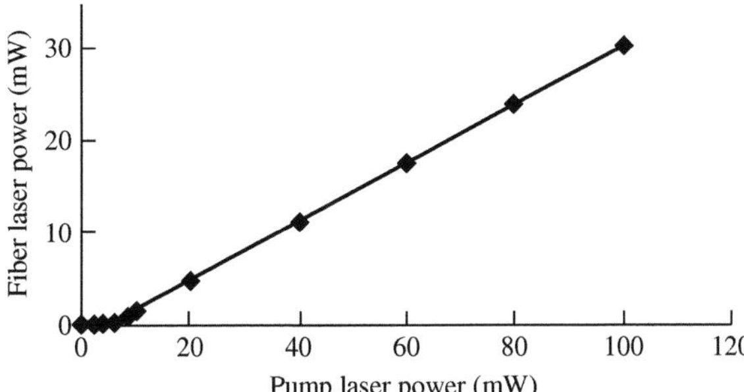

Fig. 12.9 A typical measured variation of fiber laser power with input pump power. (Adapted from Zhu et al. (2007))

the slope efficiency is about 35%. Beyond threshold the laser power increases linearly with the pump power. It is indeed possible to generate extremely high powers from fiber lasers. Figure 12.10 shows the variation of output power with pump power of a Ytterbium-doped, large-core fiber laser showing that it is possible to achieve a continuous wave output power of 1.36 kW. There are efforts to achieve even larger powers from fiber lasers and such high-power fiber lasers are expected to revolutionize the area of applications of lasers in industries. Figure 12.11 shows a photograph demonstrating breakdown in atmospheric air by a pulsed fiber laser.

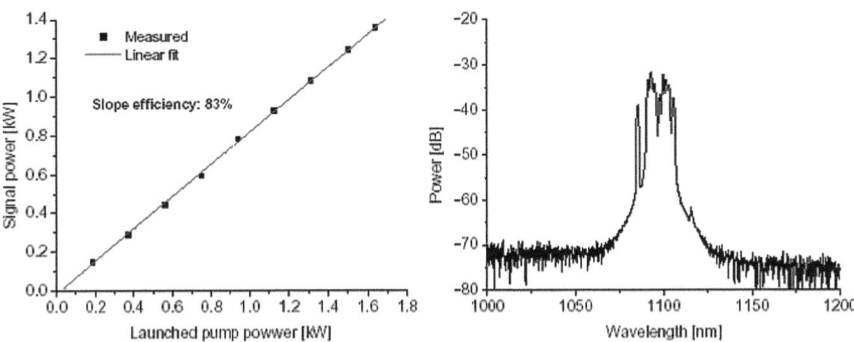

Fig. 12.10 Output laser power versus launched pump power of a very high-power fiber laser. The spectrum of the output is shown in the right curve. (Adapted with permission from Jeong et al. (2004) © 2004 OSA) Ref: Fiber output and input pump power curve and the laser output spectrum at 1.36 kW; www.optics.rochester.edu/~gweihua/hflaser.pdf

Fig. 12.11 Spark in atmospheric air produced by a focused output from a fiber laser. (Adapted with permission from Galvanauskas et al. (2007) © 2007 IEEE)

With such high optical powers, optical fibers start to exhibit non-linear effects which tend to degrade the performance of the fiber laser. Since the non-linear effects are proportional to the intensity of the propagating radiation, one way to reduce the non-linear effects is to increase the mode area so that for a given power the intensity would be less. Using photonic microstructure fibers, it is possible to achieve single-mode operation over a large wavelength range and also achieve large mode areas. Figure 12.12 shows typical examples of large mode area fibers with a microstructure cladding. These fibers have a very small numerical aperture of less than 0.03 and core diameters larger than 60 μm leading to mode field diameters of about 50 μm.

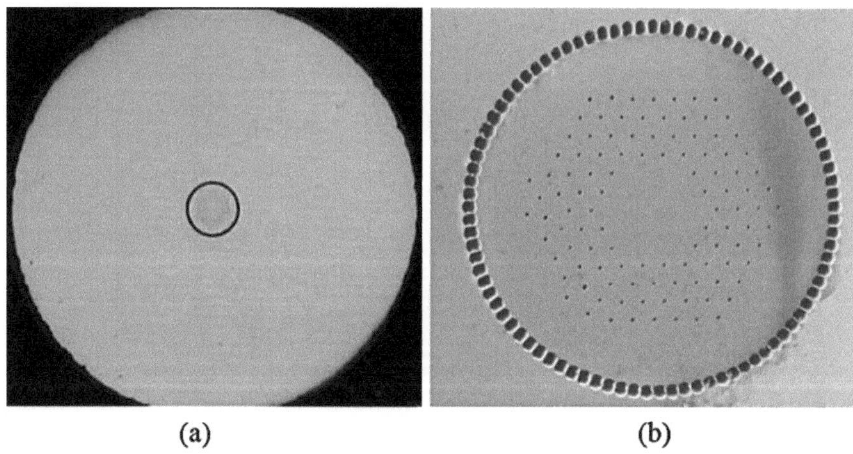

(a)   (b)

Fig. 12.12 A large mode area fiber using microstructured cladding. (Adapted with permission from Limpert et al. (2007) © 2007 IEEE)

12.4.4 Slope Efficiency

Let us consider a fiber laser cavity with $R_1 = 1$ and $R_2 = R$. Using Eqs. (12.57) and (12.59) we can obtain the following expression for the slope efficiency of the laser:

$$\eta = \frac{dP_{\text{laser}}}{dP_{\text{p}}(0)} = \frac{\nu_{\text{s}}}{\nu_{\text{p}}}\left\{1 - e^{-\alpha_{\text{p}}L}e^{(\alpha_{\text{s}}L - \frac{1}{2}\ln R)n_{\text{ss}}/n_{\text{ps}}}\right\} \tag{12.60}$$

For long doped fiber lengths, the second term in the bracket is almost equal to zero and the slope efficiency becomes $\nu_{\text{s}}\,/\,\nu_{\text{p}}$. This implies that beyond threshold, every pump photon gets converted to signal photon.

12.5 Erbium-Doped Fiber Amplifier

In the earlier sections we discussed erbium-doped fiber lasers. The doped fiber which is pumped by the laser amplifies the optical signal and the reflectors at either end of the fiber provide for feedback converting the optical amplifier to a fiber laser. In the absence of reflection from either side of the doped fiber, the pumped fiber would behave as an optical amplifier. Such optical amplifiers are playing a very important role in long-distance fiber optic communication systems (see Chapter 16). The primary characteristics of an optical amplifier are the gain, the noise figure, and the saturation behavior. The equations describing the amplifier are the same as described earlier except for the fact that unlike a laser in an amplifier, there is a signal input which propagates only along one direction through the fiber. Thus in the case of codirectional pumping, the pump light and the signal light propagate along the same direction, while in the case of contradirectional pumping, the pump and the signal propagate along opposite directions. Here we will discuss an optical amplifier operating in the codirectional configuration. Similar considerations can be had for contradirectional pumping also.

We now consider an optical fiber amplifier with codirectional pumping and assume that the pump and the signal propagate along the z-direction. Let $n_{\text{p}}(0)$ and $n_{\text{s}}(0)$ represent the input pump and signal photon fluxes at $z = 0$, the input to the amplifier. (We are not putting any superscript on the signal photon flux since the signal is propagating along only one direction.) In this case, we have $n_{\text{s}}^{-} = 0$ and $n_{\text{s}}^{+} = n_{\text{s}}$. We consider the amplifier to be operating in a steady state. Thus for the case of amplifier, Eqs. (12.43) and (12.41) become

$$n_{\text{p}}(L) = n_{\text{p}}(0)e^{-\alpha_{\text{P}}L}e^{(n_{\text{pa}}+n_{\text{sa}})/n_{\text{ps}}} \tag{12.61}$$

and

$$n_{\text{s}}(L) = n_{\text{s}}(0)e^{-\alpha_{\text{s}}L}e^{(n_{\text{pa}}+n_{\text{sa}})/n_{\text{ss}}} \tag{12.62}$$

with the same definitions of other quantities such as n_{sa}, n_{pa}, n_{ps}, and n_{ss}. Thus the change in signal and pump photon fluxes in propagating from $z = 0$ to $z = L$ is given as

$$n_{sa} = n_s(0) - n_s(L)$$
$$= n_s(0)\left[1 - e^{-\alpha_s L}e^{(n_{sa}+n_{pa})/n_{ss}}\right] \tag{12.63}$$

and

$$n_{pa} = n_p(0) - n_p(L)$$
$$= n_p(0)\left[1 - e^{-\alpha_p L}e^{(n_{sa}+n_{pa})/n_{ps}}\right] \tag{12.64}$$

Adding Eqs. (12.63) and (12.64), we obtain

$$n_{sa}+n_{pa} = n_s(0)\left[1 - e^{-\alpha_s L}e^{(n_{sa}+n_{pa})/n_{ss}}\right]+n_p(0)\left[1 - e^{-\alpha_p L}e^{(n_{sa}+n_{pa})/n_{ps}}\right] \tag{12.65}$$

Writing

$$\varsigma = n_{sa} + n_{pa} \tag{12.66}$$

Eq. (12.65) transforms to the following transcendental equation:

$$\varsigma = n_s(0)\left[1 - e^{-\alpha_s L}e^{\varsigma/n_{ss}}\right] + n_p(0)\left[1 - e^{-\alpha_p L}e^{\varsigma/n_{ps}}\right] \tag{12.67}$$

For a given doped fiber, α_s, α_p, n_{ss}, and n_{ps} are known. Thus taking a certain length L of the doped fiber and assuming a given input pump power and signal power, $n_p(0)$ and $n_s(0)$ can be found out. Equation (12.67) can then be solved for ς. Knowing the value of ς, we can immediately obtain the output signal and pump powers from the following equations:

$$n_s(L) = n_s(0)e^{-\alpha_s L}e^{\varsigma/n_{ss}}$$
$$n_p(L) = n_p(0)e^{-\alpha_p L}e^{\varsigma/n_{ps}} \tag{12.68}$$

The amplifier gain is then given as

$$G = \frac{n_s(L)}{n_s(0)} = e^{-\alpha_s L}e^{\varsigma/n_{ss}} \tag{12.69}$$

which in decibel units becomes

$$G\,(\text{dB}) = 10\log\left(\frac{n_s(L)}{n_s(0)}\right) = 10\,(\varsigma/n_{ss} - \alpha_s L)\log e \tag{12.70}$$

12.5.1 Transparency Power

From the above set of equations, we can obtain the pump power required for transparency, i.e., the pump power required so that the output signal power equals the input signal power. This will happen when

$$\varsigma = n_{\mathrm{ss}}\alpha_{\mathrm{s}}L \tag{12.71}$$

Since $n_{\mathrm{sa}}=0$, in such a case we have $\zeta = n_{\mathrm{pa}}$ and using Eq. (12.67) we obtain the pump power required for transparency as

$$P_{\mathrm{p}}(0) = h\nu_{\mathrm{p}}n_{\mathrm{p}}(0) = \frac{h\nu_{\mathrm{p}}n_{\mathrm{ss}}\alpha_{\mathrm{s}}L}{\left(1 - e^{-\alpha_{\mathrm{p}}L}e^{\alpha_{\mathrm{s}}Ln_{\mathrm{ss}}/n_{\mathrm{ps}}}\right)} \tag{12.72}$$

Example 12.9 Consider the erbium-doped fiber mentioned in Problem 12.1 for which $n_{\mathrm{ss}} = 2.35 \times 10^{15}$ s^{-1} and $n_{\mathrm{ps}} = 4.2 \times 10^{15}$ s^{-1}, $\alpha_{\mathrm{p}} = 1.4$ m^{-1}, and $\alpha_{\mathrm{s}} = 0.9$ m^{-1}. If we assume a fiber length of 10 m and a pump wavelength of 980 nm, the transparency pump power for such a doped fiber would be approximately 4.3 mW.

If the fiber length and the input pump power are such that most of the pump gets absorbed by the fiber and there is a reasonable gain in the fiber, we can assume $n_{\mathrm{p}}(L)<<n_{\mathrm{p}}(0)$, $n_{\mathrm{s}}(L)>>n_{\mathrm{s}}(0)$. In such a case

$$\varsigma = n_{\mathrm{s}}(0) - n_{\mathrm{s}}(L) + n_{\mathrm{p}}(0) - n_{\mathrm{p}}(L) \approx n_{\mathrm{p}}(0) - n_{\mathrm{s}}(L)$$

Hence using Eq. (12.62) we obtain

$$n_{\mathrm{s}}(L) = n_{\mathrm{s}}(0)\, e^{-\alpha_{\mathrm{s}}L}e^{\varsigma/n_{\mathrm{ss}}} \approx n_{\mathrm{s}}(0)\, e^{-\alpha_{\mathrm{s}}L}e^{(n_{\mathrm{p}}(0)-n_{\mathrm{s}}(L))/n_{\mathrm{ss}}}$$

which can be simplified to obtain

$$n_{\mathrm{s}}(L)e^{n_{\mathrm{s}}(L)/n_{\mathrm{ss}}} = n_{\mathrm{s}}(0)\, e^{-\alpha_{\mathrm{s}}L}e^{n_{\mathrm{p}}(0)/n_{\mathrm{ss}}} \tag{12.73}$$

which is a much simplified transcendental equation to determine $n_{\mathrm{s}}(L)$ and hence the output power for a given input signal power $P_{\mathrm{s}}(0) = h\nu_{\mathrm{s}}n_{\mathrm{s}}(0)$.

Figure 12.13 shows the variation of gain with the input pump power. As can be seen, the gain increases with the input pump power and saturates at high pump power. Pump saturation would occur when the pump has excited all the erbium ions to the excited state and there are no more erbium ions available for increasing the inversion. Figure 12.14 shows the variation of gain with the length of the fiber. As the length increases, the gain increases first and then peaks for a given length and then diminishes again. This happens due to the fact that as the pump propagates through the fiber, it gets absorbed and when the pump power reaches a value sufficient to cause equality of populations of the two levels, then at that length the gain saturates. Any doped fiber beyond this point would not have any inversion and the fiber would be absorbing. Thus for a given pump power, there is an optimum length for maximum gain.

Fig. 12.13 Simulated variation of gain with input pump power for an erbium-doped fiber amplifier

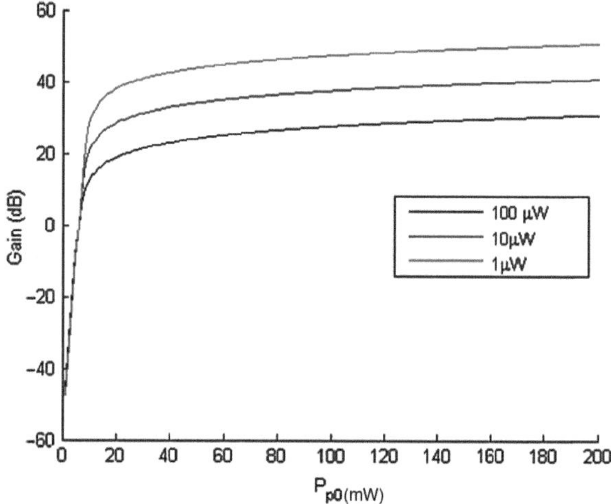

Fig. 12.14 Simulated variation of gain with fiber length of an erbium-doped fiber amplifier

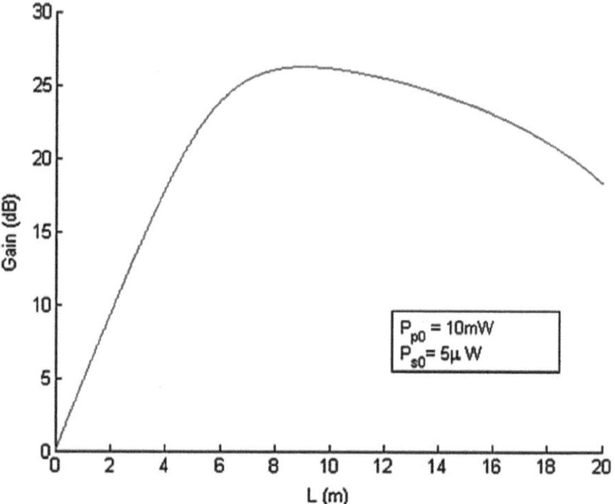

Example 12.10 Consider the fiber described in Example 12.9 and assume an input pump power of 50 mW. If the input signal power is 0.01 mW, then Eq. (12.74) can be solved to obtain an output signal power of 24.2 mW. This implies an optical gain of 10 log(24.2/.01) $\sim$34 dB.

12.6 Mode Locking in Fiber Lasers

As discussed in Chapter 7, mode locking is a technique used to generate ultrashort pulses of light. As shown in Section 7.7.3, the broader the gain bandwidth of the laser, the shorter the achievable pulse duration. With a gain bandwidth of 40 nm at 1550 nm, the corresponding spectral bandwidth is about 5 THz and if modes over this bandwidth are mode locked, then the pulse duration achievable is about 200 fs. Thus fiber lasers are very interesting candidates for generation of ultrashort pulses of a few hundred femtosecond duration.

There are many techniques used for mode locking of fiber lasers. Two important methods are based on non-linear polarization rotation (NPR) and saturable absorption. In the former the non-linearity present in the optical fiber leads to a rotation of the polarization state of the light beam and this effect is used to mode lock the laser. In the latter case, saturable absorption in a semiconductor material is used for mode locking. Here as an example we will discuss the phenomenon of non-linear polarization rotation and how it can lead to the generation of ultrashort pulses of light from fiber lasers.

12.6.1 Non-linear Polarization Rotation

This effect is a manifestation of the intrinsic non-linearity of an optical fiber. To recall we note that when a light beam propagates through a fiber, for sufficiently large intensities, the refractive index of the fiber changes due to the interaction of the electric field of the light wave with the fiber medium. This change in refractive index in turn leads to a change of phase of the propagating light wave. This leads to the phenomenon of self- phase modulation which is discussed in detail in Chapter 17. Now, if we assume that we launch an elliptically polarized light beam in an optical fiber and if the fiber has no birefringence, then the state of polarization of the light wave will remain the same as it propagates through the fiber. Such an elliptically polarized light can be considered as a superposition of two orthogonal linearly polarized waves with different amplitudes and having a phase difference of $\pi/2$ or equally as a superposition of a right circularly and a left circularly polarized light wave with different amplitudes. Considering the latter description, if we now include non-linearity in the fiber, then the amplitudes of the right circular and left circular components change with propagation distance according to the following equations:

$$\frac{\partial A_+}{\partial z} = \frac{2}{3}i\gamma \left(|A_+|^2 + 2\,|A_-|^2 \right) A_+ \tag{12.74}$$

and

$$\frac{\partial A_-}{\partial z} = \frac{2}{3}i\gamma \left(|A_-|^2 + 2\,|A_+|^2 \right) A_- \tag{12.75}$$

where A_+ and A_- represent the amplitudes of the right circular and left circular polarization components of the elliptically polarized wave and γ represents the non-linear coefficient. Equations (12.74) and (12.75) are derived in Appendix H.

In order to integrate Eqs. (12.74) and (12.75), we substitute

$$A_+(z) = \tilde{A}_+(z)e^{i\phi_+(z)} \tag{12.76}$$

and

$$A_-(z) = \tilde{A}_-(z)e^{i\phi_-(z)} \tag{12.77}$$

where $\tilde{A}_+, \tilde{A}_-, \phi_+, \phi_-$ are all real quantities. Substituting Eqs. (12.76) and (12.77) in Eqs. (12.74) and (12.75) and equating real and imaginary parts on the left-hand side and the right-hand side, we obtain the following equations:

$$\frac{\partial \tilde{A}_+}{\partial z} = 0; \tag{12.78}$$

$$\frac{\partial \tilde{A}_-}{\partial z} = 0 \tag{12.79}$$

$$\frac{d\phi_+}{dz} = \frac{2}{3}\gamma(\tilde{A}_+^2 + 2\tilde{A}_-^2) \tag{12.80}$$

and

$$\frac{d\phi_-}{dz} = \frac{2}{3}\gamma(\tilde{A}_-^2 + 2\tilde{A}_+^2) \tag{12.81}$$

The first two equations give $\tilde{A}_+ = \tilde{A}_+(0) = $ constant and $\tilde{A}_- = \tilde{A}_-(0) = $ constant using which Eqs. (12.80) and (12.81) can be solved to obtain

$$\phi_+(z) = \phi_+(0) + \frac{2}{3}\gamma(|A_+|^2 + 2|A_-|^2) \tag{12.82}$$

and

$$\phi_-(z) = \phi_-(0) + \frac{2}{3}\gamma(|A_-|^2 + 2|A_+|^2) \tag{12.83}$$

Thus due to non-linearity, the amplitudes of the right and left circular polarization components remain the same, while their phase changes in accordance with the powers of the right and left circular components. Hence the change in phase difference due to non-linearity after propagation through a length L of the fiber is

$$\Delta\phi(L) = \frac{2}{3}\gamma(|A_-|^2 - |A_+|^2) \tag{12.84}$$

Example 12.11 We first consider the propagation of a linearly polarized light. A linearly polarized light wave can be written as a superposition of a right circular and a left circular polarization having same amplitude. Thus in such a case, $|A_-|^2 = |A_+|^2$ and hence $\Delta\phi(L)=0$. Thus the polarization state would remain unaltered as it propagates through the fiber.

Example 12.12 We now consider the propagation of a right circularly polarized wave. In such a case, the wave would remain in the same polarization state and there would again be no change in the polarization state.

Example 12.13 If we consider an elliptically polarized wave and since an elliptically polarized wave is a superposition of a right circular and a left circular polarization with unequal amplitude, in this case the phase difference between the right and left circular components will change with propagation, while the amplitudes of the two components will remain the same. Since the orientation of the ellipse of the polarization state (orientation of the major and minor axes of the polarization) depends on the phase difference between the right and left polarization components, in this case, the non-linearity would result in a rotation of the orientation of the ellipse. In fact the angle of rotation is given as $\Delta\phi(L)/2$. This is referred to as non-linear polarization rotation (see Fig. 12.15).

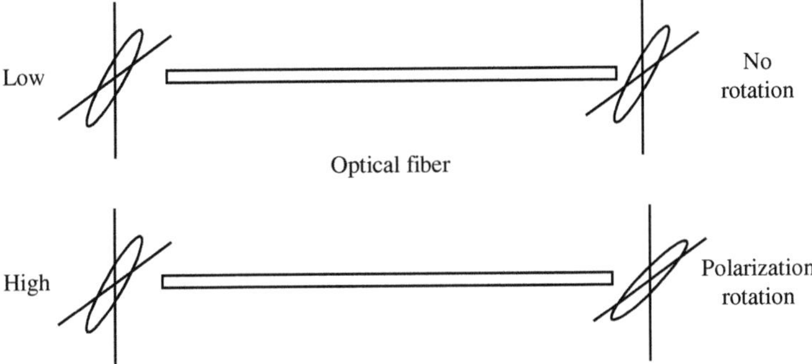

Fig. 12.15 Figure showing non-linear polarization rotation

The angle of rotation depends on the difference between the powers in the right and left circular components which in turn depends on the power in the elliptical polarization. Thus the higher the power, larger the rotation of the ellipse. It is this principle that is used in mode locking using non-linear polarization rotation; this will be discussed in the next section.

12.6.2 Mode Locking Using Non-linear Polarization Rotation

In this section we will describe how non-linear optical rotation can lead to generation of ultrashort pulses of light from the fiber laser. A typical ultrashort erbium-doped fiber laser configuration (ring laser) is shown in Fig. 12.16. It consists of a pump laser operating at 980 nm which is coupled into an erbium-doped fiber through a WDM coupler. Within the ring there are two polarization controllers and a Faraday isolator/ polarizer sandwiched in between the two polarization controllers. By adjusting the polarization controller, it is possible to achieve any output polarization state for any input polarization state. The polarizing isolator ensures that only one polarization component is able to pass through it and in only one direction. There is also an output coupler to couple out the laser light from within the ring. The allowed longitudinal modes in such a cavity are those that have a phase shift which is an integral multiple of 2π in one complete round-trip. Within the gain bandwidth

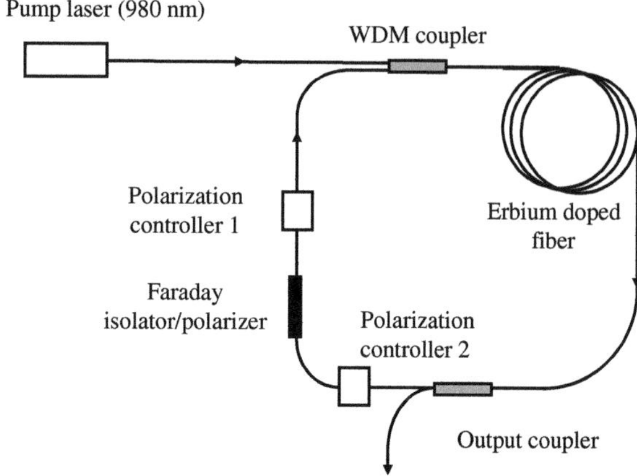

Fig. 12.16 A ring laser configuration of a fiber laser

of the erbium-doped fiber, there would be several longitudinal modes, while there would be just one transverse mode configuration, namely the LP_{01} mode of the fiber.

The pump laser creates population inversion in the fiber. As soon as erbium ions reach the excited state, some of them undergo spontaneous emission and a fraction of this is guided by the erbium-doped fiber. Due to the presence of a large number of longitudinal modes which have in general random phase relationships with each other, the light intensity within the ring will have fluctuations. If we consider a portion of the light as it starts from the polarizing isolator, it will be linearly polarized and the polarization controller 1 will change this into some elliptical polarization state. If we assume that the fibers within the ring maintain the polarization state, then as it propagates through the ring, it will get amplified and after losing a fraction of light at the output coupler enters the polarization controller 2. If the polarization controller 2 is adjusted so that it converts the incoming elliptically polarized light into a linear state which can pass through the isolator/polarizer, then in such a case, light wave corresponding to low intensity will have minimal loss in making a round-trip. If the gain in the erbium-doped fiber is sufficient to overcome the losses, then the laser would oscillate in a continuous wave. In such an arrangement, any high-intensity fluctuation will not be able to survive since the portion having high-intensity fluctuation would encounter non-linear polarization rotation with the result that after it emerges from the polarization controller 2, its polarization state would not match that of the isolator/polarizer and thus would suffer larger losses.

In the above discussion we have not considered the dispersion effects in the fiber, which are very important in analyzing propagation of very short pulses. Now consider a short pulse of high-intensity fluctuation arising out of the interference between various longitudinal modes as it starts from the exit of polarization controller 1. As it passes through the single-mode fiber, it will undergo anomalous dispersion (since the wavelength is around 1550 nm and standard single-mode fibers have anomalous dispersion at this wavelength) and get broadened and chirped with higher frequencies in the leading edge and lower frequencies in the trailing edge. If the erbium-doped fiber has normal dispersion, then as this broadened pulse propagates through the erbium-doped fiber, it will get compressed reaching a minimum pulse width around the middle of the doped fiber. The latter half of the erbium-doped fiber would again broaden the pulse which will then recompress itself as it propagates through the single-mode fiber before reaching the polarization controller 2. By adjusting the lengths of the doped fiber and the single-mode fiber, the net dispersion in the ring can be made zero.

At the same time the pulse will also undergo non-linear polarization rotation. The orientation of the two polarization controllers can be so adjusted that the high-intensity portion of the pulse recovers its linear polarization state after exiting from the polarization controller 2, which will imply low loss for the high-intensity portion. At the same time, the low-intensity portions of the pulse will not undergo polarization rotation and thus would be partially blocked by the isolator/polarizer element. Hence the ring will exhibit low losses for high-intensity portion and high losses for the low-intensity portions. This in turn will ensure that the high-intensity portions grow in intensity, while the low-intensity portions die out resulting in the generation of ultrashort pulses. Thus the phenomenon of non-linear polarization rotation can lead to the generation of ultrashort pulses of light.

The ultimate pulse width of such a pulse would depend on the gain bandwidth available. If we assume a gain bandwidth of 40 nm, then the pulse width achievable is about 200 fs. Figure 12.17 shows the output from a typical femtosecond fiber laser with output pulse width of about 160 fs.

Fig. 12.17 Output of a femtosecond fiber laser

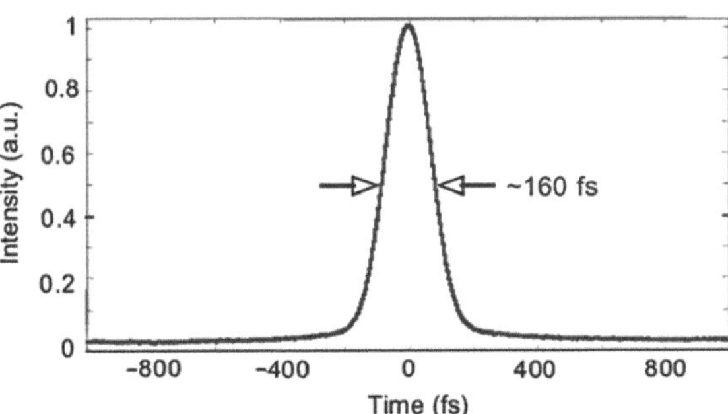

12.6.3 Semiconductor Saturable Absorbers

The other common technique used for mode locking of fiber lasers is based on saturable absorption. A saturable absorber is a medium whose absorption decreases with increase of light intensity. A saturable absorber is characterized by its wavelength range of operation (where it absorbs), its dynamic response (i.e., how fast it recovers when the incident intensity changes), and the saturation intensity and fluence (at what intensity it exhibits saturation). Semiconductor materials can absorb over a broad range of wavelengths (from the visible to the mid-infrared) and can be used as saturable absorbers.

The SESAM (semiconductor-saturable absorber mirror) is a saturable absorber that operates in reflection. In the SESAM, a highly reflecting surface is covered by the saturable absorber. At low intensities, the saturable absorber will absorb most of the incident light and the reflectivity of the mirror is small. As the incident intensity increases, the saturable absorber absorbs less and hence the reflectivity will increase. If such a device is used as a mirror in a laser cavity, then high-intensity portions will have a higher feedback compared to lower intensity portions, thus favoring high-intensity pulse formation. In the case of fiber lasers, semiconductor-based saturable absorbers based on InGaAsP or GaAs are typically used. They can be grown on top of reflecting multi-layer structures and are then referred to as semiconductor-saturable absorber mirror (SESAM). Using such saturable absorbing mirrors in the fiber laser cavity makes the design very compact and stable. The response times of such saturable absorbers are typically in the sub-picosecond range and require typically a saturation fluence of $F_s = 20$ µJ/cm^2. Assuming a fiber spot size of 5 mm, this corresponds to a saturation energy of $E_s = F_s \times \pi w^2 = 15$ pJ.

Problems

Problem 12.4 Consider an erbium-doped fiber with a core radius of 1.6 µm and a numerical aperture of 0.23.

a) Will the fiber be single moded at 980 and 1550 nm?

b) Calculate the Gaussian spot size of the fiber mode at 980 and 1550 nm.

c) Assuming that the entire core of the fiber is doped with erbium ions with a concentration of 6×10^{24} m^{-3}, obtain the absorption coefficient at 1530, 1550, and 1560 nm. The absorption cross sections at these wavelengths are 5.27, 2.57 and 1.86×10^{-25} m^2.

d) If 10 µW of power at each of these wavelengths is propagated through 5 m of the fiber, what is the output power at each wavelength?

Solution

a) *V* value at 980 nm is

$$V = \frac{2\pi}{\lambda_0} a\, NA = 2.36$$

Since for single-moded operation, $V < 2.4048$, the fiber would be single moded at 980 nm. It will also be single moded at all wavelengths larger than 980 nm.

b) Gaussian spot sizes are defined in Eq. (12.23). *V* number at 980 nm is already calculated in part (a). *V* number at 1550 nm would be 1.49. We have assumed that the numerical aperture is the same at both the wavelengths. Substituting in Eq. (12.23) we obtain $\Omega_p = 1.26$ µm and $\Omega_s = 2.04$ µm.

c) The absorption coefficient is given by Eq. (12.37) and we need to have the values of the spot size at the three wavelengths which can be obtained from Eq. (12.23). These are 1.99, 2.04, and 2.06 µm, respectively. The corresponding confinement factors of absorption coefficients are 0.476, 0.459, and 0.453, respectively. Hence the absorption coefficients are 1.5, 0.71 and 0.51 m^{-1}, respectively.

d) The output power is given as

$$P_{\text{out}} = P_{\text{in}}\, e^{-\alpha L}$$

Using the values of absorption coefficient at each of the wavelengths, we obtain output power as 0.027, 0.71, and 1.42 μW, respectively.

Problem 12.5 Consider propagation of pump and signal along the $+z$-direction. Under steady-state conditions, i.e., $d/dt = 0$, solve Eqs. (12.11)–(12.14) and show that the normalized population in level E_2 is given as

$$\tilde{N}_2 = \frac{\dfrac{P_p}{P_{p0}} + \dfrac{\sigma_{sa}}{\sigma_{se} + \sigma_{sa}}\dfrac{P_s}{P_{s0}}}{1 + \dfrac{P_p}{P_{p0}} + \dfrac{P_s}{P_{s0}}}$$

where

$$P_{p0} = \frac{h\nu_p N_t A}{\alpha_p \tau_{sp}}; \quad P_{s0} = \frac{h\nu_s N_t A}{\alpha_s(1 + \eta_s)\tau_{sp}}; \quad \eta_s = \frac{\sigma_{se}}{\sigma_{sa}}$$

Problem 12.6 Consider an EDFA operating in steady state with codirectional pumping and signal travelling along the $+z$-direction. Show that at any value of z within the doped fiber, for amplification, the pump power at that point must satisfy the following equation:

$$P_p(z) > \frac{\sigma_{sa}}{\sigma_{se}} P_{p0}$$

Problem 12.7 Using the results of Problem 12.3, calculate the value of z up to which signal amplification will take place for a pump wavelength of 980 nm, with an input pump power of 20 mW, $A = 7.1\ \mu$m^2, $\alpha_p = 1.15$ m^{-1} $N_t = 5.4 \times 10^{24}$m^{-3}, $\tau_{sp} = 12 \times 10^{-3}$s for signal wavelength of 1550 nm for which $\sigma_{sa} = 2.545 \times 10^{-25}$m^2, $\sigma_{se} = 3.41 \times 10^{-25}$m^2. [Ans: $P_{p0}= 0.56$ mW and $P_p\ (L) > 0.42$ mW, which gives $L = 3.3$ m.]

Problem 12.8 Consider the propagation of only the pump through a doped fiber. In such a case $P_s(z) = 0$ and Eq. (12.19) can be written as

$$\frac{dn_p}{n_p} = -\frac{\alpha_p}{1 + \dfrac{P_p}{P_{p0}}}\,dz = -\frac{\alpha_p}{1 + \dfrac{n_p}{n_{p0}}}\,dz$$

Obtain the solution of the above equation when $P_p \ll P_{p0}$ and $P_p \gg P_{p0}$; $P_p = n_p\, h\nu_p$ and $P_{p0} = n_{p0}h\nu_p$. [Ans: For $P_p \ll P_{p0}$, $n_p(z) = n_p(0)e^{-\alpha_p z}$; for $P_p \gg P_{p0}$, $n_p(z) = n_p(0) - \alpha_p n_{p0} z$.]

Problem 12.9 Assuming that an EDFA can be described as a two-level system, write down the equations describing the time variation of the population of the lower and upper energy levels. Under steady-state conditions, obtain the threshold pump intensity to achieve amplification at any value of z. Using the table given below, show that the threshold intensity required to achieve amplification at 1580 nm is lower than that at 1550 nm

Wavelength (nm)	σ_a(m^2)	σ_e(m^2)
1550	2.55×10^{-25}	3.41×10^{-25}
1580	0.65×10^{-25}	1.13×10^{-25}

Chapter 13
Semiconductor Lasers

13.1 Introduction

Semiconductor-based light sources such as light-emitting diodes (LED) and laser diodes have revolutionized the application of photonic components in science, engineering, and technology. They have become ubiquitous components and are found in most places, be it markets where they are used as scanners for products, at home where they are found in CD and DVD readers or laser printers, in communication systems as sources, etc. Unlike the lasers discussed earlier, laser diodes are based on semiconductors such as gallium arsenide (GaAs), gallium indium arsenide (GaInAs), gallium nitride (GaN), etc. They cover the range of wavelengths from the blue region to the infrared.

As compared to other laser systems, semiconductor lasers have some very attractive characteristics: they are very small in size, can be directly modulated by varying the drive current, are very efficient converters of electrical energy to light, can be designed to emit a broad range of wavelengths, etc.

In this chapter, we will discuss the basic principle of operation of semiconductor laser diodes and some of their important properties that lead to their widespread applications.

13.2 Some Basics of Semiconductors

The primary difference between electrons in semiconductors and other laser media is that in semiconductors, all the electrons occupy and share the entire volume of the crystal, while in the case of other laser systems such as neodymium:YAG laser and ruby laser, the lasing atoms are spaced far apart and the electrons are localized to their respective ions with very little interaction with other ions. Thus in a semiconductor, the quantum mechanical wave functions of all electrons overlap with each other and according to Pauli exclusion principle cannot occupy the same quantum state. Thus each electron in the crystal must be associated with a unique quantum state.

The atoms comprising the semiconductor when isolated have the same electron configuration. Thus electrons belonging to different atoms may be in the same

K. Thyagarajan, A. Ghatak, *Lasers*, Graduate Texts in Physics,
DOI 10.1007/978-1-4419-6442-7_13, © Springer Science+Business Media, LLC 2010

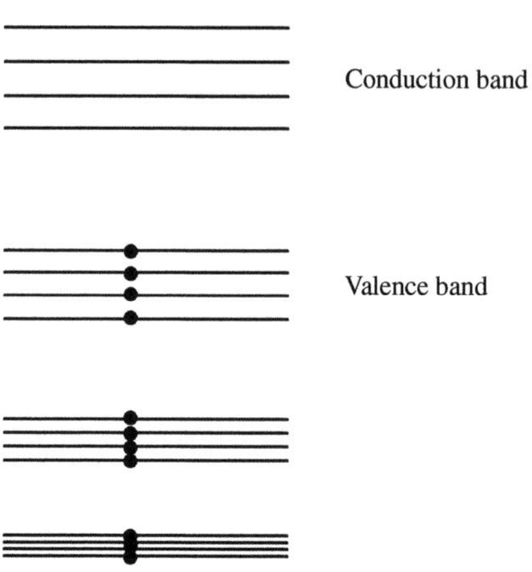

Fig. 13.1 Schematic diagram showing energy band diagram in a solid; each *horizontal line* corresponds to an energy level and *filled circles* represent electrons occupying the levels

energy state. However, when the atoms are brought close together to form the solid, interactions among the atoms lead to a splitting of the energy levels and this leads to the formation of energy bands which are separated by forbidden regions of energy. Figure 13.1 shows a schematic diagram in which each energy level is represented by a horizontal line; in each band formed by a group of energy levels, there are as many sublevels as there are atoms in the crystal. Since the number of atoms is very large, within each band, the allowed energy values are almost continuous. The highest energy band in a solid that is completely filled or occupied by electrons at 0 K is known as the valence band and the next higher band that is either vacant or partially occupied is known as the conduction band.

If the energy gap between the valence band and the conduction band is large, say > 3 eV, then thermal excitation from the valence band to the conduction band is very rare (thermal energy at room temperature of 300 K is about 25 meV). In such a case the medium behaves like an insulator. If the gap is smaller (< 2 eV), then electrons can get thermally excited from the valence band to the conduction band and they exhibit a finite electrical conductivity at temperatures higher than 0 K, which increases with temperature. Such media are referred to as semiconductors.

13.2.1 E Versus k

The wave function of an electron in a semiconductor can be written in the form of a Bloch wave function:

$$\psi(\mathbf{r}) = u_{\mathbf{k}}(\mathbf{r})\, e^{i\mathbf{k}\cdot\mathbf{r}} \tag{13.1}$$

with $u_{\mathbf{k}}(\mathbf{r})$ having the periodicity of the lattice of atoms. The solution given by Eq. (13.1) is similar to a plane wave with an amplitude function which is not constant but has space dependence with a specific periodicity. Substituting in the Schrodinger equation leads to a relationship between the energy value and $\mathbf{k}$ within the allowed energy bands. The application of periodic boundary conditions (i.e., the wave function must remain unaltered by a displacement equal to the crystal dimension along the x-, y-, and z-directions) leads to the condition that $\mathbf{k}$ cannot take arbitrary values but only a prescribed set of values. Thus if L_x, L_y, and L_z are the dimensions of the crystal along the x-, y-, and z-directions, respectively, then

$$k_x = p\frac{2\pi}{L_x}; \quad k_y = q\frac{2\pi}{L_y}; \quad k_z = r\frac{2\pi}{L_z}, \qquad p,q,r = 1,2,3,\dots . \tag{13.2}$$

where k_x, k_y, and k_z are the components of the vector $\mathbf{k}$.

Usually the energy depends not only on the magnitude of $\mathbf{k}$ but also on its direction. As a simplification, we will assume that the energy depends only on the magnitude of $\mathbf{k}$ of the electron propagation vector and is independent of its direction. We also assume that $L_x = L_y = L_z = L$.

Now for a free particle (inside a box), the relationship between the energy E and k is given as (see, e.g., Chapter 6 of Ghatak and Lokanathan (2004))

$$E = \frac{\hbar^2 k^2}{2m} = \frac{\hbar^2}{2m}\left(k_x^2 + k_y^2 + k_z^2\right)$$

where k_x, k_y, and k_z would be given by Eq. (13.2). We assume that the energy in the conduction band can be written approximately as (see Fig. 13.2)

$$E = E_\mathrm{c} + \frac{\hbar^2 k^2}{2m_\mathrm{c}} \tag{13.3}$$

where E_c is the energy at the bottom of the conduction band, $k^2 = k_x^2 + k_y^2 + k_z^2$, and m_c is referred to as the effective mass of the electron in the conduction band and is given as

$$m_\mathrm{c} = \hbar^2 \left(\frac{\mathrm{d}^2 E}{\mathrm{d}k^2}\right)^{-1}_{k=0} \tag{13.4}$$

Equation (13.3) is known as the "parabolic band approximation" and is valid because the electrons in the conduction band can be assumed to be almost free.

Similarly the relationship of energy to the electron propagation vector k in the valence band is given as (see Fig. 13.2)

$$E = E_\mathrm{v} - \frac{\hbar^2 k^2}{2m_\mathrm{v}} \tag{13.5}$$

where E_v is the energy at the top of the valence band and m_v is the effective mass of the hole in the valence band and is given as

$$m_\mathrm{v} = -\hbar^2 \left(\frac{\mathrm{d}^2 E}{\mathrm{d}k^2}\right)^{-1}_{k=0} \tag{13.6}$$

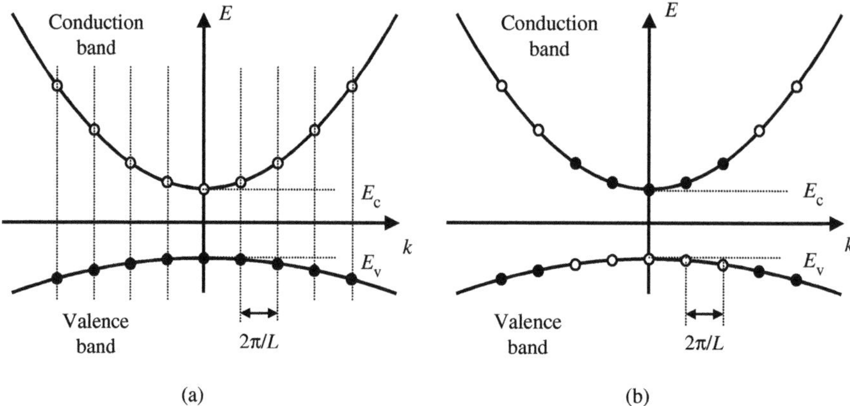

Fig. 13.2 Energy band diagram showing the variation of energy in the conduction band and the valence band with the k value (momentum) of the electron. The plot corresponds to the parabolic approximation. E_c is the energy at the bottom of the conduction band and E_v is the energy at the top of the valence band. The allowed energy states are equally separated in their k values. *Open circles* correspond to vacant states (states that are not occupied by electrons) and *filled circles* correspond to states occupied by electrons. (**a**) At 0 K in a semiconductor, all the states in the valence band are full, while all the states in the conduction band are empty. (**b**) At a finite temperature, some states at the top of the valence band are empty, while some states in the bottom of the conduction band are filled

with E given by Eq. (13.5). The effective mass of the electron in the valence band is equal to $-m_v$.

Typical values of effective mass of electrons in GaAs are $m_c = 0.067 m_0$, $m_v = 0.47 m_0$, where m_0 ($=9.109 \times 10^{-31}$ kg) is the rest mass of the electron.

The energy gap between the top of the valence band and the bottom of the conduction band is the bandgap and is given as

$$E_g = E_c - E_v \tag{13.7}$$

Figure 13.2a shows a typical energy band structure for a semiconductor such as GaAs. The dots correspond to the allowed values of k and hence by Eqs. (13.3) and (13.5) the allowed values of energy E. The filled circles represent states occupied by electrons and the open circles represent empty states. Note that the states are equally spaced along the k-axis, the spacing being $2\pi/L$, but are not equally spaced in the energy axis. The figure also shows that all states in the valence band are filled, while all states in the conduction band are empty. This situation corresponds to an intrinsic semiconductor at $T = 0$ K. If the temperature rises, then some of the electrons from the valence band can get excited to the conduction band and we will have a figure like the one shown in Fig. 13.2b with vacant states (holes) at the top of the valence band and electron filling states at the bottom of the conduction band. Note that in such media, the bottom of the conduction band and the top of the valence band occur at the same value of k. Such semiconductors are referred to as *direct bandgap* semiconductors. In semiconductors such as silicon, the bottom of the conduction band and the top of the valence band occur for different k values and such semiconductors are referred to as *indirect bandgap* semiconductors. Light emission is highly probable in direct bandgap semiconductors, while in indirect bandgap semiconductors, light emission is highly improbable.

13.3 Optical Gain in Semiconductors

In this section we will obtain the condition for achieving optical gain in a semiconductor. In order to do this, we need to introduce the concept of density of states and the occupation probability of electrons in conduction and valence bands.

13.3.1 Density of States

In Appendix E, we have calculated the number of electromagnetic modes per unit volume lying between wave vector magnitudes k and $k+ dk$ to be given as

$$p(k)\, dk = \frac{k^2}{\pi^2} dk \tag{13.8}$$

An exactly similar analysis can be performed to evaluate the number of available energy states per unit volume (density of states), with electron wavenumbers lying between k and $k + dk$ to be again given by Eq. (13.8). In the case of electromagnetic modes, we had multiplied by a factor of 2 to account for two independent states of polarization. In the case of electron waves, we have again a factor of 2 to account for the two independent spin states of the electron.

Using the relationship between energy E and k for electrons lying in the conduction band and the valence band, we can convert the density of states expression in terms of k to an expression giving the density of states lying between energy values E and $E + dE$ by using the fact that $p(E)dE = p(k)\, dk$ and thus

$$p(E) = p(k) \left(\frac{dE}{dk} \right)^{-1} \tag{13.9}$$

Using the relationships between E and k in the conduction band and the valence band [Eqs. (13.3) and (13.5)], we can evaluate $\left(\frac{dE}{dk} \right)^{-1}$ for the two bands and we get the following expressions for the density of states lying between E and $E + dE$ in the conduction band and the valence band as

$$p_c(E) = \frac{(2m_c)^{3/2}}{2\pi^2 \hbar^3} (E - E_c)^{1/2}, \qquad E > E_c \tag{13.10}$$

and

$$p_v(E) = \frac{(2m_v)^{3/2}}{2\pi^2 \hbar^3} (E_v - E)^{1/2}, \qquad E < E_v \tag{13.11}$$

The density of states is a very important quantity as it specifies the number of energy states per unit volume that are available for the electrons to occupy.

Note that in the conduction band, as E increases, the density of available energy states increases. Similarly as E decreases in the valence band, the density of available energy states increases. However, the density of states alone would not decide the electron population in the two bands; the probability of occupancy of the states along with the density of states would finally decide the electron population.

13.3.2 Probability of Occupancy of States

The density of states gives us the states that are available for occupation. In order to find the density of electrons that are actually occupying those energy states, we need to know the probability of occupancy of the states. Thus the probability that a state of energy E is occupied by an electron is given by the Fermi Dirac function (see, e.g., Saha and Srivastava (1973))

$$f(E) = \frac{1}{e^{(E-E_F)/k_B T} + 1} \tag{13.12}$$

where E_F is the Fermi energy, T is the absolute temperature, and k_B is the Boltzmann constant. It can be seen from Eq. (13.12) that at $T = 0$ K, all energy states below E_F are occupied, while all energy states above E_F are empty. At higher temperatures, the probability of having electrons above the Fermi level is finite. In fact at the energy $E = E_F$, the probability of occupation is exactly 0.5 irrespective of the temperature (in general, the Fermi energy E_F depends on the temperature). Of course it is possible that there are no energy states at E_F in which case there would be no electrons with this energy.

At thermal equilibrium, the distribution given by Eq. (13.12) describes the electron occupation probability for the conduction band as well as the valence band. When thermal equilibrium is disturbed, for example, by passing a current through a p–n junction or illuminating the semiconductor with a light beam of appropriate wavelength, then in this state we can define the probability of occupation in the conduction band and the valence band by two separate Fermi Dirac distributions, by defining two *quasi-Fermi levels* E_{Fc} and E_{Fv} (see Fig. 13.3):

Fig. 13.3 Quasi-Fermi levels in the conduction band and the valence band. At 0 K, all states below the quasi-Fermi level in the conduction band are filled, while all states above are empty. Similarly all states above the quasi-Fermi level in the valence band are empty, while all states below the level are filled

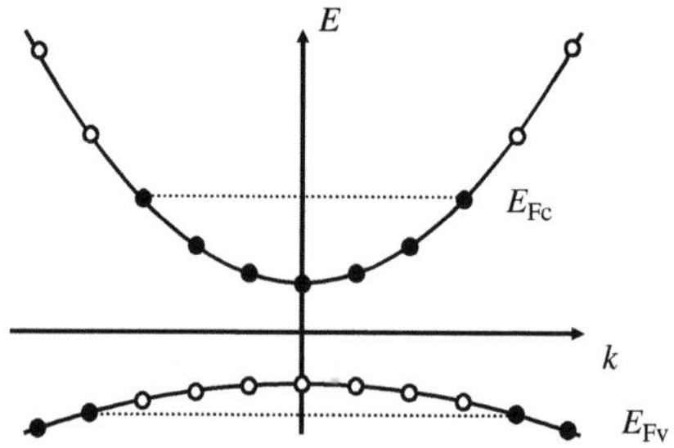

$$f_c(E) = \frac{1}{e^{(E-E_{Fc})/k_B T} + 1} \tag{13.13}$$

$$f_v(E) = \frac{1}{e^{(E-E_{Fv})/k_B T} + 1} \tag{13.14}$$

In writing Eqs. (13.13) and (13.14), it is assumed that the electrons in the conduction band and in the valence band come to a quasi-equilibrium *within the bands* very rapidly (typically within 10^{-12} s), while the transitions of electrons between the conduction band and the valence band take much longer, about 10^{-9} s.

13.3.3 Interaction with Light

Like in the case of atoms and molecules, electrons in the conduction band and holes in the valence band can interact with incident photons via three different mechanisms:

Absorption: An electron in the valence band can absorb a photon and get excited to the conduction band. Since there are no energy levels within the energy gap, the incident photon has to have a minimum amount of energy for this process to take place. If E_g $(=E_c - E_v)$ represents the energy gap, then the photon frequency must be greater than E_g/h. This process of absorption leads to the generation of electron–hole pairs (see Fig. 13.4a).

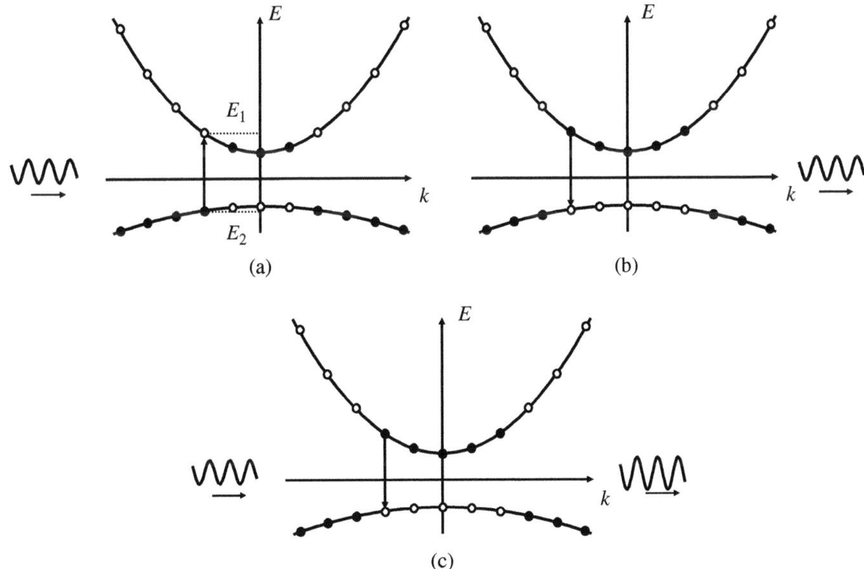

Fig. 13.4 (a) In the absorption process, an electron occupying a state in the valence band can absorb a photon of appropriate energy and get excited to a vacant state in the conduction band. (b) In the spontaneous emission process, an electron occupying a state in the conduction band can emit a photon of appropriate energy and get de-excited to a vacant state in the valence band. (c) In the case of stimulated emission, an incident photon of appropriate energy can stimulate an electron to make a transition from the conduction band to the valence band

Spontaneous emission: An electron in the conduction band can combine with a hole in the valence band (i.e., an electron can jump from the conduction band to a vacant state in the valence band) and release a photon of energy equal to the difference in the energies of the electron before and after the emission process. The photon frequency would be larger than E_g/h. This process takes place even in the absence of any photons and is termed spontaneous emission. The process of spontaneous emission is random and the emitted photon can appear in any direction. Light-emitting diodes are based on spontaneous emission arising out of electron–hole recombination (see Fig. 13.4b).

Stimulated emission: Just like in atomic systems, an incident photon having a frequency greater than E_g/h can induce a de-excitation of electron from the conduction band to the valence band (electron–hole recombination) and the emitted radiation is coherent with respect to the incident radiation. It is this process which is used in semiconductor lasers (see Fig. 13.4c).

Certain conditions are required for the above processes to take place. For absorption of an incident photon, it is essential that there be an electron available in the valence band *and a* vacant state be available in the conduction band at an energy difference corresponding to the energy of the photon (see Fig. 13.4a). Thus if v is the frequency of the incident photon, then an electron having an energy E_1 lying in the valence band can absorb this photon and get excited to a vacant energy state with energy E_2 lying in the conduction band such that

$$E_2 - E_1 = hv \tag{13.15}$$

Similarly for the spontaneous emission of a photon of energy hv, an electron occupying an energy level with energy E_2 can jump down to a vacant state (hole) with energy E_1 lying in the valence band and lead to a photon of energy $(E_2 - E_1)$. This is also termed electron–hole recombination. For stimulated emission the condition is the same as spontaneous emission with the emitted light being completely coherent with the incident light.

Apart from energy conservation described above, the processes of absorption and emission should also satisfy another condition on the wave vector $\vec{k}$ of the electron before and after the transition and the propagation vector of the photon. In fact the transition probability depends on a matrix element containing an integral over the volume of the crystal with an integrand having a term of the form $e^{i(\mathbf{k}_1 - \mathbf{k}_2 + \mathbf{k}_{op}).\mathbf{r}}$, where $\mathbf{k}_1$ represents the $\mathbf{k}$ corresponding to the electron in the valence band, $\mathbf{k}_2$ represents the $\mathbf{k}$ of the electron in the conduction band, and $\mathbf{k}_{op}$ is the propagation vector of the optical radiation interacting with the semiconductor. Since the exponential term oscillates rapidly with position $\mathbf{r}$, the integral and hence the transition probability vanishes unless $(\mathbf{k}_1 - \mathbf{k}_2 + \mathbf{k}_{op}) = 0$, i.e., the transition process needs to satisfy the following equation:

$$(\mathbf{k}_1 - \mathbf{k}_2 - \mathbf{k}_{op}) = 0 \tag{13.16}$$

The above condition can also be interpreted as a condition on the conservation of momentum in the interaction process. The momentum of the electron belonging to the conduction band is $\hbar k_2$ and if it jumps to a state in the valence band with a momentum $\hbar k_1$, then conservation of momentum implies that

$$\hbar k_2 - \hbar k_1 = \hbar k_{\mathrm{op}} = \frac{h\nu}{c} \tag{13.17}$$

where the right-hand side of Eq. (13.17) corresponds to the momentum of the emitted photon and we have replaced the vector momentum by the magnitude. The momentum of the electron in the conduction band and the valence band is much larger than that of the photon. Typically if a is the interatomic spacing, then the magnitude of the electron wave vector is comparable to $2\pi/a$, while that of the photon is of the order of $2\pi/\lambda$. Since $a \ll \lambda$, the electron possesses much larger momentum compared to the photon. Hence Eq. (13.17) essentially implies that the momentum of the electron before and after the transition must be almost equal, i.e.

$$k_2 \approx k_1 \tag{13.18}$$

This is also referred to as the k-selection rule. In the energy versus momentum diagram, this process corresponds to almost *"vertical transition"* (see Fig. 13.4).

13.3.4 Joint Density of States

Let us consider the interaction of a photon of frequency ν with an electron and a hole with the electron having an energy E_2 and lying in the conduction band and a hole having an energy E_1 and lying in the valence band. From energy conservation we have

$$E_2 - E_1 = h\nu$$

Now using Eqs. (13.3) and (13.5) we can write

$$E_2 = E_{\mathrm{c}} + \frac{\hbar^2 k^2}{2m_{\mathrm{c}}} \tag{13.19}$$

and

$$E_1 = E_{\mathrm{v}} - \frac{\hbar^2 k^2}{2m_{\mathrm{v}}} \tag{13.20}$$

where we have assumed the k value of the electron and the hole to be the same as per the k-selection rule. Hence we have

$$\begin{aligned}
h\nu &= (E_{\mathrm{c}} - E_{\mathrm{v}}) + \frac{\hbar^2 k^2}{2}\left(\frac{1}{m_{\mathrm{c}}} + \frac{1}{m_{\mathrm{v}}}\right) \\
&= E_{\mathrm{g}} + \frac{\hbar^2 k^2}{2m_{\mathrm{r}}}
\end{aligned} \tag{13.21}$$

where

$$\frac{1}{m_{\mathrm{r}}} = \left(\frac{1}{m_{\mathrm{c}}} + \frac{1}{m_{\mathrm{v}}}\right) \tag{13.22}$$

is referred to as the *reduced effective mass.*

Equation (13.21) can be rewritten as

$$k^2 = \frac{2m_{\mathrm{r}}}{\hbar^2}\left(h\nu - E_{\mathrm{g}}\right) \tag{13.23}$$

Using this expression for k^2 in Eqs. (13.19) and (13.20) we obtain

$$E_2 = E_{\mathrm{c}} + \frac{m_{\mathrm{r}}}{m_{\mathrm{c}}}\left(h\nu - E_{\mathrm{g}}\right) \tag{13.24}$$

and

$$E_1 = E_{\mathrm{v}} - \frac{m_{\mathrm{r}}}{m_{\mathrm{v}}}\left(h\nu - E_{\mathrm{g}}\right) \tag{13.25}$$

The above equations show that there is a one-to-one correspondence between the incident photon frequency ν and E_2 or ν and E_1, i.e., photons of a given frequency ν will interact primarily with electrons and holes with energy values given by Eqs. (13.24) and (13.25). Thus we can write

$$p_c(E_2)\mathrm{d}E_2 = p(\nu)\mathrm{d}\nu = p_{\mathrm{v}}(E_1)\mathrm{d}E_1 \tag{13.26}$$

where $p(\nu)$ is the joint density of states. Using the above equations we can now calculate the joint density of states as follows:

$$p(\nu) = p_{\mathrm{c}}(E_2)\frac{\mathrm{d}E_2}{\mathrm{d}\nu} = \frac{hm_{\mathrm{r}}}{m_{\mathrm{c}}}p_{\mathrm{c}}(E_2) \tag{13.27}$$

$p_{\mathrm{c}}(E_2)$ is given by Eq. (13.10) with E replaced by E_2. Using Eqs. (13.10) and (13.21) we obtain the following expression for the joint density of states:

$$p(\nu) = \frac{(2m_{\mathrm{r}})^{3/2}}{\pi\hbar^2}\left(h\nu - E_{\mathrm{g}}\right)^{1/2}, \qquad h\nu > E_{\mathrm{g}} \tag{13.28}$$

We could also have used Eq. (13.11) to obtain the same expression for the joint density of states given by Eq. (13.28). The joint density of states gives us an expression for the number of states available for an interaction (absorption or emission) to occur with a photon of energy $h\nu$. The quantity $p(\nu)\mathrm{d}\nu$ includes all possible pairs of energy states per unit volume lying in the conduction band and the valence band with an energy difference between $h\nu$ and $h(\nu+d\nu)$.

13.3.5 Absorption and Emission Rates

The rate of absorption of a photon of frequency ν would depend on the following factors:

- The probability that an electron exists in the valence band at an energy value E_1 given by Eq. (13.25) and a vacant energy state exists in the conduction band at an energy value E_2 given by Eq. (13.24).
- The joint density of states at the corresponding energy difference or equivalently at the corresponding frequency ν.

Let us assume that an electron with an energy E_1 lying in the valence band absorbs a photon and gets excited to a vacant state with an energy E_2 $(=E_1 + h\nu)$ lying in the conduction band. The probability that an electron is available at an energy E_1 in the valence band is obtained from Eq. (13.14) by replacing E by E_1:

$$f_v(E_1) = \frac{1}{e^{(E_1 - E_{Fv})/k_B T} + 1} \tag{13.29}$$

The probability that a vacant state with energy E_2 exists in the conduction band can be obtained from Eq. (13.13) by noting that since $f_c(E)$ is the probability that an electron with energy E is available in the conduction band, the probability that the energy state is not occupied (i.e., is vacant) is simply given by $(1-f_c(E))$. Hence the probability that a vacant state of energy E_2 is available in the conduction band is given as

$$1 - f_c(E_2) = 1 - \frac{1}{1 + e^{(E_2 - E_{Fc})/k_B T}} = \frac{1}{e^{-(E_2 - E_{Fc})/k_B T} + 1} \tag{13.30}$$

Hence the probability of absorption, which is equal to the probability that an electron of energy E_1 is available in the valence band and simultaneously a vacant state of energy E_2 $(= E_1 + h\nu)$ is available in the conduction band, is given as

$$f_a(\nu) = f_v(E_1)\,(1 - f_c(E_2)) \tag{13.31}$$

Similarly the emission of a photon of energy $h\nu$ depends on the availability of an electron of energy E_2 in the conduction band and simultaneously a vacant state (hole) of energy E_1 $(= E_2 - h\nu)$ in the valence band; thus the probability of emission is given as

$$f_e(\nu) = f_c(E_2)\,(1 - f_v(E_1)) \tag{13.32}$$

Let us consider the propagation of a light wave of frequency ν through the semiconductor; let ϕ_ν be the corresponding photon flux (i.e., ϕ_ν represents the number of photons crossing per unit time per unit area perpendicular to the direction of propagation). These photons will induce absorptions from the valence band to the conduction band and also stimulate emissions from the conduction band to the valence band. As discussed earlier, the transitions involving photons between the conduction and the valence bands take place primarily between electron levels that have the same value of k; the pairs of levels involved in the transition will be characterized by a specific value of k (see Fig. 13.4a). Due to inherent broadening, a group of levels with almost the same k value will be taking part in the transitions. This small range of k values would correspond to a small range of energy values within the valence and conduction bands.

Now, the rate of absorption of the photons would be proportional to (a) the photon flux ϕ_ν, (b) the probability of finding an electron in the valence band at energy E_1 and a vacant site in the conduction band with energy E_2, and (c) the joint density of states corresponding to this pair of levels. If we denote B_{vc} as the proportionality constant, then we can write the rate of absorption of incident photons as

$$R_a \, d\nu' = B_{vc}\phi_\nu p(\nu')f_v(E_1)\,(1 - f_c(E_2))\, d\nu' \tag{13.33}$$

where the range of frequency $d\nu'$ is included to account for the fact that a small range of energy states around the k value would take part in the absorption process.

Similarly the rate of stimulated emission by an electron making a transition from the conduction band to the valence band is given as

$$R_e \, d\nu' = B_{cv}\phi_\nu p(\nu')f_c(E_2)\,(1 - f_v(E_1))\, d\nu' \tag{13.34}$$

where B_{cv} is the constant of proportionality.

Exactly similar to the case of atomic systems where we found $B_{12} = B_{21}$, i.e., the constant of proportionality determining the absorption and stimulated emission was the same, here too the absorption and emission probabilities are the same, i.e., $B_{vc} = B_{cv}$.

13.3.6 Light Amplification

For light amplification we would require the rate of stimulated emission to exceed the rate of absorption. Thus from Eqs. (13.33) and (13.34) it follows that light amplification will take place if

$$f_c(E_2)\,(1 - f_v(E_1)) > f_v(E_1)\,(1 - f_c(E_2))$$

Substituting the expressions for $f_c(E_2)$ and $f_v(E_1)$ we get

$$E_{Fc} - E_{Fv} > E_2 - E_1 = h\nu \tag{13.35}$$

The above condition implies that for optical amplification by the semiconductor, the energy difference between the quasi-Fermi levels in the conduction band and the valence band must be larger than the energy of the photon. The positions of the quasi-Fermi levels in the conduction band and the valence band depend on the quasi-equilibrium population of electrons and holes in the conduction band and the valence band, respectively (see Fig. 13.3). This in turn would depend on the creation of electron and hole populations in the conduction band and the valence band by means of mechanisms like an external current source or an external illumination. The condition given by Eq. (13.35) is equivalent to the requirement of population inversion to achieve optical amplification in an atomic system.

The condition given by Eq. (13.35) can be understood graphically by considering the case at $T = 0$ K. If we assume that electrons are injected from the valence band into the conduction band, then since we are assuming $T = 0$ K, the energy states from E_c to E_{fc} will be filled in the conduction band and similarly the energy states from E_v to E_{fv} will be empty (see Fig. 13.3). Now consider a photon with an energy lying between E_g and $E_{fc} - E_{fv}$ to be incident. Due to the distribution of electrons in the conduction and the valence bands, there would be no electron in the valence band that would be capable of absorbing the photon, while electrons in the conduction band can be stimulated to make a transition to the valence band and with a consequent stimulated emission process leading to amplification. In this example, photons with frequency lying between E_g/h and $(E_{Fc} - E_{Fv})/h$ will undergo amplification, while photons with frequency greater than $(E_{Fc} - E_{Fv})/h$ will undergo absorption. What would happen to photons with frequency less than E_g/h?

Although our discussion of Eq. (13.35) has been carried out at $T = 0$ K, the condition given by Eq. (13.35) is valid at all temperatures. We recall from Chapter 4 that the rate of absorption in the case of an atomic system is given as

$$R_a = \sigma_a \phi_v N_1 \tag{13.36}$$

where N_1 is the density of atoms in the lower energy state, ϕ_v is the photon flux, and σ_a is the absorption cross section given as

$$\sigma_a = \frac{c^2}{8\pi n_0^2 v^2 t_{sp}} g(v) \tag{13.37}$$

where n_0 is the refractive index of the medium, t_{sp} is the spontaneous lifetime of the upper level, and $g(v)$ is the lineshape function. For a given pair of nondegenerate energy levels, the absorption cross section σ_a and the emission cross section σ_e are equal.

In the case of semiconductors, the quantity N_1 gets replaced by the product of the joint density of states and the probability of finding an electron in the valence band and simultaneously a vacant state in the conduction band with appropriate energy. The constant of proportionality B_{vc} is then given as

$$B_{vc} = \frac{c^2}{8\pi n_0^2 v^2 \tau_r} g(v) \tag{13.38}$$

where the spontaneous lifetime in the case of the atomic system gets replaced by τ_r, the radiative recombination time of the electron.

Thus Eq. (13.33) becomes

$$R_a \, dv' = \frac{c^2}{8\pi n_0^2 v'^2 \tau_r} g(v')\phi_v p(v')f_v(E_1)\,(1 - f_c(E_2))\, dv'$$

and the total rate of absorption is given as

$$R_{ab} = \int R_a \, dv' = \int \frac{c^2}{8\pi n_0^2 v'^2 \tau_r} g(v')\phi_v p(v')f_v(E_1)\,(1 - f_c(E_2))\, dv' \qquad (13.39)$$

The line-broadening mechanism in semiconductors that leads to the lineshape function $g(v)$ is primarily due to photon–phonon collisions. Typical collision times τ_c are about 10^{-12} s and thus this leads to a linewidth of about 300 GHz. This is very narrow compared to the frequency width of the other terms within the integral in Eq. (13.39) and hence for all practical purposes, $g(v') = \delta(v' - v)$. Thus we obtain

$$R_{ab} = \frac{c^2}{8\pi n_0^2 v^2 \tau_r} \phi_v p(v)f_v(E_1)\,(1 - f_c(E_2)) \qquad (13.40)$$

which gives the total rate of absorption of photons incident at frequency v and flux ϕ_v. In Eq. (13.40), E_2 and E_1 are given by Eqs. (13.24) and (13.25), respectively, with $E_2 - E_1 = hv$.

Similarly the rate of stimulated emission due to the incident photons is given as

$$R_{st} = \frac{c^2}{8\pi n_0^2 v^2 \tau_r} \phi_v p(v)f_c(E_2)\,(1 - f_v(E_1)) \qquad (13.41)$$

The rate of spontaneous emission depends only on the density of states and the occupation probability that an electron exists in the conduction band and a hole exists at the required energy in the valence band. Hence we can write for the rate of spontaneous emissions as

$$R_{sp} = \frac{1}{\tau_r} p(v)f_c(E_2)\,(1 - f_v(E_1)) \qquad (13.42)$$

13.4 Gain Coefficient

We will now calculate the gain coefficient of a semiconductor under quasi-equilibrium. The procedure is exactly the same as was followed in Chapter 4. As light propagates through the medium, it induces absorption and stimulated emissions. Thus if we consider a plane P_1 at z and another plane P_2 at $z + dz$ with z

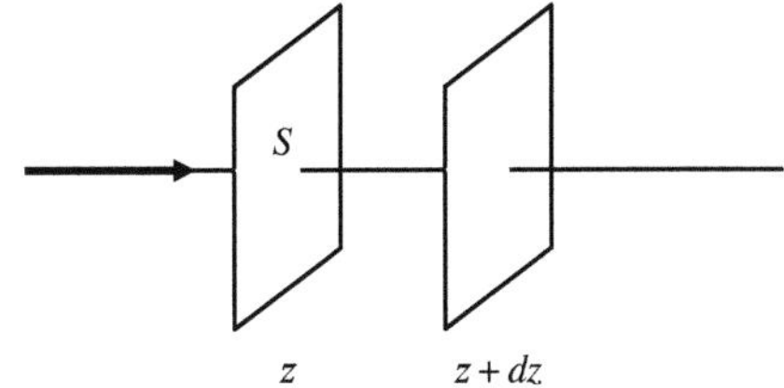

Fig. 13.5 Intensity of light entering the plane at z is $I(z)$, while that at $z + dz$ is $I(z+dz)$

being the direction of propagation of the light wave (see Fig. 13.5) and if the area of cross section is S, then the net number of photons entering the volume between P_1 and P_2 per unit time is given as

$$[\phi_v(z) - \phi_v(z + dz)]\,S = -\frac{d\phi_v}{dz}S\,dz \tag{13.43}$$

This must be equal to the net rate of absorption of photons within the volume which is given as

$$(R_{ab} - R_{st})S\,dz$$

Substituting the values of R_{ab} and R_{st} from Eqs. (13.40) and (13.41), respectively, and using Eq. (13.43) we obtain

$$\frac{d\phi_v}{dz} = -\frac{c^2}{8\pi n_0^2 v^2 \tau_r}\phi_v p(v)\left[f_v(E_1) - f_c(E_2)\right] \tag{13.44}$$

Thus if

$$E_2 - E_1 > E_{Fc} - E_{Fv}$$

then the right-hand side of Eq. (13.44) is negative and there is net absorption by the semiconductor. On the other hand if

$$E_2 - E_1 < E_{Fc} - E_{Fv}$$

then the right-hand side of Eq. (13.44) becomes positive and in such a case, there would be net amplification. The gain coefficient is given as

$$\gamma_v = \frac{c^2}{8\pi n_0^2 v^2 \tau_r}p(v)\left[f_c(E_2) - f_v(E_1)\right] \tag{13.45}$$

Hence for amplification the condition specified by Eq. (13.35) needs to be satisfied and this is usually accomplished by passing enough current through a p–n junction so as to create appropriate electron–hole population in the depletion region. This will be discussed in Section 13.4.1.

We can substitute for $p(v)$, $f_v(E_1)$, and $f_c(E_2)$ from Eqs. (13.28), (13.29), and (13.30), respectively, to obtain the following expression for the gain coefficient:

$$\gamma_v = \frac{c^2}{8\pi n_0^2 v^2 \tau_r} \frac{(2m_r)^{3/2}}{\pi \hbar^2} (hv - E_g)^{1/2} \left[\frac{1}{e^{(E_2 - E_{Fc})/k_B T} + 1} - \frac{1}{e^{(E_1 - E_{Fv})/k_B T} + 1} \right]$$

(13.46)

When electrons are injected into the conduction band from the valence band by an external mechanism such as a current or illumination, the electrons in the conduction band and the valence band come to quasi-equilibrium within the bands. Depending on the corresponding values of the quasi-Fermi energy values, for a given photon energy or frequency, the quantity γ_v will be positive or negative corresponding to amplification or attenuation. In order to find this out, we first need to estimate the electron population in the conduction band and the hole population in the valence band from which we can estimate the energy of the quasi-Fermi levels in the conduction band and the valence band. Knowing this, it is then possible to calculate the frequency dependence of gain in the semiconductor from Eq. (13.46) knowing all the other parameters. Note that the energy values E_2 and E_1 are related to the frequency through Eqs. (13.24) and (13.25).

Figure 13.6 shows a schematic variation of $p(v)$, $f_c(E_2) - f_v(E_1)$, and their product as a function of photon energy ($=E_2-E_1$) for a typical case. It can be seen that the product $p(v) [f_c(E_2) - f_v(E_1)]$ is positive only in a certain range of photon energies. In this range of photon energies, the semiconductor will exhibit gain and the figure shows a typical spectral variation of gain. If the electron–hole population is increased, then the quasi-Fermi levels would move up in the conduction band and down in the valence band, thus shifting the gain curve up and also the peak would shift to larger photon energies.

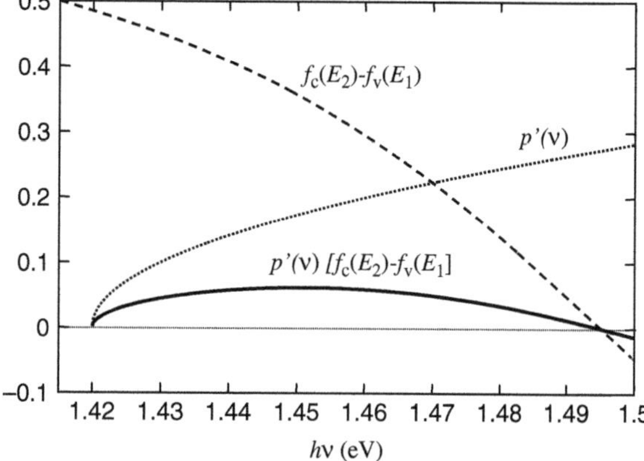

Fig. 13.6 Schematic variation of the joint density of states [$p'(v)$], the inversion factor [$f_c(E_2)-f_v(E_1)$], and their product with photon energy. Here $p'(v) = (hv - E_g)^{\frac{1}{2}}$

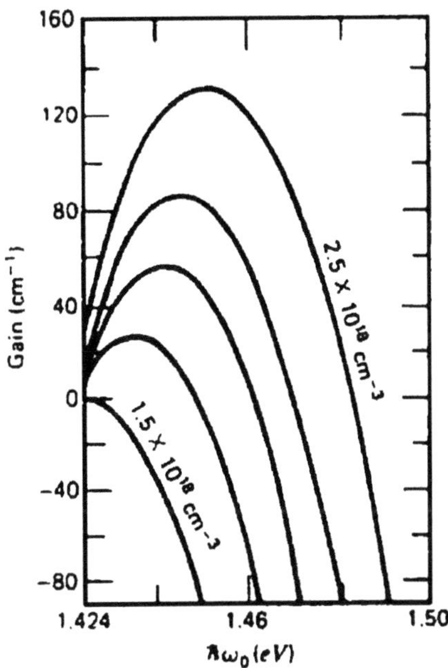

Fig. 13.7 Variation of gain coefficient with photon energy for different electron concentrations for a GaAs device. (Source: Yariv (1977), reprinted with permission)

Figure 13.7 shows a typical plot of the gain coefficient versus the energy of the photon for different injected carrier density in a GaAs device. Negative gain corresponds to absorption. Thus we see that for a given carrier concentration, the semiconductor exhibits gain over a band of photon energies or wavelengths and as the carrier concentration increases, the peak gain coefficient increases and so does the bandwidth over which gain is available. Also as the carrier concentration increases, the frequency at which the gain peaks increases, i.e., the wavelength at which the gain peaks will decrease. The increase in the bandwidth is primarily due to the shifting of the quasi-Fermi levels from the band edges with the increase in the electron–hole concentration. Note that the energy values of 1.424 and 1.50 eV correspond to photon wavelengths of 870 and 827 nm, respectively.

Figure 13.8 shows the variation of the peak gain coefficient with the carrier concentration. For a typical range of gain values, the curve can be approximated by a linear curve and we can express the gain coefficient as

$$\gamma = \sigma(n_c - n_{tr}) \tag{13.47}$$

where n_{tr} is the carrier concentration for transparency when the semiconductor becomes transparent, i.e., $\gamma = \alpha = 0$. The figure gives the required carrier concentration to achieve a certain gain coefficient.

Example 13.1 If we consider a semiconductor in overall thermal equilibrium at $T = 0$ K, then the two quasi-Fermi levels coincide with the Fermi level and if this level is within the energy gap, $f_v(E_1)=1$ and $f_c(E_2)=0$ and we would have from Eq. (13.46)

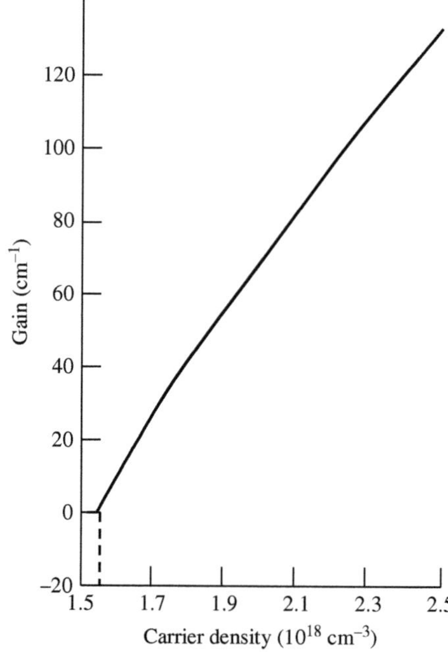

Fig. 13.8 Variation of peak gain with carrier density showing an almost linear variation (Source: Yariv (1977), reprinted with permission)

$$\gamma_v = -\alpha_v = -\frac{c^2}{8\pi n_0^2 v^2 \tau_r} \frac{(2m_r)^{3/2}}{\pi \hbar^2} \left(hv - E_g\right)^{1/2}, \qquad hv > E_g \tag{13.48}$$

Here α_v represents the absorption coefficient. The above expression gives an approximate expression for the absorption coefficient and its frequency dependence.

Example 13.2 Consider GaAs for which $m_c = 0.067m_0$, $m_v = 0.46m_0$ giving $m_r = 0.059m_0$. The bandgap energy is $E_g = 1.424$ eV, which corresponds to a frequency $v = v_0 = E_g/h = 3.43 \times 10^{14}$ Hz or a free space wavelength of 875 nm. The values of other constants are $n_0 = 3.64$ and $\tau_r = 4.6$ ns. Thus the absorption coefficient at 800 nm is approximately 8000 cm^{-1} showing a very strong absorption. If we take 10 μm of such a crystal, the fraction of light transmitted would be about 3.3×10^{-4}.

13.4.1 Electron–Hole Population and Quasi-Fermi Levels

When electrons are injected into the conduction band, they can make spontaneous transitions to the valence band if there are vacant states and in the process emit radiation. This is the radiation emitted by a light-emitting diode. In order to obtain this we need to calculate the electron and hole populations in the conduction band and the valence band, respectively. The electron population density in the conduction band between energy values E_2 and $E_2 + \mathrm{d}E_2$ is given as

$$\mathrm{d}n_c = p_c(E_2)f_c(E_2)\mathrm{d}E_2 \tag{13.49}$$

where $p_c(E_2)$ is the density of states in the conduction band and $f_c(E_2)$ is the occupation probability in the conduction band. Hence the total electron density in the conduction band is

$$n_c = \int dn_c = \int_{E_c}^{\infty} p_c(E_2) f_c(E_2) dE_2 \tag{13.50}$$

Using Eqs. (13.10) and (13.13), we obtain

$$n_c = \frac{(2m_c)^{3/2}}{2\pi^2 \hbar^3} \int_{E_c}^{\infty} (E_2 - E_c)^{1/2} \frac{1}{e^{(E_2 - E_{Fc})/k_B T} + 1} dE_2 \tag{13.51}$$

The above equation relates the electron concentration to the quasi-Fermi energy E_{Fc}.

In a similar fashion the hole concentration in the valence band and the quasi-Fermi energy in the valence band are related as

$$n_v = \frac{(2m_v)^{3/2}}{2\pi^2 \hbar^3} \int_{-\infty}^{E_v} (E_v - E_1)^{1/2} \frac{1}{e^{-(E_1 - E_{Fv})/k_B T} + 1} dE_1 \tag{13.52}$$

The integrals in Eqs. (13.51) and (13.52) have to be evaluated numerically since they cannot be analytically evaluated. A useful approximate expression relating the carrier concentration and quasi-Fermi energies is the Joyce–Dixon approximation given as

$$E_{Fc} = E_c + k_B T \left[\ln\left(\frac{n}{N_c}\right) + \frac{1}{\sqrt{8}} \frac{n}{N_c} \right];$$

$$E_{Fv} = E_v - k_B T \left[\ln\left(\frac{p}{N_v}\right) + \frac{1}{\sqrt{8}} \frac{p}{N_v} \right] \tag{13.53}$$

where n is the electron concentration and p is the hole concentration and

$$N_c = 2\left(\frac{m_c k_B T}{2\pi \hbar^2}\right)^{3/2};$$

$$N_v = 2\left(\frac{m_v k_B T}{2\pi \hbar^2}\right)^{3/2} \tag{13.54}$$

Figure 13.9 shows a normalized plot of $(E_{Fc} - E_c)/k_B T$ versus n/N_c as given by Eq. (13.53). The variation of $(E_v - E_{Fv})/k_B T$ with p/N_v will be the same. From Eq. (13.53) it follows that for $n \ll N_c$, $p \ll N_p$, we can neglect the second terms in the brackets and obtain an approximate form

Fig. 13.9 Normalized plot of $(E_{\mathrm{Fc}}-E_{\mathrm{c}})/k_{\mathrm{B}}T$ versus n/N_{c} as given by Eq. (13.53)

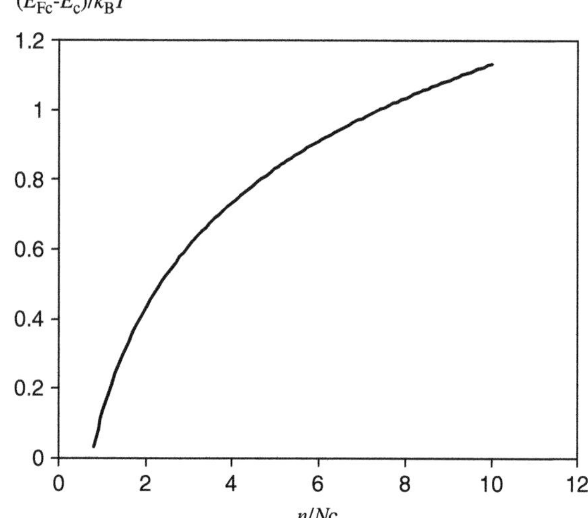

$$n = N_{\mathrm{c}} \exp\left(\frac{E_{\mathrm{Fc}} - E_{\mathrm{c}}}{k_{\mathrm{B}}T}\right); \qquad p = N_{\mathrm{v}} \exp\left(\frac{E_{\mathrm{v}} - E_{\mathrm{Fv}}}{k_{\mathrm{B}}T}\right) \tag{13.55}$$

which are referred to as Boltzmann approximations.

As an example we consider GaAs for which $m_{\mathrm{c}} = 0.0678 m_0$, $m_{\mathrm{v}} = 0.47 m_0$. Thus at 300 K, we obtain

$$N_{\mathrm{c}} \approx 4.3 \times 10^{23}\ \mathrm{m}^{-3} \text{ and } N_{\mathrm{v}} \approx 8 \times 10^{24}\ \mathrm{m}^{-3}$$

Equation (13.53) can be used to estimate the values of E_{Fc} and E_{Fv} for a given electron concentration and hole concentration or conversely estimate the electron and hole concentrations required for a given value of the quasi-Fermi energies.

Example 13.3 We now obtain the electron concentration required for transparency condition in GaAs for photon energy corresponding to the bandgap at 300 K.

For transparency, $\gamma_v = 0$ and Eq. (13.48) gives us

$$h\nu = E_2 - E_1 = E_{\mathrm{Fc}} - E_{\mathrm{Fv}}$$

We can use Eq. (13.53) for relating E_{Fc} and E_{Fv} to the electron and hole concentrations. For injection of equal number of electrons and holes as is the case with semiconductor lasers, we have

$$h\nu = E_{\mathrm{c}} - E_{\mathrm{v}} + k_{\mathrm{B}}T\left[\ln\left(\frac{n}{N_{\mathrm{c}}}\right) + \frac{1}{\sqrt{8}}\frac{n}{N_{\mathrm{c}}} + \ln\left(\frac{p}{N_{\mathrm{v}}}\right) + \frac{1}{\sqrt{8}}\frac{p}{N_{\mathrm{v}}}\right]$$

Since $E_{\mathrm{c}}-E_{\mathrm{v}}=E_{\mathrm{g}}=h\nu$ and $n = p$, the condition implies

$$\ln\left(\frac{n^2}{N_{\mathrm{c}}N_{\mathrm{v}}}\right) + \frac{n}{\sqrt{8}}\left(\frac{1}{N_{\mathrm{c}}} + \frac{1}{N_{\mathrm{v}}}\right) = 0$$

which is a transcendental equation for electron density. Using the values of N_c and N_v obtained earlier for $T = 300$ K, we can solve the above equation and obtain

$$n = p \approx 1.2 \times 10^{24}\,\mathrm{m}^{-3}$$

At this electron population density in the conduction band and an equal hole population in the valence band, the semiconductor is transparent for a photon energy equal to the bandgap energy. For photon energies larger than this (wavelengths shorter than corresponding to this wavelength), the semiconductor would be absorbing.

13.4.2 Gain in a Forward-Biased p–n Junction

Consider a p–n junction formed between a p-doped and an n-doped semiconductor as shown in Fig. 13.10a. Because of different carrier concentrations of electrons and holes in the p and n regions, electrons from the n region diffuse into the p region and holes from the p region diffuse into the n region. The diffusion of these carriers across the junction leads to a built-in potential difference between the positively charged immobile ions in the n side and the negatively charge immobile ions in

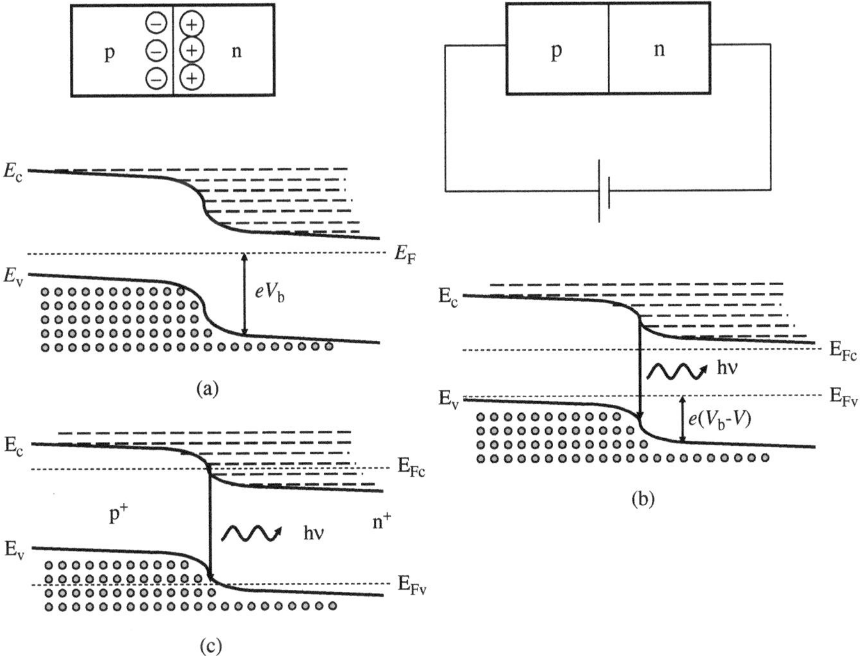

Fig. 13.10 (**a**) Unbiased p–n junction and (**b**) forward-biased p–n junction. When the p–n junction is forward biased, we can create a situation satisfying Eq. (13.35) in the depletion region and thus achieve optical amplification over a certain range of photon energies. (**c**) Forward-biased, heavily doped p–n junction

the p side of the junction. This built-in potential V_b lowers the potential energy of the electrons in the n side with respect to the potential energy of electrons in the p side, which is represented by "bending of energy bands" near the p–n junction as shown in Fig. 13.10a. Note that the Fermi levels on both sides of the p–n junction are aligned at the same energy value. This is necessary because, in the absence of any applied external energy source, the charge neutrality in the material requires that the probability of finding an electron should be the same everywhere and therefore only one Fermi function should be described the carrier distribution. In this case, there will be no net current in the medium.

If we forward bias the p–n junction by means of an external supply voltage V, then the potential energy of electrons in the n side increases and the band moves up. The band offset decreases and the Fermi levels separate out as shown in Fig. 13.10b. The increased potential energy of the carriers brings them into the depletion region, where they recombine constituting a forward current through the junction. The forward biasing leads to injection of electrons and holes into the junction region, where they recombine generating photons via spontaneous emission process. This phenomenon is also referred to as injection luminescence. Note that even though the separation between the quasi-Fermi levels is less than the bandgap energy E_g, there would be light emission because of a forward current through the device. This is the basis of operation of light-emitting diodes (LEDs) and the device therefore does not have any threshold value for the forward current to start emitting light. However, for amplification by stimulated emission, as we saw earlier, the quasi-Fermi levels have to satisfy Eq. (13.35).

It is usually not possible to satisfy Eq. (13.35) in p–n junctions formed between moderately doped p- and n-type semiconductors. However, if one starts with a p–n junction formed by highly doped p- and n-type semiconductors, in which the Fermi levels are located inside the respective bands, application of a strong bias can lead to the gain condition (see Fig. 13.10c). Indeed this is the basis of operation of injection laser diode.

As mentioned earlier, a laser diode consists basically of a forward biased p–n junction of a suitably doped direct bandgap semiconductor material. Two ends of the substrate chip are cleaved to form mirror-like end faces, while the other two ends are saw cut so that the optical resonator is formed in the direction of the cleaved ends only. The large refractive index difference at the semiconductor–air interface provides a reflectance of about 30%, which is good enough to sustain laser oscillations in most semiconductor diodes. This is primarily due to the large gain coefficients that are available in semiconductors (see Fig. 13.7).

At steady state, the rate at which excess carriers are being injected into the junction must equal the rate of recombination. At threshold this rate is just the spontaneous recombination rate which is given as

$$R_{sp} = \frac{\Delta n A d}{\tau_r} \tag{13.56}$$

where Δn is the excess carrier density, A is the area of cross section, and d is the thickness of the gain region. If J represents the current density, then the rate of injection is JA/e, where e is the electron charge. Thus we get

$$J = \frac{\Delta ned}{\tau_r} \tag{13.57}$$

Typically $d \sim 0.1$ μm (for heterostructure lasers) and $\tau_r \sim 4$ ns. Thus the threshold current density is

$$J_{th} \approx 808 \text{ A/cm}^2$$

Note that it is due to the small value of d in heterostructure lasers that we obtain a much smaller threshold current density compared to a homojunction laser wherein there is no potential step for the electrons and they diffuse away leading to a much larger value of d.

13.4.3 Laser Oscillation

We have seen earlier that the gain is approximately proportional to the excess carrier concentration [see Eq. (13.47)]. Thus we can write for the gain variation with the excess carrier concentration

$$\gamma_p = \alpha_a \left(\frac{\Delta n}{\Delta n_{tr}} - 1 \right) \tag{13.58}$$

where α_a is the absorption coefficient of the material in the absence of current injection, Δn is the excess carrier concentration in the active region due to the injection current, and Δn_{tr} is the excess carrier concentration corresponding to the condition when the gain becomes zero. At this value the medium neither absorbs nor provides gain and becomes transparent. Thus this is referred to as the transparency current density.

Using Eq. (13.57), Eq. (13.58) can be written in an alternative form in terms of current:

$$\gamma_p = \alpha_a \left(\frac{J}{J_{tr}} - 1 \right) = \alpha_a \left(\frac{I}{I_{tr}} - 1 \right) \tag{13.59}$$

where the subscript tr refers to transparency condition. Thus the medium will exhibit gain only if the current is greater than I_{tr}. However, this may not be sufficient for laser oscillation since for laser oscillation the gain should be able to compensate all losses in the cavity.

If α represents the intrinsic loss coefficient in the medium (primarily due to scattering) and if R_1 and R_2 represent the reflectivities of the ends of the cavity, then the total loss coefficient is given as [see Section 4.4]

$$\alpha_{tot} = \alpha - \frac{1}{2L}\ln(R_1 R_2) \tag{13.60}$$

where L is the length of the cavity. In order for lasing action to begin, the gain provided by the active medium must equal the loss. The gain depends on the position of the quasi-Fermi levels in the conduction band and the valence band which in turn depends on the electron–hole concentration. By injecting an external current, electron–hole population is created in the lasing region; thus a minimum current would be required for the laser action to begin.

A minimum value of excess carrier concentration Δn is needed for laser action, and for a given current, Δn depends inversely on the thickness of the active region. Typical values of d in a homojunction (a junction made up of same semiconductor material on both sides of the junction) are a micrometer or more depending on the dopant concentration on the p and n sides. The corresponding values for transparency current densities and hence threshold current density for laser oscillation are of the order of kiloampere per square centimeter. If d can be reduced to a much smaller value, then this would lead to a smaller value for the threshold current density. This is achieved in practice by using heterostructures.

13.4.4 Heterostructure Lasers

The basic laser structure shown in Fig. 13.10 is referred to as homojunction laser and was invented in 1962. In this device the p–n junction is formed by using the same semiconductor material on both sides of the junction. In such lasers the carriers drift from the junction region and occupy a larger volume and also the generated light is not confined, thus leading to reduced overlap between the electron and holes and the light radiation. Lasing in these devices can be achieved only in pulsed mode since the threshold current densities are in the range of a few amperes to tens of amperes. Such large currents can easily lead to catastrophic damage of the laser if operated continuously.

The basic configuration of the heterostructure laser is shown in Fig. 13.11. In this a thin layer of a suitable semiconductor material (such as GaAs) is sandwiched between two layers of a different semiconductor exhibiting a larger bandgap (such as $Al_xGa_{1-x}As$). Thus inner layer forms two heterojunctions at the two interfaces between the semiconductors. The bandgap energy of $Al_xGa_{1-x}As$ depends on the value of x and for $x < 0.42$, the bandgap energy can be approximated as

$$E_g = (1.424 + 1.266x)\,eV \tag{13.61}$$

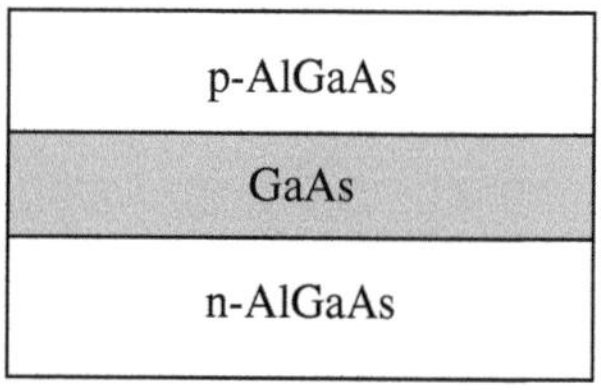

Fig. 13.11 A double-heterostructure laser with a lower bandgap semiconductor (GaAs) surrounded on either side by a higher bandgap semiconductor (AlGaAs)

p-AlGaAs

GaAs

n-AlGaAs

The increase in bandwidth with x is distributed unequally between the conduction band and the valence band according to the following approximate expressions:

$$\Delta E_{\rm c} \approx 0.67 \times 1.266x$$
$$\Delta E_{\rm v} \approx 0.33 \times 1.266x \tag{13.62}$$

Since the inner layer made up of GaAs has a smaller bandgap than the surrounding semiconducting materials on either side, carriers injected into the inner layer get confined due to the potential barrier for electrons and holes present at the two junctions. Interestingly the larger bandgap semiconductor also has a smaller refractive index than does the smaller bandgap semiconductor ($\Delta n \sim -0.7x$) and thus the inner layer of the semiconductor forms an optical waveguide confining the emitted radiation by the phenomenon of total internal reflection. The confinement of carriers by potential step and the confinement of optical radiation by refractive index steps lead to a drastic reduction in the threshold current density for laser operation in these devices.

Consider a heterostructure formed by a thin layer of GaAs sandwiched between two layers of $Al_xGa_{1-x}As$ as shown in Fig. 13.12. The bandgap of GaAs is 1.424 eV at room temperature, whereas that of $Al_xGa_{1-x}As$ increases from 1.424 eV for $x = 0$ with increasing fraction of Al as discussed earlier. Figure 13.12a shows the energy band diagram corresponding to the three regions when they are not in contact. Figure 13.12b and c shows the energy band diagrams of the composite before and after forward biasing, respectively. As can be seen from the figure, the potential barriers at the two junctions restrict the flow of electrons from the n-AlGaAs to the p-AlGaAs and of holes from the p-AlGaAs to the n-AlGaAs layers. This results in a large concentration of accumulated charge carriers in the thin GaAs layer and leads to a quasi-equilibrium with quasi-Fermi levels which satisfy the condition for gain as given by Eq. (13.35), thus leading to amplification of light. Note that the energy of the emitted photons will be around the bandgap energy of GaAs, which is smaller than that of AlGaAs, and therefore these photons will not be absorbed by the two AlGaAs layers as they have higher bandgaps.

As mentioned earlier, the lower bandgap GaAs has a higher refractive index than does AlGaAs, a typical index difference of about 0.2 around 800 nm wavelength. Thus the thin GaAs layer behaves like an optical waveguide confining the optical radiation within the region (see Fig. 13.13).

Due to the planar geometry of the thin GaAs layer, this corresponds to a planar optical waveguide. Although the structure will confine the light wave in the direction perpendicular to the junctions, there would be diffraction in the plane of the junction. Because of this spreading, the threshold currents can be large and the emission pattern may also not be stable with variation in current. To overcome these effects laterally confined semiconductor lasers have been developed. In these lasers in addition to confinement perpendicular to the junction, optical guidance is also provided in the plane of the junction by having a lower refractive index region surrounding the active region.

Fig. 13.12 Energy band diagram corresponding to the three regions of a double-heterostructure laser (**a**) when they are not in contact, (**b**) when they are in contact with no bias, and (**c**) under forward bias

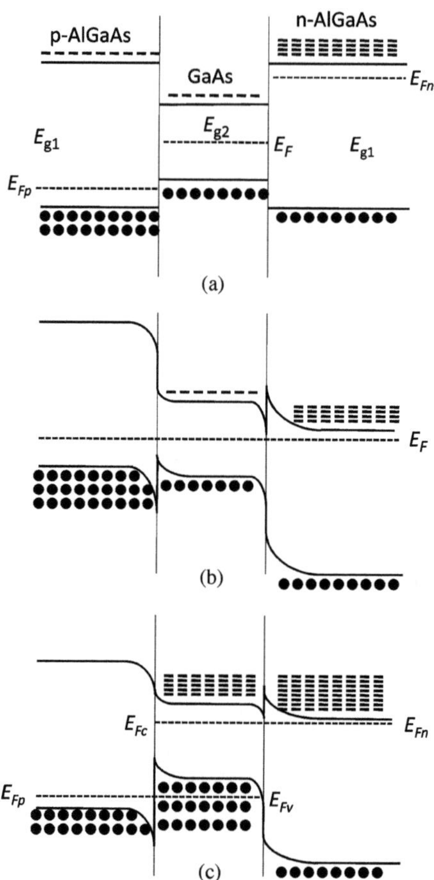

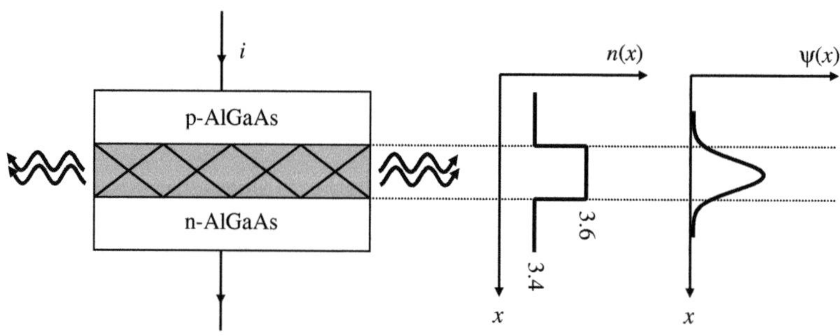

Fig. 13.13 The double heterostructure leads to optical confinement due to the formation of an optical waveguide. The lower bandgap material has a higher refractive index compared to the higher bandgap material. $n(x)$ represents the refractive index profile and $\psi(x)$ represents the electric field profile of the propagating mode

There are two types of lateral confinement; one is gain guidance and the other is index guidance. In the case of gain-guided diode lasers, the current injected into the device is restricted over a narrow stripe (see Fig. 13.14a). This can be accomplished by coating the uppermost semiconductor layer with an insulator such as silicon dioxide leaving a narrow opening for current injection. Due to this kind of injection, the carrier density is largest just under the opening and decreases as we move away from it in a direction parallel to the junction. Because of this, the gain is also a function of lateral position and this gain variation leads to confinement of optical energy in the lateral direction as well. Such lasers are hence referred to as gain-guided lasers. Since the gain distribution changes with change in current, the transverse mode profile of the laser is not very stable with changing current.

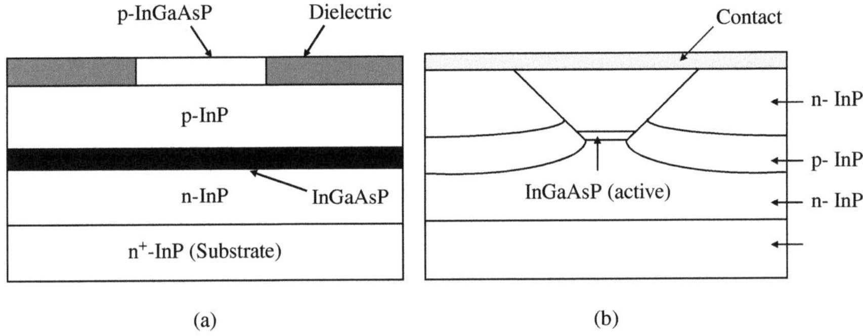

Fig. 13.14 (a) Gain-guided and (b) buried heterostructure index-guided laser structures

In the index-guided lasers, a real refractive index step is provided in the lateral direction as well. Figure 13.14b shows a typical buried heterostructure laser in which strong lateral confinement is provided by having a lower index region surrounding the gain region. In such BH lasers, the active region has typical dimensions of 0.1 μm $\times$ 1 μm with typical refractive index steps of 0.2–0.3. Because of the strong optical confinement provided by the refractive index steps, the output is a single transverse mode and is very stable with changes in current. Most fiber-optic communication systems employ BH lasers.

The combined effects of carrier confinement, optical confinement, and lower absorption losses lead to low threshold current in the range of tens of milliamperes and high overall efficiency of the laser.

13.5 Quantum Well Lasers

Quantum well semiconductor lasers consist of a very thin (about 10 nm or so) active semiconductor layer such as GaAs sandwiched between layers of a higher bandgap semiconductor such as AlGaAs. The active region may be made of only one quantum well resulting in what is referred to as a single quantum well laser or consist of multiple quantum wells separated by enough distance to avoid mutual interaction and electron tunneling, resulting in multiple quantum well lasers. The low bandgap semiconductor surrounded by higher bandgap semiconductor leads to the formation of a potential well that can trap electrons and holes. Also since the dimensions of the well are comparable to the de Broglie wavelength of electrons, quantum size effects become very important and this has a strong influence on the allowed energy values of electrons and holes in the semiconductor.

In order to calculate the density of states in a quantum well, we consider a GaAs quantum well of thickness L_z (typically 10 nm) in the z-direction surrounded by AlGaAs which has a higher bandgap (see Fig. 13.15). In order to estimate the allowed energy values of electrons in the conduction band and the valence band, we will approximate the potential well as an infinite potential well. In view of the assumed infinite potential barriers at $z = 0$ and L_z, the electron wave function should become zero at $z = 0$ and L_z. Thus the z-dependence of the electron wave function would be of the form (see Chapter 3)

$$\sin\left(\frac{p\pi z}{L_z}\right), \qquad p = 1, 2, 3, \ldots.$$

Since the width of the semiconductor along the x- and y-directions assumed to be L_x and L_y is much greater than the electron de Broglie wavelength, we can write the complete electron wave function as

$$\psi(\mathbf{r}) = u_{\mathbf{k}_\perp}(\mathbf{r}_\perp)e^{i\mathbf{k}_\perp \cdot \mathbf{r}_\perp}\sin\left(\frac{p\pi z}{L_z}\right) \tag{13.63}$$

where $\mathbf{k}_\perp = \hat{\mathbf{x}}k_x + \hat{\mathbf{y}}k_y$ is the transverse component of the $\mathbf{k}$ of the electron, $\mathbf{r}_\perp = \hat{\mathbf{x}}x + \hat{\mathbf{y}}y$ is the transverse position vector, and $u_{\mathbf{k}_\perp}(\mathbf{r}_\perp)$ is the Bloch wave func-

Fig. 13.15 The structure of a quantum well laser in which a thin (10 nm) GaAs layer is sandwiched between two higher bandgap materials. The width of the well is comparable to the de Broglie wavelength of the electron and quantum effects become very important

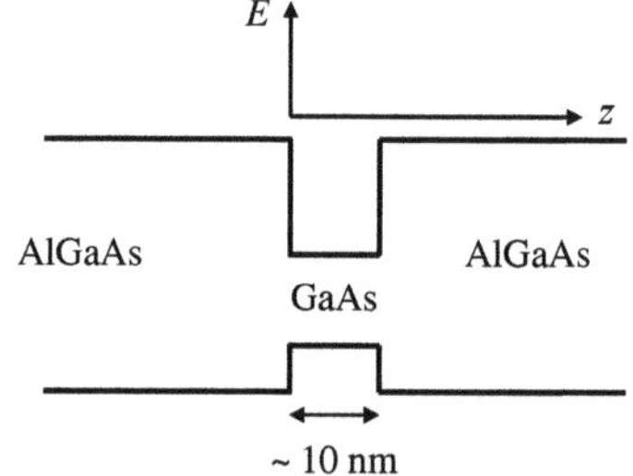

tion having the periodicity of the lattice in the x–y plane. Using periodic boundary conditions along x- and y-directions, as before, gives us

$$k_x = \frac{2\pi m}{L_x}, \qquad m = \pm 1, \pm 2, \pm 3, \ldots$$

$$k_y = \frac{2\pi n}{L_y}, \qquad n = \pm 1, \pm 2, \pm 3, \ldots . \tag{13.64}$$

In the earlier discussion on bulk semiconductors, in the parabolic approximation, the energy was taken to be proportional to $k^2 = k_x^2 + k_y^2 + k_z^2$. In the case of quantum well, the existence of boundaries at $z = 0$ and L_z modifies this relationship.

If we consider an infinite potential well of width L_z, then the allowed energy values in the conduction band due to motion along the z-direction are

$$\tilde{E}_1 = E_c + p^2 \frac{\pi^2 \hbar^2}{2m_c L_z^2}, \qquad p = 1, 2, 3, \ldots \tag{13.65}$$

where m_c is the effective mass of the hole in the conduction band and E_c is the energy at the bottom of the conduction band. Thus unlike the case of bulk semiconductor, the lowest allowed energy for the electrons in the conduction band of a quantum well is (corresponding to $p = 1$, $k_x = 0, k_y = 0$)

$$\tilde{E}_1 = E_c + \frac{\pi^2 \hbar^2}{2m_c L_z^2}$$

where the lowest energy value in the conduction band is E_c. Since the electrons are free to move in the plane perpendicular to the quantum well (i.e., along the x–y plane), the contribution to the energy from this motion would be

$$\tilde{E}_2 = \frac{\hbar^2 k_x^2}{2m_c} + \frac{\hbar^2 k_y^2}{2m_c} = \frac{\hbar^2 k_\perp^2}{2m_c} \tag{13.66}$$

Thus the total energy of the electron in the conduction band would be

$$E = E_c + \frac{\hbar^2 k_\perp^2}{2m_c} + p^2 \frac{\pi^2 \hbar^2}{2m_c L_z^2} \tag{13.67}$$

In order to calculate the density of states, i.e., the number of energy states per unit volume lying between energy values E and $E+dE$, let us first consider electrons lying in the lowest energy level corresponding to $p = 1$. Thus in this case

$$E = E_c + \frac{\hbar^2 \left(k_x^2 + k_y^2 \right)}{2m_c} + \frac{\pi^2 \hbar^2}{2m_c L_z^2} \tag{13.68}$$

The different states corresponding to the allowed values of k_x and k_y are given by Eq. (13.64); this can be represented as points in the k_x–k_y space (see Fig. 13.16). The area of each rectangle is $4\pi^2 / L_x L_y$. In the k_x–k_y space, the area occupied between

Fig. 13.16 Due to the quantum well along the z-direction, the allowed values of k_z are quantized. For a given value of k_z (corresponding to $p = 1$) in the figure, k_x and k_y can take on discrete values. Each *filled circle* corresponds to an available value of (k_x, k_y, k_z)

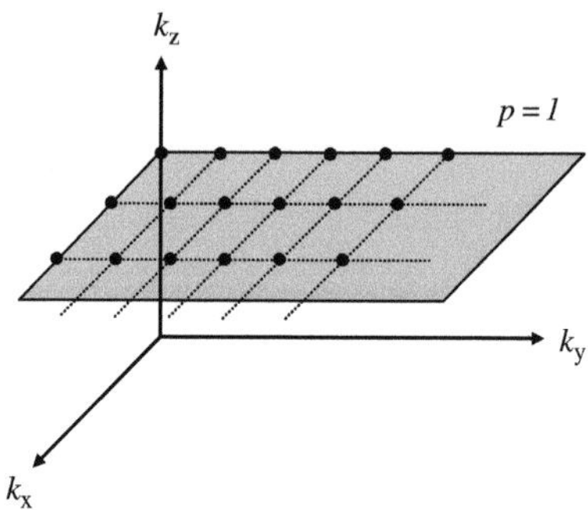

circles of radii $k_\perp = \sqrt{k_x^2 + k_y^2}$ and $k_\perp + dk_\perp = \sqrt{(k_x + dk_x)^2 + (k_y + dk_y)^2}$ is

$$\pi\,(k_\perp + dk_\perp)^2 - \pi k_\perp^2 \approx 2\pi k_\perp\,dk_\perp$$

Thus the number of states within the area would be

$$N(k_\perp)dk_\perp = \frac{2\pi k_\perp\,dk_\perp}{4\pi^2 / L_x L_y} \times 2 = \frac{L_x L_y k_\perp\,dk_\perp}{\pi} \tag{13.69}$$

where the factor 2 accounts for the two possible spin states of the electron. Now using Eq. (13.67) we have

$$dE = \frac{\hbar^2}{2m_c} 2k_\perp\,dk_\perp \tag{13.70}$$

Using this relationship in Eq. (13.69) we obtain

$$N(E)dE = N(k_\perp)dk_\perp = \frac{m_c}{\pi\hbar^2} L_x L_y\,dE \tag{13.71}$$

Hence the number of allowed energy states per unit area between E and $E + dE$ would be

$$p(E)dE = \frac{m_c}{\pi\hbar^2}\,dE \tag{13.72}$$

Since the width of the quantum well is L_z, the density of states, i.e., the number of energy states per unit volume per unit energy interval in the conduction band, would be

$$p_c(E) = \frac{m_c}{\pi\hbar^2 L_z} \tag{13.73}$$

Note that unlike the bulk case, here in the case of quantum well, the density of state is independent of energy. Of course we are considering the states lying in the lowest energy value of the quantum well.

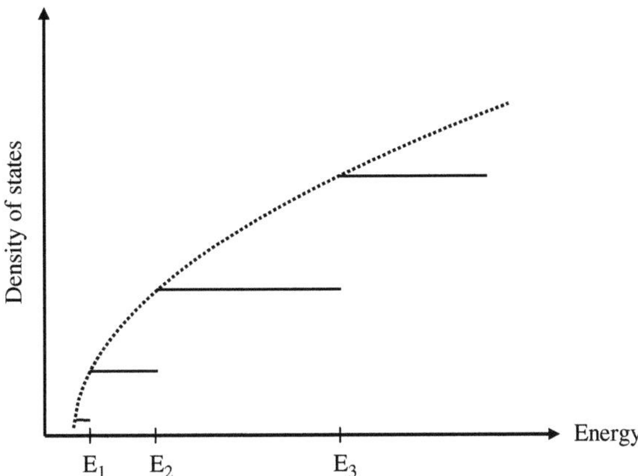

Fig. 13.17 Variation of density of states with energy for the bulk case (*dotted line*) and the quantum well (*solid line*)

As the energy increases, we will reach the second allowed energy level corresponding to $p = 2$ in Eq. (13.67). The state $p = 2$ would have the same energy density as given by Eq. (13.73) and now since the electrons can occupy either the $p = 1$ state or the $p = 2$ state, the density of states for the energy value lying between $p = 1$ and $p = 2$ would be $2m_c/\pi\hbar^2 L_z$. Every time the energy value increases to accommodate the next higher value of p, the density of states would increase by $m_c/\pi\hbar^2 L_z$. This is shown in Fig. 13.17 along with a comparison with the density of states for a bulk semiconductor. The density of states function has a jump every time the energy equals a discrete energy level of the quantum well.

In an identical fashion we can obtain the density of states in the valence band for the energy lying between the $p = 1$ and $p = 2$ quantum levels in the valence band; this would be given as

$$p_v(E) = \frac{m_v}{\pi\hbar^2 L_z} \tag{13.74}$$

where m_v is the effective mass of the hole in the valence band. Just like in the conduction band, every time the energy value crosses an energy level corresponding to the quantum well in the valence band, the density of states would increase by $m_v/\pi\hbar^2 L_z$. Note that in the case of valence band, the energy of the electrons is given as

$$E = E_v - \frac{\hbar^2\left(k_x^2 + k_y^2\right)}{2m_v} - p^2\frac{\pi^2\hbar^2}{2m_v L_z^2}, \qquad p = 1, 2, 3, \ldots \tag{13.75}$$

13.5.1 Joint Density of States

Just like for the bulk case, we can define a joint density of states between the conduction band and the valence band. Let E_1 and E_2 represent the energy of the electron in the valence band and the conduction band, respectively, that interacts with a photon of energy $h\nu$. Now

$$E_1 = E_{\mathrm{v}} - \frac{\hbar^2 k_\perp^2}{2m_{\mathrm{v}}} - \frac{\pi^2 \hbar^2}{2m_{\mathrm{v}} L_z^2} \qquad (13.76)$$

$$E_2 = E_{\mathrm{c}} + \frac{\hbar^2 k_\perp^2}{2m_{\mathrm{c}}} + \frac{\pi^2 \hbar^2}{2m_{\mathrm{c}} L_z^2} \qquad (13.77)$$

where we have assumed that the energy in the conduction and the valence band lies between the energy values corresponding to $p = 1$ and 2 states of the quantum well. Since the photon energy is $h\nu$, we have

$$h\nu = E_2 - E_1 = E_{\mathrm{g}} + \frac{\hbar^2 k_\perp^2}{2m_{\mathrm{r}}} + \frac{\pi^2 \hbar^2}{2m_{\mathrm{r}} L_z^2} \qquad (13.78)$$

where m_{r} is the reduced effective electron mass defined by Eq. (13.22). Thus

$$k_\perp^2 = \frac{2m_{\mathrm{r}}}{\hbar^2} \left(h\nu - E_{\mathrm{g}} - \frac{\pi^2 \hbar^2}{2m_{\mathrm{r}} L_z^2} \right) \qquad (13.79)$$

Using this value of $k_\perp$ in Eqs. (13.76) and (13.77), we obtain

$$\begin{aligned}
E_2 &= E_{\mathrm{c}} + \frac{m_{\mathrm{r}}}{m_{\mathrm{c}}} \left(h\nu - E_{\mathrm{g}} \right) ; \\
E_1 &= E_{\mathrm{v}} - \frac{m_{\mathrm{r}}}{m_{\mathrm{v}}} \left(h\nu - E_{\mathrm{g}} \right)
\end{aligned} \qquad (13.80)$$

Now, the joint density of states is the density of states capable of interacting with a photon of frequency ν and is given as

$$p(\nu)\mathrm{d}\nu = p(E_2)\mathrm{d}E_2$$

which can be written as

$$p(\nu) = p(E_2)\frac{\mathrm{d}E_2}{\mathrm{d}\nu} \qquad (13.81)$$

If we replace E by E_2 in Eq. (13.73), we would obtain $p(E_2)$. Using Eqs. (13.73), (13.80), and (13.81), we obtain

$$p(\nu) = \frac{4\pi m_{\mathrm{r}}}{h L_z} \qquad (13.82)$$

Note that unlike the bulk case, the joint density of states is constant in the range of energies below the second discrete level of the quantum well.

Proceeding in a manner identical to the bulk case, we obtain the following expression for the gain coefficient:

$$\gamma_v = \frac{c^2 m_r}{2n_0^2 v^2 \tau_r h L_z}\left[f_c(E_2) - f_v(E_1)\right] \tag{13.83}$$

Since the maximum value of the bracketed term is unity, the maximum gain coefficient that is achievable in a quantum well is given as

$$\gamma_0 = \frac{c^2 m_r}{2n_0^2 v^2 \tau_r h L_z} \tag{13.84}$$

For typical values $\lambda_0 = 850$ nm, $m_r = 5.37 \times 10^{-32}$ kg, $n_0 = 3.64$, $\tau_r = 4.6$ ns, and $L_z = 10$ nm, we obtain a gain coefficient of 4.8×10^4 m^{-1}.

We rewrite Eq. (13.57) relating the current density and the electron concentration as

$$J = \frac{\Delta n e d}{\tau_r}$$

The quantum well laser requires similar electron density in the conduction band and holes in the valence band as the heterostructure since the relationships of gain are very similar and the density of states in the quantum well is also very close to that of heterostructures (see Fig. 13.17). At the same time, since the value of d for a quantum well is much smaller than that for a heterostructure, the threshold current density is much smaller in the case of quantum well. If the width of the quantum well is 10 nm and that of the heterostructure is about 100 nm, the current density required for a given gain should be about 10 times smaller in the case of quantum well laser. This is usually not so since although the electrons and holes are confined to within the quantum well width of 10 nm, the optical radiation extends much beyond the quantum well. In fact the well of 10 nm width cannot efficiently guide the optical wave since the corresponding V (normalized frequency – see Chapter 16) number comes out to be about 0.2 assuming an index step of 0.3 and a well width of 10 nm. At this value of V, the optical field will spread deep into the surrounding AlGaAs region, thus reducing significantly the overlap of the optical fields with the electrons and holes. This would result in a drastic reduction of gain.

In order to overcome this, a separate structure for optical confinement (separate confinement heterostructure, SCH) is provided outside the quantum well. In particular graded refractive index SCH (GRIN-SCH), structures have provided some of the lowest threshold current densities. Figure 13.18 shows two examples in which surrounding the quantum well of width about 10 nm, two inner barrier layers of typical thickness of about 0.1 μm each are provided which are then surrounded by thick (about 1 μm) cladding layers. The inner barrier layers are chosen to have a refractive index higher than the cladding layers so that they can provide for optical confinement. For example, the refractive index of GaAs is about 3.6, that of

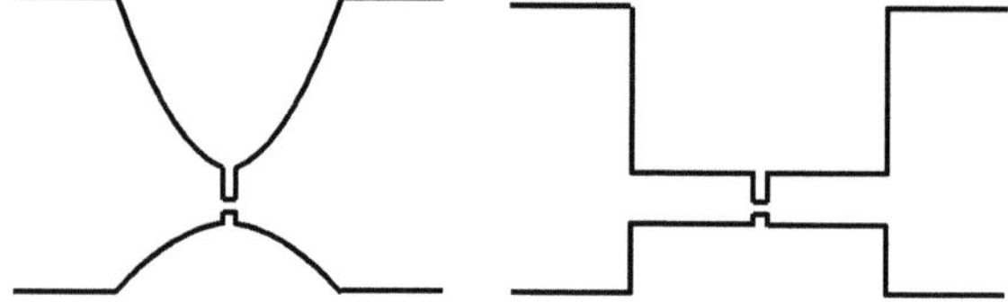

Fig. 13.18 Schematic of energy variation of the conduction band and the valence band edge with position in separate confinement heterostructures

$Al_{0.2}Ga_{0.8}As$ is about 3.48, and that of $Al_{0.6}Ga_{0.4}As$ is 3.23. The optical waveguide formed by the barrier layers has a V value of 1.8 and thus can provide for good optical guidance. Of course even in this case the optical wave almost fills the barrier region of width 0.1 μm and the quantum well occupies only a portion, 10 nm within the barrier region. The overlap between the optical mode and the electrons and holes would determine the reduction in threshold current density for achieving a gain.

Typically the required threshold current for lasing in the case of quantum well lasers is about a factor of 4 lower than that of heterostructure lasers. Thus if a heterostructure laser requires threshold currents of about 20 mA for laser action, a quantum well laser would have a threshold of a few milliamperes.

Figure 13.19 shows a typical current versus output power characteristic of a quantum well laser.

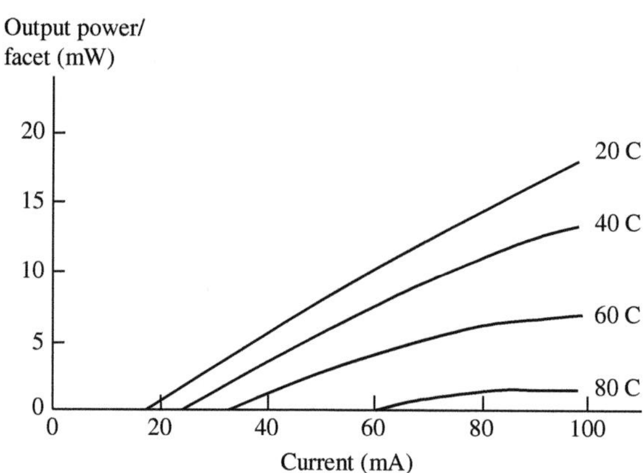

Fig. 13.19 Variation of laser output power with current in the case of a quantum well laser at different temperatures. (Source: Liu et al. (2005))

13.6 Materials

Most semiconductor lasers operate either in the 0.8–0.9 μm or in the 1–1.7 μm wavelength region. Since the wavelength of emission is determined by the bandgap, different semiconductor materials are used for the two different wavelength regions. Lasers operating in the 0.8–0.9 μm spectral region are based on gallium arsenide. By replacing a fraction of gallium atoms by aluminum, the bandgap can be increased. Thus one can form heterojunctions by proper combinations of GaAlAs and GaAs, which can provide both carrier confinement and optical wave guidance. For example, the bandgap of GaAs is 1.424 eV and that of $Ga_{0.7}Al_{0.3}As$ is ≈ 1.798 eV; the corresponding refractive index difference is about 0.19. Thus by surrounding the GaAs layer on either side with $Ga_{0.7}Al_{0.3}As$, one can achieve confinement of both carriers and light waves. For lasers operating in the 1.0–1.7 µm wavelength band, the semiconductor material is InP with gallium and arsenic used to replace fractions of indium and phosphorous, respectively, to give lasers based on InGaAsP. The above wavelength region is extremely important in connection with fiber-optic communication since silica-based optical fibers exhibit both low loss and very high bandwidth around 1.55 µm (see Chapter 16). Recently lasers have been realized in a large bandgap material like GaN ($E_g \sim 3.44$ eV). Due to the large bandgap, such lasers emit in the blue region of the spectrum and have wide applications in high-density data storage, displays, etc.

13.7 Laser Diode Characteristics

Figure 13.20 shows a typical light output versus current characteristic of a GaAs semiconductor laser. As can be seen, the output optical power starts to increase very rapidly around a threshold current, which essentially represents the beginning of laser oscillation. Below the threshold, the emission is primarily by spontaneous transitions and the emission is broadband and incoherent. On crossing the threshold, the emission becomes coherent, the linewidth reduces significantly, and the output power increases rapidly with the current.

An important property specifying a laser diode is the slope efficiency which is the slope of the light output versus current characteristic above the threshold region. Let dI represent the increase in forward current and let dP represent the corresponding

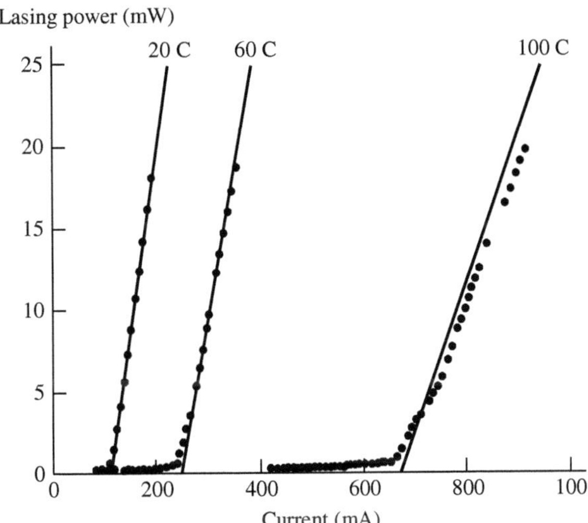

Fig. 13.20 Output power versus current characteristic of a typical laser diode. *Dots* represent experimental data and *lines* represent simulation results. (Source: Li and Piprek (2000), reprinted with permission)

increase in the output laser power. For a current increase of dI, the number of electrons being injected into the laser per unit time is dI/e, where e is the electron charge. Similarly a laser output power increase of dP would correspond to an increase in the output photon flux of d$P/h\nu$. We define the external quantum efficiency of the laser as

$$\eta_{\mathrm{D}} = \frac{\mathrm{d}P/h\nu}{\mathrm{d}I/e} = \frac{e}{h\nu}\frac{\mathrm{d}P}{\mathrm{d}I} \tag{13.85}$$

Typical continuous wave operating laser diodes have an η_{D} lying between 0.25 and 0.6. The quantity dP/dI is referred to as slope efficiency and is specified in milliwatts per milliampere.

For digital modulation of the laser diodes, they are biased at slightly above the threshold and on this bias is superposed current pulses corresponding to the digital data. Thus the electrical signal can be directly encoded into an optical signal. Biasing the laser diode near the threshold helps in turning on and off the laser at high speeds. For analog modulation the laser is usually biased above threshold and the analog signal is fed in the form of current variations. Lasers with modulation bandwidths greater than 6 GHz are commercially available.

An important characteristic of a laser diode is the spectral width of emission. In the case of laser diodes the gain bandwidth is usually quite large and thus the laser can oscillate in multiple longitudinal modes giving a multi-mode output (see Fig. 13.21).

In a multi-longitudinal mode laser, the output spectrum consists of a series of wavelengths. The oscillating wavelengths are determined by the cavity resonances given as

$$\nu = q\frac{c}{2n(\nu)L}, \qquad q = 1, 2, 3, \ldots \tag{13.86}$$

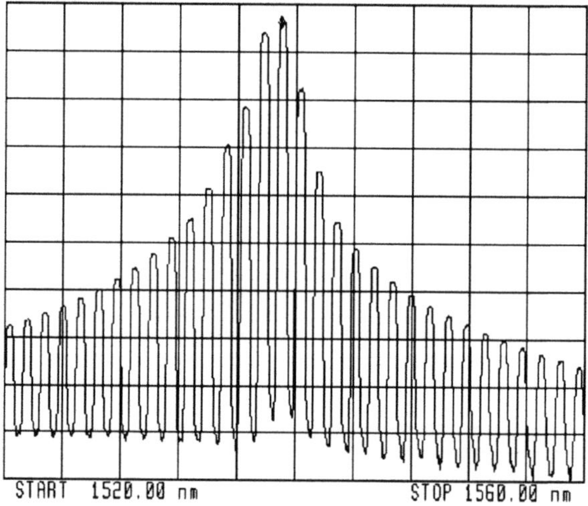

Fig. 13.21 Output spectrum of a typical semiconductor laser. The cavity is made up of two end mirrors and such lasers are referred to as Fabry–Perot lasers

where $n(\nu)$ is the refractive index of the semiconductor material at a frequency ν. If ν and $\nu + \Delta\nu$ correspond to adjacent oscillating frequencies, then we can write

$$\nu = q\frac{c}{2n(\nu)L}$$
$$\nu + \Delta\nu = (q+1)\frac{c}{2n(\nu+\Delta\nu)L} \tag{13.87}$$

For $\Delta\nu \ll \nu$, we can make a Taylor series expansion of $n(\nu+\Delta\nu)$ and obtain the following expression for the intermodal spacing:

$$\Delta\nu = \frac{c}{2n(\nu)L}\left(1 + \frac{\nu}{n}\frac{dn}{d\nu}\right)^{-1} \tag{13.88}$$

Typically $n = 3.6$, $L = 250\ \mu\text{m}$, $\frac{\nu}{n}\frac{dn}{d\nu} \approx 0.38$, and the intermode spacing comes out to be about 125 GHz. At a wavelength of 850 nm, this corresponds to a wavelength spacing of 0.3 nm. If the gain bandwidth is 3 nm, then there would be about 10 different frequencies at the output of the laser.

In some applications (for example, in fiber-optic communications) one would like to have single longitudinal mode oscillation of the laser so that its spectral width is $\Delta\lambda \ll 0.1$ nm.

One can achieve this either by using a cleaved coupled cavity configuration or by using distributed feedback. In the former case, the laser device essentially consists of two independent cavities which are optically coupled. The mode which can oscillate is the one which is a mode of either of the cavities and also has the lowest loss.

The second concept uses a periodic variation of the thickness of the layer surrounding the active region of the laser. Periodic variation of the thickness of the layer results in a feedback which is very strongly wavelength dependent. Such a periodic variation essentially acts as a Bragg reflector which is highly wavelength selective. Thus it is possible to ensure that only one of the longitudinal modes has a feedback and can oscillate. The periodic variation can be provided at the two ends of the active region and this results in what is known as distributed Bragg reflector lasers (see Fig. 13.22). If the feedback is provided throughout the cavity, then this is referred to as distributed feedback laser (see Fig. 13.23). The end facets are usually coated with anti-reflection films to avoid any formation of cavity.

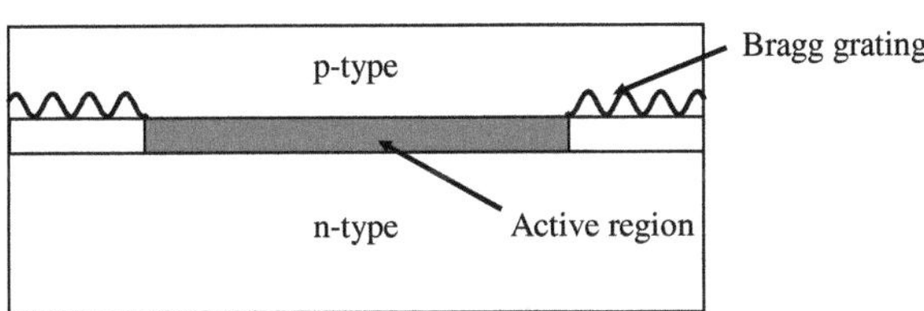

Fig. 13.22 Distributed Bragg reflector (DBR) laser

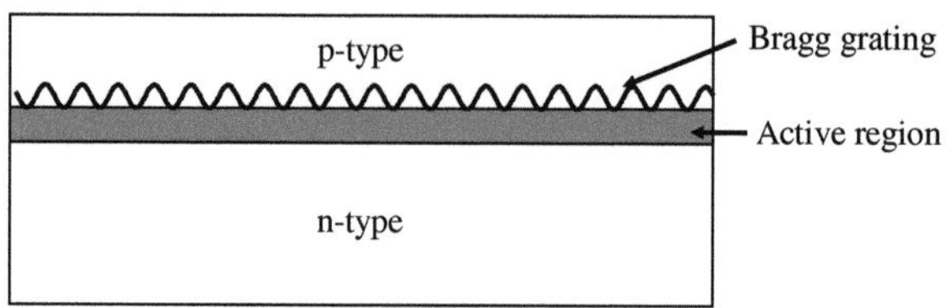

Fig. 13.23 Distributed feedback (DFB) laser

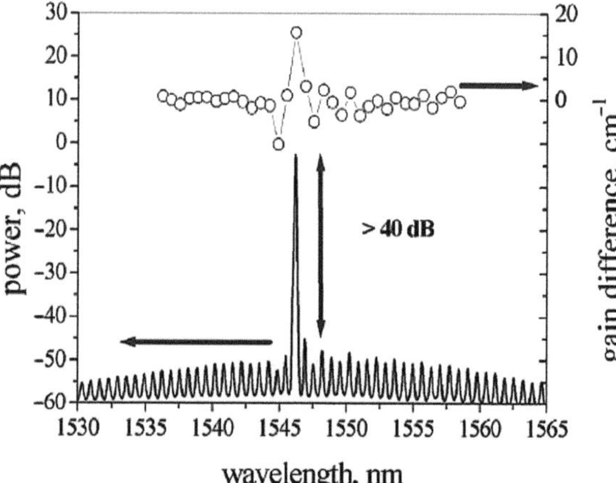

Fig. 13.24 A typical spectrum of a DFB laser. Note the very sharp spectrum. (Source: Eblana Photonics White Paper "Discrete Mode Laser Diodes with Ultra Narrow Linewidth Emission"; http://www.eblanaphotonics.com/)

If n_{eff} represents the effective index of the propagating mode in the waveguide forming the laser, then for efficient Bragg reflection we must have (see, e.g., Ghatak and Thyagarajan, 1998)

$$2\frac{2\pi}{\lambda_{\mathrm{B}}}n_{\mathrm{eff}} = \frac{2\pi}{\Lambda} \tag{13.89}$$

where Λ is the period of the grating and λ_{B} satisfying the above equation is known as the Bragg wavelength. If the periodic modulation with the period given by Eq. (13.89) is provided, then the reflectivity is strongest for the wavelength λ_{B} and the periodic structure acts like a highly wavelength-selective mirror. For a wavelength of 1550 nm, assuming $n_{\mathrm{eff}} = 3.5$, the required period comes out to be 221 nm. Gratings with short period are usually fabricated using holographic techniques.

For wavelengths away from λ_{B} the reflectivity drops very sharply and thus such gratings act as highly wavelength-selective mirrors.

Figure 13.24 shows a typical spectrum from a DFB laser operating at the telecommunication wavelength. Such lasers can provide extremely narrow emission lines and are the preferred lasers in fiber-optic communications.

13.8 Vertical Cavity Surface-Emitting Lasers (VCSELs)

The laser structures that we have discussed are all edge-emitting lasers wherein the laser beam comes out of the edge of the substrate. There is a new class of lasers

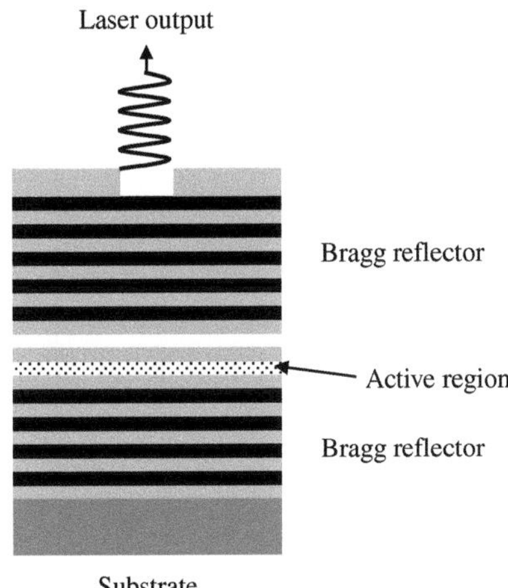

Fig. 13.25 A schematic of VCSEL structure

called vertical cavity surface-emitting lasers (VCSELs) pronounced as "vicsels" that emit light vertically from the surface. The laser cavity is vertical instead of being horizontal as in conventional laser diodes. Figure 13.25 shows a typical structure of a VCSEL wherein the active region is sandwiched between two multi-layer mirrors which could contain as many as many as 120 mirror layers. The multi-layer mirrors act like a Bragg grating and reflect only a narrow range of wavelengths back into the cavity causing emission at a single wavelength. Compared to conventional edge-emitting laser diodes it is possible to realize an array or VCSELs on a surface and they are also very small in size and have threshold currents below 1 mA.

Since the length of the laser cavity is very small (a few tens of nanometers), the inter mode spacing is very large and thus single-mode oscillation is easily possible. At the same time the gain per round-trip is also much reduced leading to increased threshold current densities. With very high reflectivity (>99.8 %) dielectric mirrors on both sides of the cavity, very high reflection into the cavity can be achieved resulting in very small losses in the cavity, thus reducing the threshold current significantly. Figure 13.26 shows a photograph of the multi-layer structure of a VCSEL and Fig. 13.27 shows a typical output power versus current characteristic.

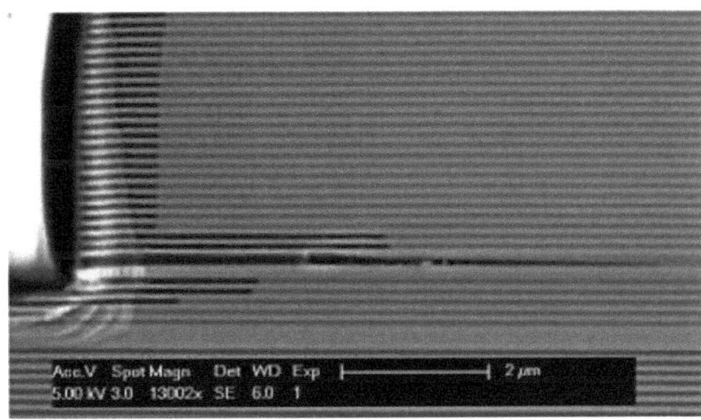

Fig. 13.26 Photograph showing the multi-layer structure of a VCSEL

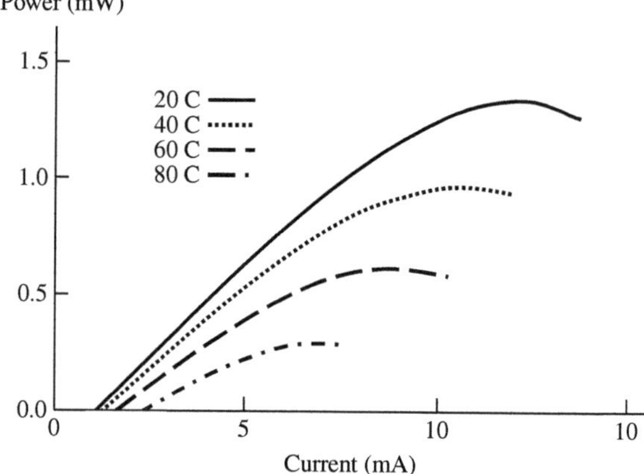

Fig. 13.27 Typical laser power versus current characteristic of a VCSEL. Note the very low threshold current. (Source: Ref. http://www.vertilas.com/pdf/Telecom_rev20.pdf)

Problems

Problem 13.1 Consider absorption of photons by GaAs whose bandgap energy is 1.424 eV. The effective electron masses in the conduction band and the valence band are $0.067m_0$ and $0.46m_0$, respectively, where m_0 is the electron rest mass. If the incident photon wavelength is 850 nm, obtain the energy of the electron and hole generated by the absorption process.

Problem 13.2 Consider GaAs and assume that electrons are excited from the valence band to the conduction band to produce an electron concentration of $1.6 \times 10^{24} \text{m}^{-3}$ in the conduction band. Obtain E_{Fc} and E_{Fv} and calculate the wavelength range over which gain can be achieved.

Problem 13.3 Consider GaAs which has a bandgap of 1.424 eV. If photons of energy 1.46 eV are incident, obtain the energy values of the electrons in the conduction band and the valence band with which the photons will interact. Also obtain the corresponding k values.

Problem 13.4 Assuming a carrier density of $2 \times 10^{24} \text{m}^{-3}$, in the conduction band in GaAs, obtain the gain coefficient of an optical wave at an energy of 1.46 eV.

Problem 13.5 Show that the joint density of states for a bulk semiconductor given by Eq. (13.28) and that for a quantum well given by Eq. (13.82) become equal at the quantized energy values corresponding to electrons and holes in the lowest level.

Problem 13.6 Consider the interaction of photons at a wavelength of 800 nm interacting with GaAs. Obtain the energy values of electrons in the conduction band and the valence band taking part in the interaction process.

Chapter 14
Optical Parametric Oscillators

14.1 Introduction

The lasers that we have discussed until now are based on amplification brought about by stimulated emission. In this scheme, population inversion is achieved between two energy levels of an atomic system and this inversion is used for amplification of light. In contrast an optical parametric oscillator (OPO) is a coherent source of light like a laser but uses the process of optical amplification brought about by the phenomenon of non-linear interaction in a crystal. Since no energy levels are involved in the amplification process it is possible to tune these lasers over a very broad range of wavelengths. In OPOs the pump is another laser which is used to pump a non-linear crystal within a resonant cavity and the non-linear interaction in the crystal leads to the conversion of the pump laser into two waves (called signal and idler) at new wavelengths. For energy conservation the sum of the frequencies of signal and idler must equal the frequency of the pump. Thus the signal and idler have wavelengths that are larger than the wavelength of the pump. The main attraction in an OPO is the possibility of achieving a tunable output. Thus starting from a laser which emits in the visible spectrum, using an OPO, it is possible to generate a tunable output in the infrared region of the spectrum (see Fig. 14.1). OPOs offer a very wide tuning range and are a primary laser source in many applications such as spectroscopy and optical amplification requiring tunability of the laser. The first successful operation of an OPO was demonstrated by Giordmaine and Miller in 1965 and ever since optical parametric processes have generated a great deal of interest and have become a powerful technique to obtain coherent sources of light with wide tunability from UV to mid-IR.

In this chapter we will discuss the process of amplification and oscillation using the non-linear characteristic of a crystal.

14.2 Optical Non-linearity

Usually the optical property of a medium is independent of the electric field amplitude of the propagating light wave. However, at large optical power densities, matter

K. Thyagarajan, A. Ghatak, *Lasers*, Graduate Texts in Physics,
DOI 10.1007/978-1-4419-6442-7_14, © Springer Science+Business Media, LLC 2010

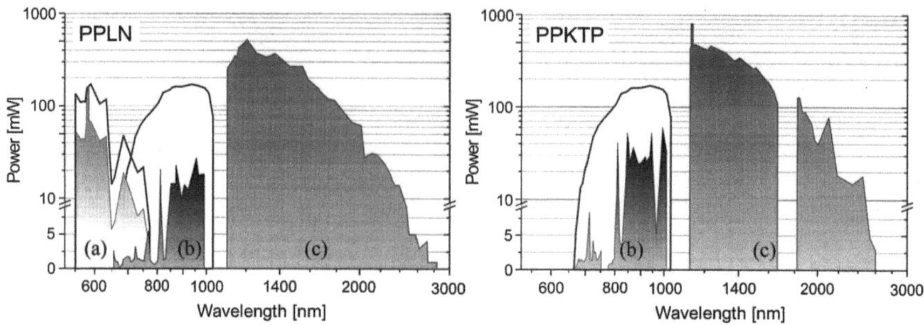

Fig. 14.1 Ultrawide tunability offered by optical parametric oscillators. (**a**) Frequency-doubled idler, (**b**) signal, and (**c**) idler power versus wavelength. *Left*, periodically poled lithium niobate (PPLN) with and without etalon; *right*, periodically poled potassium titanyl phosphate (PPKTP) without etalon. *Shaded* areas correspond to the measured powers and the curves to conservative estimates of power expected. (After Strößner et al. (2002); © 2002 OSA)

behaves in a non-linear fashion and we come across new optical phenomena such as second harmonic generation (SHG), sum and difference frequency generation, intensity-dependent refractive index and mixing of various frequencies. In SHG, an incident light beam at frequency ω interacts with the medium and generates a new light wave at frequency 2ω. Thus a red light beam (~ 800 nm) entering the crystal can get converted into a blue beam (~ 400 nm) as it comes out of the crystal (see Fig. 18.4). In sum and difference frequency generation, two incident beams at frequencies ω_1 and ω_2 mix with each other producing sum ($\omega_1 + \omega_2$) and difference ($\omega_1 - \omega_2$) frequencies at the output. In parametric fluorescence an input light wave at a frequency ω_p splits spontaneously into light waves at new frequencies ω_s and ω_i satisfying the energy conservation condition: $\omega_p = \omega_s + \omega_i$. We will now discuss the non-linear effect that leads to this phenomenon and detail how this process can be used for optical amplification and to realize a coherent source of tunable radiation.

Non-linear Polarization

In a linear medium, the electric polarization P is assumed to be a linear function of the electric field $\mathcal{E}$:

$$\mathcal{P} = \varepsilon_0 \chi \, \mathcal{E} \tag{14.1}$$

where for simplicity a scalar relation has been written. The quantity χ is termed as linear dielectric susceptibility. At high optical intensities (which corresponds to high electric fields), all media behave in a non-linear fashion. Thus Eq. (14.1) gets modified to[1]

---

[1] In actual practice electric field and polarization are vector quantities and Eq. (14.2) is a simplified scalar representation. The ith component of the polarization is given by

$$P_i = \varepsilon_0 \chi_{ij} E_j + \varepsilon_0 \chi_{ijk}^{(2)} E_j E_k + \varepsilon_0 \chi_{ijkl}^{(3)} E_j E_k E_l + \ldots$$

where χ_{ij}, $\chi_{ijk}^{(2)}$, $\chi_{ijkl}^{(3)}$, etc. are tensors and repeated indices on the right-hand side are summed over 1 to 3. For a given non-linear medium and given components of the electric fields of the interacting waves, the component equation can be used to obtain Eq. (14.2) where $\chi^{(2)}$ will be an effective second-order susceptibility.

$$\mathcal{P} = \varepsilon_0(\chi\,\mathcal{E} + \chi^{(2)}\,\mathcal{E}^2 + \chi^{(3)}\,\mathcal{E}^3 + \ldots) \tag{14.2}$$

where $\chi^{(2)}$, $\chi^{(3)}$, ... are higher order susceptibilities giving rise to the non-linear terms. The second term on the right-hand side is responsible for second harmonic generation (SHG), sum and difference frequency generation, parametric interactions, etc., while the third term is responsible for third harmonic generation, intensity-dependent refractive index, self-phase modulation, four wave mixing, etc. For media possessing inversion symmetry, $\chi^{(2)}$ is zero and there is no second-order non-linear effect. Thus silica optical fibers, which form the heart of today's communication networks, do not posses the second-order non-linearity. Second-order non-linearity is exhibited by crystals which do not possess a center of inversion symmetry like lithium niobate ($LiNbO_3$), lithium tantalite ($LiTaO_3$), potassium titanyl phosphate ($KTiPO_4$), and potassium dihydrogen phosphate (KH_2PO_4). The parametric oscillator uses the second-order non-linearity for optical amplification and oscillation.

Since the parametric process mixes waves at frequencies ω_p, ω_S, and ω_i, we first write for the electric field distributions of the waves at these frequencies as

$$\mathcal{E}_p = \frac{1}{2}\left(E_p e^{i(\omega_p t - k_p z)} + c.c\right) \tag{14.3}$$

$$\mathcal{E}_s = \frac{1}{2}\left(E_s e^{i(\omega_s t - k_s z)} + c.c\right) \tag{14.4}$$

$$\mathcal{E}_i = \frac{1}{2}\left(E_i e^{i(\omega_i t - k_i z)} + c.c\right) \tag{14.5}$$

where each of the waves is assumed to be a plane wave; E_p, E_s, and E_i are the complex electric field amplitudes of the waves; and k_p, k_s, and k_i represent the propagation constants of the waves at the pump, signal, and idler frequencies, respectively. In the above expressions c.c. implies complex conjugate of the earlier term. Since we are dealing with non-linear effects it is important to consider real electric fields and hence the complex conjugate term in each of the expressions.

At any point in the medium the existence of waves at these frequencies implies that the total electric field at any point will be given by

$$\mathcal{E} = \mathcal{E}_p + \mathcal{E}_s + \mathcal{E}_i \tag{14.6}$$

This electric field will generate a non-linear polarization given by

$$\mathcal{P}_{nl} = 2\varepsilon_0 d\mathcal{E}^2 = 2\varepsilon_0 d(\mathcal{E}_p + \mathcal{E}_s + \mathcal{E}_i)^2 \tag{14.7}$$

where we have replaced $\chi^{(2)}$ by $2d$. We now substitute the expressions for the electric fields at the pump, signal, and idler frequencies in Eq. (14.7), collect the non-linear polarization terms that have frequencies of ω_p, ω_s and ω_i and obtain

$$\mathcal{P}_{nl}^{(\omega_p)} = \frac{1}{2}\left(P_{nl}^{(\omega_p)} e^{i(\omega_p t - (k_s + k_i)z)} + c.c.\right) \tag{14.8}$$

$$\mathcal{P}_{nl}^{(\omega_s)} = \frac{1}{2}\left(P_{nl}^{(\omega_s)} e^{i(\omega_s t - (k_p - k_i)z)} + c.c.\right) \tag{14.9}$$

$$\mathcal{P}_{nl}^{(\omega_i)} = \frac{1}{2}\left(P_{nl}^{(\omega_i)} e^{i(\omega_i t - (k_p - k_s)z)} + c.c.\right) \tag{14.10}$$

where

$$P_{nl}^{(\omega_p)} = 2\varepsilon_0 d E_s E_i \tag{14.11}$$

$$P_{nl}^{(\omega_s)} = 2\varepsilon_0 d E_p E_i^* \tag{14.12}$$

$$P_{nl}^{(\omega_i)} = 2\varepsilon_0 d E_p E_s^* \tag{14.13}$$

It is the non-linear polarization that is responsible for the conversion of power from one frequency to the other.

In order to study how the non-linear polarization affects the generation and propagation of new frequencies we need to first derive the wave equation describing the propagation of the waves. We first recall Maxwell's equations from Chapter 2

$$\begin{aligned}
\nabla \cdot \mathcal{D} &= 0 \\
\nabla \cdot \mathcal{B} &= 0 \\
\nabla \times \mathcal{E} &= -\frac{\partial \mathcal{B}}{\partial t} \\
\nabla \times \mathcal{H} &= \frac{\partial \mathcal{D}}{\partial t}
\end{aligned} \tag{14.14}$$

where we have assumed absence of any free charges and free currents, and $\mathcal{E}, \mathcal{D}, \mathcal{B}$ and $\mathcal{H}$ represent, respectively, the electric field, electric displacement vector, magnetic field, and magnetic H vector. The displacement vector and the electric field vector are related through the following equation:

$$\mathcal{D} = \varepsilon_0 \mathcal{E} + \mathcal{P} = \varepsilon_0 \mathcal{E} + \mathcal{P}_1 + \mathcal{P}_{nl} = \varepsilon \mathcal{E} + \mathcal{P}_{nl} \tag{14.15}$$

where the polarization has been split into a linear part $\vec{P}_1$ and a non-linear part $\vec{P}_{nl}$ and ε represents the linear dielectric permittivity of the medium.

Taking curl of the third equation in Eq. (14.14) we get

$$\nabla \times \nabla \times \mathcal{E} = -\frac{\partial \nabla \times \mathcal{B}}{\partial t} = -\mu_0 \frac{\partial^2 \mathcal{D}}{\partial t^2}$$

or

$$\nabla(\nabla \cdot \mathcal{E}) - \nabla^2 \mathcal{E} = -\mu_0 \varepsilon \frac{\partial^2 \mathcal{E}}{\partial t^2} - \mu_0 \frac{\partial^2 \mathcal{P}_{nl}}{\partial t^2} \tag{14.16}$$

Assuming $\mathcal{E}$ is transverse to the propagation direction we can put $\nabla.\mathcal{E} = 0$ and thus Eq. (14.16) becomes

$$\nabla^2 \mathcal{E} - \mu_0\,\varepsilon\frac{\partial^2\,\mathcal{E}}{\partial t^2} = \mu_0\frac{\partial^2\,\mathcal{P}_{nl}}{\partial t^2} \tag{14.17}$$

which is the wave equation describing the propagation of the electromagnetic wave in the presence of the non-linear polarization.

In order to simplify the analysis we will assume that the directions of the electric field and the non-linear polarization are the same and consider the following scalar wave equation instead of the vector wave equation described by Eq. (14.17):

$$\nabla^2 \mathcal{E} - \mu_0\,\varepsilon\frac{\partial^2\mathcal{E}}{\partial t^2} = \mu_0\frac{\partial^2\mathcal{P}_{nl}}{\partial t^2} \tag{14.18}$$

The above equation is the wave equation describing the propagation of each of the frequencies through the non-linear medium. Thus the propagation of the wave at frequency ω_p is described by the following equation:

$$\nabla^2 \mathcal{E}_p - \mu_0\,\varepsilon(\omega_p)\frac{\partial^2\mathcal{E}_p}{\partial t^2} = \mu_0\frac{\partial^2\mathcal{P}_{nl}^{(\omega_p)}}{\partial t^2} \tag{14.19}$$

where $\varepsilon(\omega_p)$ is the dielectric permittivity of the medium at the pump frequency ω_p. Similarly the electric fields corresponding to the signal at frequency ω_s and idler at frequency ω_i will satisfy the corresponding wave equations with $\mathcal{E}_p$ replaced by $\mathcal{E}_s$ and $\mathcal{E}_i$, $\varepsilon(\omega_p)$ replaced by $\varepsilon(\omega_s)$ and $\varepsilon(\omega_i)$ and $\mathcal{P}_{nl}(\omega_p)$ replaced by $\mathcal{P}_{nl}(\omega_s)$ and $\mathcal{P}_{nl}(\omega_i)$ respectively.

In the presence of non-linearity, the electric field amplitudes E_p, E_s, and E_i will become functions of z, the propagation direction of the waves. Also since we are considering plane waves, the amplitudes are independent of the transverse coordinates x and y. Substituting for E_p and $P_{nl}(\omega_p)$ from Eqs. (14.3) and (14.8), we obtain

$$\frac{\partial^2}{\partial z^2}\left(E_p e^{i(\omega_p t - k_p z)} + c.c\right) - \mu_0\varepsilon(\omega_p)\frac{\partial^2\left(E_p e^{i(\omega_p t - k_p z)} + c.c\right)}{\partial t^2}$$
$$= \mu_0\frac{\partial^2\left(P_{nl}^{(\omega_p)} e^{i(\omega_p t - (k_s + k_i)z)} + c.c.\right)}{\partial t^2}$$

Opening up the differentials and equating the coefficients of $e^{i\omega_p t}$ on both sides we obtain

$$\left(\frac{d^2 E_p}{dz^2} - 2ik_p\frac{dE_p}{dz} - k_p^2 E_p\right) + \mu_0\omega_p^2\varepsilon(\omega_p)E_p = -\mu_0\omega_p^2 2\varepsilon_0 dE_s E_i e^{i[k_p - (k_s + k_i)]z}$$

$$\tag{14.20}$$

Assuming that the fractional change in the electric field amplitudes of the waves is negligible in distances of the order of wavelengths, we can neglect the second differential in Eq. (14.20). Also the propagation constant and the frequency are related through the following equation:

$$k_p^2 = \mu_0 \omega_p^2 \varepsilon(\omega_p) \tag{14.21}$$

Thus Eq. (14.20) becomes

$$\frac{dE_p}{dz} = -i\kappa_p E_s E_i e^{i\Delta kz} \tag{14.22}$$

where

$$\kappa_p = \frac{\omega_p d}{c n_p} \tag{14.23}$$

where we have used the fact that

$$k_p = \frac{\omega_p}{c} n_p \tag{14.24}$$

where n_p is the refractive index of the medium at the frequency ω_p and

$$\Delta k = k_p - (k_s + k_i) \tag{14.25}$$

is referred to as the phase mismatch.

Equation (14.22) describes the change in the amplitude of the electric field of the pump due to non-linear effects as it propagates through the medium. In a similar fashion we can obtain the equations describing the change of the electric field amplitudes of the signal and idler waves which are given by

$$\frac{dE_s}{dz} = -i\kappa_s E_p E_i^* e^{-i\Delta kz} \tag{14.26}$$

$$\frac{dE_i}{dz} = -i\kappa_i E_p E_s^* e^{-i\Delta kz} \tag{14.27}$$

where

$$\kappa_s = \frac{\omega_s d}{c n_s} \tag{14.28}$$

$$\kappa_i = \frac{\omega_i d}{c n_i} \tag{14.29}$$

with n_s and n_i representing the refractive indices of the medium at the signal frequency ω_s and idler frequency ω, respectively.

Equations (14.22), (14.26), and (14.27) represent the three coupled equations describing the evolution of pump, signal, and idler wave amplitudes as they propagate through the non-linear medium.

14.3 Parametric Amplification

In this section we will show that it is possible to amplify the signal wave through the non-linear interaction with the pump wave. In order to show this we consider the incidence of a strong pump wave at frequency ω_p and a weak signal wave at frequency ω_s on a non-linear medium (see Fig. 14.2). We will assume the pump to be strong and neglect the changes in the pump amplitude due to the non-linear interaction. Thus we assume the electric field $E_p(z) = E_p(0)$ at the pump frequency to be a constant and independent of z. This is also referred to as the no pump depletion approximation. Since we are assuming E_p to be a constant, we need to solve only the equations for the signal and idler [Eqs. (14.26) and (14.27)]. Using the fact that E_p is a constant we differentiate Eq. (14.26) with respect to z and get

$$\frac{d^2 E_s}{dz^2} = -i\kappa_s E_p \left(\frac{dE_i^*}{dz} - i\Delta k \, E_i^* \right) e^{-i\Delta kz}$$

which using Eq. (14.27) becomes

$$\frac{d^2 E_s}{dz^2} = -i\kappa_s E_p \left(i\kappa_i E_p^* E_s e^{i\Delta kz} - i\Delta k \left(\frac{1}{-i\kappa_s E_p} e^{i\Delta kz} \frac{dE_s}{dz} \right) \right) e^{-i\Delta kz}$$

$$= -i\Delta k \frac{dE_s}{dz} + \kappa_s \kappa_i \left| E_p \right|^2 E_s$$

or

$$\frac{d^2 E_s}{dz^2} + i\Delta k \frac{dE_s}{dz} - \kappa_s \kappa_i \left| E_p \right|^2 E_s = 0 \tag{14.30}$$

The above equation can be easily solved to obtain the following solution

$$E_s(z) = \left(Ae^{\Gamma z} + Be^{-\Gamma z} \right) e^{-i\Delta kz/2} \tag{14.31}$$

where

$$\Gamma = \left(g^2 - \frac{(\Delta k)^2}{4} \right)^{1/2} \tag{14.32}$$

$$g^2 = \kappa_s \kappa_i \left| E_p \right|^2 \tag{14.33}$$

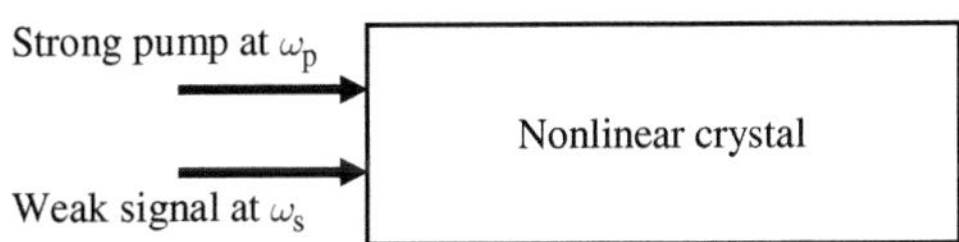

Fig. 14.2 In a parametric process, a pump at the frequency wp interacts with a signal at a frequency ws in a non-linear crystal and amplifies the signal wave if certain conditions are satisfied. This amplification process can be used to achieve a source of coherent radiation

Substituting the solution for $E_s(z)$ in Eq. (14.26) we obtain the solution for the variation of E_i with z:

$$E_i(z) = \frac{i}{\kappa_s E_p} \left(\left\{ \Gamma - i\frac{\Delta k}{2} \right\} A e^{\Gamma z} - \left\{ \Gamma + i\frac{\Delta k}{2} \right\} B e^{-\Gamma z} \right) e^{-i\Delta kz/2} \qquad (14.34)$$

Equations (14.31) and (14.34) describe the evolution of the signal and idler waves as they propagate through the non-linear medium. The constants A and B are to be determined from the initial conditions at $z = 0$.

From the solutions we can make the following observations:

1. From Eq. (14.31) we see that for signal amplification Γ should be real (if Γ is imaginary the solution would be oscillatory rather than amplifying) which implies that $g > \frac{\Delta k}{2}$. This implies

$$|E_p|^2 > \frac{(\Delta k)^2}{4\kappa_s \kappa_i} \qquad (14.35)$$

which gives the threshold value of pump electric field required to achieve amplification for a given Δk. From this we can obtain the threshold value for the intensity of the pump wave for amplification as

$$I_{p,th} = \frac{n_p}{2c\mu_0} |E_p|^2 = \frac{n_p}{2c\mu_0} \frac{(\Delta k)^2}{4\kappa_s \kappa_i} \qquad (14.36)$$

Substituting the values of κ_s and κ_i in the above equation gives us the threshold intensity for amplification as

$$I_{p,th} = \frac{cn_p n_s n_i}{8\omega_s \omega_i d^2 \mu_0} (\Delta k)^2 \qquad (14.37)$$

This implies that for a given value of Δk, there is a minimum value of pump intensity to achieve amplification of the signal. If $\Delta k = 0$ then the signal will get amplified for any non-zero value of pump intensity.

2. Maximum gain is achieved when $\Delta k = 0$ which implies

$$k_p = (k_s + k_i) \qquad (14.38)$$

The above condition is referred to as the *phase matching condition*. We shall discuss about this condition in more detail later. This equation can be written in terms of frequencies as

$$\omega_p n_p = \omega_s n_s + \omega_i n_i \qquad (14.39)$$

As we have seen earlier, the frequencies ω_p, ω_s, and ω_i also have to satisfy the following equation:

$$\omega_p = \omega_s + \omega_i \qquad (14.40)$$

For a given ω_p, Eqs. (14.39) and (14.40) give a unique set of signal and idler frequencies that would take part in the non-linear interaction process.

Another way of looking at the parametric process is to identify Eq. (14.40) as the energy conservation equation corresponding to the splitting of a photon of energy $\hbar\omega_p$ into two photons, a signal photon of energy $\hbar\omega_s$ and an idler photon of energy $\hbar\omega_i$. Similarly Eq. (14.38) can be interpreted as the momentum conservation equation for the photon splitting process. Thus in this picture the phase matching condition is nothing but the conservation of momentum for the non-linear process.

It may happen that for a given pump wavelength and a material, it may not be possible to satisfy the two conditions [Eqs. (14.39) and (14.40)] simultaneously. In this case the parametric amplification will not take place. In order to satisfy the two equations simultaneously different techniques of phase matching have been developed which includes the birefringence phase matching and quasi phase matching. In birefringence phase matching the anisotropic property of the crystal is used to satisfy the two conditions simultaneously. For example, lithium niobate is an anisotropic (uniaxial) crystal with two refractive indices, namely ordinary refractive index and extraordinary refractive index. In such a crystal for any given direction of propagation, the propagation constant of the wave depends on the polarization state of the light wave. By choosing appropriate polarization states of the pump, signal, and idler it is possible to satisfy the two conditions simultaneously. Figure 14.3 shows the dependence of the signal and idler wavelengths as a function of direction of propagation (with respect to a special direction called the optic axis) in the crystal for a pump wavelength of 1064 nm (Nd:YAG laser). For a given direction of propagation if we draw a vertical line at the corresponding angle of propagation, the line would intersect the curve at two points corresponding to the signal (lower wavelength) and the idler wavelengths. As the angle is varied, the pair of signal and idler wavelengths satisfying the two conditions varies. This gives the device tunability. Also note that below a certain angle there are no solutions to the simultaneous equations implying that this process will not take place for those propagation directions for the given pump wavelength.

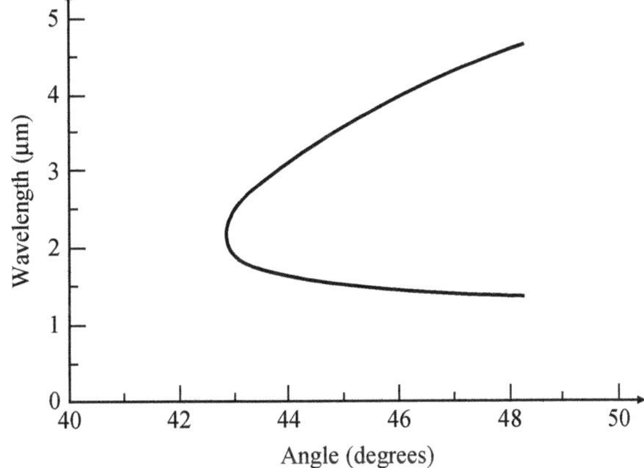

Fig. 14.3 Phase matching curve for LiNbO$_3$ for a pump wavelength of 1064 nm. [Source: http://www2.foi.se/rapp/ foir0536.pdf]

3. We now assume that phase matching condition is satisfied and obtain the gain coefficient of the amplifier. Also let us assume that at $z = 0$, the signal and idler fields are given by

$$E_s(z = 0) = E_{s0}$$
$$E_i(z = 0) = E_{i0} \tag{14.41}$$

Using these conditions in Eqs. (14.31) and (14.34), we can obtain the values of the constants A and B. Thus we get the solutions as

$$E_s(z) = E_{s0} \cosh gz - i\sqrt{\frac{\omega_s n_i}{\omega_i n_s}} E_{i0}^* \sinh gz \tag{14.42}$$

$$E_i^*(z) = i\sqrt{\frac{\omega_i n_s}{\omega_s n_i}} E_{s0} \sinh gz + E_{i0}^* \cosh gz \tag{14.43}$$

In obtaining these equation we have assumed the pump field to be real. The above two equations can be written in a matrix form as

$$\begin{pmatrix} E_s(z) \\ E_i^*(z) \end{pmatrix} = \begin{pmatrix} \cosh gz & -i\sqrt{\frac{\omega_s n_i}{\omega_i n_s}} \sinh gz \\ i\sqrt{\frac{\omega_i n_s}{\omega_s n_i}} \sinh gz & \cosh gz \end{pmatrix} \begin{pmatrix} E_{s0} \\ E_{i0}^* \end{pmatrix} \tag{14.44}$$

Equation (14.44) describes the evolution of the signal and idler electric field amplitudes due to the nonlinear interactions in the medium.

If at the input only the pump and signal fields are present, then $E_{i0} = 0$ and Eq. (14.42) gives

$$E_s(z) = E_{s0} \cosh gz \tag{14.45}$$

Since cosh is a function which increases monotonically with the increase of its argument, Eq. (14.45) implies that as the signal propagates through the crystal it gets amplified by drawing energy from the pump. For large values of gz, Eq. (14.45) can be written as

$$E_s(z) = \frac{1}{2} E_{s0}\, e^{gz} \tag{14.46}$$

showing an exponentially growing signal with g being the gain coefficient.

From Eq. (14.43) we also obtain

$$E_i^*(z) = i\sqrt{\frac{\omega_i n_s}{\omega_s n_i}} E_{s0} \sinh gz \tag{14.47}$$

showing the generation of the idler along with the amplification of the signal. Thus in this case the amplification of the signal is accompanied by the generation of the idler wave. This is also expected from the discussion above wherein we interpret this process as a process in which a pump photon splits into a signal photon and an idler photon. Every time a signal photon is generated necessarily an idler photon should have been generated. Thus amplification of the signal will be accompanied by the generation of the idler wave.

Apart from the change of amplitudes, the fields also suffer a phase change as given by the terms $e^{-ik_s z}$ and $e^{-ik_i z}$ in Eqs. (14.4) and (14.5) due to propagation over a distance z.

14.4 Singly Resonant Oscillator

We first consider a singly resonant parametric oscillator as shown in Fig. 14.4. It consists of the non-linear crystal placed inside an optical resonator. The crystal is pumped by a pump laser at the frequency ω_p. The resonator mirrors are such that both of them have high transmittivity at the pump and idler wavelengths while they have high reflectivity at the signal wavelength. Thus the resonator provides for feedback only at the signal wavelength and it is the signal wavelength which can oscillate within the cavity provided the loss in the signal wave is compensated by the gain in the signal wave due to non-linear interaction.

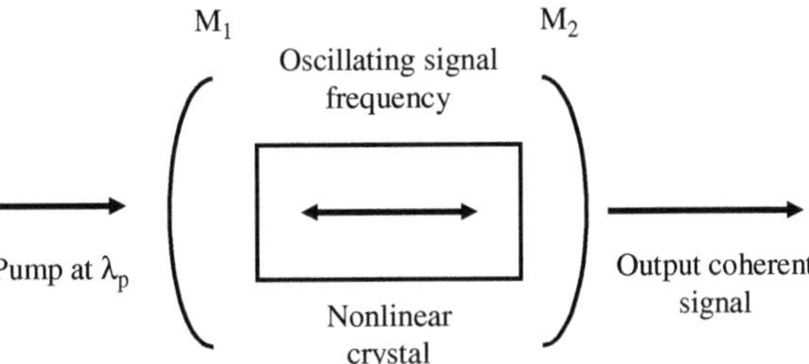

Fig. 14.4 A singly resonant oscillator (SRO) in which the resonator is formed by two mirrors which have high reflectivity at the signal wavelength and are transmitting at the pump and idler wavelengths

Now when the crystal is pumped by the external laser, then initially spontaneous parametric fluorescence (i.e., spontaneous generation of signal and idler photons from the pump photon) takes place and if phase matching condition is satisfied then this leads to spontaneous generation of light at frequencies ω_s and ω_i. This is similar to spontaneous emission in a laser cavity that initiates laser oscillation. The spontaneously emitted signal and idler waves propagate through the crystal and their amplitude changes as given by Eq. (14.44). When the waves reach the mirror, then the mirror only reflects the signal wave and transmits both the pump and the idler waves. Now when the signal is propagating in the reverse direction (opposite to the direction of the pump wave) then the phase matching condition is not satisfied and the signal suffers loss (if there are any internal losses) and no gain. When it reaches the first mirror then it gets partially reflected and the reflected signal wave again undergoes non-linear interaction and its amplitude changes. For such a singly resonant oscillator, the idler amplitude does not build up within the cavity and hence its amplitude can be neglected. Thus the signal amplitude changes within the cavity according to the formula

$$E_s(z) = E_{s0} \cosh gz \qquad (14.48)$$

Now since it is only the signal which is resonating within the cavity, only the signal wave has to satisfy the condition of standing waves within the cavity. If the cavity length is l, then assuming that the mirrors do not introduce any phase changes, the signal frequency must satisfy the following condition:

$$\omega_s = m\frac{\pi c}{n_s l}, \qquad m = 1, 2, 3\ldots \tag{14.49}$$

where we have assumed that the crystal occupies the entire length of the resonator. Thus only signal frequencies satisfying Eq. (14.49) would be able to oscillate within the cavity.

In order for oscillation, the gain per round trip must be equal to the loss per round trip. Let R_1 and R_2 represent the energy reflectivities of the two mirrors at the signal wavelength. For simplicity we will assume that there are no other losses within the cavity.

Let $E_s(0)$ be the amplitude of the signal at the mirror M_1. As it propagates to mirror M_2, the amplitude changes to

$$E_{s1}(l) = E_s(0)\cosh gl$$

A fraction of this wave is reflected by mirror M_2. Assuming that the mirrors do not introduce any phase changes, the amplitude reflectivity of the mirror M_2 would be $\sqrt{R_2}$. Thus the amplitude of the reflected signal wave on mirror M_2 and travelling toward mirror M_1 would be

$$E_{s2}(l) = \sqrt{R_2}\, E_s(0)\cosh gl \tag{14.50}$$

We neglect all other losses such as scattering loss etc. While the signal wave travels toward mirror M_1, the wave would not undergo any amplification as the phase matching condition would not be satisfied. Thus the amplitude of the signal wave as it arrives on mirror M_1 would be given by Eq. (14.50). Now mirror M_1 reflects a fraction of the signal wave and the signal wave after reflection from mirror M_1 would be

$$E_{s3}(0) = \sqrt{R_1 R_2}\, E_s(0)\cosh gl \tag{14.51}$$

For oscillation the signal amplitude should repeat itself after one round trip. Thus $E_{s3}(0) = E_s(0)$ and Eq. (14.51) gives us the following condition:

$$\sqrt{R_1 R_2}\,\cosh g_{\text{th}} l = 1 \tag{14.52}$$

where g_{th} is the threshold gain coefficient. Since in the case of parametric gain, the quantity $g_{\text{th}} l$ is very small we can expand the cosh term in Eq. (14.52) and write

$$1 + \frac{(g_{\text{th}} l)^2}{2} = \frac{1}{\sqrt{R_1 R_2}}$$

which gives us the following expression for the threshold gain coefficient required for a singly resonant parametric oscillator

$$g_{th} = \frac{\sqrt{2}}{l}\left(\frac{1}{\sqrt{R_1 R_2}} - 1\right)^{1/2} \tag{14.53}$$

If the reflectivities of the two mirrors are equal ($R_1 = R_2 = R$) and close to unity, then the above expression can be approximated by

$$g_{th} = \frac{\sqrt{2(1-R)}}{l} \tag{14.54}$$

Since the gain coefficient is related to the pump intensity [see Eq. (14.33)], Eq. (14.54) can be used to obtain an expression for the threshold pump intensity $I_{p,th}$ required for parametric oscillation. Thus

$$g_{th}^2 = \kappa_s \kappa_i \left|E_{p,th}\right|^2 = \frac{\omega_s \omega_i d^2}{c^2 n_s n_i} \frac{2c\mu_0}{n_p} I_{p,th} \tag{14.55}$$

where we have used Eqs. (14.28) and (14.29). Using Eqs. (14.55) and (14.54), we obtain an expression for the threshold pump intensity required for a singly resonant parametric oscillator as

$$I_{p,th} = \frac{c n_s n_i n_p}{\mu_0 \omega_s \omega_i d^2 l^2}(1-R) \tag{14.56}$$

As an example we take typical values corresponding to a lithium niobate non-linear crystal for which $d \sim 30 \times 10^{-12}$ m/V and $n_p \sim n_s \sim n_i = 2$. If we assume the pump wavelength to be 500 nm and the signal wavelength to be 900 nm, then using energy conservation equation [Eq. (14.40)] the corresponding idler wavelength would be 1125 nm. Assuming the reflectivity of the mirror to be 98% each at the signal wavelength and the length of the crystal to be 5 cm, substituting these values in Eq. (14.56) we obtain for the threshold pump intensity as 4.8×10^6 W/m^2. If we assume the beam to have a cross-sectional area of π mm^2 (i.e., transverse radius of 1 mm) then this corresponds to a pump power of about 14.5 W.

14.5 Doubly Resonant Oscillator

In the case of a doubly resonant oscillator, both the mirrors have high reflectivity at both the idler and the signal wavelengths and thus both are resonant within the cavity. This gives us the following two conditions:

$$\omega_s = m\frac{\pi c}{n_s l}, \qquad m = 1, 2, 3 \ldots \tag{14.57}$$

and

$$\omega_i = p\frac{\pi c}{n_i l}, \qquad p = 1, 2, 3 \ldots \tag{14.58}$$

In this case we have to consider both the equations for the signal and the idler as they would both resonate within the resonator and would both have large amplitudes.

Now, if $E_s(0)$ and $E_i(0)$ are the amplitudes of the signal and idler as they leave mirror M_1, then as they reach mirror M_2 placed at a distance l, their amplitudes would become [see Eq. (14.44)]

$$\begin{pmatrix} E_{s1}(l) \\ E_{i1}^*(l) \end{pmatrix} = \begin{pmatrix} \cosh gl & -i\sqrt{\frac{\omega_s n_i}{\omega_i n_s}}\sinh gl \\ i\sqrt{\frac{\omega_i n_s}{\omega_s n_i}}\sinh gl & \cosh gl \end{pmatrix} \begin{pmatrix} E_s(0) \\ E_i^*(0) \end{pmatrix} \tag{14.59}$$

For simplicity we assume that both mirrors are identical and have reflectivities of R_s and R_i at the signal and idler frequencies, respectively. As before we also neglect all other losses in the cavity and also assume that the mirrors do not generate any phase changes on reflection. Thus the amplitudes at the signal and idler frequencies after reflection from mirror M_2 would be

$$\begin{aligned} \begin{pmatrix} E_{s2}(l) \\ E_{i2}^*(l) \end{pmatrix} &= \begin{pmatrix} \sqrt{R_s} & 0 \\ 0 & \sqrt{R_i} \end{pmatrix} \begin{pmatrix} E_{s1}(l) \\ E_{i1}^*(l) \end{pmatrix} \\ &= \begin{pmatrix} \sqrt{R_s} & 0 \\ 0 & \sqrt{R_i} \end{pmatrix} \begin{pmatrix} \cosh gl & -i\sqrt{\frac{\omega_s n_i}{\omega_i n_s}}\sinh gl \\ i\sqrt{\frac{\omega_i n_s}{\omega_s n_i}}\sinh gl & \cosh gl \end{pmatrix} \begin{pmatrix} E_s(0) \\ E_i^*(0) \end{pmatrix} \end{aligned} \tag{14.60}$$

Now as before there would be no amplification as the waves travel from right to left and after reflection from mirror M_1 the fields would be

$$\begin{pmatrix} E_{s3}(0) \\ E_{i3}^*(0) \end{pmatrix} = \begin{pmatrix} \sqrt{R_s} & 0 \\ 0 & \sqrt{R_i} \end{pmatrix} \begin{pmatrix} E_{s2}(l) \\ E_{i2}^*(l) \end{pmatrix} \tag{14.61}$$

For oscillation the field amplitudes after one round trip must be the same as at the start. Thus we obtain the condition for oscillation as

$$\begin{pmatrix} E_{s3}(0) \\ E_{i3}^*(0) \end{pmatrix} = \begin{pmatrix} E_s(0) \\ E_i^*(0) \end{pmatrix} \tag{14.62}$$

Using Eqs. (14.60), (14.61), and (14.62), we obtain

$$\begin{aligned} \begin{pmatrix} \sqrt{R_s} & 0 \\ 0 & \sqrt{R_i} \end{pmatrix} \begin{pmatrix} \sqrt{R_s} & 0 \\ 0 & \sqrt{R_i} \end{pmatrix} & \begin{pmatrix} \cosh g_{th} l & -i\sqrt{\frac{\omega_s n_i}{\omega_i n_s}}\sinh g_{th} l \\ i\sqrt{\frac{\omega_i n_s}{\omega_s n_i}}\sinh g_{th} l & \cosh g_{th} l \end{pmatrix} \\ \begin{pmatrix} E_s(0) \\ E_i^*(0) \end{pmatrix} &= \begin{pmatrix} E_s(0) \\ E_i^*(0) \end{pmatrix} \end{aligned} \tag{14.63}$$

which gives us

$$\begin{pmatrix} R_s \cosh g_{\mathrm{th}}l - 1 & -iR_s\sqrt{\frac{\omega_s n_i}{\omega_i n_s}} \sinh g_{\mathrm{th}}l \\ iR_i\sqrt{\frac{\omega_i n_s}{\omega_s n_i}} \sinh g_{\mathrm{th}}l & R_i \cosh g_{\mathrm{th}}l - 1 \end{pmatrix} \begin{pmatrix} E_s(0) \\ E_i^*(0) \end{pmatrix} = 0 \qquad (14.64)$$

Here g_{th} is the threshold gain coefficient required for oscillation. For a non-trivial solution the following determinant

$$\begin{vmatrix} R_s \cosh g_{\mathrm{th}}l - 1 & -iR_s\sqrt{\frac{\omega_s n_i}{\omega_i n_s}} \sinh g_{\mathrm{th}}l \\ iR_i\sqrt{\frac{\omega_i n_s}{\omega_s n_i}} \sinh g_{\mathrm{th}}l & R_i \cosh g_{\mathrm{th}}l - 1 \end{vmatrix}$$

must be zero, which gives us the following condition:

$$(R_s \cosh g_{\mathrm{th}}l - 1)(R_i \cosh g_{\mathrm{th}}l - 1) - R_s R_i \sinh^2 g_{\mathrm{th}}l = 0$$

which on simplification gives us

$$\cosh g_{\mathrm{th}}l = \frac{1 + R_s R_i}{R_s + R_i} \qquad (14.65)$$

Since the reflectivities are usually quite close to unity, the right-hand side of Eq. (14.65) is almost equal to unity. This implies that $g_{\mathrm{th}}l$ is small and in such a case we can expand the cosh term and obtain the following approximate expression for the threshold gain coefficient:

$$g_{\mathrm{th}} = \frac{\sqrt{(1 - R_s)(1 - R_i)}}{l} \qquad (14.66)$$

As before using the expressions for the gain coefficient, the corresponding threshold pump intensity comes out to be

$$I_{p,\mathrm{th}} = \frac{c n_s n_i n_p}{2\mu_0 \omega_s \omega_i d^2 l^2}(1 - R_s)(1 - R_i) \qquad (14.67)$$

Compared to Eq. (14.56) which gives the threshold pump intensity in the case of singly resonant oscillator, the threshold intensity for doubly resonant oscillator is less by a factor

$$\frac{(I_{p,\mathrm{th}})_{\mathrm{dr}}}{(I_{p,\mathrm{th}})_{\mathrm{sr}}} = \frac{(1 - R_i)}{2} \qquad (14.68)$$

If $R_i = 0.98$, then the threshold pump intensity for the doubly resonant oscillator is less by a factor 0.01. If we use the same parameters as used earlier, the threshold pump intensity in this case would be 4.8×10^4 W/m^2. If the radius of the pump beam inside the crystal is 1 mm then the required threshold pump power is

$$P_{p,\mathrm{th}} = I_{p,\mathrm{th}} \times \pi \times a^2 \approx 150\,\mathrm{mW}$$

If the same crystal is made to oscillate as a singly resonant oscillator, then the threshold power would be 15 W showing the drastic reduction in threshold pump power requirement for a doubly resonant oscillator in comparison to a singly resonant oscillator.

Although the threshold pump power levels for a doubly resonant oscillator are very less compared to the singly resonant case, the need to satisfy various conditions on the signal and idler frequencies leads to instabilities in oscillation. Thus in a doubly resonant oscillator, the signal and idler have to simultaneously satisfy all the following equations:

$$\omega_p = \omega_s + \omega_i \tag{14.69}$$

$$\omega_p n_p = \omega_s n_s + \omega_i n_i \tag{14.70}$$

$$\omega_s = m \frac{\pi c}{n_s l}, \qquad m = 1, 2, 3 \ldots \tag{14.71}$$

$$\omega_i = p \frac{\pi c}{n_i l}, \qquad p = 1, 2, 3 \ldots \tag{14.72}$$

Satisfying all these conditions simultaneously puts severe requirements on the stability of the resonator. Since in singly resonant oscillators the equation for ω_i is not a requirement, the singly resonant oscillator is much more stable; of course the threshold pump power required in this case is much higher.

14.6 Frequency Tuning

One of the greatest advantages of the parametric oscillator is the ability to tune the wavelength of laser oscillation. For a given pump frequency, the signal and idler frequency that will get amplified are determined by the phase matching condition. Since the phase matching condition depends on the refractive index of the medium at the three frequencies, any parameter that can change the indices can be used to tune the frequency of oscillation. Thus by changing the temperature, or applying an external electric field which changes the indices by electro optic effect or by changing the orientation of the crystal (in the case of anisotropic crystals) if one of the waves is an extraordinary wave, it is possible to tune the frequency of oscillation.

Figure 14.5 shows a typical tuning curve of an OPO in which tuning is achieved by changing the pump wavelength. The signal and idler wavelengths span from 2 to 11 μm. Figure 14.6 shows the tunability of a commercially available OPO.

14.7 Phase Matching

For achieving efficient non-linear interaction, the phase matching condition given by Eq. (14.38) must be satisfied. Physically the condition comes about due to the

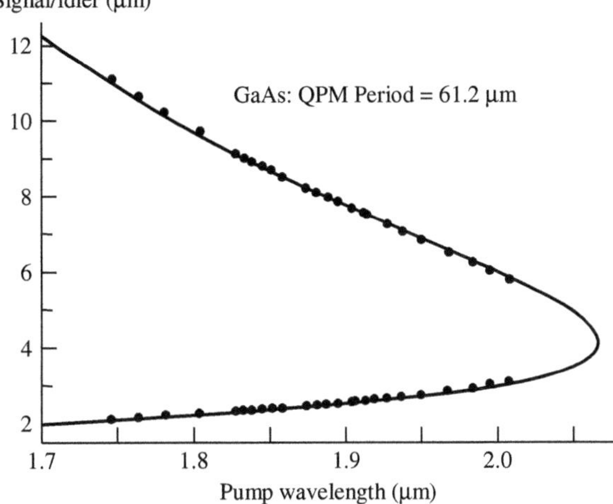

Fig. 14.5 In a tunable OPO, by changing the pump wavelength, the output signal and idler frequencies can be changed. Note that the signal and idler wavelengths together cover the range from about 2 μm to 11 μm. [After Henderson et al. (2008); reprinted with permission]

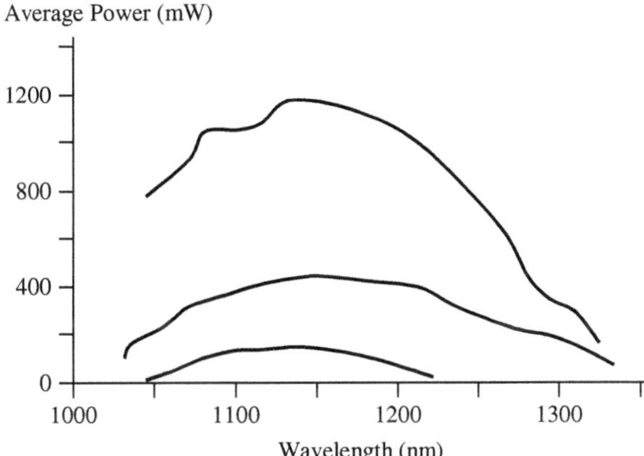

Fig. 14.6 The range of output wavelengths available from an OPO showing the wide tunability of the laser. The spectrum is of a commercially available OPO. The three curves correspond to models with different powers. (Mira OPO from Coherent, USA)

difference in speeds of the non-linear polarization that is a source of the electromagnetic wave at the new frequency that it is trying to generate and the electromagnetic wave that it is generating. Thus the non-linear polarization at the signal frequency ω_s propagates with a velocity [see Eq. (14.9)]

$$v_{\text{pol},s} = \frac{\omega_s}{k_p - k_i} \tag{14.73}$$

At the same time, the electromagnetic wave at frequency ω_s travels with a velocity [see Eq. (14.4)]

$$v_{em,s} = \frac{\omega_s}{k_s} \tag{14.74}$$

For efficient generation of the signal wave the source (non-linear polarization) and the wave that it is generating must travel at the same velocity. Thus from Eqs. (14.73) and (14.74) we see that this requires satisfying the phase matching condition.

There are two primary techniques used to achieve this. One of them is referred to as birefringence phase matching and the other is referred to as quasi phase matching.

Birefringence Phase Matching

In birefringence phase matching, the anisotropy of the crystal is used in achieving the phase matching condition. In anisotropic crystals for any given direction of propagation, there are two linearly orthogonal polarization states that travel as eigenmodes without any change in their polarization states and with different velocities. In uniaxial crystals one of the waves referred to as ordinary wave has the same velocity for all directions of propagation while the velocity of the other wave referred to as the extraordinary wave changes with the direction of propagation. Thus by appropriately choosing the polarization states of the pump, signal, and idler to correspond to ordinary or extraordinary waves it is possible to choose an appropriate direction of propagation within the crystal to achieve phase matching. If the direction of propagation is changed, then the refractive index seen by the extraordinary wave would change resulting in a change of the corresponding signal and idler frequencies satisfying the phase matching condition. Thus by changing the direction of propagation within the crystal it is possible to tune the signal and idler wavelength.

Quasi Phase Matching

In the alternative technique referred to as quasi phase matching, the non-linear coefficient of the crystal is modulated periodically along the direction of propagation. As discussed earlier when phase matching condition is not satisfied then after a distance equal to the coherence length non-linear polarization and the electromagnetic wave get out of phase. Now if the non-linear coefficient changes sign at this distance, then since the non-linear polarization is proportional to the non-linear coefficient, the phase of the non-linear coefficient would change by π bringing back the non-linear polarization and the electromagnetic wave back into phase. This would ensure that the non-linear polarization feeds energy into the signal wave in a constructive fashion. Again after propagation through a distance L_c, the non-linear polarization and the electromagnetic wave would develop a phase difference of π and if we again change the sign of the non-linear coefficient, then the non-linear polarization and the electromagnetic wave can be brought back in phase. Thus if the non-linear coefficient is periodically modulated in sign with a spatial period $2L_c$, then this would result in the growth of signal wave as it propagates along the medium. This is the basic principle of quasi phase matching.

In order to analyze parametric amplification in a periodically poled material, let us assume that the non-linear coefficient d varies sinusoidally with a period Λ. In such a case we have

$$d = d_0 \sin(Kz) \tag{14.75}$$

where d_0 is the amplitude of modulation of the non-linear coefficient and $K\ (= 2\pi/\Lambda)$ represents the spatial frequency of the periodic modulation. For easier understanding we are assuming the modulation to be sinusoidal; in general, the modulation will be periodic but not sinusoidal. Any periodic modulation can be written as a superposition of sinusoidal and cosinusoidal variations. Thus our discussion is valid for one of the Fourier components of the variation.

By using Eq. (14.76), Eqs. (14.26) and (14.27) become

$$\frac{dE_s}{dz} = -i\frac{\omega_s}{cn_s}d_0 \sin Kz\, E_p E_i^* e^{-i\Delta kz}$$
$$= -\frac{\omega_s}{2cn_s}d_0 E_p E_i^* \left(e^{-i(\Delta k - K)z} - e^{-i(\Delta k + K)z}\right) \tag{14.76}$$

$$\frac{dE_i}{dz} = -i\frac{\omega_i}{cn_i}d_0 \sin Kz\, E_p E_s^* e^{-i\Delta kz}$$
$$= -\frac{\omega_i}{2cn_i}d_0 E_p E_s^* \left(e^{-i(\Delta k - K)z} - e^{-i(\Delta k + K)z}\right) \tag{14.77}$$

Using similar arguments as earlier, it can be shown that if $\Delta k - K \approx 0$, then only the first term within the brackets in Eqs. (14.76) and (14.77) contribute to the non-linear amplification, and similarly if $\Delta k + K \approx 0$, then only the second term within the brackets contribute to the non-linear amplification.

The first condition implies that

$$k_p = k_s + k_i + K \tag{14.78}$$

If $\Lambda\ (= 2\pi/K)$ represents the spatial period of the modulation of the non-linear coefficient, and λ_0 is the wavelength of the fundamental, then the modulation period Λ required for QPM parametric amplification is

$$\Lambda = \frac{2\pi}{\left(k_p - k_s - k_i\right)}. \tag{14.79}$$

Thus any phase mismatch due to a finite value of Δk can be compensated by an appropriate value of the period Λ of the quasi phase matching.

In general, the spatial variation of the non-linear grating is not sinusoidal. In this case the efficiency of interaction would be determined by the Fourier component of the spatial variation at the spatial frequency corresponding to the period given by Eq. (14.79). It is also possible to use a higher spatial period of modulation and use one of the Fourier components for the non-linear interaction process. Thus in the case of periodic reversal of the non-linear coefficient with a spatial period given by

$$\Lambda_g = m\frac{2\pi}{\left(k_p - k_s - k_i\right)}, \qquad m = 1, 3, 5, \ldots \tag{14.80}$$

which happens to be the mth harmonic of the fundamental spatial frequency required for QPM, the corresponding non-linear coefficient that would be responsible for parametric amplification would be the Fourier amplitude at that spatial frequency. This can be taken into account by defining an effective non-linear coefficient (assuming a duty cycle of periodic reversal of 0.5):

$$d_{\mathrm{QPM}} = \frac{2d_0}{m\pi} \tag{14.81}$$

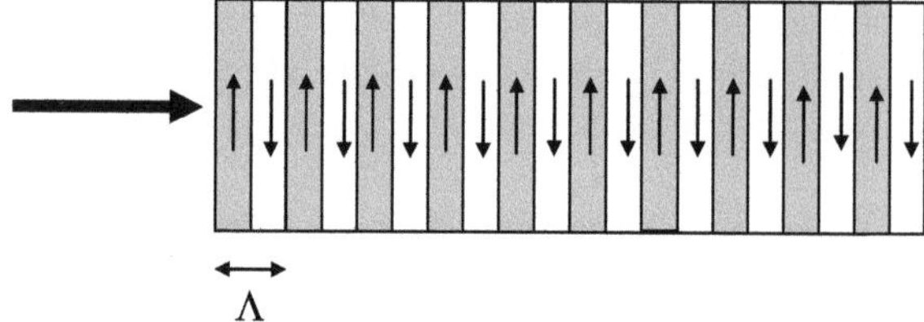

Fig. 14.7 Principle of quasi phase matching. In this method the direction of the optic axis direction is reversed every coherence length

Of course the largest effective non-linear coefficient is achieved by using the fundamental frequency with $m = 1$. Higher spatial periods are easier to fabricate but would lead to reduced non-linear efficiencies using the fundamental spatial frequency, it can be seen that the effective nonlinear coefficient is reduced by a factor of $2/\pi$ and since the efficiency depends on the square of the nonlinear coefficient, this would result in a reduction of efficiency $4/\pi^2$ as compared to the case of perfect phase matching.

In a ferroelectric material such as lithium niobate, the signs of the non-linear coefficients are linked to the direction of the spontaneous polarization. Thus a periodic reversal of the domains of the crystal can be used for achieving QPM (see Fig. 14.7). This is the currently used technique to obtain high-efficiency SHG and other non-linear interactions in $LiNbO_3$, $LiTaO_3$, and KTP. The most popular technique today to achieve periodic domain reversal in $LiNbO_3$ is the technique of electric field poling (Yamada et al. (1993), Myers and Bosenberg (1997)). In this method a high electric field pulse is applied to properly oriented lithium niobate crystal using lithographically defined electrode patterns to produce a permanent periodic domain reversed pattern. Such a periodically domain reversed $LiNbO_3$ crystal with the periodically reversed domains going through the entire depth of the crystal is also referred to as PPLN (pronounced *piplin*). For typical crystals such

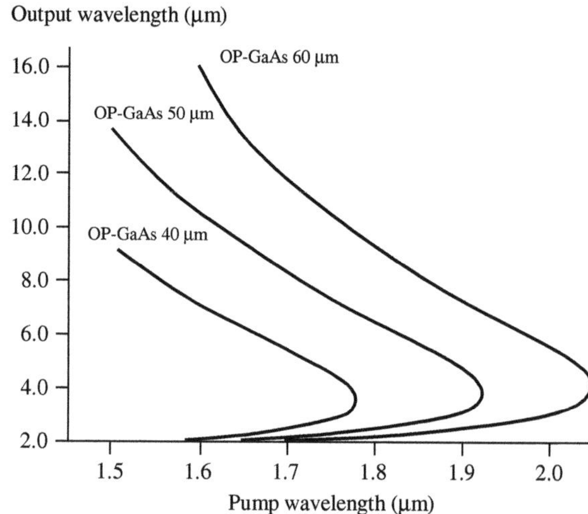

Fig. 14.8 Tunable output wavelengths achievable using quasi phase matching in GaAs (orientation patterned). (Adapted from Faye et al. (2008))

as lithium niobate the required period is of the order of 25–30 μm; such crystals are now commercially available.

Figure 14.8 shows the tuning curves of a quasi phase-matched interaction in GaAs.

Quasi phase matching offers many advantages vis a vis birefringence phase matching. Birefringence phase matching can be used only in the case of anisotropic crystals while quasi phase matching can be used even in the case of isotropic crystals. Apart from this in the case of quasi phase matching it is possible to choose the polarization states of all the interacting waves to be the same; this allows the use of the largest non-linear coefficient of the crystal. By choosing appropriate period of periodic poling quasi phase matching can be used for any set of wavelengths.

Problems

Problem 14.1 The threshold condition of a parametric oscillator is given by

$$\cosh g_{th}L = \frac{1 + R_s R_i}{R_s + R_i}$$

where symbols have their usual meaning. Show that the threshold gain required for a singly resonant OPO is much higher than that of a doubly resonant OPO.

Problem 14.2 I wish to achieve parametric amplification with signal (1.2 μm) and pump (0.8 μm) traveling along the same direction and the idler (λ_i) traveling in the reverse direction. If the refractive indices at the pump, signal and idler are 2.17, 2.15, and 2.11, respectively, calculate the period required for first-order QPM.

Problem 14.3 From the differential equations describing parametric process, show that the number of signal photons generated is equal to the number of pump photons annihilated. Assume perfect phase matching.

Problem 14.4 Consider parametric amplification of a signal at ω_s by a strong pump at ω_p. Show that even if phase matching is not exactly satisfied, it is possible to achieve amplification provided Δk satisfies some condition and obtain this condition. Neglect pump depletion.

Problem 14.5 Consider a parametric oscillator with mirrors of intensity reflection coefficients at ω_p, ω_s, and ω_i as follows:

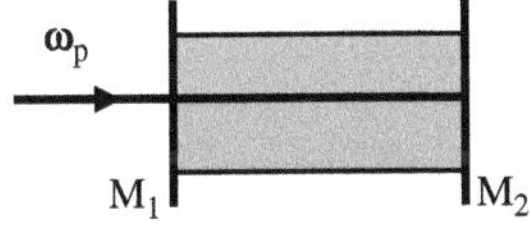

Mirror M_1 : $R_1(\omega_p) = 0$; $R_1(\omega_s) = R_s \sim 1$; $R_1(\omega_i) = 0$
Mirror M_2 : $R_2(\omega_p) = 1$; $R_2(\omega_s) = R_s \sim 1$; $R_2(\omega_i) = 0$

i.e., mirrors are highly reflecting at signal, transparent at idler, and mirror M_1 *is transparent to pump while M_2 is fully reflecting at pump. Obtain the threshold value of gain coefficient g for oscillation to begin. Neglect pump depletion.*

Problem 14.6 A parametric amplifier operates with a pump wavelength of 1 μm and a signal wavelength of 1.5 μm.

a)   Obtain the wavelength of the idler.

b)   If the input pump power is 1 W and an input signal power of 1 mW is converted to 1.5 mW at the output, obtain the output power at the idler frequency.

Problem 14.7 Consider a phase-matched parametric amplifier with pump ω_p, signal ω_s, and idler ω_i.

(a)   Starting from the coupled equations for perfectly phase-matched interaction, obtain expressions for

$$\frac{1}{\hbar\omega_p}\frac{dP_p}{dz} \quad \text{and} \quad \frac{1}{\hbar\omega_s}\frac{dP_s}{dz}$$

Compare these expressions and physically interpret their relationship.

(b)   Neglecting pump depletion and assuming perfect phase matching, obtain their solutions for the input condition

$$E_p(0) = u_{p0}, \quad E_s(0) = u_{s0}, \quad E_i(0) = +i\sqrt{\frac{\omega_i n_s}{\omega_s n_i}}\,u_{s0},$$

where u_{p0} and u_{s0} are real quantities. Determine whether the wave at ω_s gets amplified or attenuated during the non-linear interaction.

Problem 14.8 The equation describing the wavelength variation of the extraordinary refractive index of lithium niobate is given by

$$n_e^2(\lambda) = 4.5469 + \frac{0.094779}{\lambda^2 - 0.04439} - 0.026721\,\lambda^2$$

a)   Write down the phase matching condition corresponding to parametric interaction (three wave interaction) using QPM assuming the waves at pump, signal, and idler to be extraordinary waves.

b)   Assuming a QPM period of 20 μm, plot the variation of the signal wavelength (λ_s) and the idler wavelength (λ_i) as the pump wavelength (λ_p) is varied from 760 to 840 nm.

Problem 14.9 A pump wave at 1 μm and having a power of 1 W and a signal wave at 1.5 μm with a power of 1 mW are simultaneously incident on a non-linear crystal. If phase matching condition for difference frequency generation is satisfied, (a) what is the wavelength of the difference frequency and (b) if the power of the 1.5 μm wave increases to 1.1 mW what is the power exiting at the difference wavelength?

Problem 14.10 Using the expression for the parametric gain show that maximum gain is achieved at degeneracy, i.e., when the signal and idler wavelengths are equal. Neglect the frequency dependence of the refractive indices.

Problem 14.11 Two extraordinarily polarized plane waves at wavelengths of 1000 nm (power $= 1$ W) and 1500 nm (power $= 1$ mW) are incident along the y-direction in lithium niobate that is periodically poled with a spatial period of 10.34 μm. Which new wavelength will be generated most efficiently at the output of the crystal and why? Use the refractive indices of lithium niobate given below.

Extraordinary wave refractive indices of lithium niobate at different wavelengths

λ (nm)	n_e
500	2.25
600	2.21
750	2.18
1000	2.16
1500	2.14
2000	2.13
3000	2.11

Part II
Some Important Applications of Lasers

Chapter 15
Spatial Frequency Filtering and Holography

15.1 Introduction

One of the most interesting and exciting applications of lasers lies in the fields of spatial frequency filtering and holography. In this chapter, we briefly outline the principle behind spatial frequency filtering and holography and discuss their applications.

15.2 Spatial Frequency Filtering

Just as the Fourier transform of a time-varying signal gives its temporal frequency spectrum, similarly the spatial Fourier transform of a spatially varying function (like the transmittance of an object) gives the spatial frequency spectrum of the function (see Appendix F). It can indeed be shown that the field distribution produced at the back focal plane of an aberrationless converging lens is the two-dimensional Fourier transform of the field distribution in the front focal plane of the lens (see, e.g., Ghatak and Thyagarajan (1978)). Thus, if $f(x,y)$ represents the object distribution in the front focal plane of an aberrationless converging lens (see Fig. 15.1), then the field distribution in the back focal plane is given as (see Appendix F)

$$g(x,y) = \frac{i}{\lambda f} \iint f(x',y') \exp\left[\frac{2\pi i}{\lambda f}(xx'+yy')\right] dx'\,dy'$$
$$= \frac{i}{\lambda f} F\left(\frac{x}{\lambda f}, \frac{y}{\lambda f}\right) \tag{15.1}$$

where $F\left(x/\lambda f,\, y/\lambda f\right)$ represents the Fourier transform of $f(x,y)$ evaluated at the spatial frequencies $\left(x/\lambda f,\, y/\lambda f\right)$ – see Appendix G, f represents the focal length of the lens, and λ is the wavelength of illumination. (For ease of notation, in this chapter, we are representing free space wavelength by λ instead of λ_0). Thus if on the front focal plane is placed an object with a transmittance of the form

K. Thyagarajan, A. Ghatak, *Lasers*, Graduate Texts in Physics,
DOI 10.1007/978-1-4419-6442-7_15, © Springer Science+Business Media, LLC 2010

$$f(x,y) = A\cos\left(\frac{2\pi x}{a}\right) = \frac{A}{2}\left[\exp\left(\frac{2\pi ix}{a}\right) + \exp\left(-\frac{2\pi ix}{a}\right)\right] \tag{15.2}$$

then on the back focal plane one would obtain a field distribution given as

$$g(x,y) = \frac{Ai}{2\lambda f}\left[\delta\left(\frac{x}{\lambda f} + \frac{1}{a}\right) + \delta\left(\frac{x}{\lambda f} - \frac{1}{a}\right)\right]\delta(y) \tag{15.3}$$

where we have used the fact that the Fourier transform of $\exp\left(2\pi ix/a\right)$ is $\delta\left(u + 1/a\right)$, where u is the spatial frequency, which in the present case is $x/\lambda f$. Equation (15.3) represents the field corresponding to two bright dots at $\left(x = \lambda f/a, y = 0\right)$ and $\left(x = -\lambda f/a, y = 0\right)$ on the back focal plane of lens L_1 (see Fig. 15.1).

Now, the Fourier transform of the Fourier transform of a function is the original function itself except for an inversion, i.e.

$$\mathcal{F}\left[\mathcal{F}\{f(x)\}\right] = f(-x) \tag{15.4}$$

where the symbol $\mathcal{F}[\ ]$ stands for the Fourier transform of []. Thus if another converging lens L_2 is placed such that the back focal plane of the first lens L_1 is the front focal plane of the second lens (see Fig. 15.1), then on the back focal plane of L_2 one would obtain the original object distribution except for an inversion. The resultant field distribution on the plane P_3 can be controlled by suitably placing apertures on the back focal plane P_2 and thus performing operations on the spatial frequency spectrum of the object. The various apertures, stops, etc. that are placed in the plane P_2 are referred to as filters. Thus, for example, a low-pass filter would be one which allows low spatial frequencies to pass through while blocking the high

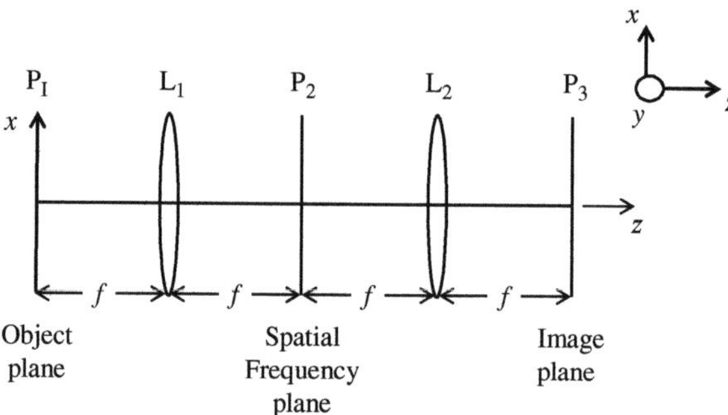

Fig. 15.1 When an object transparency is placed in the front focal plane P_1 of an aberrationless converging lens L_1 and illuminated by a parallel beam of light, then on the back focal plane P_2, one obtains a spectrum of the spatial frequency components present in the object. If a second lens L_2 is placed such that the plane P_2 coincides with its front focal plane, then in the back focal plane P_3 of L_2, one obtains the image pattern corresponding to the spatial frequency spectrum in the plane P_2. One can indeed control the spatial frequency spectrum that is responsible for forming the image on the plane P_3 by placing filters on the plane P_2

spatial frequencies. This could, for example, be a screen with a small hole at the axis of the system. Similarly, a high-pass filter would transmit all high spatial frequencies while blocking the low frequencies. One could also have complex filters which alter both the amplitude and the phase of the various spatial frequency components of the image.

As an example, let us consider an object with an amplitude variation of the form

$$f(x) = A \cos(2\pi \alpha x) + B \cos(2\pi \beta x) \tag{15.5}$$

The object represented by Eq. (15.5) has two spatial frequencies α and β. If such an object is placed in the plane P_1 and illuminated by a coherent beam of light, then in the plane P_2, one would obtain four spots at distances $x = +\lambda f\alpha, +\lambda f\beta, -\lambda f\alpha,$ and $-\lambda f\beta$ as shown in Fig. 15.2. If we do not place any obstructions on the plane P_2, then both the spatial frequencies contribute in forming the image in the plane P_3 and one obtains in the plane P_3 the same amplitude distribution as that in the plane P_1. Now, consider placing two stops at the points $x = +\lambda f\alpha$ and $x = -\lambda f\alpha$ on the y-axis in the plane P_2. Thus, no light from these points is allowed to reach the lens L_2. The lens L_2 receives light only from the spots corresponding to the spatial frequency β. Hence it follows that the image pattern in the plane P_3 will be proportional to $\cos(2\pi\beta x)$. Thus, by placing stops in the plane P_2, we have been able to filter out the frequency component α. This is the basic principle behind spatial frequency filtering. As a corollary, we may mention that if we put a stop on the axis, then it will filter out the low-frequency components. This can also be seen from the fact that if a plane wave propagating parallel to the axis (which is associated with zero spatial frequency) falls on a lens, it gets focused to the axis on the plane P_2 and if we put a small stop on the axis in the plane P_2, then there will be no light reaching the lens L_2.

Spatial frequency filtering finds widespread applications in various fields; we will discuss briefly some of these applications.

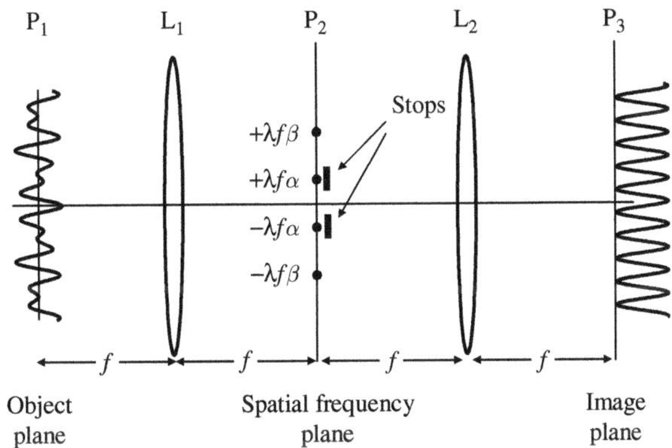

Fig. 15.2 If we place an object with a transmittance proportional to Eq. (15.5) on the plane P_1, then in the plane P_2, we will obtain four spots (on the y-axis) at $x = \pm\lambda f\alpha$ and $\pm\lambda f\beta$. If we place two stops behind the spots at $x = \pm\lambda f\alpha$, then in the plane P_3 we would obtain an image pattern represented by $\cos(2\pi\beta x)$. Thus the frequency component α has been filtered out

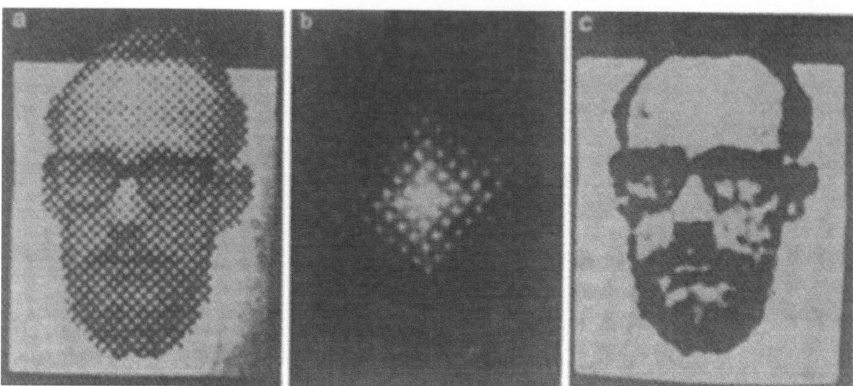

Fig. 15.3 (**a**) A photograph consisting of regularly spaced array of black and white squares. (**b**) The corresponding spatial frequency spectrum which appears on the plane P_2. If a pinhole is placed in the back focal plane P_2 to block off the high-frequency components, then an image of the form shown in (**c**) is obtained. Note that in the image, shades of grey as well as details such as the missing part of the eyeglass frame appear. (Reprinted from R.A. Philips, spatial filtering experiments for undergraduate laboratories, Am. J. Phys. *37* (1969) 536. © 1969 American Association of Physics Teachers; photographs courtesy: Dr. R.A. Philips)

If one looks closely at a newspaper photograph, one can immediately see that the image is in fact made up of a large number of closely arranged dots. These closely arranged dots represent a high spatial frequency, while the general image formed by these dots represents low-frequency components. Thus, these dots can be got rid of by spatial frequency filtering. For example, Fig. 15.3a shows a photograph which consists of regularly spaced black and white squares. The spatial frequency spectrum of the object is shown in Fig. 15.3b. If we place a screen with a small hole at the center on the back focal plane, then the image produced is devoid of these dot patterns (see Fig. 15.3c). By placing a small hole on the axis, one has essentially filtered out the high-frequency components in the object.

Another application of spatial frequency filtering is in contrast enhancement. When there is a large amount of background light in an image, the contrast in the image is poor. Since the background light represents a distribution of zero spatial frequency, if we place the object in the front focal plane and put a small stop on the axis in the back focal plane (which cuts off the low-frequency components), then since the stop removes the low-frequency, one would obtain an image with a much better contrast on the back focal plane of L_2. Such a process is termed contrast enhancement.

Spatial frequency filtering can also be used for detecting non-periodic (i.e., random) errors in a periodic structure. Thus one could either transmit all the spatial frequencies corresponding to the periodic array and stop most of the light from the random noise or block the light corresponding to the periodic array frequencies and transmit most of the light corresponding to the defects. Such techniques have indeed been used in photo mask inspection, electron tube grid inspection, etc. (see, e.g., Gagliano et al. (1969)).

Another important application of spatial frequency filtering is in character recognition problems where it is necessary to detect the presence of certain characters in an optical image. Here the filter is a complex filter produced using holographic principles and the output from the optical system is such that corresponding to the positions where the object contains the desired character, one obtains bright spots of light in the image.[1] As an example we show the identification of a fingerprint by this technique in Fig. 15.4. In the top portion, the fingerprints are matched and one obtains a bright spot of light; when the two fingerprints do not match (lower portion), there is no appearance of a bright spot but only a smear. Character recognition problems will also find application in military defense, where it might be necessary to identify certain objects of interest. For further details on optical data processing, one may look up Casasent (1978).

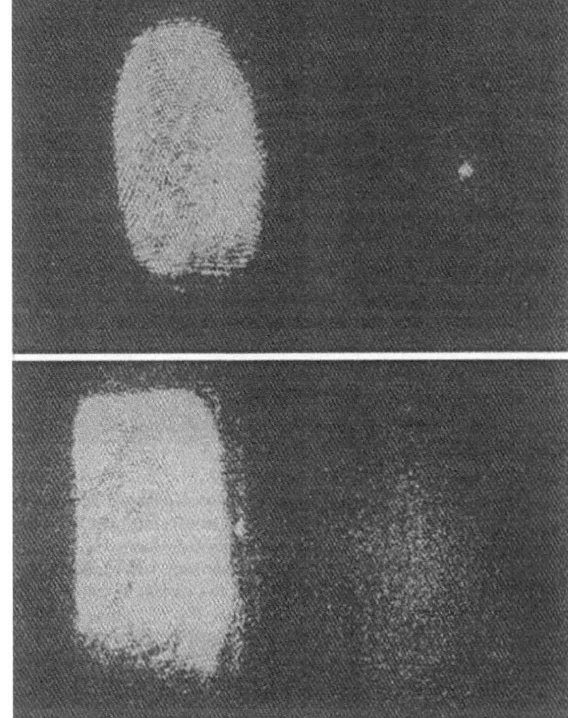

Fig. 15.4 Fingerprint identification using optical cross correlation. In the upper part, the two fingerprints are matched, which results in the appearance of a bright spot of light. In the lower part, the two fingerprints do not match and the resulting image is a smear. (Source: Tsujiuchi et al. (1971))

---

[1]For a detailed theoretical analysis of the character recognition problem, see, e.g., Ghatak and Thyagarajan (1978).

Image deblurring is another very interesting application of spatial frequency filtering. Let $f(x,y)$ represent the image of an object. If during exposure of the film, the camera moves or is out of focus, then instead of the image $f(x,y)$ we will obtain a modified image $g(x,y)$, which is the blurred image corresponding to $f(x,y)$. We shall now show how the blurring can be partially compensated, i.e., the image can be deblurred from the blurred image.

If $h(x,y)$ represents the intensity distribution of the blurred image of a point object, then the intensity distribution of the blurred image can be written as

$$
\begin{aligned}
g(x,y) &= \iint f(x',y')h(x-x',y-y')\mathrm{d}x'\mathrm{d}y' \\
&= f(x,y)^*h(x,y)
\end{aligned}
\tag{15.6}
$$

where * represents convolution. If we assume that the transmittance of the exposed film is proportional to $g(x,y)$, then the amplitude transmittance of the recorded film would be proportional to $g(x,y)$. If we place this film in the front focal plane of a lens and illuminate by a normally incident laser beam, then the amplitude distribution on the back focal plane would be the Fourier transform of $g(x,y)$. Using the fact that the Fourier transform of the convolution of two functions is the product of their Fourier transforms, the amplitude distribution on the back focal plane would be

$$
G(u,v) = F(u,v)H(u,v)
\tag{15.7}
$$

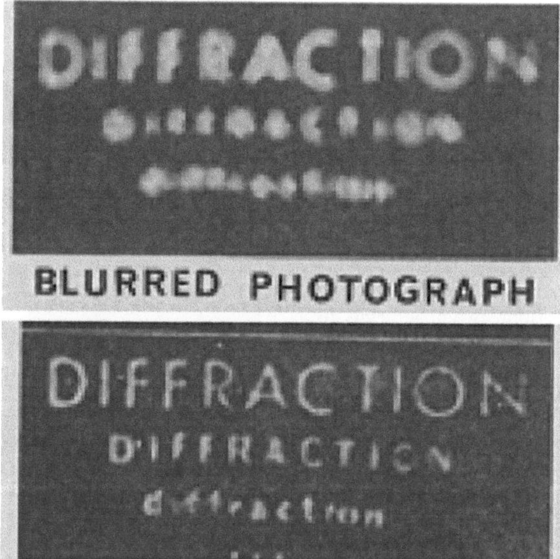

Fig. 15.5 Image deblurring using spatial frequency filtering. (**a**) shows a blurred photograph and (**b**) shows the deblurred image [After Stroke et al. (1975)]

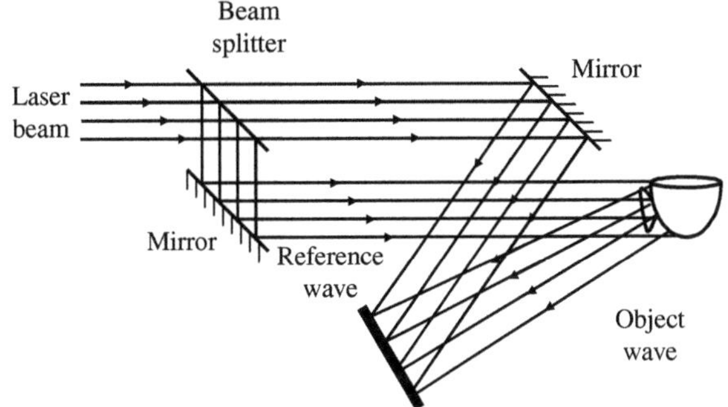

Fig. 15.6 The figure shows an arrangement for the recording of a hologram. The beam from a laser is split up into two portions: one part is used to illuminate the object and the other part is used as a reference beam. The waves scattered from the object interfere with the reference wave to form the hologram

where $u = x/\lambda f$, $v = y/\lambda f$, and G, F, and H are the Fourier transforms of g, f, and h, respectively. If we place a filter whose transmittance is proportional to $1/H(u,v)$ on the back focal plane of the lens, then the filtered spectrum would be

$$G(u, v)\,\frac{1}{H(u, v)} = F(u, v) \tag{15.8}$$

A second lens can Fourier transform the filtered spectrum further to produce the deblurred image $f(x,y)$. The filter $1/H(u,v)$ can be generated approximately using holographic principles. Figure 15.5 shows an example of image deblurring; the upper figure on the shows the blurred image and the lower figure shows the deblurred image after optical deblurring.

15.3 Holography

An ordinary photograph represents a two-dimensional recording of a three-dimensional scene. The emulsion on the photographic plate is sensitive only to the intensity variations, and hence while a photograph is recorded, the phase distribution which prevailed at the plane of the photograph is lost. Since only the intensity pattern has been recorded, the three-dimensional character (e.g., parallax) of the object scene is lost.

It was in the year 1948 that Dennis Gabor conceived of an entirely new idea and proposed a method of recording not only the amplitude but also the phase of the wave. The principle behind the method is the following: During the recording process, one superimposes on the wave (emanating from the object) another coherent wave called the reference wave (see Fig 15.6). The two waves interfere in the plane of the recording medium and produce interference fringes. This is known as the recording process. The interference fringes are characteristic of the object and the recording medium records the intensity distribution in the interference pattern.

This interference pattern has recorded in it not only the amplitude distribution but also the plane of the object wave. Thus, let

$$O\left(x,y\right) = O_0\left(x,y\right)\, e^{i\phi(x,y)} \tag{15.9}$$

represent the field produced due to the object wave at the plane of the recording medium; $O_0\left(x,y\right)$ is the amplitude part and $\phi\left(x,y\right)$ the phase part. Similarly, let

$$R\left(x,y\right) = A\, e^{i\psi(x,y)} \tag{15.10}$$

represent the field produced due to the reference wave at the recording medium. Usually the reference wave is an obliquely incident plane wave, in which case A is a constant. The total field produced at the recording medium is

$$U\left(x,y\right) = O_0\left(x,y\right)\, e^{i\phi(x,y)} + A\, e^{i\psi(x,y)} \tag{15.11}$$

and the intensity pattern recorded by the recording medium would be

$$I\left(x,y\right) = \left|U\left(x,y\right)\right|^2 = O_0^2\left(x,y\right) + A^2 + O_0\left(x,y\right)A\,\exp\left\{i\left[\phi\left(x,y\right) - \psi\right]\right\}$$
$$+ O_0\left(x,y\right)A\,\exp\left\{-i\left[\phi\left(x,y\right) - \psi\left(x,y\right)\right]\right\}$$
$$\tag{15.12}$$

where we have omitted a constant of proportionality and have carried out a time averaging.[2] It can immediately be seen from the above that the recorded intensity distribution has the phase of the object wave $\phi\left(x,y\right)$ embedded in it. Since the recorded intensity pattern has both the amplitude and the phase recorded in it, Gabor called the recording a hologram (*holos* in Greek means "whole").

This hologram has little resemblance to the object. It has in it a coded form of the object wave. The technique by which one reproduces the image is termed reconstruction. In the reconstruction process, the hologram is illuminated by a wave called the reconstruction wave; this reconstruction wave in most cases is similar to the reference wave used for recording the hologram (see Fig 15.7). When the hologram is illuminated by the reconstruction wave, various wave components emerge from the hologram, one of which is the object wave itself. In order to show this, we see that when the exposed recording medium is developed, then one, in general, gets a transparency, with a certain transmittance. Under proper conditions, the amplitude transmittance of the hologram can be made to be linearly proportional to $I\left(x,y\right)$. Thus apart from some constants, one can write for the amplitude transmittance of the hologram

---

[2] We are assuming the fields to be monochromatic with a time dependence of the form $e^{i\omega t}$. Now, for two functions f and g with time variations of the form $e^{i\omega t}$

$$\langle \mathrm{Re}f\,\mathrm{Re}\,g\rangle = \frac{1}{2}\langle \mathrm{Re}f^*g\rangle$$

where angular brackets denote time averaging and $\mathrm{Re}f$ stands for the real part of the function f.

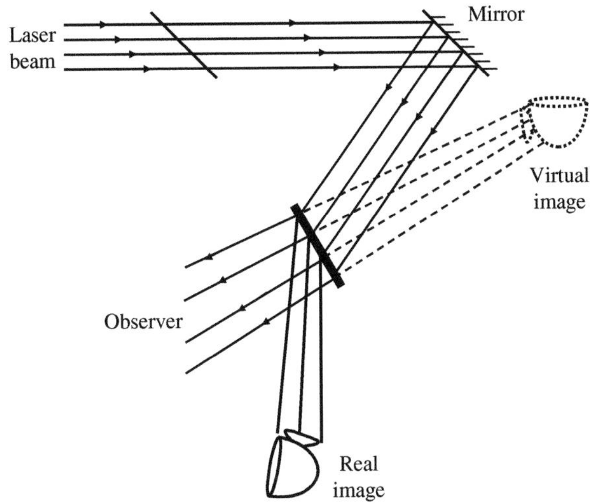

Fig. 15.7 In the reconstruction process, the hologram is illuminated by a reconstruction wave, which in most cases is identical to the reference wave used for forming the hologram. The reconstruction wave after passing through the hologram produces a real and a virtual image. The virtual image can be viewed and exhibits all the true three-dimensional characteristics like parallax and depth.

$$t\,(x, y) = I\,(x, y) \tag{15.13}$$

If this transparency is illuminated with the reconstruction wave, the emerging wave would be given as

$$
\begin{aligned}
t\,(x, y)\,A\,\exp\left[i\psi\,(x, y)\right] &= \left[O_0^2\,(x, y) + A^2\right] A\,\exp\left[i\psi\,(x, y)\right] \\
&\quad + O_0\,(x, y)\,A^2\,\exp\left[i\phi\,(x, y)\right] \\
&\quad + A^2 O_0\,(x, y)\,\exp\left\{-i\left[\phi\,(x, y) - 2\psi\,(x, y)\right]\right\}
\end{aligned}
\tag{15.14}
$$

The second term indeed represents the original object wave apart from the constant multiplicative factor A^2. The first term represents the reconstruction wave itself but which has been modulated in amplitude. The last term represents the complex conjugate of the object wave. The three wave components can be spatially separated by a proper choice of the reference wave.

The second term, which represents a reproduction of the object wave (as opposed to an image of the object), is identical to the wave that was emanating from the object when its hologram was being recorded. Thus, when one views this wave (emerging from the hologram), then one sees a reconstructed image of the object in its true three-dimensional form (see Fig. 15.7). Thus, as with the original object, one can move one's viewing position and look around the object. If the hologram has recorded in it sufficient depth of field, one has to refocus one's eyes to be able to see distinctly the objects which are far away. One can even place a lens on the path of the reconstructed wave and form an image of the object on a screen.

In addition to the virtual image, the reconstruction process generates another image, which is a real image; this is represented by the third term in Eq. (15.14).

Fig. 15.8 The in-line holography technique in which the object wave and the reference wave are traveling almost parallel to each other. During reconstruction, both the waves producing the virtual and real images are traveling approximately in the same direction; this produces some difficulties while viewing the images

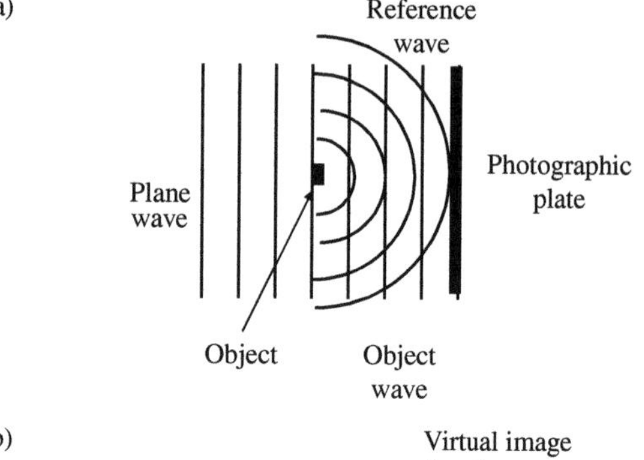

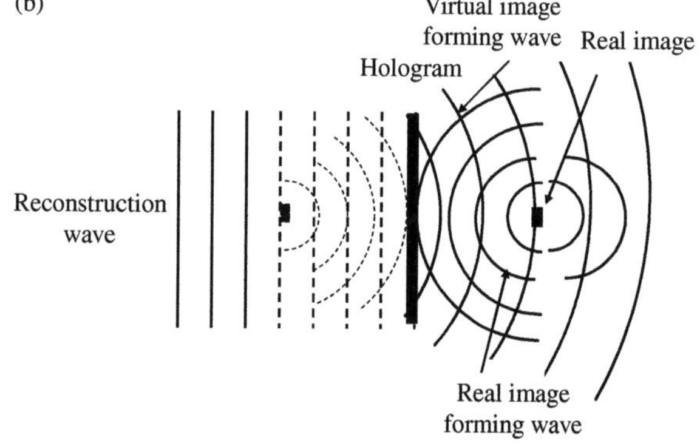

This real image can indeed be photographed by placing a suitable light-sensitive medium (like a photographic plate) at the position where the real image is formed.

Although the principle of holography was laid down by Gabor in 1948, it was not until the lasers arrived in 1960 that holography attained practical importance. Before the advent of the laser, one had to employ the method of in-line holography (as proposed by Gabor), in which the reference beam is approximately parallel to the object wave and the paths traversed by both the object wave and the reference wave are almost equal (see Fig. 15.8); this was required because the existing sources like mercury discharge lamps had only small coherence lengths.[3] The in-line technique has associated with it the disadvantage that the waves that form the virtual and real images travel along the same direction. Thus, while viewing the virtual image, one is faced with an unfocused real image and conversely. The early work on holography was in fact on the removal of this problem associated with geometry of recording.

---

[3]The high-pressure mercury arc lamp emits a green line at 5461 Å. The coherence length of this line is, in fact, only about 10 μm. (The width of the line at 5461 Å is about 5×10^{12} Hz.) On the other hand, the 6058-Å line emitted by krypton has a coherence length of $\sim$20 cm, but the power output per unit area of this source is very low. When one tries to increase the source area, one loses spatial coherence. The notion of coherence length has been discussed in Chapter 9.

It was in the year 1962 that Leith and Upatneiks introduced the technique of "off-axis holography," which overcame the difficulty associated with the in-line technique. In the technique proposed by Leith and Upatneiks, one uses a reference beam which falls obliquely on the photographic plate (see Figs. 15.6 and 15.7). Using such technique in the reconstruction process, one obtains well-separated virtual and real images. The use of such a technique was made possible by the large coherence length of the laser.[4] The importance of coherence can be seen from the fact that holography is essentially an interference phenomenon. Thus it is essential that the illuminating wave possesses sufficient spatial coherence so that the wave from every object point may interfere with the reference wave. We can obtain a wave with sufficient spatial coherence by making use of pinholes for illuminating. Alternatively, one could move the source far enough from the scene. But both the above methods essentially decrease the available power. The arrival of lasers overcame this difficulty. Further, in order that stable interference fringes be formed in the hologram of the complete object scene to be recorded, the maximum path difference between the reference wave and the wave from the object must be less than the coherence length.

Holograms exhibit very interesting properties. For example, when one records the hologram of a diffusely reflecting object, each point on the object scatters light on the complete surface of the hologram. Thus, every part of the hologram receives light from all parts of the object. Hence even if one breaks the recorded hologram into various parts, each part is capable of reconstructing the entire object; the resolution in the image decreases as the size of the hologram decreases. In fact, when one records holograms of transparencies, one often uses a ground glass screen which enables the hologram to receive light from all parts of the transparency.

The principle of holography finds applications in many diverse fields. We will discuss a few of them here.

We had observed that information about depth can also be stored in a hologram. Consider the problem of locating a transient event concerning a microscopic particle in a certain volume and studying it. If one uses an ordinary microscope, then since the event is transient, it is, in general, difficult to first locate the particle and study it. On the other hand, if one makes a holographic record of the complete volume, then one can "freeze" the event in the hologram. On reconstruction, there would emerge from the hologram the same wave with the difference that it is now not transient. If one now uses a microscope, one can easily locate the particle and study it at leisure.

One of the most important applications of holography has been in interferometry. In a technique called double-exposure holographic interferometry, one first partially exposes an object to the photographic plate with a reference wave. Now, the object is stressed, and one makes another exposure along with the same reference wave. If the resulting hologram is developed and illuminated by a reconstruction wave, then there would emerge from the hologram two object waves, one corresponding

---

[4]For further discussion on off-axis holography, the reader is referred to Collier et al. (1971), Ghatak and Thyagarajan (1978), and the Nobel Lecture by Gabor.

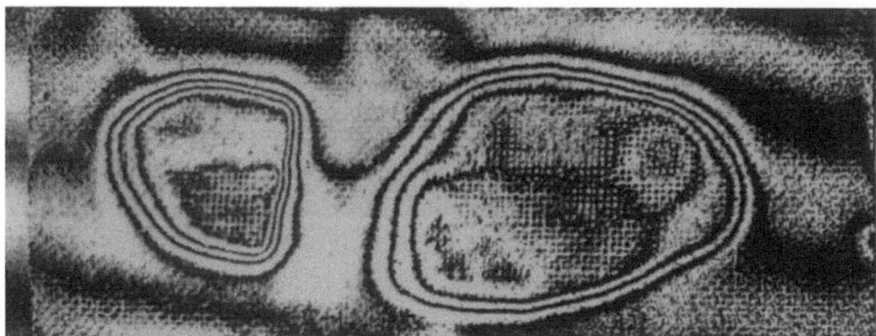

Fig. 15.9 Outline of footprint on a carpet not visible to the naked eye is revealed using holography

to the unstressed object and the other corresponding to the stressed object. These two object waves would interfere to produce interference fringes. Thus, on viewing through the hologram, one finds a reconstruction of the object superimposed with fringes. The shape and the number of fringes give one the distribution of strain in the object. One can employ the above technique in non-destructive testing of objects (see also Section 19.7).

As an example, Fig. 15.9 shows a double-exposure interferogram of a carpet on which a person had walked. Since the fibers in the carpet relax slowly, if two identical holograms are taken with a reasonable time interval between them, then the movement of the carpet surface is obvious from the interferogram. Visual inspection of the carpet would not yield any foot print!

Problems

Problem 15.1 The field variation on the front focal plane of a lens of focal length 20 cm is given as

$$g(x,y) = A + B \cos 6\pi x + C \cos 12\pi y \ (x, y \text{ in millimeters})$$

a)   What are the spatial frequencies present in the field?
b)   What pattern would you observe at the back focal plane of the lens? Assume a wavelength of 600 nm.

Problem 15.2 On plane P_1 (see figure below) the field distribution is given as $g(x) = A + B \cos 40\pi x + C \sin 25\pi x$, where x is measured in millimeters. A circular aperture of radius 1 mm is placed (with its center on the axis) on the back focal plane (P_2) of the lens. What field distribution would be obtained on the plane P_3, given that $f = 20$ cm and wavelength $= 500$ nm?

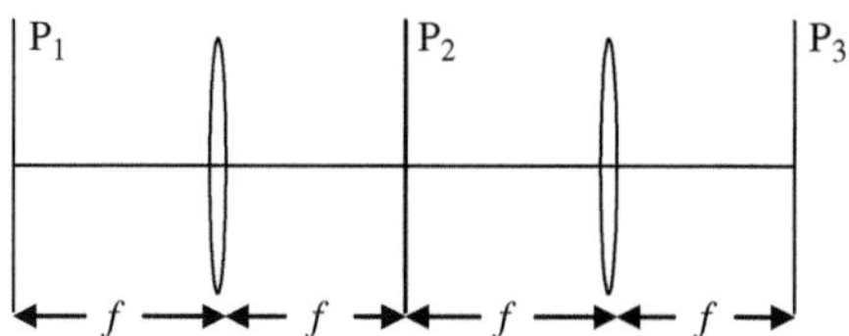

Problem 15.3 If the field distribution on the front focal plane of a lens of focal length 50 cm is given as $g(x,y) = 2 + \cos^2 10x$ (with x in centimeters), how many spots would be observed on the back focal plane? What would be their positions and relative intensities? Wavelength is 500 nm.

Problem 15.4 Consider an object distribution

$$f(x) = a + b \, \cos(20x + \pi/3) + c \, \sin(10x)$$

where x is in centimeters.

a) What are the spatial frequencies present in the object?

b) Consider an optical arrangement shown in Fig. 15.2. The amplitude distribution on plane P_1 is given as

$$g(x) = a + b \, \cos \frac{2\pi x}{a}$$

On the plane P_2 is placed a filter having a transmittance

$$T(x) = 1 \text{ for } x < 0$$
$$= 0 \text{ for } x > 0$$

Obtain the intensity distribution on the plane P_3.

Problem 15.5 Consider an object distribution of the form given below placed in the front focal plane of a lens of focal length 10 cm:

$$g(x) = (1 + 0.1 \, \cos \, (20 \, \pi \, x)]e^{-x^2}$$

where x is measured in millimeters. Plot this function. This could correspond to a case of a laser beam with spatial noise. What pattern do you expect to observe at the back focal plane of the lens? Find out how you can clean up the beam by spatial frequency filtering.

Problem 15.6 Consider an object distribution of the form

$$f(x,y) = 1 + 0.2 \, \cos \left(20x + \frac{\pi}{5}\right) + 0.3 \, \sin \left(50y + \frac{\pi}{8}\right)$$

where x and y are in millimeters. (a) What are the spatial frequencies present in the object? (b) If the above object is placed in the front focal plane of a lens of focal length 20 cm and illuminated normally by a plane wave of wavelength 1 μm, show schematically what would be observed on the back focal plane of the lens.

Problem 15.7 A circular aperture of radius a is placed (with its center on the axis) on the back focal plane of a lens of focal length f. What is the range of spatial frequencies that will be passed by the aperture?

Problem 15.8 What is the effect of placing a filter of the form $h(x) = px$, where p is a constant placed in spatial frequency plane of a spatial frequency filtering set up?

Problem 15.9 Show that a spatial frequency filter of the form $T(x) = \alpha x$ placed in the spatial frequency plane yields in the image plane a differential of the object amplitude distribution placed in the front focal plane.

Problem 15.10 Show that a spatial frequency filter of the form $T(x) = \alpha/x$ placed in the spatial frequency plane yields in the image plane an integral of the object amplitude distribution placed in the front focal plane.

Problem 15.11 Consider two plane waves travelling in the x–z-plane making angles θ_1 and θ_2 with the z-axis. A photographic plate is kept on the plane $z = 0$. Obtain the interference pattern obtained on the plane. What would be the fringe width?

Problem 15.12 Consider a plane wave propagating along the z-axis and a spherical wave emerging from a point source placed on the axis at a point $z = -d$. A photographic plate is kept on the plane $z = 0$. Obtain the shape of the fringes obtained.

Problem 15.13 The photographic plate in Problem 15.12 is developed and made into a hologram. If this is normally illuminated by a plane wave, what would be the output waves? If the hologram is illuminated by a spherical wave from a point source placed at $z = -d$, what would be the output from the hologram?

Chapter 16
Laser-Induced Fusion

16.1 Introduction

It is well known that the enormous energy released from the sun and the stars is due to thermonuclear fusion reactions, and scientists have been working for over 40 years to devise methods to generate fusion energy in a controlled manner. Once this is achieved, one will have an almost inexhaustible supply of relatively pollution-free energy. A thermonuclear reactor based on laser-induced fusion offers great promise for the future. With the tremendous effort being expended on fabrication of extremely high-power lasers, the goal appears to be not too far away, and once it is practically achieved, it would lead to the most important application of the laser.

In the next section, we discuss the basic physics behind the energy released in a fusion reaction; in Section 16.3 we discuss the laser energy requirements; and in Section 16.4 we briefly describe the laser-induced fusion reactor and some of the practical difficulties.

16.2 The Fusion Process

A nucleus of an atom consists of protons and neutrons which are nearly of the same mass. The proton has a positive electrical charge and the neutron, as the name implies, is electrically neutral. Because neutrons and protons are the essential constituents of atomic nuclei, neutrons and protons are usually referred to by the general name "nucleon." If one assumes that the forces between the nucleons are of the Coulomb type, then the nucleons would have flown apart because of the repulsion between two protons and also because no Coulomb-type forces exist between two neutrons and between a neutron and a proton. Since the nucleons are held together in the nucleus, there must be a short-range attractive force between them. Indeed, it is believed that at short distances $\left(\leq 10^{-13} \text{ cm} \right)$, very strong attractive forces exist between the nucleons[1] and this force is independent of the charge of the nucleons,

---

[1] Beyond the range of this short-range force, the forces are of Coulomb type.

K. Thyagarajan, A. Ghatak, *Lasers*, Graduate Texts in Physics, DOI 10.1007/978-1-4419-6442-7_16, © Springer Science+Business Media, LLC 2010

i.e., the force between two protons or two neutrons or between a proton and a neutron is essentially similar. Because of the strong attractive forces between the nucleons, a certain amount of energy has to be supplied to split a nucleus into its constituent nucleons; this is known as the binding energy[2] of the nucleus.

Consider a nuclear reaction in which the two deuterons react to form a tritium nucleus and a proton:

$$D + D \rightarrow T\,(1.01\,\mathrm{MeV}) + H\,(3.01\,\mathrm{MeV}) \tag{16.1}$$

The binding energy of each of the deuterium nuclei is 2.23 MeV and the total binding energy of the tritium nucleus is 8.48 MeV. Thus, there is a net gain in the binding energy, which is $8.48 - 2 \times 2.23 = 4.02$ MeV. Physically, a loosely bound system goes over to a tightly bound system resulting in the liberation of energy; this energy appears in the form of kinetic energies of tritium and proton, which are given in parentheses in Eq. (16.1). Nuclear reactions such as that represented by Eq. (16.1) in which two loosely bound light nuclei produce a heavier tightly bound nucleus are known as fusion reactions.[3]

Since both deuterium and tritium are isotopes of hydrogen with mass numbers 2 and 3, the nuclear reaction expressed by Eq. (16.1) is often written in the form[4]

$$_1H^2 + {_1H^2} \rightarrow {_1H^3} + {_1H^1} \tag{16.2}$$

---

[2]The binding energy is calculated using the famous Einstein mass–energy relation: $E = mc^2$ where $c\,(\approx 3 \times 10^{10}\,\mathrm{cm/s})$ is the speed of light in free space. If Z and N represent the number of protons and of neutrons, respectively, inside the nucleus, then the total binding energy Δ will be given as

$$\Delta = \left(Zm_{\mathrm{p}} + Nm_{\mathrm{n}} - M_{\mathrm{A}}\right)c^2$$

where m_{n}, m_{p}, and M_{A} represent the masses of the neutron, the proton, and the atomic nucleus, respectively. For example, the nucleus of the deuterium atom (which is known as the deuteron) has a mass of 2.01356 amu (1 amu $\approx 1.661 \times 10^{-24}$ g, which is equivalent to 931.5 MeV). Since deuteron consists of one proton and one neutron, one obtains

$$\begin{aligned}
\Delta &= (1.00728 + 1.00866 - 2.01356) \times 1.661 \times 10^{-24} \times \left(3 \times 10^{10}\right)^2 \mathrm{erg}\\
&= 3.56 \times 10^{-6} \times \left(1.6 \times 10^{-12}\right)^{-1} \times 10^{-6}\,\mathrm{MeV}\\
&\approx 2.23\,\mathrm{MeV}
\end{aligned}$$

which represents the binding energy of the deuteron. In the above equation, we have used 1.00728 and 1.00866 amu to represent the masses of proton and neutron, respectively.

[3]On the other hand, in a fission process, a loosely bound heavy nucleus splits into two tightly bound lighter nuclei, again resulting in the liberation of energy. For example, when a neutron is absorbed by a $_{92}U^{235}$ nucleus, the $_{92}U^{236}$ nucleus is formed in an excited state (the excitation energy is supplied by the binding energy of the absorbed neutron). This $_{92}U^{236}$ nucleus may undergo fission to form nuclei of intermediate mass numbers (like $_{56}B^{140}$ and $_{36}Kr^{93}$ along with three neutrons). The energy released in a typical fission reaction is about 200 MeV.

[4]The nuclei are identified by symbols like $_{11}Na^{23}$; the subscript (which is usually omitted) represents the number of protons in the nucleus and the superscript represents the total number of nucleons in the nucleus. Thus $_{11}Na^{23}$ represents the sodium nucleus having 11 protons and 12 neutrons. Similarly $_1H^3$ represents the tritium nucleus having 1 proton and 2 neutrons.

where $_1H^1$, $_1H^2$, and $_1H^3$ represent the nuclei of hydrogen (which is nothing but a proton), deuterium, and tritium, respectively. The following deuterium–tritium fusion reaction[5]

$$D + T \rightarrow \alpha \ (3.5\,\text{MeV}) + \text{neutron} \ (14.1\,\text{MeV}) \tag{16.3}$$

is also of considerable importance as a possible source of thermonuclear power. As indicated in Eq. (16.3), the total energy liberated is about 17.6 MeV. Tritium does not occur naturally, and one of the methods for producing it is to let the neutron [emitted in the D–T reaction – see Eq. (16.3)] interact with lithium [see Eq. (16.15)]. Even though deuterium is available in abundant quantities (it constitutes about 0.015% of natural water), one expects to use the D–T reaction in a fusion reactor, because at $T \approx 100$ million $^\circ$K,[6] the D–T reaction is about 100 times more probable than the D–D reaction[7] and the energy released in the D–T reaction is about four times that in a D–D reaction [see Eqs. (16.1) and (16.3)].

16.3 The Laser Energy Requirements

One of the difficulties associated with the fusion reaction is the requirement of a very high temperature for the fusion reactions to occur. This is due to the fact that unless the nuclei have very high kinetic energies, the Coulomb repulsion will not allow them to come sufficiently close for fusion reactions to occur.[8] The temperatures required are usually ~ 100 million K, and at such high temperatures the matter is in a fully ionized state and its confinement poses a serious problem; matter in a fully ionized state is known as a plasma. Thus, two major problems in thermonuclear fusion are (i) heating of plasmas to very high temperatures and (ii) confinement of plasmas for times long enough for substantial fusion reactions to occur.[9] For example, for the deuterium–tritium reaction [see Eq. (16.2)] at 10 keV (≈ 100 million $^\circ$K), for the fusion output energy to exceed the input energy required to heat the plasma, one must have

$$n\tau \geq 10^{14}\,\text{cm}^{-3}\,\text{s} \tag{16.4}$$

---

[5]Equation (16.3) can also be written in the form

$$_1H^2 + {_1H^3} \rightarrow {_2He^4} + {_0n^1}$$

[6]Temperatures of the order of 100 million K are required in fusion reactors; see Section 16.3.

[7]See, e.g., Booth et al. (1976).

[8]This is in contrast to fission reactions which are induced by neutrons which carry no charge. As such, even at room temperatures, there is a considerable probability for fission reactions to occur and hence it is relatively easy to construct a fission reactor. It may be mentioned that in a hydrogen bomb (where the fusion reactions are responsible for the liberation of energy), a fission bomb is first exploded to create the high temperatures required for fusion reactions to occur.

[9]In the sun (the energy of which is due to thermonuclear reactions), the plasma has a temperature of ≥ 10 million K and it is believed that the confinement is due to the gravitational forces.

where n represents the plasma density and τ is the confinement time. Equation (16.4) is known as the Lawson's criterion (see, e.g., Ribe (1975)). For $n \sim 10^{15}$ ions/cm^3, τ must be ≥ 0.1 s. Although the plasma has not yet been confined for such long times, the Russian device (known as Tokamak), using magnetic confinement, has come close to the conditions where the above equality is satisfied.

With the availability of intense laser pulses, a new idea of fusion systems has emerged. The idea is essentially compressing, heating, and confining the thermonuclear material by inertial forces which are generated when an intense laser pulse interacts with the thermonuclear material, which is usually in the form of a solid pellet. In such a confinement, it is not necessary to have a magnetic field.

For laser-induced fusion systems, instead of using the parameter $n\tau$, it is more useful to use the parameter ρR, where ρ and R represent the density of the fuel and the fuel radius, respectively. It has been shown (see, for example, Booth et al. (1976); Ribe (1975)) that if f represents the fractional burn-up of the fuel then

$$f \approx \frac{\rho R}{6 + \rho R} \tag{16.5}$$

where ρ is measured in grams per cubic centimeter and R is measured in centimeters. For $f \approx 0.05$ (i.e., 5% burn-up of the fuel), ρR must be about 0.3 g/cm^2; the higher the value of ρR, the greater the fractional burn-up of the fuel. Further, if the mass of D–T pellet is M g, then the total fusion energy released (in joules) would be given by[10]

$$E_{\text{output}} = 4.2 \times 10^{11} fM \text{ (J)} \tag{16.6}$$

Obviously

$$M = \frac{4\pi}{3} R^3 \rho = \frac{4\pi}{3} \frac{1}{\rho^2} (\rho R)^3 \tag{16.7}$$

---

[10]Since the masses of D and T nuclei are in the ratio of 2:3, the number of D nuclei will be

$$\frac{2M}{5} \frac{1}{M_{\text{d}}} = \frac{2M}{5} \frac{1}{2 \times 1.66 \times 10^{-24}}$$

where $M_{\text{d}} \left(\approx 2 \times 1.66 \times 10^{-23} \text{ g}\right)$ represents the mass of the deuteron; we have assumed equal numbers of D and T nuclei in the pellet. The energy released in a D–T reaction is 17.6 MeV [see Eq. (16.3)] and an additional 4.8 MeV is released when the neutron is absorbed by the lithium atoms in the blanket [see Eq. (16.15)] resulting in a net energy release of about 22 MeV. Thus, the output energy would be

$$E_{\text{output}} \approx f \times \frac{2M}{5} \times \frac{22 \times 1.6 \times 10^{-6}}{2 \times 1.66 \times 10^{-24}} \simeq 4.2 \times 10^{11} fM \ (J)$$

Also, to heat the D–T pellet to temperatures ($\approx$ 100 million K) at which fusion reactions will occur with high probability, the laser energy required would be[11]

$$E_{\text{laser}} \approx 4 \times 10^8 \frac{M}{\varepsilon}$$

or

$$E_{\text{laser}} \approx 4 \times 10^8 \frac{4\pi}{3} \frac{1}{\rho^2 \varepsilon} (\rho R)^3 \tag{16.8}$$

where ε represents the fraction of laser energy used for heating the pellet; usually $\varepsilon \sim 0.1$. Thus the yield ratio Y is given as

$$Y = \frac{E_{\text{output}}}{E_{\text{laser}}} \approx \frac{4 \times 10^{11} f M \varepsilon}{4 \times 10^8 M} = 10^3 \varepsilon f \tag{16.9}$$

or

$$Y \approx 100 \frac{\rho R}{6 + \rho R} \tag{16.10}$$

where we have assumed $\varepsilon \approx 0.1$. Clearly, for $Y > 1, \rho R \geq 0.1 \, \text{g/cm}^2$. Thus for a sizable burn-up and for a reasonable yield, one should at least have $\rho R \approx 0.2 \, \text{g/cm}^2$. Now, for normal (D–T) solid, $\rho \approx 0.2 \, \text{g/cm}^2$; using Eq. (16.8) one obtains

$$E_{\text{laser}} = 4 \times 10^8 \times \frac{4\pi}{3} \frac{(0.2)^3}{(0.2)^2 \times 0.1} \approx 3 \times 10^9 \, \text{J} \tag{16.11}$$

which is indeed very high. However, as is obvious from Eq. (16.8), for a given value of ρR, the laser energy requirement can be significantly decreased to the mega-joule range by increasing the value of ρ; this can be achieved by compressions to 10^3–10^4 times the normal solid density. At such high densities, the α particles that are produced in the reaction give up most of their energy to the unburned fuel before leaving the pellet; this leads to an increased fractional burning of fuel. For a typical calculation reported by Ribe (1975), for $\rho R \approx 3 \, \text{g/cm}^2$[corresponding to 30% burn-up – see Eq. (16.5)] and an ignition temperature of 10^4 eV ($\approx$ 100 million K), one obtains using Eq. (16.8)

$$E_{\text{laser}} = 4 \times 10^8 \times \frac{4\pi}{3} \frac{(3)^3}{(0.2)^2 \times 0.4} \approx 2.5 \times 10^{12} \, \text{J} \tag{16.12}$$

---

[11]For the kinetic energies to be about 10 keV ($\approx$ 100 million K), the energy imparted would be

$$2 \times \left(\frac{2M \times 10 \times 10^3 \times 1.6 \times 10^{-19}}{5 \times 2 \times 1.66 \times 10^{-24}} \right) \approx 4 \times 10^8 M \ \text{(J)}$$

where M is in grams; the quantity inside the parentheses is the energy imparted to the deuteron, and the factor of 2 outside the parentheses is due to the fact that an equal energy has also to be imparted to tritium nuclei.

where we have assumed a normal D–T solid of density $0.2\,\text{g/cm}^3$ and $\varepsilon = 0.4$. On the other hand, for $\rho = 1000\,\text{g/cm}^3$, one would obtain (for the same value of ρR)

$$E_{\text{laser}} \approx 1.2 \times 10^5\,\text{J} \tag{16.13}$$

The above numbers correspond to a pellet mass of 110 μg whose radius (before compression) is about 0.05 cm; the compressed radius of this pellet is $\sim$0.003 cm. The fusion energy yield would be given as [see Eq. (16.6)]

$$\begin{aligned} E_{\text{output}} &\approx 4.2 \times 10^{11} \times 0.3 \times 100 \times 10^{-6} \\ &\approx 14 \times 10^6\,\text{J} = 14\,\text{MJ} \end{aligned} \tag{16.14}$$

This corresponds to a yield of about 100.

16.4 The Laser-Induced Fusion Reactor

In a laser-induced fusion reactor, we take, for example, a deuterium–tritium pellet in the form of a cryogenic solid where particle densities are $\sim 4 \times 10^{22}\,\text{cm}^{-3}$ and shine laser light from all directions (see Fig. 16.1). Within a very short time, the outer surface of the pellet is heated considerably and gets converted into a very hot plasma ($T \sim 100$ million K). This hot ablation layer expands into vacuum and as a reaction gives a push to the rest of the pellet in the opposite direction. Thus,

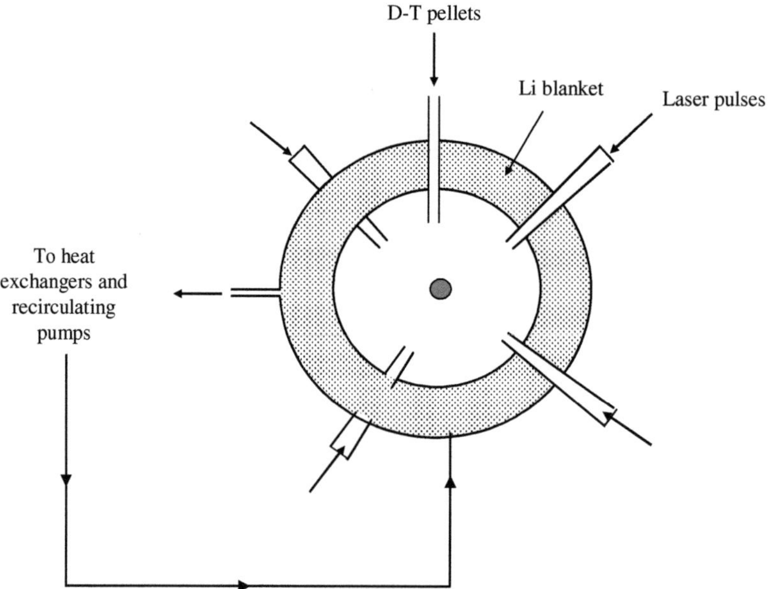

Fig. 16.1 Schematic of a D–T fusion reactor with lithium blanket

if a spherical pellet is irradiated from all sides, then a spherical implosion front travels towards the core. For a deuterium–tritium plasma, with an incident intensity of 10^{17} W/cm^2, one can get an inward pressure of $\sim 10^{12}$ atm.[12] As the implosion front accelerates toward the center, it sets up a sequence of shock waves traveling inward. Such shock waves lead to a very high compression of the core and the fusion energy is released from high compression densities[13] along with the high temperature. In order to obtain high compression densities, the time variation of the laser pulse has to be such that successive shock waves do not meet until they reach the center of the pellet. It has been shown that the time variation of the laser power should roughly be of the form $(t_0 - t)^{-2}$, where $t = t_0$ is the time when all the shocks reach the center. Thus, if the pulse duration is about 10 ns, about four-fifths of the energy goes in the last nanosecond and one-fifth in the first nine nanoseconds. A typical ideal energy profile for maximum compression in D–T pellets is shown in Fig. 16.2.

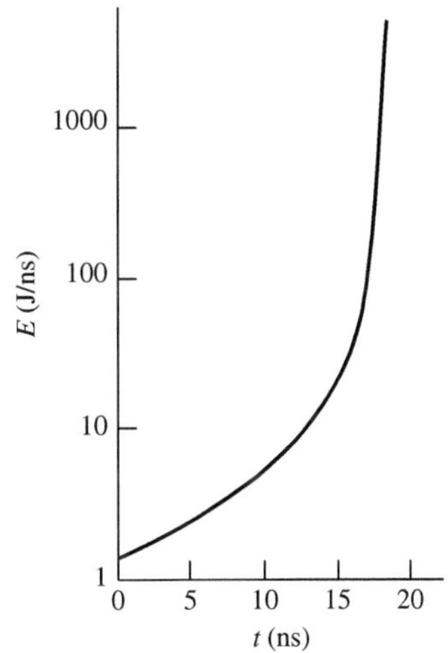

Fig. 16.2 Typical energy profile of the incident laser pulse for maximum compression in D–T pellets

According to a computer experiment by Nuckolls and his co-workers (1972), a deuterium–tritium spherical pellet of radius 0.04 cm was irradiated by a 6×10^4 J pulse of 25 ns duration[14] ($\lambda \sim 1\,\mu$m); compression densities as high as 1000 g/cm^3 were obtained and about 1.8×10^6 J of fusion energy was released in about 10^{-11} s after the compression. Thus a multiplication by a factor of about 30 was observed.

We would like to mention that laser wavelengths ($\sim 10\,\mu$m) may be too high for pellet heating. Detailed calculations show[15] that 1.06 μm radiation can heat a

---

[12] The radiation pressure corresponding to an intensity of 10^{17} W/cm^2 is only about 10^8 atm.

[13] A compression ratio of a few thousand puts the laser energy requirement in the 10^5 J range (see Section 16.3).

[14] The time variation of the incident laser pulse was assumed to be roughly of the form $(t_0 - t)^{-2}$; thus the power varied from about 10^{11} W at 10 ns to 10^{15} W at 15 ns.

[15] See, e.g., Kidder (1973).

deuterium–tritium pellet to five times the temperature in one-tenth the time as compared to 10.6 μm radiation heating the same pellet. This and other facts suggest the use of 1 μm radiation for laser-induced fusion. The 10.6 μm radiation corresponds to a CO_2 laser (see Section 11.7) and the 1.06 μm radiation corresponds to the neodymium-doped glass lasers (see Section 11.3). Further, neodymium-doped glass lasers are capable of delivering a high laser energy within a short time (see Table 16.1). One of the major drawbacks of the neodymium-doped glass laser system is the fact that the efficiency (defined as the ratio of the laser energy output to the electrical

Table 16.1 Some laser fusion facilities

Location[a]	Type	Number of beams	Total beam area (cm^2)	Maximum energy per beam	Peak power (TW)
USA					
LASL[b]	CO_2	8	9600	10 kJ	20
LASL[b]	CO_2	72	–	100 kJ	100–200
LASL[b]	CO_2	1	1200	400 J	0.4
KMS[b]	Nd:glass	2	200	200 J	0.5
NRL[b]	Nd:glass	2	70	300 J	0.2
LLL					
ARGUS[b]	Nd:glass	2	600	1 kJ	4.0
SHIVA[b]	Nd:glass	20	6300	10 kJ	30.0
NOVA	Nd:glass	100	–	300–500 kJ	300
LLE					
GDL[b]	Nd:glass	1	60	210 J	0.7
ZETA[b]	Nd:glass	6	360	1.3 kJ	3–5
OMEGA	Nd:glass	24	5500	10-14 kJ	30–40
USSR Lebedev					
UM1 35[b]	Nd:glass	32	2560	10 kJ	5–10
DELPHIN	Nd:glass	216	3430	13 kJ	10–15
UK					
Rutherford[b]	Nd:glass	2	200	200 J	0.5
France LiMeil[b] (Octal)	Nd:glass	8	500	700 J	1.0
Japan Osaka	Nd:glass	12			40

Source: Refs. [Opt. Spectra **13**(5), 30 (1979); Opt. Spectra **13**, 29 (1979); Laser Focus **15**(7), 38 (1979); Phys. Today **32**, 17 (1979); Phys. Today **32**, 20 (1979)]
[a]LASL – Los Alamos Scientific Laboratory, Los Alamos, New Mexico; KMS – KMS Fusion, Ann Arbor, Michigan; NRL – Naval Research Laboratory, Washington DC; LLL – Lawrence Livermore Laboratory, Livermore, California; LLE – Laboratory for Laser Energetics, Rochester, New York
[b]Existing facilities

Fig. 16.3 A beam of infrared light from a Nd:glass laser after passing through a 10-cm aperture potassium dihydrogen phosphate (KDP) crystal (as left) is halved in wavelength (and hence doubled in frequency), emerges as green light at 5320 Å, and is reflected by the mirror at the foreground into the target chamber at far right. The 12.8-mm-thick KDP crystals have yielded doubling efficiencies greater than 60%. The output power was 0.5 TW $\left(= 0.5 \times 10^{12}\,\text{W}\right)$, one of the highest powers in the visible region of the spectrum. (Photograph courtesy: Thomas A. Leonard, KMS Fusion, Inc.)

energy input) is extremely low ($\sim 0.2\%$). On the other hand, CO_2 lasers[16] (which operate at 10.6 µm) have efficiencies in the range of 5–7%. Although it is difficult to predict the specific laser systems which will be in operation in a laser fusion reactor, one does expect that very soon the laser technology would be sufficiently developed to meet the requirements.

Figures 16.3 and 16.4 show some photographs of the laser fusion experiments carried out at KMS Fusion, Inc., USA. Figure 16.3 shows a beam of light emerging from a neodymium:glass laser after passage through a 10-cm aperture potassium dihydrogen phosphate (KDP) crystal (at left) which doubles the frequency (harmonic generation) and hence halves the wavelength from 1.064 µm (infrared) to 5320 Å (green), which then enters the target chamber shown at the extreme right. Output powers of 0.5 TW (1 TW $= 10^{12}$ W) were measured at the output of the KDP crystals. Figure 16.4 shows a photograph taken during the laser irradiation of the cryogenic target which was a hollow spherical glass shell of 51 µm diameter with a 0.7 µm wall containing 1.3 ng of deuterium–tritium condensed into a liquid layer on the inside surface of the shell. The targets produced 7×10^7 neutrons on irradiation.

In March 2009, the National Ignition Facility (NIF) sent the first 192-beam laser shot to the center of its target chamber. The first test of the world's biggest – and

---

[16]For details of other kinds of lasers used in fusion, see, e.g., Booth et al., (1976).

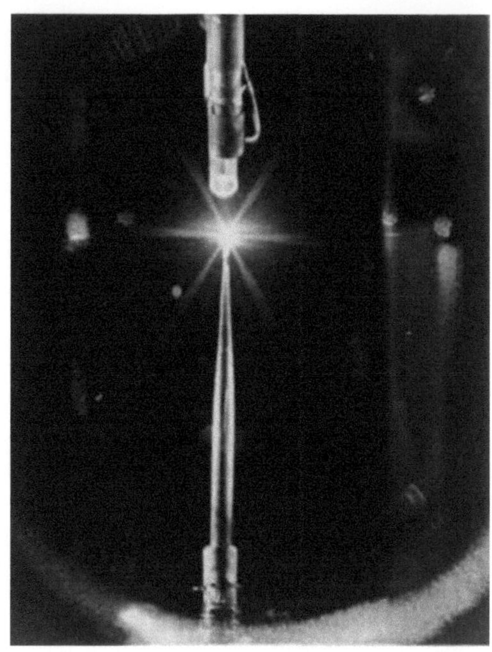

Fig. 16.4 Laser irradiation of a cryogenic target which is a hollow spherical glass shell 51 μm in diameter, with a 0.7-μm wall, containing 1.3 ng of deuterium–tritium condensed in a liquid layer on the inside surface of the shell. Two X-ray pinhole cameras are seen projecting toward the target from the top of the chamber. The targets produced 7×10^7 neutrons. (Photograph courtesy: Thomas A. Leonard, KMS Fusion, Inc.)

by up to 100 times the most energetic – laser system achieved 420 J of ultraviolet energy for each beam. Added up, the shot cycle produced 80 kJ of energy. Within 2–3 years, scientists expect to be creating fusion reactions that release more energy than it takes to produce them. If they are successful, it will be the first time this has been done in a controlled way eventually leading to fusion power plants. At the NIF located at Lawrence Livermore National Laboratory (LLNL), the 192 lasers that fire simultaneously at precisely the same point in space are designed to deliver 1.8 MJ of energy in a few nanoseconds equivalent to 500 trillion W of power. Significant results are expected sometime between 2010 and 2012. Figure 16.5 gives an aerial view of the NIF.

October 2008 marks the beginning of a 3-year *preparatory phase* of a new high-power laser energy research facility (HiPER) costing about € 1 billion. Unlike the National Ignition Facility (NIF) at the Lawrence Livermore Laboratory in the USA and the Mégajoule laboratory in France (where a single set of lasers is used to both compresses and heat the fuel), HiPER is planned to use separate laser pulses to do the compression and heating. The compression bank with its amplifiers, pulse shapers and wavelength shifters will fire 50–200 half-meter-diameter laser beams focused down to a millimeter and containing a total of 250 kJ at a wavelength of 0.35 μm over "multiple nanoseconds" creating 10^9 bars of pressure. This is expected to compresses the plasma to 300 g/cm^3 – 20 times the density of lead or even gold. Ignition is initiated by a 15-ps 70 kJ pulse, focussed to 100 μm diameter to match the size of the super-dense plasma, which heats the compressed matter to 100 million K.

Fig. 16.5 A bird's-eye view of the NIF facility shows the main 705,000 ft^2 building. The structure includes two laser bays capable of generating more than 4 MJ of infrared laser light; four capacitor bays (which store about 400 MJ of electrical energy); two switchyards, which direct all 192 laser beams into the target bay; the target bay, where experimental activities are conducted; the target diagnostic building, for assembling and maintaining diagnostic equipment for experiments; and the Optical Assembly Building (*upper left*), where optics assemblies are prepared for installation. (Source: Rick Sawicki Interview for Dartmouth Engineer Magazine, May 27, 2008)

The focussed intensities are expected to be about 10^{26} W/cm^2. Figure 16.6 gives the status of various projects on laser fusion facilities.

We end this chapter by briefly describing the laser fusion reactor (usually abbreviated as LFR). A simplified block diagram of the electricity-generating station is shown in Fig. 16.7. The reactor would roughly consist of a high-vacuum enclosure at the center of which deuterium–tritium pellets are dropped at regular intervals of time. As soon as the pellet reaches the center of the chamber, it is irradiated by synchronized pulses from an array of focused laser beams from all directions (see Fig. 16.4). The fusion energy released is absorbed by the walls of the chamber; consequently the walls get heated up, which may be used for running a steam turbine. For a commercial power station, if 100 pellets are allowed to explode per second and if in each explosion, an energy of about 10^8 J is released, one would have a 10 GW $= 10^{10}$ W power station.

One of the most important aspects of any fusion reactor is the production of tritium. Since it is not available in nature, it has to be produced through nuclear

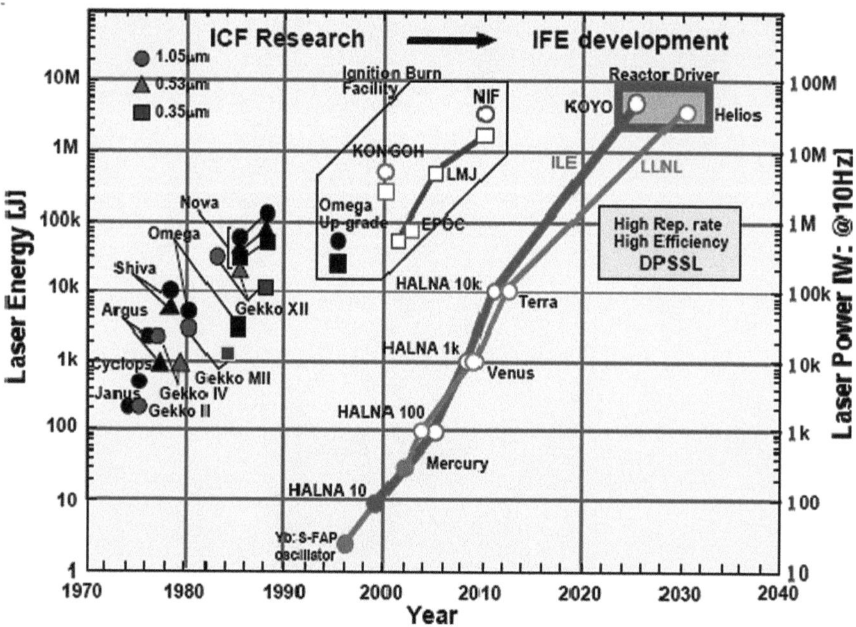

Fig. 16.6 The status of various projects on laser fusion facilities (Source: Ref. M. Dunne, HiPER: a laser fusion facility for Europe http://fsc.lle.rochester.edu/pub/workshops/FIW06/Dunne_F106. pdf)

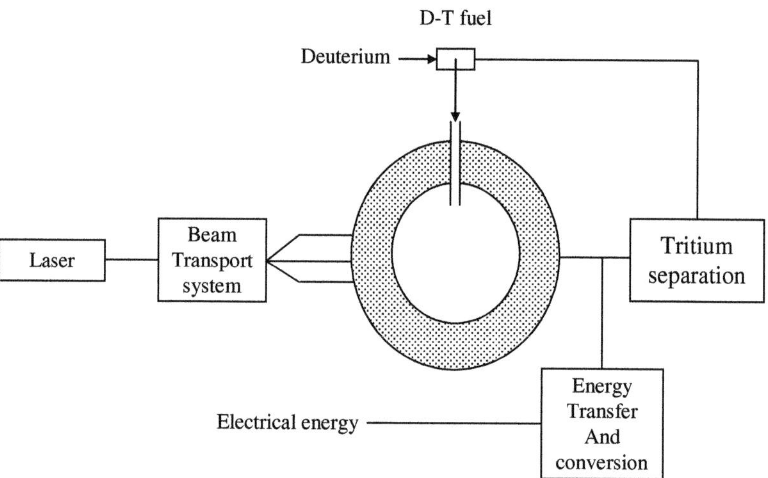

Fig. 16.7 A simplified block diagram of a laser fusion electric-generating system

reactions. This is achieved by placing lithium[17] (or its compounds) in a blanket surrounding the reactor chamber. The neutron emitted in the fusion reaction [see Eq. (16.3)] is absorbed by a lithium nucleus to give rise to tritium according to either of the following reactions:

$$_3\text{Li}^6 + {}_0\text{n}^1 \rightarrow \text{T}\left({}_1\text{H}^3\right) + \alpha\left({}_2\text{He}^4\right) + 4.79\,\text{MeV} \tag{16.15}$$

$$_3\text{Li}^7 + {}_0\text{n}^1 \rightarrow \text{T}\left({}_1\text{H}^3\right) + \alpha\left({}_2\text{He}^4\right) + {}_0\text{n}^1 - 2.46\,\text{MeV} \tag{16.16}$$

Note that in the second reaction, the neutron appears on the right-hand side also, which can again interact with a lithium nucleus to produce tritium.[18] The conceptual designs consist of liquid lithium being contained between two structural shells which enclose the reactor cavity. Liquid lithium will also be responsible for the removal of heat from the reactor and the running of a steam turbine. The heat exchangers and the lithium-processing equipment for separation of tritium are expected to be adjacent to the reactor (see Fig. 16.7).

We conclude by noting that there are many practical difficulties associated with the laser-driven fusion system, such as (a) delivery of a substantial part of the laser energy to the fuel before heating the fuel (which would, in effect, reduce the compression achieved) and before the shock wave disperses the fuel mass; (b) building of high-power lasers to a stage when the output energy from the system exceeds the input energy; and (c) design of complex targets and reliable production of such targets with extremely good surface finish, as surface irregularities of more than 1% of the thickness of the wall seem to yield very unstable compressions of thick shells. Also, projects are underway to study the feasibility of using particle beams like electron beams and ion beams as fusion drivers. It is much beyond the scope of the present book to go into the details of the various difficulties; the interested reader may look up Brueckner and Jorna (1974), Post (1973), and Stickley (1978) for further details.

---

[17]Lithium is quite abundant and has good heat transfer properties.

[18]In addition, neutron multiplication will occur in the blanket through (n, 2n) reactions; these neutrons would further produce more tritium.

Chapter 17
Light Wave Communications

17.1 Introduction

One of the most important and exciting applications of lasers lies in the field of communication. Since optical frequencies are extremely large compared to radio waves and microwaves, a light beam acting as a carrier wave is capable of carrying far more information in comparison to radio waves and microwaves. Light wave communications using hair-thin optical fibers as transmission media has become ubiquitous with the ever increasing demand for higher and higher speeds of communication. In this chapter, we will discuss briefly the concept of carrier wave communication which uses electromagnetic waves as carriers of information and then discuss some basic features of optical fibers and their application to light wave communication.

17.2 Carrier Wave Communication

Communication using electromagnetic waves is today the most reliable, economical, and fastest way of communicating information between different points. In any communication system, the information to be transmitted is generated at a source; gets transmitted through a channel such as atmosphere in radio broadcast, or electrical lines in telephone or wireless network in mobile communication, or optical fibers in a fiber-optic communication system; and finally reaches a receiver, which is the destination. Usually the channel through which the information propagates introduces loss in the signal and also distorts it to a certain extent. For a communication system to be reliable the channel must introduce minimal distortion to the signal. There should also be very little noise added by the channel so that the information can be retrieved without significant errors.

The electrical signals produced by various sources such as the telephone, computer, or video are not always suitable for transmission directly as such through the channel. These signals are made to modulate a high-frequency electromagnetic wave such as a radio wave, a microwave, or a light wave and it is this modulated electromagnetic wave that carries the information. Such a communication system is referred to as carrier wave communication.

K. Thyagarajan, A. Ghatak, *Lasers*, Graduate Texts in Physics,
DOI 10.1007/978-1-4419-6442-7_17, © Springer Science+Business Media, LLC 2010

There are different ways of modulating an electromagnetic wave in accordance with a given signal. The modulation can be either analog modulation or digital modulation. In the case of analog modulation, the amplitude, the phase, or the frequency of the carrier wave is changed in accordance with the signal amplitude, while in the case of digital modulation, the analog signal is first converted into a digital signal consisting of ones and zeroes which is then used to modulate the carrier. In the following we shall discuss these different schemes.

17.2.1 Analog Modulation

In analog modulation some characteristics of the carrier wave (amplitude, phase, or frequency) are modulated in accordance with the signal; the characteristic can take values continuously within a range.

Since the carrier wave is a sinusoidal wave, we can write the carrier wave as

$$V(t) = V_0 \sin(\omega t - \varphi) \tag{17.1}$$

where V_0 is a constant, ω represents the carrier frequency, and φ is an arbitrary phase. Here V represents either the voltage or the electric field of the electromagnetic wave. Amplitude, phase, and frequency modulations correspond to modulating the amplitude, phase, and frequency of the carrier wave.

17.2.1.1 Amplitude Modulation

In the case of amplitude modulation, the amplitude of the carrier wave is modulated in accordance with the signal to be sent. Thus we can write for an amplitude modulated wave as

$$V(t) = V_0\{1 + m(t)\}\sin(\omega t - \varphi) \tag{17.2}$$

where $m(t)$ represents the time-varying signal to be transmitted. As the signal amplitude changes, the amplitude of the modulated wave changes and thus the modulated wave carries the signal.

As an example if we consider the signal to be another sine wave with frequency Ω $(<<\omega)$, then we have the modulated wave as

$$V(t) = V_0\{1 + a \sin(\Omega t)\}\sin(\omega t - \varphi) \tag{17.3}$$

where a is a constant. Expanding the brackets and using the formulas for product of sine functions, we have

$$V(t) = V_0 \left\{ \sin(\omega t - \varphi) + \frac{1}{2}a \cos[(\omega - \Omega)t - \varphi] - \frac{1}{2}a \cos[(\omega + \Omega)t - \varphi] \right\}$$

$$\tag{17.4}$$

Hence the modulated wave now contains three frequencies ω, $\omega + \Omega$, and $\omega - \Omega$. These are the carrier frequencies, the upper side band, and the lower side band frequencies, respectively. Thus amplitude modulating a carrier wave by a sinusoidal wave generates two side bands. Since any general time-varying function can be analyzed in terms of sinusoidal functions, amplitude modulation of the carrier wave would result in the generation of an upper side band and a lower side band. If the maximum frequency of the signal is Ω_{max}, then the upper side band would lie between ω and $\omega + \Omega_{max}$ and the lower side band from $\omega - \Omega_{max}$ to ω. Both the side bands contain information on the entire signal.

Figure 17.1a shows a signal to be transmitted and Fig. 17.1b shows the carrier wave; note that the frequency of the carrier wave is much larger than the frequencies contained in the signal. Figure 17.1c shows the amplitude-modulated carrier wave. The signal now rides on the carrier as its amplitude modulation. At the receiver, the modulated carrier is demodulated and the signal can be retrieved.

As an example, let us assume that we wish to transmit speech. We first note that in order for the speech to be intelligible, it is sufficient to send signal content up to a frequency of 4000 Hz. Thus the electrical signal that comes out of the microphone into which the person speaks can be restricted to a frequency of 4000 Hz. In comparison, for sending high-fidelity music, the upper frequency is 20,000 Hz,

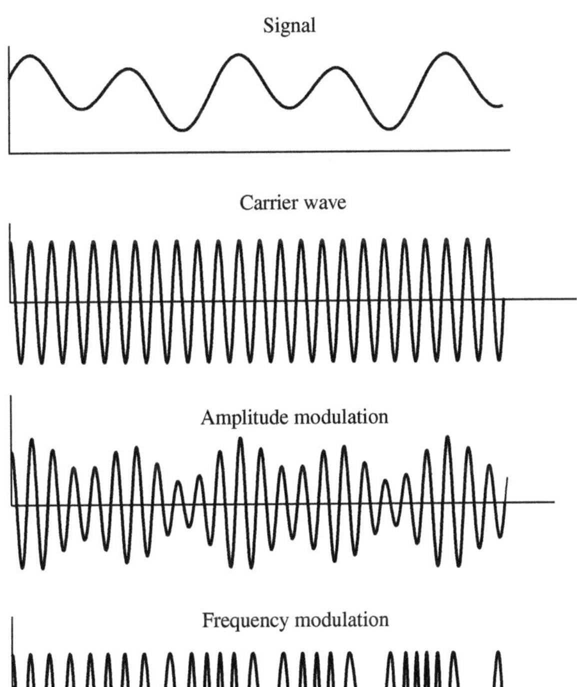

Fig. 17.1 (**a**) The signal to be transmitted; (**b**) the carrier wave using which the signal will be communicated; (**c**) amplitude-modulated carrier wave; and (**d**) frequency-modulated carrier wave

which is also the limit of frequency of human hearing. The electrical signal from the microphone could look like the one shown in Fig. 17.1a and if we are considering radio transmission, the radio wave on which the information rides will look like the one shown in Fig. 17.1b. The amplitude-modulated radio wave would then be like the one shown in Fig. 17.1c. It is this modulated radio wave that is broadcast through open space (which is the channel) and at the receiver (your radio set) it is demodulated and the signal is retrieved and you hear the speech.

An obvious question that arises is, "How can one send simultaneously more than one signal through the same channel?" In order to understand this we first note that the carrier wave shown in Fig. 17.1b is at a single frequency, while the amplitude-modulated signal shown in Fig. 17.1c has a spectrum, i.e., it has a range of frequencies. Thus if the signal occupies the frequencies up to 4000 Hz and if the carrier wave frequency is 1,000,000 Hz, then the amplitude-modulated wave has frequencies lying between 996,000 and 1,004,000 Hz (sum and difference of the carrier frequency and the maximum signal frequency). The information contained in the frequency range 996,000–1,000,000 Hz is the lower side band and the information contained in the frequency range 1,000,000–1,004,000 Hz is the upper side band and both these bands contain the entire information. Hence it is sufficient to send only one of the side bands, e.g., the components lying between 1,000,000 and 1,004,000 Hz in order for the receiver to retrieve the signal; this is referred to as upper side band transmission. Hence we see that to send one speech signal, we need to reserve the frequencies lying between 1,000,000 and 1,004,000 Hz, a band of 4000 Hz.

Now, in order to send another speech signal, we can choose a radio wave of frequency 1,004,000 Hz and send the modulated wave lying in the frequency band between 1,004,000 and 1,008,000 Hz. These frequencies lie outside the range of the frequencies of the first signal and hence will not interfere with that signal; similarly for more and more speech signals. Thus if we can use carrier frequencies over a range of say 1,000,000–3,000,000 Hz, then we can simultaneously send 2,000,000/4000 = 500 speech signals. This also makes it clear that the larger the range of frequencies of the carrier wave, the larger the number of channels that can be sent simultaneously. The range of available frequencies (which is a certain fraction of the carrier frequency) increases with the frequency of the carrier wave and this is the reason why light waves which have frequencies much higher than radio waves or microwaves can transmit much more information.

17.2.1.2 Frequency Modulation

In the case of frequency modulation, instead of modulating the amplitude of the carrier wave, its frequency is changed in accordance with the signal as shown in Fig. 17.1d. In this case, the signal information is contained in the form of the frequency of the signal. For the case of frequency-modulated signal, instead of Eq. (17.3) we would have

$$V(t) = V_0 \sin(\omega\{1 + am(t)\}t - \varphi) \tag{17.5}$$

As can be seen, in this case, the amplitude of the wave remains constant, while the frequency changes with time in accordance with the signal represented by $m(t)$. Equation (17.5) represents a wave which does not have just one frequency but many frequency components. The frequency spectrum in this case is not as simple as in the case of amplitude modulation. It can be shown that unlike the amplitude modulation case where the amplitude-modulated signal had a narrow upper and a narrow lower side bands, in the case of frequency modulation, the modulated signal contains many more frequency components. Hence an FM signal requires much larger bandwidth to transmit than does the AM signal. Thus for a given range of carrier frequencies, the number of independent channels that can be sent using frequency modulation would be smaller. In order to accommodate more channels, the carrier frequencies used in frequency modulation are much higher and fall in the range 30–300 MHz. Since the information is coded into the frequency of the carrier wave, the frequency-modulated waves are less susceptible to noise and this is quite apparent while listening to AM radio broadcast (medium-wave or short-wave channels) or FM radio broadcast.

In the above methods, simultaneous transmission of different independent signals is accomplished by reserving different carrier frequencies for different signals. This method is referred to as frequency division multiplexing. All the signals are propagating simultaneously through the transmission medium and the receiver can pick up any of the signals by filtering only the frequency band of interest to it, i.e., tune into the required signal.

17.2.2 *Digital Modulation*

The modulation scheme used in optical fiber communication is called digital modulation. The digital modulation scheme is based on the fact that an analog signal satisfying certain criteria can be represented by a digital signal. There is a theorem called the sampling theorem according to which a signal which is limited by a maximum frequency (also referred to as band-limited signal), i.e., has no frequency component above a certain frequency, is uniquely determined by its values at uniform time intervals spaced less than half of the inverse of the maximum frequency present in the signal. Thus if we consider speech which has frequencies below 4000 Hz, then the analog speech signal (like the one shown in Fig. 17.1a) can be represented uniquely by specifying the values of the signal at time intervals of less than 1/8000s. Thus if we sample the speech signal at 8000 times per second and if we are given the values of the signal at these times, then we can uniquely determine the original analog speech signal even though we are not told the value of the function at intermediate points. Figure 17.2 represents this fact; Fig. 17.2a shows again the same signal as Fig. 17.1a and Fig. 17.2b represents the sampled values at time intervals of 1/8000s. Thus instead of sending the analog signal, it is sufficient to send the values of the signal at specific times and this is sufficient to determine the signal uniquely.

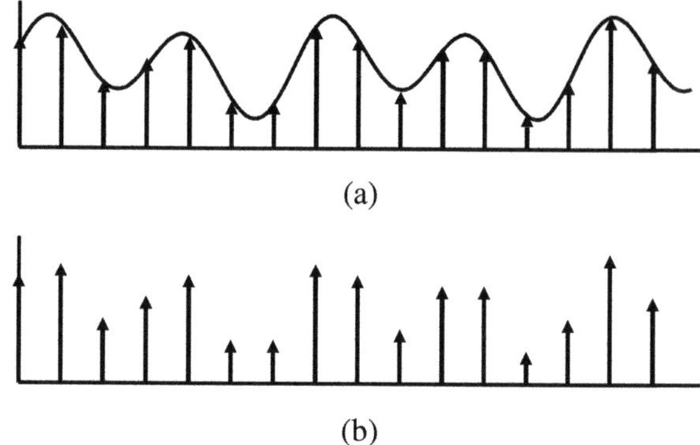

Fig. 17.2 (**a**) Sampling of the given analog signal and (**b**) sampled values of the signal. Note that even though the values of the signal between the samples are not specified, these can be uniquely determined from the sampled values

Instead of sending pulses of different amplitudes corresponding to different sampled values, it is usual to first convert the various pulse amplitudes into a binary signal which will consist of only two values of amplitudes, high amplitude referred to as 1 and low amplitude (usually 0) referred to as 0.

In the binary scheme, the rightmost place is the coefficient of 2^0, the next digit to the left is the coefficient of 2^1, the next one is the coefficient of 2^2, etc. Thus the sequence 101 in binary format, in decimal form represents the number $1 \times 2^2 + 0 \times 2^1 + 1 \times 2^0 = 5$. With three bits the largest decimal number would correspond to 111 which is 7. This number in the decimal system corresponds to the number $2^3 - 1$. Similarly the sequence of binary numbers 01100100 and 11000111 represents the numbers 100 and 199, respectively, in decimal system. Any integer between 0 and 255 $(= 2^8 - 1)$ can be represented by this sequence of 1s and 0s with eight "digits." If a number greater than 255 has to be represented, then we need to take a ninth digit before the digit corresponding to 128 and that would then correspond to the decimal number 256. Computers use the digital language for processing of information and today the binary representation is all pervasive as it is used in computer discs, digital video discs, etc.

17.2.2.1 Pulse Code Modulation

The most common modulation scheme employed in optical fiber communication is the pulse code modulation. In this, each of the amplitudes of the sampled values is represented by a binary number consisting of eight digits (see Fig. 17.3). Since the maximum decimal value with eight digits is 255 $(= 128 + 64 + 32 + 16 + 8 + 4 + 2 + 1)$, the maximum amplitude of the signal is restricted to 255 so that all integer values of signal can be represented by a sequence of 1s and 0s.

Now in pulse code modulation, the given analog signal is first sampled at an appropriate rate and then the sample values are converted to a binary form. The carrier is then modulated using the binary signal values to generate the modulated signal.

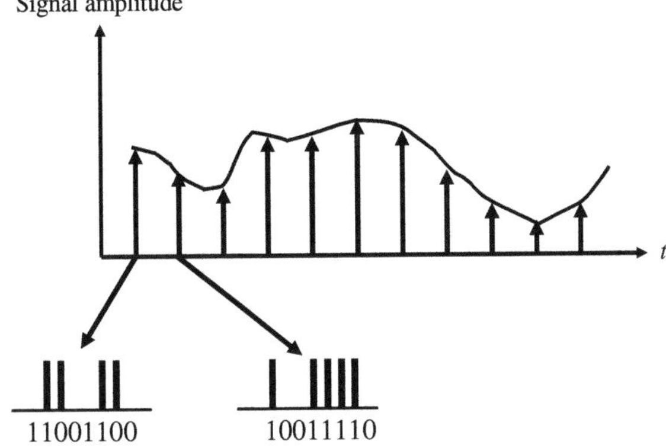

Fig. 17.3 Binary representation of each of the sampled signal values. In the figure each sampled value is represented by seven binary digits

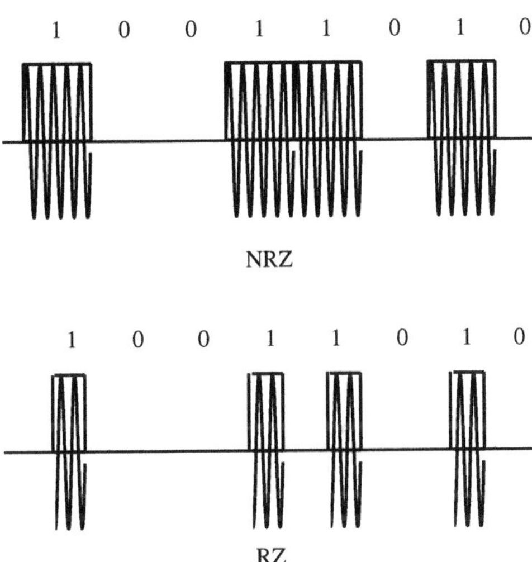

Fig. 17.4 Non-return-to-zero (NRZ) and return-to-zero (RZ) schemes

The most common scheme employed is called on–off keying (OOK). In this scheme, every digit 1 is represented by a high-amplitude value of the carrier and every digit 0 by zero amplitude of the carrier. Figure 17.4 shows the modulated wave corresponding to the binary sequence of seven digits 10011010.

In the scheme shown in Fig. 17.4a, the amplitude of the carrier does not return to zero when there are two adjacent 1s in the signal. This is referred to as non-return-to-zero (NRZ) scheme. There is another scheme called the return-to-zero (RZ) scheme in which the amplitude of the carrier returns to zero even if the adjacent digits are 1s (see Fig. 17.4b).

One of the major differences between the NRZ and RZ pulse sequences is the bandwidth requirement. To appreciate this we first note that in the NRZ scheme the fastest changes correspond to alternating sequence of 1s and 0s, while in the

Fig. 17.5 An alternating sequence of 1s and 0s corresponds to the maximum rate of change in NRZ, while a series of 1s corresponds to the maximum rate of change in RZ. The sinusoidal curves superimposed on the pulses correspond to the fundamental frequency of the pulse sequence

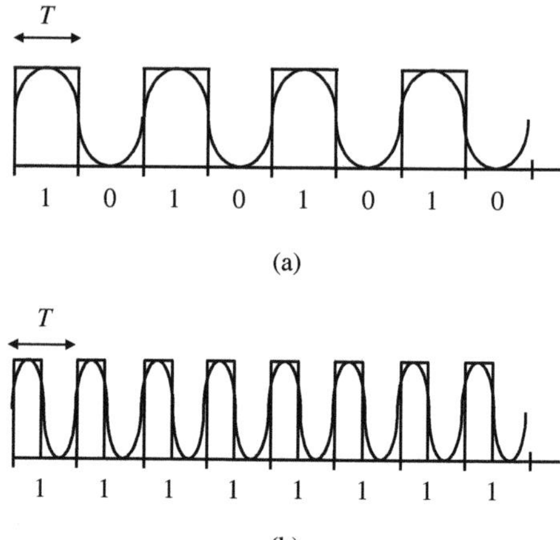

case of RZ, a sequence of 1s represents the fastest changes (see Fig. 17.5). From Fig. 17.5a it can be seen that the fundamental frequency component in the case of NRZ pulse sequence is $1/2T$, while in the case of RZ it is $1/T$. Hence for transmission without too much distortion, NRZ would require a bandwidth of at least $1/2T$, while RZ would need a bandwidth of $1/T$. If the bit rate is B, then $B = 1/T$ and thus the bandwidth requirements for NRZ and RZ are given as

$$
\begin{aligned}
\Delta f &\approx \frac{B}{2}; \quad \text{NRZ} \\
&\approx B; \quad \text{RZ}
\end{aligned}
$$

(17.6)

Thus the bandwidth requirements for RZ are more severe than those for NRZ, which is also expected since pulses in RZ format are narrower than the pulses in NRZ format. Thus, for example, a 2.5-Gb/s system would need 2.5 GHz in RZ format, while the same signal would require only 1.25 GHz in the NRZ format. Most of the fiber-optic communication systems of today use NRZ schemes for communication.

In the amplitude modulation scheme and the frequency modulation scheme, different independent signals are allocated different carrier frequencies leading to frequency division multiplexing. In digital modulation, the carrier frequency of different channels can be the same but different independent signals are multiplexed in the time domain. This concept is shown in Fig. 17.6, wherein two independent signals are sent using the pulse code modulation scheme, wherein the time slots occupied by the two signals are different. Thus the signals now overlap in the frequency domain but are sent at different times leading to what is referred to as time division multiplexing.

One of the greatest advantages of pulse code modulation scheme of communication is that the receiver has to detect only the presence or the absence of a pulse; the presence would correspond to 1, while the absence would correspond to 0. This

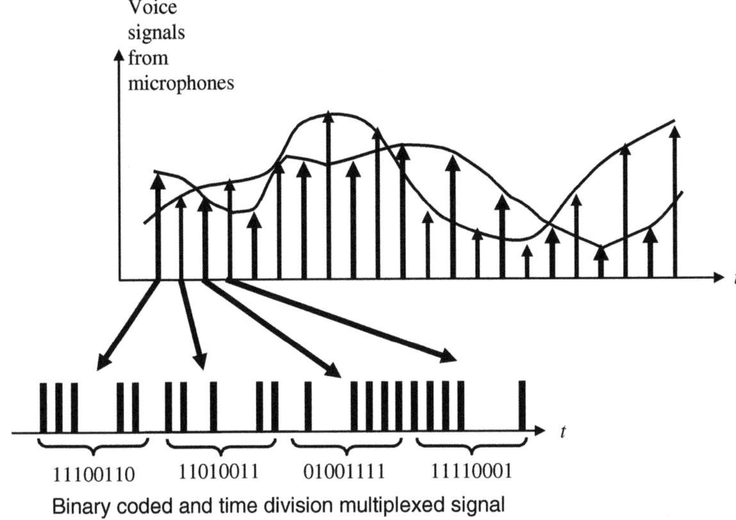

Fig. 17.6 Time division multiplexing of two independent signals

is unlike the case in amplitude modulation, wherein the receiver is supposed to precisely measure the amplitude of the signal. Since detection of the presence or the absence of a pulse is much more accurate than actual measurement of the amplitude, communication using pulse code modulation scheme suffers from much less distortion as compared to other schemes. Also for long-distance communication, the signals can be cleaned up of noise at regular intervals (at what are called as regenerators) and the accumulation of noise can be restricted. Thus for long-distance communication, pulse code modulation schemes are much preferred over any analog system.

17.2.2.2 Bit Rate Required for Speech

We have seen earlier that for transmitting speech, we need to send frequencies up to 4000 Hz and if these have to be represented by digital pulses, then we need to sample the signal at least 8000 times per second. Now each of these sampled values would be represented by the binary digit sequence. In telephony, each sampled value is represented by eight pulses (or 8 bits) and thus the number of pulses per second for each telephone channel would be 64,000. Thus speech signals require a bit rate of 64 kilobits per second (64 kbps or 64 kb/s). If we have a communication system capable of transmitting at the rate of 1 gigabits per second (1 Gb/s), then this would correspond to transmitting simultaneously (1,000,000,000/64,000) or 15,000 speech signals. Higher the bit rate, larger the capacity of information transmission. Table 17.1 gives the number of bits required for different information content per signal.

Table 17.1 Bit requirement of some common information-containing signals

Service	Bit requirement
1000 words of text	60×10^3 bits
Telephone	64×10^3 bits / s
20 volume encyclopedia	3×10^8 bits
Standard TV	100×10^6 bits / s
HDTV	1.2×10^9 bits / s

17.2.2.3 Standard Bit Rates

Table 17.2 gives the hierarchy of two common standard bit rates referred to as SDH (Synchronous Digital Hierarchy) and SONET (Synchronous Optical Network) that is used for data transmission over optical fiber networks. SDH is the international version published by the International Telecommunications Union (ITU), while SONET is the United States version of the standard published by the American National Standards Institute (ANSI).

Table 17.2 Hierarchy of digital signals in two common data rates

SDH signal	SONET signal	Bit rate (Mb/s)
STM-0	OC-1	51.840
STM-1	OC-3	155.520
STM-4	OC-12	622.080
STM-16	OC-48	2,488.320
STM-64	OC-192	9,953.280
STM-256	OC-768	39,813.120

17.3 Optical Fibers in Communication

We have just seen that light waves have a large information-carrying capacity. For long-distance communication using light waves, we need a medium of transmission and glass optical fibers are the preferred medium of transmission of information-carrying light waves. Light wave communication using glass fibers can transmit information at capacities of larger than 1 Tb/s (which is roughly equivalent to transmission of about 15 million simultaneous telephone conversations). This is certainly one of the extremely important technological achievements of the twentieth century and is one of the most important applications of lasers which is directly affecting society. The following sections provide an introduction to the propagation through optical fibers.

17.4 The Optical Fiber

At the heart of an optical communication system is the optical fiber that acts as the transmission channel carrying the light beam loaded with information. The light beam gets guided through the optical fiber due to the phenomenon of total internal reflection (often abbreviated as TIR). Figure 17.7 shows an optical fiber, which consists of a (cylindrical) central dielectric core (of refractive index n_1) cladded by a material of slightly lower refractive index n_2 ($<n_1$). The corresponding refractive index distribution (in the transverse direction) is given as

$$
\begin{aligned}
n &= n_1, & r<a \\
&= n_2, & r>a
\end{aligned}
\tag{17.7}
$$

where n_1 and n_2 ($< n_1$) represent the refractive indices of core and cladding, respectively, and a represents the radius of the core. We define a parameter Δ through the following equations:

$$
\Delta \equiv \frac{n_1^2 - n_2^2}{2n_1^2}
\tag{17.8}
$$

When $\Delta \ll 1$ (as is true for silica fibers), we may write

$$
\Delta \approx \frac{n_1 - n_2}{n_2} \approx \frac{n_1 - n_2}{n_1}
\tag{17.9}
$$

The necessity of a cladded fiber rather than a bare fiber, i.e., without a cladding, was felt because of the fact that for transmission of light from one place to another, the fiber must be supported, and supporting structures may considerably distort the fiber, thereby affecting the guidance of the light wave. This can be avoided by choosing a sufficiently thick cladding. Further, in a fiber bundle, in the absence of the cladding, light can leak through from one fiber to another leading to possible cross talk among the information carried between two different fibers. The idea of adding a second

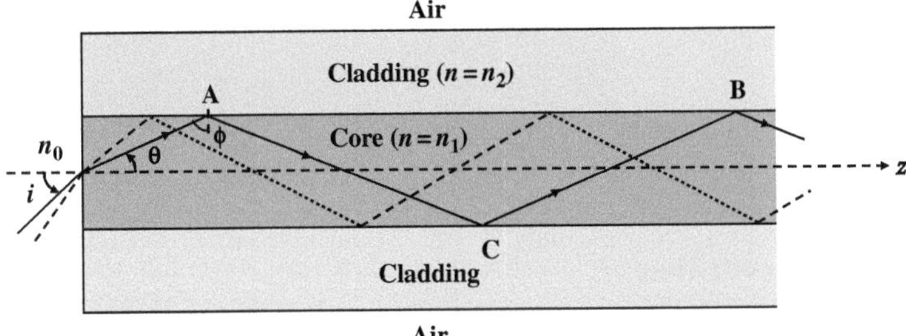

Fig. 17.7 (a) A glass fiber consists of a cylindrical central glass core cladded by a glass of slightly lower refractive index. (b) Light rays incident on the core–cladding interface at an angle greater than the critical angle are trapped inside the core of the fiber

layer of glass (namely, the cladding) came in 1955 independently from Hopkins and Kapany in the UK and van Heel in Holland. However, during that time the use of optical fibers was mainly in image transmission rather than in communications. Indeed, the early pioneering works in fiber optics (in the 1950s) by Hopkins, Kapany, and Van Heel led to the use of the fiber in optical devices.

Now, for a ray entering the fiber, if the angle of incidence ϕ at the core–cladding interface is greater than the critical angle ϕ_c [$=\sin^{-1}(n_2/n_1)$], then the ray will undergo TIR at that interface. We may mention here that the concept of rays is really valid only for multimode fibers, where the core radius a is large (≈ 25 μm or more). For a typical (multimoded) fiber, $a \approx 25$ μm, $n_2 \approx 1.45$ (pure silica), and $\Delta \approx 0.01$ giving a core index of $n_1 \approx 1.465$. The cladding is usually pure silica, while the core is usually silica doped with germanium; doping by germanium results in an increase in refractive index. Now, because of the cylindrical symmetry in the fiber structure, this ray will suffer TIR at the lower interface also and therefore get guided through the core by repeated total internal reflections. Figure 17.8 shows a photograph of light propagating through an optical fiber. The fiber is visible due to the phenomenon of Rayleigh scattering, which scatters a tiny part of the light propagating through the fiber, and makes the fiber visible. Rayleigh scattering is the same phenomenon that is responsible for the blue color of the sky and the red color of the rising or the setting sun. Even for a bent fiber, light guidance can occur through multiple total internal reflections as can be seen from Fig. 17.8.

Fig. 17.8 A long, thin optical fiber carrying a light beam. The fiber is visible due to Rayleigh scattered light

17.5 Why Glass Fibers?

Why are optical fibers made of glass? Professor W.A. Gambling, who is one of the pioneers in the field of fiber optics, quotes: "We note that glass is a remarkable material which has been in use in 'pure' form for at least 9000 years." The compositions remained relatively unchanged for millennia and its uses have been widespread. The three most important properties of glass, which makes it of unprecedented value, are the following [adapted from Gambling (1986)]:

1. First, there is a wide range of accessible temperatures where its viscosity is variable and can be well controlled unlike most materials, like water and metals, which remain liquid until they are cooled down to their freezing temperatures and then suddenly become solid. Glass, on the other hand, does not solidify at a discrete freezing temperature but gradually becomes stiffer and stiffer and eventually becoming hard. In the transition region it can be easily drawn into a thin fiber.

2. The second most important property is that highly pure silica is characterized by extremely low loss, i.e., it is highly transparent. Today in most commercially available silica fibers, 96% of the power gets transmitted after propagating through 1 km of optical fiber. This indeed represents a truly remarkable achievement.

3. The third most remarkable property is the intrinsic strength of glass. Its strength is about 2000,000 lb/in^2 so that a glass fiber of the type used in the telephone network and having a diameter (125 μm) of twice the thickness of a human hair can support a load of 40 lb.

Although for a common person, glass looks fragile, glass fibers are indeed extremely strong. It is the exposure of the glass to external atmosphere that leads to the formation of cracks, which then results in the fracture. In the case of optical fibers, these are drawn in extremely clean environment and are coated with polymers as they are being drawn. Covering the glass fiber with the polymer does not permit contact with the atmosphere and gives it the protection.

17.6 Attenuation of Optical Fibers

When light propagates through any medium, it suffers loss due to various mechanisms including those due to scattering, absorption, etc. Even the purest materials will have loss due to various intrinsic mechanisms such as scattering, absorption by the atoms and molecules forming the material, etc. In 1966 the most transparent glass available at that time had a loss of 1000 dB/km primarily due to trace amount of impurities present in the glass; a loss of 1000 dB/km (or equivalently 1 dB/m) implies that for every 10 m the power will fall by a factor of 10. Thus after propagating through 1 km of such a fiber, the output power will be negligible.

In 1966, Kao[1] and Hockham (1966) first suggested the use of optical fibers for communication and pointed out that for the optical fiber to be a viable communication medium, the losses (in the optical fiber) should be less than 20 dB/km. This

---

[1]Charles Kao was awarded the Nobel Prize in Physics in the year 2009 for *for groundbreaking achievements concerning the transmission of light in fibers for optical communication.*

suggestion triggered the beginning of serious research in removing the small amount of impurities present in the glass and developing low-loss optical fibers. Since 1966, there was a global effort to purify silica and in 1970, there was a major breakthrough: the first low-loss glass optical fibers were fabricated by Maurer, Keck, and Schultz (at Corning Glass Works, USA), with the fabricated optical fibers having a loss ~ 17 dB/km at a wavelength of 633 nm. This would imply that the power will fall by a factor of 10 in traversing through approximately 600 m length of the fiber.

Since then, the technology has been continuously improving and by late 1980s commercially available optical fibers had losses less than about 0.25 dB/km at a wavelength of about 1550 nm – a loss of 0.25 dB/km implies that the power will fall by a factor of 10 after propagating through 40 km length of the optical fiber.

Figure 17.9 shows a typical dependence of fiber attenuation (i.e., loss coefficient per unit length) as a function of wavelength of silica optical fibers that are commercially available today. It may be seen that the loss is less than 0.25 dB/km at a wavelength of about 1550 nm. The losses in optical fibers are caused due to various mechanisms such as Rayleigh scattering, absorption due to metallic impurities and water in the fiber, and due to intrinsic absorption by the silica molecule itself.

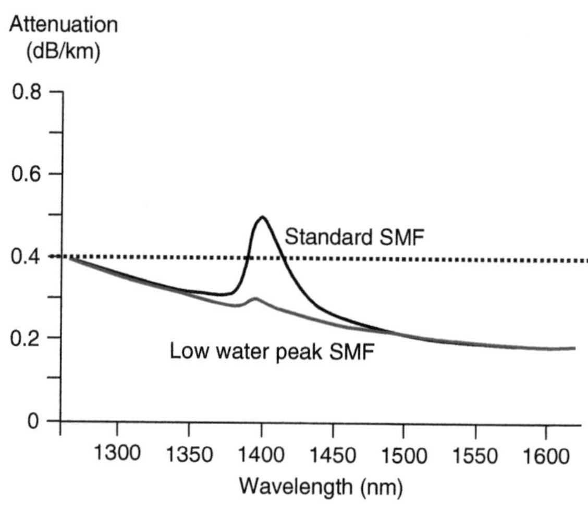

Fig. 17.9 A typical attenuation spectrum of an optical fiber SMF: Single mode optical fiber

Rayleigh scattering is a basic mechanism in which light gets scattered by very small inhomogeneities as it propagates through any medium. Rayleigh scattering loss is wavelength dependent and is such that shorter wavelengths scatter more than longer wavelengths. It is this phenomenon which is responsible for the blue color of the sky. As sunlight passes through the atmosphere, the component corresponding to blue color gets scattered more than the component corresponding to the red color (since blue wavelengths are shorter than red wavelengths). Thus more blue reaches our eye from the sky than does red and the sky looks blue. This is also the reason why the rising and setting sun looks red in color.

Rayleigh scattering causes loss of optical signals as they propagate through an optical fiber. Very small inhomogeneities present in the optical fiber scatter light out of the fiber leading to loss. In fact, this loss mechanism determines the ultimate loss of optical fibers. Since Rayleigh scattering loss decreases with increase in wavelength, optical fibers operating at higher wavelengths are expected to have lower losses if all other loss mechanisms are eliminated.

Apart from the Rayleigh scattering loss, any impurities present in the optical fiber would also cause absorption of the propagating light and would thus contribute to loss. Some of the primary impurities include metallic ions such as copper, chromium, iron, nickel, etc. Impurity levels of one part in a billion could cause increase in attenuation of 1 dB/km in the near-infrared region. Such impurities can be reduced to acceptable levels by using vapor-phase oxidation methods. Apart from impurity metal ions, one of the major contributors to loss is the presence of water dissolved in glass. An impurity level of just 1 part per million (1 ppm) of water can cause a loss of 4 dB/km at 1380 nm. This shows the level of purity that is required to achieve low-loss optical fibers.

The primary reason for the loss coefficient to decrease up to about 1550 nm is the Rayleigh scattering loss. The absorption peak around 1380 nm in Fig. 17.9 is primarily due to traces of water. If all impurities are completely removed, the absorption peak will disappear and we will have very low loss in the entire range of wavelength starting from 1250 to 1650 nm (see Fig. 17.9).

For wavelengths longer than about 1600 nm, the loss increases due to the absorption of infrared light by silica molecules themselves. This is an intrinsic property of silica and no amount of purification can remove this infrared absorption tail.

The first optical fiber communication systems operated at a wavelength around 1300 nm, where the material dispersion is negligible. However, the loss attains its absolute minimum value of about 0.2 dB/km when the wavelength is around 1550 nm – as a consequence of which the distance between two consecutive repeaters (used for amplifying and reshaping the attenuated signals) could be as large as 250 km. Furthermore, the 1550-nm window has become extremely important in view of the availability of erbium-doped fiber amplifiers (see Chapter 12).

Apart from these loss mechanisms, any deviation from straightness in the laying of the optical fiber would also result in loss. Thus when a fiber is bent, this leads to an additional loss. This additional loss can be kept to a minimum by appropriately laying the fiber cable in the link. Sharp bends are usually avoided in the laying of the fiber. Any random disturbances in the geometry of the fiber along its length would also lead to loss and thus during fabrication the uniformity of the fiber characteristics has to be maintained.

Figure 17.10 shows the loss spectrum of a typical commercially available singlemode fiber. Also shown are the various bands of wavelengths that are used in a wavelength division multiplexed (WDM) communication system. The most widely used band is the C-band lying between 1530 and 1565 nm where erbium doped fibers amplifiers operate.

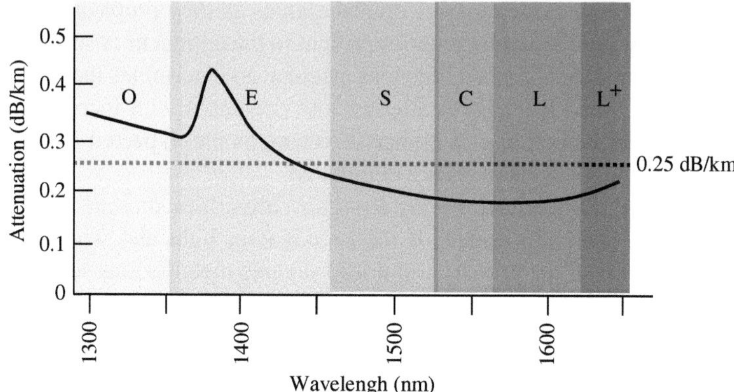

Fig. 17.10 Attenuation spectrum of a commercially available fiber from Corning Inc., USA. The low-loss spectrum is divided into various bands: O, old band (1260–1360 nm); E, extended band (1360–1460 nm); S, short band (1460–1530 nm); C, conventional band (1530–1565 nm); L, long band (1565–1625 nm)

17.7 Numerical Aperture of the Fiber

We return to Fig. 17.7 and consider a ray which is incident on the entrance aperture of the fiber making an angle i with the axis. Let the refracted ray make an angle θ with the axis. Assuming the outside medium to have a refractive index n_0 (which for most practical cases is unity), we get

$$\frac{\sin i}{\sin \theta} = \frac{n_1}{n_0} \tag{17.10}$$

Obviously if this ray has to suffer total internal reflection at the core–cladding interface

$$\sin \phi (= \cos \theta) > \frac{n_2}{n_1} \tag{17.11}$$

Thus

$$\sin \theta < \left[1 - \left(\frac{n_2}{n_1} \right)^2 \right]^{1/2} \tag{17.12}$$

and we must have $i < i_\mathrm{m}$, where

$$\sin i_\mathrm{m} = \left(n_1^2 - n_2^2 \right)^{1/2} = n_1 \sqrt{2\Delta} \tag{17.13}$$

and we have assumed $n_0 = 1$, i.e., the outside medium is assumed to be air. Thus, if a cone of light is incident on one end of the fiber, it will be guided through it provided the semiangle of the cone is less than i_m. This angle is a measure of the light-gathering power of the fiber and as such, one defines the numerical aperture

(NA) of the fiber by the following equation:

$$\text{NA} = \sin i_{\text{m}} = \sqrt{n_1^2 - n_2^2} \approx n_1\sqrt{2\Delta} \qquad (17.14)$$

Example 17.1 For a typical step-index (multimode) fiber with $n_1 \approx 1.45$ and $\Delta \approx 0.01$, we get

$$\sin i_{\text{m}} \approx 0.205 \quad \Longrightarrow \quad i_{\text{m}} \approx 12°$$

17.8 Multimode and Single-Mode Fibers

In this section we will try to understand the difference between a single-mode and a multimode fiber. Both types of fibers are used in optical communication systems. We first introduce the concept of modes: a mode is a transverse field distribution which propagates along the fiber without any change in its field distribution except for a change in phase. Mathematically, it is defined by the equation

$$\Psi(x, y, z, t) = \psi(x, y)e^{i(\omega t - \beta z)} \qquad (17.15)$$

where $\psi(x, y)$ represents the transverse field profile and β represents the propagation constant; the propagation of the mode is in the z-direction. The quantity β is similar to k for a plane wave given by $e^{i(\omega t - kz)}$. A waveguide (like an optical fiber) is characterized by a finite number of modes which are guided by the waveguide – each mode is described by a definite transverse field distribution $\psi(x, y)$ corresponding to a definite value of β. The precise form of $\psi(x, y)$ (and the corresponding value of the propagation constant β) is obtained by solving Maxwell's equations. Here we will just present results; interested readers may look up any textbook listed at the end of the book. For a step-index fiber defined by Eq. (17.7), we define a dimensionless parameter defined by the following equation:

$$V = \frac{2\pi}{\lambda_0}a\sqrt{n_1^2 - n_2^2} = \frac{2\pi}{\lambda_0}an_1\sqrt{2\Delta} \qquad (17.16)$$

λ_0 is the free space wavelength of the light beam and Δ is defined by Eq. (17.8). The parameter V (which also depends on the operating wavelength λ_0) is known as the "waveguide parameter" and is an extremely important quantity characterizing an optical fiber. For a step-index fiber if

$$V < 2.4045 \qquad (17.17)$$

the fiber is said to be a single-mode fiber. For a given fiber, the wavelength for which $V = 2.4045$ is known as the *cutoff wavelength* and is denoted by λ_c and the fiber will be single moded for $\lambda_0 > \lambda_c$ (see Example 17.2). A multimoded fiber would have a V value larger than 2.4045.

If we consider a medium with refractive index n, the propagation constant of an electromagnetic plane wave in this medium would be $k_0 n$. Thus the refractive index is the ratio of the propagation constant in the medium to k_0, which is the propagation constant of the wave in free space. In a similar manner, the quantity

$$n_{\text{eff}} = \frac{\beta}{k_0} \tag{17.18}$$

is referred to as the effective index of the mode. In an optical fiber, the effective index plays the same role as refractive index in the case of a bulk medium.

Example 17.2: Consider a step-index fiber with $n_2 = 1.447$, $\Delta = 0.003$, and $a = 4.2\,\mu\text{m}$. Thus

$$V = \frac{2\pi}{\lambda_0} \times 4.2 \times 1.447 \times \sqrt{0.006} \approx \frac{2.958}{\lambda_0}$$

where λ_0 is measured in micrometers. Thus for

$$\lambda_0 > \frac{2.958}{2.4045} \approx 1.23\,\mu\text{m}$$

the fiber will be single moded. Thus in this example, the *cutoff wavelength* $\lambda_c = 1.23\,\mu\text{m}$ and the fiber will be single moded for $\lambda_0 > 1.23\,\mu\text{m}$. If the fiber is operating at 1300 nm, then

$$V = \frac{2.958}{1.3} \approx 2.275$$

and the fiber will be single moded.

Example 17.3: Let us consider a fiber with $n_2 = 1.444$, $\Delta = 0.0075$, and $a = 2.3\,\mu\text{m}$ for which

$$V = \frac{2\pi}{\lambda_0} \times 2.3 \times 1.444 \times 0.015 \approx \frac{2.556}{\lambda_0}$$

and therefore the cutoff wavelength will be $\lambda_c = 2.556/2.4045 = 1.06\,\mu\text{m}$. If we operate at $\lambda_0 = 1.55\,\mu\text{m}$

$$V \approx 1.649$$

and the fiber will be single moded.

17.9 Single-Mode Fiber

For a step-index fiber with $0 < V < 2.4048$ we will have only one guided mode, namely the LP_{01} mode also referred to as the fundamental mode (LP stands for linearly polarized). Such a fiber is referred to as a single-mode fiber and is of tremendous importance in optical fiber communication systems.

For a single-mode step-index fiber, a convenient empirical formula for $b(V)$ is given as

$$b(V) = \left(A - \frac{B}{V}\right)^2, \quad 1.5 \lesssim V \lesssim 2.5 \tag{17.19}$$

where $A \approx 1.1428$ and $B \approx 0.996$ and b, also referred to as the normalized propagation constant, is defined through the following equation:

$$b(V) = \frac{n_{\text{eff}}^2 - n_2^2}{n_1^2 - n_2^2} \qquad (17.20)$$

17.9.1 Spot Size of the Fundamental Mode

The transverse field distribution associated with the fundamental mode of a single-mode fiber is an extremely important quantity and it determines various important parameters like splice loss at joints between fibers, launching efficiencies from sources, bending loss, etc. For a step-index fiber, one has analytical expression for the fundamental field distribution in terms of Bessel functions. For most single-mode fibers with a general transverse refractive index profile, the fundamental mode field distributions can be well approximated by a Gaussian function, which may be written in the form

$$\psi(x, y) = A\, e^{-\frac{x^2+y^2}{w^2}} = A\, e^{-\frac{r^2}{w^2}} \qquad (17.21)$$

where w is referred to as the spot size of the mode field pattern and $2w$ is called the mode field diameter (MFD). MFD is a very important characteristic of a single-mode optical fiber. For a step-index fiber one has the following empirical expression for w (see Marcuse (1978)):

$$\frac{w}{a} \approx 0.65 + \frac{1.619}{V^{3/2}} + \frac{2.879}{V^6}; \qquad 0.8 \leq V \leq 2.5 \qquad (17.22)$$

where a is the core radius. As an example, for the step-index fiber considered earlier and operating at 1300 nm, we have $V \approx 2.28$ giving $w \approx 4.8$ μm. Note that the spot size is larger than the core radius of the fiber; this is due to the penetration of the modal field into the cladding of the fiber. The same fiber will have a V value of 1.908 at $\lambda_0 = 1550$ nm giving a value of the spot size ≈ 5.5 μm. *Thus, in general, the spot size increases with wavelength.* The standard single-mode fiber designated as G.652 fiber for operation has an MFD of 9.2 ± 0.4 μm at 1310 nm and an MFD of 10.4 ± 0.8 μm at 1550 nm.

For $V \geq 10$, the number of modes (for a step-index fiber) is approximately $\frac{1}{2}V^2$ and the fiber is said to be a multimoded fiber. Different modes (in a multimoded fiber) travel with different group velocities leading to what is known as intermodal dispersion; in the language of ray optics, this is known as ray dispersion arising due to the fact that different rays take different amounts of time in propagating through the fiber. Indeed in a highly multimoded fiber, we can use ray optics to calculate pulse dispersion.

17.10 Pulse Dispersion in Optical Fibers

As discussed earlier, digital communication systems, information to be sent is first coded in the form of pulses and then these pulses of light are transmitted from the transmitter to the receiver where the information is decoded. Larger the number of pulses that can be sent per unit time and still be resolvable at the receiver end, larger would be the transmission capacity of the system. A pulse of light sent into a fiber broadens in time as it propagates through the fiber; this phenomenon is known as pulse dispersion and occurs primarily because of the following mechanisms:

1. In multimode fibers, dispersion is caused by different rays taking different times to propagate through a given length of the fiber. In the language of wave optics, this is known as *intermodal dispersion* because it arises due to different modes traveling with different group velocities.
2. Any given light source emits over a range of wavelengths and, because of the dependence of refractive index on wavelength, different wavelengths take different amounts of time to propagate along the same path. This is known as material dispersion and obviously, it is present in both single-mode and multimode fibers.
3. In single-mode fibers, since there is only one mode, there is no intermodal dispersion. However, apart from material dispersion, we have what is known as waveguide dispersion. Physically, this arises due to the fact that the spot size (of the fundamental mode) depends explicitly on the wavelength.
4. A single-mode fiber can support two orthogonally polarized LP_{01} modes. In a perfectly circular core fiber laid along a perfectly straight path, the two polarizations propagate with the same velocity. However due to small random deviations from circularity of the core or due to random bends and twists present in the fiber, the orthogonal polarizations travel with slightly different velocities and get coupled randomly along the length of the fiber. This phenomenon leads to polarization mode dispersion (PMD) which becomes important for high-speed communication systems operating at 40 Gb/s and higher.

Obviously, waveguide dispersion and polarization mode dispersion are present in multimode fibers also – however, the effects are very small and can be neglected.

17.10.1 Dispersion in Multimode Fibers

A broad class of *multimoded* graded index fibers can be described by the following refractive index distribution:

$$n^2(r) = n_1^2 \left[1 - 2\Delta \left(\frac{r}{a}\right)^q \right], \qquad 0 < r < a$$
$$= n_2^2 = n_1^2 \left(1 - 2\Delta\right), \qquad r > a$$

(17.23)

where r corresponds to a cylindrical radial coordinate, n_1 represents the value of the refractive index on the axis (i.e., at $r = 0$), and n_2 represents the refractive index of the cladding; $q = 1$, 2, and ∞ correspond to the linear, parabolic, and step-index profiles, respectively. Equation (17.23) describes what is usually referred to as a power law profile, which gives an accurate description of the refractive index variation in most multimoded fibers. The total number of modes in a highly multimoded graded index optical fiber characterized by Eq. (17.23) are approximately given as

$$N \approx \frac{q}{2\,(2+q)}V^2 \tag{17.24}$$

Thus, a parabolic index ($q = 2$) fiber with $V = 10$ will support approximately 25 modes. Similarly, a step-index ($q = \infty$) fiber with $V = 10$ will support approximately 50 modes. When the fiber supports such a large number of modes, then the continuum (ray) description gives very accurate results. For the power law profile, it is possible to calculate the pulse broadening due to the fact that different rays take different amount of time in traversing a certain length of the fiber; details can be found in many text books; see, for example, Ghatak and Thyagarajan (1998). The time taken to propagate through a length L of a multimode fiber described by a q-profile (see Eq. 17.23) is given as

$$\tau(L) = \left(A\tilde{\beta} + \frac{B}{\tilde{\beta}}\right) L \tag{17.25}$$

where

$$A = \frac{2}{c(2+q)}; B = \frac{qn_1^2}{c(2+q)} \tag{17.26}$$

and for rays guided by the fiber, $n_2 < \tilde{\beta} < n_1$. Using Eq. (17.25) we can estimate the intermodal dispersion in fibers with different q values. Thus, for $n_1 \approx 1.46$ and $\Delta \approx 0.01$, the dispersion would be 50 ns/km for a step-index fiber ($q = \infty$), 0.25 ns/km for a parabolic index fiber ($q = 2$), and 0.0625 ns/km for $q = 2\text{–}2\Delta$ (referred to as the optimum profile exhibiting minimum dispersion).

Thus we find that for a parabolic index fiber, the intermodal (or ray) dispersion is reduced by a factor of about 200 in comparison to the step-index fiber and for the optimum profile there is a further reduction by a factor of 4. It is because of this reason that first- and second-generation optical communication systems used near parabolic index fibers. In order to further decrease the pulse dispersion, it is necessary to use single-mode fibers because there will be no intermodal dispersion. However, in all fiber-optic systems we will have material dispersion which is a characteristic of the material itself and not of the waveguide; we will discuss this in the following section.

17.10.2 Material Dispersion

Above we have considered the broadening of an optical pulse due to different rays taking different amounts of time to propagate through a certain length of the fiber. However, every source of light has a certain wavelength spread which is often referred to as the *spectral width of the source*. An LED would have a spectral width of about 25 nm and a typical laser diode (LD) operating at 1300 nm would have a spectral width of about 2 nm or less. The pulse broadening (due to wavelength dependence of the refractive index) is given in terms of the material dispersion coefficient D_{m} (which is measured in picoseconds per kilometer and nanometer) and is defined as

$$D_{\mathrm{m}}\,(\mathrm{ps/km\,nm}) = -\frac{10^4}{3\lambda_0}\left[\lambda_0^2\frac{\mathrm{d}^2 n}{\mathrm{d}\lambda_0^2}\right] \tag{17.27}$$

λ_0 is measured in micrometer and the quantity inside the square brackets is dimensionless. Thus D_{m} represents the material dispersion in picoseconds per kilometer length of the fiber per nanometer spectral width of the source. At a particular wavelength, the value of D_{m} is a characteristic of the material and is (almost) the same for *all* silica fibers. When D_{m} is negative, it implies that the longer wavelengths travel faster; this is referred to as normal group velocity dispersion (GVD). Similarly, a positive value of D_{m} implies that shorter wavelengths travel faster; this is referred to as anomalous GVD.

The spectral width $\Delta\lambda_0$ of an LED operating around $\lambda_0 = 825$ nm is about 20 nm; at this wavelength for pure silica $D_m \approx 84.2$ ps/(km nm). Thus a pulse will broaden by 1.7 ns/km of fiber. It is interesting to note that for operation around $\lambda_0 \approx 1300$ nm [where $D_{\mathrm{m}} \approx 2.4$ ps/(km nm)], the resulting material dispersion is only 50 ps/km of the fiber. The very small value of $\Delta\tau_m$ is due to the fact that the group velocity is approximately constant around $\lambda_0 = 1300$ nm. Indeed the wavelength $\lambda_0 \approx 1270$ nm is usually referred to as the zero material dispersion wavelength, and it is because of such low material dispersion that the optical communication systems shifted their operation to around $\lambda_0 \approx 1300$ nm.

Optical communication systems in operation today use LDs (laser diodes) with $\lambda_0 \approx 1550$ nm having a spectral width of about 2 nm. At this wavelength the material dispersion coefficient is 21.5 ps/(km nm) and the material dispersion $\Delta\tau_m$ would be 43 ps/km.

17.10.3 Dispersion and Bit Rate

In multimode fibers, the total dispersion consists of intermodal dispersion ($\Delta\tau_i$) and material dispersion ($\Delta\tau_m$) and is given as

$$\Delta\tau = \sqrt{(\Delta\tau_i)^2 + (\Delta\tau_m)^2} \tag{17.28}$$

In the NRZ scheme the maximum permissible bit rate is approximately given as

$$B_{\text{max}} \approx \frac{0.7}{\Delta\tau} \tag{17.29}$$

Operation around 1310 nm minimizes $\Delta\tau_{\text{m}}$ and hence almost all multimode fiber systems operate at this wavelength region with optimum refractive index profiles having small values of $\Delta\tau_i$.

17.10.4 Dispersion in Single-Mode Fibers

In the case of a single-mode optical fiber, the effective index n_{eff} ($= \beta/k_0$) of the mode depends on the core and cladding refractive indices as well as the waveguide parameters (refractive index profile shape and radii of various regions). Hence n_{eff} would vary with wavelength even if the core and cladding media were assumed to be dispersionless (i.e., the refractive indices of core and cladding are assumed to be independent of wavelength). This dependence of effective index on wavelength is due to the waveguidance mechanism and is referred to as *waveguide dispersion.* Waveguide dispersion can be understood from the fact that the effective index of the mode depends on the fraction of power in the core and the cladding at a particular wavelength. As the wavelength changes, this fraction also changes. Thus even if the refractive indices of the core and the cladding are assumed to be independent of wavelength, the effective index will change with wavelength. It is this dependence of $n_{\text{eff}}(\lambda_0)$ that leads to waveguide dispersion.

Thus the total dispersion in the case of a single-mode optical fiber can be attributed to two types of dispersion, namely material dispersion and waveguide dispersion. It can be shown that the total dispersion coefficient D is given to a good accuracy by the sum of material (D_{m}) and waveguide (D_{w}) dispersion coefficients (see e.g., Ghatak and Thyagarajan, 1998). The material contribution is given by Eq. (17.27), while the waveguide contribution for a step-index fiber is given as

$$D_{\text{w}} = -\frac{n_2\Delta}{c\lambda_0}\left(V\frac{\mathrm{d}^2(bV)}{\mathrm{d}V^2}\right) \tag{17.30}$$

A simple empirical expression for waveguide dispersion for step-index fibers is

$$D_{\text{w}}\,[\text{ps/(km\,nm)}] = -\frac{n_2\Delta}{3\lambda_0} \times 10^7[0.080 + 0.549(2.834 - V)^2] \tag{17.31}$$

where λ_0 is measured in nanometers.

In the single-mode regime, the quantity within the bracket in Eq. (17.30) is usually positive; hence the waveguide dispersion is negative. Since the sign of material dispersion depends on the operating wavelength region, it is possible that the two effects, namely material and waveguide dispersions, cancel each other at a certain wavelength. Such a wavelength, which is a very important parameter of single-mode fibers, is referred to as the zero-dispersion wavelength (λ_{ZD}). For typical step-index fibers, the zero-dispersion wavelength falls in the 1310-nm-wavelength window. Since the lowest loss in an optical fiber occurs at a wavelength of 1550 nm and optical amplifiers are available in the 1550-nm window, fiber designs can be modified to shift the zero-dispersion wavelength to the 1550-nm-wavelength window. Such fibers are referred to as dispersion-shifted fibers (with zero dispersion around 1550 nm) or non-zero dispersion-shifted fibers (with finite but small dispersion around 1550 nm). With proper fiber refractive index profile design, it is also possible to have flat dispersion spectrum leading to dispersion-flattened designs. Figure 17.11 shows the total dispersion in three standard types of fibers, namely G.652, G.653, and G.655 fibers. The G.655 fibers have a small but finite dispersion around the 1550 nm wavelength. The small dispersion is required to avoid the nonlinear optical effect referred to as four wave mixing.

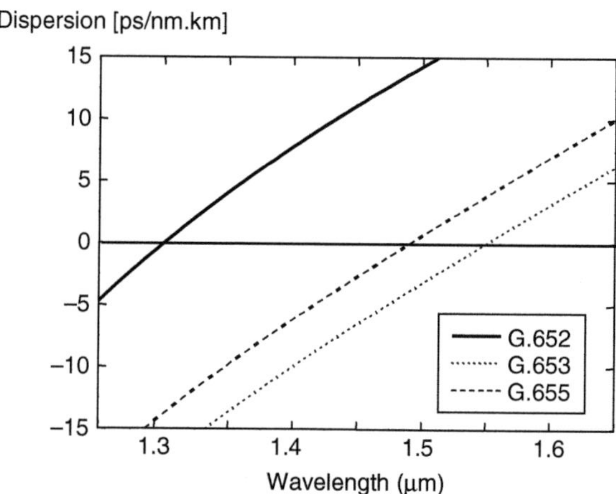

Fig. 17.11 Wavelength dependence of dispersion coefficient of single-mode fibers with different refractive index profiles

It appears that when an optical fiber is operated at the zero-dispersion wavelength, the pulses will not suffer any dispersion at all. In fact zero dispersion only signifies that the second-order dispersive effects are absent. In this case the next higher order dispersion, namely third-order dispersion, will become the dominating term in determining the dispersion. Thus in the absence of second-order dispersion, we can write for dispersion suffered by a pulse as

$$\Delta\tau = \frac{L\,(\Delta\lambda_0)^2}{2}\,\frac{\mathrm{d}D}{\mathrm{d}\lambda_0} \tag{17.32}$$

where $S = \mathrm{d}D/\,\mathrm{d}\lambda_0$ represents the dispersion slope at zero-dispersion wavelength and is measured in units of picosecond per kilometer and square nanometer. Third-order dispersion becomes important when operating close to the zero-dispersion wavelength. In the presence of only third-order dispersion, the pulse does not

Table 17.3 Values of dispersion and dispersion slope for some standard fibers at 1550 nm

Fiber type	D [ps/(km nm)]	S [ps/(km nm^2)]
Standard SMF (G.652)	17	0.058
LEAF (Corning)	4.2	0.085
Truewave-reduced slope (OFS)	4.5	0.045
TeraLight (Alcatel)	8.0	0.057

remain symmetric. Table 17.3 lists values of D and S for some standard fibers at 1550 nm.

17.10.5 Dispersion and Maximum Bit Rate in Single-Mode Fibers

In a digital communication system employing light pulses, pulse broadening would result in an overlap of adjacent pulses, resulting in intersymbol interference leading to errors in detection. Apart from this, since the energy in the pulse gets reduced within the time slot, the corresponding signal-to-noise ratio (SNR) will decrease. One can offset this by increasing the power in the pulses. This additional power requirement is termed as *dispersion power penalty*. Increased dispersion would imply increased power penalty.

In order to keep the interference between adjacent bits below a specified level, the root mean square width of the dispersed pulse needs to be kept below a certain fraction ε of the bit period. For a 2 dB power penalty, $\varepsilon \sim 0.491$. Using this condition we can estimate the maximum bit rate B for a given link length L and dispersion coefficient D operating at 1550 nm as

$$B^2DL < 1.9 \times 10^5 \, \text{Gb}^2\text{ps/nm} \tag{17.33}$$

where B is measured in gigabits per second, D in picoseconds per kilometer and nanometer and L in kilometer. Thus for a bit rate of 2.5 Gb/s the maximum allowed dispersion $(D.L)$ is approximately 30,400 ps/nm, while for a bit rate of 10 Gb/s the maximum allowed dispersion is 1900 ps/nm.

Problems

Problem 17.1 Consider an analog signal given as

$$f(t) = 10 + 2\sin 5\pi t + \cos 8\pi t + 3\sin\left(15\pi t + \frac{\pi}{9}\right)$$

where t is measured in seconds. At what rate would you sample so that the analog signal can be retrieved from the digitized signal?

Solution The frequencies present in the given signal are 0, 2.5, 4, and 7.5 Hz. Hence the minimum sampling rate for the signal is 15 samples per second.

Problem 17.2 If in the above problem each sample is represented by 8 bits, then what would be the bit rate?

Solution The bit rate would be $15 \times 8 = 120$ bits/s.

Problem 17.3 Consider a communication system having a rise time of 20 ns. What bit rates can it support when operated in NRZ and RZ formats?

Solution Using Eq. (17.6) we can obtain the bit rates that can be supported by the system as 35 and 17.5 MHz.

Problem 17.4 A typical digital TV channel required bit rates of 10 Mb/s. Can a system having a rise time of 50 ns support this transmission?

Solution For sending 10 Mb/s signal the rise time should be smaller than 35 ns for RZ coding and 70 ns for NRZ coding. Hence the system can support the transmission only if NRZ coding is used.

Problem 17.5 High-definition television (HDTV) requires higher bit rates than does standard TV. Typical bit rates would be about 16 Mb/s. What is the rise time requirement of the system if I need to send 10 HDTV channels through the same link?

Solution The bit rate for 10 HDTV channels would be 160 Mb/s. Thus the rise time requirement in the NRZ scheme would be about 4.3 ns.

Problem 17.6 Consider an RC circuit and assume that a step voltage V_0 is applied at $t = 0$. Relate the rise time of the RC circuit to its bandwidth.

Solution It is well known that when a step voltage is applied to an RC circuit, then the voltage variation with time is given as

$$V_c(t) = V_0 \left(1 - e^{-t/RC} \right)$$

Thus the voltage rises to 10% of the peak value V_0 in a time given as

$$0.1 V_0 = V_0 \left(1 - e^{-t_{0.1}/RC} \right)$$

or

$$t_{0.1} = RC \ln \left(\frac{1}{0.9} \right)$$

Similarly the time taken to rise to 90% of the peak value is given as

$$t_{0.9} = RC \ln \left(\frac{1}{0.1} \right)$$

Hence the rise time is

$$T_r = (t_{0.9} - t_{0.1}) = RC \ln (9) \approx 2.2RC$$

The 3-dB bandwidth of an RC circuit is given as

$$\Delta f = \frac{1}{2\pi RC}$$

Hence from the above equations we get

$$T_r \approx \frac{0.35}{\Delta f}$$

Problem 17.7 Consider a Gaussian function given as

$$f(t) = \frac{1}{\sqrt{2\pi}\,\sigma} e^{-t^2/2\sigma^2}$$

where σ is called the root mean square (RMS) pulse width. Relate the full width at half maximum (FWHM) of this function to its bandwidth.

Solution The 50% rise time $t_{0.5}$ is obtained from the below equation:

$$f(t_{0.5}) = 0.5f(0)$$

which gives

$$t_{0.5} = \sqrt{2\ln 2}\,\sigma$$

Thus the FWHM t_{FWHM} is $2t_{0.5}$.
The Fourier transform of $f(t)$ is given as

$$F(\omega) = \frac{1}{\sqrt{2\pi}} e^{-\omega^2\sigma^2/2}$$

Thus the 3-dB bandwidth is defined as the frequency $f_{0.5}$ ($=2\pi/\omega$) at which the function has fallen to a value half of that at zero frequency. This can be obtained as

$$f_{0.5} = \frac{\sqrt{2\ln 2}}{2\pi\sigma}$$

Hence we get

$$f_{0.5} = \frac{0.44}{t_{\text{FWHM}}}$$

Problem 17.8 Using Eqs. (17.19) and (17.30) obtain an expression for waveguide dispersion. Obtain the value of waveguide dispersion at 1310 and 1550 nm for a standard single-mode fiber.

Problem 17.9 Using Eq. (17.22) plot the variation of w with wavelength in the range of wavelength 1100–1600 nm for a standard single-mode fiber.

Chapter 18
Lasers in Science

18.1 Introduction

In this chapter, we discuss some experiments in physics and chemistry (and related areas) which have become possible only because of the availability of highly coherent and intense laser beams. In Sections 18.2, 18.3, and 18.4 we briefly discuss second-harmonic generation, stimulated Raman emission, and the self-focusing phenomenon, respectively; all these phenomena have opened up new avenues of research after the discovery of the laser. Second-harmonic generation is a process similar to what we had discussed in Chapter 14, namely the parametric process of three-wave interaction and the optical parametric oscillator that is a very important tunable source of coherent radiation. In Section 18.5, we outline how laser beams can be used for triggering chemical and photochemical reactions. In Sections 18.6, 18.7, and 18.8, we discuss some experiments which can be carried out with extreme precision with the help of lasers. The Nobel Lecture by Ted Hansch at the end of the book discusses very nicely the application of lasers to precision clocks. Finally, in Section 18.9, we give a fairly detailed account of isotope separation using lasers. We would like to point out that there are in fact innumerable experiments that have become possible only with the availability of laser beams; here we discuss only a few of them.

18.2 Second-Harmonic Generation

With the availability of high-power laser beams, there has been a considerable amount of work on the non-linear interaction of optical beams with matter. One of the most striking non-linear effects is the phenomenon of second-harmonic generation, in which one generates an optical beam of frequency 2ω from the interaction of a high-power laser beam (of frequency ω) with a suitable crystal. The first demonstration of the second-harmonic generation was made by Franken et al. (1961) by focusing a 3-kW ruby laser pulse ($\lambda = 6943\ \text{Å}$) on a quartz crystal (see Fig. 18.1).

K. Thyagarajan, A. Ghatak, *Lasers*, Graduate Texts in Physics,
DOI 10.1007/978-1-4419-6442-7_18, © Springer Science+Business Media, LLC 2010

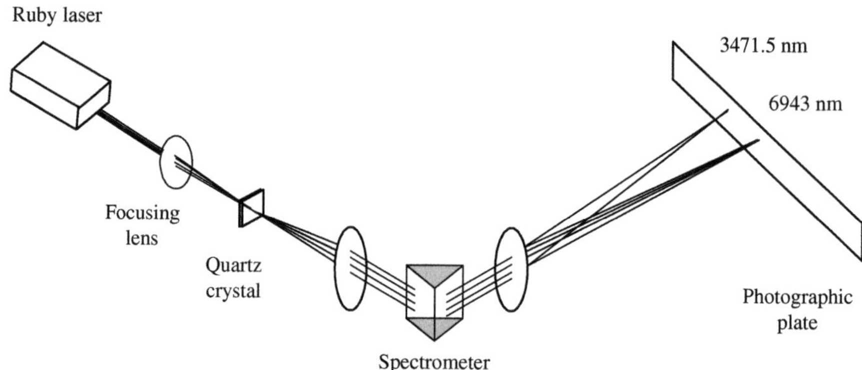

Fig. 18.1 Schematic of the experimental setup used by Franken and his co-workers for the generation and observation of the second-harmonic light. The beam from the ruby laser is focused by a lens into a quartz crystal which converts a very small portion of the incident light energy (which is at a wavelength of 6943 Å) into its second harmonic, which has a wavelength of 3471.5 Å (this wavelength falls in the ultraviolet part of the spectrum). The spectrometer arrangement shown in the figure disperses the components of light at wavelengths 6943 and 3461.5 Å, which can be photographed

In order to understand second-harmonic generation we consider the non-linear dependence of the electric polarization[1] on the electric field. It is well known that an atom consists of a positively charged nucleus surrounded by a number of electrons. When an electric field is applied, the electron cloud gets displaced and each atom behaves as an electric dipole. This leads to the medium being polarized. For weak intensities of the light beam, the polarization induced in the crystal is proportional to the oscillating electric field associated with the light beam. However, for an intense focused beam, the electric fields in the focal region could be very high ($\geq 10^7$ V/cm) and for such strong fields the response of the crystal is no more linear and may be of the type shown in Fig. 18.2. From the figure one can see that the electric field in a particular direction can be more effective in polarizing the material compared to a field in the opposite direction. This indeed happens in crystals which have no center of inversion symmetry. These crystals are usually piezoelectric in nature.[2] In Fig. 18.2 we show typical linear and non-linear dependences of the optical polarization on the electric field which induces the polarization. Note that in the absence of any non-linearities, the curve between the induced polarization and the applied electric field is a straight line, which implies that if the field is increased (or decreased) by a factor α, the polarization would also increase or decrease by the same factor. On the other hand, in the presence of non-linearities, only for weak fields is the curve approximately a straight line.

---

[1] By "polarization" we imply here the dipole moment induced by the electric field. This induced dipole moment is due to the relative displacement of the center of the negative charge from that of the nucleus.

[2] A piezoelectric material converts mechanical energy into electrical energy.

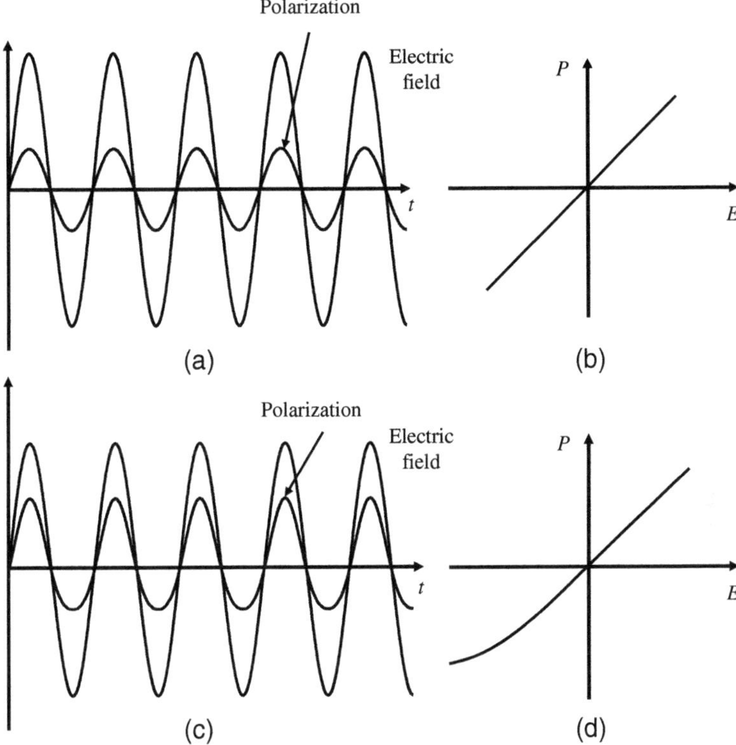

Fig. 18.2 The curves show the time variation of the electric field of the light wave incident on the crystal and the induced polarization. A typical linear dependence between the induced polarization wave and the applied electric field of the incident light wave is shown in (**a**), and (**c**) corresponds to a typical non-linear dependence. The distorted nature of the induced polarization wave in (**c**) leads to the production of harmonics of the incident light (see Fig. 18.3). (**b**) and (**d**) show the dependence of the polarization on the applied electric field for a linear and a non-linear response of a crystal

For weak electric fields, media behave linearly and the polarization P induced in the medium is directly proportional to the electric field E of the light wave. Thus we have

$$P = \varepsilon_0 \chi E \tag{18.1}$$

Now when the electric field strength increases and if the medium does not possess a center of inversion symmetry, then the polarization is no more related linearly to the electric field. In this case instead of Eq. (18.1) we have

$$P = \varepsilon_0 \chi E + 2\varepsilon_0 d E^2 \tag{18.2}$$

where d represents the non-linear coefficient and gives the strength of the non-linearity of the medium. Equation (18.2) is a simplified form of equation; polarization and electric fields are vectors and thus actually the quantity d is a tensor of

Fig. 18.3 By making a
Fourier analysis of the wave
shown in (**a**), it can be shown
that it consists of three
components, namely a
zero-frequency component
shown in (**b**), a component
with the fundamental
frequency ω shown in (**c**), and
a second-harmonic
component with a frequency
2ω shown in (**d**). Thus a
non-linear polarization wave
leads to the production of a
second harmonic of the
fundamental frequency

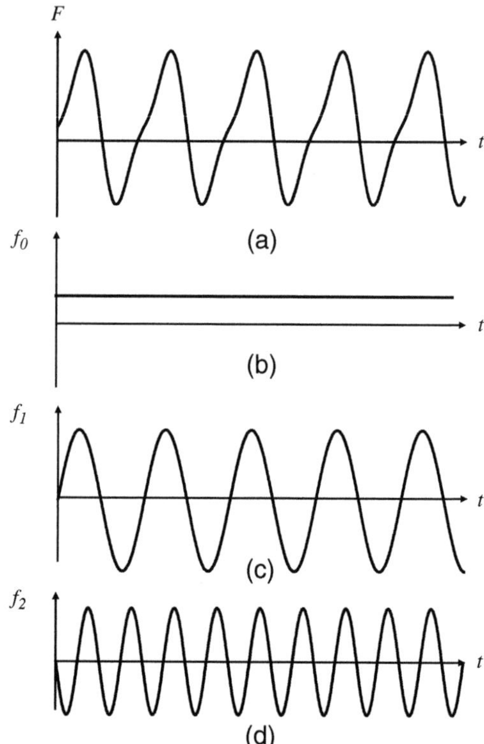

rank 3 with 27 elements. For a given medium and given states of polarization of the
incident electric field and polarization, we can write Eq. (18.2) with d representing
an effective non-linear coefficient.

If we assume the incident light wave to have an electric field variation given as

$$E = E_0 \cos (\omega t - kz) \tag{18.3}$$

then the induced polarization is given as

$$P = \varepsilon_0 \chi E_0 \cos (\omega t - kz) + 2\varepsilon_0 d E_0^2 \cos^2 (\omega t - kz) \tag{18.4}$$

Using standard trigonometric identity we can write Eq. (18.4) as

$$P = \varepsilon_0 \chi E_0 \cos (\omega t - kz) + \varepsilon_0 d E_0^2 + \varepsilon_0 d E_0^2 \cos 2 (\omega t - kz) \tag{18.5}$$

Thus in this case in addition to the term oscillating at frequency ω, there is a
term which is constant and independent of time and the other term oscillates at a frequency 2ω. Figure 18.3 shows the sine wave components of the polarization as given
by Eq. (18.5). Since polarization is the source of electromagnetic radiation, the term

Fig. 18.4 When a ruby laser beam (red light) is passed through a crystal of potassium dihydrogen phosphate, a portion of the energy gets converted into blue light, which is the second harmonic. The remaining ruby red light is suppressed by using a glass filter after the crystal. The two beams are made visible by means of smoke-filled glass troughs. (Photograph courtesy: Dr. R.W. Terhune)

oscillating at a frequency 2ω will lead to the generation of electromagnetic waves at frequency 2ω. In order that the radiation emitted by individual dipoles of the medium be added constructively, we need to satisfy the following phase matching condition:

$$k(2\omega) = 2\,k(\omega) \tag{18.6}$$

where $k(2\omega)$ and $k(\omega)$ represent the propagation constants of the medium at frequency 2ω and ω, respectively. Equation (18.6) can be written as

$$n(2\omega) = n(\omega) \tag{18.7}$$

i.e., the refractive index of the medium must be equal at frequencies ω and 2ω. This is accomplished using various techniques such as birefringence phase matching, quasi-phase matching. Chapter 14 discusses the concept of phase matching in more detail for the case of difference frequency generation. In fact, Eq. (18.7) is a momentum conservation equation for the interaction process. Second-harmonic generation can be considered to be a process in which two photons at frequency ω merge into a single photon at frequency 2ω. For maximum efficiency of this process, we need to conserve momentum and since the momentum of a photon is represented by $\hbar k$, Eq. (18.6) is nothing but momentum conservation equation.

The constant term in Eq. (18.5) is referred to as dc polarization and corresponds to zero frequency; this dc polarization was also observed by Franken et al. (1961).

Figure 18.4 shows how a ruby laser beam (red light) when passed through a crystal of ammonium dihydrogen phosphate gets partially converted into its second harmonic which lies near the ultraviolet region. The remaining ruby red light has been filtered by using a glass filter after the crystal.

There has been an extensive amount of work in the field of harmonic generation as it provides a coherent beam at frequency 2ω at the output of the crystal.

Since efficient lasers in the infrared wavelengths and red wavelengths are available, SHG can be used to achieve coherent sources at the ultraviolet and blue wavelengths for various applications. For example green laser wavelengths can be generated by pumping an yttrium orthovanadate ($Nd:YVO_4$) crystal by a laser diode at 808 nm resulting in emission at 1064 nm. Using a crystal such as potassium titanyl phosphate (KTP) as a frequency doubler, the laser generates 532 nm (green) wavelength. Output powers of up to 5 W can be generated using such techniques at 532 nm wavelength. Such lasers are finding applications in DNA sequencing, flow cytometry, cell sorting, holography, laser printing, etc.

The optical parametric process discussed in Chapter 14 is based on the same non-linearity that leads to second-harmonic generation. Both these processes have very important applications in science as well as technology. In fact the parametric downconversion process discussed in Chapter 14 is used for generation of squeezed light which has many applications including in gravitational wave detection.

18.3 Stimulated Raman Emission

Another important application of the laser light is in stimulated Raman emission. Raman scattering can be understood by considering light of frequency ω to consist of photons of energy $\hbar\omega$. When a monochromatic light beam gets scattered by a transparent substance, one of the following may occur:

1. Over 99% of the scattered radiation has the same frequency as that of the incident light beam (see Fig. 18.5); this is known as Rayleigh scattering. The sky looks blue because of Rayleigh scattering and the light that comes out from the side of the optical fiber (see Fig. 17.8) is also due to Rayleigh scattering.
2. A very small portion of the scattered radiation has a frequency different from that of the incident beam – this may arise due to one of the following three processes:

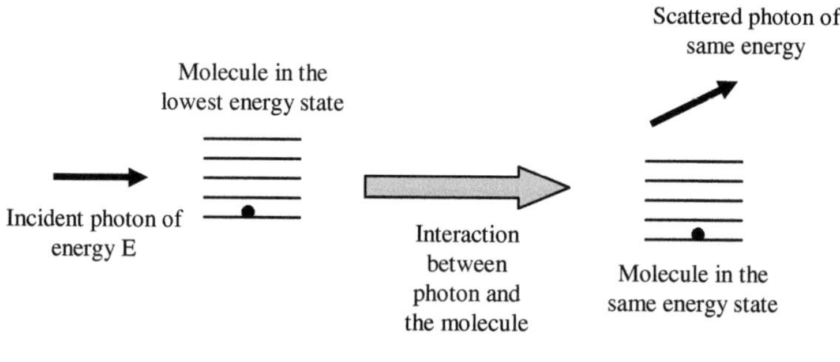

Fig. 18.5 Rayleigh scattering in which the frequency of the scattered photon is the same as that of the incident photon

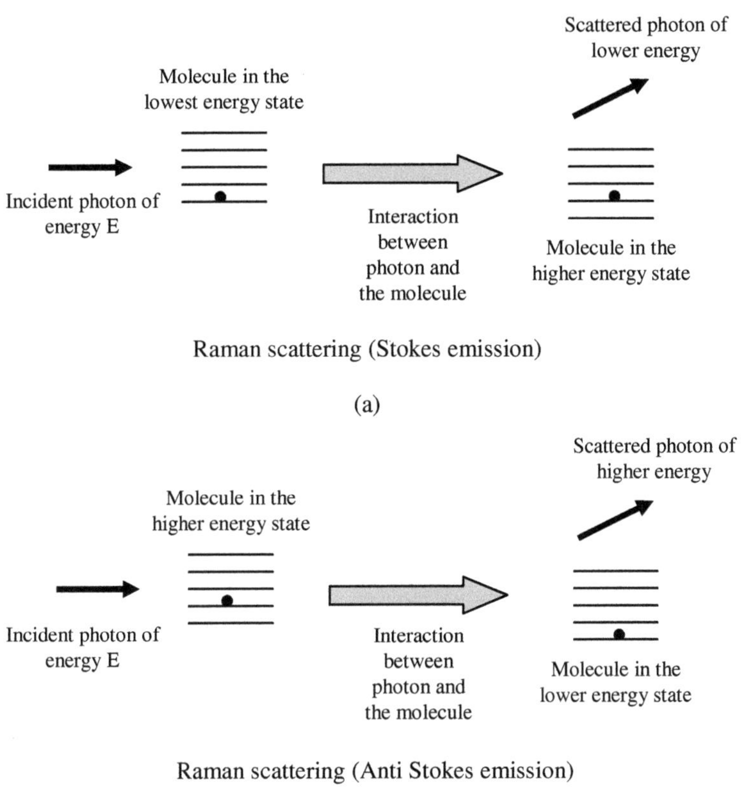

Fig. 18.6 (**a**) Stokes emission in which the molecule takes away some energy from the incident photon and the scattered photon has a lower frequency as compared to the incident photon. (**b**) Anti-Stokes emission in which the molecule gives some energy to the incident photon and the scattered photon has a higher frequency as compared to the incident photon

i)   A part of the energy of the incident photon may be absorbed to generate translatory motion of the molecules – this would result in the scattered light having a very small shift of frequency, typically in the 10–20 GHz range. This scattering process is known as Brillouin scattering.[3]

ii) A part of the energy $\hbar\omega$ of the incident photon is taken over by the individual molecule in the form of rotational (or vibrational) energy and the scattered photon has a smaller energy $\hbar\omega'$ with $\omega' < \omega$. The resulting spectral lines are known as Raman Stokes lines (see Fig. 18.6a).

iii) On the other hand, the photon can undergo scattering by a molecule which is already in an excited vibrational or rotational state. The molecule can

---

[3]The shift in frequency is usually measured in wavenumber units, which is defined later in this section. In Brillouin scattering, the shift is $\lesssim 0.1$ cm^{-1}. On the other hand, in Raman scattering, the shift is $\lesssim 10^4$ cm^{-1}.

de-excite to one of the lower energy states and in the process, the incident photon can take up this excess energy and come out with a higher frequency ($\omega' > \omega$). This leads to the scattered light having a larger energy and hence larger frequency. The resulting spectral lines are known as Raman anti-Stokes lines (see Fig. 18.6b).

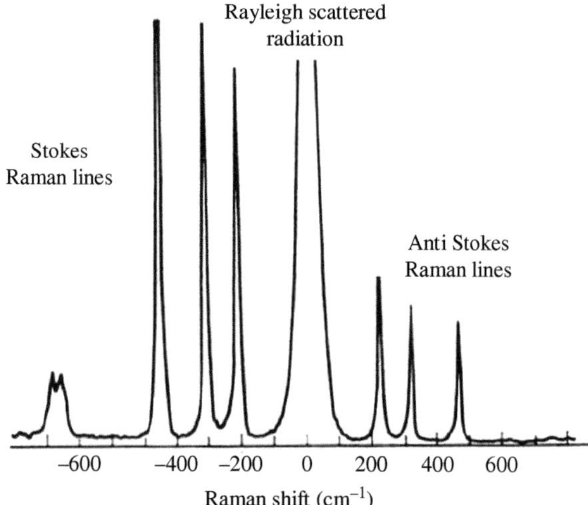

Fig. 18.7 Scattered spectrum showing Rayleigh scattering and Raman scattering

The difference energy, which is $\hbar\left(\omega' - \omega\right)$ for the Raman Stokes line and $\hbar\left(\omega - \omega'\right)$ for the Raman anti-Stokes line, would therefore correspond to the energy difference between the rotational (or vibrational) energy levels of the molecule and would therefore be a characteristic of the molecule itself.

The quantity $\hbar\left(\omega' - \omega\right)$ or $\hbar\left(\omega - \omega'\right)$ is usually referred to as the "Raman shift" (see Fig. 18.7) and is independent of the frequency of the incident radiation. Through a careful analysis of the Raman spectra, one can determine the structure of molecules, and there lies the tremendous importance of Raman effect.

Since the probability of finding molecules is higher in the lower energy state as compared to the higher energy state, the Stokes emission (into a lower energy photon) is much more intense than the anti-Stokes emission (into higher energy). Figure 18.7 shows a typical spectrum of scattered light observed showing the Rayleigh scattering at the incident frequency, Stokes lines at lower frequency and much weaker anti-Stokes line at a higher frequency. Note that the frequency shift in Stokes and anti-Stokes is equal; both of them involve the same set of energy levels of the molecule and the anti-Stokes emission is much weaker than the Stokes emission.

Figure 18.8a shows the Raman scattered spectrum from a mixture of hydrogen and deuterium when the mixture is illuminated by a laser beam at 488 nm wavelength, the corresponding frequency being approximately 615 THz ($= 6.15 \times$

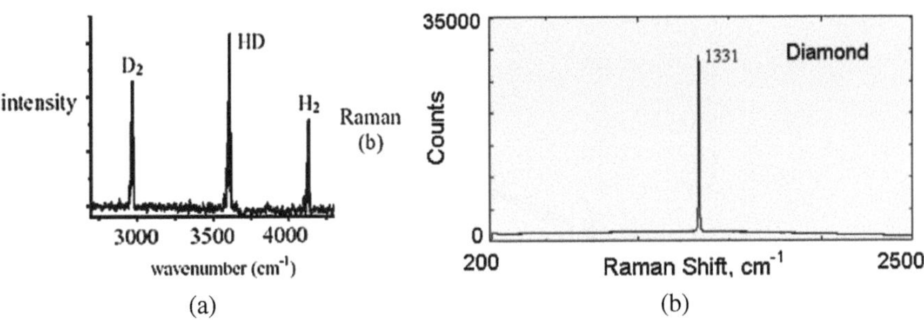

Fig. 18.8 (**a**) Raman scattered spectrum of hydrogen and it isotopes. (**b**) Raman spectrum of diamond

10^{14} Hz). Now, there is an energy level corresponding to vibration of the hydrogen molecule which is separated from the lowest level by 125 THz; thus the Stokes line should appear at a (lower) frequency of 490 THz (690 nm wavelength) and the very weak anti-Stokes line should appear at a (higher) frequency of 740 THz (405 nm). In spectroscopy the frequency shift is measured in wavenumber units (i.e., inverse wavelength) and the units are usually cm^{-1} (read as centimeter inverse)

$$\Delta T\,(cm^{-1}) = \frac{\Delta E}{hc}$$

where ΔE is measured in ergs (1 erg $= 10^{-7}$ J), $h \approx 6.634 \times 10^{-27}$ erg s is the Planck's constant, and $c \approx 3 \times 10^{10}$ cm/s is the speed of light in free space. Thus the frequency shift in hertz will be given as

$$\Delta v\,(Hz) = \frac{\Delta E}{h} = \Delta T\,(cm^{-1}) \times 3 \times 10^{10}$$

Thus a frequency shift of 4155 cm^{-1} in wavenumbers corresponds to 125 THz, which is consistent with the value given above. In Fig. 18.8a, the Raman shift due to D_2 and HD molecules are also shown; once again the frequency shift is a "signature" of the molecule and is determined by the energy difference of the vibrational levels of the molecule. Since deuterium is heavier than hydrogen, the vibrational frequency of HD is smaller than that of hydrogen and that of D_2 is even smaller. This is clearly seen in Fig. 18.8a, where the frequency shift of HD is smaller than that of H_2 and that of D_2 is even smaller. Similarly, Fig. 18.8b shows the Raman scattered spectrum of diamond showing a Raman shift of 1331 cm^{-1} which for a pump wavelength of 514 nm would correspond to a wavelength of 620 nm.

The Raman effect is extremely weak, which means that for proper observation, it requires a very bright light source. In fact, before the invention of the laser, there was little interest in using Raman scattering as a general spectroscopic tool. Raman scattering using light sources such as the sodium lamp was so weak that only ideal samples yielded good spectra. The laser dramatically changed that situation, and

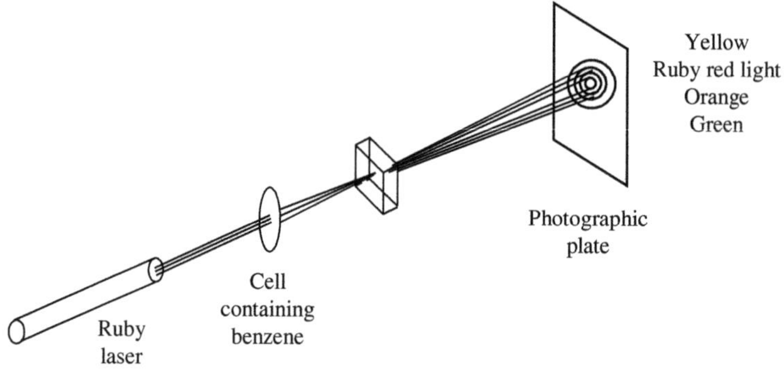

Fig. 18.9 Schematic of the experimental setup used for the observation of stimulated Raman emission. The light emerging from the ruby laser is focused by a lens into a cell containing benzene. The Stokes lines occur in the infrared and the anti-Stokes lines occur in the visible region. The various anti-Stokes lines appearing in the red, orange, yellow, and green regions of the spectrum form rings around the central ruby laser spot (see Fig. 18.10)

today all Raman instruments use laser excitation. The Raman instruments available today use the many technological advances in the areas of light sources, fiber-optic guides, good spectrometers. etc. making Raman scattering a very valuable tool in many applications including industrial applications such as in chemical processing and quality control. Lasers have made available virtually any wavelength from the ultraviolet to near-infrared for analysis allowing for optimization of Raman studies. An example of Raman scattering studies is the in situ growth monitoring of diamond films in plasma reactors at temperatures as high as 1000°C using a UV Raman system operating at 244 nm. With the advent of the laser, Raman scattering became an important tool in research laboratories as well as in industrial applications.

In the case of stimulated Raman emission, the photons emitted in the spontaneous Raman effect are made to stimulate further Raman emissions. A layout of an experimental arrangement for observing stimulated Raman emission from Raman laser material (e.g., nitrobenzene) is shown in Fig. 18.9. An intense laser beam is focused on the Raman laser material. The photons corresponding to the Stokes lines have a wavelength in the infrared region which does not appear on the color film. In the anti-Stokes emission, the frequency of the emitted photon is higher, which results in orange, yellow, and green rings (see Fig. 18.10).[4]

The stimulated Raman effect has been observed in a large number of materials, which provides us with hundreds of coherent sources of light from ultraviolet to infrared. Stimulated Raman scattering and continuous wave Raman oscillation have also been observed in optical fibers. Stimulated Raman scattering can be used

---

[4]That the emitted photons of a particular frequency should appear in well-defined cones about the direction of the incident photons follows from the conservation of momentum. The direction of the emitted photon can be found by using the fact that a photon of frequency v has a momentum equal to hv/c

Fig. 18.10 The stimulated Raman effect obtained by focusing a ruby laser beam in a cell filled with benzene and photographing the scattered radiation in a setup similar to that shown in Fig. 18.5. (Photograph courtesy: Dr. R.W. Terhune)

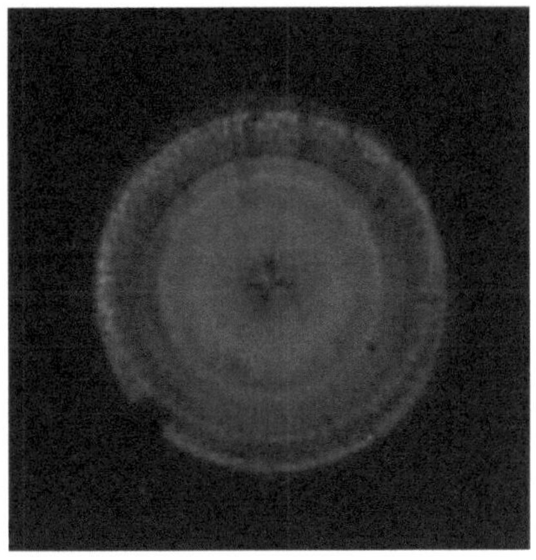

for amplification of light waves at frequency corresponding to the Raman shift. Stimulated Raman amplifiers are very important components of today's optical fiber communication systems as they are capable of providing optical amplification at any signal wavelength.

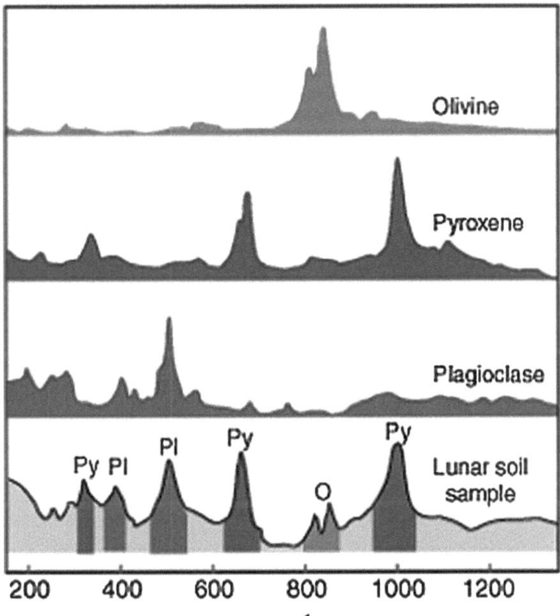

Fig. 18.11 Raman spectrum from various minerals including that from a sample of lunar soil. (Source: Ref. http://hyperphysics.phy-astr.gsu.edu/Hbase/molecule/raman2.html)

Raman amplification and oscillation in optical fiber waveguides requires low pumping powers because of the strong transverse confinement of the optical energy in the waveguide and the very long interaction lengths (hundreds of meters to tens of kilometers) that are possible with the use of optical fibers. The pumping source is usually the 1.064-μm radiation of the Nd:YAG laser (see Section 9.5) or high-power 1450-nm diode laser. The broad Raman gain bandwidth of silica glass fibers has indeed been used in making a tunable continuous wave fiber Raman oscillator (Lin et al. 1977a, Stolen et al. 1977). Such Raman oscillators can indeed be used as tunable light sources and are especially useful for fiber-optic communication studies as the region of operation at 1.3 μm wavelength is indeed becoming very important due to the existence of a region of zero dispersion in single-mode glass fibers. [see e.g. Thyagarajan and Ghatak (2007)].

Raman scattering is a very useful tool for the identification of minerals. Raman spectra for different minerals tend to have sharp unique patterns and hence serving as "fingerprints" for the minerals. Since Raman spectra can be collected remotely, they show great promise for planetary exploration. Figure 18.11 shows spectra of the common silicate minerals olivine, pyroxene, and plagioclase that are compared to a Raman spectrum of a lunar soil sample identified as 71501 (Ref. http://hyperphysics.phy-astr.gsu.edu/Hbase/molecule/raman2.html).

18.4 Intensity-Dependent Refractive Index

Another interesting non-linear effect is the intensity dependence of refractive index of a medium which leads to phenomena such as self-focusing and soliton formation. The intensity dependence of refractive index arises because of the third-order non-linearity in which the electric polarization depends on the cube of the electric field instead of square of the electric field as described by Eq. (18.2). Due to the third-order non-linearity, the dependence of refractive index of a medium on intensity can be written in the following form:

$$n = n_0 + n_2 I \tag{18.8}$$

where n_0 is the refractive index of the medium at low intensities, n_2 is a constant, and I represents the intensity of the laser beam. Thus, if the intensity[5] of the beam is maximum on the axis and decreases radially, and if n_2 is positive, then the beam would get focused (see Fig. 18.12). This can be qualitatively understood from the fact that the velocity (which is inversely proportional to the refractive index) will be minimum on the axis, and by simple Huygens' construction, one can show that a plane wave front would become converging. This results in what is known as the self-focusing phenomenon. Since a light beam usually suffers diffraction due to the

---

[5]This is indeed the case for a laser beam where one usually has a Gaussian variation of intensity along the wave front.

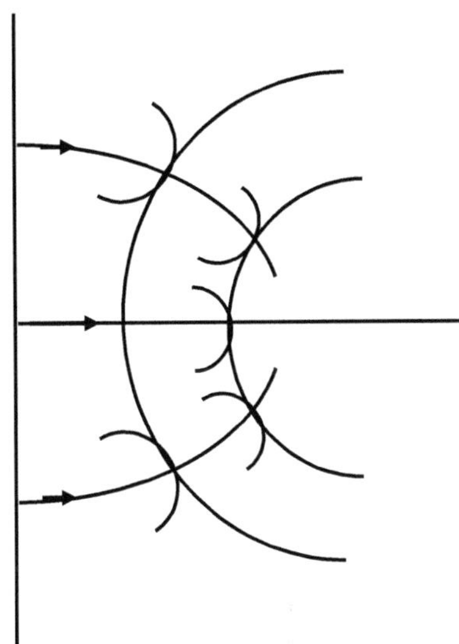

Fig. 18.12 Self-focusing of the beam occurs when a high-power laser beam with an intensity distribution which decreases away from the center passes through a medium with a positive value of n_2

finite spatial extent of the beam, it is possible to have beams in which the diffraction effects are compensated by the focusing generated by the non-linearity. In such a case the beam will propagate with no change in the spatial profile and such beams are referred to as spatial solitons. Since diffraction effects are governed by the spatial dimension of the beam and the non-linear effects are controlled by the intensity of the beam, there is a critical power at which the beam (with a characteristic spatial intensity profile) will behave as a spatial soliton.

The same intensity-dependent refractive index generates some very interesting effects in the temporal domain. Let us consider a pulse of light such as a Gaussian pulse (see Fig. 18.13); the intensity is maximum at the center and falls off on either side. Considering the intensity-dependent refractive index, for the center of the pulse, the refractive index of the medium will be larger than that for the leading and trailing edges of the pulse. Since the speed of propagation depends on the refractive index, this would imply that the center of the pulse would travel slower as compared to the leading and trailing edges of the pulse. This leads to a crowding of the waves toward the trailing edge of the pulse and an opposite effect in the leading edge, resulting in a chirping of the pulse, i.e., a pulse with a frequency which changes

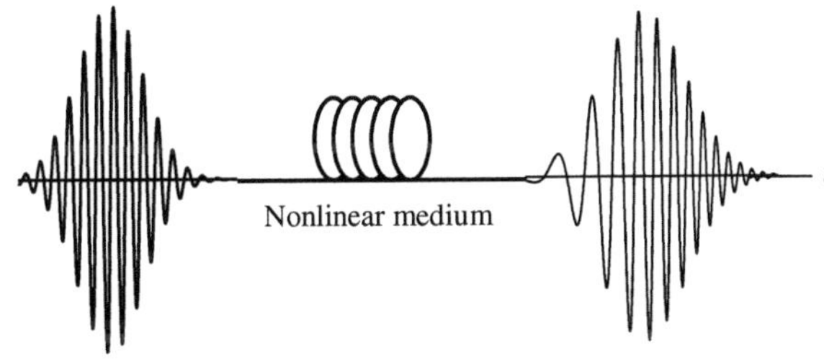

Fig. 18.13 Chirping in an optical pulse due to the non-linear dependence of the refractive index of the material of the optical fiber

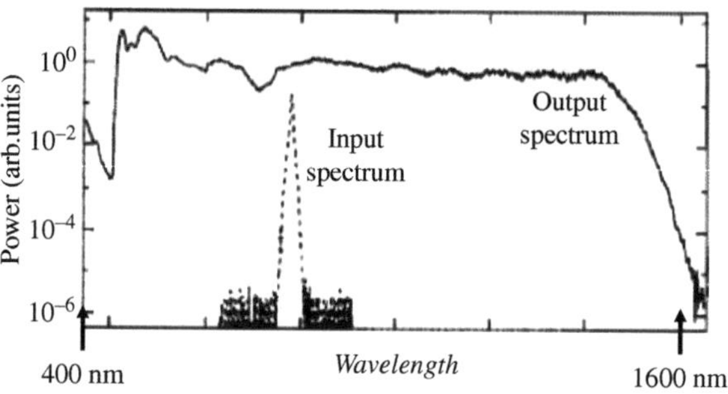

Fig. 18.14 Supercontinuum generation from an optical fiber. The incident light is a light beam at about 800 nm and the output contains frequencies from 400 to 1600 nm [Reprinted from Ranka et al. (2000). © 2000 OSA]

within the duration of the pulse (see Fig. 18.13). It is known from Fourier transform theory that the spectral width of a chirped pulse is greater than an unchirped pulse of the same duration. Thus this non-linearity leads to a spectral broadening of the pulse. Since the minimum possible temporal width of a pulse depends inversely on the spectral width, the spectrally broadened pulse can be compressed using dispersive elements to a pulse of shorter duration than the incident pulse. This is a very important technique for pulse compression. Using optical fibers, the increase in spectral width can be extremely large. Thus it is possible to generate spectra from 400 to 1600 nm starting from a few hundred femtosecond pulse at about 800 nm (see Fig. 18.14). This phenomenon is referred to as supercontinuum generation and finds wide application in realizing broadband sources for many applications such as spectroscopy, characterization, optical clocks (see Nobel Lecture by Hansch at the end of the book), etc.

18.5 Lasers in Chemistry

Lasers are expected to find important applications in chemistry. Because of the extremely large temperatures obtainable at the focus of a laser beam, the laser is an excellent tool for triggering chemical and photochemical reactions.[6] Electric fields larger than 10^9 V/cm are obtainable at the focus; such fields are larger than the fields that hold the valence electrons to the atoms.

Lasers attached to microscopes are used as microprobes in microanalysis. Such microanalysis can give information regarding the presence of trace metals in various tissues. The technique essentially involves sending a giant pulse to a preselected area and vaporizing some of the target material. The vapor so produced may itself emit a spectrum of wavelengths or one may pass a discharge through it to produce a characteristic spectrum. An analysis of the spectrum gives the presence of various elements.

---

[6]The use of lasers in photophysics and photochemistry has been discussed at a popular level by Letokhov (1977).

It has been demonstrated that molecules that have been excited by an infrared laser react faster than the molecules that are in the ground state. The extremely high monochromaticity of the laser allows one to selectively excite different bands of a molecule and thus leads to the possibility of producing some new chemical products.

With the generation of intense laser pulses lasting for a few picoseconds,[7] one can now study ultrafast physical and chemical processes. This gives an opportunity for understanding the most fundamental processes with unprecedented time resolution. The technique is essentially to excite the sample with an intense pulse and then to study the behavior of a certain characteristic parameter of the sample (e.g., absorption or scattering) as a function of the delay time after excitation. Such studies have indeed been successfully applied to study various processes such as the redistribution of light energy absorbed by chlorophyll in the photosynthetic process, to observe ultrafast chemical reactions, to obtain the vibrational and rotational decay constants of molecules, to study photovisual processes, and others. For a review of some of these applications, the reader is referred to the articles by Alfano and Shapiro (1975) and Busch and Rentzepis (1976).

With the availability of laser pulses with durations in the tens of femtoseconds, it is possible to observe in real time chemical reactions in which chemical bonds break, form, or geometrically change with extreme rapidity involving motion of electrons and atomic nuclei. The field of femtochemistry is expected to permit femtosecond time resolution in the observation of chemical reactions. Today femtochemistry finds applications in studies on various types of bonds and addresses increasingly complex molecular systems, from diatomics to proteins to DNA, leading to the new branch of femtobiology. For a nice review of femtochemistry, readers are referred to the excellent article by Zewail (2000).

18.6 Lasers and Ether Drift

Lasers have made possible an experiment to test the presence of ether drift with an accuracy a thousand times better than could be obtained before. A setup of the experiment is shown in Fig. 18.15. The beams from two lasers oscillating at slightly different frequencies are combined with a beam splitter and detected by a photomultiplier. The slight difference in frequency between the two lasers produces beats at a frequency equal to the difference between the two frequencies. The oscillation frequency of the laser depends on the length of the resonant cavity and also on the speed of light in the cavity. Thus a rotation of the apparatus must change the frequency of oscillation of the lasers and hence the beat frequency, if there was any ether drift. When the experiment was performed, no change in beat frequency was observed. The apparatus was sensitive enough to detect a velocity change as small as 0.03 mm/s.

---

[7] $1 \text{ ps} = 10^{-12}$ s.

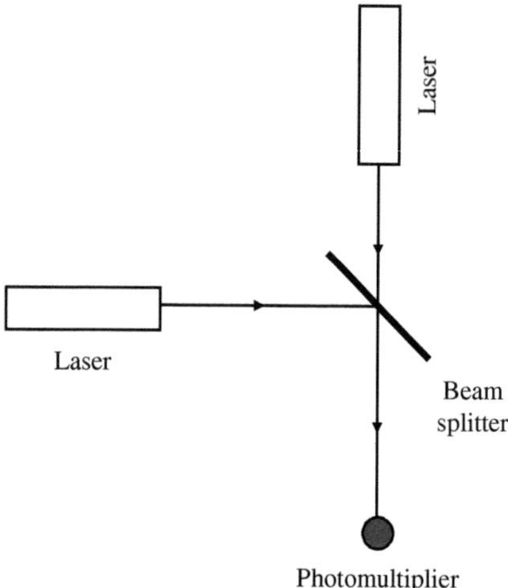

Fig. 18.15 An experimental setup for detecting the presence of ether drift

18.7 Lasers and Gravitational Waves

Einstein's general theory of relativity predicted the existence of gravitational waves which are ripples in the fabric of space–time. These waves are very weak (even for events such as supernova explosions) and supposed to be produced when massive objects accelerate through space. Scientists have been working on using the principles of interferometry to detect the existence of these waves. Gravitational waves

Fig. 18.16 The LIGO interferometer being built to detect gravitational waves. Squeezed light is expected to be used in the interferometer for increasing the sensitivity of the sensor. (Source: Ref. http://physicsworld.com/cws/article/news/33755)

passing through a Michelson interferometer (with arm lengths of a few kilometers) are supposed to stretch one arm and compress the other leading to a change of phase of the interference pattern. However the change of length is extremely tiny, about 10^{-18} m. Thus the expected fringe shift is extremely small and the interferometer to detect such effects needs to have very large arm lengths. Figure 18.16 shows a photograph of The LIGO (Laser Interferometer Gravitational Observatory) being built to detect gravitational waves. Current detectors are not sensitive enough to measure such small changes. The ultimate sensitivity is determined by quantum noise in the detector. We had discussed in Chapter 9 about squeezed states which exhibit noise level below that of vacuum state in one quadrature. Such squeezed states are expected to enable detection of gravitational waves. Such squeezed light is proposed to be produced using non-linear effects in crystals and experiments on squeezed light have demonstrated noise levels below the vacuum state.

18.8 Rotation of the Earth

Lasers have been used to detect the absolute rotation of the Earth. If light is made to rotate in both clockwise and anti-clockwise directions around a square with the help of mirrors as shown in Fig. 18.17, then if the square is at rest, the time taken for light to travel around the square in both the clockwise and the anti-clockwise directions would be the same. But if the square is rotated about an axis which is normal to the plane of the square, then the time taken for light to travel along one direction will be different from that taken along the other direction. Thus if one could measure this difference, one could obtain information about the rotation of the square.[8]

An experiment to detect the rotation of the Earth by using such a method with ordinary light sources was performed by Sagnac in 1914 and then by Michelson and Gale in 1925. With the use of lasers, one can do similar experiments with much more precision and sensitivity. The sides of the square are gas discharge tubes containing helium and neon. The corners of the square are occupied by mirrors (see Fig. 18.17). Light beams traveling along either direction would undergo amplification. Thus one would have two beams, one propagating in the clockwise direction and the other in the anti-clockwise direction. The frequency of oscillation of the laser would depend on the path length along the square. Since the path lengths along the two directions are different when the system is rotating in the plane, two different frequencies are obtained. By mixing the light beams of the two frequencies, one can detect the beat frequency of the beams and hence the rate of rotation of the square. For example, at New York (which is at a latitude of $40°40'$N) the effective speed of rotation is about one-sixth of a degree per minute; this would correspond to a beat frequency of 40 Hz.

---

[8]The difference in path length between the two paths is extremely small; thus only a shift of a hundred-thousandth of a wavelength would be produced when the square is of side 3 m and is kept at a latitude of $40°$ on the surface of the earth.

Fig. 18.17 A ring laser for detecting the absolute rotation of the earth. When the system is at rest, both the clockwise and anti-clockwise rotating beams have the same frequency. When the system is rotated about an axis normal to the plane containing the system, the clockwise rotating and the anti-clockwise rotating beams have slightly different frequencies. When they are combined, then beats are produced which can easily be detected

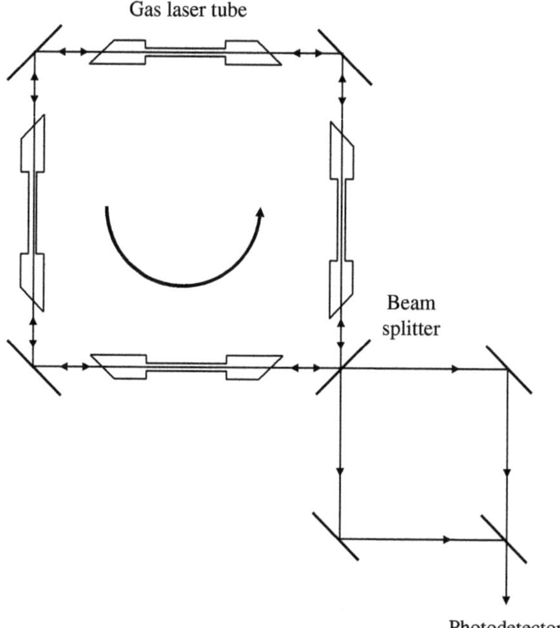

In the ring interferometer that we have discussed, one can show that the phase difference introduced by the counter propagating beams is (see, e.g., Post (1967))

$$\Delta\phi = \frac{8\pi}{c\lambda}\mathbf{\Omega} \cdot \mathbf{A} \tag{18.9}$$

where $\mathbf{\Omega}$ is the rotation vector and $\mathbf{A}$ is the area enclosed by the optical path. This phase change is generally too small for direct measurement in the range of rotation speeds encountered in inertial navigation. For example, a rotation rate of $5 \times 10^{-7}\,\mathrm{rad/s}(0.1°/h)$ over an area of $0.1\,\mathrm{m^2}$ at $\lambda = 0.6\,\mu\mathrm{m}$ yields a phase shift of $7 \times 10^{-9}\,\mathrm{rad}$ ($\sim 10^{-9}$ of a fringe!). The availability of extremely low-loss, single-mode optical fibers makes it possible to increase $\Delta\varphi$ by three to four orders of magnitude by increasing the effective area A by having the light beams propagate through a large number of turns of an optical fiber. In addition to this increase in area, it is possible to have both the clockwise and counterclockwise beams follow identical paths, thus stabilizing the differential path lengths.

Fiber-optic gyroscope is a very versatile device for measurement of rotation and has been in production by many industries in the world. Low-cost fiber-optic gyros are also finding applications in vehicles such as cars.

18.9 Photon Statistics

Let us consider an experiment in which a beam of light from a source is allowed to fall on a photodetector for a specific time interval T by having a shutter open in front of the detector for the time T; one then registers the number of photoelectrons so liberated. Then the shutter is again opened for an equal time interval T after a certain time delay which is longer than the coherence time of the source and the number of photoelectrons counted during the interval is again registered. This experiment is repeated a large number of times ($\sim 10^5$) and the number of photoelectrons produced in equal intervals of time is counted. The results of such an experiment give one the probability distribution $p(n, T)$ of counting n photons in a time T. Here it is assumed that the light source is stationary, i.e., the long time average of the intensity is fixed and independent or the particular long time period chosen for measuring it. The above-obtained probability distribution contains information regarding the statistical properties of the source, and such studies have applications in spectroscopy, stellar interferometry, etc.

It can be shown that for a polarized thermal source, for counting times T much smaller than the coherence time T_c, the photoelectron counting distribution $p(n)$ is given by[9] the following formula (see, e.g., Loudon (1973)):

$$p(n) = \frac{\langle n \rangle^n}{[1 + \langle n \rangle]^{n+1}} \tag{18.10}$$

where $\langle n \rangle$ represents the mean number of counts in time T. Figure 18.18a shows typical plots of the above distribution for $\langle n \rangle = 4$ and $\langle n \rangle = 10$.

---

[9]It is much beyond the scope of this book to go into the derivation of Eqs. (18.10) and (18.11). However, it may be worthwhile to mention that the probability distribution $p(n, t, T)$ of registering n photo-electrons by an ideal detector in a time interval t to $t + T$ is given by (see, e.g., Mandel (1959))

$$p(n, t, T) = \int_0^\infty \frac{(\alpha w)^n}{n!} e^{-\alpha w} P(w)\, dw$$

where α is the quantum efficiency of the detector,

$$w = \int_t^{t+T} I(t')\, dt'$$

is the integrated light intensity, and $P(w)$ is the probability distribution corresponding to the variable w. Thus the photoelectron counting distribution given by the above equation depends on the particular form of the probability distribution $P(w)$. Now, for a polarized thermal source, it can be shown that

$$P(w) = \left(1/\langle w \rangle\right) e^{-w/\langle w \rangle}, \qquad \text{for } T << T_c$$
$$= \delta(w - \langle w \rangle), \qquad \text{for } T >> T_c$$

where $\langle w \rangle = T \langle I \rangle$ and angular brackets denote averaging. Further, for an ideal laser, the beam would not exhibit any intensity fluctuations and $P(w) = \delta(w - \langle w \rangle)$. On substituting the above equations for $P(w)$ in the equation for $p(n, t, T)$, one gets Eqs. (18.10) and (18.11).

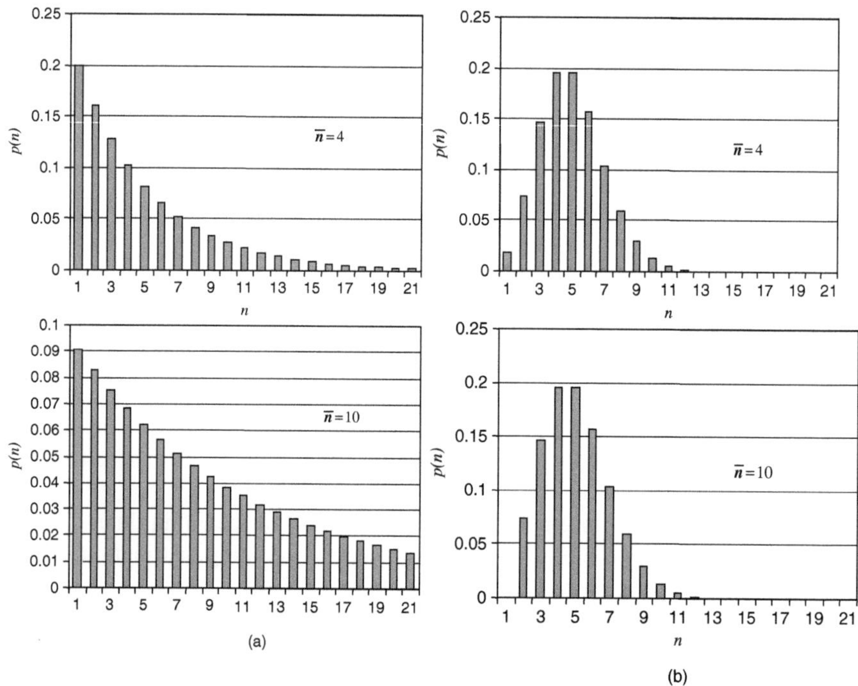

Fig. 18.18 (**a**) The probability distribution for the arrival of photons from a thermal source. (**b**) The probability distribution for the arrival of photons from a laser source corresponds to the Poisson distribution

On the other hand, for very large values of T as compared to T_c, all the fluctuations in the intensity may be expected to be averaged out during the counting period. For such a case, the photon counting distribution is given as

$$p(n) = \frac{\langle n \rangle^n}{n!} e^{-\langle n \rangle} \tag{18.11}$$

which is a Poisson distribution.

When the source is an ideal laser, the beam would not exhibit any intensity fluctuations and the counting distribution would be again given by Eq. (18.11).

Figure 18.18b shows the probability distribution given by Eq. (18.11) for $\langle n \rangle = 4$ and 10. As can be observed from the figure, even for a beam of constant intensity the fluctuations still exist.

Experiments on photoelectron counting were first carried out by Arecchi et al. (1966a, b), in which they have shown that for a laser operating much beyond threshold, the counting distribution is indeed Poissonian. The results on the counting distribution from a laser source near threshold, from thermal sources and from a mixture of a laser and a thermal source, have also been discussed by Arecchi et al.

(1966a, b). For further details, the reader is referred to Arecchi (1976), Mandel and Wolf (1970), and Mehta (1970).

18.10 Lasers in Isotope Separation

A new application of lasers (in particular, tunable lasers), which has been much discussed in recent years, is in isotope separation.[10] The technique seems to promise efficient separation processes that may perhaps revolutionize the economics of the separation and the use of isotopes. The major interest in the so-called LIS (laser isotope separation) process would be its possibility for large-scale enrichment of uranium for use in nuclear power reactors. There are, however, other applications for pure isotopes in medicine, science, and technology, if they could be produced economically. We shall, in the following, briefly discuss the principle of LIS and the variety of options which have been considered and already demonstrated on the laboratory scale.

Isotopes are atoms that have the same number of protons and electrons but which differ in the number of neutrons. Since most chemical properties are determined by the electrons surrounding the nucleus, the isotopes of an element behave in almost indistinguishable ways. Thus one of the common methods of separating an element from a mixture (by making use of its chemical properties) becomes cumbersome when an isotope is to be separated from a mixture with another isotope of the same element. Light elements lying below oxygen in the Periodic Table may be separated using repeated chemical extraction. Isotopes of heavier atoms may be separated using physical methods. For example, because of the differences in masses of the isotopes, they diffuse at different rates through a porous barrier and repeated passage through various stages leads to separation of required concentration of the isotope.

Isotope separation using a laser beam is a fundamentally different technique where one makes use of the slight differences in the energy levels of the atoms of the isotopes due to the difference in nuclear mass. This difference is termed the isotope shift. Thus, one isotope may absorb light of a certain wavelength, while the other isotope of the element may not absorb it.[11] Since the light emerging from a laser is extremely monochromatic, one may shine laser light on a mixture of two isotopes and excite the atoms of only one of the isotopes, thus earmarking it for subsequent separation. A block diagram of a typical LIS process is shown in Fig. 18.19. It may be of interest to note here that the basic physical idea of isotope separation by light was conceived more than 70 years ago. The first successful separation of mercury

---

[10]The material in this section was kindly contributed by Dr. S.V. Lawande of Bhabha Atomic Research Centre, Mumbai.

[11]When an atom absorbs light, it jumps from one energy level to another, the difference in energy between the two levels being just equal to the energy of the incident photon. Since the energy levels are slightly different for two different isotopes, their absorption properties are also different. The isotope shift between hydrogen and deuterium is given as $\Delta v/v \approx 2.7 \times 10^{-4}$. For uranium, the isotope shift is given as $\Delta v/v \approx 0.6 \times 10^{-4}$.

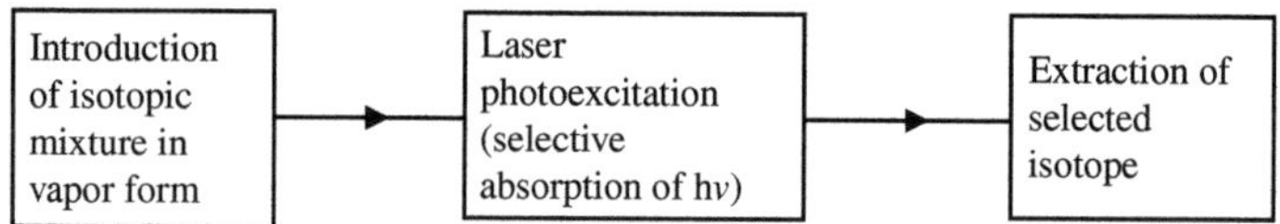

Fig. 18.19 Block diagram of a typical laser isotope separation process. A laser excites one of the isotopes from the isotopic mixture through selective absorption and the excited isotope atoms are separated using one of the many techniques

isotope (^{202}Hg) was reported by Zuber (1935), who irradiated a cell containing natural mercury vapor with the light from a mercury lamp. Since the invention of tunable lasers, it has become possible to revive the interest in photochemical separation processes for large-scale isotope separation.

In addition to the high monochromaticity, the high intensity of the laser is also responsible for its application in isotope separation because with low-intensity beams the separation rate would be too little for practical use.

The basic principle behind the laser isotope separation process is to first selectively excite the atoms of the isotope by irradiating a stream of the atoms by a laser beam and then separate the excited atoms from the mixture. Various techniques exist for separation. We will discuss a few of them; for more details the reader is referred to Zare (1977).

18.10.1 Separation Using Radiation Pressure

An interesting method of laser isotope separation is the deflection of free atoms or molecules by radiation pressure. A photon of energy hv carries with it a momentum of hv/c. When this photon is absorbed by an atom, conservation of momentum requires that the atom acquire this momentum. Thus the absorption tends to push the atom in the direction of travel of the incident photon. The momentum acquired in a single absorption is very small; hence for the atom to gain sufficient momentum, it must absorb many photons. This requires that the atoms have a short lifetime in the excited state before dropping back to the ground state. It should be noted that every time an atom emits a photon, it acquires a momentum equal and opposite to that it gained during absorption. Since the emissions occur in all random directions, the net effect of many absorptions and emissions is to push the atoms along the laser beam. In the present technique, a laser beam is allowed to impinge on an atomic beam at right angles (see Fig. 18.20) and the atoms of the isotope (which absorb the radiation) are deflected by the laser beam.

To give some idea of the numbers involved, the velocity resulting from the momentum transfer is about 3 cm/s in the case of sodium atoms. After an excitation event the atom/molecule remains in the excited state for a certain time $\tau \sim 10^{-8}$ s. Thus a sodium atom traveling with a thermal velocity $v \approx 10^5$ cm/s through an interaction zone of length 1 cm will be subjected to 10^3 absorption events and gain a net velocity of 3×10^3 cm/s in the direction of the laser beam. The resulting deflection of

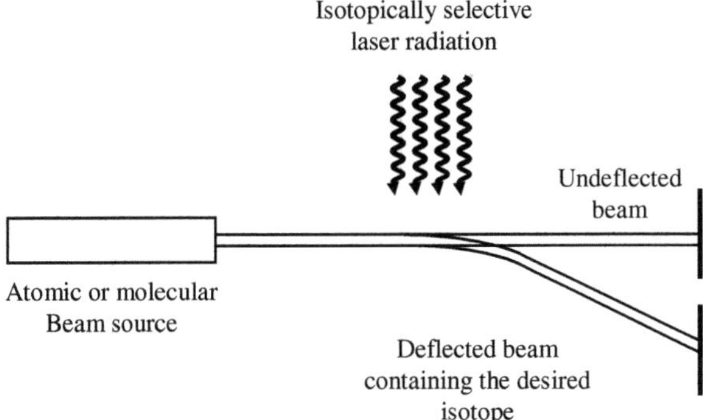

Fig. 18.20 Separation of isotope by deflection caused by selective absorption. The atomic beam emerging from the source is impinged by a laser beam tuned to excite atoms of the isotope to be separated. The absorption causes the atoms to acquire a momentum, and by repeated absorption they gain enough kinetic energy to get deflected from the main beam

30 mrad will be sufficient for separating the sodium atom. Such a scheme has been used to separate the isotopes of barium (Bernhardt et al. 1974). More sophisticated modifications of this scheme which improve the photon economy have also been conceived.

18.10.2 Separation by Selective Photoionization or Photodissociation

The most popular and perhaps universally applicable scheme of isotope separation is the two-step photoionization of atoms or the two-step dissociation of molecules (Fig. 18.21). The first step causes the selective excitation; this is followed by a second excitation which ionizes the excited atoms or dissociates the excited molecules. In the case of atoms the separation can be carried out by extracting the ions by means of electric fields. In the case of molecules the dissociation products must be separated from the other molecules. This may be carried out directly or by means of chemical reactions. It must be mentioned here that the two-step photoionization was used to demonstrate the feasibility of LIS for uranium at the Lawrence Livermore Laboratory in the USA.[12] In this experiment an atomic beam of uranium, generated in a furnace at a temperature of about 2100°C, was excited by the light of a dye laser (isotope-selective excitation) and then ionized by the light of a high-pressure mercury lamp and the ^{235}U isotope, which is present in natural uranium in a concentration of 0.71%, was enriched to 60%.

---

[12]Reported in Physics Today **27**(9), 17 (1974).

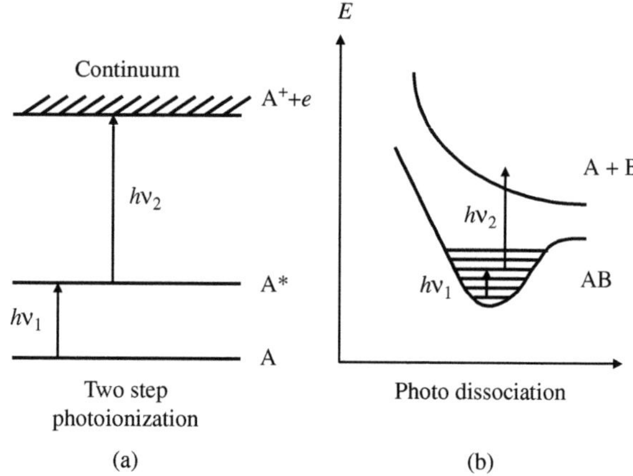

Fig. 18.21 (a) Two-step photoionization of atoms; (b) two-step photodissociation of molecules. The ionized atoms can be separated by application of an electric field. The dissociation products from a molecule may be separated by chemical reactions

The application of two-step excitation of molecules is described in an experiment on the separation of the isotopes[13] ^{10}B and ^{11}B. In this experiment, ^{11}BC$_3$ isotope was selectively excited by the light of a CO$_2$ laser which emits lines corresponding to vibrational transitions of ^{11}BC$_3$. The molecules excited in this manner were dissociated by light with a wavelength between 2130 and 2150 Å. The fragments generated by this dissociation, originating mainly from ^{11}BC$_3$, were bound by reaction with O$_2$. It was found that with five light pulses of the CO$_2$ laser radiation, a 14% isotopic enrichment of a 5-μg sample could be obtained.

18.10.3 Photochemical Separation

Another possible way of separating selectively excited atoms/molecules from those in the ground state is by means of a chemical reaction. The reaction must be so chosen that it takes place only with atoms or molecules in the excited state but not with those in the ground state. The isotope of interest can be separated from the other isotopes present by the chemical separation of the reaction products. The basic idea is illustrated in Fig. 18.22. An example of a separation using this scheme is the enrichment of deuterium using an HF laser (Mayer et al. 1970). Some lines of the HF laser coincide with strong transitions of methanol but not with the corresponding lines of deuteromethanol. The excitation activates the reaction

$$CH_3OH + Br_2 \rightarrow 2HBr + HCOOH$$

A one-to-one gas mixture of CH$_3$OH + CD$_3$OD can be converted under irradiation for 60 s with a 90-W HF laser in the presence of Br$_2$ into a mixture containing 95% CD$_3$OD.

---

[13]Reported in Physics Today **27**(9), 17 (1974).

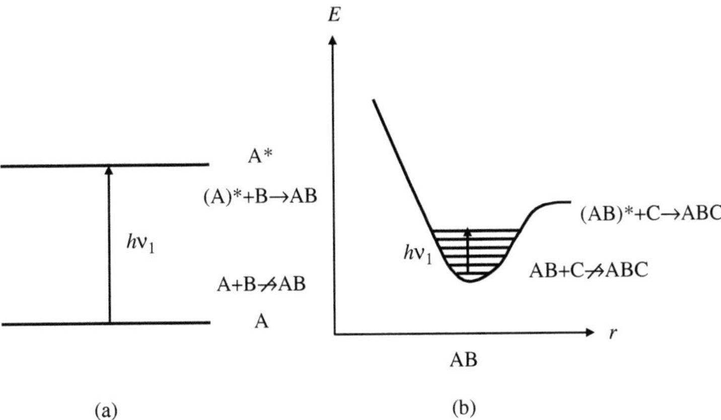

Fig. 18.22 Chemical reactions that take place only with excited atoms or molecules may be employed for the separation of the excited isotope atoms or molecules

One of the most important fields in which the laser isotope separation process would find application is in the nuclear power industry, which requires uranium enriched with the isotope of mass number 235. The present method of enrichment is through gaseous diffusion through a number of stages. This process is quite costly; the cost of obtaining uranium 235 of 90% purity is about 2.3 cents per milligram (Zare 1977). Similarly, the cost of other isotopes of such a concentration is also high. In addition to the nuclear power industry, the isotope separation process would also help obtain isotopes used as tracers in medicine, agriculture, research, industry, etc.

Problems

Problem 18.1 From Eq. (18.5) estimate the velocity of the non-linear polarization at 2ω and compare with the velocity of the electromagnetic wave at frequency ω.

Problem 18.2 The peak Raman scattering in silica appears at about a frequency shift of 13 THz. If the pump wavelength is 1450 nm, at what wavelength would you expect the maximum Raman scattering to take place?

Problem 18.3 The value of n_2 for pure silica is about $3 \times 10^{-20}\,\mathrm{m^2/W}$. Consider a light wave carrying a power of 100 mW propagating through an optical fiber in which the mode occupies a transverse area of 30 $\mu\mathrm{m}^2$. Calculate the non-linear change in refractive index. If the light beam propagates through 1 km of the fiber, then calculate the change in phase of the light beam due to non-linearity. Assume a wavelength of 1500 nm and neglect attenuation in the fiber.

Problem 18.4 Consider a 1-ps pulse at 1500 nm. What is the spectral width of the pulse? If the spectrum increases by 50% due to non-linear effects as it propagates in a medium, to what minimum pulse width can the output pulse be compressed?

Problem 18.5 Consider a fiber-optic gyroscope with 500 m of fiber wound on a coil of radius 10 cm. What is the phase difference between the clockwise and anti-clockwise propagating beams for a rotation rate corresponding to the earth's rotation? Assume a wavelength of 633 nm.

Problem 18.6 Consider a laser beam obeying Poisson statistics and assume that the average number of photons is 1. What is the probability of detecting no photons?

Problem 18.7 What is the change in phase at the output of a 1-km-long fiber when the power is changed from 10 μW to 100 mW. Assume the area of the light beam propagating in the fiber to be 50 μm^2 and $n_2 = 3 \times 10^{-20}$ m^2/W. Neglect fiber loss.

Chapter 19
Lasers in Industry

19.1 Introduction

In Chapter 10 we discussed the special properties possessed by laser light, namely its extreme directionality, its extreme monochromaticity, and the large intensity associated with some laser systems. In the present chapter, we briefly discuss the various industrial applications of the laser.

The beam coming out of a laser is usually a few millimeters (or more) in diameter and hence, for most material processing applications, one must use focusing elements (like lenses) to increase the intensity of the beam. The beam from a laser has a well-defined wave front, which is either plane or spherical. When such a beam passes through a lens, then according to geometrical optics, the beam should get focused to a point. In actual practice, however, diffraction effects have to be taken into consideration (see Chapter 2), and one can show that if λ is the wavelength of the laser light, a is the radius of the beam, and f is the focal length of the lens, then the incoming beam will get focused into a region of radius (see Fig. 19.1)[1]

$$b \approx \lambda f / a \qquad (19.1)$$

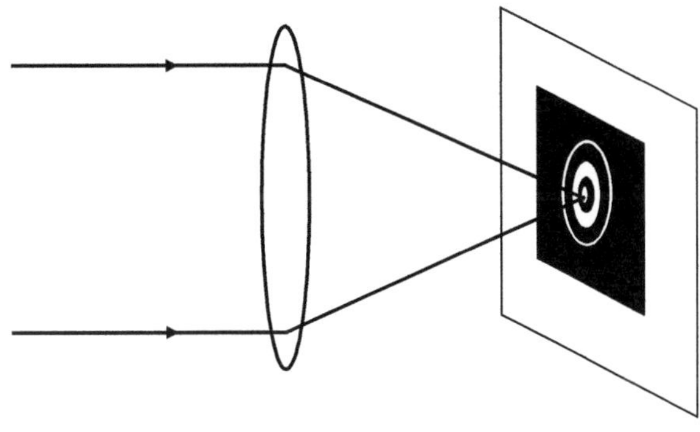

Fig. 19.1 When a plane wave of wavelength λ falls on a lens of radius a, then at the focal plane F of the lens, one obtains an intensity distribution of the type shown in the figure. About 84% of the total energy is confined within a region of radius $\lambda f / a$

---

[1] Here we have assumed that the aperture of the lens is greater than the width of the beam. If the converse is true, then a would represent the radius of the aperture of the lens.

K. Thyagarajan, A. Ghatak, *Lasers*, Graduate Texts in Physics, DOI 10.1007/978-1-4419-6442-7_19, © Springer Science+Business Media, LLC 2010

As can be seen, the dimension of this region[2] is directly proportional of f and λ (the smaller the value of λ, the smaller the size of the focused spot) and inversely proportional to the radius a. If P represents the power of the laser beam, then the intensity I, obtained at the focused region, would be given as

$$I \approx \frac{P}{\pi b^2} \approx \frac{Pa^2}{\pi \lambda^2 f^2} \tag{19.2}$$

Thus if we focus a 1-W laser beam (with $\lambda = 1.06 \ \mu$m and having a beam radius of about 1 cm)[3] by a lens of focal length 2 cm, then the intensity obtained at the focused spot would be given as

$$I \approx \frac{1}{3.14 \times \left(1.06 \times 10^4 \times 2\right)^2} \text{W} / \text{cm}^2 \tag{19.3}$$

$$\approx 7 \times 10^6 \text{W} / \text{cm}^2$$

Figure 10.4 shows the spark created in air at the focus of a 3-MW peak power giant pulsed ruby laser. The electric field strengths produced at the focus are of the order of 10^9 V / m. Note here that such large intensities are produced in an extremely small region whose radius is $\sim 2 \times 10^{-6}$ m. Further, as can be seen from Eq. (19.2), the larger the value of a, the greater the intensity; as such, one often uses a beam expander to increase the diameter of the beam; a beam expander usually consists of a set of two convex lenses as shown in Fig. 19.2.

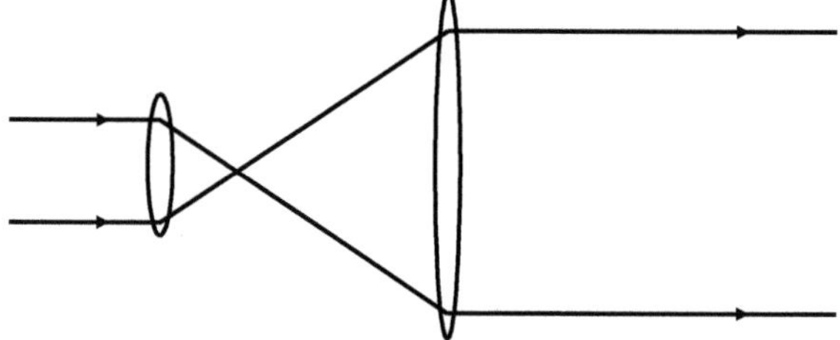

Fig. 19.2 A beam expander consisting of two convex lenses

It may be noted that when one produces such small focused laser spots, the beam has a large divergence, and hence near the focused region, the beam expands again within a very short distance. This distance (which may be defined as the distance over which the intensity of the beam drops to some percentage of that at the focus) defines the depth of focus. Thus, smaller focused spots lead to a smaller depth of

---

[2]The dimension of the focused region is usually larger than that given by Eq. (19.1) due to the multimode emission of the laser. We are also assuming here that the lenses are aberrationless. In general, aberrations increase the spot dimension, resulting in lower intensities.

[3]The 1.06-μm radiation is emitted from the neodymium-doped YAG or glass laser (see Chapter 11).

focus. This must also be kept in mind while choosing the parameters in a laser processing application.

We now discuss in the next few sections some of the important applications of the laser in industry.

19.2   Applications in Material Processing

Since laser beams have high power and can be focused to very small areas, they can generate very high intensities in the region of focus. The intensity levels at the focus can be adjusted by controlling the power and the focused area of the beam. This property of lasers is used in many industrial applications. The primary lasers used for such applications are the Nd:YAG laser emitting at 1060 nm (infrared) with typical powers of 5 kW, carbon dioxide laser emitting at 10.6 μm (far infrared) with powers of up to 50 kW, excimer lasers emitting at 157–350 nm (ultraviolet range) with powers of up to about 500 W. Depending on the application, both continuous wave and pulsed lasers are used. The applications include welding, cutting, hole drilling, micromachining, marking, photolithography, etc.

19.2.1   Laser Welding

One of the simplest applications is in welding wherein high temperature is required to melt and join materials such as steel. High-power lasers have found many important applications in the area of welding. For example, carbon dioxide lasers emitting a wavelength of 10.6 μm and with a power of 6 kW of power are used in welding of $\frac{1}{4}$-in.-thick stainless steel. Lasers are routinely used in the manufacture of automobiles. Figure 19.3 shows welding of car parts using a laser.

Pulsed ruby lasers have also been used in welding. For example, a pulsed ruby laser beam having an energy of 5 J with pulse duration of about 5 ns was used in

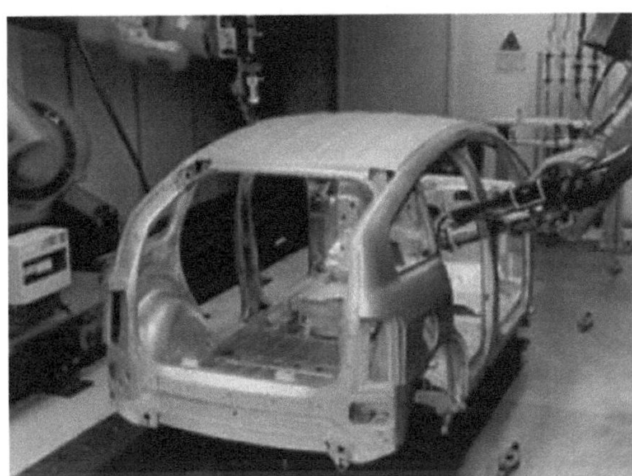

Fig. 19.3 Welding of auto parts by a high-power laser

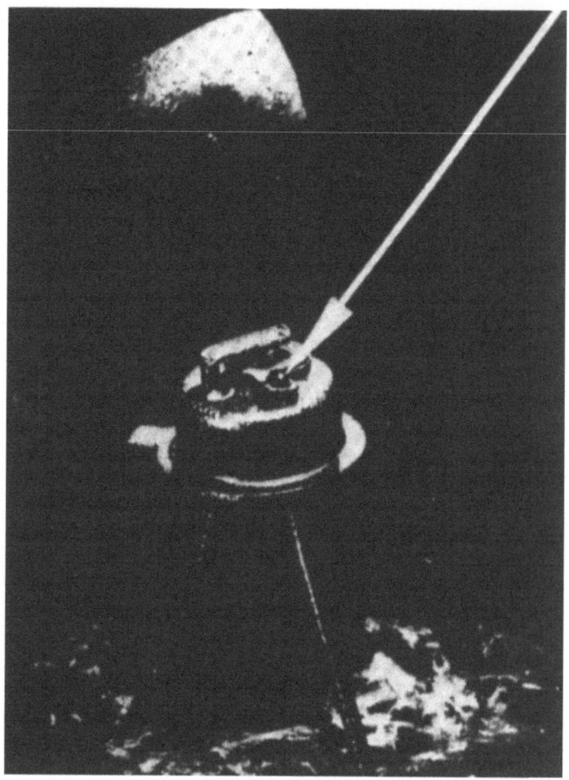

Fig. 19.4 Welding of parts on a transistor; the *arrow* shows the position of the weld. (Adapted from Gagliano et al. (1969))

welding 0.18-mm-thick stainless steel. The weld was made using overlapping spots and the laser was pulsed at a rate of 20 pulses/min. The focused spot was about 1 mm in diameter and the associated power density was $\sim 6 \times 10^5$ W/cm^2.

Laser welding has found important applications in the fields of electronics and microelectronics which require precise welding of very thin wires (as small as 10 μm) or welding of two thin films together. In this field, the laser offers some unique advantages. Thus, because of the extremely short times associated with the laser welding process, welding can be done in regions adjacent to heat-sensitive areas without affecting these elements. Figure 19.4 shows a weld performed with a laser on a transistor unit. Further, welding in otherwise inaccessible areas (like inside a glass envelope) can also be done using a laser beam. Figure 19.5 shows such an example in which a 0.03-in. wire was welded to a 0.01-in.-thick steel tab without breaking the vacuum seal. In laser welding of two wires, one may have an effective weld even without the removal of the insulation.

Laser welds can easily be performed between two dissimilar metals. Thus, a thermocouple may easily be welded to a substrate without much damage to adjacent material. One can indeed simultaneously form the junction and attach the junction to the substrate. This method has been used in attaching measuring probes

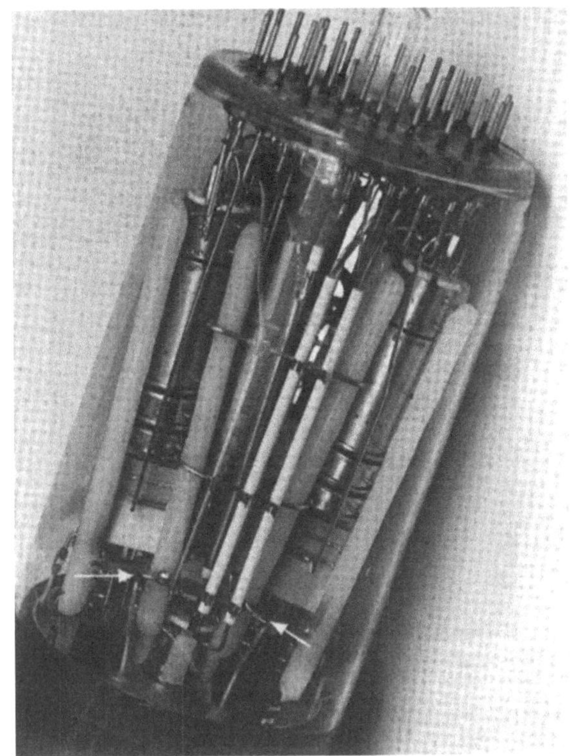

Fig. 19.5 Laser welding in inaccessible areas. The photograph shows the welding of a 0.03-in. wire to a 0.01-in.-thick steel tab inside a vacuum tube without breaking the vacuum seal. The *arrows* point toward the repaired connections. (Adapted from Weaver (1971); photograph courtesy: Dr. Weaver)

to transistors, turbine blades, etc. Laser weld not only achieves welding between dissimilar metals but also allows precise location of the weld.

In welding, material is added to join the two components. Thus the laser power must not be too high to evaporate the material; removal of material leads, in general, to bad welds. Thus the laser used in welding processes must have a high average power rather than high peak power. The neodymium:YAG lasers and carbon dioxide lasers are two important kinds of lasers that find wide-ranging applications in welding.

19.2.2 Hole Drilling

Drilling of holes in various substances is another interesting application of the laser.[4] For example, a laser pulse having a pulse width of about one hundredth of a second and an energy of approximately 0.05 J can burn through a 1-mm-thick steel plate

---

[4]In the early 1960s, the power of a focused laser beam was measured by the number of razor blades that the beam could burn through simultaneously, the "Gillette" being the unit of measurement of power per blade burnt through.

Fig. 19.6 Hole drilled in a 1-mm-thick stainless steel

leaving behind a hole of radius about 0.1 mm. Further, one can use a laser beam for the drilling of diamond dies used for drawing wires. Drilling holes less than about 250 μm in diameter by using metal bits becomes very difficult and is also accompanied by frequent breakage of drill bits. With laser one can easily drill holes as small as 10 μm through the hardest of substances. Figure 19.6 shows a typical laser-drilled hole in a 1-mm-thick stainless steel. The Swiss watch industry in Europe has been using flash-pumped neodymium:YAG laser to drill ruby stones used in timepieces. In addition to the absence of problems like drill breakage, laser hole drilling has the advantage of precise location of the hole.

Figure 19.7 shows drilling through a piece of rock using an Nd:YAG laser emitting at 1.06 μm. Laser drilling can indeed reduce drilling time by more than a factor of 10 and hence reduce cost dramatically in oil exploration applications. Typical drilling speed of 1 cm/s is possible by using different types of lasers.

Due to the extremely small areas to which the laser beams can be focused, they are used in the area of micromachining. Figure 19.8 shows how it is possible to write on a human hair using lasers. Lasers are also being used in the removal of microscopic quantities of material from balance wheels while in motion. They have also been used in trimming resistors to accuracies of 0.1%. Such micromachining processes find widespread use in semiconductor circuit processing. The advantages offered by a system employing lasers for such purposes include the small size of the focused image with a precise control of energy, the absence of any contamination, accuracy of positioning, and ease of automation.

19.2.3 Laser Cutting

Lasers also find application in cutting materials. The most common laser that is used in cutting processes is the carbon dioxide laser due to its high output power.

Fig. 19.7 Drilling through a piece of rock using laser. (Adapted from Ref. http://www.ne.anl.gov/facilities/lal/laser_drilling.html)

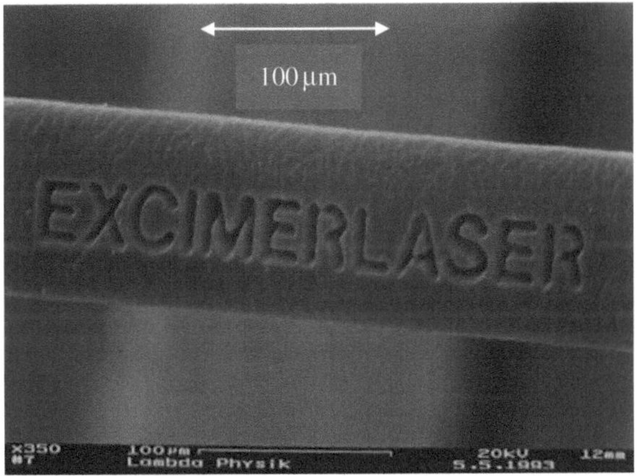

Fig. 19.8 Micromachining in a piece of hair using a laser. (Adapted from Lambda Physik, Germany)

In the cutting process, one essentially removes the materials along the cut. When cuts are obtained using pulsed lasers, then the repetition frequency of the pulse and the motion of the laser across the material are adjusted so that a series of partially overlapping holes are produced. The width of the cut should be as small as possible with due allowance to avoid any rewelding of the cut material. The efficiency of laser cutting can be increased by making use of a gas jet coaxial with the laser (see Fig. 19.9). In some cases one uses a highly reactive gas like oxygen so that when the laser heats up the material, it interacts with the gas and gets burnt. The gas jet also helps in expelling molten materials. Such a method has been used to cut materials like stainless steel, low-carbon steel, and titanium. For example, a 0.13-cm-thick

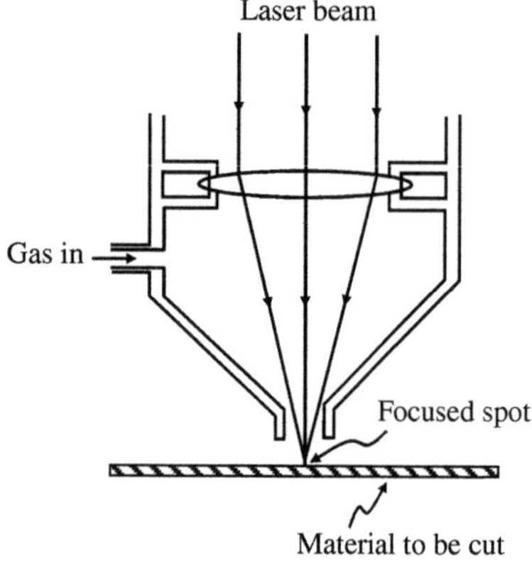

Fig. 19.9 Cutting using a focused laser spot

Fig. 19.10 Use of carbon dioxide laser to cut wood. (Photograph courtesy: Ferranti Ltd.)

stainless steel plate was cut at the rate of 0.8 m/min using a 190-W carbon dioxide laser using oxygen jet.

In some methods, one uses inert gasses (like nitrogen or argon) in place of oxygen. Such a gas jet helps in expelling molten materials. Such a technique would be very efficient with materials which absorb most radiation at the laser wavelength. Wood, paper, plastic, etc. have been cut using such a method. A gas jet-assisted CO_2 laser can be used for obtaining parallel cuts of up to 50 mm depth in wood products. At the cut edges, carbonization occurs, but it is usually limited to a small depth (about a few tens of micrometers) of the material. This causes a discoloration only and can be decreased by increasing the cutting speed. Figure 19.10 shows how a carbon dioxide laser is used (with a gas jet) in cutting wood. Laser cutting of stainless steel, nickel alloys, and other metals finds widespread application in the aircraft and automobile industries.

19.2.4 Other Applications

Lasers also find applications in vaporizing materials for subsequent deposition on a substrate. Some unique advantages offered by the laser in such a scheme include the fact that no contamination occurs, some preselected areas of the source material may be evaporated, and the evaporant may be located very close to the substrate.

An interesting application of laser is in the opening of oysters. A laser beam is focused on that point on the shell where the muscle is attached. This results in detachment of the muscle, the opening of the shell, and leaving the raw oyster alive in the half shell (see Fig. 19.11).

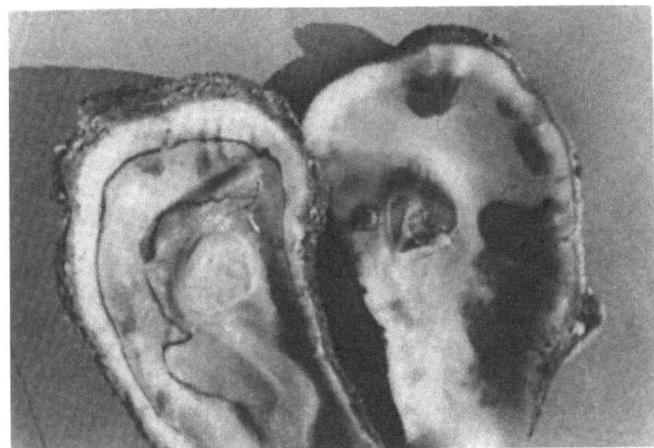

Fig. 19.11 The photograph shows an oyster opened with a CO_2 laser, which neatly detaches adductor muscle from the shell and leaves the raw oyster alive in the half shell. (Photograph courtesy: Professor Gurbax Singh of the University of Maryland)

19.3 Laser Tracking

By tracking we imply either determining the trajectory of a moving object like an aircraft or a rocket, or determining the daily positions of a heavenly object (like the Moon) or an artificial satellite; a nice review on laser tracking systems has been given by Lehr (1974). The basic principle of laser tracking is essentially the same as that used in microwave radar systems. In this technique, one usually measures the time taken to travel to and fro for a sharp laser pulse sent by the observer to be reflected by the object and received back by the observer (see Fig. 19.12); suitably modulated continuous wave (CW) lasers can also be used for tracking.

One of the main advantages of a laser tracking system over a microwave radar system is the fact that not only a laser tracking system has a smaller size but also its cost is usually much less. Further, in many cases one can use a retroreflector on the object; in a retroreflector, the incident and reflected rays are parallel and travel in opposite directions. A cube corner is often used to act as a retroreflector (see Fig. 19.13). For example, on the surface of the moon, or on a satellite, one can have a retroreflector to reflect back the incident radiation. For a laser tracking system, the size of the retroreflector is much smaller than the corresponding microwave reflector

Fig. 19.12 Light detection
and ranging (LIDAR)

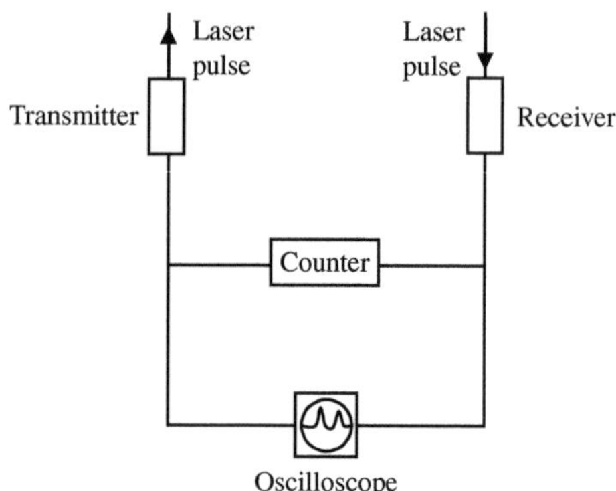

Fig. 19.13 Cube corner as a
retroreflector

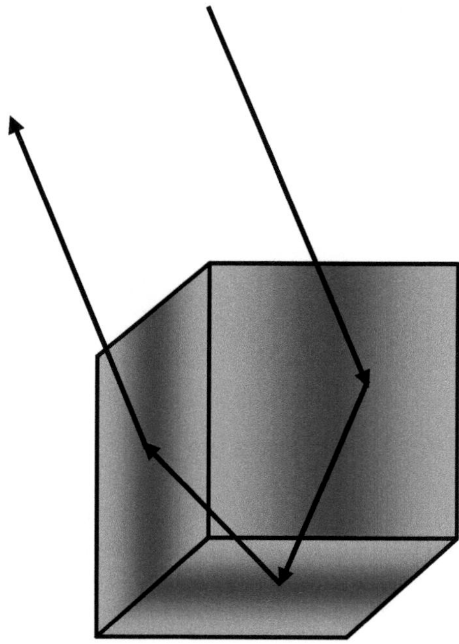

owing to the smaller wavelength of the optical beam and hence the reflector can be
more conveniently mounted in the system involving lasers.

In a microwave radar system, one has to incorporate corrections because of the
presence of the ionosphere and also because of the presence of water vapor in the
troposphere. These corrections are much easier to incorporate in the case of an optical beam. As compared to a microwave radar system, the laser radar offers much
higher spatial resolution.

On the other hand, there are some disadvantages in using a laser tracking system. For example, when fog and snow are present in the atmosphere, it is extremely

Table 19.1 Characteristics of pulsed lasers used in tracking systems

Type	Wavelength (μm)	Efficiency (%)	Energy (J)	Pulse duration (ns)	Pulse repetition rate	Spectral width (nm)
Nd:YAG	1.06	0.1	0.02	10–25	100 s^{-1}	0.5
GaAs	0.9	4	10^{-4}	100	100 s^{-1}	2
Ruby	0.694	0.013	7	3	20 min^{-1}	0.04
Nd:glass	0.530	0.04	20	20	12 h^{-1}	0.9

Source: Adapted from Lehr (1974)

Table 19.2 Typical ranges and velocities

Object	Distance (m)	Angular velocity (arcsec/s)
Moon	3.8×10^8	14.5
Near-Earth satellite	10^6	10^3
Aircraft (DC-10)	2×10^4	500
Rocket (at launch)	5×10^3	10^5

Source: Adapted from Lehr (1974)

difficult to work at optical frequencies. Further, during daytime there is a large background noise. The losses in the transmitter and the receiver are also considerably larger in laser systems.

In Table 19.1 we have tabulated some of the typical lasers that have been used in tracking systems. Typical ranges and velocities of various objects measured by a laser tracking system are tabulated in Table 19.2. One can see that the distance that can be covered range from 5000 m to hundreds of megameters.

The transmitter which is pointed toward the object may simply consist of a beam expander as shown in Fig. 19.2. For tracking a moving object, both the laser and the telescope may be moved. One could alternatively fix the laser and bend the laser beam by means of mirrors. There are other ways of directing the laser beam toward the object; for further details, the reader is referred to the review article by Lehr (1974) and the references therein. The receiver which is also pointed toward the object may consist of a reflector or a combination of mirrors and lenses. The detector may simply be a photomultiplier.

Figure 19.14 gives a block diagram of a laser radar system for tracking of a satellite. A portion of the pulse that is sent is collected and is made to start an electronic counter. The counter stops counting as soon as the reflected pulse is received back. The counter may be directly calibrated in units of distance.

National Aeronautics and Space Administration, USA, had launched an aluminum sphere called the Laser Geodynamic Satellite (LAGEOS) into orbit at an altitude of 5800 km for studying the movements in the Earth's surface, which would be of great help in predicting earthquakes. Figure 19.15 shows the satellite, which is 60 cm in diameter, weighs 411 kg, and has 426 retroreflectors which return the

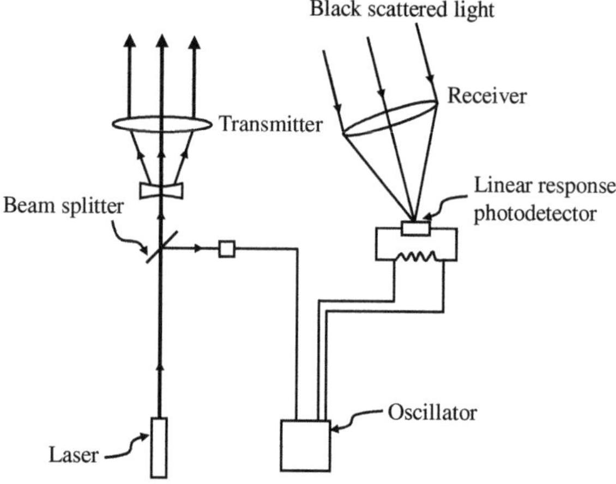

Fig. 19.14 Block diagram of a typical pulsed LIDAR system

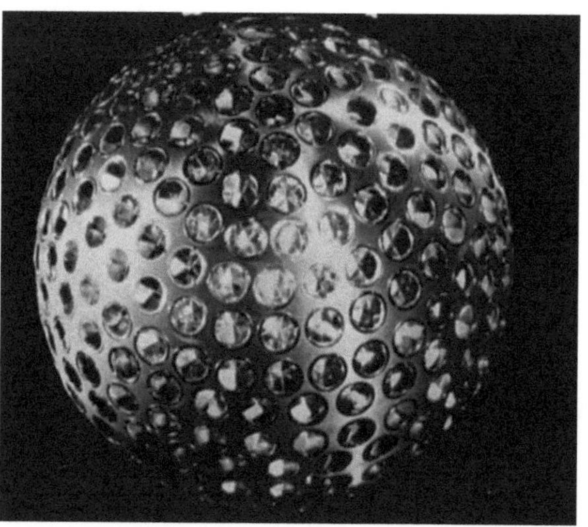

Fig. 19.15 The Laser Geodynamic Satellite (LAGEOS) put into orbit by the National Aeronautics and Space Administration, USA, for measuring minute movements of the Earth's crust, which would be helpful in predicting earthquakes. The satellite is 60 cm in diameter, weighs 411 kg, and is studded with 426 retroreflectors, which return the incident laser pulses to their origin on the surface of the Earth. Minute movements of the Earth's crust are detected by measuring the flight time of a light pulse to the satellite and back. (Photograph courtesy: United States Information Services, New Delhi)

laser pulses exactly back to the point of origin on the Earth. Accurate measurements of the time of flight of laser pulses to the satellite and back should help scientists in measuring minute movements of the Earth's crust. Figure 19.16 shows scientists performing the prelaunch testing of the satellite.

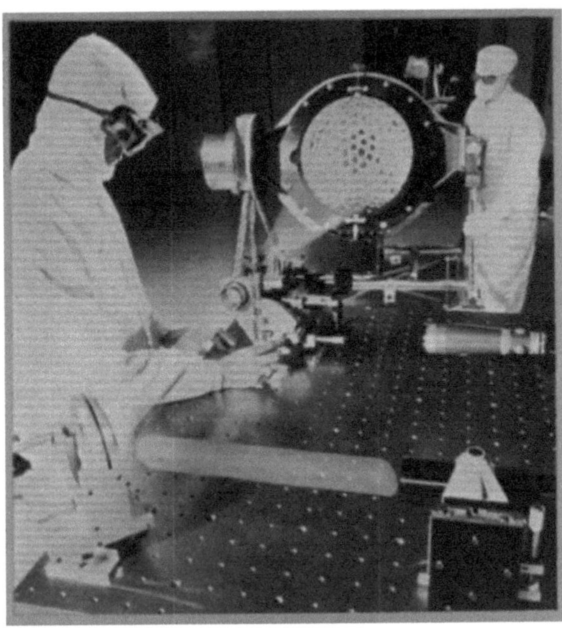

Fig. 19.16 LAGEOS undergoing prelaunch testing in the laboratory. (Photograph courtesy: United States Information Service, New Delhi)

19.4 Lidar

Laser systems have also been used for monitoring the environment. Such systems are called LIDARs (acronym for light detection and ranging) and they essentially study the laser beam scattered from the atmosphere. It may be mentioned that studies of the atmosphere using an optical beam had been carried out even before the advent of the laser; for example, using a searchlight, Hulbert in 1937 studied atmospheric turbidity to a height of 28 km. The arrival of the laser on the scene revolutionized the atmospheric study using coherent laser beams.

Pulses of laser light are sent and the radiation that is scattered by various particles present in the atmosphere is picked up by the receiver. The background sunlight is removed by using filters. This scattered light gives information regarding the particles present in the atmosphere with a sensitivity that is much more than that obtainable from microwave radars.

In Fig. 19.14 we have given a block diagram of a pulsed LIDAR system to study the nature of aerosols present in the atmosphere. One usually measures the time dependence of the intensity of the backscattered laser light using a photodetector. The time variation can be easily converted into the height from which the laser beam has been backscattered. A typical time dependence of the backscattered laser radiance is plotted in Fig. 19.17a which corresponds to an atmosphere which has no aerosols, i.e., the backscattering is by pure molecular gases such as N_2, O_2, and Ar. On the other hand, if the atmosphere contained aerosols, then the time dependence of the backscattered laser radiance would of the form shown in Fig. 19.17b.

Fig. 19.17 Backscattered
radiation from (**a**) a clear
atmosphere and (**b**)
atmosphere containing
aerosols

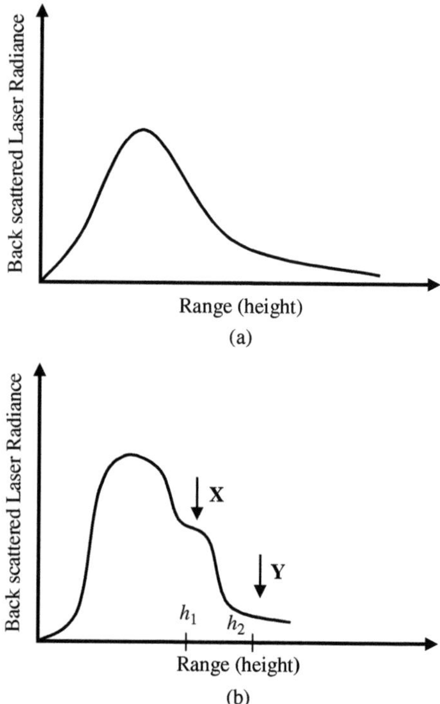

Note the kinks that appear in the curve at the points marked X and Y; these are
due to the fact that between the heights h_1 and h_2 there are aerosols which are
responsible for a greater intensity (compared to that for a clear atmosphere) of the
backscattered laser light. Thus a curve like that shown in Fig. 19.17b implies a
haze which exists between the heights h_1 and h_2. It may be seen that corresponding
to the heights h_2 the intensity is roughly the same as that from a pure molecular atmosphere. Thus, beyond the height h_2, one does not expect the presence of
any aerosols. With the LIDAR one can also study the concentrations and sizes of
various particles present in the atmosphere, which are of extreme importance in pol-
lution studies. Small particles are difficult to detect with the microwave radar; the
microwave radar can detect the presence of rain, hail, or snow in the atmosphere.
This difference arises essentially due to the larger amount of scattering that occurs
at optical wavelengths. In addition, a LIDAR can also be used to study the visibility
of the atmosphere, the diffusion of particulate materials (or gases released at a point)
in the atmosphere, and also the presence of clouds, fog, etc.; the study of turbulence
and winds and the probing of the stratosphere have also been carried out by LIDAR
systems. For further details on the use of laser systems for monitoring the environment, the reader is referred to the review article by Hall (1974) and the references
therein.

19.5 Lasers in Medicine

Perhaps the most important use of lasers in the field of medicine is in eye surgery. Hundreds of successful eye operations have already been performed using lasers. The tremendous use of the laser in eye surgery is primarily due to the fact that the outer transparent regions of the eye allow light at suitable wavelengths to pass through for subsequent absorption by the tissues at the back of the eye.

As is well known, the eye is roughly spherical and consists of an outer transparent wall called the cornea, which is followed by the iris (which can adjust its opening to control the amount of light entering the eye), and a lens. Between the cornea and the lens is the aqueous humor. The back part of the eye contains the light-sensitive element, namely the retina. Light falling on the eye is focused by the lens on the retina, and the photosensitive pigment-containing cells present in the retina convert the light energy into electrical signals, which are carried by the optic nerve to the brain, resulting in the process of seeing.

As a result of some disease or heavy impact, the retinal layer may get detached from the underlying tissue, creating a partial blindness in the affected area. Earlier, a xenon arc lamp was used for welding together the detached portion of the retina. But the long exposure times of this source required administering anesthesia for safety. Also it cannot be focused sharply. The unique advantages of using a laser beam for welding a detached retina are that since it can be focused to an extremely small spot, precise location of the weld can be made and also the welds are much smaller in size. The spot size of a typical xenon arc beam on the retina when focused by the eye lens is about 500–1000 μm in diameter; this is much larger than the typical diameter (about 50 μm) obtainable using a laser beam. The time involved in laser beam welding is so short that the eye does not need any clamping. Pulses of light from a ruby laser lasting for about 300 μs at levels below 1 J are used for retinal attachment.

Lasers are also expected to be used extensively in the treatment of cancer. In an experiment reported in the USSR, amelanotic melanoma was inculcated from human beings on nine animals. These animals were irradiated with a ruby laser beam and it was reported that within 1 month the tumors completely disappeared. The power associated with the ruby laser beam was about 100 MW with a total energy of about 200 J. Successful skin cancer treatment with lasers has also been reported on human beings.

Lasers can also be used for correction of focusing defects of the eye. In the method referred to as LASIK (*la*ser *i*n *s*itu *k*eratomileusis), the cornea of the eye can be crafted to adjust the curvature so that the focusing by the eye lens takes place on the retina (see Fig. 19.18). This method can correct for eye defects requiring high lens powers and is a very popular technique.

It is impossible to list all the applications of lasers in the field of medicine. Extensive use of lasers is anticipated in surgery, dentistry, and dermatology. For further details and other applications of lasers in medicine, the reader is referred to the recent article by Peng et al. (2008).

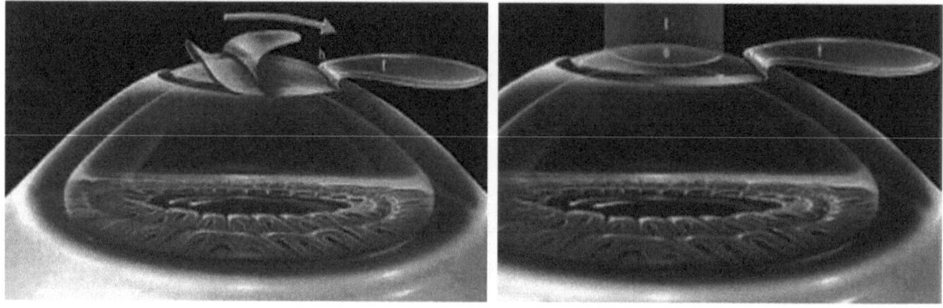

Fig. 19.18   Application of lasers in LASIK

19.6 Precision Length Measurement

The large coherence length and high output intensity coupled with a low divergence enables the laser to find applications in precision length measurements using interferometric techniques. The method essentially consists of dividing the beam from the laser by a beam splitter into two portions and then making them interfere after traversing two different paths (see Fig. 19.19). One of the beams emerging from the beam splitter is reflected by a fixed reflector and the other usually by a retroreflector[5] mounted on the surface whose position is to be monitored. The two reflected beams interfere to produce either constructive or destructive interference. Thus, as the reflecting surface is moved, one would obtain alternatively constructive

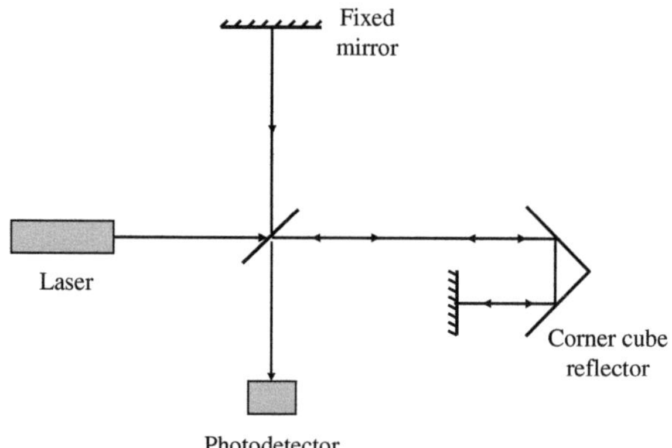

Fig. 19.19   Laser interferometer arrangement for precision length measurements

---

[5]As mentioned earlier, a retroreflector reflects an incident beam in a direction exactly opposite to that of an incident beam (see Fig. 19.11), and it is characterized by the property that minor misalignments of the moving surface do not cause any significant errors.

and destructive interferences, which can be detected with the help of a photodetector. Since the change from a constructive to a destructive interference corresponds to a change of a distance of half a wavelength, one can measure the distance traversed by the surface on which the reflector is mounted by counting the number of fringes which have crossed the photodetector. Accuracies up to 0.1 μm can be obtained by using such a technique.

This technique is being used for accurate positioning of aircraft components on a machine tool, for calibration and testing of machine tools, for comparison with standards, and many other precision measurements; for further details, see, e.g., Harry (1974), Chapter 5 and the references therein. The conventional cadmium light source can be used only over path differences of about 20 cm. With the laser one can make very accurate measurements over very long distances because of the large coherence length. The most common type of laser used in such applications is the helium laser and since the distance measurement is being made in terms of wavelength, in these measurements, a high wavelength stability of the laser output must be maintained.

19.7 Laser Interferometry and Speckle Metrology

The phenomenon of interference, which was briefly discussed in Chapter 2, is a widely used technique for many extremely accurate measurements in science, technology and engineering. The field which uses interference phenomena for such measurements is referred to as *interferometry*. To achieve interference between two beams of light, an interferometer divides a light beam into two or more parts, which are made to travel different paths after which they are united to produce an interference pattern. The interference pattern exhibits the effect of the paths travelled by the beams. Since the wavelength of light is very small ($\sim$ 500 nm), interference principles are capable of resolving changes in distance to the order of a few tens of nanometers.

As discussed in Chapter 2, good interference between waves requires the waves to be coherent and before the advent of the laser, such spatially and temporally coherent sources were realized by using a pinhole to have a point source leading to spatial coherence and a wavelength filter to achieve temporal coherence (see Chapter 10 for a detailed discussion on coherence). This led to a drastic reduction in the intensity of the light that is available for the interference phenomena. The appearance of the laser provided interferometry with an intense spatially and temporally coherent source and laser interferometry has become a very important tool in the hands of the scientists and engineers for measurement purposes. Laser interferometers are used for high-precision measurements from a few nanometers to about 100 m for measuring distances, angles, flatness, straightness, velocity, acceleration, vibrations, etc. Accurate measurement of displacement is very crucial in many industries such as machine tool industries. Some of the highest demands for accurate measurement come from integrated circuit manufacturing industries where interferometers are used to control wafer steppers. Some of the more common lasers that are used in interferometry are the He–Ne laser, Argon ion laser, Nd:YAG laser, and diode laser. Among these, diode lasers are the most compact with low power consumption and are available over a very broad range of wavelengths. They are also tunable over a limited wavelength range. The beam from a diode laser is usually highly divergent and does not have a circular cross section. Additional optics is usually used to produce a collimated beam.

In Chapter 18 we also discussed about applications of laser in the detection of gravitational waves. Here we shall discuss some applications of laser interferometry in detection of small displacements and vibrations. In Section 19.7.4 we shall discuss speckle metrology which has very important applications.

19.7.1 Homodyne and Heterodyne Interferometry

One of the most common interferometers used in laser interferometry is the Michelson interferometer arrangement shown in Fig. 19.20. Beam from a light source, here a laser, is directed toward a beam splitter which reflects half the incident light toward a fixed mirror and reflects the other half toward a movable mirror (see Fig. 19.20). The returning beams interfere after getting transmitted and reflected at the beam splitter. The intensity in the transmitted arm of the interferometer changes with the displacement of the mirror through the following expression:

$$I(d) = I_1 + I_2 + 2I_1I_2 \cos(2\pi d/\lambda) \tag{19.4}$$

where I_1 and I_2 are the intensities of the beams returning via paths 1 and 2, respectively, the last term is the interference term and $2d$ is the path difference between the two arms. As d changes, the intensity received in the transmitted arm changes.

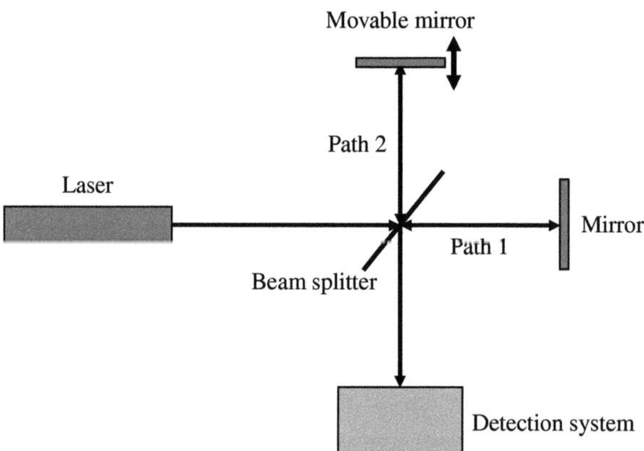

Fig. 19.20 A schematic of the Michelson interferometer setup

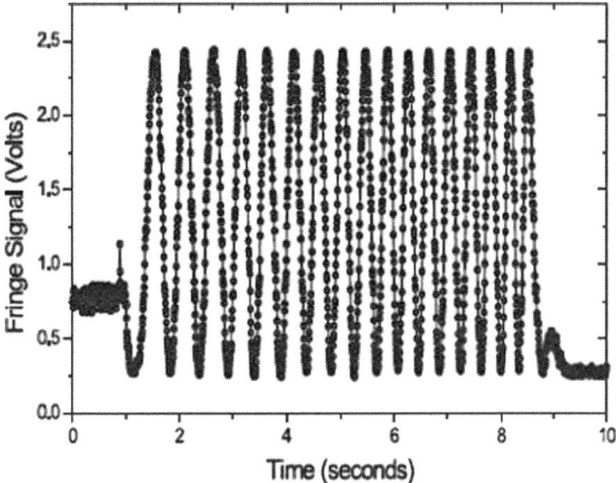

Fig. 19.21 The fringe signal from the detector when the mirror is moved. For every movement by one half wavelength of the mirror, the intensity goes through one cycle. (Source: Teach Spin)

A change of d by half a wavelength would cause the intensity to go over one full period of interference (see Fig. 19.21). Thus a precise measurement of the intensity change can lead to precise measurement of the change in d.

Movement of the mirror by half a wavelength changes the phase of the interfering beams by one full wavelength resulting in passage through a full interference fringe. Thus starting from a minimum position, moving of the mirror by one half wavelength passes the interference to go from this minimum to the next minimum. Since a laser has a large coherence length, a measurement such as this can be made over a large distance and due to the monochromaticity, the measurements are very accurate. The intensity measurements are usually carried out by photodetectors to obtain an electronic signal for further processing. Although the principle looks very simple, there are many possible errors that can contribute to an inaccuracy in the measurement. For example, the laser wavelength needs to be known precisely and should not vary during the measurement, there could be problems due to varying atmospheric conditions etc. As an example it may be noted that a change of temperature of air by 1°C or pressure by 2.5 mmHg or 80% change in humidity will cause an error of one part in a million.

In the above discussion, the laser is assumed to emit a single wavelength; thus the two interfering beams have the same frequency and the output intensity remains constant as long as the mirrors are stationary. This is also referred to as *homodyne interferometry*. In contrast, in *heterodyne interferometry*, the laser is made to emit two closely lying wavelengths or frequencies. The two frequencies from the laser can be generated, for example, using an acousto-optic modulator or by using external fields such as magnetic field across a He–Ne laser tube which creates two closely lying energy levels via the Zeeman effect. Using the Zeeman effect a maximum frequency difference of about 4 MHz can be generated, while using acousto-optic modulators, it is possible to generate frequency shits of 20 MHz or more. In such an interferometer instead of a polarization-independent beam splitter, one uses a

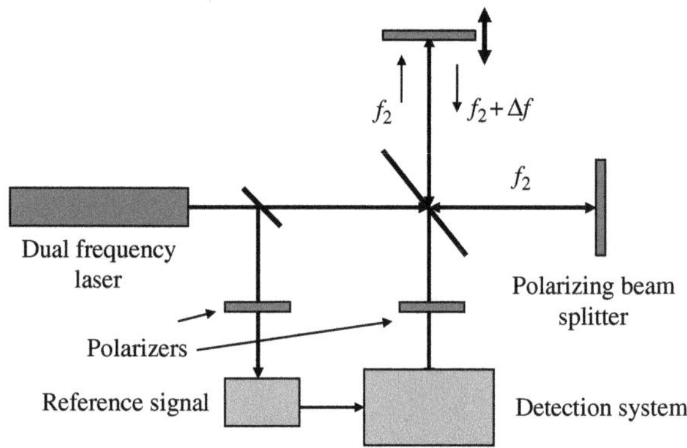

Fig. 19.22 A schematic of the laser heterodyne interferometer

polarizing beam splitter (see Fig. 19.22). Assuming that the two frequencies f_1 and f_2 are in orthogonal linear polarization states, light at these frequencies incident on the polarizing beam splitter splits and takes two different paths. In the interferometer, a part of the light emitted by the laser is reflected to a photodetector and mixed producing a current that is modulated at the beat frequency f_1–f_2. The beams reflected by the mirror in the reference arm and the mirror in the measurement arm return to the beam splitter and are made to interfere in another photodetector. When both the mirrors are stationary, the beat frequency measured is f_1–f_2. When the movable mirror on which the wave at frequency f_2 is incident, moves, the reflected wave undergoes Doppler shift and thus the frequency that mixes with the other wave changes from f_2 to $f_2 + \Delta f$ or f_2–Δf depending on the direction of motion of the mirror; here the Δf term is due to the Doppler shift. This signal is then electronically "compared" with the reference signal f_1–f_2. A phase detector is then used to measure the phase between the reference and the measured signals, which is used to get precise information on the position and velocity of the movable interferometer arm.

If the mirror moves with a velocity v, then the Doppler shift Δf is given as

$$\Delta f = \frac{2vf_2}{c} \tag{19.5}$$

where c is the velocity of light in free space. The change of phase of the signal for a movement between times t_1 and t_2 is given as

$$\varphi = \int_{t_1}^{t_2} 2\pi \, \Delta f \, dt = \frac{2\pi f_2}{c} \Delta l \tag{19.6}$$

where Δl is the distance moved by the movable mirror. The displacement is obtained by a measurement of the difference in phase between the reference beat signal and the measured signal.

A few aspects of the heterodyne interferometer make it superior to the homodyne interferometer. Since the displacement information is carried on an ac signal rather than a dc signal, it is less sensitive to laser power fluctuations and other noise-like ambient light, etc. It also requires only a single detector; however it requires a very highly stable dual-frequency laser source and the signal processing is also more complex.

19.7.2 Holographic Interferometry

In an earlier chapter we had discussed the basic principles of holography. Here we outline a very important application of holography, namely holographic interferometry. The technique of holographic interferometry was first discovered in 1965 and ever since has been widely used for many applications. Here we discuss in brief the following: double-exposure interferometry, real-time interferometry, and time-average interferometry.

19.7.2.1 Double-Exposure Interferometry

This technique is used to determine minute deformations (in the scale of wavelength of light) in an object from which information on the quality of the object can be obtained. Due to the basic nature of the holographic principle, rough surfaces can be studied with interferometric precision. The object is holographically recorded twice on the same photosensitive device, once before and once after introducing deformation in the object. As discussed in Chapter 15, the hologram reconstructs the two *object waves* simultaneously and since the object has had distortions between the two exposures, the two reconstructions are not identical, leading to an interference between the two waves. The resulting interference pattern contains information on the deformation of the object.

In order to analyze this, let the object wave emerging before the deformation be represented by $O(x,y)$ and the object wave after the deformation be represented by $O'(x,y)$. If the deformation is small and if we assume that it leads only to changes in phase, then we can write

$$O'(x, y) = O(x, y)\exp(-i\phi(x, y)) \tag{19.7}$$

where $\phi(x,y)$ is the change of phase due to the deformation. During the reconstruction after double exposure, the object waves $O(x,y)$ and $O'(x,y)$ are simultaneously generated and what we observe is the interference between the two waves. This leads to an intensity distribution given as

$$I(x, y) = K([O(x, y) + O'(x, y)])^2 = 2K(O(x, y))^2(1 + \cos\varphi(x, y)) \tag{19.8}$$

where K is a constant. The term within the brackets on the right-hand side describes the interference pattern that would be observed superimposed on the object intensity

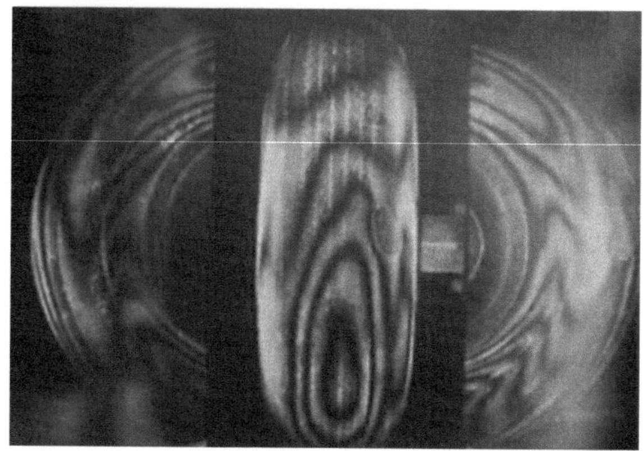

Fig. 19.23 Double-exposure interferogram of a tire; the two exposures correspond to different levels of air filling. (Source: Ref. www.holophile.com/history.htm)

pattern defined by $(O(x, y))^2$. The fringes will have the shape given by the curves $\varphi(x, y) = $ constant.

This principle of double-exposure holographic interferometry is used in non-destructive testing of objects. Thus if the deformation corresponds to application of stress on the object or change of temperature between the two exposures, then the fringe pattern will be a direct indication of the deformation caused in the object due to the applied stress or the temperature change. Defects in the object become immediately visible due to the strain being different at points where the defect is present in the object as compared to other regions of the object. Figure 19.23 shows a typical example of a double-exposure interferogram showing clearly the fringes formed due to distortion of the object. The contours of the fringes clearly indicate the strain contours of the object.

19.7.2.2 Real-Time Interferometry

This is similar to double-exposure interferometry, except that the hologram is recorded only once and is placed exactly at the same position and illuminated with the reconstruction wave. Thus the hologram produces the object wave at the time of recording. Now any distortion in the object generates an object wave which will interfere with the reconstructed object wave to produce interference fringes. If the distortions vary with time, the interference pattern will vary with time, thus helping visualization of time variation of distortion of the object.

19.7.2.3 Time-Average Interferometry

This technique is used in the case of steady-state vibration problems. The technique gives information on the vibration modes, amplitude distribution, etc. Consider an object which is vibrating in one of its normal modes. If the period of vibration is much shorter than the recording time, then the recording is an overlap of several recordings. Thus when the hologram is reconstructed, the wave fronts stored in the hologram corresponding to various positions of the object are reconstructed and they form an interference pattern.

Let the vibration of the object be described as $A \cos \omega t$. If we assume that the vibrations are of small amplitude, as the object vibrates, only the phase of the object wave can be assumed to change and if ω is the frequency of vibration, we can assume that the phase of the object wave changes according to the following equation:

$$\Delta = F \cos \omega t \tag{19.9}$$

where F is the maximum value of the phase change due to the vibration. If the recording time is large compared to the time period of vibration, then the hologram records a continuous distribution of images of the object corresponding to one cycle of vibration. During reconstruction, the total object wave reconstructed will be the sum of all the object waves recorded during the recording process. Hence the reconstruction process generates a reconstructed wave with an intensity given as

$$I = I_0 \left(\int e^{(iF \cos \omega t)} dt \right)^2 = I_0 J_0^2(F) \tag{19.10}$$

where I_0 is a constant corresponding to the reconstruction when the object is stationary $(F = 0)$ and J_0 is the Bessel function of first kind of order zero. Thus the intensity distribution in the reconstruction corresponds to the object image modulated by the Bessel function term. The brightest image point corresponds to $F = 0$, which corresponds to the nodal point of the vibration pattern. Zeroes of the Bessel function correspond to dark fringes and subsequent maxima of the Bessel function J_0 correspond to bright fringes. The fringe pattern gives the vibration pattern of the object (see Fig. 19.24). Knowing the positions of maxima and minima of the Bessel function, it is possible to analyze the vibration characteristics of the object.

19.7.3 Laser Interferometry Lithography

In recent years, laser interferometry has been used for nanolithography to produce periodic and quasi-periodic nanostructures for various applications. In this a periodic interference pattern is produced using two, three, or four laser beams propagating in appropriate directions. The periodic interference pattern is produced on photoresist and the pattern produced on the photoresist is then transferred using conventional photolithographic techniques to the underlying layer. Another recent technique involves using high-power laser beams for interference and directly writing the pattern into recording materials to produce nanostructured surfaces and devices. The advantages of this technique vis-a-vis other lithography techniques such as electron beam lithography is its high efficiency and the lower cost.

If we assume N laser beams propagating in different directions and interfering, then the electric field distribution in the interference pattern is given as

Fig. 19.24 Time-average holograms of a vibrating guitar. (Adapted from Jansson (1969))

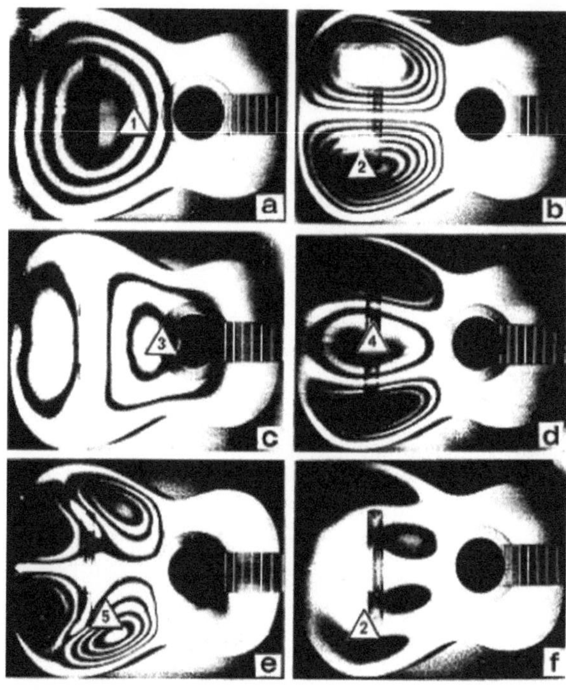

$$E = \sum A_n \exp(\omega t - \mathbf{k}_n.\mathbf{r} + \phi_n) \qquad (19.11)$$

where A_n is the amplitude, ϕ_n the phase, and $\mathbf{k}_n$ the corresponding propagation vector of the nth beam. By choosing appropriate values of the various parameters, it is possible to generate various interference patterns which can then be transferred to the material. Lasers used for this purpose should have an appropriate wavelength, high degree of spatial and temporal coherences, and enough power. Typical lasers used for this include excimer lasers and solid-state lasers. The lasers may require beam shaping before interfering in order to achieve a good uniform interference pattern. Since the fringe width in interference depends on the wavelength, the smaller the wavelength, the smaller the fringe width; short wavelength lasers are interesting to generate finer features. Thus using 157- or 193-nm UV lasers, pattern sizes of about 50 nm are possible. Figure 19.25 shows a typical pattern on silicon surface obtained by using direct writing laser interference lithography.

19.7.4 Speckle Metrology

The granular pattern that is observed when a highly coherent beam of light such as the output from a laser undergoes diffuse reflection from a rough surface is referred

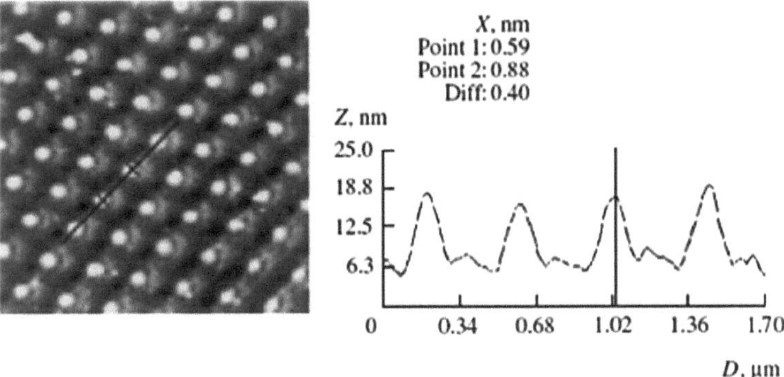

Fig. 19.25 Nanolithography using interference among various laser beams

Fig. 19.26 An arrangement to see speckles. When a coherent beam from a laser illuminates a rough surface, a speckle pattern is seen on a screen placed in front

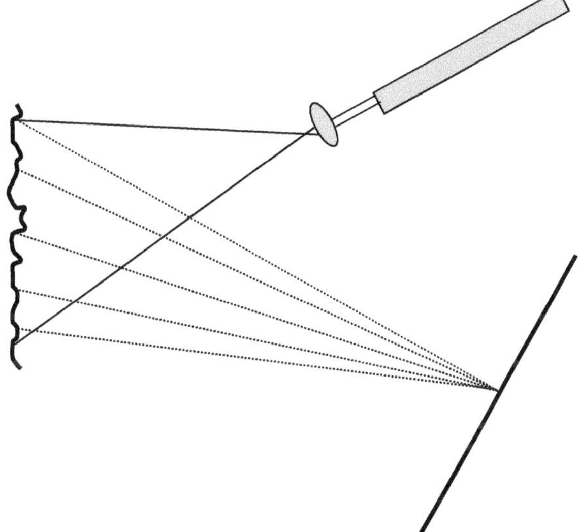

to as *laser speckle*. When a rough surface is illuminated by coherent light, then light gets scattered from different points and the light reaching any point on a screen consists of these various scattered waves (see Fig. 19.26). Due to the nature of the surface, the phases of the various waves reaching the given point on the screen may lie anywhere between 0 and π. When waves with these random phases are added, the resultant could lie anywhere between a maximum and a minimum value. At a nearby point, the waves may add to generate a different intensity value. In such a circumstance, what we observe on the screen is a speckle pattern; Figure 19.27 shows a typical speckle pattern observed on a screen. The mean speckle diameter is approximately given as

$$s \approx 1.22\frac{\lambda L}{d} \tag{19.12}$$

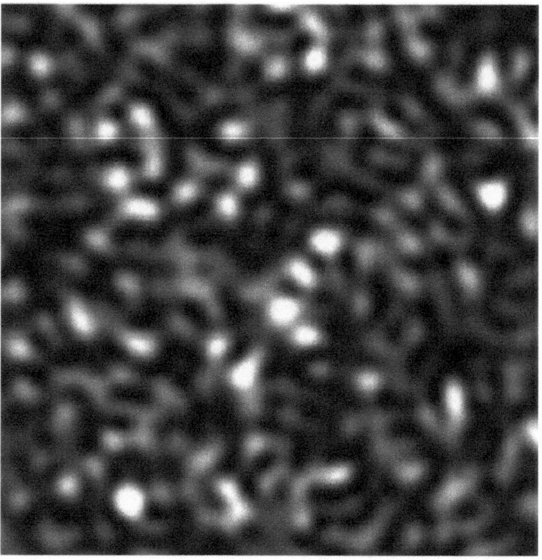

Fig. 19.27 A typical speckle pattern as observed on a screen

where λ is the wavelength of illumination, L is the distance between the screen and the rough surface, and d is the diameter of region of illumination of the object. Thus if we assume $d = 2$ cm, $\lambda = 500$ nm, and $L = 1$ m, we obtain $s \sim 30$ μm.

Figure 19.28 shows another geometry where speckles are observed; this corresponds to an imaging geometry. In this case, the imaging system images each point on the object into a diffraction spot and the random interference among the various diffraction images leads to a speckle pattern. In this case the mean speckle diameter is approximately given by the following relation:

$$s \approx 1.22\lambda F(1 + M) \tag{19.13}$$

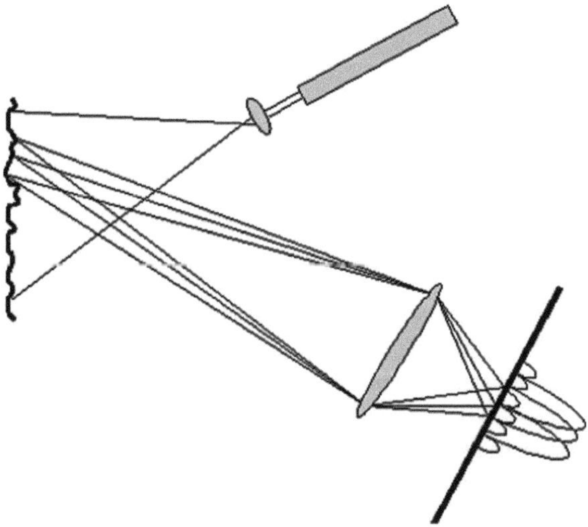

Fig. 19.28 An imaging geometry in which a lens images the rough surface. Interference among the various diffraction spots gives rise to speckles

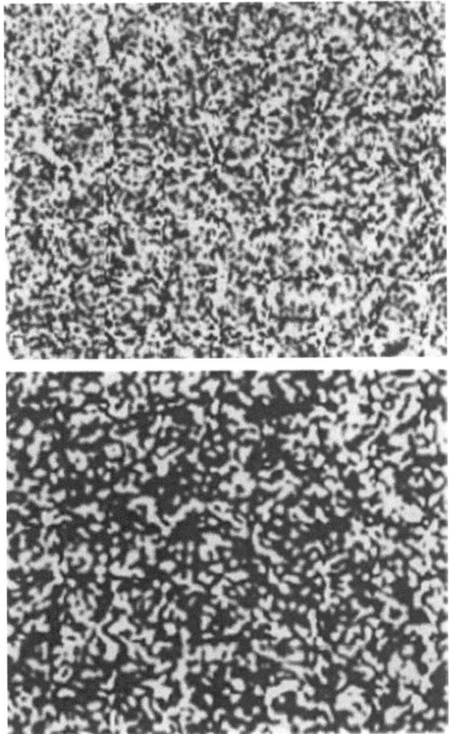

Fig. 19.29 The size of speckles depends on the aperture size. The *upper figure* corresponds to a larger aperture than the lower figure. (Source: Bates et al. (1986))

where F is the F number of the lens (focal length divided by diameter of the lens) and M is the magnification.

As an example if we are observing a rough surface and if we assume a pupil diameter of 4 mm, then we have an F number of 6 (assuming an eye lens having a focal length of 24 mm). If the object is at a distance of 25 cm from the eye, then we can assume $M \ll 1$ and we obtain for the approximate size of the speckle as formed on the retina as $s \sim 3.6$ μm. Note that the size of the observed speckles depends on the resolution of the optical system. Thus observing the speckles through a pinhole placed in front of the eye will lead to increase in size since this will lead to an increase in the F number which in turn leads to a reduction in the resolution of the eye. Thus if we put a pinhole of diameter 1 mm in front of the eye, then the speckle size would increase by a factor of 4 to about 14 μm. Figure 19.29 shows the speckles with two different apertures showing that the speckle size increases with the decrease in the aperture.

The contrast of the speckles is zero for a perfectly reflecting surface and it increases as the surface roughness increases. However, even when the roughness is well within the wavelength of the light, the contrast becomes unity and remains so for any higher roughness.

It is interesting to note that when laser light is coupled into a multi-mode fiber, the output from the fiber also exhibits a speckle pattern due to interference among the various modes of propagation of the fiber. Any disturbance of the fiber in terms of vibration or change of temperature leads to a change in the phases of the interfering modes and this results in the movement of the speckle pattern. There are indeed some fiber-optic sensors based on this phenomenon to detect movement.

Speckle contrast measurement has proved to be a powerful tool for the non-destructive testing of small surface roughness within the light wavelength. Other important applications include displacement or motion analysis and relevant non-destructive testing. To illustrate, we discuss the simple example of the well-known laser speckle photography for lateral (in-plane) displacement analysis. Historically, the method involving a single illuminating beam is called photography, whereas with two beams it is called interferometry, although both are interferometric methods. For this, the surface to be studied is illuminated by a divergent laser beam (Fig. 19.28). Laser speckles are formed on the camera focused on the film plane. The smallest speckle size S_0 on the object plane will thus be governed by the lens resolution, given by Eq. (19.13). Obviously we assume that the film resolution is capable of recording the pattern.

If the object is displaced slightly from its original position, the speckle pattern would in general change. For a small lateral displacement of the object, the nature of the random structure on the image plane can be assumed to remain unchanged with only the position of the speckles suffering a displacement along the direction of the displacement of the object. If a second exposure of the speckles with object displaced from its original position is recorded on the same film, we get a double exposure of the same speckle structure but one of them laterally displaced with respect to the other. We will show that the analysis of the transparency thus formed will give the displacement of the object.

Let the intensity distribution of the speckle during the first exposure be denoted by $S(x,y)$, where x and y are measured on the plane of the recording medium. The second exposure of the speckle with the displaced position of the object would then be represented by $S(x-x_0,y)$, where x_0 is the displacement of the object assumed to be along the x-direction. Thus the total exposure of the recording medium will be the sum of the two exposures and if we assume that the transmittance of the recorded negative is proportional to the total exposure, then we can write for the transmittance of the recorded negative as

$$T(x, y) = A - B\left[S(x, y) + S(x - x_0, y)\right] \qquad (19.14)$$

where A and B are constants. Equation (19.14) can be written in an alternative form as

$$T(x, y) = A - BS(x, y) \otimes \left[\delta(x, y) + \delta(x - x_0, y)\right] \qquad (19.15)$$

where $\otimes$ represents convolution operation.

We now place the negative on the front focal plane of a lens and illuminate it with a coherent beam of wavelength λ. On the back focal plane of the lens we would observe the Fourier transform of the amplitude distribution on the front focal plane (see Chapter 15) and thus the amplitude distribution on the back focal plane would be proportional to

$$\tilde{T}(u,v) = \mathcal{F}.\mathcal{T}.\left[T(x,y)\right] = A\delta(u,v) - B\tilde{S}(u,v)\left[1 + e^{2\pi i u x_0}\right] \tag{19.16}$$

where

$$u = \frac{x}{\lambda f}; \quad v = \frac{y}{\lambda f} \tag{19.17}$$

with f representing the focal length of the lens and x and y refer here to the coordinates on the back focal plane of the lens and tilde representing the Fourier transform of the corresponding variables.

The first term in Eq. (19.16) represents a bright spot on the axis, while the second term represents the pattern that would be observed on the screen. Concentrating on the second term, we note that it consists of the Fourier transform of the speckle pattern modulated by the interference term. The intensity pattern produced by the second term would be given as

$$I(x,y) = \left|\tilde{S}(u,v)\right|^2 \left|1 + e^{2\pi i u x_0}\right|^2 = 4\left|\tilde{S}(u,v)\right|^2 \cos^2\left(\frac{\pi x_0 x}{\lambda f}\right) \tag{19.18}$$

Equation (19.18) shows that on the back focal plane, we would have a diffused illumination represented by $\left|S(\tilde{u},v)\right|^2$, which is modulated by a fringe pattern represented by the function

$$\cos^2\left(\frac{\pi x_0 x}{\lambda f}\right)$$

which is similar to Young's interference fringes. The separation between two consecutive bright or dark fringes would be $\lambda f / x_0$. Thus knowing the wavelength and the focal length of the lens, it is possible to estimate the lateral shift in the object. Figure 19.30 shows a typical fringe pattern observed on the back focal plane of a lens showing clearly the interference fringes and the direction of displacement.

There is another method for the analysis of the double-exposure speckle pattern, namely by pointwise scanning of the photograph (see Fig. 19.31). This method is very convenient due to its simplicity. A point of the recorded transparency is simply illuminated by a laser beam. A diffraction halo with a set of Young's fringes is observed at a distance on the screen. The fringes are the measure of the object displacement D given as

$$D = \frac{\lambda L}{md} \tag{19.19}$$

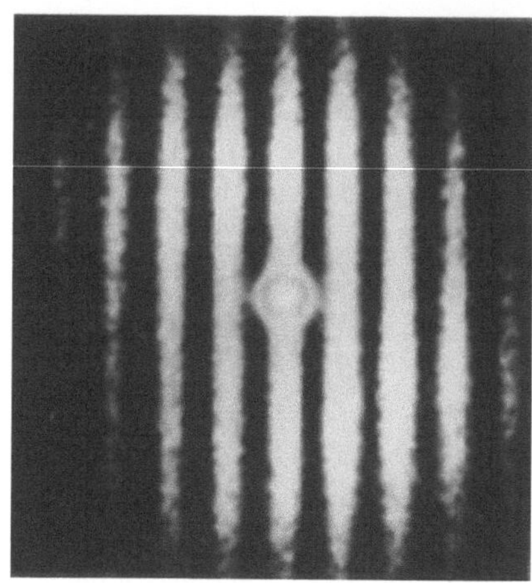

Fig. 19.30 Young's type interference fringes produced on the focal plane of a lens when a double-exposure speckle photograph is placed on the front focal plane and illuminated by a laser beam. (Source: Bates et al. (1986))

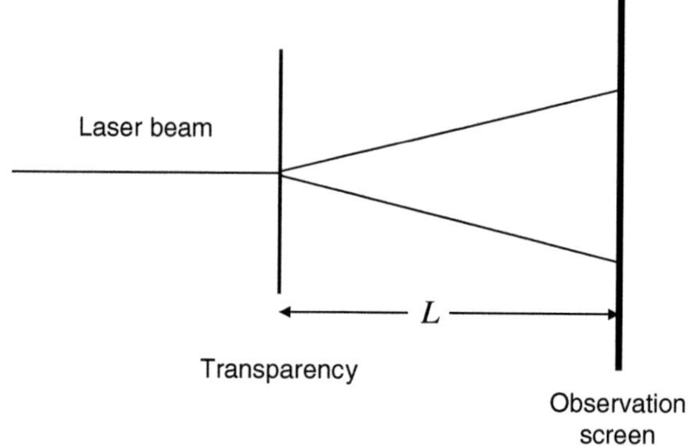

Fig. 19.31 Young's fringes (pointwise scanning) method of the analysis of speckle photographs

where d is the fringe spacing, L is the distance from the screen to the transparency, and m is the magnification. The orientation of the fringes is perpendicular to the direction of the displacement.

There is yet another method. This involves illuminating the object by two beams, particularly to reduce the measurement range and to observe the changes in real time. There are methods to process the image electronically (no photographic recording) using a TV vidicon under the subject *electronic speckle pattern interferometry*.

Attempts have also been made to eliminate the need for wet chemical processing and still use the conventional manner of analysis. These are replacing the usual photographic film by instant films, thermoplastic photographic materials, liquid-crystal light valves, BSO crystals, photographic diffusers, real-time heterodyne approach, etc.; there is a method of magnifying the speckles using lenses and a TV camera–monitor system to observe these movements directly. However, a detailed discussion of these techniques is beyond the scope of this text.

19.8 Velocity Measurement

It is well known that when a light beam gets scattered by a moving object, the frequency of the scattered wave is different from that of the incident wave; the shift in the frequency depends on the velocity of the object. Indeed, if v represents the light frequency and υ represents the velocity of the moving object which is moving at an angle θ with respect to the incident light beam (see Fig. 19.32), then the change in frequency Δv between the incident and the reflected beams is given as

$$\frac{\Delta v}{v} = \frac{2\upsilon}{c} \cos \theta \qquad (19.20)$$

where c represents the velocity of light in free space. Thus the change in frequency Δv is directly proportional to the velocity υ of the moving object; this is known as the Doppler shift. Thus, by measuring the change in frequency suffered by a beam when scattered by a moving object, one can determine the velocity of the object. This method has been successfully used for velocity determination of many types of materials from about 10 mm/min to about 150 m/min (Harry 1974). Further, using the above principle, portable velocity-measuring meters have been fabricated which measure speeds in the range of 10–80 miles/h; these have been used by traffic police. Laser Doppler velocimeters have also been used for measuring fluid flow rates.

The basic arrangement for velocity measurements is the following: the beam from a CW laser (usually a helium–neon laser – see Section 9.4) is split by a beam splitter; one of the components is reflected back from a fixed mirror and the other component undergoes scattering from the moving object. The two beams are then combined and made to interfere as shown in Fig. 19.32, and because of the difference in frequency between the two beams, beating occurs. The beat frequency is a direct measure of the velocity of motion of the object.

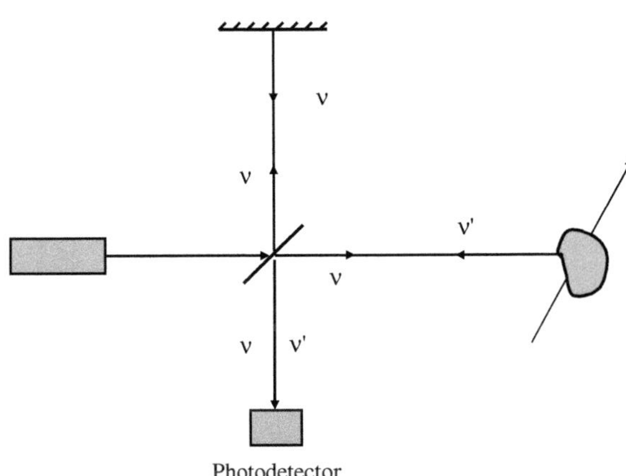

Fig. 19.32 Schematic of an arrangement for measuring the velocity of a moving object using Doppler shift

19.8.1 Lasers in Information Storage

Lasers find widespread applications in the storage, transmission, and processing of information. As we have discussed earlier, combined with optical fibers, they have revolutionized the field of transmission of information. An extremely important application of lasers is in the field of information storage. We are all familiar with compact discs (CDs) storing data, music, pictures, videos, etc. With progress in lasers and materials, the capacity of information storage has been steadily rising and today CDs are used to routinely store gigabytes of information.

Compact discs store information in digital form. Any form of information is first converted into digital form with just a sequence of 1s and 0s. In a CD these 1s and 0s are recorded in the form of pits or depressions along a spiral track on a plastic material with a metal coating (see Fig. 19.33). The total length of the track would be about 6 km! The usual coding is such that any transition from pit to land (flat area) or land to pit is read as 1s, while the duration in the pit or in the land is read as 0s (see Fig. 19.34). In CDs the radial distance between adjacent tracks, which is the track pitch, is 1.6 μm, while the length of data marks is about 0.6 μm. In order to write on the CD, the data stream is used to generate pulses of light corresponding

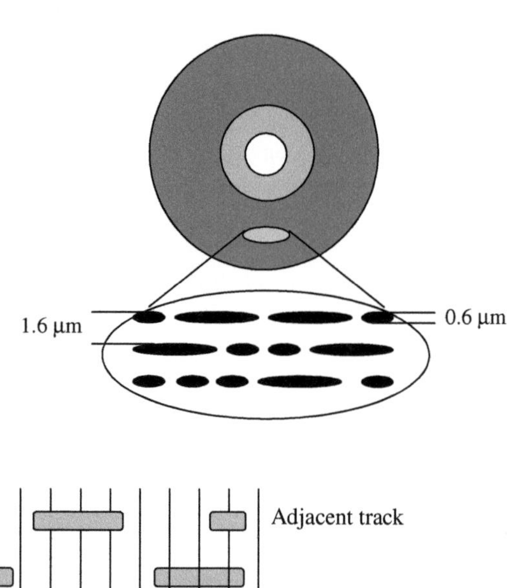

Fig. 19.33 In a compact disc (CD), information is stored in the form of pits along a spiral track

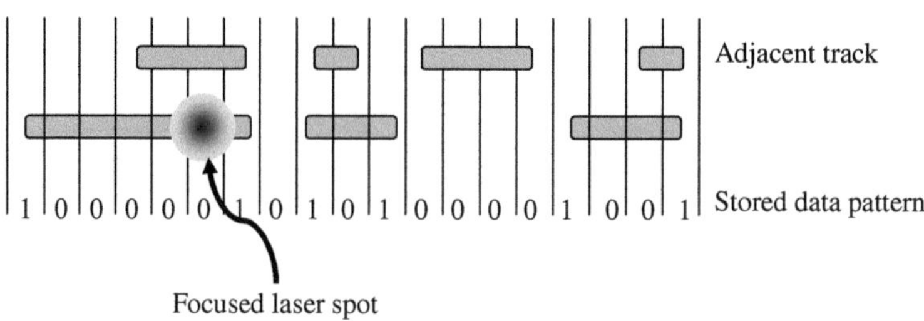

Fig. 19.34 In a CD, transition from land to pit or pit to land is read as "1s," while the duration within the pit or the land is read as "0s"

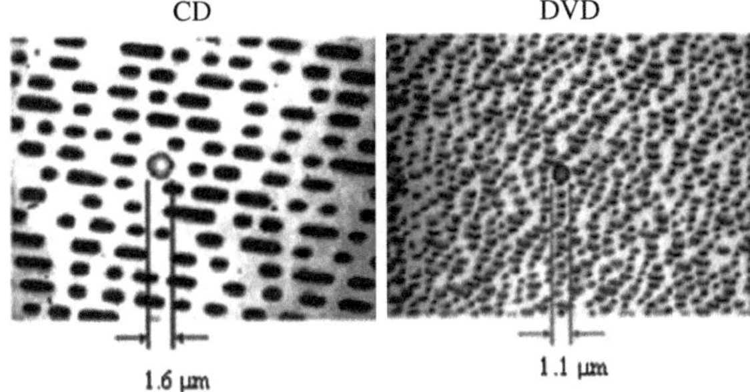

Fig. 19.35 Comparison between the pit sizes in a CD and a DVD

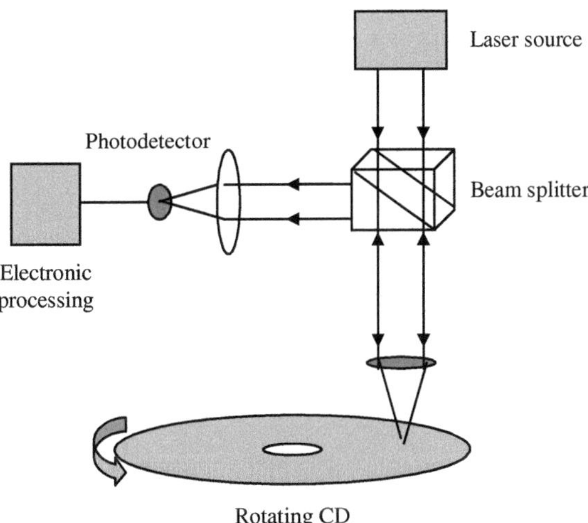

Fig. 19.36 A laser beam is focused on the CD and the reflected intensity is read and converted into the signal

to the data stream. The laser which emits an intense beam of light is focused on the surface of the CD using an objective. As the CD rotates under the laser spot, a small region heats up whenever the laser beam hits and changes the reflectivity of the surface. Digital video discs (DVDs) use the same principle except that the track pitch is about 0.74 µm instead of 1.6 µm and the data marks are narrower and the focused laser spot is also smaller (see Fig. 19.35). The data are written on the CD along a spiral track on the surface. To read the information stored in the CD, a laser beam is focused through a beam splitter on the disc and the spot size of the focused laser is about the track width (see Fig. 19.36). As the CD moves under the focused laser spot, it leads to a modulation of the reflected intensity, which is then directed by the beam splitter to a photodetector which converts the intensity variations to electric current variations for further processing.

It is clear that the smaller the data points and the smaller the track pitch, the larger the amount of data that can be stored per unit area in the disc and hence the larger the capacity of the disc. For writing and reading of the data, we use lasers and the minimum spot to which the laser can be focused will determine the size

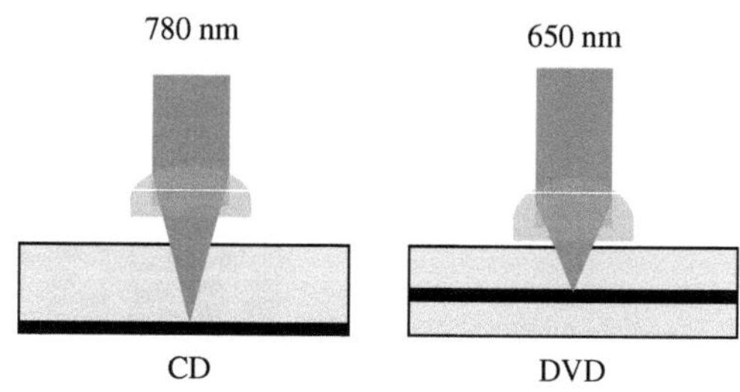

Fig. 19.37 Reading of a CD at a wavelength of 780 nm and a DVD at a wavelength of 650 nm

Table 19.3 Comparison of various parameters of CD, DVD, and Blue Ray DVD

Parameter	CD	DVD	Blue Ray DVD
Laser wavelength (μm)	0.78	0.65	0.405
Track to track spacing (μm)	1.6	0.74	0.32
Spot size of focused spot (μm)	1.6	1.1	0.48
User capacity (Gb)	0.68	9	50

Source: Milster (2005)

of the data points so that the readout can be precise. The fundamental limitation to the size of the focused spot of a laser beam arises due to diffraction. Smaller spot sizes can be achieved using smaller wavelengths and smaller focal length lenses. Since the CD is covered by a protective layer, the focusing needs to be carried out through the protective layer and this determines the smallest focal length that can be used. A CD uses a 1.2-mm clear substrate and data are recorded on the recordable layer through the clear substrate. The substrate also acts as a protective layer for the data. The reading wavelength is typically 780 nm. In contrast, in a DVD, two clear substrates each of 0.6 mm thick are bonded together and data are recorded on the bond side of each substrate. The reading wavelength in DVDs is typically 650 nm (see Fig. 19.37). Thus DVDs can store much more data than do CDs. Recent developments of blue lasers emitting a wavelength of 405 nm have triggered development of DVDs with much higher capacities since they operate with smaller wavelength and hence can be focused to smaller spot sizes. Table 19.3 lists a comparison of CDs, DVDs, and Blue Ray DVDs with regard to some important characteristics.

Further increases in data storage capacity are possible with new technologies such as holographic discs. In holographic storage, information is stored within the entire volume of the recording medium rather than on a surface. Thus holographic storage offers orders of magnitude increase of storage capacity. It is in principle possible to store 1 Tb of information per cubic centimeter of the medium using a wavelength of 500 nm. In holographic data storage, the required data are transferred to and from the storage medium as two-dimensional images composed of thousands of pixels (picture elements). The data which need to be stored are first presented to the recording system as pixels on a device called the spatial light modulator

(SLM). This is a planar device which encodes the data into small checkerboard pattern of light and dark pixels with the data arranged as an array on the page; each pixel is a small shutter which can either stop (corresponding to bit 0) or pass (corresponding to bit 1) a light beam. Commercial devices containing 1000 × 1000 pixels are available. Light from a laser is split into two parts and one part is used as a reference beam, while the other part illuminates the SLM. Interference occurring between the two beams is recorded in the storage medium as the data pattern. Multiple pages are recorded by recording the holograms with reference waves incident along different directions. Readout is performed by using a laser beam at the appropriate angle when the entire page is read out as a single bit. The reconstructed beam is detected by a detector array and converted into electronic data. By changing the angle of the beam, one can read one entire page at a time. Thus apart from the ability to store large amounts of data, holographic data storage also promises extremely fast readouts of about 1 Gb/s as compared to DVDs, wherein the readouts are about hundred-fold smaller. There is intense research activity to realize efficient recordable media and holographic data storage devices are expected to become commercially available within the next few years.

19.8.2 Bar Code Scanner

The technology associated with identification of all types of products using bar codes is one of the very important developments of the past century. The Universal Product Code (UPC) was introduced in the USA in 1973 and the European Article Numbering (EAN) system was developed in Europe in 1978 and is presently the most widely used bar code scheme used in the world. A special form of the EAN code is the International Standard Book Numbering (ISBN) system and is used for identification of books. A bar code consists of a series of strips of dark and white bands (see Fig. 19.38). Each strip has a width of about 0.3 mm and the total width of the bar code is about 3 cm. Information such as the country of origin, manufacturer of the product, the direction of scan, price, reading error checking, weight of the product, and expiry date can be stored in the pattern of dark and white strips. By a simple scanning, complete information regarding the product can be obtained.

Fig. 19.38 A bar code consisting of series of strips of dark and white bands

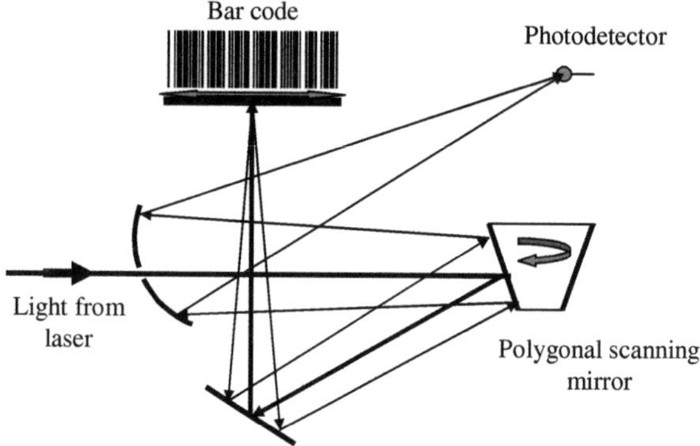

Fig. 19.39 A laser beam is scanned across the bar code and the scattered light is focused on a detector which converts the code into information

The primary purpose of the laser in this application is the optical reading of the bar code. In the bar code scanner, a low-power ($\sim$ 0.5 mW) laser beam is deflected by a rotating polygon mirror to scan along a line (see Fig. 19.39). Typical scanning speeds are about 200 m/s. Such high speeds are chosen to ensure that even if the product is moving while it is getting scanned, the scanned object does not move significantly while getting scanned. When the laser beam hits the bars, the amount of scattered light depends on whether the strip is black or white. As the laser beam scans across the black and white strips at a certain speed, the variation of scattering with time contains the information of the bar code. The scattered light is focused on a photodetector which converts the optical signal to an electrical signal for further processing. In order to be able to scan the product in any arbitrary direction for ease of scanning, the laser beam is made to scan in multiple directions by using multiple mirrors with the rotating polygon.

Problems

Problem 19.1 Consider a retroreflector shown in Fig. 19.13. Assuming the rays to be described by vectors, show that the three mirror system reflects any incident wave in exactly the reverse direction.

Problem 19.2 In the Michelson interferometer shown in Fig. 19.20, by what distance would one have to move the mirror for the output on the photodetector to change from one interference maximum to the next one?

Problem 19.3 Calculate the change in frequency of an incident light at 633 nm when it gets scattered by an object moving at the speed of 100 km/h away from it.

Problem 19.4 In the interferometer shown in Fig. 19.20, if the polarization state of the light returning from the object is different from the one reflected by the fixed mirror, what would happen to the signal at the photodetector?

Problem 19.5 A laser operating at 450 nm is focused by a lens having an NA of 0.6. What would be the approximate area of the focused spot?

Problem 19.6 A Gaussian beam is incident on a converging lens of focal length f with its waist at the front focal plane of the lens. What will be the intensity distribution on the back focal plane of the lens? Assume the lens diameter to be large.

The Nobel Lectures

CHARLES H. TOWNES

Production of coherent radiation by atoms and molecules

Nobel Lecture, December 11, 1964

From the time when man first saw the sunlight until very recently, the light which he has used has come dominantly from spontaneous emission, like the random emission of incandescent sources. So have most other types of electromagnetic radiation - infrared, ultraviolet, or gamma rays. The maximum radiation intensities, or specifically the power radiated per unit area per unit solid angle per unit frequency bandwidth, have been controlled by Planck's black-body law for radiation from hot objects. This sets an upper limit on radiation intensity - a limit which increases with increasing temperature, but we have had available temperatures of only a few tens of thousands or possibly a few millions of degrees.

Radio waves have been different. And, perhaps without our realizing it, even much of our thinking about radio waves has been different, in spite of Maxwell's demonstration before their discovery that the equations governing radio waves are identical with those for light. The black-body law made radio waves so weak that emission from hot objects could not, for a long time, have been even detected. Hence their discovery by Hertz and the great use of radio waves depended on the availability of quite different types of sources - oscillators and amplifiers for which the idea of temperature and black-body radiation even seems rather out of place. For example, if we express the radiation intensity of a modern electronic oscillator in terms of temperature, it will typically be in the range 10^{10} to 10^{30} degrees Kelvin.

These two regimes, radio electronics and optics, have now come much closer together in the field known as quantum electronics, and have lent each other interesting insights and powerful techniques.

The development of radar stimulated many important applications of electronics to scientific problems, and what occupied me in particular during the late 1940's was microwave spectroscopy, the study of interactions between microwaves and molecules. From this research, considerable information could be obtained about molecular, atomic, and nuclear structure. For its success, coherent microwave oscillators were crucial in allowing a powerful

K. Thyagarajan, A. Ghatak, *Lasers*, Graduate Texts in Physics,
DOI 10.1007/978-1-4419-6442, © Springer Science+Business Media, LLC 2010

high-resolution technique. Consequently it was important for spectroscopy, as well as for some other purposes, to extend their range of operation to wavelengths shorter than the known limit of electronic oscillators, which was near 1 millimeter. Harmonic generation and some special techniques allowed interesting, though rather slow, progress. The basic problem with electronic amplifiers or oscillators seemed to be that inevitably some part of the device which required careful and controlled construction had to be about as small as the wavelength generated. This set a limit to construction of operable devices[1]. It was this experimental difficulty which seemed inevitably to separate the techniques which were applicable in the radio region from those applicable to the shorter waves of infrared or optical radiation.

Why not use the atomic and molecular oscillators already built for us by nature? This had been one recurring theme which was repeatedly rejected. Thermodynamic arguments tell us, in addition to the black-body law of radiation, that the interaction between electromagnetic waves and matter at any temperature* cannot produce amplification, for radiation at the temperature of matter cannot be made more intense by interaction of the two without violating the second law. But already by 1917, Einstein had followed thermodynamic arguments further to examine in some detail the nature of interactions between electromagnetic waves and a quantum-mechanical system. And a review of his conclusions almost immediately suggests a way in which atoms or molecules can in fact amplify.

The rate of change of electromagnetic energy confined in a region where it interacts with a group of molecules must, from Einstein's work, have the form

$$\frac{dI}{dt} = A N_b - B I N_a + B' I N_b \tag{1}$$

where N_a and N_b are the numbers of molecules in the upper and lower of two quantum states, which we assume for simplicity to be nondegenerate (that is, single). A and B are constants, and thus the first and second terms represent spontaneous emission and absorption, respectively. The third term represents emission from the upper state produced by the presence of a radiation intensity I, and is hence called stimulated emission.

At equilibrium, when

$$\frac{dI}{dt} = 0, \quad I = \frac{A N_b}{B N_a - B' N_b}$$

---

* Strictly speaking, at any *positive* temperature. Negative absolute temperatures can be defined as will be noted below.

Rather simple further thermodynamic reasoning shows that $B' = B$ and gives the ratio A/B. While Boltzmann's law $N_b = N_a e^{-W/kT}$ requires $N_b < N_a$ at any temperature T, it is immediately clear from Eqn. 1 that if $N_b > N_a$, dI/dt will always be positive and thus the radiation amplified. This condition is of course one of nonequilibrium for the group of molecules, and it hence successfully obviates the limits set by black-body radiation. The condition $N_b > N_a$ is also sometimes described as population inversion, or as a negative temperature[2], since in Boltzmann's law it may be obtained by assuming a negative absolute temperature.

Thermodynamic equilibrium between two states of a group of atoms requires not only a Boltzmann relation $N_b = N_a e^{-W/kT}$ but also a randomness of phases of the wave functions for the atoms. In classical terms, this means that, if the atomic electrons are oscillating in each atom, there must not be a correlation in their phases if the entire group can be described as in temperature equilibrium. Einstein's relation (Eqn. 1) in fact assumed that the phases are random. And, if they are not, we have another condition which will allow the atoms to amplify electromagnetic waves, even when $N_b < N_a$. This represents a second type of loophole in the limits set by the black-body law and thermodynamic equilibrium, and one which can also be used alone or in conjunction with the first in order to produce amplification.

Thermodynamic arguments can be pushed further to show that stimulated emission (or absorption) is coherent with the stimulating radiation. That is, the energy delivered by the molecular systems has the same field distribution and frequency as the stimulating radiation and hence a constant (possibly zero) phase difference. This can also be shown somewhat more explicitly by a quantum-mechanical calculation of the transition process.

Stimulated emission received little attention from experimentalists during the 1920's and 1930's when atomic and molecular spectroscopy were of central interest to many physicists.

Later, in the 1940's, experiments to demonstrate stimulated emission were at least discussed informally and were on the minds of several radio spectroscopists, including myself. But they seemed only rather difficult demonstrations and not quite worth while. In the beautiful 1950 paper of Lamb and Retherford on the fine structure of hydrogen[3] there is a specific brief note about "negative absorption" with reversal of population. And a year later Purcell and Pound[4] published their striking demonstration of population inversion and stimulated emission. As a matter of fact, population inversion and its effects on radiation had already shown up in a somewhat less accented form in the resonance experiments of Bloch[5] and others. But all these effects were so small that any amplification was swamped by losses due to other competing processes, and their use for amplification seems not to have been seriously considered until the work of Basov and Prokhorov[6], Weber[7], and of Gordon, Zeiger, and Townes[8,9] in the early 1950's.

My own particular interest came about from the realization that probably only through the use of molecular or atomic resonances could coherent oscillators for very short waves be made, and the sudden discovery in 1951 of a particular scheme* which seemed to really offer the possibility of substantial generation of short waves by molecular amplification.

Basic Maser Principles

The crucial requirement for generation, which was also recognized by Basov and Prokhorov, was to produce positive feedback by some resonant circuit and to ensure that the gain in energy afforded the wave by stimulated molecular transitions was greater than the circuit losses. Consider a resonant microwave cavity with conducting walls, a volume V, and a quality factor Q. The latter is defined by the fact that power lost because of resistance in the walls is

$$\frac{\overline{E^2} V v}{4Q}$$

where $\overline{E^2}$ is the electric field strength in the mode averaged over the volume and v is the frequency. If a molecule in an excited state is placed in a particular field of strength E, the rate of transfer of energy to the field is

$$\left(\frac{\overline{E}\mu}{\hbar}\right)^2 \frac{hv}{3\Delta v}$$

when the field's frequency coincides with the resonance frequency v between the two molecular states. Here   is a dipole matrix element for the molecular transition and   is the width of the molecular resonance at half maximum (if a Lorentz line shape is assumed). Hence for N_b molecules in the upper state and N_a in the lower state the power given the field in the cavity is

$$(N_b - N_a)\left(\frac{\overline{E}\mu}{\hbar}\right)^2 \frac{hv}{3\Delta v}$$

If the molecules are distributed uniformly throughout the cavity, E^2 must be averaged over the volume. For the net power gain to be positive, then,

$$(N_b - N_a)\left(\frac{\overline{E\mu}}{\hbar}\right)^2 \frac{h\nu}{3\Delta\nu} \geq \frac{\overline{E^2}V\nu}{4Q}$$

This gives the threshold condition for buildup of oscillations in the cavity

$$(N_b - N_a) \geq \frac{3h\,V\Delta\nu}{16\pi^2 Q\mu^2} \tag{2}$$

There is by now an enormous variety of ways in which the threshold condition can be met, and some of them are strikingly simple. But the system which first seemed to give an immediate hope of such an oscillator involved a beam of ammonia molecules entering a resonant cavity, as shown in Fig. 1. The

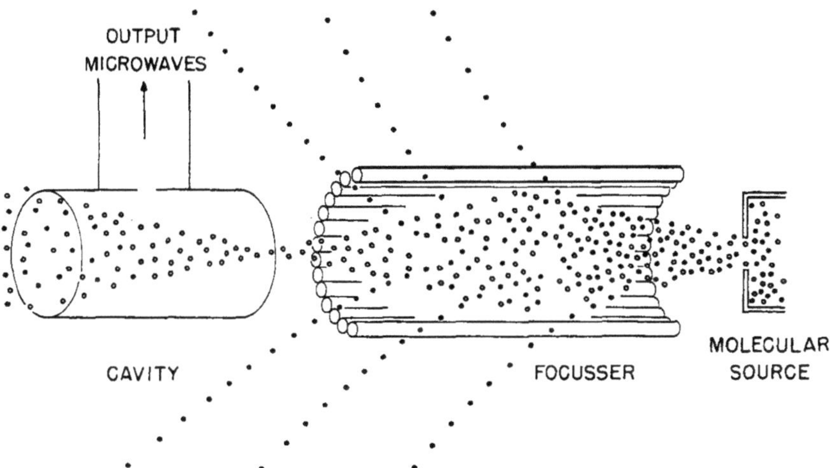

Fig.1. The ammonia (beam-type) maser. Molecules diffuse from the source into a focusser where the excited molecules (open circles) are focused into a cavity and molecules in the ground state (solid circles) are rejected. A sufficient number of excited molecules will initiate an oscillating electromagnetic field in the cavity, which is emitted as the output microwaves. Because of energy given to the field, some molecules return to the ground state toward the end of their transit through the cavity.

transition used was the well-known inversion transition of ammonia at 23,870 Mc/sec. A "focuser", involving inhomogeneous electric fields, tends to remove molecules in the ground state from the beam and to focus molecules in the excited state along the axis of the beam and into the cavity, thus ensuring that $N_b >> N_a$. J. P. Gordon played a crucial role in making operable the first such system in 1954, after 2,5 years of experimental work[9,10], and H. J. Zeiger was a valuable colleague in the first year of work and early designs. We called this general type of system the maser, an acronym for *m*icrowave *a*mplication by stimulated *e*mission of *r*adiation. The idea has been successfully extended to such a variety of devices and frequencies that it is probably well to generalize the name - perhaps to mean *m*olecular *a*mplification by stimulated *e*mission of *r*adiation. But in the radio-frequency range it is sometimes called the *r*aser, and for light the term *l*aser is convenient and commonly used. Maser amplification is the key process in the new field known as quantum electronics - that is, electronics in which phenomena of a specifically quantum-mechanical nature play a prominent role.

It is well known that an amplifier can usually be made into an oscillator, or *vice versa,* with relatively minor modifications. But it was only after experimental work on the maser was started that we realized this type of amplifier is exceedingly noise-free. The general reason for low noise can be stated simply. The molecules themselves are uncharged so that their motions, in contrast to motions of electrons through vacuum tube amplifiers, produce no unwanted electromagnetic signals. Hence a signal introduced into the resonant cavity competes only with whatever thermal noise is in the cavity as the result of thermal radiation from the cavity walls, and with spontaneous emission from the excited molecules. Spontaneous emission can be regarded for this purpose as that stimulated by a fluctuating field of energy $h\nu$. Since $kT \approx 200\,h\nu$ for microwaves in a cavity at room temperature, the thermal radiation kT in the cavity is much more important than spontaneous emission. It is then only the thermal radiation present which sets the limit to background noise, since it is amplified precisely as is the signal.

The above discussion also shows that, if the cavity is at $0°K$ and no extraneous noise enters the cavity with the input signal, the limiting noise fluctuation is determined by the spontaneous emission, which is equivalent to only one quantum of energy in the cavity. It can be shown, in fact, that masers can yield the most perfect amplification allowed by the uncertainty principle.

The motion of an electromagnetic wave is analogous to that of a mechanical harmonic oscillator, the electric and magnetic fields corresponding to position and momentum of the oscillator. Hence the quantum-mechanical uncertainty principle produces an uncertainty in the simultaneous determination of the electric and magnetic fields in a wave, or equivalently in determination of the total energy and phase of the wave. Thus one can show that, to the extent that phase of an electromagnetic wave can be defined by a quantum- mechanical operator, there is an uncertainty relation[11]

$$\Delta n \Delta \phi \geq {}^{1}/_{2} \tag{3}$$

Here n is the uncertainty in the number of photons in the wave, and is the uncertainty in phase measured in radians.

Any amplifier which gives some representation of the phase and energy of an input wave in its output must, then, necessarily involve uncertainties or fluctuations in intensity. Consider, for example, an ideal maser amplifier composed of a large number of molecules in the upper state interacting with an initial electromagnetic wave, which is considered the signal. After some period of time, the electromagnetic wave will have grown to such magnitude that it contains a very large number of quanta and hence its phase and energy can be measured by classical means. By using the expected or average gain and phase relation between the final ελεχτρομαγνετιχ wave and the initial signal the maser amplifier thus allows a measurement of the initial wave.

A calculation by well-established quantum-mechanical techniques of the relation between input and output waves shows that this measurement of the input wave leaves an uncertainty just equal to the minimum required by the uncertainty principle[11]. Furthermore, the product H of uncertainties in the electric and magnetic fields has the minimum value allowed while at the same time $(\Delta E)^2 + (\Delta H)^2$ is minimized. The uncertainty in number n of quanta in the initial wave is

$$\Delta n = \sqrt{n+1}$$

and in phase it is

$$\Delta \phi = \frac{1}{2\sqrt{n}}$$

so that

$$\Delta n \Delta \phi = {}^{1}/_{2} \frac{\sqrt{n+1}}{n}$$

The phase has real meaning, however, only when there are as many as several quanta, in which cased $n\Delta\phi \rightarrow \tfrac{1}{2}$, the minimum allowed by Eqn. 3. The background noise, which is present with no input signal (n = o), is seen to be equivalent to a single quantum ($\Delta n = 1$) of input signal.

A somewhat less ideal maser might be made of N_a and N_b molecules in the upper and lower states, respectively, all interacting with the input signal. In this case fluctuations are increased by the ratio $N_b/(N_b - N_a)$. If the amplifier has a continuous input signal, a continuous amplified output, and a bandwidth for amplification , the noise power output can be shown to be equivalent to that produced by an input signal[12]

$$N = \frac{h\nu\Delta\nu}{1 - \dfrac{N_a}{N_b}}$$

The noise power N is customarily described in terms of the noise temperature T_n of the amplifier, defined by $N = kT_n\Delta\nu$. Thus the minimum noise temperature allowed by quantum mechanics is that for a maser with $(N_a/N_b) << 1$, which is

$$T_n = \frac{h\nu}{k} \tag{4}$$

This is equivalent to the minimum energy uncertainty indicated above of one quantum ($\Delta n = 1$). In the microwave region, T_n given by Eqn.4 is approximately $1°$, whereas the best other microwave amplifiers when maser amplifiers were first being developed had noise fluctuations about 1000 times greater.

It is interesting to compare an ideal maser as a detector with a perfect photodetector, such as a y-ray counter. The y-ray counter can detect a single photon with almost no false signals, whereas a maser must always have a possible false signal of about one photon. But the photodetector gives no information about the phase of the signal; it only counts quanta, which is why the uncertainty principle allows $\Delta n \rightarrow 0$. Unfortunately, there are no perfect photodetectors in the microwave or radio regions, so that the maser is our best available detector for these waves.

The same freedom from noise which makes the maser a good amplifier helps make it a strikingly good source of monochromatic radiation since, when the threshold condition is fulfilled and the maser oscillates, the low noise implies a minimum of random frequency fluctuations.

Consider now a maser oscillator consisting of a group of excited molecules in a resonant cavity. Let the molecular transition frequency be ν_m, its half width at half-maximum intensity $\Delta\nu_m$, and the resonant-cavity frequency be ν_c with a half width $\Delta\nu_c$. If ν_m and ν_c differ by much less than $\Delta\nu_m + \Delta\nu_c$, the radiation produced by the oscillation can be shown to occur at a frequency[13]

$$\nu = \frac{\nu_m Q_m + \nu_c Q_c}{Q_m + Q_c} \tag{5}$$

where the quality factors Q_m and Q_c are $\nu_m/\Delta\nu_c$ and $\nu_m/\Delta\nu_c$ respectively. Thus if the molecular resonance is much sharper than that of the cavity, as in the ammonia-beam maser $(Q_m >> Q_c)$, the frequency of oscillation is[10]

$$\nu = \nu_m + (\nu_c - \nu_m)\frac{Q_c}{Q_m} \qquad (6)$$

If the cavity is tuned so that $\nu_c - \nu_m$ is small, then the frequency of oscillation coincides very closely with the natural molecular frequency ν_m, and one has an almost constant frequency oscillator based on a molecular motion, a so-called atomic clock.

The frequency ν is not precisely defined or measurable because of noise fluctuations, which produce random phase fluctuations of the wave. In fact, the maser is essentially like a positive feedback amplifier which amplifies whatever noise source happens to be present and thereby produces a more or less steady oscillation. If Q_m or Q_c is high, and the amplifier gain is very large, then the bandwidth of the system becomes exceedingly small. But it is never zero, nor is the frequency ever precisely defined. The average deviation in frequency from Eqn. 5 which these phase fluctuations produce when averaged over a time t is[14]

$$\varepsilon = \Delta\nu \left(\frac{W_n}{Pt}\right)^{\frac{1}{2}} \qquad (7)$$

where

$$\Delta\nu = \frac{\Delta\nu_c \, \Delta\nu_m}{\Delta\nu_c + \Delta\nu_m}$$

P is the power generated by the oscillator, and W_n is the effective energy in the source of fluctuations. Where $kT >> h\nu$ in a cavity at temperature T and resonant frequency ν, the effective energy comes from thermal noise and $W_n = kT$. If the noise fluctuations come from spontaneous emission, as they do when $kT << h\nu$, then $W_n = h\nu$.

It is also useful to state the spectral width of the radiation emitted from a maser oscillator, as well as the precision to which the frequency can be determined. The half width of the spectral distribution is again determined by the same noise fluctuation and is given by[10,15,16]

$$\delta = \frac{2\pi W_n}{P}(\Delta\nu)^2 \qquad (8)$$

where δ, W_n, and P are the same as in Eqn. 7. This widths is typically so small in maser oscillators that they provide by far the most monochromatic sources of radiation available at their frequencies.

Maser Clocks and Amplifiers

Although the ammonia beam-type maser was able to demonstrate the low-noise amplification which was predicted[17], its extremely narrow band- width makes it and other beam-type masers more useful as a very monochromatic source of electromagnetic waves than as an amplifier. For the original maser, the power output P was about 10^{-9} Watt, and the resonance width δ about 2 kilocycles, as determined by the length of time required for the beam of molecules to pass through the cavity. Since the frequency of oscillation ν is 23,874 megacylces, the fractional spectral width, according to Eqn. 8, is $\delta/\nu \approx 10^{-14}$. In a time $t = 100$ seconds, Eqn. 7 shows that the frequency can be specified to a fractional precision $\delta\nu/\nu = 2 \cdot 10^{-14}$, and of course the precision increases for longer times proportionally to $\mathbf{1/t^{1/2}}$.

As a constant-frequency oscillator or precise atomic clock, however, the ammonia maser has an additional problem which is not so fundamental, but which sets a limit on long- term stability. This comes from long-term drifts, particularly of the cavity temperature, which vary ν,. These variations can be seen, from Eqn. 6, to « pull » ν. Variations of this type have limited the long-term stability'* of ammonia masers to fractional variations of about 10^{-11}; this still represents a remarkably good clock.

A beam-type maser using the hyperfine structure transition in the ground state of hydrogen, which is at 1420 megacycles, has recently been developed by Goldenberg, Kleppner, and Ramsey[19]. In this case, the excited atoms bounce many times from glass walls in the cavity, and thereby a resonance width as small as 1 cycle per second is achieved. Present designs of the hydrogen maser yield an oscillator with long-term fractional variations no larger than about 10^{-13}. This system seems likely to produce our best available clock or time standard.

Masers of reasonably wide utility as amplifiers came into view with the realization that certain solids containing paramagnetic impurities allowed attainment of the maser threshold condition[20]. Microwave resonances of paramagnetic atoms in solids, or in liquids, had been studied for some time, and many of their properties were already well known. The widths of these resonances vary with materials and with impurity concentration from a small fraction of a megacycle to many hundreds of megacycles, and their frequencies depend on applied magnetic field strengths, so that they are easily tunable. Thus they offer for maser amplifiers a choice of a considerable range of bandwidth, and a continuous range of frequencies.

A paramagnetic atom of spin ½ has two energy levels which, when placed in a magnetic field, are separated by an amount usually of about $v = 2.8\,H$ Mc. Here H is the field in gauss, and from this it is clear that most of the microwave frequency range can be covered by magnetic fields of normal magnitudes. The first paramagnetic masers suggested involved impurity atoms of this type in crystals of silicon or germanium. Relaxation between the two states was slow enough in these cases that a sufficient population inversion could be achieved[20]. However, before very long a very much more convenient scheme for using paramagnetic resonances was proposed by Bloembergen[21], the so-called three-level solid-state maser. This system allowed continuous inversion of population, and hence continuous amplification, which was very awkward to obtain in the previous two-level system.

Paramagnetic atoms with an angular momentum due to electron spin S greater than $1/2$ have $2S + 1$ levels which are degenerate when the atom is in free space. But these levels may be split by "crystalline fields", or interaction

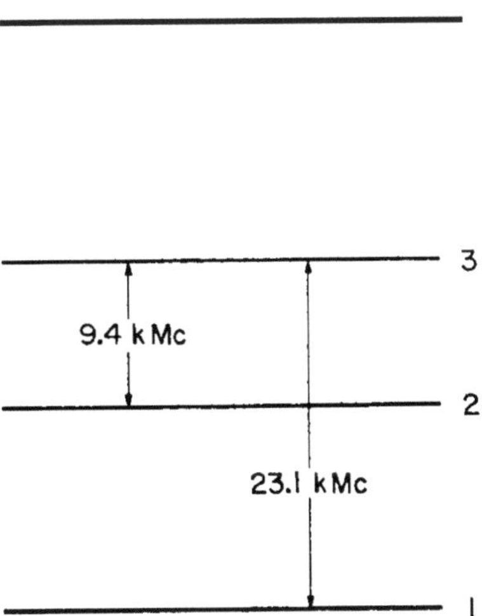

Fig. 2. Energy levels of Cr^{2+} in ruby with a particular crystalline orientation in a magnetic field of 3900 oersteds. For a three-level maser, 23.1 kMc ($23.1\,10^3$ Mc) is the frequency of the pumping field and 9.4 kMc is the frequency of amplification or oscillation.

with neighboring atoms if the atoms are imbedded in a solid, and frequently the splittings lie in the microwave range. The energy levels of such a system, involving a spin of 3/2 and four levels, can be as indicated in Fig. 2 when the system is in a magnetic field. If a sufficiently large electromagnetic wave of frequency ν_{13} (the transition frequency between levels 1 and 3) is applied, the population of these two levels can be equalized or « saturated ». In this case, the ratio of the population of level 2 to that of level 1 or 3 under steady conditions is

$$\frac{n_2}{n_1} = \frac{\dfrac{1}{T_{12}} e^{-\frac{h\nu_{12}}{kT}} + \dfrac{1}{T_{23}}}{\dfrac{1}{T_{12}} + \dfrac{1}{T_{23}} e^{-\frac{h\nu_{23}}{kT}}}$$

Here T is the temperature of the crystal containing the impurities, and T_{12} and T_{23} are the times for relaxation between the states 1 and 2 or 2 and 3, respectively. For $h\nu_{12} >> kT$ and $h\nu_{23} >> kT$, as occurs at very low temperatures or at ordinary temperatures if the levels are separated by optical frequencies,

$$\frac{n_2}{n_1} = \frac{T_{12}}{T_{23}}$$

When $h\nu_{12} << kT$ and $h\nu_{23} << kT$, which is more commonly the case for microwaves,

$$\frac{n_2}{n_1} = 1 + \frac{h}{kT} \frac{\dfrac{\nu_{12}}{T_{12}} - \dfrac{\nu_{23}}{T_{23}}}{\dfrac{1}{T_{12}} + \dfrac{1}{T_{23}}} \tag{9}$$

Thus if

$$\frac{\nu_{12}}{T_{12}} > \frac{\nu_{23}}{T_{23}}$$

there is an inversion of population between levels 2 and 1, or if,

$$\frac{\nu_{12}}{T_{12}} < \frac{\nu_{23}}{T_{23}}$$

there is an inversion of population between levels 3 and 2, since the populations n_3 and n have been equalized by the « pumping » radiation. Equation 9 is essentially the result obtained by Bloembergen[21], who also suggested several promising paramagnetic materials which might be used. Basov and Prokhorov had already proposed a rather similar three-level « pumping » scheme for application to a molecular beam system[22].

The first successful paramagnetic maser of this general type was obtained by Scovil *et al.*[23], using a rare-earth ion in a water-soluble crystal. But, before long, other more suitable crystals such as ruby[24] (chromium ions in Al_2O_3) became more or less standard and have provided amplifiers of remarkable sensitivity for radio astronomy, for satellite communication, and for communication with space probes[25]. They have considerably improved the potentialities of radio astronomy, and have already led to some new discoveries[26,27]. These systems generally require cooling with liquid helium, which is a technological difficulty that some day may be obviated. But otherwise they represent rather serviceable and convenient amplifiers.

A maser amplifier of microwaves can rather easily be built which has a theoretical noise temperature as low as 1° or 2°K, and experimental measurements have confirmed this figure[28]. However, such a low noise level is not easy to measure because almost any measurement involves attachment of input and output circuits which are at temperatures much higher than 1°K, and which radiate some additional noise into the amplifier. The lowest overall noise temperature so far reported for an entire receiving system[29] using a maser amplifier is about 10°K. This represents about 100 times the sensitivity of microwave amplifiers built before invention of the maser. But masers have stimulated other amplifier work, and some parametric amplifiers, using more or less classical properties of materials rather than quantum electronics, now have sensitivities within a factor of about 5 of this figure.

Optical ad Infrared Masers, or Lasers

Until about 1957, the coherent generation of frequencies higher than those which could be obtained from electronic oscillators still had not been directly attacked, although several schemes using molecular-beam masers for the far-infrared were examined from time to time. This lack of attention to what had been an original goal of the maser came about partly because the preliminary stages, including microwave oscillators, low-noise amplifiers, and their use in various scientific experiments, had proved so interesting that they distracted attention from the high-frequency possibilities.

But joint work with A. L. Schawlow[30], beginning at about this time, helped open the way for fairly rapid and interesting development of maser oscillators in the far-infrared, optical, and ultraviolet regions - as much as 1000 times higher in frequency than any coherent sources of radiation previously available. It is masers in these regions of the spectrum, frequently called lasers (light amplification by stimulated emission of radiation), which have perhaps provided the most striking new scientific tools and results. Important aspects of this work were clear demonstrations that there are practical systems which can meet the threshold condition of oscillation, and that particular resonator designs allow the oscillations to be confined to certain specific and desirable modes. The resonator analyzed was composed simply of two parallel mirrors-the well-known Fabry-Perot interferometer, but ofspecial dimensions.

For light waves, the wavelength is so short that any macroscopic resonator constructed must have dimensions that are large compared with the wavelength. In this case, the electromagnetic field may to some reasonable approximation be considered to travel in straight lines and be reflected from the walls of the resonator. The threshold condition may be written

$$\left(\frac{\mu E}{\hbar}\right)^2 \frac{h\nu\,(N_b - N_a)}{12\pi\,\Delta\nu} \geq \frac{E^2}{8\pi}\frac{V}{t} \tag{10}$$

where t is the decay time for the light in a cavity ofreflecting walls and volume V. If the light has a random path in the cavity, the decay time can be expressed generally in terms of the reflection coefficient r of the walls, the volume V, the wall area A, and the velocity of light c,

$$t = \frac{6V}{(1-r)Ac}$$

Hence Eqn. 10 becomes[30]

$$N_b - N_a \geq \frac{\Delta\nu}{\nu}\frac{h\,(1-r)Ac}{16\pi^2\,\mu^2}\frac{1}{V} \tag{11}$$

It can be seen that this critical condition is almost independent of frequency if the fractional line width $\Delta\nu/\nu$ does not change with frequency (as, for example, in the Doppler effect). The reflection coefficient and dipole moment matrix element   are not particularly dependent on frequency over the range in question. Hence, if the critical condition can be met for one frequency, it can probably be met over the entire range from the far-infrared to the ultraviolet.

There is a problem with a resonator which is large compared to a wavelength in that there are many modes. Hence, unless the modes in which oscillations occur are successfully controlled, the electromagnetic field may build up simultaneously in many modes and at many frequencies. The total number of modes in a cavity with frequencies which lie within the line width   of the atomic molecular resonance is

$$p = \frac{8\pi^2 \, V v^2 \, \Delta v}{c^3}$$

or about 10^9 for a cavity volume of $1 \, cm^3$, a frequency in the optical region, and ordinary atomic line widths. But fortunately the possibility of oscillation can be eliminated for most of these modes.

Two small parallel mirrors separated by a distance much larger than their diameter will allow a beam of light traveling along the axis joining them to travel back and forth many times. For such a beam, the decay time t is L/c $(1 - r)$, where L is the mirror separation and r the reflectivity. Hence the threshold condition is

$$N_b - N_a \geq \frac{3\Delta v}{8\pi^2 v} \, \frac{hc \, (1-r)}{\mu^2 L}$$

This assumes that diffraction losses are negligible. A beam of light which is not traveling in a direction parallel to the axis will disappear from the volume between the mirrors much more rapidly. Hence the threshold condition for off-axis beams will require appreciably more excited atoms than that for axial beams, and the condition for oscillation can be met for the latter without a build-up of energy in off-axis light waves.

Many features of the modes for the electromagnetic wave between two square, plane, parallel mirrors of dimension D and separation L can be approximately described as those in a rectangular box of these dimensions, although the boundary conditions on the enclosed sides of the « box » are of course somewhat different. The resonant wavelengths of such a region for waves traveling back and forth in a nearly axial direction are[30]

$$\lambda = \frac{2L}{q} \left[1 - \tfrac{1}{2} \left(\frac{Lr}{Dq}\right)^2 - \tfrac{1}{2} \left(\frac{Ls}{Dq}\right)^2 \right] \tag{12}$$

where q, r, and s are integers, and $r << q$, $s << q$. More precise examination of the modes requires detailed numerical calculation[31]. For a precisely axial direction, $r = s = 0$, and the modes are separated by a frequency $c/2L$. If this frequency is somewhat greater than the atomic line width   , then only one axial mode can oscillate at a time. The axial wave has an angular width due to diffraction of about λ/D, and, if this is comparable with the angle D/L, then all off-axis modes (r or s $\neq$ o) are appreciably more lossy than are the axial ones, and their oscillations are suppressed.

If one of the mirrors is partially transparent, some of the light escapes from the axial mode in an approximately plane wave and with an angular divergence, approximately λD, determined by diffraction.

A number of modified resonator designs have been popular and useful in optical masers, in particular ones based on the confocal Fabry-Perot interferometer. However, the plane-parallel case seems to offer the simplest means of selecting an individual mode.

Although a number of types of atomic systems and excitation seemed promising in 1958 as bases for optical masers, optical excitation of the alkali vapors lent itself to the most complete analysis and planning for an operable oscillator. One such system has been shown to oscillate as expected[32]; but the alkali vapors are no longer of great interest, because other systems which were at the time much less predictable have turned out to be considerably more useful.

The first operating laser, a system involving optical excitation of the chromium ions in ruby and yielding red light, was demonstrated by Maiman in 1960[33]. He took what seemed at first a rather difficult route of inverting the population between the ground state and excited states of the chromium ion. This technique requires that at least half of the very large number of atoms in the ground state must be excited in order to have the possibility of a population inversion. In the case of two normally unpopulated atomic states, the total amount of excitation required is much less. However, Maiman succeeded handsomely in exciting more than half the chromium ions in a ruby with chromium concentration of about 1/2000 by applying a very intense pulse of light from a flash tube. This type of system is illustrated schematically in Fig. 3. Success immediately yielded a very-high-energy maser oscillation because, to get population inversion at all, a large amount of energy must be stored in the excited atomic states. Surfaces of the ruby served as the reflecting mirrors. Collins et al.[34] quickly demonstrated that the ruby laser showed many of the characteristics predicted for such an oscillator.

The ruby laser is operated normally only in pulses, because of the high power required to reach threshold, and emits intense bursts of red light at power levels between about 1 kilowatt and 100 megawatts. It has given rise to a whole family of lasers involving impurities in various crystals of glasses, and covering frequencies from the near infrared into the optical region.

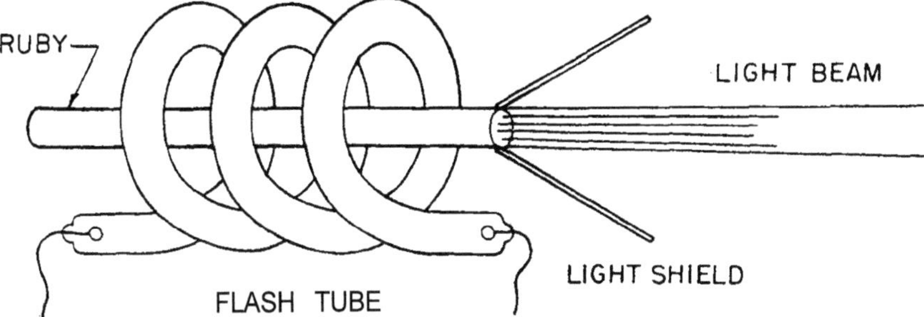

Fig. 3. Schematic diagram of a ruby (optically excited solid-state) laser. When the gas
flash tube is activated, electromagnetic oscillations occur within the ruby rod, and some
of these light waves are emitted in a beam through one partially reflecting end of the rod.

Not very long after the ruby laser was developed, Javan, Bennett and Herriott[35] obtained maser oscillations from neon atoms excited by collisions of
the second kind with metastable helium, in accordance with an idea previously put forward by Javan[36]. This system, illustrated in Fig. 4, requires only a
gaseous discharge through a tube containing a mixture of helium and neon
at low pressure, and two reflectors at the ends of the tube. It oscillates at the

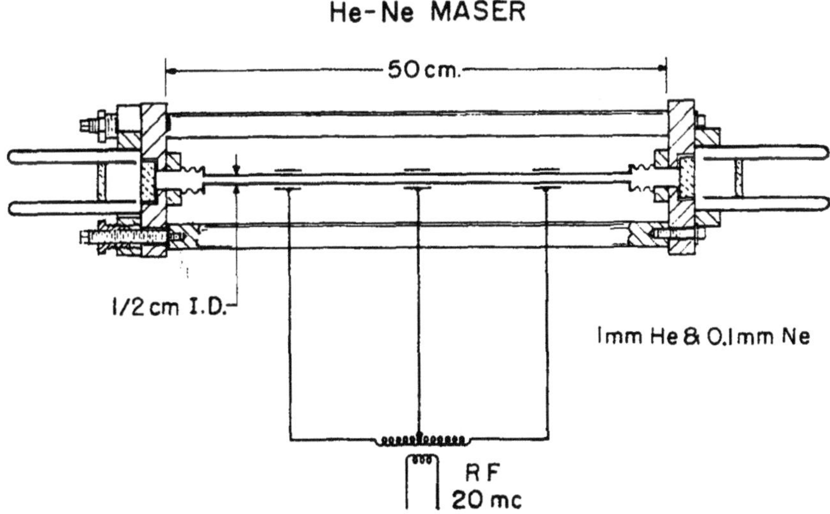

Fig.4. Schematic diagram of a helium-neon (gas discharge) laser. Electrical excitation
can initiate a steady maser oscillation, resulting in an emitted light beam from either end
of the gas discharge, where there are reflecting mirrors.

relatively low power of about 1 milliwatt, but approaches ideal conditions much more closely than the ruby system, and affords a continuous source of infrared radiation of great purity and directivity.

The technique of gaseous excitation by electrical discharge has also led to a large family of lasers, producing hundreds of different frequencies from many different gases which range from wavelengths as long as a few tenths of 1 millimeter down into the ultraviolet. For some systems, a heavy discharge pulse in the gas is needed. Others, particularly some of the infrared frequencies in rare gases, oscillate so readily that it seems probable that we have had lasers accidentally all along. Very likely some neon or other rare-gas electric signs have been producing maser oscillations at infrared wavelengths, which have gone unnoticed because the infrared could not escape from the glass neon tubes. Some of these oscillation frequencies represent atomic transitions which were previously undetected; for others, the transition has not yet even been identified.

Another class of lasers was initiated through the discovery[37] that a *p-n* junction of the semiconductor gallium arsenide through which a current is passed can emit near-infrared light from recombination processes with very high efficiency. Hall *et al.*[38] obtained the first maser oscillations with such a system, with light traveling parallel to the junction and reflected back and forth between the faces of the small gallium arsenide crystal. His results were paralleled or followed immediately, however, by similar work in two other laboratories[39,40]. This type of laser, illustrated in Fig. 5, is of the general size and cost of a transistor. It can be made to oscillate simply by passage of an electric current, and in some cases the radiation emitted represents more than 50 percent of the input electrical energy - an efficiency greater than that of other man-made light sources.

There quickly developed a large family of semiconductor lasers, some involving junctions and, recently, some using excitation by an external beam of electrons[41]. They range in wavelength from about 10 microns, in the infrared, to the center of the visible region.

Normal Raman scattering can be regarded as spontaneous emission from a virtual state, as indicated in Fig. 6. Associated with any such spontaneous emission there must be, in accordance with Einstein's relations, a stimulated emission. Javan showed[42] the principles involved in using this stimulated emission for a Raman maser. What is required is simply a large enough number of molecular systems which are sufficiently strongly excited by radiation of frequency greater than some Raman-allowed transition.

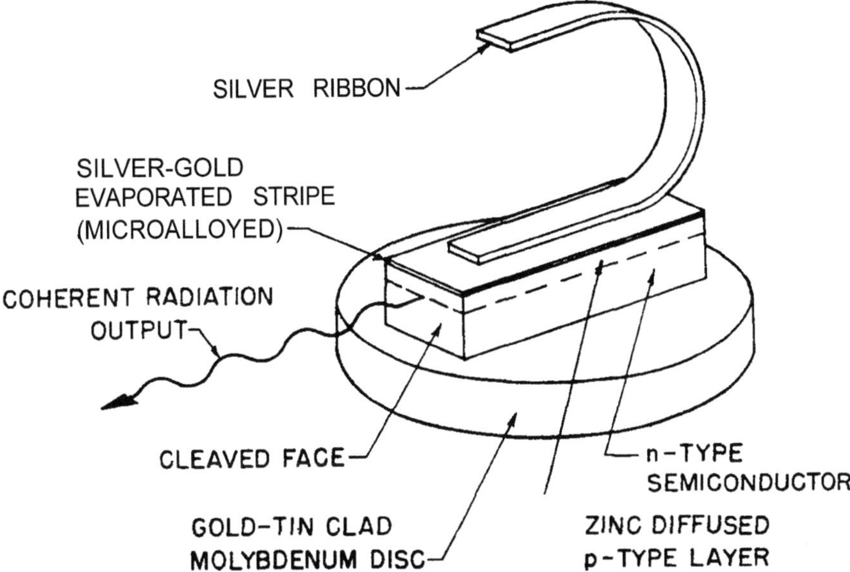

Fig. 5. Schematic diagram of a gallium arsenide (injection, or semiconductor) laser. A small voltage applied between the silver ribbon and the molybdenum disc can produce maser oscillations with resulting emission of coherent infrared radiation.

One might consider the population of the virtual level in a Raman maser (see Fig. 6) to be greater than that of the first excited state, so that there is no population inversion. On the other hand the initial state, which is the ground state, needs to be more populated than the first excited state. One can quite properly consider the amplification process as a parametric one with the molecular frequency as idler, or as due to a mixture of ground and excited states in which there is phase coherence between the various molecules. This is the second type of loophole through the black-body radiation law mentioned earlier. The ammonia-beam maser itself illustrates the case of amplification without the necessity of population inversion. As the ammonia molecules progress through the cavity and become predominantly in the ground state rather than the excited state, they continue to amplify because their oscillations are correlated in phase with each other, and have the appropriate phase with respect to the electromagnetic wave.

Raman masers were first demonstrated by Woodbury and Ng[43] as the result of excitation of various liquid molecules with a very intense beam from a pulsed laser. They too have now many versions, giving frequencies which

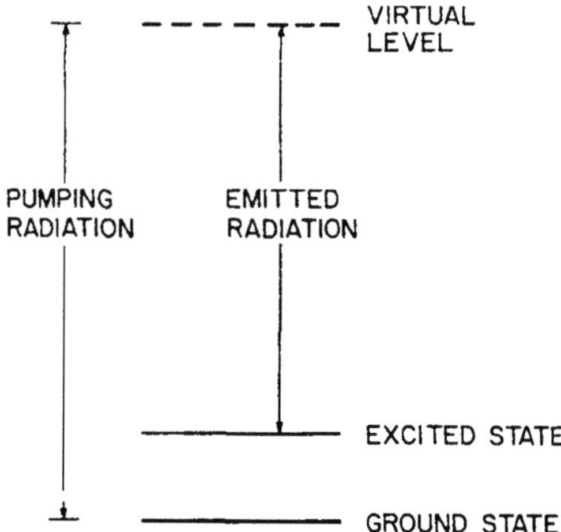

Fig. 6. Representation of energy levels in a Raman maser. This system resembles qualitatively a three-level maser, one of the levels being « virtually », or not characteristic of the molecule when no field is present.

differ from the original driving maser beam by some small integer times a molecular-vibrational frequency. Their action has been considerably extended by Terhune[44] and treated in a number of theoretical papers[42,45].

Present Performance of Lasers

Where now do we stand in achieving the various theoretical expectations for performance of masers?

First, consider the general extension of the frequency range where we have coherent amplifiers and oscillators. This has been increased by a factor of somewhat more than 1000; there are still additional spectral regions where such techniques need to be developed, but the pace has been quite rapid in the last few years. Maser oscillations in the infrared, optical, and ultraviolet regions have now been obtained in many ways and appear easy; new excitation mechanisms and systems are continually turning up. There are still two frequency regions, however, where such sources of radiation are rare or nonexistent. One is in the submillimeter region or far-infrared. The region has, in a sense, now been crossed and conquered by maser oscillators. But techniques in this spectral region are still rudimentary, and the frequency coverage with masers is spotty. Presumably further work will allow interesting explorations in this region and a very fruitful, high-resolution spectroscopy.

Another region where coherent oscillators have not yet been developed is that of still shorter wavelengths stretching indefinitely beyond the near-ultra-violet, where the first such oscillators are now available. It can be shown that a rather severe and fundamental limitation exists as one proceeds to shorter wavelengths because of the continually increasing number of electromagnetic modes in a given volume and the faster and faster dissipation of energy into them by spontaneous emission.

Consider a cavity resonator of fixed volume, fixed-wall reflectivity, and fixed-fractional frequency-width $\Delta v/v$. Meeting the threshold condition (Eqn. II) in such a resonator requires that there is power which increases as v^4, radiated by spontaneous emission into all modes of the system[30]. In the optical region this dissipated power for typical conditions, whereas at 50 ångströms, in the soft X-ray region, it would be about 10^5 watts. The threshold condition would then be very difficult to maintain. But, by the same token, if it is maintained, the coherent X-ray beam produced would contain many kilowatts of power. It seems reasonable to expect, on this basis, that masers will be developed to wavelengths somewhat below 1 000 ångströms, but that maser oscillations in the X-ray region will be very much more difficult.

Secondly, let us examine the monochromaticity which has been achieved. For the ammonia-beam maser, the variation of microwave oscillations was shown experimentally to agree with the theoretical expression (Eqn. 7) within the experimental precision of about 50 percent. This was done by beating two independent ammonia oscillators together and examining their relative phase variations[46]. A similar procedure can be carried out for two optical oscillators by mixing their two light beams together in a photocell and detecting the beat frequency. However, the technical difficulties in obtaining theoretical performance are rather more demanding than in the case of the ammonia maser. Equation 8 for a typical helium-neon laser predicts a frequency spread of about 10^{-2} cycle per second, or a fraction $3\cdot10^{-17}$ of the oscillation frequency of $3\cdot10^{14}$ cycles per second.

Almost all masers so far oscillating in the optical or near-infrared region require a sharper resonance, or higher Q, of the cavity than of the atomic resonance. Hence the frequency of oscillation is primarily determined, from Eqn. 5 by the cavity resonance. The frequency of oscillation thus depends on the separation L between mirrors, since from Eqn. 12 $v=qc/2L,$ where q is some integer. If, then, the radiated frequencies are to have a fractional band-width of about $3 \cdot 10^{-17}$, such as would come from fundamental noise according to Eqn. 8, the mirror separation must not vary by more than this fractional amount. For a mirror separation of 1 meter, the motion allowed would be less than $3 \cdot 10^{-13}$ centimeter - a demanding requirement!

If the mirror separation is held constant by cylindrical rods, L must still vary as a result of thermal excitation of the lowest frequency-stretching modes of the rods. This gives an additional fluctuation which is usually larger than that from spontaneous emission. It produces a fractional motion[47]

$$\left(\frac{2kT}{YV}\right)^{\frac{1}{2}}$$

where T is the temperature, V the volume of the separators, and Y their Young's modulus.

In order to examine the monochromaticity of lasers, two helium-neon systems were carefully shock-mounted in an acoustically insulated wine cellar of an unoccupied and isolated house so that acoustic vibrations would be minimized[47]. Their pairs of mirrors were separated by heavy invar rods about 60 centimeters long. For this case, the limiting theoretical fluctuations set by thermal motions of the rods corresponded to fractional frequency variations of $5 \cdot 10^{-15}$, or a frequency fluctuation of 2 cycles per second. Light from each laser was sent into a photodetector, and the beat frequency examined electronically. Under good conditions free from acoustic disturbances or thermal transients in the invar spacers, this experiment showed that variations in the laser frequencies over periods of a few seconds were less than 20 cycles per second, or about one part in 10^{13}. This was ten times the limit of thermal fluctuations, but corresponded to detection of motions of the two mirrors as small as $5 \cdot 10^{-12}$ centimeter, a dimension comparable with nuclear diameters. Presumably, with great care, one can obtain results still nearer to the theoretical values.

The narrowest atomic spectral lines have widths of the order of 10^8 cycles per second, so that the laser measured was more monochromatic than earlier light sources by a factor of about 10^6. Light of this type can interfere with itself after traveling a distance of about 10 000 kilometers. Hence it could in principle measure changes in such a large distance to a precision of one wavelength of light, if there were any optical path so constant. Interference work has been done in several laboratories with laser light over distances of a few hundred meters, which does not require quite such special elimination of acoustic or other disturbances.

A third property of laser light which is of interest is its directivity, or the spacial coherence across the beam. As indicated above, certain modes of oscillation should represent approximately a plane wave of cross section comparable with the mirror diameter D. The helium-neon maser seems to easily allow adjustment so that such a mode of oscillation occurs, and its beam has been shown[35,48] to have nearly the expected divergence λ/D due to diffraction.

The spacial coherence or planarity of a laser beam implies that the entire beam can be focused by a microscope to a region as small as about $\lambda/2$, or the resolving power of the microscope. Similarly, it may be transmitted through a telescope in a beam whose angular width is simply determined by the angular resolution of the telescope, and hence much less than the angular divergence λ/D as the beam emerges from a small laser. The entire energy is originally created in the ideal laser in a single mode; it can be transmitted into other single modes by optical systems without violating the well-known brightness laws of optics.

This brings us to a fourth important property, the intensity or brightness which can be achieved by maser techniques. As indicated initially, once one has the possibility of coherent amplification, there is no firm limit to intensity because equilibrium thermodynamics and Planck's law no longer are controlling. The only limit is set by the available energy input, heat dissipation, and size of the apparatus used.

If only the 1 milliwatt of power emitted by a helium-neon laser is focused by a good lens, the power density becomes high because the cross-sectional area of the focused spot would be only about $\lambda^2/4$. This gives a power density of $4·10^5$ Watt/cm^2. The effective temperature of such a beam, because of its monochromaticity, is also rather high - about 10^{19} °K for the light of 20-cycle per-second bandwidth.

The pulsed systems, such as ruby lasers in particular, emit much greater power, although they do not quite approach the limits of coherence which the gaseous systems do. Ruby lasers emit a few tenths of a joule to a few hundred joules of energy in pulses from about 10^{-3} second to 10^{-8} second in length. The power can thus be as great as 10^9 Watts or more. Effective temperatures of the radiation are of the order 10^{23} °K. The actual limit of power density will generally be set by the limit of light intensity optical materials can stand without breakage or ionization. Power of 10^9 Watts focused to a spot 10^{-2} millimeter in diameter produces an electric field strength in the optical wave of about 10^9 Volt/cm, which is in the range of fields by which valence electrons are held in atoms. Hence this power ionizes and disrupts all material.

The radiation pressure also becomes large, being about 10^{12} dyne/cm^2, or 10^6 atmospheres, at such a focal point.

Some Applications of Lasers

It is clear that light in more ideal and in more intense form, which maser techniques have produced, can be expected to find application in wide and numerous areas of technology and of science simply because we find our present techniques of producing and controlling light already so widely applied. Most of these applications are still ahead of us, and there is not time to treat here even those which are already beginning to develop. I shall only mention that in technology lasers have been put to work in such diverse areas as radar, surgery, welding, surveying, and microscopy. A little more space will be devoted here to discussing three broad areas of science to which optical, infrared, and ultraviolet masers are expected to contribute.

Masers seem to provide the most precise techniques for measurement of the two fundamental dimensions of time and length. Over short periods of time maser oscillators clearly give the most oscillations; for longer times the hydrogen maser also seems to provide the most precise clock yet available. Light from optical masers allows new precision in the measurement of distance, and already seems capable of improving our standard of length. This new precision suggests interesting experiments on certain fundamental properties of our space, as well as the application of higher precision to a variety of physical effects. So far, experiments have been done to improve the precision with which the Lorentz transformation can be experimentally verified[49,50]. It appears that improved precision in measurement of the speed of light can also be expected. If we look some distance in the future, it seems clear that the techniques of quantum electronics will allow direct measurement of the frequency of light, rather than only its wavelength. This can be accomplished by generation of harmonics of a radio frequency, amplification of the new frequency, and further generation of harmonics until the radio region is linked with optical frequencies. This should eventually allow measurement of the velocity of light, c, to whatever precision we define time and length. Or, it will allow the elimination of separate standards of time and of length because c times a standard time will define a standard length with more precision than we can now achieve.

The power of spectroscopy should be considerably increased by use of masers. In particular, these very monochromatic sources can very much improve spectroscopic resolution and thus allow more detailed examination of the structure of atoms, molecules, or solids. This advance can be particularly striking in the infrared and far-infrared, where present resolution is far less than the widths of atomic or molecular lines. Already some high-resolution spectroscopy has been done with lasers[51, 52] and still more interesting work of this general type can be expected before long.

A third interesting field for which lasers are important has emerged as a field almost entirely because of the existence of lasers, and is the area where scientific research has so far been most active. This is what is usually called nonlinear optics[53, 54] although it includes some phenomena which might not previously have been described in this way. We have been accustomed in the past to discussing the progress of light through a passive optical material of more-ore-less fixed properties. But, in the intense laser beams now available, interactions between the light and the optical medium are sufficiently large that properties of the medium can no longer be regarded as fixed. The medium distorts, it molecules vibrate, and polarization of electrons in its atoms no longer responds linearly to the applied field. One must now also consider the dynamics medium, and interactions between their two motions. Some of the new phenomena observed are multiple-quanta absorption, which makes absorption depend on intensity[55, 56], harmonic generation in optical materials and mixing of light frequencies[57-60], excitation of coherent molecular vibrations and stimulated Raman effects[42-45], and stimulated Brillouin scattering[61, 62]. Only the last two of these will be discussed, partly because they bear on still another kind of maser, one which generates phonons.

The Phonon Maser

Acoustic waves follow equations that are of the same general form as the equations of light and manifest many of the same phenomena. An acoustic wave can produce an atomic or molecular excitation, or receive energy from it by either spontaneous or stimulated emission. Hence, one may expect maser action for acoustic waves if a system can be found in which molecules are sufficiently coupled to an acoustic field and appropriate excitation can be obtained to meet the threshold condition. The first such systems suggested involved inversion of the spin states of impurities in a crystal in ways similar to those used for solid-state electromagnetic masers[63]. A system of this type has been shown to operate as expected[64]. However, a more generally applicable technique seems to be Brillouin scattering and its close associate Raman scattering, which utilize phase correlation rather than population inversion to produce amplification. This process can also be viewed as parametric amplification.

Light may be scattered by the train of crests and troughs in an acoustic wave much as by a grating. Since the wave is moving, the scattering involves a Doppler shift. The net result, first analyzed by Brillouin[65], is that the scattered light is shifted in frequency from the frequency v_0 of the original beam by an amount

$$v = 2\,v_0 v / c \sin \theta/2 \tag{13}$$

where v and c are the phase velocities of sound and of light, respectively, in the medium, and θ is the scattering angle. The energy lost, hv, is given to the scattering acoustic wave of frequency v. If the light is of sufficient intensity, it can thus give energy to the acoustic field faster than it is lost and fulfill a threshold condition which allows the acoustic energy to build up steadily.

For the very high acoustic frequencies (10^9 to 10^{10} c/ sec) implied by Eqn.13 when θ is not very small, the losses are usually so large that interesting amplification cannot be achieved with ordinary light. But, with laser beams of hundreds of megawatts per square centimeter, it is quite feasible to produce an intense build-up of acoustic waves by this process of stimulated Brillouin scattering[61, 62] - so intense, in fact, that the acoustic energy can crack glass or quartz. This gives a method of producing and studying the behavior of very-high-frequency acoustic waves in almost any material which will transmit light - a possibility which was previously not so clearly available.

Brillouin scattering by spontaneous emission has been studied for some time. But the intense monochromatic light of lasers allows now much greater precision in work with this technique[52] and it too is yielding interesting information on the propagation of hypersonic waves in materials.

There is no firm limit to the acoustic frequencies which can be produced by stimulated emission, even though Eqn.13 indicates a kind of limit, for $\theta = \pi$, of $2v_0 v / c$. But in the optical branch of acoustic waves the phase velocity v can be very high. In fact, stimulated Raman scattering, or the Raman maser mentioned briefly above, represents excitation of the optical branches of acoustic spectra, and generates coherent molecular oscillations. Quantum-electronic techniques can thus allow interesting new ways to generate and explore most of the acoustic spectrum as well as much of the electromagnetic domain.

Concluding　Remarks

In a few years this brief report will no longer be of much interest because it will be outdated and superseded, except for some matters of general principle or of historical interest. But, happily, it will be replaced by further striking progress and improved results. We can look forward to another decade of rapid development in the field of quantum electronics - new devices and unsuspected facets of the field, improved range and performance of masers, and extensive application to science and to technology. It seems about time now for masers and lasers to become everyday tools of science, and for the exploratory work which has demonstrated so many new possibilities to be increasingly repla-ced by much more finished, more systematic, and more penetrating appli-cations. It is this stage of quantum electronics which should yield the real benefits made available by the new methods of dealing with radiation.

1. J. R. Pierce, *Phys. Today*, 3, No.11 (1950) 24.
2. N. F. Ramsey, *Phys. Rev.*, 103 (1956) 20.
3. W.E. Lamb and R.C. Retherford, *Phys. Rev., 79 (1950) 549.*
4. E.M. Pyrcell and R.V. Pound, *Phys. Rev.*, 81 (1951) 279.
5. Bloch, Hansen and Packard, *Phys. Rev.*, 70 (1946) 474.
6. N.G. Basov and A.M. Prokhorov, *J. Exptl. Thoeret. Phys. (U.S.S.R.), 27 (1954) 431.*
7. J. Weber, *Inst. Electron. Elec. Engrs., Trans. Electron Devices, 3 (1953)* I.
8. *Columbia Radiation Lab.Quart.Progr. Rept.,* (Dec.1951); C.H.Townes, *J. Inst.Elec. Commun.Eng. (Japan), 36 (1953) 650* [in Japanese].
9. J.P. Gordon, H. J. Zeiger and C.H. Townes, *Phys. Rev., 95 (1954)* 282.
10. J. P. Gordon, H. J. Zeiger and C. H. Townes, *Phys. Rev., 99 (1955)* 1264.
11. R. Serber and C. H. Townes, in C. H. Townes (Ed.), *Quantum Electronics,* Columbia University Press, New York, 1960, p. 233 ; see also H. Friedburg, *ibid.,* p. 227.
12. Shimoda, Takahashi and Townes. J. *Phys. Soc. (Japan),* 12 (1957) 686; M. W. Muller, *Phys. Rev.,* 106 (1957) 8; M.W.P.Strandberg, *ibid.,* 106 (1957) 617; R.V. Pound, *Ann. Phys. (N. Y.),* I(1957) 24.
13. C. H. Townes, in J. R. Singer (Ed.), *Advances in Quantum Electronics,* Columbia University Press, New York, 1961, p. 3.
14. Shimoda, Wang and Townes, *Phys. Rev.,* 102 (1956) 1308.
15. A.L. Schawlow and C.H. Townes, *Phys. Rev.,* 112 (1958) 1940.
16. A correction of a factor of 2 to δ given by Gordon, Zeiger and Townes[10], and Schawlow and Townes[15], is demonstrated by P. Grivet and A. Blaquiere, *Proc. Symp. Optical Masers,* Polytechnic Inst., Brooklyn, 1963.
17. Alsop, Giordmaine, Townes and Wang, *Phys. Rev.,* 107 (1957) 1450; J.P. Gordon and L.D. White, *ibid.,* 107 (1957) 1728.

18. J. de Prins and P. Kartaschoff, *in Proc. Course XVII, Enrico Fermi Intern. School of Physics,* Academic Press, New York, 1962, p. 88.

19. Goldenberg, Kleppner and Ramsey, *Phys. Rev. Letters, 5 (1960)* 361; Phys. Rev., 123 (1961) 530.

20. Combrisson, Honig and Townes, *Compt. Rend.,* 242 (1956) 2451; Feher, Gordon, Buehler, Gere and Thurmond, *Phys. Rev., 109 (1958)* 221.

21. N. Bloembergen, *Phys. Rev.,* 104 (1956) 324.

22. N. G. Basov and A.M. Prokhorov, *J. Exptl. Theoret. Phys. (U. S. S. R.), 28 (1955) 249.*

23. Scovil, Feher and Seidel, *Phys. Rev.,* 105 (1957) 762.

24. Makhov, Kikuchi, Lambe and Terhune, *Phys. Rev., 109 (1958) 1399.*

25. A. E. Siegman, *Microwave Solid State Masers,* McGraw-Hill, New York, 1964.

26. Alsop, Giordmaine, Mayer and Townes, *Astron.J.,* 63 (1958) 301; *Proc. IEEE (Inst. Electron. Elec. Engrs.),* 47 (1959) 1062.

27. E. Epstein, *Astron.]., 69 (1964) 490.*

28. R. W. DeGrasse and H.E.D. Scovil, *J. Appl. Phys.,* 31 (1960) 443.

29. DeGrasse, Hogg, Ohm and Scovil, *J. Appl. Phys.,* 30(1959) 2013.

30. A.L. Schawlow and C.H.Townes, *Phys. Rev.,* 112 (1958) 1940.

31. A.G. Fox and T. Li, *Bell System Tech. J.,* 40 (1961) 453.

32. Jacobs, Gould and Rabinowitz, *Phys. Rev. Letters,* 7 (1961) 415.

33. T.H. Maiman, *Nature,* 187 (1960) 493 ; Maiman, Hoskins, D'Haenens, Asawa and Evtuhov, *Phys. Rev.,* 123 (1961) 1151.

34. Collins, Nelson, Schawlow, Bond, Garrett and Kaiser, *Phys. Rev. Letters, 5 (1960)* 303.

35. Javan, Bennett and Herriott, *Phys. Rev. Letters, 6 (1961)* 106.

36. A. Javan, *Phys. Rev. Letters, 3 (1959)* 87.

37. R. J. Keyes and T.M. Quist, *Proc. IEEE (Inst. Electron. Elec. Engrs.), 50 (1962)* 1822.

38. Hall, Fenner, *Kingsley,* Soltys and Carlson, *Phys. Rev. Letters, 9 (1962) 366.*

39. Nathan, Dumke, Burns, Dill and Lasher, *Appl. Phys. Letters,* I(1962) 62.

40. Quist, Rediker, Keyes, Krag, Lax, McWhorter and Zeiger, *Appl. Phys. Letters,* I (1962) 91.

41. Basov, Bogdankevich and Devyatkov, *Dokl. Akad. Nauk S. S. S. R.,* 155 (1964) 783 ; *Proc. Symp. Radiative Recombination in Semiconductors, Paris,* 1964, Dunod Paris, 1965 ; C.Benoit à la Guillaume and T. M. Debever, *ibid.*

42. A. Javan, *J. Phys.Radium,* 19 (1958) 806; *Bull. Am. Phys. Soc.,* 3 (1958) 213; *Proc. Course XXXI, Enrico Fermi Intern. School of Physics,* Academic Press, New York, 1964.

43. E. J. Woodbury and W.K. Ng, *Proc. IEEE (Inst. Electron. Elec. Engrs.), 50 (1962)* 2367; Eckhardt, Hellwarth, McClung, Schwarz, Weiner and Woodbury, *Phys. Rev. Letters, 9 (1962) 455.*

44. R. W.Terhune, *Solid State* Design, 4 (1963) 38 ; Minck, Terhune and Rado, *Appl. Phys. Letters, 3 (1963)* 181; see also B.Stoicheff, *Proc. Course XXXI, Enrico Fermi Intern. School of Physics,* Academic Press, New York, 1964.

45. R.W. Hellwarth, *Phys. Rev.,* 130 (1963) 1850; H. J. Zeiger and P.E. Tannenwald, in P. Grivet and N. Bloembergen (Eds.), *Quantum Electronics: Proc. 3rd Intern. Congr.,* Columbia University Press, New York, 1964, p.1589; Garmire, Pandarese and Townes, *Phys. Rev. Letters,* II (1963) 160.

46. E.L. Tolnas, *Bull. Am. Phys. Soc.*, 5 (1960) 342.

47. Jaseja, Javan and Townes, *Phys. Letters*, 10 (1963) 165.

48. D. Herriott, *J. Opt. Soc. Am.*, 52 (1962) 31.

49. J.P. Cedarholm and C.H. Townes, *Nature*, 184 (1959) 1350.

50. Jaseja, Javan, Murray and Townes, *Phys. Rev.*, 133 (1964) A1221.

51. A. Szöke and A. Javan, *Phys. Rev. Letters*, 10 (1963) 521.

52. R. Chiao and B. Stoicheff, *J. Opt. Soc. Am.*, 54 (1964) 1286; Benedek, Lastovka, Fritsch and Greytak, *ibid.*, 54 (1964) 1284; Masch, Starunov and Fabelinskii, *J. Exptl. Theoret. Phys. (U. S. S. R.)*, 47 (1964) 783.

53. C. A. Akhmanov and R. B. Khoklov, *Problems in Nonlinear Optics, 1962-1963*, Academy of Sciences, U.S.S.R., Moscow, 1964.

54. N. Bloembergen, *Nonlinear Optics*, Benjamin, New York, 1965.

55. I. Abella, *Phys. Rev. Letters, 9 (1962) 453.*

56. J.A. Giordmaine and J.A. Howe, *Phys. Rev. Letters*, II (1963) 207.

57. Franken, Hill, Peters and Weinreich, *Phys. Rev. Letters*, 7 (1961) 118; P.A. Franken and J.F. Ward, *Rev. Mod. Phys.*, 35 (1963) 23.

58. J.A. Giordmaine, *Phys. Rev. Letters*, 8 (1962) 19; Terhune, Maker and Savage, *Appl. Phys. Letters*, 2 (1963) 54.

59. J. Ducuing and N. Bloembergen, *Phys. Rev. Letters*, 10 (1963) 474.

60. Armstrong, Bloembergen, Ducuing and Pershan, *Phys. Rev.*, 127 (1962) 1918; N. Bloembergen and P. S. Pershan, *ibid.*, 128 (1962) 606.

61. Chiao, Garmire and Townes, *Proc. Course XXXI Enrico Fermi Intern. School of Physics*, Academic Press, New York, 1964.

62. Chiao, Townes and Stoicheff, *Phys. Rev. Letters*, 12 (1964) 592.

63. C.H.Townes, in C.H. Townes (Ed.), *Quantum Electronics*, Columbia University Press, New York, 1960, p-402; C. Kittel, *Phys. Rev. Letters, 6 (1961) 449.*

64. E.B. Tucker, *Phys. Rev. Letters, 6 (1961) 547.*

65. L. Brillouin, *Ann. Phys. (Paris)*, 17 (1922) 88; M. Born and K. Wang, *Dynamical Theory of Crystal Lattices*, Oxford University Press, New York, 1964.

A.M. PROCHOROV

Quantum electronics

Nobel Lecture, December 11, 1964

One may assume as generally accepted that quantum electronics started to exist at the end of 1954 - beginning of 1955[1,2]. Just by that time theoretical grounds had been created, and the first device, - a molecular oscillator - had been designed, and constructed. A basis for quantum electronics as a whole is the phenomenon of an induced radiation, predicted by A. Einstein in 1917. However, quantum electronics was developed considerably later. And it is quite natural to ask the questions: why did it happen so? What reasons put obstacles for the creation of quantum devices considerably earlier, for instance, in the period of 1930-1940?

In order to try to answer these questions I should like to say some words about the bases on which quantum electronics is founded.

As I have already noted, the phenomenon of an induced radiation was predicted by Einstein. It is well-known that an atom being in an excited state may give off its energy in the form of emission of radiation (quantum) in two ways. The first way is a spontaneous emission of radiation, *i.e.* when an atom emits energy without any external causation. All usual light sources (filament lamps, gas-discharge tubes, etc.) produce light by way of such spontaneous radiation. It means that scientist engaged in the field of optical spectroscopy were well acquainted with that type of emission many years ago.

The second way for an atom to give off its energy is through stimulated emission of radiation. That phenomenon was noted by Einstein to be necessary in order to describe thermodynamic equilibrium between an electromagnetic field and atoms. The phenomenon of stimulated emission occurs when an excited atom emits due to interaction with an external field (quantum). Then two quanta are involved: one is the external one, the other is emitted by the atom itself. Those two quanta are indistinguishable, *i.e.* their frequency and directivity coincide. This very significant characteristic of an induced radiation (which was, apparently, first pointed at by Dirac in 1927) made it possible to build quantum electronic devices.

In order to observe a stimulated emission, it is necessary, firstly, to have excited atoms and, secondly, that the probability of an induced radiation must be greater than that of a spontaneous emission. If atoms are in a thermal equilibrium, optical levels are not populated. If the atoms became excited they make a transition to the lower level due to spontaneous emission. This happens because the probability of a stimulated emission radiation is small at usual densities of the light energy. Therefore scientists engaged in the field of spectroscopy did not take into account the stimulated radiation and some of them, apparently, considered that phenomenon as a « Kunststück » of a theorist necessary only for the theory.

It is absolutely clear that if all atoms are in an excited state, such a system of atoms will amplify the radiation, and many scientists understood this already before 1940, but none of them pointed to the possibility of creating light oscillators in this way. It may seem strange because, in principle, optical quantum oscillators (lasers) could have been made even before 1940. But definite fundamental results were necessary. They appeared after Second World War, when radiospectroscopy started to develop rapidly. And just the scientists engaged in the field of radiospectroscopy laid down foundations for quantum electronics[1,2].

How should one explain this? There were some favourable circumstances which had not been available to the scientists working in the field of optical spectroscopy.

First of all, since for systems in a thermal equilibrium, the excited levels in the radio range, contrary to the optical ones may have a large population and of course one should then take into account induced radiation. Indeed, if the concentration of particles on the lower level equals n_1, and on the excited level n_2, one may write down a net absorption coefficient for an electromagnetic wave in the form

$$\alpha = \frac{1}{v}\, hv\, (n_1 B_{12} - n_2 B_{21})$$

(1)

the value of B_{12} characterizes the probability of an absorption act, and B_{21} characterizes the probability of an induced radiation act. If the levels are not degenerated, $B_{12} = B_{21}$, then will take the form

$$\alpha = \frac{1}{v}\, hv\, (n_1 - n_2)\, B_{12}$$

(2)

For a frequency as in the optical range, under usual conditions of thermal equilibrium, one may put with a high accuracy n_2 equal to zero, and then the absorption coefficient will become

$$a = \frac{1}{v} \, hv \, n_1 B_{12} \qquad (3)$$

Therefore, for the optical range the absorption coefficient depends only on the population of the lower level. For as in the radio range, as a rule, $hv << kT$. In that case

$$n_2 = n_1 e^{-\frac{hv}{kT}} \approx n_1\left(1 - \frac{hv}{kT}\right)$$

Then the value of   will be

$$a = \frac{1}{v} \, hv \, n_1 B_{12} \frac{hv}{kT} \qquad (4)$$

As is seen from Eqn. 4, due to stimulated emission, the value of the absorption coefficient becomes reduced by a factor kT/hv compared to what it would be without the presence of induced emission. Therefore, all scientists engaged in radiospectroscopy have to take into account the effect of induced radiation. Moreover, for increasing an absorption coefficient one has to lower the temperature in order to decrease the population of the upper level and to weaken, in this way, the influence of stimulated radiation. It follows from Eqn. 2 that for systems that are not in thermal equilibrium, but have $n_2 > n_1$, the net absorption coefficient becomes negative, *i.e.* such a system will amplify radiation. In principle, such systems were known to physicists long time ago for the radio range. If we pass molecular or atomic beams through inhomogeneous magnetic or electric fields, we can separate out molecules in definite state. In particular, one may obtain molecular beams containing molecules in the upper state only. Actually physicists engaged in the field of a micro-wave radiospectroscopy started to think about application of molecular beams for increasing the resolving power of radio spectroscopes. In order to gain a maximum absorption in such beams, one must have molecules either in the lower state only or in the upper states only, *i.e.* one must separate them using inhomogeneous electric or magnetic fields. If molecules are in the upper state, they will amplify a radiation.

As is well-known from radio engineering, any system able to amplify can be made to oscillate. For this purpose a feedback coupling is necessary. A theory for ordinary tube oscillators is well developed in the radio range. For description of those oscillators, the idea of a negative resistance or conductance is introduced, *i. e.* an element in which so-called negative losses take place. In the case of a quantum oscillator the medium with a negative absorption factor is that « element ». Therefore the condition of self-excitation for the quantum oscillator should be written in the similar way as for a tube oscillator. According to the analogy with usual tube oscillators, it is quite natural to expect that for a quantum oscillator the oscillations will also be quite monochromatic.

Finally, the resonator system is a very significant element of a quantum oscillator (maser or laser) as well as in any other oscillator with sinusoidal oscillations. However, resonator systems were well worked out for the radio range, and just those resonators operating in the radio range were used for masers. Thus, a very important element - a cavity - was also well-known to the scientists engaged in radiospectroscopy. Therefore all elements of masers really existed separately but it was necessary to do a very important step of synthesis in order to construct the maser. First two papers - one of which was published in the U.S.S.R. and the other in the U.S.A. - appeared independently; and they both were connected directly with the construction of radio spectroscopes with a high resolving power, using molecular beams. As is easily seen from the aforesaid, this result is quite natural.

Those two papers initiated the development of quantum electronics, and the first successes in this new field of physics stimulated its further progress. Already in 1955, there was proposed a new method - the method of pumping for gaining a negative absorption[3]. That method was further developed and applied for the construction of new types of quantum devices. In particular, the method of pumping was developed and applied for designing and building quantum amplifiers for the radio range on the basis of an electronic paramagnetic resonance[4,5]. Quantum devices according to the suggestion of Prof. Townes were called masers. One might think that after the successful construction of masers in the radio range, there would soon be made quantum oscillators (lasers) in the optical range as well. However, this did not occur. Those oscillators were constructed only 5-6 years later. What caused such a delay?

There were two difficulties. One of them was as follows: at that time no resonators for the optical wavelength range were available. The second difficulty was that no methods were immediately available for gaining an inverse population in the optical wavelength range.

Let us consider firstly the question of resonators. It is well-known that radio engineering started its development from the region of long waves where resonators were used in the form of self-inductance coils combined with condensers. In that case the size of the resonator is much less than one wavelength.

With development towards short waves the cavity resonators were used. They are closed volume cavities. The size of those cavities was comparable with a wavelength. It is quite clear that with the help of such cavities it is impossible to advance into the region of very short waves. In particular, it would be impossible to reach the optical range.

In 1958 there was proposed the so-called open type of cavities for masers and lasers in the region of very short waves[6,7]. Practically speaking this is Fabry-Perot's interferometer; however, a « radio engineering » approach made it possible to suggest using such a system as a resonator. Afterwards, spherical mirrors were used together with plane mirrors. The size of these resonators is much more than that wavelength.

At present open cavities are widely utilized for lasers.

There were also systems suggested for the production of a negative absorption in the submillimeter (far-infrared) wavelength range[6], the infrared and optical wave ranges[7-10]. Those works stimulated a further advancement in the region of shorter waves and, in particular, into the optical range. However, the first quantum-optical oscillator was made as late as 1960[11]. It was a ruby laser. After carrying out investigations in the optical range, many scientists started to think about further extension into the X-ray field. In that wavelength range the same difficulties arise as in the optical wavelength range. It was necessary to suggest new types of resonators and to find also the proper system that would produce negative absorption. As is known X-ray quantum oscillators have not yet been constructed. We have also considered this problem* and we have found that there are essential difficulties.

Indeed in the X-ray region the lifetime of an excited level state is small and one may assume that the line width is determined by that lifetime only. Then the absorption coefficient may be written in a very simple form

$$\alpha = \frac{\lambda^2}{4\pi} (n_1 - n_2) \qquad (5)$$

where λ is the wavelength and n_1 and n_2 are the densities of particles in the lower and upper level stated respectively. As seen from Eqn. 5 the absorption coefficient decreases sharply as the wavelength becomes shorter. This is an extremely unpleasant circumstance. Indeed for the laser operation the value of α should be of the order of one inverse cm. If $\lambda = 1\,\text{Å}$, the density of particles on the upper level must be not less than $10^{17}\,\text{cm}^{-3}$. The lifetime in the upper

---

* This problem was also considered in ref. 1.

level τ is of the order of 10^{-16} sec. Therefore, 10^{33} particles/cm^3 per second must be excited. In order to fulfil this condition one has to overcome essential experimental difficulties.

Nevertheless, the successes of quantum electronics are enormous even without the construction of laser in the X-ray region.

At present the range in which lasers and masers operate is extremely wide. Recently a far-infrared range had not been available but now investigations in this region are carried out with a great success. In practice with the help of masers and lasers one may produce emission from the the lowest radio frequencies to the ultraviolet region.

Operation of all masers and lasers is based on the fact that in media with a negative absorption the processes of induced emission dominate due to a large field intensity over spontaneous or non-radiative transitions. Moreover, at present, for instance, one may produce with the help of a ruby laser such radiation energy densities at which the probability of multiquantum processes becomes comparable with the probability of one quantum process or even exceeds it. This is a new qualitative jump which leads to interesting results of several kinds.

First of all one may estimate[12] the maximum power which a ruby laser is able to give per cm^2. That power equals 10^{11} Watt/cm^2 that is one hundred gigaWatts / cm^2. At that power the probability of a simultaneous absorption of three quanta of red light with transition of an electron to the conduction band is so great that a further growth of the field stops. For three-quantum processes the losses grow in proportion to the cube of the energy density, i. e. a very strong dependence on the field takes place.

Large electric fields available in a laser beam may carry out ionization and dissociation of molecules and breakdown in a solid as well.

Multiquantum processes do not always have a bad effect (for instance, restriction of the maximum density given by laser) but they open up new possibilities for a further development of quantum electronics. This interesting and principally new direction is connected with the construction of lasers which utilize two-quantum transitions. It was pointed out in 1963 in the U. S. S. R.[13] and somewhat later but independently in the U. S. A.[14,15] that construction of these oscillators should be possible. The idea of this laser is that if there is an inverse population between two levels with the energy difference $E_2 - E_1 = h\nu$, generation of two frequencies ν_1 and ν_2 is possible in such a way that

$$\nu = \nu_1 + \nu_2 \tag{6}$$

In particular, frequencies ν_1 and ν_2 may coincide. However, frequencies ν_1 and v_2 may have any value as long as only condition 6 is fulfilled.

Operation of such an oscillator, as it was mentioned above, is connected with two-quantum transitions, the probability of which is rather great, if the field density is considerable. For self-excitation of these oscillators it is necessary to have another oscillator of a sufficiently large initial energy density with frequencies ν_1 or ν_2, and one may remove the external field only after self-excitation of the two quantum oscillator. Such two quantum oscillators have two possibilities : *(1)* Faster growth of the field density than in the case of usual lasers. *(2)* Possibility of producing any frequency within the framework of the relation 6.

Construction of an oscillator for any given radiation frequency will greatly extend the region of application of lasers. It is clear that if we make a laser with a sweep frequency, we apparently shall be able to influence a molecule in such a way that definite bonds will be excited and, thus, chemical reactions will take place in certain directions.

However, this problem will not be simple even after design of the appropriate lasers. But one thing is clear: the problem is extremely interesting and perhaps its solution will be able to make a revolution in a series of branches of chemical industry.

1. N. G. Basov and A.M. Prochorov, *Zh. Eksperim. i Teor. Fiz.*, 27 *(1954) 431*.
2. J.P. Gordon, H. J. Zeiger and C. H. Townes, *Phys. Rev.*, 95 (1954) 282.
3. N. G. Basov and A. M. Prochorov, *Zh. Eksperim, i Teor. Fiz.*, 28 (1955) 249.
4. N. Bloembergen, *Phys. Rev.*, 104 (1956) 324.
5. H.E.D. Scovil, G. Feiher and H. Seidel, *Phys. Rev.*, 105 (1957) 762.
6. A.M. Prochorov, *Zh. Eksperim. i Teor. Fiz.*, 34 *(1958)* 1658.
7. A.L. Schawlow and C.H. Townes, *Phys. Rev.*, 112 (1958) 1940.
8. N.G. Basov, B.M. Vul and Yu. M. Popov, *Zh. Eksperim. i Teor. Fiz.*, 37 *(1959)* 587.
9. A. Javan, *Phys. Rev. Letters*, 3 (1959) 87.
10. F. A. Butaeva and V. A. Fabrikant, *O srede s otritsatelnym koeffitsentom pogloshcheniya. Issledovaniya po eksperimentalnoy i teoreticheskoy fizike.* Sbornik pamyati G. S. Landsberga, Izdatelstvo Akad. Nauk S. S. S.R., 1959, p.62.
11. T.H. Maiman, *Brit. Commun. Electron.*, 1(1960) 674.
12. F. V. Bunkin and A. M. Prochorov, in preparation.
13. A.M. Prochorov and A.C. Selivanenko, *U.S.S.R. Patent Application* No. 872, 303, priority 24 Dec. 1963.
14. P.P. Sorokin and N. Braslau, *IBM J. Res. Develop.*, 8 (1964) 177.
15. R.Z. Gorwin, *IBM J. Res. Develop.*, 8 (1964) 338.

N IKOLAI G . BA SOV

Semiconductor lasers

Nobel Lecture, December 11, 1964

In modern physics, and perhaps this was true earlier, there are two different trends. One group of physicists has the aim of investigating new regularities and solving existing contradictions. They believe the result of their work to be a theory; in particular, the creation of the mathematical apparatus of modern physics. As a by-product there appear new principles for constructing devices, physical devices.

The other group, on the contrary, seeks to create physical devices using new physical principles. They try to avoid the inevitable difficulties and contradictions on the way to achieving that purpose. This group considers various hypotheses and theories to be the by-product of their activity.

Both groups have made outstanding achievements. Each group creates a nutrient medium for the other and therefore they are unable to exist without one another; although, their attitude towards each other is often rather critical. The first group calls the second « inventors », while the second group accuses the first of abstractness or sometimes of aimlessness. One may think at first sight that we are speaking about experimentors and theoreticians. However, this is not so, because both groups include these two kinds of physicists.

At present this division into two groups has become so pronounced that one may easily attribute whole branches of science to the first or to the second group, although there are some fields of physics where both groups work together.

Included in the first group are most research workers in such fields as quantum electrodynamics, the theory of elementary particles, many branches of nuclear physics, gravitation, cosmology, and solid-state physics.

Striking examples of the second group are physicists engaged in thermonuclear research, and in the fields of quantum and semiconductor electronics.

Despite the fact that the second group of physicists strives to create a physical device, their work is usually characterized by preliminary theoretical analysis. Thus, in quantum electronics, there was predicted theoretically the possibility of creating quantum oscillators: in general, also, there were predicted the high monochromaticity and stability of the frequency of masers, the high sensitivity of quantum amplifiers, and there was investigated the possibility of the creation of various types of lasers.

This lecture is devoted to the youngest branch of quantum electronics - semiconductor lasers, which was created only two years ago, although a theoretical analysis started already in 1957 preceded the creation of lasers[1].

However, before starting to discuss the principles of operation of semiconductor lasers we would like to make some remarks of the theoretical « by-products » of quantum electronics. There are many of them but we shall consider only three:

(1) The creation of quantum frequency oscillators of high stability and the transition to atomic standards of time made it possible to raise the question of solving the problem of the properties of atomic time.

Dicke[2] in his paper at the first conference on quantum electronics pointed out the possibility of an experimental check of the hypothesis on the variation of fundamental physical constants with time on the basis of studying changes in frequencies of different quantum standards with time. There arises the question about the maximum accuracy of atomic and molecular clocks depending on the nature of quantum of emission, especially about the accuracy of the measurement of short time intervals.

(2) Due to quantum electronics there was started an intensive investigation of a new « super non-equilibrium state of matter » - the state with negative temperature, which in its extreme state of negative zero is close in its properties to the absolute ordering intrinsic for the temperature of absolute zero. It is just this property of high ordering of a system with negative temperature which makes it possible to produce high-coherent emission in quantum oscillators, to produce high sensitive quantum amplifiers, and to separate the energy stored in the state with negative temperature in a very short time, of the order of the reciprocal of the emission frequency.

(3) Quantum electronics gives examples of systems in which there occurs radiation with a very small value of entropy. For instance, spontaneous low temperature radiation from flash tubes, distributed through very large number of degrees of freedom is converted with the help of a system in a state of negative temperature (quantum oscillators) into high-coherent laser emission, the temperature of which in present experiments already attains a value of 10^{20} degrees.

Apparently, the regularities established by quantum electronics for radiation may be generalized for other natural phenomena. The possibility of obtaining a high degree of organization with the help of feed-back systems may be of interest for chemical and biological research, and for cosmology. The question arises as to whether or not the maser principle is used in Nature.

We believe that the above questions need attention from physicists of the first group, because these questions go beyond the limits of the theory of oscillations, the theory of radiation and usual optics which form the basis of modern quantum electronics.

I. Conditions for the Production of Negative Temperature in Semiconductors

Investigations of semiconductor quantum oscillators were a direct continuation of research on molecular oscillators and paramagnetic amplifiers. One should note that at the beginning of research on semiconductor lasers, due to investigations in the field of semiconductor electronics, there became known the physical characteristics of semiconductors, which were essential in the development and practical realization of lasers; such as, optical and electric characteristics, structure of energy bands, and relaxation time.

Various pure and alloyed semiconductors were made, and the technique of measurements of their various properties and the technology of making *p-n* junctions were worked out. All of this considerably simplified investigations of semiconductor lasers. Semiconductors were very intriguing because of the possibility of using them for making oscillators with a frequency range from the far-infrared region to the optical or even to the ultraviolet range, as well as because of the variety of methods by means of which states with negative temperatures may be obtained within them and because of their large factor of absorption (amplification). As the following studies have shown, semiconductor lasers may have extremely high efficiency, in some cases approximating 100 percent.

In contrast to an isolated atom, in semiconductors there do not exist separate energy levels, but rather there exist groups of energy levels arranged very close to one another, which are called bands (Fig. 1). The upper group of levels, called the conduction band, and a lower group of excited levels, called the valent band, are divided by a band of forbidden energy (Fig. 1).

The distribution of electrons on energy levels is described by the Fermi function: each level is occupied by two electrons, the electrons being distributed in the energy range of the order of the energy of kT thermal motion; and, the probability of finding an electron beyond the kT interval sharply decreases when the energy level increases. If the energy of thermal motion is

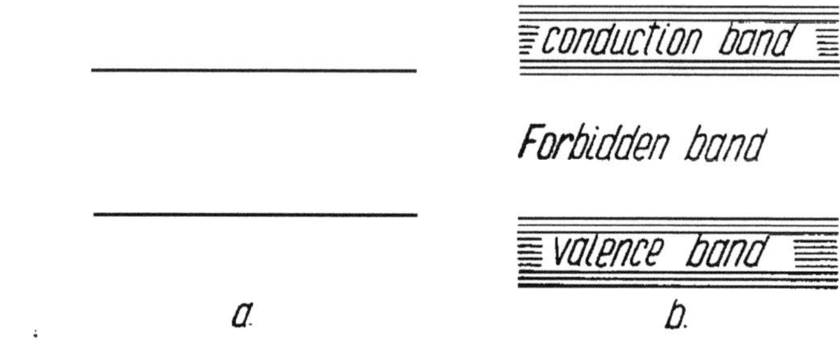

Fig.1. Energy-level diagram. (a) For atoms with two energy levels; (b) for semiconductors.

significantly less than the energy difference between the conduction and valent bands, then practically all electrons will be found in the valent band, filling its levels, while practically all levels of the conduction band remain free (Fig. 2a). In such a state the semiconductor cannot conduct electric current and becomes an insulator, since the electric field applied to the semiconductor is unable to change the motion of the electrons in the valent band (all energy levels are occupied).

If the energy of thermal motion is sufficient, then a part of the electrons are thrown into the conduction band. Such a system may serve as a conductor of electric current. Current is able to flow both due to variation of the electron energy under the action of the external field, as well as due to changes in the electron distribution within the valent and conduction bands. Current within the valent band behaves as if those places free from electrons (holes) moved in a direction opposite to that of the electrons. A vacant place or « hole » is entirely equivalent to a positively charged particle (Fig. 2b).

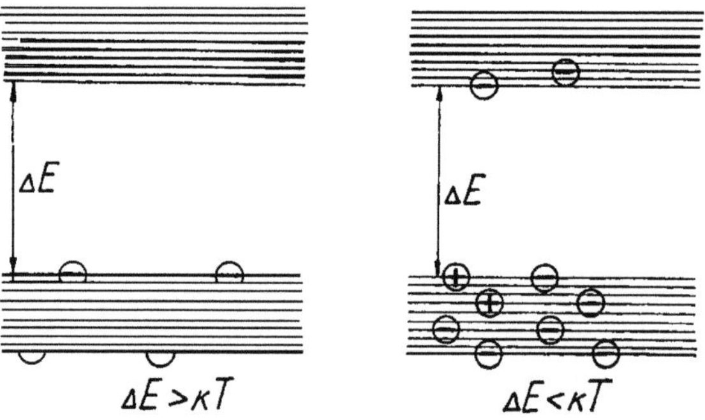

Fig. 2. Distribution of the electrons on energy levels.

During interaction with light, a semiconductor, similar to an isolated atom, may undergo three processes:

(1) A light quantum may be absorbed by the semiconductor: and, in this case an electron-hole pair is produced. The difference in energy between the electron and the hole is equal to the quantum energy. This process is connected with the decrease in energy of the electromagnetic field and is called *resonance absorption*   (Fig. 3a).

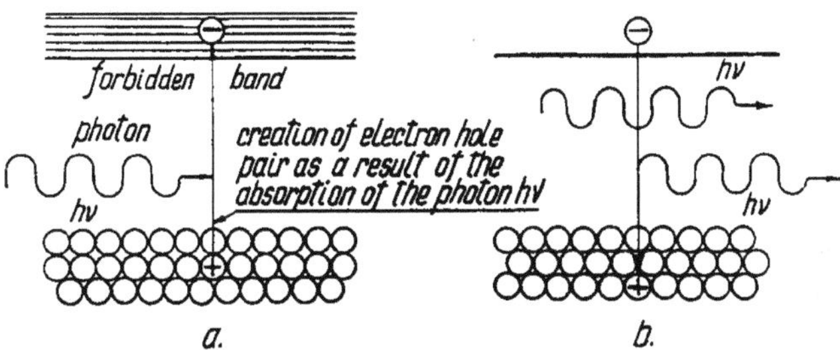

Fig. 3. Processes of the interaction with light. (a) Resonance absorption; (b) stimulated emission.

(2) Under the influence of a quantum, an electron may be transferred from the conduction band to a vacant place (hole) on the valent band. Such a transfer will be accompanied by the emission of a light quantum identical in frequency, direction of propagation and polarization to the quantum which produced the emission. This process is connected with an increase of the field energy and is called *stimulated emission* (Fig. 3b). We recall that stimulated emission was discovered by A. Einstein in 1917 during an investigation of thermodynamical equilibrium between the radiation field and atoms.

(3) Besides resonance absorption and stimulated emission, a third process may take place - *spontaneous emission*. An electron may move over to a vacant place-hole (recombine with the hole) in the absence of any radiation quanta.

Since the probabilities of stimulated radiation and resonance absorption are exactly equal to one another, a semiconductor in an equilibrium state at any temperature may only absorb light quanta, because the probability of finding electrons at high levels decreases as the energy increases. In order to make the semiconductor amplify electromagnetic radiation, one must disturb the equilibrium of the distribution of electrons within the levels and artificially produce a distribution where the probability of finding electrons on higher energy levels is greater than that of finding them on the lower levels[1,3]. It is very difficult to disturb the distribution inside a band because of the strong interaction between the electrons and the lattice of the semiconductor: it is restored in 10^{-10} to 10^{-12} sec. It is much simpler to disturb the equilibrium between the bands, since the lifetime of electrons and holes is considerably greater in the bands. It depends on the semiconductor material and lies in the interval of 10^{-3} to 10^{-9} sec.

Due to the fact that electrons and holes move in semiconductors, in addition to the law of the conservation of energy, the law of the conservation of momentum should be fulfilled during emission. Since the photon impulse is extremely small, the law of the conservation of momentum, approximately speaking, requires that the electrons and holes must have the same velocity during the emission (or absorption) of a light quantum. Fig. 4 shows graphi-

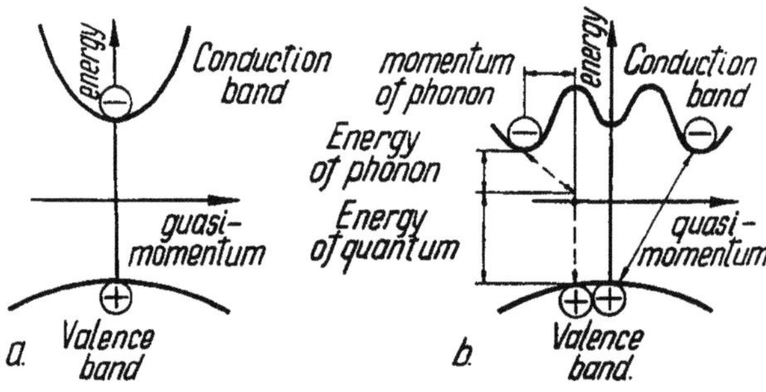

Fig. 4. Diagram of the electron-hole energy dependence on the quasi-momentum. (a) Direct transitions; (b) indirect transitions.

cally the dependence of energy on momentum. There are two types of semiconductors. For one group of semiconductors, the minimum of electron energy in the conduction band is exactly equal to the maximum of hole energy in the valent band. In such semiconductors there may take place so called « direct transitions ». An electron having minimum energy may recombine with a hole having maximum energy. For another group of semiconductors, the minimum energy in the conduction band does not coincide with the maximum energy in the valent band. In this case the process of emission or absorption of a light quantum should be accompanied by a change in the amplitude of the oscillatory state of the crystal lattice, that is by the emission or absorption of a phonon which should compensate for the change in momentum. Such processes are called indirect transitions. The probability of indirect transitions is usually less than that of direct transitions.

In order to make a semiconductor amplify incident radiation under inter-band transitions, one should distinguish two cases:

(a) *In the case of direct transitions*

It is necessary to fill more than half of the levels in the band of the order of kT near the band's edge with electrons and holes. Such states, both for atoms and molecules, came to be called states with inverse populations, or states with negative temperature. The distribution of electrons when all levels in the kT zone of the conduction band are occupied by electrons, and in the valent band - by holes, corresponds to the temperature minus zero degrees. In this state (in contrast to the state of plus zero degrees), the semiconductor is only able to emit (stimulated and spontaneous) light quanta and is unable to absorb emissions.

The state of a semiconductor when most levels in a certain energy band are occupied by electrons or holes was named the *degenerated state.*

Thus, for the creation of negative temperature there must occur degeneration of electrons and holes in the semiconductor. With a given number of electrons and holes it is always possible to produce degeneration by means of lowering the semiconductor's temperature; since, as the temperature decreases the energy band width occupied by the electrons also decreases. At the temperature of liquid nitrogen for degeneration to take place it is necessary to have an electron concentration[3] of 10^{17}-10^{18} $1/\text{cm}^3$.

(b) *In the case of indirect transitions*

Degeneration is not necessary for the creation of negative temperature. This is connected with the fact that when indirect transitions occur, the probability of quantum-stimulated emission may not be equal to the probability of resonance absorption,

Consider, for instance, an indirect transition in which a quantum and a phonon are emitted simultaneously. The process of the simultaneous absorption of a quantum and a phonon is the inverse of that process.

The probability of absorption is proportional to the number of phonons in the crystal lattice. The number of phonons decreases with a lowering of temperature. At low temperature phonons are absent. Therefore, by means of lowering the temperature of the sample one may make the probability of emission much greater than the probability of absorption. This means that with indirect transitions negative temperature may be attained with a considerably lower concentration of electrons and holes[4].

One should note that the absorption and emission of quanta during transitions within a band also takes place due to indirect transitions. When negative temperature is created between bands, the distribution of electrons (and holes) within a band corresponds to a positive temperature and leads to the absorption of emission.

In the case of direct transitions, when the probability of interband transitions is much greater than that of innerband transitions, one may neglect the innerband transitions; that is, states with negative temperature can be used for the amplification of emission.

In the case of indirect transitions for amplification to take place it is not sufficient to attain negative temperature. It is necessary that the probability of interband transitions be greater than that of innerband transitions. The necessity of fulfilling this condition makes it difficult to utilize indirect transitions. According to Dumke's estimate, this condition cannot be fulfilled for germanium[5]. However, it may be fulfilled for other semiconductors[6].

In a number of cases in semiconductors, an electron and a hole form an interconnected state something like an atom-exciton. The excitons may recombine, producing an emission. They may be also used to obtain quantum amplifiers, but we shall not consider this in detail.

We have studied conditions for the production of negative temperature in semiconductors possessing an ideal lattice. In a non-ideal crystal there occur additional energy levels connected with various disturbances in the crystalline lattice (impurities, vacancies, dislocations, etc.). As a rule, these states are localized near the corresponding centre (for instance, near an impurity atom) and in this they differ from those states in the valent and conduction bands which belong to the crystal as a whole.

In an ideal crystal the number of electrons in the conduction band is exactly equal to the number of holes in the valent band. However, in an actual crystal the number of current carriers - electrons and holes - is determined, mainly, by the existence of impurities (Fig. 5).

There are two kinds of impurities: one type has energy levels arranged near the conduction band and creates excess electrons due to thermal ionization. These are called « donor » impurities. Other impurities having energy levels near the valent band are able of removing electrons from the valent band and thus producing an excess number of holes in it. These impurities are called « acceptors ».

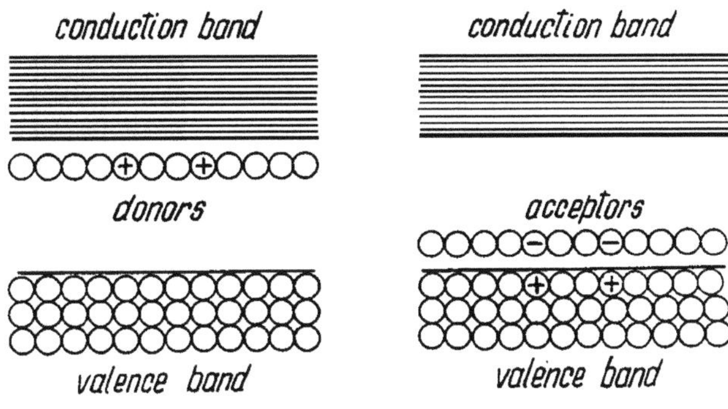

Fig. 5. Donor and acceptor levels.

One should note that a semiconductor with an equal number of donor and acceptor impurities behaves as if it were a pure semiconductor, since the holes produced by acceptors recombine with the electrons produced by donors.

In a number of cases, transitions of electrons between bands, between impurity atoms or between other levels may also be accompanied by emission of photons. One may likewise use these transitions for the creation of negative temperature. However, because of time limitations, we shall not discuss this question.

II. Methods of Obtaining States with Negative Temperature in Semiconductors

(a) The method of optical pumping

In the case of semiconductors one may utilize the « three-level » scheme[7] which has been used successfully for paramagnetic quantum amplifiers[8] and optical generators based on luminescent crystals and glasses[9] (Fig. 6).

Since the relaxation time of electrons and holes in the band levels[10] is much less than the lifetime of electrons and holes in the corresponding bands, one may obtain an inverse population by means of optical pumping.

Semiconductors have a very large absorption index which sharply increases as the radiation frequency increases. Therefore, to obtain an inverse population in samples of relatively large thickness, it is reasonable to use monochromatic radiation with a frequency close to that of the interband transitions[11]. In the case when the frequency of the exciting radiation is greater than the width of the forbidden band, a state with negative temperature is produced in a narrow band, several microns deep (on the order of the electrons' diffusion length)

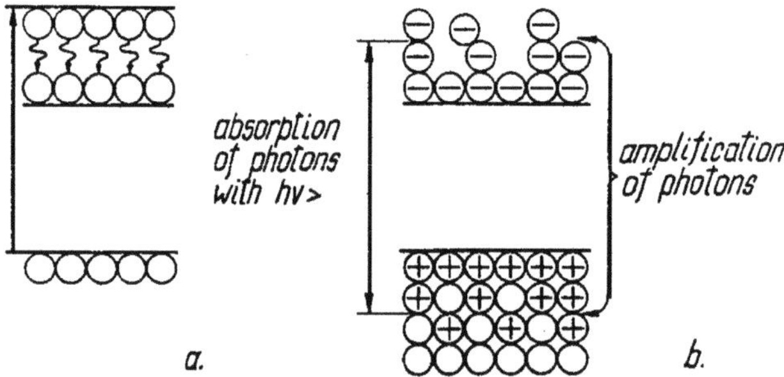

Fig. 6. Optical pumping. (a) Three levels diagram for atoms; (b) for semiconductors.

near the surface of the sample. As a source of radiation one may use the light from other types of lasers: gas lasers, lasers based on luminescent crystals or lasers based on *p-n* junctions[11].

(b) The excitation of semiconductors by a beam of fast electrons

If a beam of fast electrons is directed into the surface of a semiconductor, the electrons easily penetrate into the semiconductor. On their way the electrons collide with the atoms of the crystal and create electron-hole pairs. Calculations and experiments[12,13] have shown that an amount of energy approximately three times greater than the minimum energy difference between the bands is spent on the production of one electron-hole pair (Fig. 7a). The electrons and holes obtained give their excess energy to the atoms of the lattice and accumulate in the levels near the edges of the corresponding bands. In this case a state with negative temperature may be created[14,15]. The higher the electron energy, the deeper they will penetrate. However, there exists a

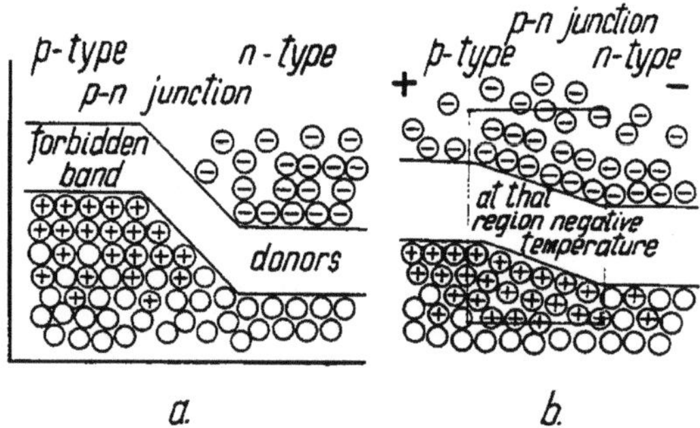

Fig. 7. (a) *p-n* junctional equilibrium; (b) *p-n* junction in the external electrical field.

certain threshold energy, beginning with which the electrons will produce defects in the crystal; that is, will destroy the crystalline lattice. This threshold energy depends upon the binding energy of the atoms in the crystals and is usually equal to about several hundred KeV. Experiments have shown that electrons with energy in the range of 200-500 KeV are not yet capable of noticeably harming the lattice.

The current density of fast electrons at which negative temperature is produced strongly depends upon the lifetime of electrons and holes. For semiconductors with a lifetime of 10^7 sec at the temperature of liquid nitrogen, the threshold of the current density has the order of one Ampere per cm^2. Since in the presence of such large currents it is difficult to remove the energy released in the semiconductor, the impulse method of excitation with a short impulse duration is usually employed.

(c) The injection of electrons and holes through p-n junctions

As it was noted above, a specific characteristic of semiconductors is that their energy levels may be filled with electrons or holes by introducing into the crystals special types of impurity atoms. However, the simultaneous introduction of donor and acceptor impurities does not result in the production of states with negative temperature. Therefore, in order to obtain an inverse population one does as follows: take two pieces of a semiconductor, inject donor impurities into one of them, and inject acceptor impurities into the other. If one then connects one piece to the other, a *p-n* junction will be created. On the boundary between the semiconductors there arises a potential difference which does not allow electrons to penetrate into the crystal having holes and likewise does not allow holes to penetrate into the crystal having electrons (Fig. 7a). As it was pointed out above, a large concentration of electrons and holes is necessary for the production of an inverse population (more than half of the levels in a certain energy band should be occupied), that is the semiconductor must have a large number of impurities.

If one applies an external voltage to a *p-n* junction, removing the potential difference between the two pieces of the semiconductor, the equilibrium of the distribution of electrons will be disturbed, and current will flow through the semiconductor. In this case electrons appear to flow into the region with a large concentration of holes, and holes - into the region with a large concentration of electrons. An inverse population arises in a narrow region near the *p-n* junction at a distance of several microns. Thus, there is obtained a layer of the semiconductor which is able to amplify electromagnetic waves by means of the stimulated emission of quanta during the transition of electrons from the conduction band to the valent band[16] (Fig. 7b).

Many methods for the production of *p-n* junctions were worked out during research on semiconductors. At the present time two methods of making *p-n* junctions are used for the creation of lasers: the diffusion method[17,18] and the method of dopping with different impurities during the process of growing a crystal[19].

III. Semiconductor Lasers

In order to carry out generation on the basis of systems with negative temperature, one must introduce feedback coupling into the system. This feedback coupling is carried out with the aid of cavities. The simpliest type of cavity in the optical range is a cavity with plane-parallel mirrors[20,21]. Light quanta reflecting from the mirrors will pass many times through the amplifying medium. If a light quantum, before its absorption by the mirrors or inside the sample, has time to cause stimulated emission of more than one quantum (that is, if the condition of self-excitation is fulfilled in the system), that system will operate as a laser (Fig. 8). If one maintains a certain negative

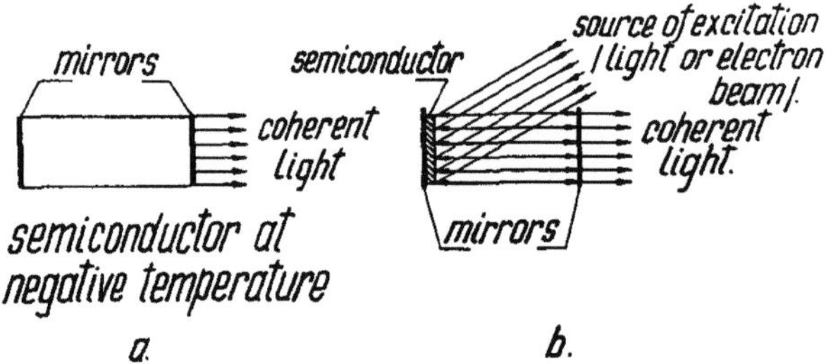

Fig. 8. Diagram of semiconductor lasers. (a) Usual; (b) with radiative mirrors.

temperature in the sample with the help of an external energy source, the number of quanta in the cavity will increase until the quantity of electrons excited per time unit becomes equal to the number of emitted quanta.

It should be especially noted, that when a quantum system with feed-back coupling operates as a laser, its emission has a very narrow frequency band. This characteristic makes laser emission different from all other light sources:

filament lamps, luminescent lamps and light sources with very narrow atomic and molecular spectral lines.

The monochromatic emission of lasers is a result of the properties of stimulated radiation: the quantum frequency of the stimulated radiation equals the frequency of the quantum which produced the radiation. The initial line width in semiconductors is usually about several hundred ångstroms. At the present time it has been shown that the line width in lasers which use a *p-n* junction in GaAs is less than fifty megacycles[22,24]. The minimum value of the line width in lasers is connected with the phenomenon of spontaneous emission.

Spatial directivity of the emission arises together with change in the spectral composition of the oscillation regime. It is connected also with the nature of stimulated radiation: during stimulated radiation, a light quantum has the same direction of propagation as the quantum which produced it.

Usually in semiconductor lasers, the sample itself serves as a cavity; since semiconductor crystals have a large dielectric constant, and, since the polished boundary of the division between the air and the dielectric is able of reflecting about 30% of the radiation.

The first semiconductor lasers were made utilizing *p-n* junctions in crystals of GaAs[17,18]. Some time later, lasers were made under excitation by an electron beam[15], and, recently, under excitation by a light beam[23]. In Table I different semiconducting materials are shown with which lasers have been made, and the methods of excitation are given.

With the help of semiconductors it has already become possible to cover a large frequency range from 0.5μ to 8.5μ. In a number of cases it is possible to continuously overlap a very large frequency range, since the variation of the concentration of components in three-component semiconductors units results in changes in the distances between the bands, that is, allows one to continuously change the emission frequency. For instance, variation of composition in the system In As-In P results[25] in frequency changes from 0.9μ to 3.2μ.

At present, the highest degree of development has been obtained with lasers utilizing *p-n* junctions in GaAs. Impulse and continuous regimes were obtained with an average power of several Watts, and peak power of up to 100 Watts, with an efficiency[24] of about 30%.

The most interesting characteristic of semiconductor lasers is their high efficiency.

Since a direct transformation of electric current into coherent emission takes place in lasers utilizing *p-n* junctions, their efficiency may approach uni-

Table I
Semiconductor lasers

The semi‑conductor material	The wave range of the radiation (in microns)	The method of the excitation	References
CdS	0.5	high speed electron beam	15
CdTe	0.8	high speed electron beam	30
GaAs	0.85	*p-n* junction	17,18
		high speed electron beam	29
		optical excitation	23
InP	0.9	*p-n* junction	31
GaSb	1.6	*p-n* junction	32
		high speed electron beam	33
InAs	3.2	*p-n* junction	34
		high speed electron beam	35
InSb	5.3	*p-n* junction	36
		high speed electron beam	6
PbTe	6.5	*p-n* junction	24
PbSe	8.5	*p-n* junction	24
GaAs-GaP	0.65-0.9	*p-n* junction	37
InAs-InP	0.9-3.2	*p-n* junction	25
GaAs-InAs	0.85-3.2	*p-n* junction	38

ty. Even now it has become possible to make diodes with an efficiency[26] of 70-80%.

Lasers with monochromatic optical pumping should also have a very high efficiency, since the pumping frequency may be made close to the emission frequency[11].

The efficiency of lasers with electron excitation cannot be higher[12] than about 30%, since two thirds of the energy is spent on heating of the lattice during the production of electron-hole pairs. However, such lasers may be rather powerful. This type of excitation will evidently make it possible to create sources of coherent emission working in the far ultraviolet range.

Another characteristic of semiconductors is a high coefficient of amplification, attaining a value of several thousands of reverse centimeters, which makes possible to construct lasers with dimensions measured in microns, that is, with cavity dimensions close to the length of the emission wave. Such cavities should have a very short setup time, of the order of 10^{-12}-10^{-13} sec, which opens the way for the control of high frequencies by using the oscillations in semiconductor lasers, and for the creation of superfast-operating circuits on the basis of lasers, such as components for superfast-operating electronic computers. Q-switch lasers giving very short light pulses may be built out of semiconducting materials.

The small dimensions of semiconductor lasers make it possible to construct quantum amplifiers with an extremely high sensitivity, since sensitivity increases with a decrease in the number of modes of oscillation which may be excited in the cavity. For the first time light amplifiers with an amplification index of about 2000 cm^{-1} have been produced[28].

The high amplification index in semiconductor lasers makes it possible to create for them a new type of cavity - the cavity with emitting mirrors (Fig. 8)[27].

A silver mirror is covered by a thin semiconductor film which is then covered by a transparent film. If one produces in the semiconducting film a state with negative temperature which can compensate for the mirror losses, such a mirror may be used in the construction of a laser. As in the case of a gas laser, one may expect to observe very high monochromaticity and spatial coherence in the emission. A significant advantage of such a system is the simplicity of removing heat from the thin semiconducting film, which indicates that it should be possible to obtain considerable power.

In order to produce negative temperature in a semiconducting film, one may use electronic excitation or optical pumping. The utilization of semiconductor lasers with *p-n* junctions for optical pumping makes it possible to attain high efficiency in the system as a whole.

The question as to the maximum power which may be obtained using semiconductor lasers is not quite clear at present. However, the employment of emitting mirrors of sufficiently large area will make it possible, apparently, to utilize a considerable quantity of semiconducting material. The maximum value of a mirror's cross-section is determined by such factors as the precision of its manufacture and the homogeneity of its semiconducting layer. Various deviations from optical homogeneity will produce the highest modes of oscillations.

Among the disadvantages of semiconductor lasers are their relatively small power, their large spatial divergence and their insufficiently high monochromaticity.

However, in speaking about those disadvantages one should keep in mind that the field of semiconductor quantum electronics is still in its infancy. Furthermore, the means of overcoming these disadvantages are already in sight. It is quite clear in what directions to proceed in order to develop semiconductor quantum electronics, and to increase the sphere of application of semiconductor lasers. All of this gives reason to hope that semiconductor quantum electronics will continue to play a fundamental role in the development of lasers.

1. N.G. Basov, B.M Vul and Yu.M. Popov, *Soviet JETP (U.S.S.R.)*, 37(1959) 585.
2. R.H. Dicke, *Quantum Electronics*, Columbia University Press, New York, 1960, p. 572.
3. N.G. Basov, O.N. Krokhin and Yu.M. Popov, *Usp. Fiz. Nauk*, 72 (1960) 161.
4. N.G. Basov, O.N. Krokhin and Yu.M. Popov, *Soviet JETP (U.S.S.R.), 39 (1960)* 1001.
5. W.P. Dumke, *Phys. Rev.*, 127 (1962) 1559.
6. C. Benoit à la Guillaume and J. M. Debever, *Proc. Symp. Radiative Recombination in Semiconductors*, Paris, 1964, Dunod, Paris, 1965.
7. N.G. Basov and A.M. Prochorov, *Soviet JETP (U.S.S.R.)*, 28 (1955) 249.
8. N. Bloembergen, *Phys. Rev.*, 104 (1956) 324.
9. T.H. Maiman, *Nature*, 187 (1960) 493.
10. O.N. Krokhin and Yu.M. Popov, *Soviet JETP (U.S.S.R.)*, 38 (1960) 1589.
11. N.G. Basov and O.N. Krokhin, *Soviet JETP (U.S.S.R.)*, 46 (1964) 1508.
12. Yu.M. Popov, *Proc. FIAN, 23 (1963) 67.*
13. V.S. Vavilov, *Usp. Fiz. Nauk*, 25 (1961) 263.
14. N.G. Basov, O.N. Krokhin and Yu.M. Popov, *Advan. Quant. Elec.*, (1961) 496.
15. N.G. Basov, O.V. Bogdankevich and A. Devyatkov, *Dokl. Akad. Nauk (S.S.S.R.)*, 155 (1964) 783.
16. N.G. Basov, O.N. Krokhin and Yu.M. Popov, *Soviet JETP (U.S.S.R.), 40 (1961)* 1897.
17. R.N. Hall, G.E. Fenner, J.D. Kingsley, T.J. Soltys and R.O. Carlson, *Phys. Rev. Letters*, 9 (1962) 366.
18. M.I. Nathan, W.P. Dumke, G. Bums, H.F. Dill and G.J. Lasher, *Appl. Phys. Letters*, 1 (1962) 62.
19. M. Bernard, personal communication.
20. A.M. Prochorov, *Soviet JETP (U.R.S.S.)*, 34 (1959) 1658.
21. A.L. Schawlow and C.H. Townes, *Phys. Rev.*, 112 (1958) 1940.
22. J.A. Armstrong and A.W. Smith, *Appl. Phys. Letters*, 4 (1964) 196.
23. N.G. Basov, A.Z. Grasjuk and V.A. Katulin, *Dokl. Akad. Nauk (S.S.S.R.)*, 161 (1965) 1306.
24. *C.* Hilsurn, *Lasers and their Applications*, London, 1964.
25. F.B. Alexander, *Appl. Phys. Letters*, 4 (1964) 13.
26. M.I. Nathan, *Proc. Electron. Elec. Engrs.*, 52 (1964) 770.
27. N.G. Basov and O.V. Bogdankevich, *Proc. Symp. Radiative Recombination in Semiconductors, Paris, 1964*, Dunod, Paris, 1965.

28. J.W. Crowe and R.W. Craig, *Appl. Phys. Letters,* **4** (1964) 57.

29. C.E. Hurwitz and R.J. Keyes, *Appl. Phys. Letters,* **5** (1964) 139.

30. V.S. Vavilov, E.L. Nolle and V.D. Egorov, *FTT (U.S.S.R.), 7 (1965)934.*

31. G. Bums, R.S. Levitt, M.I. Nathan and K. Weiser, *Proc. Electron. Elec. Engrs.,* **51** (1963) 1148.

32. T. Deutsch *et al., Phys. Stat. solids,* 3 (1963) 1001.

33. C. Benoit à la Guillaume and J.M. Debever, *Compt.* Rend., 259 (1964) 2200.

34. I. Melngilis, *Appl. Phys. Letters,* **2** (1963) 176.

35. C. Benoit à la Guillaume and J.M. Debever, Solid *State* Commun., 2 (1964) 145.

36. R.J. Phelan, A.R. Calawa, R.H. Rediker, R. J. Keyes and B. Lax, *Appl. Phys. Letters, 3 (1963) 143.*

37. N. Holonyak Jr. and S.F. Bevacqua, *Appl. Phys. Letters,* 1 (1962) 82.

38. T.M. Quist, R.H. Rediker, R.J. Keyes and W.E. Prag, *Bull. Am. Phys. Soc.,* (1963) 88.

PASSION FOR PRECISION

Nobel Lecture, December 8, 2005

by

THEODOR W. HÄNSCH

Max-Planck-Institut für Quantenoptik, Garching, and
Department of Physics, Ludwig-Maximilians-Universität München, Germany.

ABSTRACT

Optical frequency combs from mode-locked femtosecond lasers have revolutionized the art of counting the frequency of light. They can link optical and microwave frequencies in a single step, and they provide the long missing clockwork for optical atomic clocks. By extending the limits of time and frequency metrology, they enable new tests of fundamental physics laws. Precise comparisons of optical resonance frequencies of atomic hydrogen and other atoms with the microwave frequency of a cesium atomic clock are establishing sensitive limits for possible slow variations of fundamental constants. Optical high harmonic generation is extending frequency comb techniques into the extreme ultraviolet, opening a new spectral territory to precision laser spectroscopy. Frequency comb techniques are also providing a key to attosecond science by offering control of the electric field of ultrafast laser pulses. In our laboratories at Stanford and Garching, the development of new instruments and techniques for precision laser spectroscopy has long been motivated by the goal of ever higher resolution and measurement accuracy in optical spectroscopy of the simple hydrogen atom which permits unique confrontations between experiment and fundamental theory. This lecture recounts these adventures and the evolution of laser frequency comb techniques from my personal perspective.

INTRODUCTION

In our highly complex and ever changing world it is reassuring to know that certain physical quantities can be measured and predicted with very high precision. Precision measurements have always appealed to me as one of the most beautiful aspects of physics. With better measuring tools, one can look where no one has looked before. More than once, seemingly minute differences between measurement and theory have led to major advances in fundamental knowledge. The birth of modern science itself is intimately linked to the art of accurate measurements.

Since Galileo Galilei and Christiaan Huygens invented the pendulum clock, time and frequency have been the quantities that we can measure with the highest precision. Today, it is often a good strategy to transform other quantities such as length or voltage into a frequency in order to make accurate measurement. This is what my friend and mentor Arthur Schawlow at Stanford University had in mind when he advised his students: "Never measure anything but frequency!" Measuring a frequency, i.e. counting the number of cycles during a given time interval, is intrinsically a digital procedure that is immune to many sources of noise. Electronic counters that work up to microwave frequencies have long been available. In 1967, the Conference Generale des Poids et Mesures (CGPM) has defined the second, our unit of time, as the period during which a cesium-133 atom oscillates 9 192 631 770 times on a hyperfine clock transition in the atomic ground state. Today, after 50 years of continuous refinement, microwave cesium atomic clocks reach a precision of 15 decimal digits [1].

Even much higher precision is expected from future optical atomic clocks which use atoms or ions oscillating at the frequency of light as the "pendulum". By slicing time into a hundred thousand times finer intervals, such clocks will greatly extend the frontiers of time and frequency metrology. The long missing clockwork mechanism can now be realized with a femtosecond laser frequency comb, an ultra-precise measuring tool that can link and compare optical frequencies and microwave frequencies phase coherently in a single step. Laser frequency combs provide powerful tools for new tests of fundamental physics laws. Precise comparisons of optical resonance frequencies of atomic hydrogen and other atoms with the microwave frequency of a cesium atomic clock are already establishing sensitive limits for possible slow variations of fundamental constants. Optical high harmonic generation is extending frequency comb techniques into the extreme ultraviolet, opening a new spectral territory to precision laser spectroscopy. Frequency comb techniques are also providing a key to attosecond science by offering control of the electric field of ultrafast laser pulses.

Femtosecond laser frequency combs have been highlighted in the citation for the 2005 Nobel Prize in Physics. Although perfected only about seven years ago, they have already become standard tools for precision spectroscopy and optical frequency metrology in laboratories around the world. Commercial instruments have quickly moved to the market, and extensive review articles and books have been written on frequency comb techniques [2 - 4]. In this lecture I will try to give my personal perspective on the evolution of these intriguing measuring tools for time and frequency. Far from attempting a comprehensive review, I have selected references that helped guide my own insights along a winding path.

THE DAWN OF DOPPLER-FREE LASER SPECTROSCOPY

High resolution laser spectroscopy and precise spectroscopic measurements have appealed to me since I was a graduate student at the University of Heidelberg. For my diploma and thesis research I worked with helium-neon gas lasers in the group of Peter Toschek at the Institute of Applied Physics, headed by Christoph Schmelzer. I was intrigued by the central narrow Lamb dip that Abraham Szöke and Ali Javan had first observed while scanning the frequency of a single-mode gas laser across the Doppler-broadened gain profile [5]. Such a dip had been predicted by Willis Lamb in his semiclassical laser theory [6]. Bill Bennett was the first to give a simple explanation in terms of saturation and spectral hole burning the two counter propagating waves inside the standing wave laser cavity [7]. Other researchers such as John Hall, Veniamin Chebotaev, or Christian Bordé soon explored "inverted Lamb dips" by placing some absorbing molecular gas inside the laser cavity [8]. With resonances of unprecedented spectral resolution, one could almost smell the revolution in laser spectroscopy that would unfold within the next few years. At that time, however, such Doppler free spectroscopy remained limited to the study of gas laser transitions or of a few molecular absorption lines in accidental coincidence. In my own work with Peter Toschek, I studied quantum interference effects in coupled atomic three-level systems [9, 10], demonstrating phenomena which have recently been recognized as important, such as lasing without inversion or electromagnetically induced transparency. They are also essential to understand slow light.

In 1970, I joined Arthur L. Schawlow at Stanford University as a postdoc. Collaborating in separate experiments with Peter Smith, then at Berkeley [11], and with Marc Levenson at Stanford [12], I perfected a new method of Doppler-free saturation spectroscopy that did not require the sample to be placed inside a laser cavity. Soon afterwards, I succeeded in making a nitrogen-laser-pumped widely tunable pulsed dye laser so highly monochromatic that we could apply Doppler-free saturation spectroscopy to arbitrarily chosen atomic resonance lines [13, 14]. Broadly tunable laser action in liquid solutions of organic dyes had been discovered in 1966, independently by Fritz Schäfer [15] and Peter Sorokin [16].

LASER SPECTROSCOPY OF ATOMIC HYDROGEN

Arthur Schawlow at Stanford suggested to apply our technique to the red Balmer-α line of atomic hydrogen that had been at the center of attention of atomic spectroscopists in the 1930s because of suspected discrepancies between the observed line profile and the predictions of Dirac's relativistic quantum theory [17]. In those days, spectroscopists could only observe a blend of unresolved fine structure components because Doppler broadening is particularly large for the light hydrogen atoms. Spectroscopy of the simple hydrogen atom has long played a central role in the history of atomic physics. The visible Balmer spectrum was the Rosetta stone that allowed us to decipher the laws of quantum physics. It has inspired the path breaking discoveries of Niels Bohr, Arnold Sommerfeld, Louis De Broglie, Erwin Schrödinger, Paul Dirac, and even Willis Lamb at the origin of modern quantum electrodynamics.

In 1972, graduate student Issa Shahin and myself were proud to present to Arthur Schawlow a Doppler-free saturation spectrum of the red hydrogen

Balmer-α line, recorded with our pulsed tunable dye laser [18]. The 2S Lamb shift, i.e. the splitting between the $2S_{1/2}$ and $2P_{1/2}$ states that should be degenerate according to the Dirac theory, appeared clearly resolved in the optical spectrum. This was the beginning of a long adventure in precision spectroscopy of the simple hydrogen atom, which permits unique confrontations between experiment and theory. This quest continues today. It has inspired many advances in spectroscopic techniques, including the first proposal for laser cooling of atomic gases [19], and, most recently, the femtosecond laser frequency comb.

Fig. 1 illustrates how the accuracy of optical spectroscopy of atomic hydrogen has advanced over time [20]. Classical spectroscopists remained limited to about six or seven digits of precision by the large Doppler broadening of hydrogen spectral lines. In 1971, our group at Stanford overcame this barrier by nonlinear spectroscopy with a tunable dye laser. Other groups, notably in New Haven, Oxford, and Paris, soon joined in to improve the accuracy by three orders of magnitude over the next two decades. Around 1990, a new barrier appeared: the limits of optical wavelength metrology due to unavoidable geometric wave front errors. Progress beyond a few parts in 10^{10} has been achieved only because we have learned increasingly well how to measure the frequency of light rather than its wavelength. In 2003, the accuracy has reached 1.4 parts in 10^{14} [21]. Further progress is becoming difficult, because we are again approaching a barrier: the limits of how well we know our unit of time, the second. Cesium atomic clocks have been continually refined over the past 50 years [1], as shown by the dashed line in Fig. 1, but the potential for further improvements seems almost exhausted. However, our optical frequency counting techniques make it now feasible to develop optical atomic clocks, based on sharp optical resonances in laser-cooled trapped ions, neutral atoms or molecules. With such clocks future spectroscopic measurements may reach accuracies of parts in 10^{18} and beyond.

In atomic hydrogen, the highest resolution can be achieved on the ultraviolet 1S-2S two-photon resonance with a natural line width of only 1 Hz. First order Doppler shifts cancel if this transition is excited with two counterpropagating laser waves, as was first pointed out by Veniamin Chebotaev [22]. The first Doppler-free spectra have been recorded in our laboratory at Stanford in 1975 [23]. At Garching, we observe this resonance by collinear excitation of a cold hydrogen atomic beam [21]. Starting in 1986, many generations of graduate students and postdocs have made important contributions to advance the state of the art.

Today, the hydrogen atoms are produced by microwave dissociation of molecules and cooled to a temperature of about 6 K by collisions with the walls of a nozzle mounted to a helium cryostat. A collinear standing wave field at 243 nm for Doppler-free two-photon excitation is produced by coupling the frequency-doubled output of a dye laser into a buildup cavity inside the vacuum chamber. Atoms excited to the 2S metastable state after traveling along a path of about 10 cm are detected by applying a quenching electric field and counting the emitted vacuum ultraviolet Lyman-α photons. The

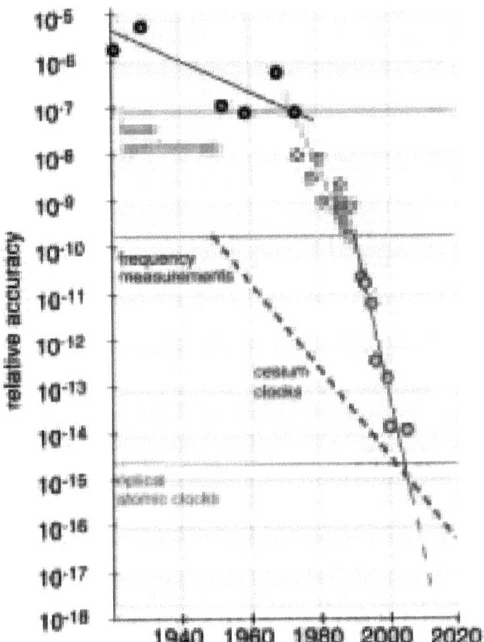

Figure 1. The relative accuracy in optical spectroscopy of atomic hydrogen is charted over eight decades. Major barriers have been overcome in the early 1970s with the advent of Doppler-free laser spectroscopy and in the early 1990s with the introduction of optical frequency measurements. The accuracy of such measurement will soon be limited by the performance of cesium atomic clocks. Dramatic future advances are expected from the development optical atomic clocks.

laser light is periodically blocked by a chopper, and the photon counts are sorted into bins corresponding to different delay times. With slow atoms selected by a delay time of 1.3 ms, the line width is now reduced to about 530 Hz at 243 nm corresponding to a resolution of 4.3 parts in 10^{13}. We would have to reach an accuracy of 5 parts in 10^{15} in order to measure the line position to 1% of this width.

MEASURING OPTICAL FREQUENCIES

The observation of sharp optical resonances by nonlinear laser spectroscopy with a resolution much beyond the measurement limits of wavelength interferometry had long created a strong need for methods to measure the frequency rather than the wavelength of light. The quest for an optical frequency counter is almost as old as the laser itself. Ali Javan, the co-inventor of the helium-neon laser, was the first to superimpose the beams from two different lasers with a beam splitter on a photo detector to observe a beat note, similar to the interference of the sound waves from two tuning forks [24]. This was an extraordinary result, because it proved that laser light can behave like a classical radio wave. A coherent laser wave can have a well defined phase and amplitude, so that it must be possible to count the ripples of such a light wave. However, at a frequency near 500 000 billion oscillations per second, there are no electronic detectors and circuits fast enough to build an optical frequency counter.

At MIT in the early 1960s, Ali Javan started a research project, aimed at extending microwave frequency counting techniques into the optical spectral region. He experimented with whisker-like metal-insulator-metal point contacts as antennas, detectors and mixers for infrared laser waves. Such elements were later used by John Hall and Ken Evenson at NBS (now NIST) in Boulder to realize the first harmonic laser frequency chain, that was used to determine the speed of light by measuring both the wavelength and the frequency of a methane-stabilized 3.39 μm helium-neon gas lasers [25]. Harmonic laser frequency chains were highly complex systems, engineered to measure just one particular optical frequency, and only a handful of these chains have ever been constructed at a number of well-equipped national metrology laboratories. In the early 1980s, a chain at NBS in Boulder had been perfected so that it could measure the frequencies of some iodine-stabilized visible helium-neon lasers to 10 decimal digits. This demonstration led the Conference Generale des Poids et Mesures in 1983, to redefine the meter by defining the speed of light in vacuum c as exactly 299 792 458 meters per second. One meter is then the distance traveled by light during the time of 1/299 792 458 seconds. From now on, one could determine the precise wavelength of a laser in vacuum, λ, by simply measuring the frequency f, since f · λ = c.

Unfortunately, the complex NBS frequency chain had to be abandoned soon after this definition was in the books, and for the next decade there was not a single laboratory in the U.S. that could have followed this prescription. A number of European laboratories did better, notably the Observatoire in Paris (now BNM SYRTE) and the Physikalisch-Technische Bundesanstalt (PTB) in Braunschweig. In an article published in early 1996 [26], a team from the PTB laid claim to the first phase coherent frequency measurement of visible radiation. An elaborate frequency chain filling three large laboratories spread over two separate buildings was assembled to compare the frequency of the red intercombination line of atomic calcium with the microwave frequency of a cesium atomic clock. To reach sufficient phase stability, the clock frequency was first reduced to the 100 MHz of a stable quartz oscillator. From here, the chain traversed the entire electromagnetic spectrum in discrete steps, always generating some harmonic frequency in a suitable nonlinear element and producing enough power for the next step with a phase-locked transfer oscillator. A tricky puzzle had to be solved to reach the desired final frequency with the help of several auxiliary oscillators.

It was obvious that we could not afford to assemble such a harmonic laser frequency chain for our hydrogen experiments at Garching. As a simpler alternative, I proposed a frequency interval divider chain in 1988, that worked with frequency differences rather than the frequencies themselves so that one could stay in a convenient region of the electromagnetic spectrum such as the near infrared where compact diode laser sources are available [27]. The basic building block is an interval divider stage with a laser that is servo-controlled to oscillate at the precise mid-point of two input frequencies. To this end, the second harmonic frequency of the central laser is compared to the sum of the two input frequencies, as generated in a nonlinear optical crystal. With a cascaded chain of n such interval dividers, a large frequency interval can be divided by 2^n. To measure an absolute laser frequency f, one could start with the interval between f and its second harmonic 2f, which is just equal to the frequency f. After repeatedly cutting this interval in half with perhaps 15 stages, the remaining frequency gap is small enough that it can be observed as a beat note with a fast photo-detector and measured with a microwave frequency counter. With Harald Telle, who had joined us from the PTB, and Dieter Meschede, we demonstrated the first working interval divider in 1990 [28].

We never assembled a complete optical frequency counter, but we constructed a chain of four interval dividers to measure a frequency interval of 1 THz that we encountered when comparing our hydrogen 1S-2S frequency with the infrared frequency of a methane-stabilized infrared helium-neon laser at 3.39 μm as the starting point of our own short harmonic laser frequency chain [29]. This intermediate frequency reference had to be repeatedly shuttled to Braunschweig to be calibrated with the PTB frequency chain against a cesium clock. In 1997, we established a new record in optical frequency metrology [29] by determining the ultraviolet 1S-2S frequency to within 3.7 parts in 10^{13}. From this and other spectroscopic measurement in hydrogen, we were able to derive a new value of the Rydberg constant, the scaling factor for any spectroscopic transition, and the most precisely known of the fundamental constants. We could also derive the Lamb shift of the 1S ground state accurately enough to provide a stringent new test of bound-state QED. Assuming that QED is correct we could, moreover, determine new values for the rms charge radius of the proton and the structure radius of the deuteron [29, 30]. We were rather proud that the accuracy achieved with our table-top experiments exceeded that of electron scattering experiments with large accelerators by an order of magnitude.

Soon, a number of metrology laboratories set out to build optical frequency counters based on optical interval division. At Garching, we were also experimenting with electro-optic frequency comb generators kindly provided by Motonobu Kourogi, that could produce an evenly spaced comb of modulation sidebands extending over several THz [31]. A frequency counter could have been realized with only six or seven interval divider stages, if the final frequency gap was bridged with such an electro-optic comb generator. During an extended visit to Garching, Motonobu Kourogi showed us how to observe even feeble comb lines by heterodyne detection, improving the signal to noise ratio with optical balanced receivers and variable beam splitters. We soon verified the accuracy of his frequency comb generator and our frequency interval divider chain in a direct comparison [32].

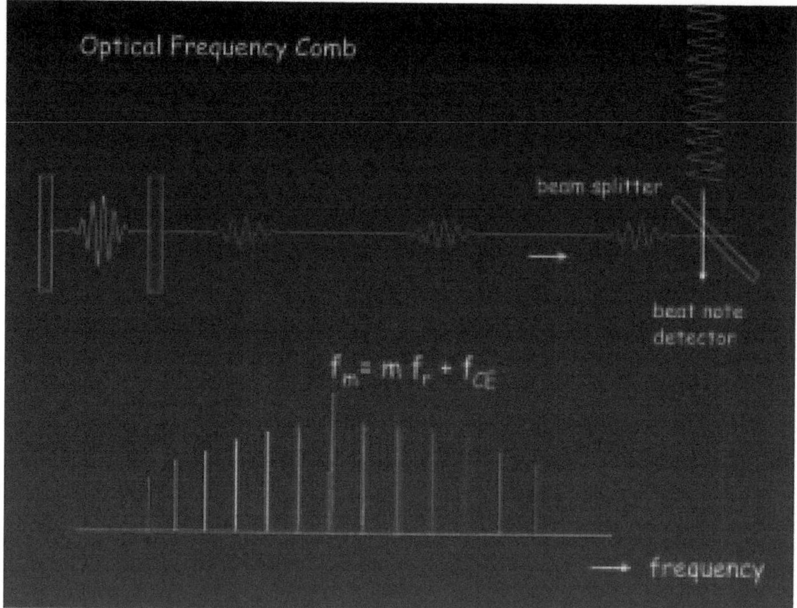

Figure 2. Scheme of a femtosecond laser frequency comb synthesizer.

Many other alternatives have been explored in the long quest for precise optical frequency measurements, including interferometry with modulated laser waves [33, 34] or frequency division with phase-locked optical parametric oscillators [35]. David Wineland has proposed to synchronize the cyclotron motion of a single electron to a laser wave [36]. In the meantime, all these approaches have become obsolete. Since 1998, optical frequency measurements have been enormously simplified with the advent of femtosecond laser optical frequency comb synthesizers [2,3].

FEMTOSECOND LASER OPTICAL FREQUENCY COMBS

The scheme of a frequency comb synthesizer is rather simple, as illustrated in Fig. 2. At the heart is a single mode-locked femtosecond laser which maintains a soliton-like short pulse circulating inside the optical cavity. This laser can be compared to Einstein's Gedanken light clock. With each roundtrip, an attenuated copy of the light pulse escapes so that the laser emits a regular train of ultrashort pulses. To measure the unknown frequency of a laser wave, the beam and the pulse train are superimposed with a beam splitter, and a photo detector registers an interference signal. In the idealized case of a perfectly periodic pulse train, we would expect a low frequency beat note whenever the laser frequency comes close to a value where an integer number of oscillations fits in the time interval between two pulses. To give an example, if we know that the laser emits precisely one billion pulses per second and if we can be sure that the laser wave oscillates precisely 500 000 times during the pulse repetition period, then we know that the optical frequency must be 500 000 billion cycles per second.

In the frequency domain, we can argue that the coupled longitudinal modes of the pulsed laser form an evenly spaced comb of spectral lines. A low frequency beat note is expected whenever the unknown laser frequency approaches one of these comb lines. The origin of the comb spectrum is well explained by Antony E. Siegman in his classic textbook [37]. Consider an arbitrary optical waveform circulating inside an optical cavity. During each roundtrip, an attenuated copy escapes through a partly transmitting mirror. A single copy will have a broad and more or less complicated spectrum. However, two identical copies end-to-end will produce interference fringes in the spectrum, somewhat reminiscent of Young's double slit experiment. Three copies produce a spectrum that resembles the interference pattern of a triple-slit, and an infinite series of copies produces sharp lines which can be identified with the modes of the cavity. Mathematically, an ideal periodic pulse train can be described in terms of a Fourier series, and the comb lines correspond to the elements of this series.

The separation between two modes or comb lines is just equal to the repetition frequency f_r. This remains true even if the pulses are not identical replicas but if we allow for a (reproducible) slip of the phase of the electromagnetic "carrier"-wave relative to the pulse envelope from pulse to pulse [38, 2, 3]. Such phase slips are unavoidable in a real laser because of dispersion in the cavity. The entire comb will then be shifted relative to the integer harmonics of the repetition frequency f_r by a carrier-envelope offset frequency f_{CE}, that equals the net phase slip modulo 2π per pulse interval. The frequency of a comb line with integer mode number m is then given by

$$f_m = m\, f_r + f_{CE}.$$

Such a comb acts like a ruler in frequency space that can be used to measure a large separation between two different optical frequencies in terms of the pulse repetition rate f_r. If these two frequencies are known multiples or fractions of the same laser frequency f, such a measurement reveals the optical frequency f itself. With a known repetition frequency f_r, the beat signal between a known optical frequency f and the nearest comb line reveals the previously unknown offset frequency f_{CE}. The frequency of any comb line can be calculated from the two radio-frequencies f_r and f_{CE} together with the integer mode number.

It has been surprising to most experts how far this frequency comb approach can be pushed. The frequency spectrum of a femtosecond laser oscillator can be broadened in a nonlinear optical medium to span more than an optical octave without destroying the integrity of the comb lines. In a now common implementation, the pulse train from a Kerr-lens mode-locked Ti:sapphire laser is sent through a micro structured silica fiber, with a small solid fiber core surrounded by air-filled holes [39, 40]. The large change in refractive index at the silica-air interface permits guiding by total internal reflection even if the incoming beam is tightly focused to a high intensity. Since part of the light travels as an evanescent wave in air, an additional engineering parameter is available in such a fiber to reduce the spreading of an injected pulse due to group velocity dispersion. Inside the fiber, the pulse spectrum is broadened by self-phase modulation due to the intensity dependent refractive index, soliton splitting, shock wave formation, and other nonlinear optical processes. The emerging white light can be dispersed with a grating to form a rainbow of colors. However, this is not ordinary white light. Remarkably, the processes generating the white light can be so highly reproducible that successive pulses are still correlated in their phases and can interfere in the spectrum to form a comb of several hundred thousand sharp spectral lines.

By now, it has been confirmed in many experiments that the line spacing is very precisely equal to the repetition frequency f_r. With a comb spanning more than an octave, it is particularly simple to measure the carrier envelope offset frequency f_{CE}. One can simply select a few thousand comb lines from the red end of the spectrum and send the light pulses through a frequency doubling crystal, so that new comb lines are generated which are now displaced by twice the offset frequency f_{CE}. A collective beat note with the corresponding original comb lines near the blue end of the spectrum directly reveals the shift f_{CE}. Once this frequency can be measured, it can be controlled, for instance by adjusting the dispersion in the laser cavity or simply by changing the pump power. One can even set f_{CE} to zero so that the frequencies of the comb lines become precise integer multiples of the laser repetition frequency f_r.

So far, we have treated all light waves as classical electromagnetic waves. The quantum optical aspects of frequency combs, i.e. expected correlations in the noise due to photons and their entanglement, have not yet been explored. Such studies may lead to a rich new field of research.

A laser frequency comb provides a direct link between optical frequencies and microwave frequencies. This link can be used in either direction. We can measure or control the repetition frequency f_r with a cesium atomic clock to synthesize several hundred thousand sharp optical reference frequencies which are precisely known in terms of the primary standard of time. Any unknown frequency can then be determined by first making a wavelength measurement with a conventional wave meter that is sufficiently accurate to determine the integer order number m of the nearest comb line. The precise distance from this reference line is then measured by feeding a beat signal to a microwave counter. In the reverse direction, we can start with a sharp optical reference line in some cold trapped ion, cold atoms, or slow molecules, and lock a nearby comb line to this optical reference. All the other comb line frequencies are then rational multiples of the optical reference frequency, and the repetition frequency becomes a precisely known fraction.

Frequency comb synthesizers act as if we had several hundred thousand ultra-stable and precisely tuned lasers operating at once. With the help of nonlinear sum and difference frequency generation, they can precisely measure any frequency from radio waves to the near ultraviolet. They provide the long-missing clockwork for optical atomic clocks. They can even generate microwaves with extreme phase stability [41]. As sources of phase-stabilized femtosecond pulses they have even given us a key to the intriguing field of attosecond science [42]. With their electronic servo controls, frequency-comb synthesizers can be relatively simple, robust, and increasingly user-friendly devices.

THIS IS A SIMPLE IDEA! WHAT TOOK SO LONG?

In hindsight, the ideas behind the optical frequency comb look rather simple and almost obvious. Why did all the experts, including our own laboratory, struggle for so long with much more cumbersome harmonic laser frequency chains?

The main reason may be that nobody believed seriously that such an approach could actually work. There were good arguments why it should be impossible to bridge the gap between radio frequencies and optical frequencies in a single step. The phase noise of even the best quartz oscillator is so large that any comb structure would be completely washed out if one could somehow multiply its frequency up into the visible region. In a harmonic frequency chain, the intermediate transfer oscillators act as phase noise filters and electromagnetic "flywheels" to overcome this "coherence collapse" [43].

Another reason may be that two separate scientific communities have evolved since the early days of laser science. People interested in precise high resolution spectroscopy used their ingenuity to perfect the frequency stability of continuous wave lasers. On the other side, there were people who invented clever techniques to produce ever shorter pulses with mode-locked lasers. They applied their spectrally broad light flashes to the study of ultrafast phenomena in semiconductors, liquids, or in chemical reaction dynamics or to generate ever higher peak intensities for experiments in plasma physics. These two communities went to their own separate conferences, and neither side felt a strong need to keep track of the other frontier.

For our own work, I cannot hide behind this latter excuse. I knew since the early experiments with multi-mode helium-neon lasers [44] that the longitudinal modes of a laser are well defined and their phases can be coupled so as to produce a short light pulse circulating inside the cavity [45, 46]. Much shorter pulses were produced some years later with broadband dye lasers by locking their axial modes with the help of a saturable absorber or by synchronous pumping with a modulated argon laser [47]. At Stanford in the mid-seventies, I became intrigued by the idea of high resolution spectroscopy of atomic resonance lines by Ramsey-like excitation with a coherent train of multiple light pulses [48]. Resonant excitation with separated light pulses has also been explored at the time by Michael Salour at MIT [49] and by Veniamin Chebotaev at Novosibirsk [50]. After initial encouraging experiments with a dye laser pulse injected into a passive cavity [48], our group at Stanford

with graduate student Jim Eckstein and visiting Lindemann Fellow Allister
Ferguson demonstrated that a synchronously pumped mode-locked picosecond dye laser could produce a stable phase coherent pulse train which we
used for Doppler-free two-photon excitation of atomic sodium [51]. The
comb lines served as a frequency ruler to measure some atomic fine structure
intervals. To improve the precision, we replaced the original radio frequency
driver for the modulator of our argon pump laser with a high quality frequency synthesizer. Much to our delight, the performance of the dye laser improved so much that we were the first to generate sub-picosecond pulses directly from a synchronously pumped dye laser [52]. At that time we should
have learned an important lesson: what is good for frequency stability is also
good for the generation of ultrashort light pulses! We later explored
Doppler-free polarization spectroscopy with our frequency comb [53], as well
as two-photon spectroscopy with the frequency comb of an FM mode-locked
laser where the phases of the modes adjust so that the intensity remains constant but the frequency is sweeping back and forth periodically [54].

   During the Stanford experiments, we were painfully aware that we could
not know the absolute positions of our comb lines because the dispersion inside the laser resonator would lead to unknown phase slips of the carrier
wave relative to the pulse envelope. Such phase slips shift the entire comb
spectrum by an unknown amount f_{CE}, as worked out in considerable detail in
the 1978 Stanford Ph.D. thesis of Jim Eckstein [38]. With a comb spectrum
spanning only 800 GHz, we had no means to observe and measure the offset
frequency f_{CE}. Therefore, we did not know how to measure absolute optical
frequencies with our laser frequency combs in the late seventies.

   The idea of somehow generating much broader frequency combs surfaced
again in my mind after I had moved back to Germany in 1986. In 1990, I published a proposal for a synthesizer of sub-femtosecond pulses that would
superimpose the frequencies from separate phase-locked cw laser oscillators
to form a very wide comb [55]. In the early 1990s, the technology of ultrafast
lasers advanced dramatically with the discovery of Kerr-lens mode locking by
Wilson Sibbett at the University of St. Andrews [56]. Soon, commercial
Ti:sapphire femtosecond lasers became available, that made the generation
of ultrashort light pulses much easier. Intrigued by these new sources I discussed with Peter Lambropoulos at Garching the possibility of finding some
highly nonlinear effect, such as above-threshold-ionization (ATI) , that would
depend on the phase of the electric field relative to the pulse envelope and
could reveal the offset frequency f_{CE} of the laser comb lines. Calculations
soon showed that such effects would be observable only for pulses lasting at
most a few optical cycles [57]. Today, such sources have become available,
and the phase dependence of ATI has since been demonstrated by Gerhard
Paulus and Herbert Walther [58]. In 1994, I also discussed the problem of
the carrier-envelope phase slips with Ferenc Krausz at the Technical
University of Vienna. His group was the first to observe such pulse-to-pulse
phase slips in an interferometric correlation experiment in 1996 [59].

   I remember a trade show in 1994, when I was captivated by an exhibit of a

(Coherent Mira) mode-locked Ti:sapphire femtosecond laser with regenerative amplifier. The laser beam was focused into a glass slide to produce a white light continuum which a prism dispersed into a rainbow of colors. Such white light pulses are produced by a combination of self-focusing, self phase modulation, and other nonlinear processes, and they have been used for a long time in ultrafast pump-probe experiments [60]. A striking feature was the laser-like speckle pattern in the rainbow colors which indicated a high degree of spatial coherence. It occurred to me that such a system might produce an octave-spanning frequency comb if the phases of successive pulses were sufficiently correlated. Such a wide comb could then be used as a ruler to measure the large interval between a laser frequency and its second harmonic, which must be equal to the laser frequency itself. Even though the pulse repetition frequency of a few hundred kHz remained inconveniently low for frequency comb experiments, I felt sufficiently intrigued to acquire such a system for our frequency metrology laboratory at Garching in 1994. In the back of my mind I had the hope that it might somehow be possible to produce white light directly with the pulses from the laser oscillator, without the regenerative amplifier with its much lower pulse rate, by sending the pulse train into a small waveguide made from some highly nonlinear optical material, so that it would not be necessary to reach the threshold power for self-focusing.

We did not pursue the femtosecond laser approach seriously right away, because we had come quite far in perfecting our alternative scheme of optical interval division. An accurate measurement of the 1S-2S frequency seemed almost within reach. We also felt that we would need an independent tool to verify any measurement with a femtosecond laser frequency comb, since the frequency metrology community would otherwise distrust our results. The hydrogen measurements were finally completed in 1997 [29, 30].

In February 1997, I visited the European Laboratory for Nonlinear Spectroscopy, LENS, in Florence, Italy. There, Marco Bellini was working with an amplified Ti:sapphire femtosecond laser producing pulses of 1 mJ energy at a rate of 1 kHz. As is common in many ultrafast laboratories, he produced a white light continuum for pump-probe experiments by focusing part of the laser beam into a thin plate of CaF_2. I asked what would happen if we split the laser beam in two parts and focus these beams at two spatially separate spots. Would the two white light pulses interfere?

In an earlier joint experiment at the Lund Laser Center, we had investigated the same question for the generation of high harmonic radiation in a gas jet [61]. From that time, Marco Bellini still had a Michelson interferometer on his shelve that we could quickly place into the laser beam, somewhat misaligned so that two beams would escape in two slightly different directions. By adjusting the length of one arm we could make sure that the two focused pulses arrived on the CaF_2 plate at precisely the same time. I felt electrified when we observed stable interference fringes of high contrast for all the colors that I could record with my handheld camcorder electronic notebook, as shown in Fig. 3 [62]. The white light pulses had to be phase-locked to the

Figure 3. Interference fringes between two white light pulses [Ref. 62].

driving laser field! No matter how complicated the process of white light continuum generation might be, the process was reproducible. If such pulses were separated in time rather than in space, they would interfere in the spectrum to produce a very broad frequency comb.

By March 30, 1997, I had written a confidential 6-page proposal for a universal optical frequency comb synthesizer *"...which produces a wide comb of absolutely known equidistant marker frequencies throughout the infrared, visible, and ultraviolet spectral range. To this end, a white light continuum with a pulse repetition rate* f_r *is produced by focusing the output of a mode-locked femtosecond laser into an optical fiber or bulk medium with a third order nonlinear susceptibility. The rate of phase slippage of the laser carrier relative to the pulse envelope,* f_{CE}, *is monitored by observing a beat signal between the white light continuum and the second harmonic of the laser."* The envisioned self-referencing scheme could find the carrier-envelope offset frequency f_{CE} without any auxiliary laser. I asked Thomas Udem and Martin Weitz in our laboratory to witness and sign every page on April 4, 1997, since this might become important for later patent applications.

At Garching, we soon started a serious experimental effort towards optical frequency measurements with femtosecond laser frequency combs. Even if we did not yet know how to broaden the spectrum of our Mira laser oscillator to more than an optical octave, we could always follow the 1988 proposal [27] and measure the frequency of the dye laser in our hydrogen spectrometer by implementing a short chain of two or three interval divider stages, using small semiconductor lasers, to arrive at a frequency gap that could be bridged with our laser comb. Propelled by such visions, Thomas Udem and Jörg Reichert investigated the frequency comb spectrum of the Mira femtosecond laser. They were later joined by Ronald Holzwarth. By that time, hundreds of such lasers were in use in laboratories around the world, but they were mostly used to study ultrafast phenomena. Nobody had ever looked for any comb lines, as far as we could tell. With a repetition frequency of 76.5 MHz, the comb spectrum of our femtosecond laser was so densely spaced that no spectrometer in our laboratory could resolve the comb lines. Therefore, we resorted to heterodyne detection, employing a cw diode laser as a local oscillator. The diode laser beam and the pulse train were superimposed with a beam splitter, and a beat signal was detected with an avalanche photodiode after some spectral filtering. After paying attention to the mechanical stability of the femtosecond laser, we could observe stable comb lines. Next, we investigated the spacing of these lines. We phase-locked two diode lasers to two arbitrarily chosen comb lines and used an optical interval divider stage to produce a new frequency precisely at the center. A beat note with the nearest comb line confirmed, much to our delight, that the comb lines were perfectly evenly spaced, way out into the wings of the emission spectrum, within a few parts in 10^{17} [63].

It was now certain that the frequency comb of such a mode-locked femtosecond laser did not suffer from "coherence collapse" and could serve as a ruler in frequency space to measure large optical frequency intervals. In a first demonstration of an optical frequency measurement with a femtosecond laser comb, we determined the frequency interval between the cesium D1 resonance line and the fourth harmonic of a transportable CH_4-stabilized 3.39 µm He-Ne-laser, which had been calibrated with a harmonic laser frequency chain at the PTB Braunschweig [64]. The optical cesium frequency was needed for a determination of the fine structure constant α from the atomic recoil energy as measured by atom interferometry in the group of Steve Chu at Stanford. Soon these experiments found a considerable resonance in newspapers and journals. They demonstrated to the optical frequency metrology community that femtosecond laser frequency combs could be powerful tools to measure the frequency of light.

We next turned to the more ambitious goal of measuring the absolute optical frequency of our 486 nm dye laser in the hydrogen 1S-2S spectrometer. At that time, we had broadened the frequency comb of our mode-locked laser oscillator by self-phase-modulation in a short length of ordinary optical fiber to span 60 or 70 THz. Rather than building a few new interval divider stages, we recognized a more expedient approach for a first proof-of-principle experiment. Our hydrogen spectrometer [29] required only minor modifications to produce two different fractional subharmonics, 4/7 and 1/2, of the dye laser frequency, which we could bridge with our femtosecond laser comb [65, 66]. The CH_4-stabilized helium neon laser served now as part of an interval divider and no longer as an intermediate reference standard. The primary reference for our first absolute frequency measurements was a commercial HP cesium atomic beam clock which we used to determine the pulse repetition rate f_r and the carrier-envelope offset frequency f_{CE} [65]. Once the optical frequency gap was measured, we knew the absolute frequency of the dye laser as well as the absolute frequencies of all the comb lines. To control the position of the comb lines, we learned how to change the frequency f_{CE} of our Mira Ti:sapphire laser by tilting an end mirror where the spectrum is slightly dispersed by a prism pair inside the cavity. In this way, we were the first to generate femtosecond laser pulses with controlled slips of the carrier-envelope phase.

In October 1998 we proudly showed our experiment to Norman Ramsey, who had come to the Max-Planck-Institute as a member of the Scientific Advisory Board. For the first time, we could compare the hydrogen 1S-2S frequency directly with a cesium atomic clock in our own laboratory, without involving any large harmonic laser frequency chain. Later in the same year, we demonstrated our experiment to John Hall who had come to Munich to attend a meeting commemorating our mutual friend Veniamin Chebotaev. John soon became an ardent evangelist for "this goofy technique, that makes everything obsolete that we have worked on for so long." He started to assemble a powerful professional team in Boulder to advance research on femtosecond laser frequency combs, and he persuaded his colleague Steve Cundiff at JILA, an expert on femtosecond lasers from Bell Laboratories, to visit our laboratory in the spring of 1999. An increasingly heated competition did much to accelerate the development of the new tools in the coming months and to ignite a firework of novel applications in the years to follow [3].

Until the summer of 1999, we had felt as if the new femto-comb playing ground belonged just to us. Thomas Udem had reported on our experiments at a conference in Perth, Australia, in late 1998 [43]. But we had delayed publications such as a first article on "measuring the frequency of light with mode-locked lasers" [67] to appear after the filing of our first patent application in March 1999 because, according to German law, an invention can no longer be patented once it has been published.

In June 1999, we could directly compare the hydrogen frequency with a highly accurate transportable cesium fountain clock (PHARAO), built at the LPTF (now BNM SYRTE) in Paris [66]. This measurement yielded a new value of the hydrogen 1S-2S frequency accurate to 1.8 parts in 10^{14}, surpassing all earlier optical frequency measurements by more than an order of magnitude. By now, the compelling advantages of laser frequency combs had been clearly demonstrated. A number of different possible approaches for carrier-envelope offset phase control were soon proposed [68].

As a next step in our own work, we wanted to drastically simplify our setup. Around that time, a new tool had appeared on the horizon that made it likely that we would no longer need any optical interval dividers. At the CLEO conference at Baltimore, MD, held in May 1999, researchers from Bell Laboratories had reported on a novel micro-structured "rainbow fiber" that could broaden the spectrum of pulses of a Ti:sapphire femtosecond laser oscillator without further amplification to a rainbow of colors [40]. After the white light interference experiments in Florence [62], I felt rather confident that this magic fiber would preserve the phase coherence of successive pulses and produce comb lines with a desirable large frequency spacing.

In June 1999, John Hall came to Germany to participate in the annual retreat of our research group at the Ringberg Castle near the Tegernsee south of Munich. Together, we phoned many of our old friends at Bell Laboratories, to try and obtain a sample of the magic fiber. We were hoping that we could demonstrate an octave-spanning frequency comb while John Hall was still in Germany. Unfortunately, this plan was foiled by the lawyers at Lucent Technologies who did not allow the fiber to leave Bell Laboratories. Ronald Holzwarth traveled to Bell Laboratories at Holmdel, NJ, during his 1999 summer vacation, but he had to leave without a piece of the fiber. John Hall's team in Boulder experienced similar difficulties at first, but in October 1999, they could demonstrate the first octave-spanning self-referencing laser frequency comb after finally securing some of the holey fiber [69, 70]. At Garching we realized a similar comb system a few weeks later [71], after we had received some "photonic crystal fiber" from the group of Philip Russell at the University of Bath in the UK. We had found out too late that these British researchers had actually pioneered micro-structured silica fibers some years earlier [39]. Both laboratories submitted their first short publications on octave-spanning frequency combs on the same day (Nov. 12, 1999) to the CLEO/QELS 2000 conference in San Francisco.

Similar to the Boulder experiments, we used a commercial small Ti:sapphire ring laser for our first octave-spanning frequency comb, producing pulses of about 25 fs duration at a repetition frequency of 625 MHz. Launching about 170 mW into a 30 cm length of photonic crystal fiber, we immediately produced a frequency comb spanning more than an octave. The spectrum showed a complicated structure, with valleys and peaks, but it offered useable comb lines everywhere. Together with a nonlinear interferometer for control of the offset frequency f_{CE}, the entire optical setup did easily fit on a single breadboard. While the traditional harmonic frequency chains with their factory halls full of lasers could measure just one single optical frequency, our new system was ready to measure any frequency throughout the visible and near infrared.

Since then, Ti:sapphire femtosecond lasers have been developed that produce an octave-spanning spectrum directly from the oscillator, without any need for external spectral broadening [72]. Octave spanning combs can also be generated with erbium-doped fiber lasers [73], pumped by very reliable and robust laser diodes developed for telecommunications. As turn-key instruments, such fiber comb generators can run for months without human attention.

In a first stringent test, Ronald Holzwarth has compared an octave spanning frequency comb synthesizer with the more complex frequency synthesizer used in the 1999 hydrogen frequency measurement [71]. By starting with a common 10 MHz radiofrequency reference and comparing comb lines near 350 THz, he could verify agreement within a few parts in 10^{16}, limited by Doppler shifts due to air pressure changes or thermal expansion of the optical tables. In 2002, a group at the PTB in Braunschweig demonstrated how a femtosecond laser frequency comb generator can be used as a transfer oscillator to precisely measure optical frequency ratios [74]. As a test case, they measured the frequency ratio between the second harmonic of a Nd:YAG laser and the fundamental frequency, verifying the expected value of 2 with an uncertainty of 7 parts in 10^{19}. More recently, Marcus Zimmermann in our laboratory has pushed a related experiment to an uncertainty of 6 parts in 10^{21} [75]. In 2004, researchers in Boulder compared four different frequency combs from different laboratories, finding agreement between neighboring comb lines at an uncertainty level of 10^{-19} [76]. So far, no systematic error has been identified which would limit the potential accuracy of future precision spectroscopy or optical atomic clocks.

NEW FREQUENCY MEASUREMENT OF HYDROGEN 1S-2S IN 2003: ARE THE FUNDAMENTAL CONSTANTS CONSTANT?

In February 2003, we used an octave spanning comb synthesizer in a new measurement of the hydrogen 1S-2S transition frequency [21]. Marc Fischer and Nikolai Kolachevsky had implemented many improvements in the hydrogen spectrometer. Light from the dye laser was sent through a fiber into the frequency metrology laboratory, where an octave-spanning Ti:sapphire femtosecond laser frequency comb synthesizer was used to compare the optical frequency to the radio frequency of the Paris PHARAO atomic fountain clock, which had again been brought to Garching.

With such immediate absolute frequency calibration, the hydrogen spectroscopy could be performed with much confidence. Compared to the 1999 measurements, the statistical error of the data recorded on a given day was much reduced. Nonetheless, the day-to-day fluctuations of our measurements remained of similar magnitude as before. They indicate some uncontrolled systematic errors. After careful further experiments probing for possible causes of systematic line shifts and after a thorough statistical analysis of all recorded data, we believe that the fluctuations are caused by some residual small first order Doppler shifts. Such shifts are expected if the two counter-propagating wave fronts of the exciting 243 nm radiation do not match perfectly. A mismatch may be caused by imperfect mode-matching of the frequency-doubled dye laser beam that enters the build-up cavity inside the atomic beam apparatus through a 2% input coupling mirror. Another cause may be the accumulation of frozen molecular hydrogen on the walls of the cold copper nozzle for the hydrogen atomic beam, which can grow until it distorts the optical wave fronts by vignetting and diffraction. A second cause of systematic line shifts in two-photon spectroscopy with frequency-doubled laser light may be unwanted correlations between amplitude noise and phase noise that could be caused by imperfect servo-locks of the enhancement cavities. For the future, we are preparing an ultraviolet build-up cavity of higher finesse, and we are working towards an all-solid-state laser source with a line width of only a few Hz.

From the 2003 measurement, we find a frequency of 2 466 061 102 474 851 ± 34 Hz for the F=1 to F'=1 hyperfine component of the hydrogen 1S-2S frequency, with a relative uncertainty of 1.4 parts in 10^{14}. The new results agrees within the error limits with the 1999 measurement of 2 466 061 102 474 880 ± 46 Hz. A difference of 29 ± 57 Hz in 44 months corresponds to a relative drift of the 1S-2S transition frequency of $(3.2 \pm 6.3) \times 10^{-15}$ per year, i.e. it is compatible with zero drift.

This experiment has attracted some attention because it can be considered as a test for a possible slow variation of the electromagnetic fine structure constant α. During the 2003 experiment, theorist Harald Fritzsch called frequently to get some preliminary results, because he had predicted an observable drift of the microwave frequency of the cesium clock relative to the hydrogen frequency [77]. Starting point for his arguments were astronomical observations of spectral lines in the light of distant quasars performed at the Keck Observatory [78]. Differential red shifts seemed to suggest that the electromagnetic fine structure constant α in the early universe was somewhat smaller than today. Making the simplest assumption of a linear drift, the data would indicate a drift of $(6.4 \pm 1.35) \times 10^{-16}$ per year, too small to be observable in our laboratory experiment. However, Fritzsch had argued with ideas from grand unification and quantum chromodynamics that α cannot change simply by itself. If all known forces are to remain unified at very high energies, other coupling constants must change as well. As a result, the masses and magnetic moments of hadrons (in units of the Bohr magneton) should change relative to those of the electron. Fritzsch pointed out a possible magnifying effect, that could change the hyperfine transition frequency of the cesium atomic clock about 20 times faster than the optical hydrogen frequency. So far, we have not found any evidence for such a drift. There are also other more recent observations of quasar spectra that do not support the evidence for a changing fine structure constant [79].

Regardless of such speculations, we have to admit that the hydrogen measurements of 1999 and 2003 do not strictly rule out a changing fine structure constant α. It is conceivable that the magnetic moment of the cesium nucleus also changes at just the right rate to give a null result in our experiment. Fortunately, such measurements are not limited to hydrogen. One can also measure transition frequencies in heavier atoms with stronger relativistic effects that respond in a different way to changes in α. One such candidate is the clock transition in a single cold Hg^+ ion, that has been compared by Jim Bergquist and his team at Boulder with a cesium atomic clock in 2000 and 2002, also using a laser frequency comb [80]. In addition, Eckhard Peik and his team at the PTB have made two separate measurements of the clock transition in a single Yb^+ ion [81]. Together, these laboratory experiments give now some upper limits for the possible rates of change of the fine structure constant α and the cesium nuclear magnetic moment μ_{Cs} on the order of $(-0.3 \pm 2.0) \times 10^{-15}$ per year and $(2.4 \pm 6.8) \times 10^{-15}$ per year, respectively.

Within the next few years we can expect much more stringent experimental limits on possible variations of physical constants from such laboratory experiments. If we found such changes it would not have any consequences for our everyday lives, but it would give reason for fascinating speculations about the nature of the universe.

OPTICAL ATOMIC CLOCKS

Sensitive limits for the variations of fundamental constants will be established in the comparison of different types of optical atomic clocks that are now being developed by strong professional teams in many industrialized countries. The perfection of optical frequency standards has been progressing at a much more rapid pace than that of microwave cesium clocks [82]. With the femtosecond laser comb now available as a perfect clockwork mechanism, the efforts must concentrate on ever more perfect laser frequency stabilization and on the management of systematic line shifts in precision spectroscopy of narrow optical resonances that serve as the "pendulum" of optical clocks. Much progress has already been made in experiments with cold trapped ions, notably Hg^+, Yb^+, In^+, and Sr^+. Cold neutral atoms such as H, Ca, or Sr are also appealing candidates because many atoms can be observed simultaneously without disturbing Coulomb repulsion, improving the signal-to-noise ratio and the speed with which a resonance frequency can be established. A particularly promising approach has been proposed by Hidetoshi Katori [83]. In his neutral atom clock, many cold neutral Sr atoms are captured in the microscopic dipole force potential wells of an optical lattice. Light shifts are minimized by choosing a proper "magic" wavelength of the lattice field. In 2005, the accuracy of the best optical frequency standards has become comparable to that of the best cesium fountain clocks. But even if they are not yet more accurate, optical frequency standards offer one important advantage already. It takes hours or days to compare two cesium clocks to a part in 10^{15}. Two optical frequencies can be compared to this level within just seconds.

It is interesting to look at the historical evolution of the accuracy of clocks. The clocks in medieval church towers were only good to about 20 minutes per day. In the 18th century, the nautical clock H4 of the legendary watchmaker John Harrison reached an accuracy of some 100 msec per day. The best primary cesium fountain clocks of today can be accurate to within 100 psec per day. Some experts hope that optical atomic clocks will reach a hundred or thousand fold higher accuracy within the next decade.

Better atomic clocks will be enabling tools for many scientific and technical applications, so that this pursuit will be worthwhile even if we do not discover any changes of fundamental constants. They can extend the frontiers of precision spectroscopy and of time and frequency metrology. They will make it possible to precisely synchronize clocks over large distances. In astronomy, such synchronized clocks may allow an extension of large baseline interferometry to infrared and optical wavelengths. Better clocks can improve the performance of satellite navigation systems and the tracking of probes in deep space. Accurate clocks are also needed to synchronize optical telecommunication networks. In fundamental physics, more accurate clocks will permit more stringent tests of special and general relativity, as well as other fundamental laws.

TOWARDS FREQUENCY COMBS IN THE EXTREME ULTRAVIOLET

So far, we have not discovered any fundamental limits for the potential accuracy of future clocks. It should even be possible to extend frequency comb techniques into the extreme ultraviolet and soft X-ray spectral regions, so that we would be able to slice time into still much finer intervals. High harmonic radiation that is generated when an intense femtosecond laser pulse is focused into a gas jet can be harnessed to produce coherent pulse trains at such wavelengths.

Since the pioneering work of Charlie Rhodes [84] and Anne L'Huiller [85] in the late 1980s, high harmonic generation has been studied in many laboratories. In a simple model first proposed by Paul Corkum [86], gas atoms are field ionized and the electrons are accelerated by the strong laser field until the light field reverses direction. Dependent on the time of escape, such electrons can return to the ion core with substantial kinetic energy which they can radiate in the form of energetic photons, with one burst emitted during each half cycle of the driving laser wave.

The mutual phase coherence of short pulses was much on my mind when I visited the high harmonic experiments of Anne L'Huiller and Claes Göran Wahlström at the Lund Laser Center in 1995. I wondered if two successive high harmonic pulses would be mutually phase coherent. As a test I proposed to split the driving laser beam into two parts that could be focused into the gas jet at separate spots, and to look for interference fringes in the high harmonic radiation, similar to the later white light experiments at Florence [62]. At first, there seemed to be good reasons why this was not to be expected, because the phase of the harmonic radiation should depend strongly on the varying intensity of the driving laser pulse. But after I left, graduate student Raoul Zerne tried the experiment and observed some fleeting interference fringes. With much excitement, we scheduled some serious joint experiments at Lund. Marco Bellini from LENS agreed to participate and to construct a stable Michelson interferometer so that the timing of the two laser pulses could be finely adjusted. We soon observed clean interference fringes of high contrast up to the 15th harmonic and beyond [61]. In a subsequent experiment we even discovered a regime where the harmonic beam was surrounded by a divergent halo beam of very short coherence length. This behavior could be explained in terms of two different electron trajectories that can contribute to a given harmonic photon energy [87].

These results demonstrate that high harmonic pulses can be mutually phase coherent so that a regular train of such pulses could form a frequency comb in the extreme ultraviolet. However, the necessary peak intensities of the order of 10^{14} W/cm^2 could only be produced with amplified femtosecond laser systems of low repetition frequency. Very recently, Christoph Gohle and Thomas Udem at Garching have succeeded in producing high harmonic radiation down to wavelengths of 60 nm at a repetition frequency of 112 MHz [88]. To this end, they stacked the pulses from a mode-locked Ti:sapphire laser oscillator in a dispersion-compensated passive build-up cavity and placed a xenon gas jet at an intra-cavity focus. The high harmonic radiation is coupled out by external reflection from a thin sapphire Brewster plate, that has a refractive index smaller than 1 in the extreme ultraviolet. Similar experiments have also been reported by Jun Ye in Boulder [89].

In a future ambitious project, we plan to apply frequency combs in the XUV directly to precision spectroscopy of sharp resonances in laser-cooled trapped ions. The hydrogen-like helium ion with a 1S-2S two-photon transition near 60 nm is a particularly interesting candidate. In one envisioned scenario, helium ions will be sympathetically cooled by laser-cooled magnesium ions in the same trap, and the signal might be detected via the production of doubly charged helium ions due to photo ionization.

CONCLUSIONS

Spectroscopy of the simple hydrogen atom has sparked off the cross fertilization of two seemingly unrelated frontiers, precise optical spectroscopy and the study of ultrafast phenomena. Femtosecond frequency combs are revolutionizing precision measurements of time and frequency. Future optical atomic clocks will find important applications in many areas of science and technology. Ultraprecise optical spectroscopy can be harnessed for new tests of fundamental physics laws. However, many other spectroscopic applications of laser frequency combs can be envisioned, such as massively parallel ultra-sensitive cavity ring-down spectroscopy [90] or broadband spectral interferometry. At the same time, frequency comb techniques are also offering powerful new tools for ultrafast physics. By controlling the phase of the electric field of intense light pulses lasting for only a few cycles, they make it possible to study ultrafast electronic processes in light matter interactions, such as the production of single sub-femtosecond pulses of soft X-rays in high harmonic generation [42]. Only the future can show what we will discover with such exquisite new instruments.

ACKNOWLEDGEMENT

More than ninety students, postdocs, visiting scholars, and senior colleagues have made essential contributions to our research on precision laser spectroscopy during more than 4 decades. I owe particular gratitude to Thomas Udem and Ronald Holzwarth who played a key role in making the frequency comb synthesizer become reality. I am also very grateful to John L. Hall, who has long shared his precious insights on how to make lasers unbelievably stable.

REFERENCES

[1]   C. Audoin, and G. Bernard, *The Measurement of Time: Time, Frequency, and the Atomic Clock*, Cambridge: Cambridge University Press, 2001

[2]   Th. Udem, R. Holzwarth, and T.W. Hänsch, Nature, **416,** 233 (2002)

[3]   Jun Ye, and S. Cundiff, eds., *Femtosecond Optical Frequency Comb: Principle, Operation and Applications,* Springer Verlag, New York, 2005

[4]   P. Hannaford, ed., *Femtosecond Laser Spectroscopy,* Springer Verlag, New York, 2005

[5]   A. Szöke, and A. Javan, Phys. Rev. Lett. **10,** 521 (1963)

[6]   W.E. Lamb, Phys. Rev. **134,** 1429 (1964)

[7]   W.R. Bennett, Phys. Rev. **126,** 580 (1962)

[8]   V.S. Letokhov, and V.P. Chebotaev, *Nonlinear Laser Spectroscopy,* Springer Series in Optical Sciences, Vol. 4, Springer Verlag, New York, 1977

[9]   T.W. Hänsch, R. Keil, A. Schabert, Ch. Schmelzer, and P. Toschek, Z. Physik, **226,** 293 (1969)

[10] T.W. Hänsch, and P. Toschek, Z. Physik **236,** 213 (1970)

[11] P.W. Smith, and T.W. Hänsch, Phys. Rev. Lett. **26,** 740 (1971)

[12] T.W. Hänsch, M.D. Levenson, and A.L. Schawlow, Phys. Rev. Lett. **26,** 949 (1971)

[13] T.W. Hänsch, Appl. Opt. **11,** 895 (1972)

[14] T.W. Hänsch, I.S. Shahin, and A.L. Schawlow, Phys. Rev. Lett. **27,** 707 (1971)

[15] P.P. Sorokin, and J.R. Lankard, IBM J. Res. Dev. **10,** 162 (1966)

[16] F.P. Schäfer, W. Schmidt, and J. Volze, Appl. Phys. Lett. **9,** 306 (1966)

[17] G.W. Series, *Spectrum of Atomic Hydrogen,* Oxford University Press, Oxford, 1957

[18] T.W. Hänsch, I.S. Shahin, and A.L. Schawlow, Nature **235,** 63 (1972)

[19] T.W. Hänsch, and A.L. Schawlow, Opt. Comm. **13,** 68 (1975)

[20] S.G. Karshenboim, F.S. Pavone, F. Bassani, M. Inguscio, and T.W. Hänsch, eds., *The Hydrogen Atom, Precision Physics of Simple Atomic Systems,* Springer Verlag, Lecture Notes in Physics, New York, 2001

[21] M. Fischer, N. Kolachevsky, M. Zimmermann, R. Holzwarth, Th. Udem, T.W. Hänsch, M. Abgrall, J. Grünert, I. Maksimovic, S. Bize, H. Marion, F. Pereira Dos Santos, P. Lemonde, G .Santarelli, P. Laurent, A. Clairon, and C. Salomon, M. Haas, U. D. Jentschura, and C. H. Keitel, Phys. Rev. Lett. **92,** 230802 (2004)

[22] E.V. Baklanov, and V.P. Chebotaev, Opt. Comm. **12,** 312 (1974)

[23] T.W. Hänsch, S.A. Lee, R. Wallenstein, and C. Wieman, Phys. Rev. Lett. **34,** 307 (1975)

[24] A. Javan, E.A. Ballik, and W.L. Bond, J. Opt. Soc. Am. 7, 553 (1962)

[25] K. M. Evenson, J. S. Wells, F. R. Petersen, B. L. Danielson, G. W. Day, R. L. Barger, and J. L. Hall, Phys. Rev. Lett. **29,** 1346 (1972)

[26] H. Schnatz, B. Lipphardt, J. Helmcke, F. Riehle, and G. Zinner, Phys. Rev. Lett. **76,** 18, (1996)

[27] T.W. Hänsch, in *The Hydrogen Atom, Proceedings of the Symposium held in Pisa, Italy, June 1988,* G.F. Bassani, M. Inguscio, and T.W. Hänsch, eds., Springer Verlag, New York, 1989, pp. 93

[28] H.R. Telle, D. Meschede, and T.W. Hänsch, Opt. Lett. **15,** 532 (1990)

[29] T. Udem, A. Huber, B. Gross, J. Reichert, M. Prevedelli, M. Weitz, and T.W. Hänsch, Phys. Rev. Lett. **79,** 2646 (1997)

[30] A. Huber, T. Udem, B. Gross, J. Reichert, M. Kourogi, K. Pachucki, M. Weitz, and T.W. Hänsch, Phys. Rev. Lett. **80,** 468 (1998)

[31] M. Kourogi, K. Nakagawa, and M. Ohtsu, IEEE J. Quant. El. 29, 2693 (1993)

[32] T. Udem, J. Reichert, M. Kourogi, and T.W. Hänsch, Optics Letters, **23,** 1387 (1998)

[33] Z. Bay, G.G. Luther, and J.A. White, Phys. Rev. Lett. **29,** 189 (1972)

[34] R. G. DeVoe, C. Fabre, K. Jungmann, J. Hoffnagle, and R. G. Brewer, Phys. Rev. A 37, 1802 (1988)

[35] N.C. Wong, Opt. Lett. **17,** 13 (1992)

[36] D.J. Wineland, J. Appl. Phys. **50,** 2528 (1979)

[37] A.E. Siegmann, *Lasers*, University Science Books, Mill Valley, 1986

[38] J.N. Eckstein, *Ph.D. Thesis*, Stanford University, 1978

[39] T.A. Birks, P.J. Roberts, P. St. J. Russell, D.M. Atkin, and T.J. Shepherd, Electron. Lett. **31**, 1941 (1995)

[40] J.K. Ranka, R.S. Windeler, and A.J. Stentz, Opt. Lett. 25, 25 (2000)

[41] J.J. McFerran, E.N. Ivanov, A. Bartels, G. Wilpers, C.W. Oates, S.A. Diddams, and L. Hollberg, Electron. Lett. **41**, 650 (2005)

[42] A. Baltuska, Th. Udem, M. Uiberacker, M. Hentschel, E. Goulielmakis, Ch. Gohle, R. Holzwarth, V.S. Yakovlev, A. Scrinzi, T. W. Hänsch, and F. Krausz) Nature **421**, 611 (2003)

[43] A.N. Luiten, ed., *Frequency Measurement and Control: Advanced Techniques and Future Trends*, Topics in Applied Physics Vol. 79, Springer Verlag, New York, 2001

[44] L.E. Hargrove, R.L. Fork, and M.A. Pollack, Appl. Phys. Lett. 5, 4 (1964)

[45] A. Yariv, J. Appl. Phys. 36, 388 (1965)

[46] O.P. McDuff and S.E. Harris, IEEE J. Quantum Electron. **3**, 101 (1967)

[47] C.V. Shank, and E.P. Ippen, in *Dye Lasers*, F.P. Schäfer. Ed., Topics in Applied Physics, Vol. 1, Springer Verlag, New York, 3rd edition, 1990, pp 139

[48] R. Teets, J. Eckstein, and T.W. Hänsch, Phys. Rev. Lett. **38**, 760 (1977)

[49] M.M. Salour, and C. Cohen-Tannoudji, Phys. Rev. Lett. **38**, 757 (1977)

[50] E.V. Baklanov and V.P. Chebotaev, Kvantovaya Elektronika **4**, 2189 (1977)

[51] J.N. Eckstein, A.I. Fergusons, and T.W. Hänsch, Phys. Rev. Lett. **40**, 847 (1978)

[52] A.I. Ferguson, J.N. Eckstein, and T.W. Hänsch, J. Appl. Phys. **49**, 5389 (1978)

[53] A.I. Ferguson, J.N. Eckstein, and T.W. Hänsch, Appl. Phys. **18**, 257 (1979)

[54] T.W. Hänsch, and N.C. Wong, Metrologia **16**, 101 (1980)

[55] T.W. Hänsch, Opt. Comm. **80**, 71 (1990)

[56] D.E. Spencer, P.N. Kean, and W. Sibbett, Opt. Lett. **16**, 42 (1991)

[57] T. Nakajima and P. Lambropoulos, Phys. Rev. A **50**, 595 (1994)

[58] G.G. Paulus, F. Lindner, H. Walther, A. Baltuska, E. Gouielmakis, M. Lezius, and F. Krausz, Phys. Rev. Lett. 91, 253004 (2003)

[59] L. Xu, C. Spielmann, A. Poppe, T. Brabec, F. Krausz, and T.W. Hänsch, Opt. Lett. **21**, 2008 (1996)

[60] R.L. Fork, C.V. Shank, C. Hirlimann, R. Yen, and W.J. Tomlison, Opt. Lett. **8**, 1 (1983)

[61] R. Zerne, C. Altucci, M.B. Gaarde, A. L'Huillier, C. Lynga, C.-G. Wahlström, M. Bellini, and T.W. Hänsch, Phys. Rev. Lett., **79**, 1006 (1997)

[62] M. Bellini and T.W. Hänsch, Opt. Lett. **25**, 1049 (2000)

[63] Th. Udem, J. Reichert, R. Holzwarth, and T.W. Hänsch, Opt. Lett. **24**, 881 (1999)

[64] Th. Udem, J. Reichert, R. Holzwarth, and T.W. Hänsch, Phys. Rev. Lett. **82**, 3568 (1999)

[65] J. Reichert, M. Niering, R. Holzwarth, M. Weitz, Th. Udem, and T. W. Hänsch, Phys. Rev. Lett. **84**, 3232 (2000)

[66] M. Niering, R. Holzwarth, J. Reichert, P. Pokasov, Th. Udem, M. Weitz, T. W. Hänsch, P. Lemonde, G. Santarelli, M. Abgrall, P. Laurent, C. Salomon, and A. Clairon, Phys. Rev. Lett. **84**, 5496 (2000)

[67] J. Reichert, R. Holzwarth, Th. Udem, and T. W. Hänsch, Opt. Comm., 172, 59 (1999)

[68] H.R. Telle, G. Steinmeyer. A.E. Dunlop, J. Stenger, D.H. Sutter, and U. Keller, Appl. Phys. B **69**, 327 (1999)

[69] S. A. Diddams, D. J. Jones, J. Ye, S. Cundiff, J.L. Hall, J. K. Ranka, R. Windeler, R. Holzwarth, Th. Udem, and T. W. Hänsch, Phys. Rev. Lett. **84**, 5102 (2000)

[70] D.J. Jones, S.A. Diddams, J.K. Ranka, A. Stentz, R.S. Windeler, J.L. Hall, and S. Cundiff, Science **288**, 635 (2000)

[71] R. Holzwarth, Th. Udem, T.W. Hänsch, J. C. Knight, W. J. Wadsworth, and P.St.J. Russell, Phys. Rev. Lett. **85**, 2264 (2000)

[72] L. Matos, D. Kleppner, O. Kuzucu, T. R. Schibli, J. Kim, E. P. Ippen, and F.X. Kaertner, Opt. Lett. **29**, 1683 (2004)

[73] B.R. Washburn, S.A. Diddams, N.R. Newbury, J.W. Nicholson, M.F. Yan, and C.G. Jorgensen, Opt. Lett. **29**, 250 (2004)

[74] J. Stenger, H. Schnatz, C. Tamm, and H.R. Telle, Phys. Rev. Lett. **88**, 073601 (2002)

[75] M. Zimmermann, Ch. Gohle, R. Holzwarth, Th. Udem, and T.W. Hänsch, Opt. Lett. **29**, 310-312 (2004)

[76] L.S. Ma, B. Zhiyi, A. Bartels, L. Robertsson, M. Zucco, R.S. Windeler, G. Wilpers, C. Oates, L. Hollberg, and S.A. Diddams, Science **303**, 1843 (2004)

[77] X. Calmet and H. Fritzsch, Phys. Lett. B **540**, 173 (2002)

[78] M.T. Murphy, J.K. Webb, and V.V. Flambaum, MNRAS 345, 609 (2003)

[79] S. Srianand, H. Chand, P. Petitjean, and B. Aracil, Phys. Rev. Lett. **92**, 121302 (2004)

[80] S. Bize, S.A. Diddams, U. Tanaka, C.E. Tanner, W.H. Oskay, R.E. Drullinger, T.E. Parker, T.P. Heavner, S.R. Jefferts, L. Hollberg, W.M. Itano, and J.C. Bergquist, Phys. Rev. Lett. 90, 150802 (2003)

[81] E. Peik, B. Lipphardt, H. Schnatz, T. Schneider, C. Tamm, and S.G. Karshenboim, Phys. Rev. Lett. **93**, 170801 (2004)

[82] L. Hollberg, C.W. Oates, G. Wilpers, C.W. Hoyt, Z.W. Barber, S.A. Diddams, W.H. Oskay, and J.C. Bergquist, J. Phys. B **38**, S469 (2005)

[83] M. Takamoto, F.L. Hong, R. Higashi, and H. Katori, Nature **435**, 321 (2005)

[84] A. McPherson, G. Gibson, H. Jara, U. Johann, T.S. Luk, I. McIntyre, K. Boyer, and C.H. Rhodes, J. Opt. Soc. Am, B 4, 595 (1987)

[85] M. Ferray, A. L'Huiller, X.F. Li, A. Lompré, G. Mainfray, and C. Manus, J. Phys. B **21**, L31 (1988)

[86] P. Corkum, Phys. Rev. Lett. **71**, 1994 (1993)

[87] M. Bellini, C. Lynga, A. Tozzi, M.B. Gaarde, A. L'Huillier, C.-G. Wahlström, and T.W. Hänsch, Phys. Rev. Lett., **81**, 297 (1998)

[88] Ch. Gohle, Th. Udem, J. Rauschenberger, R. Holzwarth, M. Herrmann, H.A. Schüssler, F. Krausz, and T.W. Hänsch, Nature, **436**, 234 (2005)

[89] R. J. Jones, K. D. Moll, M. J. Thorpe, and J. Ye, Phys. Rev. Lett. **94**, 193201 (2005)

[90] Jun Ye, private communication

Appendix A
Solution for the Harmonic Oscillator Equation

In this appendix we will show that for the solution of the following equation

$$\frac{d^2\psi}{d\xi^2} + \left[\Lambda - \xi^2\right]\psi(\xi) = 0 \tag{A.1}$$

to be well behaved we must have $\Lambda = 1,3,5,7,\ldots$; i.e., Λ must be an odd integer. These are the eigenvalues of Eq. (A.1). We introduce the variable

$$\eta = \xi^2 \tag{A.2}$$

Thus

$$\frac{d\psi}{d\xi} = \frac{d\psi}{d\eta}\frac{d\eta}{d\xi} = \frac{d\psi}{d\eta}2\xi \tag{A.3}$$

and

$$\frac{d^2\psi}{d\xi^2} = 4\eta\frac{d^2\psi}{d\eta^2} + 2\frac{d\psi}{d\eta} \tag{A.4}$$

Substituting in Eq. (A.1), we obtain

$$\frac{d^2\psi}{d\eta^2} + \frac{1}{2\eta}\frac{d\psi}{d\eta} + \left[\frac{\Lambda}{4\eta} - \frac{1}{4}\right]\psi(\eta) = 0 \tag{A.5}$$

In order to determine the asymptotic form, we let $\eta \to \infty$ so that the above equation takes the form

$$\frac{d^2\psi}{d\eta^2} - \frac{1}{4}\psi(\eta) = 0$$

the solution of which would be $e^{\pm\frac{1}{2}\eta}$. This suggests that we try out the following solution

$$\psi(\eta) = y(\eta)e^{-\frac{1}{2}\eta} \tag{A.6}$$

Thus

$$\frac{d\psi}{d\eta} = \left[\frac{dy}{d\eta} - \frac{1}{2}y\right]e^{-\frac{1}{2}\eta} \tag{A.7}$$

and

$$\frac{d^2\psi}{d\eta^2} = \left[\frac{d^2y}{d\eta^2} - \frac{dy}{d\eta} + \frac{1}{4}y(\eta)\right]e^{-\frac{1}{2}\eta} \tag{A.8}$$

Substituting Eqs. (A.7) and (A.8) in Eq. (A.5) we get

$$\eta\frac{d^2y}{d\eta^2} + \left(\frac{1}{2} - \eta\right)\frac{dy}{d\eta} + \frac{\Lambda - 1}{4}y(\eta) = 0 \tag{A.9}$$

Now the confluent hypergeometric equation is given by (see, e.g., Refs. Alfano and Shapiro (1975), Arditty et al. (1980))

$$x\frac{d^2y}{dx^2} + (c - x)\frac{dy}{dx} - ay(x) = 0 \tag{A.10}$$

where a and c are constants. For $c \neq 0, \pm 1, \pm 2, \pm 3, \pm 4$, the two independent solutions of the above equation are

$$y_1(x) = {}_1F_1(a, c, x) \tag{A.11}$$

and

$$y_2(x) = x^{1-c} \, {}_1F_1(a - c + 1, 2 - c, x) \tag{A.12}$$

where ${}_1F_1(a, c, x)$ is known as the confluent hypergeometric function and is defined by the following equation:

$$_1F_1(a, c, x) = 1 + \frac{a}{c}\frac{x}{1!} + \frac{a(a+1)}{c(c+1)}\frac{x^2}{2!} + \frac{a(a+1)(a+2)}{c(c+1)(c+2)}\frac{x^3}{3!} + \cdots \tag{A.13}$$

Obviously, for $a = c$ we will have

$$_1F_1(a, a, x) = 1 + \frac{x}{1!} + \frac{x^2}{2!} + \frac{x^3}{3!} + \ldots = e^x \tag{A.14}$$

Thus although the series given by Eqs. (A.13) and (A.14) is convergent for all values of x, they would blow up at infinity. Indeed the asymptotic form of ${}_1F_1(a, c, x)$ is given as

$$_1F_1(a, c, x) \underset{x\to\infty}{\to} \frac{\Gamma(c)}{\Gamma(a)}x^{a-c}\, e^x \tag{A.15}$$

The confluent hypergeometric series ${}_1F_1(a, c, x)$ is very easy to remember and its asymptotic form is easy to understand. Returning to Eq. (A.9), we find that $y(\eta)$ satisfies the confluent hypergeometric equation with

$$a = \frac{1 - \Lambda}{4} \quad \text{and} \quad c = \frac{1}{2} \tag{A.16}$$

Thus the two independent solutions of Eq. (A.1) are

$$\psi_1(\eta) = {}_1F_1\left(\frac{1 - \Lambda}{4}, \frac{1}{2}, \eta\right) e^{-\frac{1}{2}\eta} \tag{A.17}$$

and

$$\psi_2(\eta) = \sqrt{\eta}\,{}_1F_1\left(\frac{3 - \Lambda}{4}, \frac{3}{2}, \eta\right) e^{-\frac{1}{2}\eta} \tag{A.18}$$

We must remember that $\eta = \xi^2$. Using the asymptotic form of the confluent hypergeometric function [Eq. (A.15)], one can readily see that if the series does not become a polynomial, then as $\eta \to \infty$, $\psi(\eta)$ will blow up as $e^{\frac{1}{2}\eta}$. In order to avoid this, the series must become a polynomial. Now $\psi_1(\eta)$ becomes a polynomial for $\Lambda = 1, 5, 9, 13, \ldots$, and $\psi_2(\eta)$ becomes a polynomial for $\Lambda = 3, 7, 11, 15$. Thus only when

$$\Lambda = 1, 3, 5, 7, 9, \ldots \tag{A.19}$$

we will have a well-behaved solution of Eq. (A.1) – these are the eigenvalues of Eq. (A.1). The corresponding wave functions are the Hermite–Gauss functions:

$$\psi(\xi) = NH_m(\xi) \exp\left(-\frac{1}{2}\xi^2\right), \qquad m = 0, 1, 2, 3, \ldots \tag{A.20}$$

Indeed

$$H_n(\xi) = (-1)^{n/2} \frac{n!}{\left(\frac{n}{2}\right)!} {}_1F_1\left(-\frac{n}{2}, \frac{1}{2}, \xi^2\right) \quad \text{for} \quad n = 0, 2, 4, \ldots \tag{A.21}$$

and

$$H_n(\xi) = (-1)^{(n-1)/2} \frac{n!}{\left(\frac{n-1}{2}\right)!} 2\xi\, {}_1F_1\left(-\frac{n-1}{2}, \frac{3}{2}, \xi^2\right) \quad \text{for} \quad n = 1, 3, 5, \ldots$$

$$\tag{A.22}$$

Appendix B
The Solution of the Radial Part of the Schrödinger Equation

For a particle in a spherically symmetric potential $V = V(r)$, the radial part of the Schrödinger equation is given as

$$\frac{1}{r^2}\frac{d}{dr}\left(r^2\frac{dR}{dr}\right) + \frac{2\mu}{\hbar^2}\left[E - V(r) - \frac{l(l+1)\hbar^2}{2\mu r^2}\right]R(r) = 0 \qquad (B.1)$$

If we define a new radial function

$$u(r) = rR(r) \qquad (B.2)$$

we would get

$$r^2\frac{dR}{dr} = r^2\frac{d}{dr}\left[\frac{u(r)}{r}\right]$$

$$= r\frac{du}{dr} - u(r)$$

Thus

$$\frac{1}{r^2}\frac{d}{dr}\left(r^2\frac{dR}{dr}\right) = \frac{1}{r}\frac{d^2u}{dr^2}$$

and Eq. (B.1) would become

$$\frac{d^2u(r)}{dr^2} + \frac{2\mu}{\hbar^2}\left[E - V(r) - \frac{l(l+1)\hbar^2}{2\mu r^2}\right]u(r) = 0$$

For the hydrogen-like atom problem,

$$V(r) = -\frac{Zq^2}{4\pi\varepsilon_0 r}$$

and therefore

$$\frac{d^2u}{dr^2} + \frac{2\mu}{\hbar^2}\left[E + \frac{Zq^2}{4\pi\varepsilon_0 r} - \frac{l(l+1)\hbar^2}{2\mu r^2}\right]u(r) = 0 \qquad (B.3)$$

In order to solve the above equation, we introduce the dimensionless variable

$$\rho = \gamma r \tag{B.4}$$

where the parameter γ is to be conveniently chosen later. Since

$$\frac{\mathrm{d}^2 u}{\mathrm{d}r^2} = \gamma^2 \frac{\mathrm{d}^2 u(\rho)}{\mathrm{d}\rho^2}$$

we readily get

$$\frac{\mathrm{d}^2 u}{\mathrm{d}\rho^2} + \left[\frac{2\mu E}{\hbar^2 \gamma^2} + \frac{2\mu}{\gamma \hbar^2} \frac{Zq^2}{4\pi \varepsilon_0 \rho} - \frac{l(l+1)}{\rho^2} \right] u(\rho) = 0 \tag{B.5}$$

We choose[1]

$$\gamma^2 = \frac{8\mu |E|}{\hbar^2} = -\frac{8\mu E}{\hbar^2} \tag{B.6}$$

and set

$$\lambda = \frac{2\mu Z}{\hbar^2 \gamma} \left(\frac{q^2}{4\pi \varepsilon_0} \right) = Zc\alpha \left(\frac{\mu}{2|E|} \right)^{1/2} \tag{B.7}$$

Where

$$\alpha = \frac{q^2}{4\pi \varepsilon_0 \hbar c} \approx \frac{1}{137}$$

is the fine structure constant. Thus Eq. (B.5) becomes

$$\frac{\mathrm{d}^2 u}{\mathrm{d}\rho^2} + \left[-\frac{1}{4} + \frac{\lambda}{\rho} - \frac{l(l+1)}{\rho^2} \right] u(\rho) = 0 \tag{B.8}$$

Now as $\rho \to \infty$, the first term inside the square bracket is the most dominating term and we may approximately write

$$\frac{\mathrm{d}^2 u}{\mathrm{d}\rho^2} - \frac{1}{4} u(\rho) = 0$$

the solution of which is given as

$$u(\rho) \sim e^{\pm \rho/2}$$

We reject the exponentially amplifying solution and write

---

[1] We are looking for energy levels corresponding to bound states; therefore $E < 0$.

$$u(\rho) \sim e^{-\rho/2} \quad \text{as} \quad \rho \to \infty \tag{B.9}$$

Now as $\rho \to 0$, the third term inside the square brackets is the most dominating term and we may approximately write

$$\frac{d^2 u}{d\rho^2} - \frac{l(l+1)}{\rho^2} u(\rho) = 0$$

If we write

$$u(\rho) \sim \rho^g$$

then we readily obtain

$$g(g-1) = l(l+1)$$

giving

$$g = -l \quad \text{or} \quad (l+1)$$

We reject the $g = -l$ solution because it will diverge for $\rho \to 0$ so that we may write

$$u(\rho) \sim \rho^{l+1} \text{ as } \rho \to 0 \tag{B.10}$$

The above equations suggest that we try to solve Eq. (B.8) by defining $y(\rho)$ through the following equation:

$$u(\rho) = \rho^{l+1} e^{-\rho/2} y(\rho) \tag{B.11}$$

Simple manipulations will show that $y(\rho)$ satisfies the following equation:

$$\rho \frac{d^2 y}{d\rho^2} + (c - \rho)\frac{dy}{d\rho} - ay(\rho) = 0 \tag{B.12}$$

where

$$a = l + 1 - \lambda; \qquad\qquad \tag{B.13}$$
$$c = 2l + 2$$

Equation (B.12) is the confluent hypergeometric equation (see Appendix A) and the solution which is finite at the origin is given as

$$y(\rho) = {}_1F_1(a, c, \rho) = 1 + \frac{a}{c}\frac{\rho}{1!} + \frac{a(a+1)}{c(c+1)}\frac{\rho^2}{2!} + \cdots \tag{B.14}$$

which is known as the confluent hypergeometric function. Thus the complete solution of Eq. (B.8) which is well behaved at $\rho = 0$ is given as

$$u(\rho) = N\rho^{l+1} e^{-\rho/2} {}_1F_1(a, c, \rho) \tag{B.15}$$

where N is the normalization constant so that

$$\int_0^\infty u^2(r)\mathrm{d}r = \int_0^\infty R^2(r)r^2\,\mathrm{d}r = 1 \tag{B.16}$$

The value of N is given by Eq. (3.152) We may mention the following:

(i) The solution given by Eq. (B.15) is a rigorously correct solution of Eq. (B.5). Writing the solution in the form of Eq. (66) does *not* involve *any* approximation.

(ii) The infinite series given by Eq. (B.14) is convergent for *all* values of ρ in the domain $0 < \rho < \infty$.

(iii) For $a = c$, the infinite series given in Eq. (B.14) is simply e^ρ; thus as $\rho \to \infty$, $u(\rho)$ will diverge as $\rho^{l+1} e^{\rho/2}$.

(iv) Indeed

$$ {}_1F_1(a, c, \rho) \xrightarrow[\rho \to \infty]{} \rho^{a-c} e^\rho$$

Thus

$$u(\rho) \xrightarrow[\rho \to \infty]{} \rho^{l+1} \rho^{a-c} e^{\rho/2}$$

and will blow up as $\rho \to \infty$ [although the infinite series given by Eq. (B.14) is always convergent!]. In order to avoid this we must make the infinite series a polynomial which can happen only if a is a negative integer:

$$a = -n_\mathrm{r}, \qquad n_\mathrm{r} = 0, 1, 2, \ldots \tag{B.15}$$

Thus

$$\lambda = l + 1 + n_\mathrm{r} = n, \qquad n = 1, 2, \ldots \tag{B.16}$$

The quantities n_r and n are usually referred to as the radial quantum number and the total quantum number, respectively. Now, $\lambda = n$ implies [see Eq. (62)]

$$\frac{Z^2 c^2 \alpha^2 \mu}{2|E|} = n^2$$

or

$$E = E_n = -\frac{Z^2 c^2 \alpha^2 \mu}{2n^2} \tag{B.17}$$

which represent the energy eigenvalues of a hydrogen-like atom. The corresponding normalized wave functions are given by Eqs. (3.151) and (3.152).

Appendix C
The Fourier Transform

According to Fourier's theorem, a periodic function can be expressed as a sum of sine and cosine functions whose frequencies increase in the ratio of natural numbers. Thus, a periodic function with period α, i.e.

$$f(x + n\alpha) = f(x), \qquad n = 0, \pm 1, \pm 2, \pm 3, \ldots \tag{C.1}$$

can be expanded in the form of

$$f(x) = \frac{1}{2}a_0 + \sum_{n=1}^{\infty} [a_n \cos(nkx) + b_n \sin(nkx)] \tag{C.2}$$

where

$$k = \frac{2\pi}{\alpha} \tag{C.3}$$

The coefficients a_n and b_n can be easily determined by using the following properties of the trigonometric functions:

$$\frac{2}{\alpha} \int_{x_0}^{x_0+\alpha} \cos(nkx) \cos(mkx) \mathrm{d}x = \delta_{mn} \tag{C.4}$$

$$\frac{2}{\alpha} \int_{x_0}^{x_0+\alpha} \sin(nkx) \sin(mkx) \mathrm{d}x = \delta_{mn} \tag{C.5}$$

$$\frac{2}{\alpha} \int_{x_0}^{x_0+\alpha} \sin(nkx) \cos(mkx) \mathrm{d}x = 0 \tag{C.6}$$

where x_0 is arbitrary and δ_{mn} is the Kronecker delta function defined through the following equation:

$$\delta_{mn} = \begin{cases} 0, & \text{if } m \neq n \\ 1, & \text{if } m = n \end{cases} \tag{C.7}$$

If we multiply Eq. (C.1) by $\cos(mkx)$ and integrate from x_0 to $x_0 + \alpha$, we would obtain

$$a_n = \int_{x_0}^{x_0+\alpha} f(x)\cos(nkx)dx, \quad n = 0, 1, 2, \ldots \tag{C.8}$$

Similarly

$$b_n = \int_{x_0}^{x_0+\alpha} f(x)\sin(nkx)dx, \quad n = 1, 2, 3, \ldots \tag{C.9}$$

For the sake of convenience we choose $x_0 = -\alpha/2$. If we now substitute the above expressions for a_n and b_n in Eq. (C.2), we would obtain

$$
\begin{aligned}
f(x) = \frac{1}{\alpha} \int_{-\alpha/2}^{\alpha/2} f(x')dx' + \sum_{n=1}^{\infty} \Bigg[\frac{2}{\alpha} \cos(nkx) \int_{-\alpha/2}^{\alpha/2} f(x') \cos\left(nkx'\right) dx' \\
+ \frac{2}{\alpha} \sin\left(nkx\right) \int_{-\alpha/2}^{+\alpha/2} f(x') \sin\left(nkx'\right) dx' \Bigg]
\end{aligned}
\tag{C.10}
$$

or

$$f(x) = \frac{1}{2\pi}\Delta s \int_{-\pi/\Delta s}^{+\pi/\Delta s} f(x')dx' + \sum_{n=1}^{\infty} \left\{ \frac{\Delta s}{\pi} \int_{-\pi/\Delta s}^{+\pi/\Delta s} f(x') \cos\left[n\,\Delta s\,(x'-x)\right] dx' \right\} \tag{C.11}$$

where

$$\Delta s = \frac{2\pi}{\alpha} = k \tag{C.12}$$

We now let $\alpha \to \infty$ so that $\Delta s \to 0$. Thus, if the integral

$$\int_{-\infty}^{+\infty} \left| f(x') \right| dx'$$

exists (i.e., it has a finite value), then the first term on the right-hand side of Eq. (C.11) would go to zero. Further, since

$$\int_0^{\infty} F(s)ds = \lim_{\Delta s \to 0} \sum_{n=1}^{\infty} F(n\,\Delta s)\,\Delta s \tag{C.13}$$

we have, in the limit $\Delta s \to 0$

$$f(x) = \frac{1}{\pi} \int_0^\infty \left\{ \int_{-\infty}^{+\infty} f(x') \cos\left[s\left(x' - x\right)\right] dx' \right\} ds \tag{C.14}$$

Equation (C.14) is known as the Fourier integral. Since the cosine function inside the integral is an even function of s, we may write

$$f(x) = \frac{1}{2\pi} \int_{-\infty}^{+\infty} \left\{ \int_{-\infty}^{+\infty} f(x') \cos\left[s\left(x' - x\right)\right] dx' \right\} ds \tag{C.15}$$

Further, since $\sin\left[s\left(x' - x\right)\right]$ is an odd function of s

$$0 = \frac{i}{2\pi} \int_{-\infty}^{+\infty} \left\{ \int_{-\infty}^{+\infty} f(x') \sin\left[s\left(x' - x\right)\right] dx' \right\} ds \tag{C.16}$$

Adding (or subtracting) Eqs. (C.15) and (C.16), we get

$$f(x) = \frac{1}{2\pi} \int_{-\infty}^{+\infty} \int_{-\infty}^{+\infty} f(x') e^{\pm is(x'-x)} dx'\, ds \tag{C.17}$$

Thus, if

$$F(s) = \frac{1}{(2\pi)^{1/2}} \int_{-\infty}^{\infty} f(x') e^{isx'}\, dx' \tag{C.18}$$

then

$$f(x) = \frac{1}{(2\pi)^{1/2}} \int_{-\infty}^{+\infty} F(s) e^{-isx}\, ds \tag{C.19}$$

The function $F(s)$ defined by Eq. (C.18) is known as the Fourier transform of $f(x)$ and conversely.

Since the Dirac delta function $\delta\left(x - x'\right)$ is defined by the equation

$$f(x) = \int_{-\infty}^{+\infty} f(x') \delta\left(x - x'\right) dx' \tag{C.20}$$

we obtain, comparing Eqs. (C.17) and (C.20), the following representation of the delta function:

$$\delta\left(x' - x\right) = \frac{1}{2\pi} \int_{-\infty}^{+\infty} e^{\pm is(x'-x)}\, dx \tag{C.21}$$

For a time-dependent function, we can write the Fourier transform in the following form:

$$F(\omega) = \frac{1}{(2\pi)^{1/2}} \int_{-\infty}^{+\infty} f(t') e^{i\omega t'}\, dt' \tag{C.22}$$

$$f(t) = \frac{1}{(2\pi)^{1/2}} \int_{-\infty}^{+\infty} F(\omega)e^{-i\omega t}d\omega \tag{C.23}$$

where ω represents the angular frequency. As an example, we consider a time-dependent pulse of the form

$$f(t) = A\,e^{-t^2/2\tau^2}e^{-i\omega_0 t} \tag{C.24}$$

Thus

$$F(\omega) = \frac{A}{(2\pi)^{1/2}} \int_{-\infty}^{+\infty} \exp\left(-\frac{t^2}{2\tau^2}\right)e^{i(\omega-\omega_0)t}dt$$

$$= A\tau \exp\left[-\frac{(\omega-\omega_0)^2\,\tau^2}{2}\right] \tag{C.25}$$

where we have used the following result:

$$\int_{-\infty}^{+\infty} e^{-\alpha x^2+\beta x}\,dx = e^{\beta^2/4\alpha} \int_{-\infty}^{+\infty} \exp\left[-\alpha\left(x-\frac{\beta}{2\alpha}\right)^2\right]dx$$
$$= (\pi/\alpha)^{1/2}\exp\left(\beta^2/4\alpha\right) \tag{C.26}$$

From Eqs. (C.24) and (C.25) it can be immediately seen that a temporal pulse of duration $\sim\tau$ has a frequency spread of $\Delta\omega \sim 1/\tau$. Thus, one obtains

$$\tau\Delta\omega \sim 1 \tag{C.27}$$

If $\tau \to \infty$, the pulse becomes almost monochromatic and $\Delta\omega \to 0$. The result expressed by Eq. (C.27) is quite general in the sense that it is independent of the shape of the pulse. For example, for a pulse of the form

$$f(t) = \begin{cases} A\,e^{-i\omega_0 t} & |t| < \tau/2 \\ 0 & \text{everywhere else} \end{cases} \tag{C.28}$$

one would obtain

$$F(\omega) = A\left(\frac{2}{\pi}\right)^{1/2}\frac{\sin(\omega-\omega_0)\tau/2}{\omega-\omega_0} \tag{C.29}$$

which also has a spread $\sim 1/\tau$.

If we introduce the variable

$$u = s/2\pi \tag{C.30}$$

then Eq. (C.17) becomes

$$f(x) = \int_{-\infty}^{+\infty} \int_{-\infty}^{+\infty} f\left(x'\right) e^{2\pi iu(x'-x)} \, dx' \, du \qquad (C.31)$$

Thus, if we write

$$F(u) = \int_{-\infty}^{+\infty} f\left(x'\right) e^{2\pi iux'} \, dx' = \mathcal{F}\left[f\left(x\right)\right] \qquad (C.32)$$

then

$$f(x) = \int_{-\infty}^{+\infty} F(u) \, e^{-2\pi iux} \, du \qquad (C.33)$$

where the symbol $F[\ ]$ stands for the Fourier transform of the quantity inside the brackets. The quantity u is termed the spatial frequency (see Section 10.2).

Now

$$\mathcal{F}\left[\mathcal{F}\left[f(x)\right]\right] = \mathcal{F}\left[F(u)\right]$$
$$= \int_{-\infty}^{+\infty} F(u) \, e^{2\pi iux} \, dx \qquad (C.34)$$
$$= f(-x)$$

Thus the Fourier transform of the Fourier transform of a function is the original function itself except for an inversion. We also see that

$$\mathcal{F}\left[\exp\left(\frac{2\pi ix}{a}\right)\right] = \int_{-\infty}^{+\infty} \exp\left[2\pi ix\left(\frac{1}{a} + u\right)\right] dx \qquad (C.35)$$
$$= \delta\left(u + 1/a\right)$$

where we have used Eq. (C.21). In addition

$$\mathcal{F}\left[\delta(t - t_0)\right] = \int_{-\infty}^{\infty} \delta\left(t - t_0\right) e^{2\pi ivt} \, dt \qquad (C.36)$$
$$= e^{2\pi ivt_0}$$

Convolution theorem. The convolution of two functions $f(t)$ and $g(t)$ is defined by the relation

$$f(t) * g(t) = \int_{-\infty}^{\infty} f\left(t'\right) g\left(t - t'\right) dt' = g\left(t\right) * f\left(t\right) \qquad (C.37)$$

We will now show that the Fourier transform of the convolution of two functions is the product of their Fourier transforms. This can be seen as follows:

$$\mathcal{F}\left[f(t)*g(t)\right] = \int_{-\infty}^{\infty} \int_{-\infty}^{\infty} f\left(t'\right) g\left(t - t'\right) e^{2\pi i v t}\, dt'\, dt$$

$$= \int_{-\infty}^{\infty} dt'\, f\left(t'\right) e^{2\pi i v t} \int_{-\infty}^{\infty} g\left(t - t'\right) e^{2\pi i v (t-t')}\, dt \qquad \text{(C.38)}$$

$$= F(v)\, G(v)$$

where $F(v) = \mathcal{F}\left[f(t)\right]$ and $G(v) = \mathcal{F}\left[g(t)\right]$. Similarly

$$\mathcal{F}\left[f(t)g(t)\right] = \int_{-\infty}^{\infty} f\left(t'\right) g\left(t'\right) e^{+2\pi i v t'}\, dt'$$

$$= \int_{-\infty}^{\infty} f\left(t'\right) \int_{-\infty}^{\infty} G\left(v'\right)^{-2\pi i v' t'}\, dv'\, e^{2\pi i v t'}\, dt'$$

$$= \int_{-\infty}^{\infty} G\left(v'\right) \int_{-\infty}^{\infty} f\left(t'\right) e^{2\pi i (v-v') t'}\, dt'\, dv' \qquad \text{(C.39)}$$

$$= \int_{-\infty}^{\infty} G\left(v'\right) F\left(v - v'\right)\, dv'$$

$$= F(v)*G(v)$$

i.e., Fourier transform of the product of two functions is the convolution of their Fourier transforms.

We will now consider some examples. The convolution of a function $f(t)$ with a delta function $\delta(t - t_0)$ is

$$f(t)*\delta(t - t_0) = \int_{-\infty}^{\infty} f\left(t'\right) \delta\left(t - t' - t_0\right) dt' \qquad \text{(C.40)}$$

$$= f\left(t - t_0\right)$$

Thus the convolution of a function with a delta function yields the same function but with a shift in origin.

Similarly, the convolution of two Gaussian functions can again be shown to be another Gaussian function:

$$\exp\left(-\frac{x^2}{\alpha^2}\right) * \exp\left(-\frac{x^2}{\beta^2}\right) = \frac{\alpha\beta\,(\pi)^{1/2}}{\left(\alpha^2 + \beta^2\right)^{1/2}} \exp\left(-\frac{x^2}{\alpha^2 + \beta^2}\right) \qquad \text{(C.41)}$$

Sampling theorem. We will now derive the sampling theorem according to which *a band-limited function, i.e., a function which has no spectral components beyond a certain frequency, say v_m, is uniquely determined by its value at uniform intervals less than $(1 / 2v_m)$ seconds.*

A band-limited function having a maximum spectral component v_m implies that its Fourier transform is zero beyond v_m (see Fig. C.1). Thus if $f(t)$ is a band-limited function, then

Fig. C.1 (**a**) A band-limited function having no spectral components beyond a frequency v_m and (**b**) the corresponding Fourier spectrum

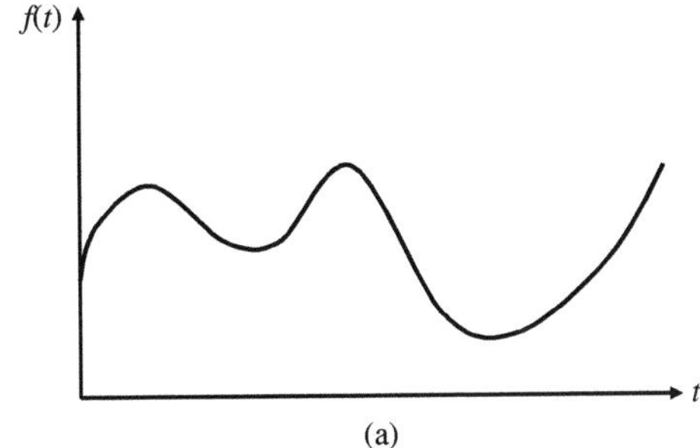

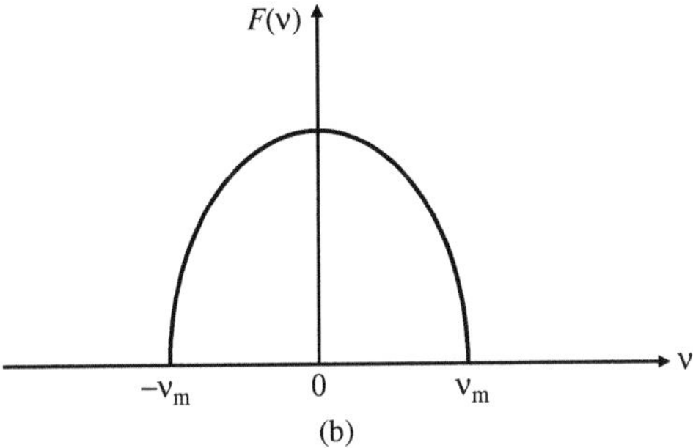

$$\mathcal{F}\left[f(t)\right] = F(v) = 0, \quad \text{for } v > v_m \tag{C.42}$$

We now consider the sampled function which consists of impulses every T seconds with the strength of each impulse being equal to the value of the function at that time. Thus the sampled function can be written as

$$f_s(t) = \sum_{n=-\infty}^{\infty} f(t)\delta(t - nT) = f(t) \sum_{n=-\infty}^{\infty} \delta(t - nT) \tag{C.43}$$

The sampled function $f_s(t)$ is shown in Fig. C.2a. Here T represents the sampling interval and $\delta(t - nT)$ represents the Dirac delta function.

In order to obtain the Fourier spectrum of the sampled function, we have to first find the Fourier transform of the function $\sum_{n=-\infty}^{\infty} \delta(t - nT)$. Using Eq. (C.36) we have

$$\mathcal{F}\left[\sum_{n=-\infty}^{\infty} \delta(t - nT)\right] = \int_{-\infty}^{\infty} \sum_{n=-\infty}^{\infty} \delta(t - nT)\, e^{2\pi i v t}\, \mathrm{d}t$$

$$= \sum_{n=-\infty}^{\infty} e^{2\pi i n v T} \tag{C.44}$$

Fig. C.2 (**a**) The function shown here is obtained by periodically sampling the original function depicted in Fig. C.1 at uniform intervals by impulses separated in time by T; the strength of the sampled signal at the sampling times is equal to the value of the original function at that time. (**b**) The Fourier spectrum corresponding to the sampled function when the sampling interval T is smaller than $1/2v_m$, the Nyquist interval. Observe that in such a case the various repetitively appearing spectra do not overlap and it is possible to use a filter to retrieve $F(v)$ and hence $f(t)$

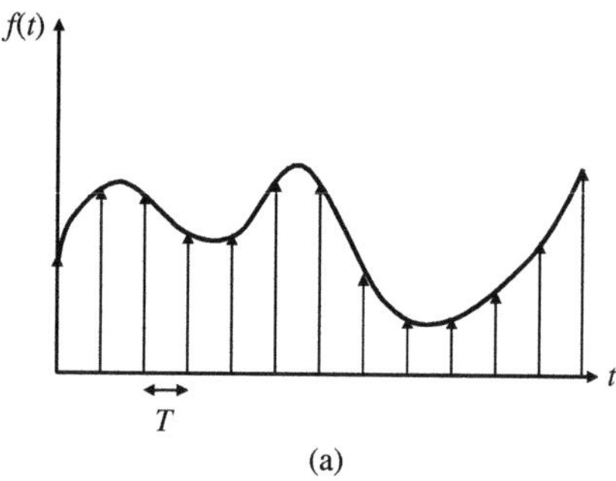

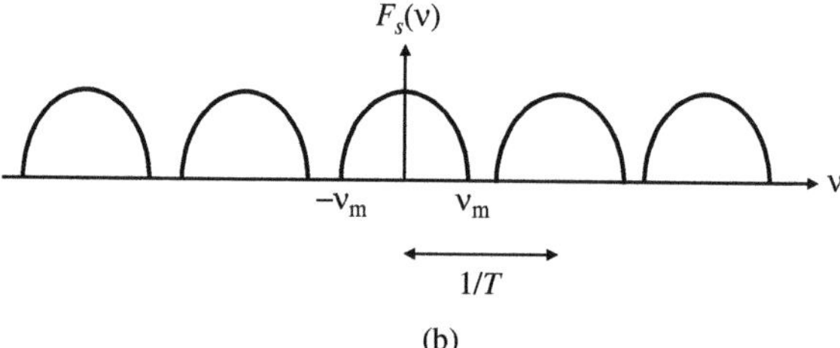

In order to evaluate the sum of the right-hand side of Eq. (C.44), we consider the following finite sum:

$$\sum_{n=-N}^{N} e^{2\pi i v n T} = e^{-2\pi i n N T}\left(\frac{1 - e^{2\pi i v T(2N+1)}}{1 - e^{2\pi i v T}}\right) = \frac{\sin\left[\pi v T\,(2N+1)\right]}{\sin\left(\pi v T\right)} \quad \text{(C.45)}$$

which is a periodic function with a period $1/T$. If we restrict ourselves to the region $-1/2T < v < 1/2T$, then for a large value of N, the above function is sharply peaked around $v = 0$ and we may write

$$\lim_{N \to \infty} \frac{\sin \pi v T\,(2N+1)}{\sin \pi v T} = \lim_{g \to \infty} \frac{\sin g v T}{\pi v T} \quad \text{(C.46)}$$

In order to determine the value of the limit on the right-hand side of Eq. (C.46), we have from Eq. (C.21)

$$\delta(t) = \frac{1}{2\pi} \int_{-\infty}^{\infty} e^{ivt}\, dv$$

$$= \lim_{N \to \infty} \frac{1}{2\pi} \int_{-N}^{N} e^{ivt}\, dv \quad \text{(C.47)}$$

$$= \lim_{N \to \infty} \frac{\sin Nt}{\pi t}$$

Thus we get from Eq. (C.46)

$$\lim_{N\to\infty} \frac{\sin \pi vT\,(2N+1)}{\sin \pi vT} = \delta\,(vT) \tag{C.48}$$

and from Eqs. (C.44) and (C.45)

$$\mathcal{F}\left[\sum_{n=-\infty}^{\infty} \delta(t-nT)\right] = \frac{1}{T} \sum_{n=-\infty}^{\infty} \delta\left(v-\frac{n}{T}\right) \tag{C.49}$$

Hence the Fourier transform of the sampled function is given as

$$\begin{aligned}
\mathcal{F}\left[f_{\mathrm{s}}(t)\right] &= \mathcal{F}\left[f(t)\sum_{n=-\infty}^{\infty}\delta(t-nT)\right] \\
&= \mathcal{F}\left[f(t)\right] * \mathcal{F}\left[\sum_{n=-\infty}^{\infty}\delta\,(t-nT)\right] \qquad \text{using Eq. (C.39)} \\
&= F(v)*\frac{1}{T}\sum_{n=-\infty}^{\infty}\delta\left(v-\frac{n}{T}\right) \qquad \text{using Eq. (C.49)} \\
&= \frac{1}{T}\sum_{n=-\infty}^{\infty} F(v)*\delta\left(v-\frac{n}{T}\right) \\
&= \frac{1}{T}\sum_{n=-\infty}^{\infty} F\left(v-\frac{n}{T}\right) \qquad \text{using Eq. (C.40)}
\end{aligned} \tag{C.50}$$

Here

$$F(v) = \mathcal{F}\left[f(t)\right] = \int_{-\infty}^{\infty} f(t)\, e^{2\pi ivt}\, dt \tag{C.51}$$

represents the Fourier transform of the function $f(t)$. Hence from above it follows that the Fourier transform of the sampled function consists of an infinite sum of repetitively appearing Fourier transforms of the original function (see Fig. C.2b), i.e., if the function $F(v)$ is band limited up to v_m, then in order that the various repetitively appearing Fourier spectra in the sampled function do not overlap, one must have

$$2v_m \le 1/T$$

or

$$T \le 1/2v_m \tag{C.52}$$

Hence as long as we sample $f(t)$ at regular intervals less than $1/2v_m$ seconds apart, the Fourier spectrum of the sampled function is a periodic replica of $F(v)$. One can thus recover $f(t)$ completely by allowing the sampled signal to pass through a low-pass filter which attenuates all frequencies beyond v_m and which passes without distortion the frequency components below v_m. On the other hand, if $T > 1/2v_m$, then the various spectra $F(v)$ overlap and it is not possible to recover $f(t)$ from the sampled values (see Fig. C.3). Thus in order to completely preserve the information content of a signal in a retrievable form, the sampling interval should not be more than $1/2v_m$; this maximum sampling interval is also referred to as the Nyquist interval.

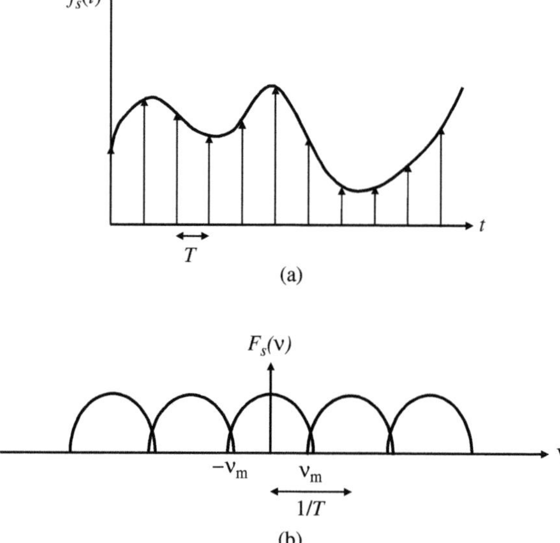

Fig. C.3 (**a**) The same as shown in Fig. C.2(a) but here the sampling interval is larger than $\frac{1}{2v_m}$ (**b**) The corresponding Fourier spectrum which shows overlap of the various repetitively appearing spectra. In such a case it is not possible to retrieve the signal

Hence from the above theorem it follows that complete information of a band-limited signal can be transmitted by just sending the discrete sampled values. This is the basic principle behind digital transmission systems (see Section 17.2).

It may be mentioned here that no signal is strictly band limited. But for most practical situations, the energy content in high frequencies is so small to be negligible. Thus, one can consider such signals to be essentially band limited.

Appendix D
Planck's Law

In Section 7.2 we solved Maxwell's equations in a rectangular cavity and obtained the allowed frequencies of oscillation of the field in the cavity [see Eq. (7.16)]. In Appendix E, we will calculate the density of such modes and will show that the number of modes per unit volume in a frequency interval $d\omega$ will be given as [see Eq. (E.10)]

$$p(\omega)\,d\omega = \frac{1}{\pi^2 c^3}\omega^2\,d\omega \qquad (D.1)$$

In Section 8.2, we have shown that quantum mechanically, we can visualize the radiation field (inside a cavity) as consisting of an infinite number of simple harmonic oscillators (each oscillator corresponding to a particular mode of the cavity) and that the energy of each oscillator can take only the discrete values

$$E_{n_\lambda} = \left(n_\lambda + \frac{1}{2}\right)\hbar\omega_\lambda, \quad n_\lambda = 0, 1, 2, \ldots \qquad (D.2)$$

[see Eq. (9.59)]. Now, according to Boltzmann's law, if E_{n_λ} is the energy of the λth mode in the nth excited state, then the probability p_{n_λ} of the system being in this state (at thermal equilibrium) is proportional to

$$\exp\left(-E_{n_\lambda}\,/\,k_B T\right)$$

where k_B is the Boltzmann's constant and T is the absolute temperature. Because the λth mode must be in one of the states, we must have

$$\sum_{n_\lambda=0}^{\infty} P_{n_\lambda} = 1 \qquad (D.3)$$

To ensure this condition, we must have

$$P_{n_\lambda} = \frac{\exp\left(-E_{n_\lambda}\,/\,k_B T\right)}{\sum_{n_\lambda=0}^{\infty}\exp\left(-E_{n_\lambda}\,/\,k_B T\right)} \qquad (D.4)$$

Substituting for E_{n_λ} from Eq. (D.2), we have

$$P_{n_\lambda} = \frac{\exp(-n_\lambda x)}{\sum_{n_\lambda=0}^{\infty} \exp[-n_\lambda x]} \tag{D.5}$$

where

$$x = \frac{\hbar\omega_\lambda}{k_B T} \tag{D.6}$$

Carrying out the summation of the geometrical series in the denominator, we obtain

$$P_{n_\lambda} = \left[1 - e^{-x}\right] e^{-n_\lambda x} \tag{D.7}$$

The mean number of photons $\bar{n}_\lambda$ associated with the λth mode is given as

$$\bar{n}_\lambda = \sum_{n_\lambda=0}^{\infty} P_{n_\lambda} n_\lambda$$

$$= \left(1 - e^{-x}\right) \sum_{n_\lambda=0}^{\infty} n_\lambda e^{-n_\lambda x}$$

$$= -\left(1 - e^{-x}\right) \frac{d}{dx} \sum_{n_\lambda=0}^{\infty} e^{-n_\lambda x}$$

$$= -\left(1 - e^{-x}\right) \frac{d}{dx} \left(\frac{1}{1 - e^{-x}}\right)$$

or

$$\bar{n}_\lambda = \frac{1}{e^x - 1} = \frac{1}{e^{\hbar\omega_\lambda/k_B T} - 1} \tag{D.8}$$

Thus the mean energy associated with each mode will be

$$\bar{n}_\lambda \hbar\omega_\lambda = \frac{\hbar\omega_\lambda}{e^{\hbar\omega_\lambda/k_B T} - 1} \tag{D.9}$$

Consequently, if $u(\omega)$ represents the energy of the radiation field (per unit volume) in the frequency interval $d\omega$, then

$$u(\omega)d\omega = \text{(energy associated with a mode)} \times \text{(number of}$$
$$\text{modes per unit volume in the frequency interval } d\,\omega)$$
$$= \frac{\hbar\omega}{e^{\hbar\omega/k_B T} - 1} \times \frac{1}{\pi^2 c^3} \omega^2 d\omega$$

or

$$u(\omega) = \frac{\hbar\omega^3}{\pi^2 c^3} \frac{1}{e^{\hbar\omega/k_B T} - 1} \tag{D.10}$$

which is the famous Planck's law. For $k_B T >> \hbar\omega$, one obtains

$$u(\omega) \approx \frac{\omega^2}{\pi^2 c^3} k_B T \tag{D.11}$$

which is known as the Rayleigh–Jeans law. On the other hand, at low temperatures, where $k_B T << \hbar\omega$, one obtains

$$u(\omega) \approx \frac{\hbar\omega^3}{\pi^2 c^3} \exp\left(-\frac{\hbar\omega}{k_B T}\right) \tag{D.12}$$

which is known as Wien's law.

Appendix E
The Density of States

In Section 7.2, we had solved Maxwell's equations in a rectangular cavity and had shown that the electric field inside the cavity is given as

$$E_x = E_{0x} \cos k_x x \, \sin k_y y \, \sin k_z z \qquad \text{(E.1)}$$

with similar expressions for E_y and E_z [see Eq. (7.14)]. By imposing proper boundary conditions, the following allowed values of k_x, k_y, and k_z are obtained [see Eqs. (7.12) and (7.13)]:

$$k_x = \frac{m\pi}{2a}, \quad m = 0, 1, 2, \ldots \qquad \text{(E.2)}$$

$$k_y = \frac{n\pi}{2b}, \quad n = 0, 1, 2, \ldots \qquad \text{(E.3)}$$

$$k_z = \frac{q\pi}{d}, \quad q = 0, 1, 2, \ldots \qquad \text{(E.4)}$$

Now, the number of modes whose x component of $\mathbf{k}$ lies between k_x and $k_x + dk_x$ would simply be the number of integers lying between $(2a/\pi)\, k_x$ and $(2a/\pi)\,(k_x + dk_x)$. This number would be approximately equal to $(2a/\pi)\, dk_x$. Similarly, the number of modes whose y and z components of $\mathbf{k}$ lie between k_y and $k_y + dk_y$ and k_z and $k_z + dk_z$ would, respectively, be

$$\left(\frac{2b}{\pi} dk_y\right) \text{ and } \frac{d}{\pi} dk_z$$

Thus, there will be

$$\left(\frac{2a}{\pi} dk_x\right) \left(\frac{2b}{\pi} dk_y\right) \left(\frac{d}{\pi} dk_z\right) = \frac{V}{\pi^3} dk_x \, dk_y \, dk_z \qquad \text{(E.5)}$$

modes in the range $dk_x \, dk_y \, dk_z$ of $\mathbf{k}$; here $V\,(= 2a \times 2b \times d)$ represents the volume of the cavity. Thus the number of modes per unit volume in the $\mathbf{k}$ space would be

$$\frac{V}{\pi^3}$$

If $P(k)dk$ represents the number of modes whose $|k|$ lies between k and $k + dk$, then

$$P(k)dk = 2 \times \frac{1}{8} \times \frac{V}{\pi^3} 4\pi k^2 \, dk \tag{E.6}$$

where the factor $4\pi k^2 \, dk$ represents the volume element (in the $\mathbf{k}$ space) lying between k and $k + dk$ and the factor $1/8$ is due to the fact that k_x, k_y and k_z can take only positive values [see Eqs. (E.2)–(E.4)] so that while counting the modes in the $\mathbf{k}$ space we must consider only the positive octant; the factor of 2 corresponds to the fact that corresponding to a particular value of $\mathbf{k}$, there are two independent modes of polarization. If $p(k)dk$ represents the corresponding number of modes per unit volume, then

$$p(k)dk = \frac{1}{\pi} k^2 \, dk \tag{E.7}$$

Now, if $p(\nu)d\nu$ represents the number of modes (per unit volume) in the frequency interval $d\nu$, then

$$p(\nu) \, d\nu = p(k) \, dk = \frac{1}{\pi^2} k^2 \, dk = \frac{1}{\pi^2} \left(\frac{2\pi \nu}{c} \right)^2 \frac{2\pi}{c} d\nu \tag{E.8}$$

or

$$p(\nu) \, d\nu = \frac{8\pi \nu^2}{c^3} d\nu \tag{E.9}$$

which is identical to Eq. (7.18). In deriving Eq. (E.9) we have used the relation $k = 2\pi \nu / c$. Transforming to the ω space, we obtain

$$p(\omega)d\omega = \frac{1}{\pi^2 c^3} \omega^2 \, d\omega \tag{E.10}$$

It should be mentioned that in the derivation of the above formula we have assumed $|\mathbf{k}|$ to lie between k and $k + dk$ and have integrated over all the directions of $\mathbf{k}$. If, however, we are interested in the number of modes for which $\mathbf{k}$ lies between k and $k + dk$ but the direction of $\mathbf{k}$ lies in the solid angle $d\Omega$, then the number of such modes would be given as

$$N(k) \, dk \, d\Omega = \frac{1}{8} \times \frac{V}{\pi^3} \times 4\pi k^2 \, dk \times \frac{d\Omega}{4\pi} \tag{E.11}$$

and we have not taken into account the two independent states of polarization. Thus

$$N(\omega)\mathrm{d}\omega\,\mathrm{d}\Omega = N(k)\mathrm{d}k\,\mathrm{d}\Omega = \frac{V}{8\pi^3} \times k^2\,\mathrm{d}k\,\mathrm{d}\Omega$$

or

$$N(\omega)\,\mathrm{d}\omega\,\mathrm{d}\Omega = \frac{V\omega^2}{8\pi^3 c^3}\mathrm{d}\omega\,\mathrm{d}\Omega \tag{E.12}$$

which is identical to Eq. (9.176).

Appendix F
Fourier Transforming Property of a Lens

In this appendix, we will show that the amplitude distribution at the back focal plane of a lens is the spatial Fourier transform of the object field distribution at the front focal plane. In order to show this, we must first determine the effect of a lens on an incident field.

Consider an object point O which is at a distance d_1 from an aberrationless thin lens of focal length f (see Fig. F.1). We know that under geometrical optics approximation, the lens images the point O at the point I which is at a distance d_2 from the lens where

$$\frac{1}{d_2} = \frac{1}{f} - \frac{1}{d_1} \tag{F.1}$$

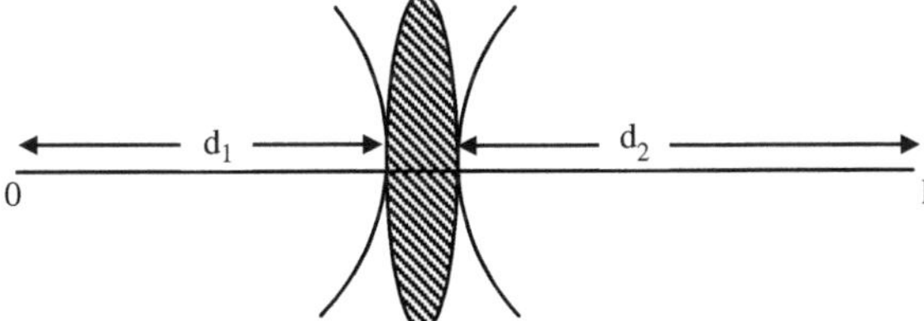

Fig. F.1 A spherical wave emanating from an object point which is at a distance d_1 from a lens is converted into a converging spherical wave converging toward the point I at a distance d_2 satisfying Eq. (F.1)

Thus, the lens transforms an incident diverging spherical wave emanating from O into a converging spherical wave converging to the point I. The phase of the incident diverging spherical wave from O can be written as

$$
\begin{aligned}
\exp(-ikr) &= \exp\left[-ik\left(x^2 + y^2 + d_1^2\right)^{1/2} \right] \\
&= \exp(-ikd_1)\exp\left(-ik\frac{x^2+y^2}{2d_1}\right)
\end{aligned}
\tag{F.2}
$$

where we have assumed a time dependence of the form $e^{i\omega t}$ so that $\exp[i(\omega t - kr)]$ and $\exp[i(\omega t + kr)]$ represent, respectively, a diverging and a converging spherical wave and in writing in last expression, we have assumed that $x, y \ll d_1$. The wave after passing through the lens becomes a converging spherical wave given by

$$\exp\left(+ikd_2\right)\exp\left(+ik\frac{x^2+y^2}{2d_2}\right) \tag{F.3}$$

where we have assumed $x,\ y\ \ll\ d_2$ and the positive sign refers to the fact that the wave is a converging spherical wave. Hence, if we represent by p_L the effect of the lens on the incident field distribution, then

$$\exp\left(-ikd_1\right)\exp\left[-\frac{ik}{2d_1}\left(x^2+y^2\right)\right]p_L=\exp\left(+ikd_2\right)\exp\left[+\frac{ik}{2d_2}\left(x^2+y^2\right)\right]$$

or

$$p_L=\exp\left[ik\left(d_2+d_1\right)\right]\exp\left[\frac{ik}{2}\left(\frac{1}{d_1}+\frac{1}{d_2}\right)\left(x^2+y^2\right)\right] \tag{F.4}$$

Using Eq. (F.1) and neglecting the first factor on the right-hand side of Eq. (F.4) as it is a constant phase factor, one obtains

$$p_L=\exp\left[+\frac{ik}{2f}\left(x^2+y^2\right)\right] \tag{F.5}$$

Thus, the effect of a thin lens of focal length f is to multiply the incident field distribution by the factor p_L given by Eq. (F.5).

In order to obtain the field distribution at the back focal plane, we would also require the effect of propagation through space; this is indeed given by Eq. (7.72), namely,

$$g\left(x,y,z\right)\approx\frac{i}{\lambda z}e^{-ikz}\iint f\left(x',y'\right)\exp\left\{-\frac{ik}{2z}\left[\left(x-x'\right)^2+\left(y-y'\right)^2\right]\right\}dx'dy' \tag{F.6}$$

where $f\left(x',y'\right)$ is the field distribution on the plane $z=0$ and $g(x,y,z)$ is the field at the point x,y,z.

Let us now consider a field distribution $f\left(x,y\right)$ on the front focal plane (P_1) of a lens of focal length f (see Fig. F.2). The field on the plane P_3 would be given by Eq. (F.6) with z replaced by f, i.e.,

$$\frac{i}{\lambda f}e^{-ikf}\iint f(\xi,\eta)\exp\left\{-\frac{ik}{2f}\left[\left(x'-\xi\right)^2+\left(y'-\eta\right)^2\right]\right\}d\xi\,d\eta$$

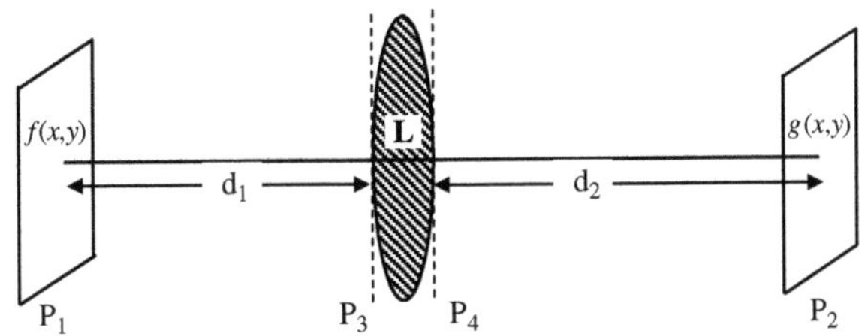

Fig. F.2 The field distribution on the back focal plane of a lens of focal length f is the Fourier transform of the field distribution on the front focal plane P_1

This field on passing through the lens becomes

$$\frac{i}{\lambda f} e^{-ikf} \iint (\xi, \eta) \exp\left\{ -\frac{ik}{2f} \left[(x' - \xi)^2 + (y' - \eta)^2 \right] \right\} d\xi d\eta$$

$$\times \exp\left[\frac{ik}{2f} \left(x'^2 + y'^2 \right) \right]$$

Thus, the field on the back focal plane P_2 would be

$$g(x, y) = \left(\frac{i}{\lambda f} \right)^2 e^{-2ikf} \iint \iint f(\xi, \eta) \exp\left\{ -\frac{ik}{2f} \left[(x' - \xi)^2 + (y' - \eta)^2 \right] \right\}$$

$$\times \exp\left\{ -\frac{ik}{2f} \left[(x + x')^2 + (y + y')^2 \right] \right\} \times \exp\left[\frac{ik}{2f} \left(x'^2 + y'^2 \right) \right] d\xi d\eta dx' dy' \tag{F.7}$$

which on simplification gives

$$g(x, y) = -\frac{1}{\lambda^2 f^2} e^{-2ikf} \exp\left[-\frac{ik}{2f} \left(x^2 + y^2 \right) \right] \iint d\xi d\eta f(\xi, \eta)$$

$$\times \exp\left[-\frac{ik}{2f} \left(\xi^2 + \eta^2 \right) \right] \int \exp\left\{ -\frac{ik}{2f} \left[x'^2 - 2x' (x + \xi) \right] \right\} dx'$$

$$\times \int \exp\left\{ -\frac{ik}{2f} \left[y'^2 - 2y' (y + \eta) \right] \right\} dy'$$

$$= -\frac{1}{\lambda^2 f^2} \frac{\pi 2f \lambda}{2\pi i} e^{-2ikf} \exp\left[-\frac{ik}{2f} \left(x^2 + y^2 \right) \right] \iint d\xi d\eta f(\xi, \eta) \tag{F.8}$$

$$\times \exp\left[-\frac{ik}{2f} \left(\xi^2 + \eta^2 \right) \right]$$

$$\times \exp\left[+\frac{ik}{2f} \left(x^2 + \xi^2 + 2x\xi + y^2 + \eta^2 + 2y\eta \right) \right]$$

$$= \frac{i}{\lambda f} e^{-2ikf} \iint d\xi d\eta \, f(\xi, \eta) \exp\left[2\pi i \left(\frac{x\xi}{\lambda f} + \frac{y\eta}{\lambda f} \right) \right]$$

where we have used the following result:

$$\int_{-\infty}^{\infty} e^{-px^2+qx}\,\mathrm{d}x = \left(\frac{\pi}{p}\right)^{1/2} \exp\left(\frac{q^2}{4p}\right) \tag{F.9}$$

Thus, apart from a constant phase factor, the amplitude distribution at the back focal plane of a lens is nothing but the Fourier transform of the amplitude distribution at the front focal plane evaluated at spatial frequencies $x/\lambda f$ and $y/\lambda f$.

Appendix G
The Natural Lineshape Function

A spontaneous transition from a state b to a lower energy state a does not give radiation at a single frequency ω_{ba}. A finite lifetime τ of the excited state gives it an energy width of the order $\hbar/\tau$ so that the emitted radiation has a frequency distribution. This argument may be made more precise by considering the simple case of an atom with only two states a and b ($E_b > E_a$) undergoing spontaneous transition from b and a. In this section we will calculate the frequency distribution of the emitted radiation from such a spontaneous transition; the analysis is based on the treatment given by Heitler (1954), Section 18, and is known as Weisskopf-Wigner theory of the natural line width.

It is assumed that at $t = 0$, the atom is in the excited state b and the radiation field has no photons. We denote this state by

$$|1\rangle = |b; 0, 0, 0, \ldots\rangle \tag{G.1}$$

[compare with Eqs. (9.174)]. The atom makes a transition to the state a emitting a photon of frequency ω_λ in the mode characterized by λ [see Eq. (9.175)]. We denote this state by

$$|a\lambda\rangle = |a; 0, 0, \ldots 1_\lambda, 0, \ldots\rangle \tag{G.2}$$

If we denote the corresponding probability amplitudes by

$$C_1(t) \quad \text{and} \quad C_{a\lambda}(t)$$

then

$$\begin{aligned} C_1(0) &= 1 \\ C_{a\lambda}(0) &= 0 \end{aligned} \tag{G.3}$$

Further [see Eq. (9.159)]

$$\begin{aligned} i\hbar \frac{dC_1}{dt} &= \sum_\lambda \langle 1|H\prime|a\lambda\rangle\, e^{i(W_1 - W_{a\lambda})t/\hbar}\, C_{a\lambda}(t) \\ &= \sum_\lambda \langle 1|H'|a\lambda\rangle\, e^{i(\omega_0 - \omega_\lambda)t}\, C_{a\lambda}(t) \end{aligned} \tag{G.4}$$

and

$$ i\hbar \frac{dC_{a\lambda}}{dt} = \langle a\lambda | H' | 1 \rangle \, e^{-i(\omega_0 - \omega_\lambda)t} \, C_1(t) \tag{G.5} $$

where

$$ \hbar\omega_0 = E_b - E_a \tag{G.6} $$

and

$$ W_1 - W_{a\lambda} = E_b - (E_a + \hbar\omega_\lambda) = \hbar(\omega_0 - \omega_\lambda) \tag{G.7} $$

We try to solve Eqs. (G.4) and (G.5) by assuming[2]

$$ C_1(t) = e^{-\gamma t/2} \tag{G.8} $$

which satisfies the condition $C_1(0) = 1$. It may be noted that since $|C_1(t)|^2$ represents the probability of finding the atom in the upper state and the quantity $1/\gamma$ represents the mean lifetime of the state. If we substitute for $C_1(t)$ in Eq. (G.5) and carry out the integration we get

$$ C_{a\lambda}(t) = \frac{\langle a\lambda | H' | 1 \rangle \exp\left\{ -\left[i(\omega_0 - \omega_\lambda) + \gamma/2 \right] t \right\} - 1}{(\omega_0 - \omega_\lambda) - i\gamma/2} \tag{G.9} $$

We substitute Eqs. (G.8) and (G.9) in (G.4) to obtain

$$ -i\hbar\frac{\gamma}{2} = \sum_\lambda \frac{\left|\left| H' \right\rangle\right|^2}{\hbar} \frac{1 - \exp\left[i(\omega_0 - \omega_\lambda)t + \gamma t/2 \right]}{(\omega_0 - \omega_\lambda) - i\gamma/2} \tag{G.10} $$

where $\langle H' \rangle \equiv \langle a\lambda | H' | 1 \rangle$. The summation on the right-hand side of the above equation is over a number of states λ with very nearly the same frequency. Under these circumstances, the summation can be replaced by an integral

$$ \sum_\lambda \rightarrow \iint N(\omega) d\omega \, d\Omega \tag{G.11} $$

where $N(\omega)$ represents the density of states [see Eq. (9.176)]. Equation (H.10) then becomes

$$ \gamma = \frac{2i}{\hbar}\left[\frac{1}{\hbar} \iint |\langle H \rangle|^2 N(\omega) J(\omega) \, d\omega \, d\Omega \right] \tag{G.12} $$

---

[2]It should be mentioned that the present treatment is consistent with the treatment given in Chapter 9. Indeed, in Chapter 9 we have shown that A [as given by Eq. (9.182)] represents the probability per unit time for the spontaneous emission to occur; the decay given by Eq. (G.8) follows with $\gamma = A$; this is explicitly shown later in this section.

where

$$J(\omega) = \frac{1 - \exp\left[i\left(\omega_0 - \omega\right)t\right] e^{\gamma t/2}}{\left(\omega_0 - \omega\right) - i\gamma/2} \tag{G.13}$$

We will assume that $\gamma \ll \omega_0$; i.e., the inverse of the lifetime $\left(\leq 10^9 \text{s}^{-1}\right)$ is much smaller than the characteristic frequencies ($\sim 10^{15}$ s^{-1}); this will indeed follow from the final result. Under this assumption γ can be neglected in Eq. (G.13) to obtain

$$\begin{aligned}
J &\approx \frac{1 - e^{i(\omega_0 - \omega)t}}{\omega_0 - \omega} \\
&\approx \frac{1 - \cos\left[\left(\omega_0 - \omega\right)t\right]}{\left(\omega_0 - \omega\right)} - i\frac{\sin\left[\left(\omega_0 - \omega\right)t\right]}{\left(\omega_0 - \omega\right)}
\end{aligned} \tag{G.14}$$

The first term has very rapid variation around $\omega \approx \omega_0$ and gives negligible contribution to any integral over ω except around $\omega \approx \omega_0$, where the function itself vanishes. Thus this term will lead to a small imaginary value of γ. We are in any case interested in the real part of γ (which would give the lifetime, etc.) which will be given by

$$\gamma \approx \frac{2}{\hbar}\left[\frac{1}{\hbar}\int |\langle H'\rangle|^2 \, N\left(\omega_0\right) \mathrm{d}\Omega \int_{-\infty}^{\infty} \frac{\sin\left(\omega_0 - \omega\right)t}{\left(\omega_0 - \omega\right)}\mathrm{d}\omega\right] \tag{G.15}$$

We have made two approximations here. First, we have pulled out factors which are essentially constant in the neighborhood of $\omega \approx \omega_0$. Second, the limits of the integral have been taken from $-\infty$ to $+\infty$ since in any case the contribution vanishes except around ω_0. Thus

$$\begin{aligned}
\gamma &\approx \frac{2\pi}{\hbar}\left[\frac{1}{\hbar}\int |\langle H'\rangle|^2 \, N(\omega_0)\mathrm{d}\Omega\right] \\
&= \frac{2\pi}{\hbar}\frac{1}{\hbar}\frac{V\omega_0^2}{8\pi^3 c^3}\int |\langle H'\rangle|^2 \, \mathrm{d}\Omega
\end{aligned} \tag{G.16}$$

Now

$$\begin{aligned}
\langle H'\rangle &= \langle a\lambda|\, H'\, |1\rangle \\
&= \langle a; 0, 0, \ldots 1_\lambda, 0, \ldots|\,(-ie) \\
&\quad \times \sum_\lambda \left(\frac{\hbar\omega_\lambda}{2\varepsilon_0 V}\right)^{1/2}\left[\hat{a}_\lambda e^{i\mathbf{k}_\lambda \cdot \mathbf{r}} - \hat{a}_\lambda^\dagger e^{-i\mathbf{k}_\lambda \cdot \mathbf{r}}\right]\hat{\mathbf{e}}_\lambda \cdot \mathbf{r}\,|b; 0, 0, \ldots, 0_\lambda, \ldots\rangle \\
&= (ie)\left(\frac{\hbar\omega_\lambda}{2\varepsilon_0 V}\right)^{1/2}\langle a|\, e^{i\mathbf{k}_\lambda \cdot \mathbf{r}}\mathbf{r}\,|b\rangle \cdot \hat{\mathbf{e}}_\lambda \\
&= (ie)\left(\frac{\hbar\omega_\lambda}{2\varepsilon_0 V}\right)^{1/2}\langle a|\, \mathbf{r}\,|b\rangle \cdot \hat{\mathbf{e}}_\lambda
\end{aligned} \tag{G.17}$$

where in the last step we have used the dipole approximation. On substitution in Eq. (G.16), we get

$$\gamma \approx \frac{1}{2\pi} \left[\frac{e^2}{4\pi\varepsilon_0\hbar c} \right] \frac{\omega_0^3}{c^2} \int \left| \hat{\mathbf{e}}_\lambda \cdot \langle a| \mathbf{r} |b\rangle \right|^2 d\Omega \tag{G.18}$$

[compare Eq. (9.180)]. On carrying out the integration and summing over the two states of polarization we get [see Eq. (9.182)]

$$\gamma = A \tag{G.19}$$

i.e., γ is simply the Einstein A coefficient corresponding to spontaneous emissions as it indeed should be!

Now the probability of a photon being emitted in the mode λ is given by [see Eq. (G.9)]

$$|C_{a\lambda}(t=\infty)|^2 = \frac{\left|\langle H'\rangle\right|^2}{\hbar^2} \frac{1}{(\omega_0 - \omega_\lambda)^2 + \gamma^2/4} \tag{G.20}$$

If we multiply the above equation by $N(\omega)\, d\omega d\Omega$, which gives the number of modes in the frequency interval ω to $\omega + d\omega$ and in the solid angle $d\Omega$, and carry out the integration over $d\Omega$ and sum over the two states of polarization, we would get

$$g(\omega)\, d\omega = \frac{\gamma}{2\pi} \frac{d\omega}{(\omega_0 - \omega)^2 + \gamma^2/4} \tag{G.21}$$

where we have used Eq. (G.18). The above equation indeed gives us the probability that the spontaneously emitted photon has its frequency between ω and $\omega + d\omega$, which is nothing but the Lorentzian line shape; notice that the above expression is normalized, which implies that the probability that the atom makes a spontaneous transition from the upper state to the lower state is unity (as $t \to \infty$), as it indeed should be.

Appendix H
Nonlinear polarization in optical fibers

At low optical intensities, all media behave as linear media, i.e., the electric polarization P developed in the medium is linearly related to the electric field E of the lightwave:

$$P = \varepsilon_0 \chi E \qquad (H.1)$$

When the intensity of the light wave increases, then the amplitude of the electric field associated with the light wave also increases and in such a situation the relationship described by Eq. (H.1) is no more valid. In such a situation the polarization is not linearly related to the electric field and we can write for the total polarization as

$$P = \varepsilon_0 \chi E + \varepsilon_0 \chi^{(3)} E^3 \qquad (H.2)$$

where $\varepsilon_0 \chi^{(3)} E^3$ represents the nonlinear polarization term and $\chi^{(3)}$ is referred to as the third-order nonlinear susceptibility. Note that there is no term proportional to E^2 the quadratic term is present only in media not possessing a center of inversion symmetry like many crystals. The second term in Eq. (H.2) is responsible for many nonlinear effects in optical fibers; these include self phase modulation, cross phase modulation, and four wave mixing. Here we will give a brief outline of the derivation of Eq. (H.2) which results in nonlinear polarization rotation. Such an effect is used in the generation of ultrashort pulses from fiber lasers (see Chapter 12).

Equation (H.2) is written in scalar form but since both electric fields and polarization are vector fields we have to write a vector equation. Thus the nonlinear polarization in vector form is given by

$$P_i^{\text{NL}} = \varepsilon_0 \chi_{ijkl}^{(3)} E_j E_k E_l \qquad (H.3)$$

where the subscripts correspond to various Cartesian components of the fields and $\chi_{ijkl}^{(3)}$ is called the third-order nonlinear optical susceptibility tensor. It is characterized by 81 elements and depending on the symmetry present in the medium some of them may be zero, some may be equal to some other elements, etc. For an optical fiber, the only nonzero elements are

$$\chi_{1122}^{(3)} = \chi_{1212}^{(3)} = \chi_{1221}^{(3)} = \chi_{2211}^{(3)} = \chi_{2121}^{(3)} = \chi_{2112}^{(3)} = \chi_{2233}^{(3)} = \chi_{2323}^{(3)}$$
$$= \chi_{2332}^{(3)} = \chi_{3322}^{(3)} = \chi_{3232}^{(3)} = \chi_{3223}^{(3)} = \chi_{3311}^{(3)} = \chi_{3131}^{(3)} \tag{H.4}$$
$$= \chi_{3113}^{(3)} = \chi_{1133}^{(3)} = \chi_{1313}^{(3)} = \chi_{1331}^{(3)} = \frac{1}{3}\chi_{1111}^{(3)} = \frac{1}{3}\chi_{2222}^{(3)} = \frac{1}{3}\chi_{3333}^{3}$$

All other elements of the nonlinear susceptibility tensor are zero.

Let the propagation direction along the fiber be designated as z-direction and the transverse coordinates are assumed to be x and y. If we assume the incident field to have both x and y components then the electric field at any point along the fiber would be given by

$$\mathbf{E}(x,y,z,t) = \frac{1}{2}[(\hat{\mathbf{x}}E_x(z) + \hat{\mathbf{y}}E_y(z))e^{i(\omega t - \beta z)} + \text{c.c}] \tag{H.5}$$

where c.c represents complex conjugate, E_x and E_y are the complex amplitudes of the x-component and y-component of the electric field of the light wave, β represents the propagation constant, and ω the frequency of the wave. For simplicity we are neglecting the transverse dependence of the fields within the fiber; taking them into account will result in an overlap integral which would primarily lead to a reduction of the nonlinear effects.

For an electric field E, the power carried is given by

$$F = \frac{n}{2c\mu_0}|E|^2 S \tag{H.6}$$

where S is the area of the beam and n represents the refractive index of the medium. We now define

$$A_x = \sqrt{\frac{nS}{2c\mu_0}}E_x; \; A_y = \sqrt{\frac{nS}{2c\mu_0}}E_y \tag{H.7}$$

The constants are chosen so that the powers carried by the x-component and y-component of the wave are directly given by $|A_x|^2$ and $|A_y|^2$.

Substituting from Eqs. (H.4), (H.5) into Eq. (H.3), and collecting terms at frequency ω, we get for the x-component of nonlinear polarization to be

$$P_x^{(NL)} = \frac{3}{8}\varepsilon_0\chi_{1111}\left[\left\{|E_x|^2E_x + \frac{2}{3}|E_y|^2E_x + \frac{1}{3}E_x^*E_y^2\right\}e^{i}(\omega t - \beta z) + \text{c.c.}\right] \tag{H.8}$$

The first term within the bracket on the right-hand side leads to self phase modulation while the second terms are referred to as cross phase modulation.

The wave equation describing the propagation of a wave in the presence of nonlinear polarization is given by [see Eq. (14.18)]

$$\frac{\partial^2 E_x}{\partial x^2} + \frac{\partial^2 E_x}{\partial y^2} + \frac{\partial^2 E_x}{\partial z^2} - \varepsilon\mu_0\frac{\partial^2 E_x}{\partial t^2} = \mu_0\frac{\partial^2 P_x^{(NL)}}{\partial t^2} \tag{H.9}$$

Substituting from Eq. (H.5) and (H.8) into Eq. (H.9) and neglecting the second differential of E_x with respect to z, we obtain the following equation for A_x:

$$\frac{\partial A_x}{\partial z} = -i\gamma \left[\left(|A+x|^2 + |A_y|^2 \right) A_x + \frac{1}{3} A_x^* A_y^2 \right] \tag{H.10}$$

where γ is the nonlinear coefficient defined by

$$\gamma = \frac{3}{4} \frac{\omega \chi_{1111}^{(3)}}{c^2 \varepsilon_0 n^2 S} = \frac{\omega n_2}{cS} \tag{H.11}$$

where

$$n_2 = \frac{3}{4} \frac{\chi_{1111}^{(3)}}{c \varepsilon_0 n^2} \tag{H.12}$$

Using a similar procedure we can derive the following equation satisfied by A_y:

$$\frac{\partial A_y}{\partial z} = -i\gamma \left[\left(|A_y|^2 + \frac{2}{3} |A_x|^2 \right) A_y + \frac{1}{3} A_y^* A_x^2 \right] \tag{H.13}$$

These equations are written in terms of the two linearly polarized components of the propagating wave. The terms $A_x^* A_y^*$ and $A_x^* A_y^*$ can be eliminated by describing the wave in terms of circular polarization components rather than linear polarization components. For this we define the following:

$$A+ = \frac{1}{\sqrt{2}}(A_x + iA_y); \quad A- = \frac{1}{\sqrt{2}}(A_x - iA_y) \tag{H.14}$$

A_+ represents the amplitude of the right circular component and A_- represents the amplitude of the left circular component. Substituting from Eq. (H.14) into Eqs. (H.10) and (H.13), we obtain the following equations:

$$\frac{\partial A_+}{\partial z} = \frac{2}{3} i\gamma (|A_+|^2 + 2|A_-|^2) A_+ \tag{H.15}$$

and

$$\frac{\partial A_-}{\partial z} = \frac{2}{3} i\gamma (|A_-|^2 + 2|A_+|^2) A_- \tag{H.16}$$

These equations are used in Section to describe nonlinear polarization rotation.

It is interesting to note that Eqs. (H.15) and (H.16) satisfy the following equations:

$$\frac{\partial |A_+|^2}{\partial z} = 0; \quad \frac{\partial |A_-|^2}{\partial z} = 0 \tag{H.17}$$

showing that the powers in the right circular and left circular polarizations do not change with propagation in spite of the nonlinear interaction. Thus circular polarization modes are the modes of propagation in such a case. The nonlinearity introduces only an additional phase difference other than due to linear propagation. The phase shift depends on the amplitude of the right and left circular components.

References and Suggested Reading

Alfano, R. R., and Shapiro, S. L. (1975), Ultrashort phenomena, *Phys. Today* **31**(7), 30.

Arditty, H., Papuchon, M., Puech, C., and Thyagarajan, K. (1980), Recent developments in guided wave optical rotation sensors, presented at the Conference on Integrated and Guided-Wave Optics, Nevada, January 28–30.

Arecchi, F. T. (1976), Quantum optics and photon statistics, in *Lasers and Their Applications* (A. Sona, ed.), p. 497, Gordon and Breach, New York.

Arecchi, F. T., Berne, A., and Burlamacchi, P. (1966a), High-order fluctuations in a single mode laser field, *Phys. Rev. Lett.* **16**, 32.

Arecchi, F. T., Gatti, E., and Sona, A. (1966b), Time distribution of photons from coherent and Gaussian sources, *Phys, Lett.* **20**, 27.

Baldwin, G. C. (1969), *An Introduction to Nonlinear Optics*, Plenum Press, New York.

Barnoski, M. (1976), *Fundamentals of Optical Fiber Communications*, Academic Press, New York.

Bates, B., Murphy, A. A., and Tweedie, M. (1986), Laser speckle photography: some simple experiments for the undergraduate laboratory, *Phys. Educ.* **21**, 54.

Baym, G. (1969), *Lectures on Quantum Mechanics*, W.A. Benjamin, Inc., New York.

Bennett, W. R. (1962), Hole burning effects in a He–Ne optical Maser, *Phys. Lett.* **25**, 617.

Bernhardt, A. F., Duerre, D. E., Simpson, J. R., and Wood, L. L. (1974), Separation of isotopes by laser deflection of atomic beam, I. Barium, *Appl. Phys. Lett.* **25**, 617.

Bleuler, E., and Goldsmith, G. J. (1952), *Experimental Nucleonics*, Holt, Rinehart and Winston, New York.

Booth, L. A., Freiwald, D. A., Frank, T. G., and Finch, F. T. (1976), Prospects of generating power with laser driven fusion, *Proc. IEEE* **64**, 1460.

Born, M., and Wolf, E. (1999), *Principles of Optics*, 7th Edition, Cambridge University Press, Cambridge.

Boyd, G. D., and Gordon, J. P. (1961), Confocal multimode resonator for millimeter through optical wavelength masers, *Bell Syst. Tech. J.* **40**, 489.

Brouwer, W. (1964), *Matrix Methods in Optical Instrumental Design*, W.A. Benjamin, Inc., New York.

Brueckner, K. A., and Jorna, S. (1974), Laser driven fusion, *Rev.Mod. Phys.* **46**(2), 325.

Busch, G. E., and Rentzepis, P. M. (1976), Picosecond chemistry, *Science* **194**, 276.

Casasent, D. (ed.) (1978), *Optical Data Processing*, Springer, Berlin.

Cattermole, W. (1969), *Principles of Pulse Code Modulation*, Iliffe Books Ltd., London.

Charschan, S. S. (1972), *Lasers in Industry*, Van Nostrand Reinhold Co., New York.

Chynoweth, A. G. (1976), Lightwave communications: the fiber lightguide, *Phys. Today* **29**(5), 28.

Collier, R. J., Burckhardt, C. B., and Lin, L. H. (1971), *Optical Holography*, Academic Press, New York.

Condon, E. U., and Shortley, G. H. (1935), *Theory of Atomic Spectra*, Cambridge University Press, Cambridge.

Dirac, P. A. M. (1958a), *The Principles of Quantum Mechanics*, Oxford University Press, London.

Dirac, P. A. M. (1958b), *Quantum Theory of Emission and Absorption in Quantum Electrodynamics* (J. Schwinger, ed.), Dover Publications, New York.

Dunne, M. (2010), *Fusion's bright new down, Physics World*, May issue, p. 29.

Einstein, A. (1917), On the quantum theory of radiation, *Phys. Z.* **18**, 121. (This paper has been reprinted in *Laser Theory* by F. S. Barnes, IEEE Press, New York, 1972.)

Faye, D., Grisard, A., Lallier, E., Gérard, B., Kieleck, C., and Hirth, A. (2008), Orientation-patterned gallium arsenide: engineered materials for infrared sources, SPIE, Micro/Nano Lithography and Fabrication, DOI: 10.1117/2.1200806.1164.

Fox, A. G., and Li, T. (1961), Resonant modes in a maser interferometer, *Bell. Syst. Tech. J.* **40**, 453.

Fox, A. G., Schwarz, S. E., and Smith, P. W. (1969), Use of neon as a nonlinear absorber for mode locking a He-Ne laser, *Appl. Phys. Lett.* **12**, 371.

Franken, P. A., Hill, A. E., Peters, C. W., and Weinreich, G. (1961), Generation of optical harmonics, *Phys. Rev. Lett.* **7**, 118.

French, P. M. W., Kelly, S. M. J., and Taylor, J. R. (1990), Mode locking of a continuous-wave titanium-doped sapphire laser using a linear external cavity, *Opt. Lett.* **15**, 378.

Gagliano, F. P., Lumley, R. M., and Watkins, L. S. (1969), Lasers in industry, *Proc. IEEE* **57**, 114.

Gallawa, R. (1979), Optical systems: a review I. U. S. and Canada: initial results reported, *IEEE Spectrum* **16**(10), 71.

Galvanauskas, A., Cheng, M.-Y., Hou, K.-C., and Liao, K.-H. (2007), High peak power pulse amplification in large-core Yb-doped fiber amplifiers, *IEEE J. Sel. Topics Quant. Electron.* **13**, 559.

Gambling, W. A. (1986), Glass, light and the information revolution, Ninth W.E.S. Turner Lecture, *Glass Technology*, **27**(6), 179.

Gerard, J. M. (2003), Boosting photon storage, *Nat. Mater.* **2**, 140.

Gerrard, A., and Burch, J. M. (1975), *An Introduction to Matrix Methods in Optics*, Wiley, New York.

Gerstner, E. (2007), Extreme light, *Nature* **446**, 17.

Ghatak, A. K. (2009), *Optics*, McGraw-Hill, New York.

Ghatak, A. K., Goyal, I. C., and Chua, S. J. (1985), *Mathematical Physics*, Macmillan India, New Delhi.

Ghatak, A. K., and Lokanathan, S. (2004), *Quantum Mechanics,* Macmillan, New Delhi.

Ghatak, A. K., and Thyagarajan, K. (1978), *Contemporary Optics*, Plenum Press, New York.

Ghatak, A. K., and Thyagarajan, K. (1980), Graded index optical waveguides: a review, in *Progress in Optics* (E. Wolf, ed.), Vol. XVIII, North-Holland Publishing Company, Amsterdam.

Ghatak, A. K., and Thyagarajan, K. (1989), *Optical Electronics*, Cambridge University Press, Cambridge, UK [Reprinted by Foundation Books, New Delhi].

Ghatak, A., and Thyagarajan, K. (1998), *Introduction to Fiber Optics*. Cambridge University Press, UK.

Giordmaine, J. A., and Miller, R. C. (1965), Tunable coherent parametric oscillation in LiNbo3 at optical frequencies, *Phys. Rev. Lett.* **14**, 973.

Gloge, D. (1976), *Optical Fiber Technology*, IEEE Press, New York.

Gloge, D. (1979), Inside the new region: second generation fiber systems, *Opt. Spectra* **131**(11), 50.

Goldman, L., and Rockwell, R. J. (1971), *Lasers in Medicine*, Gordon and Breach Science Publishers Inc., New York.

Goldstein, H. (1950), *Classical Mechanics,* Addison-Wesley, Reading, Massachusetts.

Goldstein, R., and Goss, W. G. (1979), Fiber optic rotation sensor (FORS), Laboratory performance evaluation, *Opt. Eng.* (July–Aug.) **18**(4), 381.

Gopal, E. S. R. (1974), *Statistical Mechanics and Properties of Matter*, Wiley, New York.

Gordon, J. P., Zeiger, H. J., and Townes, C. H. (1954), Molecular microwave oscillator and new hyperfine structure in the microwave spectrum of NH3, *Phys. Rev.* **95**, 282.

Gordon, J. P., Zeiger, H. J., and Townes, C. H. (1955), The Maser–New type of microwave amplifier, frequency standard and spectrometer, *Phys. Rev.* **99**, 1264.

Gupta, J. A., Barrios, P. J., Aers, G. C., and Lapointe, J. (2008), 1550 nm GaInAsSb distributed feedback laser diodes on GaAs, *Electron. Lett.* **44**, 578.

Hall, F. F. (1974), Laser systems for monitoring the environment, in *Laser Applications* (M. Ross, ed.), Vol. 2, Academic Press, New York.

Harry, J. E. (1974), *Industrial Lasers and Their Applications*, McGraw Hill, London.

Heitler, W. (1954), *Quantum Theory of Radiation*, Oxford University Press, London.

Henderson, A., Staffort, R., and Hofman, P. (2008), High power CW OPOs span the spectrum, *Laser Focus World*, October, 65.

Hong, C. K., Ou, Z. Y., and Mandel, L. (1987), Measurement of subpicosecond time intervals between two photons by interference, *Phys. Rev. Lett.* **59**, 2044.

Irving, J., and Mullineux, N. (1959), *Mathematics in Physics and Engineering*, Academic Press, New York.

Jacobs, S. F. (1979), How monochromatic is laser light? *Am. J. Phys.* **47**, 597.

Jacobs, I., and Miller, S. E. (1977), Optical transmission of voice and data, *IEEE Spectr.* **14**(2), 32.

Jansson, E. V., (1969), A comparison of acoustical measurements and hologram interferometry measurements of the vibrations of a guitar top plate, *STL-QPSR*, **10**, 36.

Javan, A., Bennet, W. R., and Herriott, D. R. (1961), Population inversion and continuous optical maser oscillation in a gas discharge containing a He-Ne mixture, *Phys. Rev. Letts.* **6**, 106.

Jenkins, F. A., and White, H. E. (1981), *Fundamentals of Optics*, 4th Edition, McGraw-Hill, New York.

Jeong, Y., Sahu, J. K., Payne, D. N., Nilsson, J. (2004), Ytterbium-doped large-core fiber laser with 1.36 kW, continuous-wave output power, *Opt. Express*, **12**, 6088–6092.

Kao, C. K., and Hockham, G. A. (1966), Dielectric fiber surface waveguides for optical frequencies, *Proc. IEE*, **113**, 1151.

Khokhlov, R. V., Akhmanov, S. A., and Sukhorukov, A. P. (1976), Self-focusing, self-defocusing and self-modulation of light beams in nonlinear media, in *Laser Handbook* (F. T. Arecchi and E. O. Schulz-Dubois, eds.), North-Holland Publishing Company, Amsterdam.

Kidder, R. E. (1973), Some aspects of controlled fusion by use of lasers, in *Fundamental and Applied Laser Physics* (M. S. Feld, A. Javan, and N. A. Kurnit, eds.), Wiley, New York.

Koechner, W. (1976), *Solid state laser engineering*, Springer, New York.

Kogelnik, H., and Li, T. (1966), Laser beams and resonators, *Appl. Opt.* **5**, 1550.

Kressel, H. (1979), Semiconductor diode lasers, in *Quantum Electronics, Part A: Methods of Experimental Physics* (C. L. Tang, ed.), Vol. 15, Academic Press, New York.

Kressel, H., and Butler, J. K. (1977), *Semiconductor Lasers Heterojunction LEDs*. Academic Press, New York.

Kressel, H., Ladany, I., Ettenberg, M., and Lockwood H. F. (1976), Lightwave communications: light sources, *Phys. Today* **29**(5), 38.

Lamb, W. E. (1964), Theory of an optical maser, *Phys. Rev.* **134**, A1429.

Lehr, C. G. (1974), Laser tracking systems, in *Laser Applications* (M. Ross, ed.), Vol. 2, Academic Press, New York.

Letokhov, V. S. (1977), Lasers in research: photophysics and photochemistry, *Phys. Today* **30**(5), 23.

Limpert, J., Roser, F., Klingebiel, S., Schreiber, T., Wirth, C., Peschel, T., Eberhardt, R., and Tünnermann, A. (2007), The rising power of fiber lasers and amplifiers, *IEEE J. Sel. Topics Quant. Electron.* **13**, 537.

Lin, C., Stolen, R. H., and Cohen, L. G. (1977a), A tunable 1.1 μm fiber Raman oscillator, *Appl. Phys. Lett.* **2**, 97.

Lin, C., Stolen, R. H., French, W. G., and Malone, T. G. (1977b), A CW tunable near infrared (1.085–1.175 μm) Raman oscillator, *Opt. Lett.* **1**, 96.

Liu, C. Y., Yoon, S. F., Fan, W. J., Ronnie Teo, J. W., and Yuan, S. (2005), Low threshold current density and high characteristic temperature narrow-stripe native oxide-confined 1.3-μm InGaAsN triple quantum well lasers, *Opt. Express* **13**, 9045.

Loudon, R. (1973), *The Quantum theory of Light,* Clarendon Press, Oxford.

Maiman, T. H. (1960), Stimulated optical radiation in ruby, *Nature,* **187,** 493.

Maitland, A., and Dunn, M. H. (1969), *Laser Physics,* North-Holland Publishing Company, Amsterdam.

Mandel, L. (1958), Fluctuations of photon beams and their correlations, *Proc. Phys. Soc.* **72,** 1037.

Mandel, L. (1959), Fluctuations of photon beams: the distribution of the photo-electrons, *Proc. Phys. Soc.* **74,** 233.

Mandel, L., and Wolf, E. (1970), *Selected Papers on Coherence and Fluctuations of Light,* Vols. I and II, Dover Publications, New York.

Marcuse, D. (1978), Gaussian approximation of the fundamental modes of graded-index fibers, *J. Opt. Soc. Am.* **68,** 103–109.

Mayer, S. W., Kwok, M. A., Gross, R. W. F., and Spencer, D. J. (1970), Isotope separation with the CW hydrogen fluoride laser, *Appl. Phys. Lett.* **17,** 516.

McFarlane, R. A., Bennet, W. R., and Lamb, W. E. (1963), Single mode tuning dip in the power output of an He–Ne Optical maser, *Appl. Phys. Lett.* **2,** 189.

Mehta, C. L. (1970), Theory of photoelectron counting, in *Progress in Optics* (E. Wolf, ed.), Vol. VIII, p. 375, North-Holland Publishing Company, Amsterdam.

Melchoir, H. (1977), Detectors for light wave communication, *Phys. Today* **30**(11), 32.

Midwinter, J. E. (1979a), *Optical Fibers for Transmission,* Wiley, New York.

Midwinter, J. E. (1979b), Optical systems a review: III, Europe business benefits, *IEEE Spectr.* **16**(10), 76.

Milster, T. D. (2005), Horizons for optical data storage, *Opt. Photonics News,* March, 28.

Minck, R. W., Terhune, R. W., and Wang, C. C. (1966), Nonlinear optics, *Appl. Opt.* **5,** 1595.

Miya, T., Terunuma, Y., Hosaka, T., and Miyashita, T. (1979), Ultra low loss single-mode fibers at 1.55 μm, *Rev. Electr. Commun. Lab.* **27**(7–8), 497.

Mooradian, A. (1985), Laser linewidth, *Phys. Today,* May, 43.

Mourou, G. A., and Yanovsky, V. (2004), Relativistic optics; a gateway to attosecond physics, *Opt. Photonics News,* May, 40.

Myers, L. E., and Bosenberg, W. R. (1997), Periodically poled lithium niobate and quasi phase matched optical parametric oscillators, *IEEE J. Quant. Electron.* **33,** 1663.

Nelson, D. F., and Collins, R. J. (1961), Spatial coherence in the output of an optical maser, *J. Appl. Phys.* **32,** 739.

Nuckolls, J., Wood, L., Thiessen, A., and Zimmerman, G. (1972), Laser compression of matter to super high densities: thermonuclear applications, *Nature* **239,** 139.

Nuese, C. J., Olsen, G. H., Ettenberg, M., Gannon, J. J., and Zamarowsky, T. J. (1976), CW room temperature $In_xGa_{1-x}As/In_yGa_{1-y}P1.06$ μm laser, *Appl. Phys. Lett.* **29,** 807.

Nussbaum, A. (1968), *Geometric Optics: An Introduction,* Addison-Wesley, Reading, Massachusetts.

Pal, B. P. (1979), Optical communication fiber waveguide fabrication: a review, *Fiber Int. Opt.* **2**(2), 195.

Patel, C. K. N. (1966), Optical harmonic generation in the infrared using a CO_2 laser, *Phys. Rev. Lett.* **16,** 613.

Patel, C. K. N., Slusher, R. E., and Flurry, P. A. (1966), Optical nonlinearities due to mobile carriers in semiconductors, *Phys. Rev. Lett.* **17,** 1011.

Peng, Q., Juzeniene, A., Chen, J., Svaansand, L. O., Sarloe, T., Giercksky, K., and Moan, J. (2008), Lasers in medicine, *Rep. Prog. Phys.* **71,** 056701.

Personick, S. D. (1976), Photodetectors for fiber systems, in *Fundamentals of Optical Fiber Communications* (M. K. Barnoski, ed.), p. 155, Academic Press, New York.

Philips, R. A. (1969), Spatial filtering experiments for undergraduate laboratories, Am. J. Phys. **37,** 536.

Post, E. J. (1967), Sagnac effect, *Rev. Mod. Phys.* **39,** 475.

Post, R. F. (1973), Prospects for fusion power, *Phys. Today* **26**(4), 30.

Powell, J. L., and Craseman, B. (1961), *Quantum Mechanics*, Addison-Wesley, Reading, Massachusetts.

Ranka, J. K., Windeler, R. S., and Stentz, A. J. (2000), Visible continuum generation in air-silica microstructure optical fibers with anomalous dispersion at 800 nm, *Opt. Lett.* **25**, 25.

Ribe, F. L. (1975), Fusion reactor systems, *Rev. Mod. Phys.* **47**, 7.

Sacchi, C. A., and Svelto, O. (1965), Spiking behavior of a multimode ruby laser, *IEEE J. Quant. Electron.* **1**, 398.

Saha, M. N., and Srivastava, B. N. (1973), *Treatise on Heat*, 5th Edition. The Indian Press Pvt. Ltd. Allahabad.

Salzman, W. R. (1971), Time evolution of simple quantum mechanical systems: II. Two state system in intense fields, Phys.*Rev. Lett.* **26**, 220.

Sawicki, R. (2008), Interview for Dartmouth Engineer Magazine, May 27, 2008.

Schafer, F. P. (1973), *Dye Lasers*, Springer, Berlin.

Schawlow, A. L., and Townes, C. H. (1958). Infrared and optical masers, *Phys. Rev.* **112**, 1940.

Shapiro, S. L. (1977), *Ultrashort Light Pulses: Picosecond Techniques and Applications*, Springer, Berlin.

Shimada, S. (1979), Optical systems: a review, II. Japan: unusual applications, *IEEE Spectr.* **16**(10), 74.

Simon Li, Z., and Piprek, J. (2000), Software and computing: simulation software gives laser designers insight, *Laser Focus World*, January.

Slepian, D., and Pollack, H. O. (1961), Prolate spheroidal wave functions–Fourier analysis and uncertainty–I, *Bell Syst. Tech. J.* **40**, 43.

Smith, P. W. (1972), Mode selection in lasers, *Proc. IEEE* **60**, 422.

Sodha, M. S., and Ghatak, A. K. (1977), *Inhomogeneous Optical Waveguides*, Plenum Press, New York.

Sodha, M. S., Ghatak, A. K., and Tripathi, V. K. (1974), *Self-focusing of laser beams in Dielectrics, Plasmas and Semiconductors*, Tata McGraw Hill, New Delhi.

Sodha, M. S., Ghatak, A. K., and Tripathi, V. K. (1976), *Self-focusing of Laser Beams in Plasmas and Semiconductors, Progress in Optics* (E. Wolf, ed.), Vol. XIII, North-Holland Publishing Company, Amsterdam.

Steensma, P. D., and Mondrick, A. (1979), Battlefield fiber network, *Laser Focus* **15**(7), 52.

Stickley, C. M. (1978), Laser fusion, *Phys. Today* **31**(5),50.

Stolen, R. H., Ippen, E. P., and Tynes, A. R. (1972), Raman oscillation in glass optical waveguides, *Appl. Phys. Lett.* **20**, 62.

Stolen, R. H., Lin, C., and Jain, R. K. (1977), A time dispersion tuned fiber Raman oscillator, *Appl. Phys. Lett.* **30**, 340.

Strößner, U., Meyn, J.-P., Wallenstein, R., Urenski, P., Arie, A., Rosenman, G., Mlynek, J., Schiller, S., Peters, A. (2002), Single-frequency continuous-wave optical parametric oscillator system with an ultrawide tuning range of 550 to 2830 nm, *J. Opt. Soc. Am. B* **19**, 1419.

Stroke, G. W., Halioua, M., and Srinivasan, N. (1975), Holographic image restoration using fourier spectrum analysis of blurred photographs in computer-aided synthesis of Wiener filters, *Phys. Lett.* **51**A, 383.

Susskind, L., and Glogower, J. (1964), Quantum mechanical phase and time operator, *Physics* **1**, 49.

Svelto, O. (1975), Self focusing, self trapping and self modulation of laser beams, *Progress in Optics* (E. Wolf, ed.), Vol. XII, North-Holland Publishing Company, Amsterdam.

Tait, J. H. (1964), *Neutron Transport Theory*, Longmans, London.

Thyagarajan, K., and Ghatak, A. (2007), *Fiber optic Essentials*, John Wiley, NY, 2007

Tolansky, S. (1955), *An Introduction to Interferometry*, Longmans Green and Co., London.

Tsujiuchi, J., Matsuda, K., and Takeja, N. (1971), Correlation techniques by holography and its application to fingerprint identification, in *Applications of Holography* (E. S. Barrekette, W. E. Kock, T. Ose, J. Tsujiuchi, and G. W. Stroke, eds.), p. 247, Plenum Press, New York.

Turitsyn, S. K., Ania-Castanon, J. D., Babin, S. A., Karalekas, V., Harper, P., Churkin, D., Kablukov, S. I., El-Taher, A. E., Podivilov, E. V., and Mezentsev, V. K. (2009), 270-km ultralong Raman fiber laser, *Phys. Rev. Lett.* **103**, 133901-1.

Vahala, K. J. (2003), Optical microcavities, *Nature* **424**, 839.

Vali, V., and Shorthill, R. W. (1977), Ring interferometer 950 m long, *Appl. Opt.* **16**, 290.

Weaver, L. A. (1971), Machining and welding applications, in *Laser Applications* (M. Ross, ed.), Vol. I, p. 201, Academic Press, New York.

Yamada, M., Nada, N., Saitoh, M., and Watanabe, K. (1993), First order quasi phase matched LiNbO3 by applying an external field for efficient blue second harmonic generation, *Appl. Phys. Lett.* **62**, 435.

Yariv, A. (1977), *Optical Electronics in Modern Communications*, 5th Edition, Oxford University Press.

Zare, R. N. (1977), Laser separation of isotopes, *Sci. Am.* **236**(2), 86.

Zernike, F., and Midwinter, J. E. (1973), *Applied Nonlinear Optics*, Wiley, New York.

Zewail, A. H. (2000), Femtochemistry. Past, present, and future, *Pure Appl. Chem.*, **72**, 2219–2231.

Zhu, W. Qian, L., Helmy, A. S. (2007), Implementation of three functional devices using erbium-doped fibers: an advanced photonics lab, ETOP Proceedings. June 3–5, 2007, Ottawa, Canada.

Zuber, K. (1935), *Nature* **136**, 796.

Index

MIX
Papier aus verantwortungsvollen Quellen
Paper from responsible sources
FSC® C105338

If you have any concerns about our products,
you can contact us on
ProductSafety@springernature.com

In case Publisher is established outside the EU,
the EU authorized representative is:
Springer Nature Customer Service Center GmbH
Europaplatz 3, 69115 Heidelberg, Germany

Printed by Libri Plureos GmbH
in Hamburg, Germany